NOTES ON

ACI 318-89

BUILDING CODE REQUIREMENTS FOR REINFORCED CONCRETE

with Design Applications

Edited by: S. K. Ghosh and Basile G. Rabbat

PORTLAND CEMENT  ASSOCIATION

An organization of cement manufacturers to improve and extend the uses of portland cement and concrete through market development, engineering, research, education, and public affairs work.

5420 Old Orchard Road, Skokie, Illinois 60077-1083

© 1990 Portland Cement Association

Fifth edition, First printing, 1990

Printed in U.S.A.

Library of Congress catalog card number 89-64412

ISBN 0-89312-089-8

About the building on the cover

Cover art: line representation derived from posterized photograph
Water Tower Place, Chicago, Illinois
Architects—Loebel Schlossman, Bennett and Dart, Chicago, Illinois
Engineers—C. F. Murphy Associates, Chicago, Illinois

Preface

The first edition of this reference manual was prepared to aid users in applying the provisions of the 1971 edition of "Building Code Requirements for Reinforced Concrete (ACI 318-71)." The second, third, and fourth editions updated the material in conformity with the provisions of the 1977 Code Edition, the 1980 Code Supplement, and the 1983 Code edition, respectively. Through four editions, much of the initial material has been revised to better emphasize the subject matter and new chapters added to assist the engineer and designer in proper application of the ACI 318 design provisions.

This fifth edition addresses the 1989 edition of "Building Code Requirements for Reinforced Concrete (ACI 318-89)." This edition, for the first time, has been produced using desktop publishing, to give the publication an enhanced professional look and to make it more readable. Although considerable new material has been added, the fifth edition is slightly more compact than the previous edition. Also for the first time in this edition, column and and slab design examples have been worked out using PCA's computer software (programs PCACOL and ADOSS) as well as manually. Readers familiar with the fourth edition of the "Notes" may be interested to note that Part 1 of the current edition (General Requirements) is new; Part 2 through 13 correspond to Parts 1 through 12, respectively, of the previous edition; Part 14 combines the previous Parts 13 (Shear) and 19 (Horizontal Shear in Composite Flexural Member); Parts 15 through 19 correspond to Parts 14 through 18, respectively of the previous edition; Part 20 is new; and Parts 21 through 31 correspond to Parts 20 thorugh 30, respectively, of the fourth edition. Some of the previous text and design examples have been revised to reflect, where possible, comments received from users of the "Notes" who suggested improvements in wording, identified errors, and recommended items for inclusion or deletion.

The primary purpose for publishing this manual is to assist the engineer and designer in the proper application of the ACI 318-89 design standard. The emphasis is placed on "how-to-use" the Code. For complete background information on the development of the Code provisions, the reader is referred to the "Commentary on Building Code Requirements for Reinforced Concrete (ACI 318R-89)" which, for the first time, has been published together with the Code itself under the same cover.

This manual can also be a valuable aid to educators, architects, contractors, materials and products manufacturers, building code authorities, inspectors, and others involved in the design, construction, and regulation of concrete structures.

Although every attempt has been made to impart editorial consistency to the thirty-one chapters, many inconsistencies probably still remain. Some errors, in all probablility, are also to be found. PCA would be grateful to any reader who would bring such errors and inconsistencies to our attention. Other suggestions for improvement are also genuinely welcome.

S.K. Ghosh
Director, Engineered Structures and Codes

Acknowledgments

Gerald B. Neville, now with the International Conference of Building Officials, did the preliminary updating of the fourth edition to ACI 318-89. Sincere thanks are due him for his most conscientious effort.

The publication of this fifth edition would have been impossible without the wholehearted cooperation of the following members of PCA's Engineered Structures and Codes/Computer Programs Marketing staff: Basile G. Rabbat who served as co-editor and was fully or partly responsible for most phases of the project; Vytenis P. Markevicius who did preliminary editing of a number of chapters, and revised and/or produced most of the figures; August W. Domel, Jr. who co-ordinated the later stages of production; Joseph "Jim" Messersmith who did preliminary editing on a number of chapters; Sonya Farmer who produced a part of the manuscript on the desktop publishing system and revised and/or produced a number of figures; and James P. Hurst who collaborated in the production of the final chapter and the table of contents. The dedicated effort of Lisa Vlamis of PCA's Publication Services, who single-handedly produced a major part of the complex manuscript, needs special acknowledgment. Her cheerful devotion to the project was exemplary. Thanks are also due to Susan P. Heyn of Engineered Structues and Codes for her overall supporting role; To Mark Kluver, Jim Messersmith, and Steve Skalko of Engineered Structures and Codes for their review of Part 1, and to Steven H. Kosmatka of Construction Information Services for his review of Part 2. Mary Louise Carr of Computer Programs Marketing helped with the table of contents. Not enough can be said about the thorough professionalism of these fine men and women who performed at their best through periods of pressure that was often intense.

Anthony L. Felder of the Concrete Reinforcing Steel Institute was kind enough to review Parts 3 and 4. Alex Aswad of Penn State, Harrisburg was simiarly kind, and served as a reviewer for Parts 26 and 27. Donald W. Musser of the Tilt-Up Concrete Association graciously reviewed several other chapters. Grateful acknowledgment is due all three of them.

Last but not the least, sincere gratitude must be expressed to the authors of the various parts of the first, second, third and fourth editions of the "Notes", whose initial work is carried over into this edition, although their names are no longer separately identified with the various parts.

S.K. Ghosh

Contents

PART 3 - DETAILS OF REINFORCEMENT

PART 4 - Development And Splices of Reinforcement

PART 5 - DESIGN METHODS AND STRENGTH REQUIREMENTS

PART 6 - GENERAL PRINCIPLES OF STRENGTH DESIGN

PART 7 - DISTRIBUTION OF FLEXURAL REINFORCEMENT

PART 8 - DEFLECTIONS

PART 9 - MOMENT REDISTRIBUTION

PART 10 - DESIGN FOR FLEXURE

PART 14 - SHEAR

PART 15 - TORSION

MISSING

1

General
Requirements

1.1 SCOPE

As the name implies, *Building Code Requirements for Reinforced Concrete* (ACI 318-89) is meant to be adopted by reference in a general building code, to regulate the design and construction of buildings of reinforced concrete. For the '89 Code, Section 1.1.1 has been revised to further emphasize the intent and format of the ACI 318 document and its status as part of a legally adopted general building code. The ACI 318 Code has no legal status unless adopted by a state or local jurisdiction having the police power to regulate building design and construction by a legally appointed Building Official. It is also recognized that when the ACI Code is made part of a legally adopted general building code, that general building code may modify some provisions of ACI 318 to reflect local conditions and requirements. For areas where there is no general building code, there is no law to make ACI 318 the "code." In such cases, the ACI Code may serve as a standard of accepted practice even though it has no legal status.

1.1.6* Structural Plain Concrete

With publication of the 1983 edition of the ACI Code, ACI 318.1 *Building Code Requirements for Structural Plain Concrete*, addressing design and construction of structural members of plain concrete, was incorporated as part of ACI 318. By definition, structural plain concrete is concrete in members that are either unreinforced or contain less reinforcement than the minimum amount specified in ACI 318 for reinforced concrete. ACI 318.1 contains specific design provisions for structural plain concrete walls, footings, and pedestals. The designer should especially note Section 1.2.2; since the structural integrity of plain concrete members depends solely on the properties of the concrete, use of plain concrete should be limited to members that are principally in a state of compression, that is, members that are continuously supported by soil or supported by other structural members capable of providing continuous vertical support.

---

* Section numbers correspond to those of ACI 318-89

1.1.7 Special Provisions for Earthquake Resistance

Prior to publication of the 1989 Code, the special provisions for seismic design were contained in Appendix A to the Code. The appendix location for the seismic design provisions implied that they were not applicable unless specifically adopted as part of the general building code. When local jurisdictions adopt one of the model codes which adopts ACI 318 by reference, they invariably adopt the main body of the ACI 318 Code; adoption of the ACI 318 appendices usually requires additional procedures. For the '89 Code, Appendix A is moved to the main body of the Code as Chapter 21, ensuring adoption of the special seismic design provisions when a jurisdiction adopts the ACI Code by reference as part of its general building code.

1.1.7.1 Regions of Low Seismic Risk—For concrete structures located in regions of low seismic risk (no or minor risk of damage), no special design or detailing is required; thus, the general requirements of the Code, excluding Chapter 21, apply. Concrete structures proportioned by the general requirements of the Code are considered to have a level of toughness adequate for low earthquake intensities.

The designer should be aware that several design provisions of the Code are included specifically to improve toughness, in order to increase resistance of concrete structures to earthquakes and other catastrophic or abnormal loads. For example, when a beam is part of a primary lateral load resisting system, a portion of the positive moment reinforcement must be anchored at supports to develop yield (Section 12.11.2). Similarly, hoop reinforcement must be provided in certain types of beam-column connections (Section 11.11.2). Since publication of the '71 Code, other design provisions, such as those requiring minimum shear reinforcement (Section 11.5.5) and improvements in bar splicing and anchorage details (Chapter 12), also increased toughness and the ability of concrete structures to withstand reversing loads due to earthquakes. For the '89 Code, a new provision addressing special reinforcement for structural integrity (Section 7.13) has been added to enhance the overall integrity of concrete structures in the event of damage to a major supporting element or abnormal loading.

1.1.7.2 Regions of Moderate or High Seismic Risk—For concrete structures located in regions of moderate seismic risk (moderate risk of damage), Section 21.9 includes special reinforcing details that are applicable to reinforced concrete moment frames (beam-column or slab-column framing systems) required to resist earthquake effects. The special reinforcing details will serve to accommodate a suitable level of inelastic behavior, if the frame is subjected to an earthquake of such magnitude as to require it to perform inelastically. There are no special design or detailing requirements for other structural or nonstructural components of buildings in regions of moderate seismic risk, including structural walls (shearwalls) provided to resist earthquake effects. Structural walls proportioned by the general requirements of the Code are considered to have sufficient toughness at drift levels anticipated in regions of moderate seismicity.

The type of framing system provided for earthquake resistance in regions of moderate seismicity will dictate whether the special reinforcing details need to be included in the design.

If the lateral load resisting system consists of moment frames, the special details must be provided. Note that even if the combination of dead, live, and wind loads (Section 9.2.2) govern design over the combination of dead, live, and earthquake loads (Section 9.2.3), the special reinforcing details must still be provided to ensure a limited level of toughness in the moment resisting frames. Whether or not the specified earthquake loads govern design, the frames are the only defense against the effects of an earthquake.

For a combination frame-shearwall structural system, inclusion of the special details will depend on how the earthquake loads are "assigned" to the shearwalls and the frames. If the total earthquake forces are assigned to the shearwalls, the special detailing is not required for the frames. If frame-shearwall interaction is considered in the analysis, with some of the earthquake forces being resisted by the frames, then special details are required to toughen up the frame portion of the dual framing system.

If structural walls resist total vertical and lateral load effects of dead, live, and wind or earthquake loads (shearwall buildings), no special details are required; the general requirements of the Code apply.

For concrete structures located in regions of high seismic risk (major risk of damage), all building components, structural and nonstructural, must satisfy the special proportioning and detailing requirements of Chapter 21. If, for purposes of analysis, some of the frame members are not considered as part of the lateral load resisting system, special consideration is still required in the design and detailing of these frame members (Section 21.8). The special provisions for seismic design of Chapter 21 are intended to provide a monolithic reinforced concrete structure with adequate toughness to respond inelastically under severe earthquake motions.

1.1.7.3 Seismic Risk Level Specified in General Building Code—The Code addresses levels of seismic risk as "low," "moderate," or "high." The Commentary indicates that definitions of low, moderate, and high seismic risk are not precise. Seismic risk levels are under the jurisdiction of general building codes, and are usually designated by zones (i.e., areas of equal probability of risk of damage, related to intensity of ground shaking). The model codes specify which sections of Chapter 21 must be satisfied, based on the seismic zone. Also, in some cases, the model codes require that important structures in lower risk areas satisfy provisions normally applied in higher risk areas. As a guide, in the absence of specific requirements in the general building code, seismic risk levels and seismic zones generally correlate as follows:

Seismic Risk Level	Seismic Zone
Low	0 and 1
Moderate	2
High	3 and 4

In the absence of a general building code that addresses earthquake loads and seismic zoning, it is intended that local authorities (engineers and geologists) should decide on the need and proper application of the special provisions of Chapter 21 for seismic design. The model code predominately used in the area may be used as a guide.

With publication of the 1982 edition of ANSI A58.1, *Minimum Design Loads for Buildings and Other Structures,*[1.1] a new seismic zoning map was adopted. The new zoning map, based on a uniform probability of earthquake occurrence not considered in earlier zoning maps, placed some of the major metropolitan areas of the U.S. in different seismic zones, as compared with the previous ANSI (1972) map. Some major U.S. cities that are located in Seismic Zone 2, indicating a potential for moderate earthquake damage, according to the ANSI A58.1-1982 seismic map, are listed in Table 1-1.

Table 1-1 – Cities Located in Seismic Zone 2
(ANSI A58.1-1982)

Albany[1]	Hartford[1]
Atlanta[2]	Kansas City[2]
Boston[1]	New York[2]
Buffalo[3]	Providence[1]
Charlotte[2]	St. Louis[2]

Many cities or metropolitan areas are located near boundary lines of the ANSI A58.1 seismic zones. A listing of some major cities located along the seismic zone lines is given in Table 1-2.

Table 1-2 – Cities Along Seismic Zone Boundaries
(ANSI A58.1-1982)

City, State	Seismic Zones	City, State	Seismic Zone
Atlanta, GA	1,2[4]	New York, NY	1,2
Baltimore, MD	0,1	Norfolk, VA	0,1
Bangor, ME	1,2	Oklahoma City, OK	1,2
Burlington, VT	1,2	Omaha, NE	1,2
Charlotte, NC	1,2	Philadelphia, PA	1,2
Chicago, IL	0,1	Richmond, VA	0,1
Dallas, TX	0,1	Roanoke, VA	1,2
Kansas City, MO	1,2	Savannah, GA	1,2
Knoxville, TN	1,2	St. Louis, MO	1,2
Lincoln, NE	1,2	Tulsa, OK	1,2
Memphis, TN	2,3	Washington, DC	0,1

The above listed metropolitan areas represent high concentrations of buildings and structures, with the impact of higher seismic zone assignment on design being potentially significant. The designer will need to refer to the local building code for seismic risk levels that need to be considered in design of reinforced concrete structures.

1.3 INSPECTION

The ACI Code requires that concrete construction be inspected as required by the legally adopted general building code. Clarification that the minimum level of inspection required is determined by the legally adopted building code was added to the '89 Code.

In the absence of any inspection requirements, the provisions of Section 1.3 may serve as a guide to providing an acceptable level of inspection and inspection responsibility for concrete construction.

---

1. Located in Zone 2 of ANSI A58.1-1972 map.
2. Located in Zone 1 of ANSI A58.1-1972 map.
3. Located in Zone 3 of ANSI A58.1-1972 map.
4. Underlining indicates zone in which the city is located.

The three model building codes[1.2-1.4], adopted extensively in the U.S. to regulate building design and construction, require inspections of concrete construction in varying degrees. However, administrative provisions such as these are frequently amended when the model code is adopted locally. The engineer should refer to the specific inspection requirements contained in the local building code having jurisdiction over the construction.

In addition to periodic inspections by the code official, special inspections of concrete structures by a special inspector may be required. The engineer should check the local building code or with the local building official to ascertain if special inspection requirements exist within a specific jurisdiction. Degree of inspection and inspection responsibility should be set forth in the contract documents.

1.3.4 Records of Inspection

Inspectors and inspection agencies will need to be aware of the wording of Section 1.3.4. Records of inspection must be preserved for two years after completion of a project, or longer if required by the legally adopted building code. Preservation of inspection records for a minimum two-year period after completion of a project is to ensure that records are available, should disputes or discrepancies arise subsequent to owner acceptance or issuance of certificate of occupancy, concerning workmanship or any violations of the approved plans, specifications, or the general building code.

1.3.5 Special Inspections

For the '89 Code, continuous inspection by a specially qualified inspector is required for placement of all reinforcement and concrete for moment frames (beam and column framing systems resisting earthquake-induced forces) located in regions of high seismic risk (Seismic Zones 3 and 4) and designed according to the special provisions of Chapter 21 for seismic design. In addition, the specially qualified inspector must be under the supervision of the engineer or architect responsible for the design.

The new special inspection requirement is patterned after a similar provision contained in the *Uniform Building Code*. According to the UBC, the specially qualified inspector must "demonstrate competence for inspection of the particular type of construction requiring special inspection." Duties and responsibilities of the special inspector are further outlined as follows:

1. Observe the work for conformance with the design drawings and specifications.

2. Furnish inspection reports to the building official, the engineer or architect of record, and other designated persons.

3. Submit a final inspection report indicating whether the work was in conformance with the design drawings and specifications and acceptable workmanship.

REFERENCES

1.1 *Minimum Design Loads for Buildings and Other Structures,* (ANSI A58.1-1982), American National Standards Institute, New York, 1982, and (ANSI/ASCE 7-88), American Society of Civil Engineers, New York, 1989.

1.2 *The BOCA National Building Code,* 1990 edition, Building Officials and Code Administrators International, Inc., Country Club Hills, IL.

1.3 *Standard Building Code,* 1988 edition (including 1989 Revisions), Southern Building Code Congress International, Birmingham, AL.

1.4 *Uniform Building Code,* 1988 edition (including 1989 Supplement), International Conference of Building Officials, Whittier, CA.

Materials, Concrete Quality

CHAPTER 3 — MATERIALS

3.1   TESTS OF MATERIALS

Testing agencies will need to be aware of the wording of Section 3.1.3. Records of tests of materials and of concrete must be preserved for two years after completion of a project, or longer if required by the legally adopted general building code. Preservation of test records for a minimum two-year period after completion of a project is to ensure that records are available, should questions arise (subsequent to owner acceptance or issuance of certificate of occupancy) concerning quality of materials and of concrete, or concerning any violations of the approved plans and specifications or of the general building code.

3.2   CEMENTS

The cement used in the work must correspond to the type upon which the selection of concrete proportions for strength and other properties has been based. This may simply mean the same type of cement or it may mean cement from the same source. In the case of a plant that has determined the standard deviation from tests involving cements from several sources, the former would apply. The latter would be the case if the standard deviation of strength tests used in establishing the required target strength was based on one particular type of cement from one particular source.

3.3   AGGREGATES

The nominal maximum aggregate size is limited to (i) one-fifth the narrowest dimension between sides of forms, (ii) one-third the depth of the slab, and (iii) three-quarters the minimum clear spacing between reinforcing bars or prestressing tendons or ducts. Note that the limitations on nominal maximum aggregate size may be waived if, in the judgment of the engineer, the workability and methods of consolidation of the concrete are such that the concrete can be placed without honeycomb or void. The engineer in charge of inspection must decide whether the limitations on maximum size of aggregate may be waived.

3.4 WATER

Water used in reinforced concrete (prestressed or nonprestressed) or in concrete that is to have aluminum embedments or is to be cast against stay-in-place galvanized metal forms should not contain deleterious amounts of chlorine ion. Concern over a high chloride content in mixing water is chiefly due to the possible effect of chloride ions on the corrosion of embedded reinforcing steel or prestressing tendons. Limits on chloride ion content contributed from the ingredients including water, aggregates, cement, and admixtures are given in Chapter 4, Table 4.3.1. An in-depth discussion of this and many other topics covered in Chapters 3 through 5 of the Code may be found in Ref. 2.1.

Seawater is not suitable for use in making reinforced concrete, due to risk of corrosion of the reinforcement. Such corrosion is accelerated in warm and humid environments.

3.5 METAL REINFORCEMENT

3.5.2 Welding of Reinforcement

Guidelines on welding to existing reinforcing bars are given in Commentary Section 3.5.2. For the '89 Code, additional guidelines on field welding of cold drawn wire are added. The welding of wire is not covered in AWS D1.4. The welding of cold drawn wire may significantly reduce its yield strength and ductility. Cold drawn wire is used as spiral reinforcement, and wires or wire fabric may occasionally be welded for various reasons. The engineer will need to specify procedures or performance criteria if welding of wire or wire fabric is to be required on a project.

3.5.3 Deformed Reinforcement

Only deformed reinforcement as defined in Chapter 2 may be used for nonprestressed reinforcement, except that plain bars and plain wire may be used for spiral reinforcement. Welded plain wire fabric is included under the Code definition of deformed reinforcement. Plain wire fabric bonds to concrete by positive mechanical anchorage at each wire intersection. Deformed wire fabric utilizes wire deformations plus welded intersections for bond and anchorage. This difference in bond and anchorage for the plain vs. deformed fabric is reflected in the development and lap splice provisions of Chapter 12.

Reinforcing bars rolled to ASTM A615 specifications (billet steel) are the most commonly specified for construction. Rail and axle steels (ASTM A616 and A617) are not generally available, except in a few areas of the country.

ASTM A706 covers low-alloy steel deformed bars (Grade 60 only) intended for special applications where welding or bending or both are of importance. Reinforcing bars conforming to A706 should be specified wherever critical or extensive welding of reinforcement is required, and for use in reinforced concrete structures located in regions of high seismic risk where more bendability and controlled ductility are required. The special provisions of Chapter 21 for seismic design require that reinforcement resisting earthquake-induced flexural and axial forces in frame members and in wall boundary elements forming parts of structures located in regions of high seismic risk must comply with ASTM A706 (Section 21.2.5). Before specifying A706 reinforcement, local availability should be investigated. Most rebar producers can make A706 bars, but generally not in quantities less than one heat of steel for each bar size ordered. A heat of steel varies from 50 and 200 tons, depending on the mill. A706 in lesser quantities of single bar sizes may not be immediately available from any single producer.

Section 9.4 permits designs based on a yield strength of reinforcement up to a maximum of 80,000 psi. Currently there is no ASTM specification for a Grade 80 reinforcement. However, deformed reinforcing bars #11, #14, and #18 with a yield strength of 75,000 psi (Grade 75) are included in the 1987 edition of the ASTM A615 specification. The provisions in the A615-87 specification for Grade 75 bars are compatible with the ACI Code requirements, including the requirement that yield strength correspond to a strain of 0.35 percent (Section 3.5.3.2). The designer may simply specify Grade 75 to meet A615-87 and no special requirements, or he may delineate exceptions. Certified mill test reports should be obtained from the supplier when Grade 75 bars are used. Before specifying Grade 75, local availability should be investigated. The higher yield strength #11, #14, and #18 bars are intended primarily as column reinforcement. They are used in conjunction with higher strength concrete to reduce the size of columns in high-rise buildings, and in other applications where high capacity columns are required. For welded wire fabric, yield strength above 60,000 psi is available; however, the Code assigns a yield strength value of 60,000 psi, while making provisions for use of higher yield strengths, provided the stress used in design corresponds to a strain of 0.35 percent.

Since publication of the 1977 ACI Code, two exceptions to the ASTM deformed reinforcing bar specifications have been prescribed:

1. Yield strength of bars must correspond to that determined by tests on full-size bars. Tests of reduced-section specimens may indicate higher strength than tests on actual full-size bars;

2. Bend tests of bars must be made on full-size bars bent around a pin generally smaller than the minimum bend diameters permitted in construction (Section 7.2).

Current editions of the ASTM deformed reinforcing bar specifications referenced in Section 3.5.3.1 cover the original ACI Code exceptions as follows:

For A615 (billet steel), the two exceptions were incorporated into the main body of ASTM A615-85. Thus, the more restrictive requirements for bend tests, and the requirement that tensile tests be performed on full-size bars are now an integral part of A615. The specification also requires all billet steel reinforcing bars to be marked with the letter "S." For code conformance, the inspector need only check for the "S" marking on bars supplied to the job site.

For A616 (rail steel), the exception requiring tighter bend tests is contained in Supplementary Requirement S1. The exception requiring that yield strength be based on tests of full-size bars is part of A616. Note: The S1 requirement is optional and applies only when specified by the purchaser. Thus, to comply with the ACI Code A616 reinforcing bars must be specified as conforming to ASTM Specification A616 plus Supplementary Requirement S1 — ASTM A616-85 (S1). Bars meeting the S1 requirement will be marked with the letter "R," in addition to the rail symbol. For code conformance, the inspector need only check for the "R" marking on the bars supplied to the job site.

For A617 (axle steel), the two exceptions are included directly in the main body of the specification.

For A706 (low-alloy steel), both exceptions are satisfied; full-size bar specimens are required for tension testing, and the required bend test pin diameters are compatible with the minimum bend diameters specified in Section 7.2.

3.5.3.7 Coated Reinforcement—Appropriate references to the ASTM specifications for coated reinforcement, A767 (galvanized) and A775 (epoxy-coated), are included in the Code to reflect increased usage of coated, especially epoxy-coated, reinforcement for corrosion protection. Epoxy-coated reinforcing bars provide a viable corrosion protection system for reinforced concrete structures. In recent years, usage of epoxy-coated bars has spread to many types of reinforced concrete structures such as parking garages (exposed to deicing salts), wastewater treatment plants, marine structures, and other facilities located near coastal areas where the risk of corrosion of reinforcement is high because of exposure to seawater—particularly if the climate is warm and humid.

Designers specifying epoxy-coated reinforcing bars should clearly outline in the project specifications special hardware and handling methods to minimize damage to the epoxy coating during handling, transporting, and placing coated bars, and placing of concrete:[2.2]

1. Nylon lifting slings should be used, or wire rope slings should be padded.

2. Spreader bars should be used for lifting bar bundles, or bundles should be lifted at the third points with nylon or padded slings. Bundling bands should be made of nylon, or be padded.

3. Coated bars should be stored on padded or wooden cribbing.

4. Coated bars should not be dragged over the ground, or over other bars.

5. Walking on coated bars during or after placing should be held to a minimum, and tools or other construction materials should not be dropped on bars in place.

6. Bar supports should be of an organic material or wire bar supports should be coated with an organic material such as epoxy or vinyl compatible with concrete.

7. Epoxy- or plastic-coated tie wire, or nylon-coated tie wire should be used to minimize damage or cutting into the bar coating.

8. Concrete conveying and placing equipment should be set up, supported, and moved carefully to minimize damage to the bar coating.

Project specifications should also address field touch-up of the epoxy coating after bar placement. Permissible coating damage and repair are included in the A775 specification and in Ref. 2.3. Reference 2.4 contains suggested project specification provisions for epoxy-coated reinforcing bars.

The designer should be aware that epoxy-coated reinforcement requires increased development and splice lengths for bars in tension (Section 12.2.4.3).

CHAPTER 4 – DURABILITY REQUIREMENTS

UPDATE FOR THE '89 CODE

Chapters 4 and 5 have been completely reformatted for the '89 Code to emphasize the importance of the special exposure requirements on concrete durability. The special exposure requirements contained in Section 4.5 of the '83 Code are relocated to a new Chapter 4—Durability Requirements. The code provisions for concrete proportioning and strength evaluation, Chapter 4 in the '83 Code, are included in a new Chapter 5—Concrete Quality, Mixing And Placing. As in previous code editions, selection of concrete proportions must be established to provide:

(a) resistance to special exposures as required by Chapter 4, and

(b) conformance with the strength requirements of Chapter 5.

Resistance to special exposures are addressed in Section 4.1—Freeze-thaw exposure, Section 4.2—Sulphate exposure, and Section 4.3—Corrosion of reinforcement. Conformance with strength test requirements is addressed in Section 5.6—Evaluation and acceptance of concrete. Depending on design and exposure requirements, the lower of the water-cement ratios required for concrete durability (Chapter 4) and for concrete strength (Chapter 5) must be specified (see Example 2.2).

With the new Chapter 4, greater emphasis is placed on the special exposure requirements for improved concrete durability. Unacceptable deterioration of concrete structures in many areas due to severe exposure to freezing and thawing, to deicing salts used for snow and ice removal, to sulphate in soil and water, and to chloride exposure, warranted a stronger code emphasis on the special exposure requirements. The new Chapter 4 directs special attention to the need for considering concrete durability, in addition to concrete strength.

In the context of the Code, durability refers to the ability of concrete to resist deterioration from the environment or service in which it is placed. Properly designed and constructed concrete should serve its intended function without significant distress throughout its service life. The Code, however, does not include provisions for especially severe exposures such as to acids or high temperatures, nor is it concerned with aesthetic considerations such as surface finishes. Items like these, which are beyond the scope of the Code, must be covered specifically in the project specifications. Concrete ingredients and proportions must be selected to meet the minimum requirements stated in the Code and the additional requirements of the contract documents.

In addition to the proper selection of cement, adequate air entrainment, maximum water-cement ratio, and limiting chloride ion content of the materials, as set forth in Chapter 4, other requirements for durable concrete exposed to adverse environments such as low slump, adequate consolidation, uniformity, adequate cover of reinforcement, and sufficient moist curing to develop the potential properties of the concrete, are essential.

4.1 FREEZING AND THAWING EXPOSURES

For concrete that will be exposed to moist freezing and thawing or deicer salts, air-entrained concrete must be specified with minimum air contents as set forth in Table 4.1.1 for severe and moderate exposures. Severe exposure is an environment in which concrete is exposed to wet freeze-thaw conditions or to deicing salts, or other aggressive agents. Moderate exposure is an environment in which the concrete is exposed to freezing, but will not be continually moist, nor exposed to water for long periods before freezing, and will not be in contact with deicers or aggressive chemicals. Project specifications should allow the air content of the delivered concrete to be within (−1.5) and (+1.5) percentage points of Table 4.1.1 target values.

Use of intentionally entrained air significantly improves resistance of hardened concrete to freezing when exposed to water and deicing salts. Concrete that is dry or contains only a small amount of moisture is essentially not affected by even a large number of cycles of freezing and thawing. Sulfate resistance is also improved by air entrainment.

The entrainment of air in concrete can be accomplished by adding an air-entraining admixture at the mixer, by using an air-entraining cement, or by a combination of both. Air-entraining admixtures, added at the mixer, must conform to ASTM C260 (Code Section 3.6.4); air-entraining cements must comply with the specifications in ASTM C150 and C595 (Code Section 3.2.1). Air-entraining cements are sometimes difficult to obtain; and their use has been decreasing as the popularity of air-entraining admixtures has increased. ASTM C94, Standard Specifications for Ready Mixed Concrete, which is adopted by reference by the ACI Code, requires that air content tests by conducted. The frequency of these tests is the same as required for strength evaluation. Samples must be obtained and tested in accordance with ASTM C172.

Freeze-thaw resistance is also significantly improved when concrete has a low water-cement ratio and a minimum cement content. Concrete that will be exposed to wet freeze-thaw conditions and exposed to deicing salts must be proportioned so that the water-cement ratio as set forth in Table 4.1.2 for the exposure conditions indicated is not exceeded. A minimum strength is specified for lightweight aggregate concrete due to the variable absorption characteristics of lightweight aggregates, which makes calculation of water-cement ratio impractical. A minimum cement content of 520 lb per cu yd must also be specified for concrete exposed to freezing and thawing in the presence of deicers. The minimum cement content is a new requirement for the '89 Code.

Concrete used in water-retaining structures or exposed to weather or other severe exposure conditions must be virtually impermeable or watertight. Low permeability not only improves freezing and thawing resistance, especially in the presence of deicing salts, but also improves the resistance of concrete to chloride ion penetration. Concrete that is intended to have low permeability to water must be proportioned so that the water-cement ratio does not exceed 0.50. If concrete is to be exposed to freezing and thawing in a moist condition, the specified water-cement ratio must be no more than 0.45. Also, for corrosion protection of reinforcement in concrete exposed to deicing salts (wet freeze-thaw conditions), and in concrete exposed to seawater (including seawater spray, particularly in warm and humid climates), the concrete must be proportioned so that the water-cement ratio does not exceed 0.40.

4.2 SULFATE EXPOSURES

Sulfate attack of concrete can occur when it is exposed to soil, seawater, or groundwater having a high sulfate content, unless measures to reduce sulfate attack, such as the use of sulfate-resistant cement, have been taken. The susceptibility to sulfate attack is greater for concrete exposed to moisture, such as in foundations and slabs on ground, and in structures directly exposed to seawater. For concrete that will be exposed to sulfate attack from soil and water, sulfate-resisting cement must be specified. Table 4.2.1 lists the appropriate types of sulfate-resisting cements and maximum water-cement ratios for various exposure conditions. Degree of exposure is based on the amount of water-soluble sulfate concentration in soil or on the amount of sulfate concentration in water. Note that Table 4.2.1 lists seawater under "moderate exposure," even though it generally contains more than 1500 ppm of sulfate concentration. The reason is that the presence of chlorides in seawater inhibits the expansive reaction that is characteristic of sulfate attack.[2.1]

In selecting a cement type for sulfate resistance, the principal consideration is the tricalcium aluminate (C_3A) content. Cements with low percentages of C_3A are especially resistant to soils and waters containing sulfates. Where precaution against moderate sulfate attack is important, as in drainage structures where sulfate concentrations in groundwater are higher than normal but not necessarily severe (0.10 – 0.20 percent), Type II portland cement (maximum C_3A content of eight percent under ASTM C150) should be specified.

Type V portland cement should be specified for concrete exposed to severe sulfate attack — principally where soils or groundwaters have a high sulfate content. The high sulfate resistance of Type V cement is attributed to its low tricalcium aluminate content (maximum C_3A content of five percent).

Certain blended cements provide sulfate resistance. Other types of cement produced with low C_3A contents are usable in cases of moderate to severe sulfate exposure. Sulfate resistance also increases with air-entrainment and increasing cement contents (low water-cement ratios).

Before specifying a sulfate resisting cement, its availability should be checked. Type II cement is usually available, especially in areas where moderate sulfate resistance is needed. Type V cement is available only in particular areas where it is needed to resist high sulfate environments. Blended cements may not be available in many areas.

4.3 CORROSION OF REINFORCEMENT

Chlorides can be introduced into concrete through its ingredients: mixing water, aggregates, cement, and admixtures, or through exposure to deicing salts, seawater, or salt-laden air in coastal environments. The chloride ion content limitations of Table 4.3.1 are to be applied to the chlorides contributed by the concrete ingredients, not to chlorides from the environment surrounding the concrete (chloride ion ingress).

In addition to a high chloride content, oxygen and moisture must be present to induce the corrosion process. The availability of oxygen and moisture adjacent to the embedded steel will vary with service exposure from one structure to another, and between different parts of the same structure.

If significant amounts of chlorides may be introduced into the hardened concrete from the concrete materials to be used, the individual concrete ingredients, including water, aggregates, cement, and any admixtures, must be tested to ensure that the total chloride ion concentration contributed from the ingredients does not exceed the limits of Table 4.3.1. These limits have been established to provide a threshold level, to avoid corrosion of the embedded reinforcement at the outset, prior to service exposure. Chloride limits for corrosion protection also depend upon the type of construction and the environment to which the concrete is exposed during its service life, as indicated in Table 4.3.1.

Chlorides are present in variable amounts in all of the ingredients of concrete. Both water soluble and insoluble chlorides exist; however, only water soluble chlorides induce corrosion. Tests are available for determining either the water soluble chloride content or the total (soluble plus insoluble) chloride content. The test for soluble chloride is more time-consuming and difficult to control, and is therefore more expensive than the test for total chloride. An initial evaluation of chloride content may be obtained by testing the individual concrete ingredients for total (soluble plus insoluble) chloride content. If the total chloride ion content is less than that permitted by Table 4.3.1, obviously water-soluble chloride need not be determined. If the total chloride content exceeds the permitted value, testing of samples of the hardened concrete for water-soluble chloride content will need to be performed for direct comparison with Table 4.3.1 values. Some of the soluble chlorides in the ingredients will react with the cement during hydration and become insoluble, further reducing the soluble chloride ion content, the corrosion-inducing culprit. Of the total chloride ion content in hardened concrete, only about 50 to 85 percent is water soluble; the rest is insoluble. Note that hardened concrete should be at least 28 days of age before sampling.

Chlorides are among the more abundant materials on earth, and are present in variable amounts in all of the ingredients of concrete. Potentially high chloride-inducing materials and conditions might be: use of seawater as mixing water or as washwater for aggregates, since seawater contains significant amounts of sulfates and chlorides; use of marine-dredged aggregates, since such aggregates often contain salt from the seawater; use of aggregates that have been contaminated by salt-laden air in coastal areas; use of admixtures containing chloride, such as calcium chloride; and use of deicing salts where salts may be tracked onto parking structures by vehicles. The engineer will need to be cognizant of the potential hazard of chlorides to concrete in marine environments or other exposures to soluble salts. Research has shown that the threshold value for a water soluble chloride content of concrete necessary for corrosion of embedded steel can be as low as 0.15 percent by weight of cement. When chloride content is above this threshold value, corrosion is likely if moisture and oxygen are readily available. If chloride content is below the threshold value, the risk of corrosion is low.

Depending on the type of construction and the environment to which it is exposed during its service life, and the amount and extent of protection provided to limit chloride ion ingress, the chloride level in concrete may increase with age and exposure. Protection against chloride ion ingress from the environment is addressed in Section 4.3.2, with reference to Table 4.1.2. A maximum water-cement ratio of 0.40 must be provided for corrosion protection of "concrete exposed to deicing salts, brackish water, seawater or spray from these sources." Resistance to corrosion of embedded steel is also improved with an increase in the thickness of concrete cover. Code Commentary Section 7.7.5 recommends a minimum concrete cover of 2 in. for cast-in-place walls and slabs, and 2½ in. for other members, when concrete will be exposed to external sources of chlorides in service. For plant-produced precast members, the corresponding recommended minimum concrete covers are 1½ in. and 2 in., respectively.

Other methods of reducing environmentally caused corrosion include the use of epoxy-coated reinforcing steel, corrosion-inhibiting admixtures, surface treatments, and cathodic protection. Epoxy coating of reinforcement prevents chloride ions from reaching the steel. Corrosion-inhibiting admixtures attempt to chemically arrest the corrosive reaction. Surface treatments attempt to stop or reduce chloride ion penetration at the concrete surface. Cathodic protection methods reverse the corrosion current flow through the concrete and reinforcing steel.

CHAPTER 5 – CONCRETE QUALITY, MIXING, AND PLACING

The following discussion of Code Chapter 5 addresses the selection of concrete mixture proportions for strength, based on probabilistic concepts.

5.1.1 Concrete Proportions for Strength

It is emphasized in Section 5.1.1 that the average strength of concrete produced must always exceed the specified value of f_c' that was used in the structural design phase. This is based on probabilistic concepts, and is intended to ensure that adequate concrete strength will be developed in the structure.

5.1.3 Test Age for Strength of Concrete

Section 5.1.3 permits f_c' to be based on tests at ages other than the customary 28 days. If other than 28 days, the test age for f_c' must be indicated on the design drawings or specifications. Higher strength concretes, exceeding 6,000 psi compression strength, are often used in tall buildings; and concretes having strengths of 20,000 psi have been used. Test ages for the higher strength concretes can justifiably be higher than the customary 28 days. In high-rise structures requiring high-strength concrete, the process of construction is such that the columns of the lower floors are not fully loaded until a year or more elapses after commencement of construction. For this reason, compressive strengths based on 56- or 90-day test results are commonly specified.

5.2 SELECTION OF CONCRETE PROPORTIONS

Recommendations for proportioning concrete mixtures are given in detail in *Design and Control of Concrete Mixtures*.[2.1] Recommendations for selecting proportions for concrete are also given in detail in "Standard Practice for Selecting Proportions for Normal, Heavyweight, and Mass Concrete" (ACI 211.1)[2.5] and "Standard Practice for Selecting Proportions for Structural Lightweight Concrete" (ACI 211.2).[2.6]

The selected water-cement ratio must be low enough, or the compressive strength high enough (for lightweight concrete), to satisfy both the strength criteria (Sections 5.3 or 5.4) and the special exposure requirements (Chapter 4).

The Code emphasizes the use of field experience or laboratory trial batches (Section 5.2) as the preferred method for selecting concrete mixture proportions. When no prior experience or trial batch data are available, permission may be granted by the project engineer to base concrete proportions on water-cement ratio limits prescribed in Section 5.4.

5.3 PROPORTIONING ON THE BASIS OF FIELD EXPERIENCE AND/OR TRIAL MIXTURES

5.3.1 Standard Deviation

For establishing concrete mixture proportions, emphasis is placed on the use of laboratory trial batches or field experience as the basis for selecting the required water-cement ratio. The Code emphasizes a statistical approach to establishing the target strength f_{cr}' required to ensure attainment of the specified compressive strength, f_c', used in the structural design. If an applicable standard deviation, s, from strength tests of the concrete is known, this establishes the target strength level for which the concrete must be proportioned. Otherwise, the proportions must be selected to produce a conservative target strength sufficient to allow for a high degree of variability in strength test results. For background information on statistics as it relates to concrete, see "Recommended Practice for Evaluation of Compression Test Results of Concrete"[2.7] and "Statistical Product Control."[2.8]

Concrete used in background tests to determine standard deviation is considered to have been "similar" to that specified, if it was made with the same general types of ingredients, under no more restrictive conditions of control over material quality and production methods than are specified to exist on the proposed work, and if its specified strength did not deviate by more than 1000 psi from the f_c' specified. A change in the type of concrete or a significant increase in the strength level may increase the standard deviation. Such a situation might occur with a change in the type of aggregate; i.e., from natural aggregate to lightweight aggregate or vice versa, or with a change from non-air-entrained concrete to air-entrained concrete. Also, there may be an increase in standard deviation when the average strength level is raised by a significant amount, although the increment in standard deviation should be somewhat less than directly proportional to the strength increase. When there is reasonable doubt as to its reliability, any estimated standard deviation used to calculate the required average strength should always be on the conservative (high) side.

Statistical methods provide valuable tools for assessing the results of strength tests. It is important that concrete technicians understand the basic language of statistics and be capable of effectively utilizing its tools to evaluate strength test results.

Figure 2-1 illustrates several fundamental statistical concepts. Data points represent six (6) strength test results[1] from consecutive tests on a given class of concrete. The horizontal line represents the average of tests which is designated $\overline{X}$. The average is computed by adding all test values and dividing by the number of values summed; i.e., in Fig. 2-1:

---

1. A strength test result is the average of the strengths of two cylinders made from the same batch of concrete and tested at the same time.

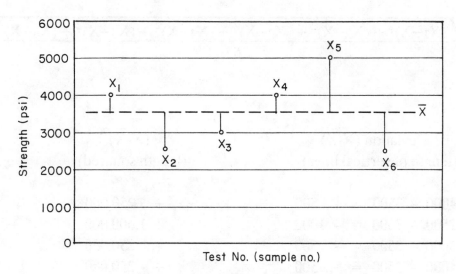

Fig. 2-1   Illustration of Statistical Terms

$$\overline{X} = (4000 + 2500 + 3000 + 4000 + 5000 + 2500)/6 = 3500 \text{ psi}$$

The average $\overline{X}$ gives an indication of the overall strength level of the concrete tested.

It would also be informative to have a single number which would represent the variability of the data about the average. The up and down deviations from the average (3500 psi) are given as vertical lines in Fig. 2-1. If one were to accumulate the total length of the vertical lines without regard to whether they are up or down, and divide that total length by the number of tests, the results would be the average length, or the average distance from the average strength: (500 + 1000 + 500 + 500 + 1500 + 1000)/6 = 833 psi. This is one measure of variability. If concrete test results were quite variable, the vertical lines would be long. On the other hand, if the test results were close, the lines would be short.

In order to emphasize the impact of a few very high or very low test values, statisticians recommend the use of the square of the vertical line lengths. The square root of the sum of the squared lengths divided by one less than the number of tests (some texts use the number of tests) is known as the standard deviation. This measure of variability is commonly designated by the letter s. Mathematically, s is expressed as:

$$s = \sqrt{\frac{\Sigma (X - \overline{X})^2}{n-1}}$$

where: s = standard deviation, psi
           Σ indicates summation
           X = an individual strength test result, psi
           $\overline{X}$ = average strength, psi
           n = number of tests

For example, for the data in Fig. 2-1, the standard deviation would be:

$$s = \sqrt{\frac{(X_1-\overline{X})^2 + (X_2-\overline{X})^2 + (X_3-\overline{X})^2 + (X_4-\overline{X})^2 + (X_5-\overline{X})^2 + (X_6-\overline{X})^2}{6-1}}$$

or:

Deviation $(X-\overline{X})$ (length of vertical lines)	$(X-\overline{X})^2$ (length squared)
$4000 - 3500 = + 500$	$+ \quad 250,000$
$2500 - 3500 = - 1000$	$+ 1,000,000$
$3000 - 3500 = - 500$	$+ \quad 250,000$
$4000 - 3500 = + 500$	$+ \quad 250,000$
$5000 - 3500 = + 1500$	$+ 2,250,000$
$2500 - 3500 = - 1000$	$+ 1,000,000$
	Total $\quad + 5,000,000$

$$s = \sqrt{\frac{5,000,000}{5}} = 1,000 \text{ psi (a very large value)}$$

Obviously, it would be time consuming to actually calculate s in the manner described above. There are many short-cut methods that can be used. Example 2.2 illustrates a relatively easy method.

The coefficient of variation V is simply the standard deviation expressed as a percentage of the average value. The mathematical formula is:

$$V = \frac{s}{\overline{X}} \times 100\%$$

For the test results of Fig. 2-1:

$$V = \frac{1000}{3500} \times 100 = 29\%$$

Standard deviation may be computed either from a single group of successive tests of a given class of concrete or from two groups of such tests. In the latter case, a **statistical average** value of standard deviation is to be used, calculated by usual statistical methods as follows:

$$s_3 = \sqrt{\frac{(n_1 - 1)(s_1)^2 + (n_2 - 1)(s_2)^2}{n_{total} - 2}}$$

where: n_1 = number of samples in group 1

n_2 = number of samples in group 2

s_1 or s_2 is calculated as follows:

$$s = \sqrt{\frac{(X_1 - \overline{X})^2 + (X_2 - \overline{X})^2 + \ldots + (X_n - \overline{X})^2}{n - 1}}$$

For ease of computation,

$$s = \sqrt{\frac{X_1^2 + X_2^2 + X_3^2 + \ldots + X_n^2 - n\,\overline{X}^2}{n - 1}}$$

or $$s = \sqrt{\frac{\left(X_1^2 + X_2^3 + X_3^2 + \ldots + X_n^2\right) - \dfrac{(X_1 + X_2 + X_3 + \ldots + X_n)^2}{n}}{n - 1}}$$

where $X_1, X_2, X_3, \ldots X_n$ are the strength results of individual specimens and n is the total number of specimens tested.

5.3.2   Required Average Strength

Where the concrete production facility has a record based on at least 30 consecutive strength tests representing materials and conditions similar to those expected (or a record based on 15 to 29 consecutive tests with the calculated standard deviation modified by the applicable factor from Table 5.3.1.2), the strength used as the basis for selecting concrete proportions must be the larger of:

$$f'_{cr} = f'_c + 1.34s \tag{5-1}$$

or $$f'_{cr} = f'_c + 2.33s - 500 \tag{5-2}$$

If the standard deviation is unknown, the required average strength f'_{cr} used as the basis for selecting concrete proportions must be determined from Table 5.3.2.2:

For f'_c less than 3000 psi $f'_{cr} = f'_c + 1000$ psi
      between 3000 and 5000 psi $f'_{cr} = f'_c + 1200$ psi
      greater than 5000 psi $f'_{cr} = f'_c + 1400$ psi

Formulas for calculating the required target strengths are based on the following criteria:

(1) A probability of 1 in 100 that an average of 3 consecutive strength tests will be below the specified strength, f'_c ($f'_{cr} = f'_c + 1.34s$), and

(2) A probability of 1 in 100 that an individual strength test will be more than 500 psi below the specified strength, f'_c ($f'_{cr} = f'_c + 2.33s - 500$).

Criterion (1) will produce a higher required target strength than Criterion (2) for low to moderate standard deviations, up to 500 psi. For higher standard deviations, however, Criterion (2) will govern.

The indicated average strength levels are intended to reduce the probability of concrete strength being questioned on the following usual bases: (1) strength averaging below specified f'_c for an appreciable period (three consecutive tests); or (2) an individual test being disturbingly low (more than 500 psi below specified f'_c).

5.3.3 Documentation of Average Strength

Mix approval procedures are necessary to ensure that the concrete furnished will actually meet the strength requirements. The steps in a mix approval procedure can be outlined as follows:

1. Determine the expected standard deviation from past experience.

   (a) This is done by submitting a record of 30 consecutive tests made on a similar mix.

   (b) If it is difficult to find a job with 30 tests, the standard deviation can be computed from two jobs, if the total number of tests exceeds 30. The standard deviations are computed separately and then averaged by the statistical averaging method already described.

2. Use the standard deviation to select the appropriate target strength from the larger of Eqs. (5-1) and (5-2).

   (a) For example, if the standard deviation is 450 psi, then overdesign must be by the larger of 1.34(450) = 603 psi or 2.33(450) − 500 = 549 psi. Thus, for a 3000 psi specified strength, the average strength used as a basis for selecting concrete mixture proportions must be 3600 psi.

   (b) Note that if no acceptable test record is available, the average strength must be 1200 psi greater than f'_c (i.e., 4200 psi average for a specified 3000 psi concrete), see Table 5.3.2.2.

3. Furnish data to document that the mix proposed for use will give the average strength needed. This may consist of:

   (a) A record of 30 tests of field concrete. This would generally be the same test record that was used to document the standard deviation, but it could be a different set of 30 results; or

   (b) Laboratory data obtained from a series of trial batches.

Section 5.3.3.2(c) permits tolerances on slump and air content when proportioning by laboratory trial batches. The tolerance limits are stated at maximum permitted values because most specifications, regardless of form, will permit establishing a maximum value for slump or air content. The wording also makes it clear that these tolerances on slump and air content are to be applied only to laboratory trial batches and not a record of tests on field concrete.

5.4 PROPORTIONING BY WATER-CEMENT RATIO

Table 5.4 can be used to select a water-cement ratio, only with the permission of the project engineer, when no field or trial mixture data are available. Note that a single table relating concrete strength to water-cement ratio must, of necessity, be conservative. In the interest of economy of materials, the use of Table 5.4 should be limited to relatively small projects where the added cost of trial mixture data are not warranted. Note also that Table 5.4 applies only for concrete strengths up to 4500 psi for non-air-entrained concrete, and 4000 psi for air-entrained concrete. For higher concrete strengths, proportioning by field experience or by trial mixture data is required.

Proportioning concrete by the water-cement ratios of Table 5.4 is not permissible for IS or IP cements carrying a MH or LH suffix, or for Type II cement for which the optional moderate heat of hydration requirements have been used. Because of the conservative water-cement ratio limits, Table 5.4 is considered usable for all of the cement types in Section 5.4.2, despite the fact that the minimum specified 3-, 7-, and 28-day strengths of these cements vary from one cement to another. Typically, cement strengths exceed the ASTM minimum requirements by significant but different amounts. The following recommendations are given as guidelines to aid in the use of blended hydraulic cements:

(1) The cement used in the work should correspond to that on which the selection of concrete proportions was based.

(2) When Types V and P cements are used, proper recognition should be given to the effects of slower strength gain and lower heat of hydration on concrete proportioning and construction practices.

(3) In concretes made with blended hydraulic cements, if fly ash or other pozzolans are used as admixtures resulting in dilution of the cement component, proper recognition should be given to changes in the properties of concrete, such as strength, durability, and protection against corrosion of reinforcement.

5.6 EVALUATION AND ACCEPTANCE OF CONCRETE

5.6.2.3 Strength Test Criteria—Once a mix is approved for the proposed work, the tests made on job concretes must meet **both** the following two criteria for the concrete to be considered acceptable:

(1) No single test strength (the average of the strengths of two cylinders from a batch) shall be more than 500 psi below the specified compressive strength f_c'; i.e., 2500 psi for a specified 3000 psi concrete.

(2) The average of any three consecutive test strengths must equal or exceed the specified compressive strength, f_c'.

REFERENCES

2.1 *Design and Control of Concrete Mixtures,* 13th Edition, Publication EB001T, Portland Cement Association, Skokie, IL, 1988.

2.2 "Epoxy-Coated Reinforcing Bars," Engineering Data Report No. 14, Concrete Reinforcing Steel Institute, Schaumburg, IL.

2.3 "Guidelines for Inspection and Acceptance of Epoxy-Coated Reinforcing Bars at the Job Site," Concrete Reinforcing Steel Institute, Schaumburg, IL, 1986.

2.4 "Suggested Project Specifications Provisions for Epoxy-Coated Reinforcing Bars," Engineering Data Report No. 19, Concrete Reinforcing Steel Institute, Schaumburg, IL, 1984.

2.5 ACI Committee 211, "Standard Practice for Selecting Proportions for Normal, Heavyweight, and Mass Concrete (ACI 211.1-81, Revised 1985)," American Concrete Institute, Detroit, 1985.

2.6 ACI Committee 211, "Standard Practice for Selecting Proportions for Structural Lightweight Concrete (ACI 211.2-81)," American Concrete Institute, Detroit, 1981.

2.7 ACI Committee 214, "Recommended Practice for Evaluation of Compression Test Results of Concrete (ACI 214-77, Reapproved 1983)," American Concrete Institute, Detroit, 1983.

2.8 "Statistical Product Control," Publication IS172T, Portland Cement Association, Skokie, IL, 1970.

Example 2.1—Simplified Method for Calculating Standard Deviation

The averages of strength results for 46 pairs of cylinders sampled from a particular class of concrete delivered to a project are as follows:

Average Strengths for 46 Pairs of Cylinders

3395	2975	4220	3395	3045
3555	3200	3820	2965	2665
3545	3120	3995	3655	3485
3110	3055	3675	3815	4035
3260	3500	3220	4480	3500
3575	3840	3455	3650	3025
3975	3055	2980	3385	3435
3775	2815	3195	3595	3600
3670	3410	3260	3250	3515
3835				

The mean and standard deviation of the given set of test results is required.

Calculations and Discussion

Table 2-1 shows a simplified method of calculating the mean or average, $\overline{X}$, and standard deviation, s, for the above set of test results.

Table 2-1—Simplified Statistical Computations

1	2	3	4	5	6	7	8
Cell Boundaries	Tally Count	Mid-cell Values	Column 3 −3500	Column 4 ÷ 200	Frequency, Column 2	Column 5 × Column 6	Column 5 × Column 7
2600–2799	1	2700	−800	−4	1	−4	16
2800–2999	4	2900	−600	−3	4	−12	36
3000–3199	7	3100	−400	−2	7	−14	28
3200–3399	8	3300	−200	−1	8	−8	8
3400–3599	11	3500	0	0	11	0	0
3600–3799	6	3700	200	1	6	6	6
3800–3999	6	3900	400	2	6	12	24
4000–4199	1	4100	600	3	1	3	9
4200–4399	1	4300	800	4	1	4	16
4400–4599	1	4500	1000	5	1	5	25
Totals					46	−8	168
Squares of Totals					2116	64	

Example 2.1—Continued

Calculations and Discussion

$$\frac{-8 \, (\text{ Column 7 total })}{46 \, (\text{ Column 6 total })} \times 200 = -35$$

Mean, $\overline{X} = -35 + 3500 = 3465$ psi, rounded up to 3470 psi.

46 (column 6 total) × 168 (column 8 total) = 7728.

$8 \times 8 = 64$ (square of column 7 total).

7728 – 64 = 7664.

7664 ÷ 2116 (square of column 6 total) = 3.62.

Standard deviation, $s = 200 \times \sqrt{3.62} = 380$ psi, or
directly from Table 2-2, s = 380 psi.

Steps in developing Table 2-1 are as follows:*

Step 1 — Group the results into cells (groups) of 200 psi intervals and tally them in increasing order (Columns 1, 2).

Step 2 — Tabulate the mid-cell values (Column 3) rounded to the nearest 100 psi.

Step 3 — Examine Column 3 to determine the mid-cell value with the highest frequency (highest count). Call this value the Central Value. In our example, this value is 3,500 psi. Subtract it from each of the mid-cell values and tabulate the results in Column 4.

Step 4 — Divide all values in Column 4 by 200 and tabulate the results in Column 5.

Step 5 — Tabulate in Column 6 the frequencies tallied in Column 2.

Step 6 — Multiply each of the values in Column 5 by the corresponding frequencies in Column 6 and tabulate the results in Column 7 (some values will be positive and some will be negative).

Step 7 — Multiply each of the values in Column 5 by the corresponding values in Column 7 and tabulate the results in Column 8 (all values will be positive).

Step 8 — Add Columns 6, 7, and 8. (The total of Column 6 is the total number of tests.)

---

*To simplify this discussion, equations and derivations of formulas have been omitted. Interested readers should refer to ACI 214,[2.7] or to any textbook on statistics.

Example 2.1 — Continued

Calculations and Discussion

Step 9 — Divide the total of Column 7 by the total of Column 6 (if the sum of Column 7 is negative, then this result will also be negative). Do not calculate beyond two figures after the decimal point. Multiply this result by 200.

Step 10 — To obtain the mean, $\overline{X}$, add (or subtract) the result of Step 9 to (from) the Central Value. Round out the result to the nearest 10 psi.

Step 11 — Multiply the total of Column 6 by the total of Column 8.

Step 12 — Square the total of Column 7 (the result will always be positive).

Step 13 — Subtract the result of Step 12 from that of Step 11.

Step 14 — Divide the result of Step 13 by the square of the total of Column 6. Do not calculate beyond two figures after the decimal point.

Step 15 — Calculate the square root of the result of Step 14 rounded to two figures after the decimal point, and multiply it by 200. This is the standard deviation, s. Table 2-2 may be used to determine s directly from the results of Step 14 (intermediate values will have to be estimated).

Other methods of computation may yield more accurate results. However, the accuracy of the values obtained by the above described method is adequate for general product control purposes.

Table 2-2 — Standard Deviation, s*

Result of Step 14	s	Result of Step 14	s	Result of Step 14	s	Result of Step 14	s
1.00	200	4.00	400	9.00	600	16.00	800
1.21	220	4.41	420	9.61	620	16.81	820
1.44	240	4.84	440	10.24	640	17.64	840
1.69	260	5.29	460	10.89	660	18.49	860
1.96	280	5.76	480	11.56	680	19.36	880
2.25	300	6.25	500	12.25	700	20.25	900
2.56	320	6.76	520	12.96	720	21.16	920
2.89	340	7.29	540	13.69	740	22.09	940
3.24	360	7.84	560	14.44	760	23.04	960
3.61	380	8.41	580	15.21	780	24.01	980

*Only applicable for cell intervals of 200 psi.

Example 2.2 — Selection of Water-Cement Ratio for Strength and Durability

The first step in proportioning a concrete mixture is the selection of an appropriate water-cement ratio (w/c) for the durability and strength needed. The water-cement ratio selected for mix design must be the lowest value required to meet the design exposure conditions. When durability does not control, the water-cement ratio must be selected on the basis of the specified compressive strength of the concrete f'_c to satisfy the structural design requirements.

Design conditions: Concrete is required for a loading-dock slab that will be exposed to moisture in a severe freeze-thaw climate, but not subject to deicers. A specified compressive strength f'_c of 3000 psi is required. Assume the proposed work will use similar materials and conditions of control as that resulting in the test records of Example 2.1, with an established standard deviation of 380 psi. Assume Type I cement with ¾ in. maximum size aggregate.

Calculations and Discussion	Code Reference

1. Determine the maximum allowable w/c ratio based on specified compressive strength of the concrete.

   Selection of water-cement ratio for required strength should be based on trial mixtures or field data made with actual job materials to determine the relationship between water-cement ratio and strength. Typical trial mixture or field data strength curves are given in Ref. 2.1. For a standard deviation of 380 psi, the required average compressive strength to be used as the basis for selection of concrete proportions must be the larger of   **5.3.2**

   $$f'_{cr} = f'_c + 1.34s = 3000 + 1.34(380) = 3500 \text{ psi}$$ Eq. 5-1
   or
   $$f'_{cr} = f'_c + 2.33s - 500 = 3000 + 2.33(380) - 500 = 3400 \text{ psi}$$ Eq. 5-2

   Therefore $f'_{cr} = 3500$ psi

   Note: The average strength required for the mix design should equal the specified strength plus an allowance to account for variations in materials; variations in methods of mixing, transporting, and placing the concrete; and variations in making, curing, and testing concrete cylinder specimens. For this example, with a standard deviation of 380 psi, an allowance of 500 psi for any variations is required.

   Using the field data strength curve, Fig. 7-1 of Ref. 2.1, reproduced in Fig. 2-2, the required water-cement ratio for air-entrained concrete is 0.53 for an f'_{cr} of 3500 psi.

2. Determine the maximum allowable w/c ratio based on exposure conditions.

   Concrete exposed to freezing and thawing must be air-entrained, with air content indicated in Table 4.1.1. For concrete in a cold climate and exposed to wet freeze-thaw conditions, an air content of 6% is required for a ¾-in. maximum size aggregate.   **4.1.1**

Example 2-2 — Continued

	Code
Calculations and Discussion	**Reference**

For concrete exposed to freezing and thawing in a moist condition, Table 4.1.2 requires a maximum water-cement ratio of 0.45.   4.1.2

Since the slab will not be exposed to deicing salts, the minimum cement content required by Section 4.1.3 does not apply.   4.1.3

If significant amounts of chloride are suspected from the concrete materials to be used, Table 4.3.1 would require a maximum chloride-ion content of 0.30% by weight of cement, for corrosion protection of the reinforcement.   4.3

3. Determine the maximum allowable w/c ratio for strength and durability.

Since the water-cement ratio of 0.53 required for strength is greater than the 0.45 or less   5.5
required for exposure conditions, the exposure requirements govern. A water-cement ratio of 0.45 or less must be used to establish the mixture proportions, even though this may produce strengths higher than needed to satisfy structural requirements. Note that the specified compressive strength, $f_c' = 3000$ psi, is the strength that is expected to be equalled or exceeded by the average of any set of three consecutive strength tests, with no individual test more than 500 psi below the specified strength. The higher strength resulting from using a water- cement ratio of 0.45 to satisfy exposure conditions will enhance the "probability" of satisfying the strength test requirements of Section 5.6. Referring to Fig. 2-2, with a required w/c ratio of 0.45, a strength level approximating 4200 psi can be expected for air-entrained concrete.

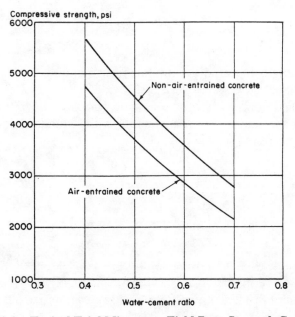

Fig. 2-2   Typical Trial Mixture or Field Data Strength Curves

2-21

Details of Reinforcement

UPDATE FOR THE '89 CODE

For ACI 318-89, a new Section 7.13 has been added addressing requirements for structural integrity. Experience has shown that structures designed in accordance with previous editions of the ACI Code possess adequate overall integrity, and have provided the required safety to the public. The intent of new Section 7.13 is to enhance the overall integrity of a structure through "minor changes in detailing of reinforcement." The additional reinforcement details are intended to improve the redundancy and ductility in structures, so that in the event of damage to a major supporting element or an abnormal loading incident, the resulting damage may be confined to a relatively small area. The structure then will have a better chance to maintain overall stability.

Two additional provisions of Code Chapter 7 dealing with details of reinforcement have been revised for the '89 Code; new guidance on field bending of reinforcing bars has been added to Commentary Section 7.3.2, and the vertical spacer requirements for installation of column spirals has been removed from Code Section 7.10. See discussion on Code Sections 7.3.2 and 7.10.4.

GENERAL CONSIDERATIONS

Good reinforcing details are vital to satisfactory performance of reinforced concrete structures. Standard practice for reinforcement details has developed gradually. The Building Code Committee (ACI 318) continually collects reports of research and practice with reinforcing materials, suggests new research needed, receives reports on new research, and translates the results into specific code provisions for details of reinforcement.

The ACI Detailing Manual, Ref. 3.1, provides recommended methods and standards for preparing design drawings, typical details, and drawings for fabrication and placing of reinforcing steel in reinforced concrete structures. Separate sections of the manual define responsibilities of both the engineer and the reinforcing bar detailer. The CRSI Manual of Standard Practice, Ref. 3.2, provides recommended industry practices for reinforcing steel. As an aid to designers, Recommended Industry Practices for Estimating, Detailing, Fabrication, and Field Erection of Reinforcing Materials are included in Ref. 3.2, for direct reference in project drawings and specifications. The WRI *Structural Welded Wire Fabric Detailing Manual*,[3.3] provides information on detailing welded wire fabric reinforcement systems.

7.1 STANDARD HOOKS

Requirements for standard hooks and minimum finished inside bend diameters for reinforcing bars are illustrated in Tables 3-1 and 3-2. The standard hook details for stirrups and ties apply only to #8 and smaller bar sizes.

Table 3-1—Standard Hooks for Primary Reinforcement*

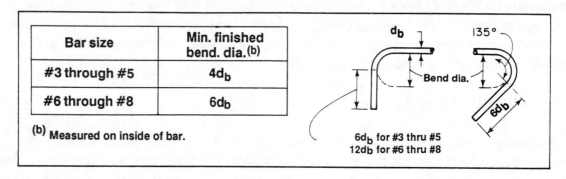

Bar size	Min. finished bend. dia.(a)
#3 through #8	$6d_b$
#9, #10, #11	$8d_b$
#14 and #18	$10d_b$

(a) Measured on inside of bar.

Table 3-2—Standard Hooks for Stirrups and Tie Reinforcement*

Bar size	Min. finished bend. dia.(b)
#3 through #5	$4d_b$
#6 through #8	$6d_b$

(b) Measured on inside of bar.

$6d_b$ for #3 thru #5
$12d_b$ for #6 thru #8

7.2 MINIMUM BEND DIAMETERS

Minimum bend diameter for a reinforcing bar is defined as "the diameter of bend measured on the **inside** of the bar." Minimum bend diameters are dependent on bar size and multiples of bar diameters; for #3 to #8, the minimum bend diameter is 6 bar diameters; for #9 to #11, the minimum bend diameter is 8 bar diameters; and for #14 and #18, the minimum bend diameter is 10 bar diameters. Exceptions to these provisions are:

---

*Table 1, Part C of Reference 3.1 provides actual bar dimensions for end hooks, and stirrup and tie hooks.

(1) For stirrups and ties in sizes #5 and smaller, the minimum bend diameter is 4 bar diameters. For #6 through #8 stirrups and ties, the minimum bend diameter is 6 bar diameters.

(2) For welded wire fabric used for stirrups and ties, inside diameter of bend must not be less than four wire diameters for deformed wire larger than D6 and two wire diameters for all other wire. Welded intersections must be at least four wire diameters away from bends with inside diameters of less than eight wire diameters.

7.3 BENDING

All reinforcement must be bent cold unless otherwise permitted by the engineer. For unusual bends, special fabrication including heating may be required and the engineer must give approval to the techniques used.

Reinforcing bars partially embedded in concrete must not be field bent without authorization of the engineer. Code Commentary Section 7.3.2 provides guidelines for field bending, and heating if necessary, of bars partially embedded in concrete. For the '89 Code, new guidance on field bending is added. Recent tests have shown that billet steel bars (A615) can be cold bent up to 90° and straightened at or near the minimum bend diameters permitted in Section 7.2.

7.5 PLACING REINFORCEMENT

Supports for reinforcement are required to adequately secure the reinforcement against displacement during casting, but the supports are not required to be of any specific material or type. Use of epoxy-coated reinforcing bars will, however, require bar supports made of dielectric material, or wire supports coated with a dielectric material such as epoxy or vinyl compatible with concrete. See discussion on Code Section 3.5.3.7 concerning special hardware and handling to minimize damage to the epoxy coating during handling, transporting, and placing epoxy-coated bars.

Note that welding of cross bars (tack welding) for assembly of reinforcement is prohibited except as specifically authorized by the engineer.

Tolerances for placing reinforcement are given for minimum concrete cover and for effective depth, d. Since both dimensions are components of the total depth, tolerances on these dimensions are directly related. The amount of tolerance allowed is dependent on the size of the member expressed as a function of the effective depth d. These tolerances are illustrated in Table 3-3. Exceptions to these provisions are:

(1) Tolerance for clear distance to formed soffits must be minus $\frac{1}{4}$ in.

(2) Tolerance for cover must not exceed minus one-third the minimum concrete cover required in the design drawings and specifications.

For ends of bars and longitudinal location of bends, the tolerance is $\pm$ 2 in., except at discontinuous ends of members where the tolerance is $\pm$ $\frac{1}{2}$ in. These tolerances are illustrated in Fig. 3-1.

Table 3-3—Critical Dimensional Tolerances for Locating Reinforcement

Effective Depth d	Tolerance on d	Tolerance on Min. Cover	
d ≤ 8 in.	± 3/8 in.	– 3/8 in.	
d > 8 in.	± 1/2 in.	– 1/2 in.	

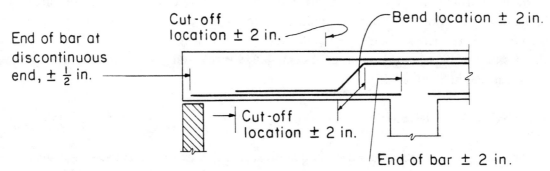

Fig. 3-1   Tolerances for Bar Bend and Cutoff Locations

7.6   SPACING LIMITS FOR REINFORCEMENT

Spacing (clear distance) between bars must be as follows:

Minimum Spacing

For members with parallel bars in a layer, the clear spacing between bars must be at least one bar diameter but not less than 1 in.; and for reinforcement in two or more layers, bars in the upper layers must be directly above bars in the bottom layer, with at least 1 in. clear vertically between layers. For spirally reinforced and tied reinforced compression members, the clear distance between longitudinal bars must be at least 1½ bar diameters, but not less than 1½ in.   These spacing requirements also apply to clear distance between contact-lap-spliced single or bundled bars. Section 3.3.3, which contains spacing requirements based on maximum nominal aggregate size, may be applicable. Clear distances between bars are illustrated in Table 3-4.

Maximum Spacing

In walls and slabs other than concrete joists, primary flexural reinforcement must not be spaced greater than 3 times the wall or slab thickness nor 18 in.

Table 3-4—Clear Distances Between Bars, Bundles, or Tendons

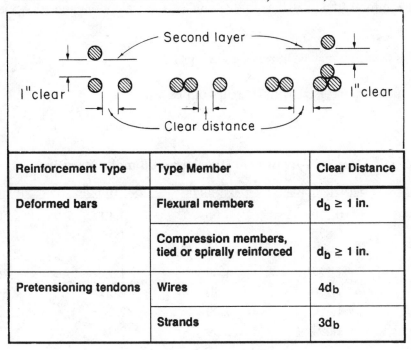

Reinforcement Type	Type Member	Clear Distance
Deformed bars	Flexural members	$d_b \geq 1$ in.
	Compression members, tied or spirally reinforced	$d_b \geq 1$ in.
Pretensioning tendons	Wires	$4d_b$
	Strands	$3d_b$

Notes: (1) Clear distance must also be greater than $\frac{4}{3}$ of the maximum nominal aggregate size used (Section 3.3.3).
(2) For bundled bars, diameter of a single bar of equivalent total area must be used, see Table 3-5.
(3) Closer vertical spacing and bundling of pretensioning tendons may be allowed in the middle portion of a member (Section 7.6.7.1).

7.6.6 Bundled Bars

For isolated situations requiring heavy concentration of reinforcement, bundles of standard bar sizes can save space and reduce congestion for placement and consolidation of concrete. As a design consideration, bundling of bars in columns is a means to better locating and orienting the reinforcement for increased column capacity; also, less lateral ties will be required if column bars are bundled.

Bundling of bars (parallel reinforcing bars in contact, assumed to act as a unit) is permitted, but only if such bundles are enclosed by lateral ties or stirrups. Some limitations are placed on the use of bundled bars as follows:

(1) #14 and #18 bars cannot be bundled in beams.

(2) If individual bars in a bundle are cut off within the span of beams, such cutoff points must be staggered at least 40 bar diameters.

(3) A maximum of two bundled bars in any one plane is implied (three or four adjacent bars in one plane are not considered as bundled bars).

(4) For spacing and concrete cover, a unit of bundled bars must be treated as a single bar with an area equivalent to the total area of all bars in the bundle. Equivalent diameters of bundled bars are given in Table 3-5.

(5) A maximum of four bars may be bundled (See Fig. 3-2).

(6) Bundled bars must be enclosed within stirrups or ties.

Table 3-5—Equivalent Diameters of Bundled Bars, in.

Bar Size	Bar Diameter	2-Bar Bundle	3-Bar Bundle	4-Bar Bundle
# 6	0.750	1.06	1.30	1.50
# 7	0.875	1.24	1.51	1.75
# 8	1.000	1.42	1.74	2.01
# 9	1.128	1.60	1.95	2.26
#10	1.270	1.80	2.20	2.54
#11	1.410	1.99	2.44	2.82
#14	1.693	2.39	2.93	3.39

Fig. 3-2   Possible Reinforcing Bar Bundling Arrangements

7.6.7 Prestressing Tendons and Ducts

Clear distance between pretensioning tendons at ends of members is handled separately and is limited to 4-wire diameters for individual wires, or 3-strand diameters. Closer vertical spacing or bundling of tendons is permitted in the middle portion of the span if special care in design and fabrication is employed. Post-tensioning ducts may be bundled if concrete can be satisfactorily placed and if provision is made to prevent the tendons from breaking through the duct when tensioned. Spacing requirements for tendons are illustrated in Table 3-4.

7.7 CONCRETE PROTECTION FOR REINFORCEMENT

Concrete cover or protection requirements are specified for members cast against earth, in contact with earth or weather, and for interior members not exposed to weather. Slightly reduced cover or protection is permitted under these same conditions for precast concrete manufactured under plant control, and other values are given for prestressed concrete. The term "manufactured under plant controlled conditions" does not necessarily mean that precast members must be manufactured in a plant. Structural elements precast at the job site will also qualify for the lesser cover if the control of form dimensions, placing of reinforcement, quality of concrete, and curing procedure are equivalent to those normally expected in a plant operation. Larger diameter bars and bundled bars require slightly greater cover. Corrosive environments or fire protection may also warrant increased cover. The designer should take special note of the Commentary recommendations (Section 7.7.5) for increased cover when concrete will be exposed to external sources of chlorides in service, such as deicing salts and seawater.

7.8 SPECIAL REINFORCING DETAILS FOR COLUMNS

Section 7.8 covers the special detailing requirements for offset bent longitudinal bars and steel cores of composite columns.

When column offsets are necessary, longitudinal bars may be bent, subject to the following limitations:

(1) Slope of inclined portion of an offset bar with axis of column must not exceed 1 in 6 (see Fig. 3-3).

(2) Portions of bar above and below an offset must be parallel to axis of column.

(3) Horizontal support at offset bends must be provided by lateral ties, spirals, or parts of the floor construction; lateral ties or spirals, if used, shall be placed not more than 6 in. from points of bend (see Fig. 3-3). Horizontal support provided must be designed to resist 1½ times the horizontal component of the computed force in the inclined portion of an offset bar.

(4) Offset bars must be bent before placement in the forms.

(5) When a column face is offset 3 in. or more, longitudinal column bars parallel to and near that face must not be offset bent. Separate dowels, lap spliced with the longitudinal bars adjacent to the offset column faces, must be provided (see Fig. 3-3). In some cases, a column might be offset 3 in. or more on some faces, and less than 3 in. on the remaining faces, which could possibly result in some offset bent longitudinal column bars and some separate dowels being used in the same column.

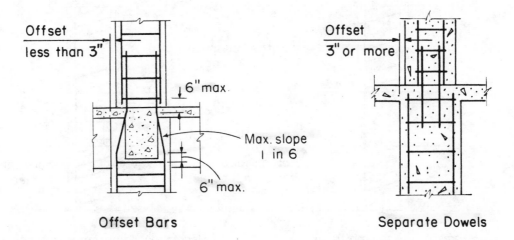

Fig. 3-3 Special Column Details

Steel cores in composite columns may be detailed to allow transfer of up to 50 percent of the compressive load in the core by direct bearing. The remainder of the load must be transferred by welds, dowels, splice plates, etc. This should ensure a minimum tensile capacity similar to that of a more common reinforced concrete column.

7.9 CONNECTIONS

Enclosures must be provided for splices of continuing reinforcement, and for end anchorage of reinforcement terminating at beam and column connections. This confinement may be provided by surrounding concrete or internal closed ties, spirals, or stirrups.

7.10 LATERAL REINFORCEMENT FOR COMPRESSION MEMBERS

7.10.4 Spirals

Minimum diameter of spiral reinforcement in cast-in-place construction is ⅜ in. and the clear spacing must be between the limits of 1 in. and 3 in. This requirement does not preclude the use of a smaller minimum spiral diameter for precast units. Splices in spirals must be welds or tension lap splices of at least 48 spiral bar or wire diameters but not less than 12 in. Anchorage of spiral reinforcement must be provided by 1½ extra turns at each end of a spiral unit.

Spiral reinforcement must extend from the top of footing or slab in any story to the level of the lowest horizontal reinforcement in slabs, drop panels, or beams above. If beams or brackets do not frame into all sides of the column, ties must extend above the top of the spiral to the bottom of the slab or drop panel (see Fig. 3-4). In columns with capitals, spirals must extend to a level where the diameter or width of capital is twice that of the column.

Spirals must be held firmly in place, at proper pitch and alignment, to prevent displacement during concrete placement. For ACI 318-89, the vertical spacer requirements for installation of column spirals has been

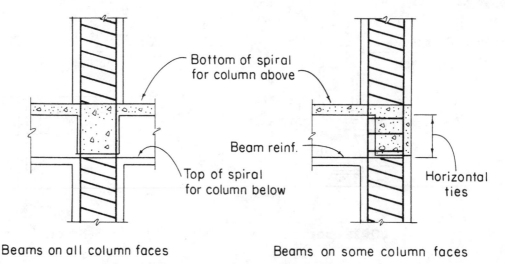

Fig. 3-4 Termination of Spirals

removed from the Code. Section 7.10.4.9 simply states that "spirals must be held firmly in place and true to line." This new performance provision will now permit alternative methods, such as field tying, to hold the fabricated spiral cage in place during construction, which is current practice in most areas where spirals are used. The original vertical spacer requirements are now in the Code Commentary as guidance where spacers are used for spiral installation. Note that the project specifications should cover the vertical spacer requirements (if used) or the field tying of the spiral reinforcement.

7.10.5 Ties

In tied reinforced columns, ties must be located no more than half a tie spacing above floor or footing and no more than half a tie spacing below the lowest horizontal reinforcement in the slab or drop panel above. If beams or brackets frame from four directions into a column, ties may be terminated not more than 3 in. below the lowest horizontal reinforcement in the shallowest of such beams or brackets (see Fig. 3-5). Minimum size of lateral ties in tied reinforced columns is related to the size of the longitudinal bars. Minimum tie sizes are #3 for longitudinal bars #10 and smaller, and #4 for #11 longitudinal bars and larger and for bundled bars. The following conditions also apply: spacing must not exceed 16 longitudinal bar diameters, 48 tie bar diameters, or the least dimension of the column; every corner bar and alternate bar must have lateral support provided by the corner of a tie with an included angle of not more than 135°. No unsupported bar shall be farther than 6 in. from a supported bar (see Fig. 3-6). Note that the 6-in. clear distance is measured along the tie.

Welded wire reinforcement of equivalent area may be used for ties. When main reinforcement is arranged in a circular pattern, it is permissible to use complete circular ties at the specified spacing. This provision allows the use of circular ties at a spacing greater than that specified for spirals in spirally reinforced columns.

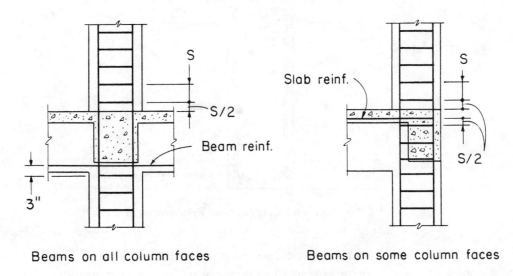

Beams on all column faces   Beams on some column faces

Fig. 3-5   Termination of Column Ties

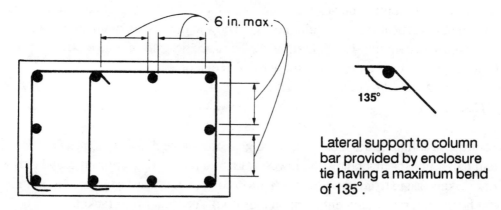

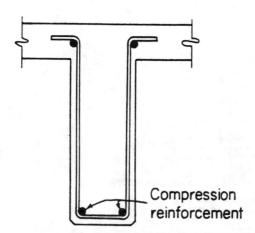

135°

Lateral support to column
bar provided by enclosure
tie having a maximum bend
of 135°.

Fig. 3-6   Lateral Support of Column Bars by Ties

7.11   LATERAL REINFORCEMENT FOR FLEXURAL MEMBERS

Where compression reinforcement is used to increase the flexural strength of a member (Section 10.3.4), or to control long-term deflection [Eq. (9-10)], Section 7.11.1 requires that such reinforcement be enclosed by ties or stirrups. Requirements for size and spacing of the ties or stirrups are the same as for ties in tied columns. Welded wire fabric of equivalent area may be used. The ties or stirrups must extend throughout the distance where the compression reinforcement is required for flexural strength or deflection control. Section 7.11.1 is interpreted not to apply to reinforcement located in a compression zone to help assemble the reinforcing cage and hold the web reinforcement in place during concrete placement.

Enclosing reinforcement required by Section 7.11.1 is illustrated by the U-shaped stirrup in Fig. 3-7; the

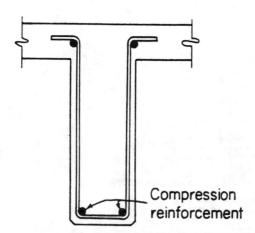

Compression
reinforcement

Fig. 3-7   Enclosed Compression Reinforcement

continuous bottom portion of the stirrup satisfies the enclosure intent of Section 7.11.1 for the two bottom bars shown. A completely closed stirrup is ordinarily not necessary, except in cases of high moment reversal, where reversal conditions require that both top and bottom longitudinal reinforcement be designed as compression reinforcement.

Torsion reinforcement, where required, must consist of completely closed stirrups, closed ties, or spirals, as required by Section 11.6.7.3.

7.11.3 Closed Stirrups

According to Section 7.11.3, a closed stirrup is formed either in one piece with overlapping 90° or 135° end hooks around a longitudinal bar, or in one or two pieces with a Class B lap splice, as illustrated in Fig. 3-8. The one-piece closed stirrup with overlapping end hooks is not practical for placement. Neither of the closed stirrups shown in Fig. 3-8 is considered effective for members subject to high torsion. Tests have shown that,

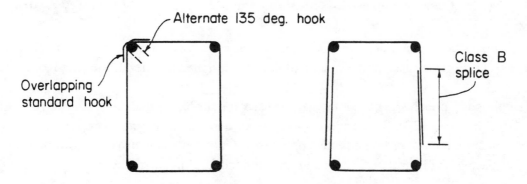

Fig. 3-8   Code Definition of Closed Tie or Stirrup

with high torsion, loss of concrete cover and subsequent loss of anchorage result if the 90° hook and lap splice details are used where confinement by external concrete is limited. See Fig. 3-9. The ACI Detailing Manual, Reference 3.1, recommends the details illustrated in Fig. 3-10 for closed stirrups used as torsional reinforcement.

7.12   SHRINKAGE AND TEMPERATURE REINFORCEMENT

Minimum shrinkage and temperature reinforcement normal to primary flexural reinforcement is required for structural floor and roof slabs (not slabs on ground) where the flexural reinforcement extends in one direction only. Minimum steel ratios, based on the gross concrete area, are:

(1)   0.0020 for Grade 40 and 50 deformed bars;

(2)   0.0018 for Grade 60 deformed bars or welded wire fabric;

(3) $0.0018 \times 60,000/f_y$ for reinforcement with a yield strength greater than 60,000 psi; but not less than 0.0014.

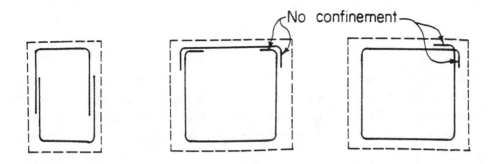

These details are not considered effective for members subject to high torsion. Note lack of confinement when compared to similar members with confinement shown in Fig. 3-10.

Fig. 3-9   Closed Stirrup Details Not Recommended for Members Subject to High Torsion

Spacing of shrinkage and temperature reinforcement must not exceed 5 times the slab thickness nor 18 in. Splices and end anchorages of such reinforcement must be designed for the full specified yield strength.

Bonded or unbonded prestressing tendons may be used for shrinkage and temperature reinforcement in structural slabs (Section 7.12.3). The tendons must provide a minimum average compressive stress of 100 psi on the gross concrete area, based on effective prestress after losses. Spacing of tendons must not exceed 6 ft. When the spacing is greater than 54 in., additional bonded reinforcement must be provided at slab edges.

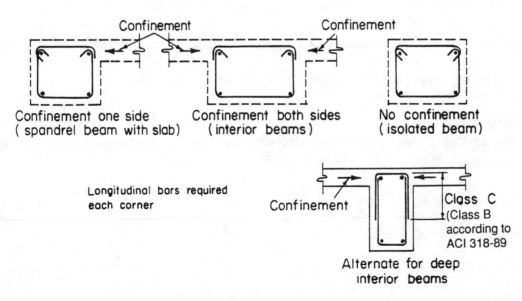

Fig. 3-10   Recommended Two-Piece Closed Stirrup Details[3.1]

7.13 REQUIREMENTS FOR STRUCTURAL INTEGRITY

Structures capable of safely supporting all conventional design loads may suffer local damage from severe local abnormal loads, such as explosions due to gas or industrial liquids; vehicle impact; impact of falling objects; and local effects of very high winds such as tornadoes. Generally, such abnormal loads or events are not ordinary design considerations. The overall integrity of a reinforced concrete structure to withstand such abnormal loads can be substantially enhanced by providing relatively minor changes in the detailing of the reinforcement. The intent of the new Section 7.13 is to improve the redundancy and ductility of structures. This is achieved by providing, as a minimum, some continuity reinforcement or tie between horizontal framing members. In the event of damage to a major supporting element or an abnormal loading event, the integrity reinforcement is intended to confine any resulting damage to a relatively small area. Therefore, the structure will have a better chance to maintain overall stability.

It is not the intent of Section 7.13 that a structure be designed to resist general collapse caused by gross misuse or to resist severe abnormal loads acting directly on a large portion of the structure. General collapse of a structure as the result of abnormal events such as wartime bombing, landslides, and floods are beyond the scope of any practical design.

7.13.1 General Structural Integrity

Since accidents and misuse are normally unforseeable events, they cannot be defined precisely; likewise, providing general structural integrity to a structure is a requirement that cannot be stated in simple terms. The performance provision..."members of a structure shall be effectively tied together to improve integrity of the overall structure," will require a level of judgment on the part of the design engineer, and will generate certain differing opinions among engineers as to how to effectively provide a general structural integrity solution for a particular framing system. It is obvious that all conditions that might be encountered in design cannot be specified in the Code. The Code, however, does set forth specific examples of certain reinforcing details for cast in-place joists, beams, and two-way slab construction.

With damage to a support, top reinforcement which is continuous over the support, but not confined by stirrups, will tend to tear out of the concrete and will not provide the catenary action needed to bridge the damaged support. By making a portion of the bottom reinforcement in beams continuous over supports, some catenary action can be provided. By providing some continuous top and bottom reinforcement in edge or perimeter beams, an entire structure can be tied together; also, the continuous tie provided to perimeter beams of a structure will toughen the exterior portion of a structure, should an exterior column be severely damaged. Other examples of ways of detailing for required integrity of a framing system to carry loads around a severely damaged member can be cited. The design engineer will need to evaluate his particular design for specific ways of handling the problem. The concept of providing general structural integrity was first introduced in the 1982 edition of ANSI A58.1, *Minimum Design Loads for Buildings and Other Structures.*[3.4] The reader is referred to that document for further discussion of design concepts and details for providing general structural integrity.

Figures 3-11 through 3-13 illustrate the required reinforcing details for the general case of cast-in-place joists and beams.

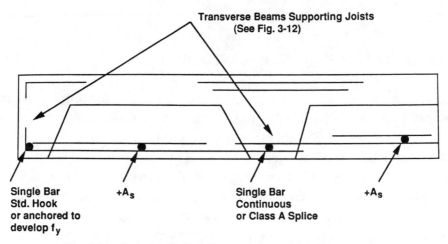

Fig. 3-11   Continuity Reinforcement for Joist Construction

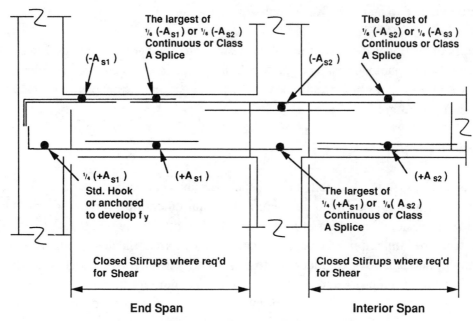

Fig. 3-12   Continuity Reinforcement for Perimeter Beams

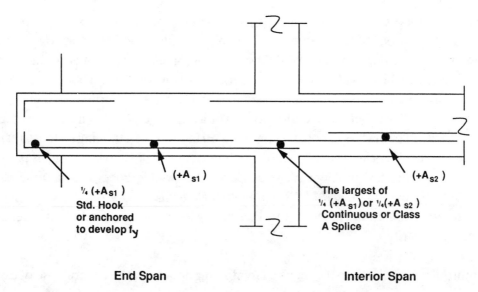

Fig. 3-13   Continuity Reinforcement for Beams without Closed Stirrups

While the new requirements for structural integrity are prescriptive for cast-in-place construction, the '89 Code provides only performance requirements for precast construction. Precast structures can be built in a lot of different ways. The Code requires tension ties for precast concrete buildings of all heights. Connections that rely solely on friction due to gravity forces are not permitted.

The general requirement for structural integrity (7.13.1) states that "...members of a structure shall be effectively tied together...". The '89 Commentary cautions that for precast concrete construction, connection details should be arranged so as to minimize the potential for cracking due to restrained creep, shrinkage, and temperature movements. Ref. 3.5 contains information on industry practice for connections and detailing requirements. Prescriptive requirements recommended by the PCI for precast concrete bearing wall buildings are given in Ref. 3.6.

REFERENCES

3.1 ACI Committee 315, *ACI Detailing Manual – 1988,* Publication SP-66(88), American Concrete Institute, Detroit, 1988.

3.2 *Manual of Standard Practice,* 24th edition, Concrete Reinforcing Steel Institute, Schaumburg, IL, 1986.

3.3 *Structural Welded Wire Fabric Detailing Manual,* 1st edition, Wire Reinforcement Institute, McLean, VA, 1983.

3.4 *Minimum Design Loads for Buildings and Other Structures,* (ANSI A58.1 - 1982), American National Standards Institute, New York, 1982.

3.5 *Design and Typical Details of Connections for Precast and Prestressed Concrete,* Publication MNL-123-88, Prestressed Concrete Institute, Chicago, 1988.

3.6 PCI Building Code Committee, "Proposed Design Requirements for Precast Concrete," *PCI Journal,* Vol. 31, No. 6, Nov.-Dec. 1986, pp.32-47.

Development and Splices of Reinforcement

UPDATE FOR THE '89 CODE

To reflect research results and provide proper development for closely spaced bars and for bars with minimal cover, the development lengths for deformed bars and deformed wire in tension (Section 12.2) have been extensively revised for ACI 318-89. The '89 Code provisions increase the development lengths for closely spaced bars and for bars with minimal cover. For an in-depth discussion of the various parameters and conditions that affect bar development, the reader is referred to Commentary Section 12.2 and Commentary Refs. 12.2 and 12.3, and Refs. 12.6 through 12.9.

As a result of the new provisions for development of bars in tension, tension lap splice lengths, which are a multiple of the tension development length ℓ_d, will also be affected by the new modification factors. In consideration of the multitude of new factors to account for closely spaced bars and bars with minimal cover, the Class C tension lap splice has been eliminated.

Also, for the '89 Code, the special splice requirements for columns (Section 12.17) have been greatly simplified. Designer confusion and misinterpretation of the splice provisions for columns prompted the ACI Committee 318 to review the special requirements. New Code Section 12.17 is also reorganized to define more clearly the design requirements for the different types of bar splices permitted in columns. See discussion of Section 12.17.

Until further research is completed and to ensure ductility and safety of structures built with high strength concrete, the term $\sqrt{f_c'}$ has been limited to 100 psi. Existing design equations for development of straight bars in tension and compression, and standard hooks in tension, are all a function of $\sqrt{f_c'}$. These equations were developed from results of tests on bars embedded in concrete with compressive strengths of 3000 to 6000 psi. ACI Committee 318 was prudent in limiting $\sqrt{f_c'}$ at 100 psi pending completion of tests to verify applicability of current design equations to bars in high strength concrete.

GENERAL CONSIDERATIONS

The development length concept for anchorage of reinforcement, such as for deformed bars and deformed wire, is based on the attainable average bond stress over the length of embedment of the reinforcement. In application, the development length concept requires the specified minimum lengths or extensions of reinforcement beyond all locations of peak stress in the reinforcement. Such peak stresses generally occur in flexural members at the locations of maximum stress and where adjacent reinforcement terminates or is bent.

The strength reduction factor φ is not used in Chapter 12 since the specified development lengths already include an allowance for understrength. The required development and lap splice lengths are the same for either the Strength Design Method or the Alternate Design Method of Appendix A (Section A.4).

12.1 DEVELOPMENT OF REINFORCEMENT—GENERAL

Development length or anchorage of reinforcement is required on both sides of a location of peak stress at each section of a reinforced concrete member. In continuous members, for example, reinforcement typically continues for a considerable distance on one side of a critical stress location so that detailed calculations are usually required only for the side where the reinforcement is terminated.

12.2 DEVELOPMENT OF DEFORMED BARS AND DEFORMED WIRE IN TENSION

12.2.1 Development Length

Development of reinforcement in tension involves calculation of a basic development length, ℓ_{db}, as a function of bar size and yield strength, and compressive strength of concrete, modified by factors to reflect influence of bar spacing and cover, enclosing transverse reinforcement, location of bar (top bar effect), type of aggregate, epoxy coating, and ratio of required to provided area of reinforcement to be developed. The applicable modification factors are given in Sections 12.2.3 through 12.2.5 and are summarized in Table 4-1.

Section 12.2.1 requires that the development length ℓ_d (including all applicable modification factors) must not be less than 12 in. The product of ℓ_{db} and the modification factors for bar spacing and cover, and enclosing transverse reinforcement (Items 1, 2 and 3 of Table 4-1) must not be less than a minimum development length specified in Section 12.2.3.6.

12.2.2 Basic Development Length

Requirements for basic tension development length ℓ_{db} of deformed bars and deformed wire are given in Section 12.2.2:

For bar sizes #3 to #11, and deformed wire ....$\ell_{db} = 0.04\ A_b f_y/\sqrt{f_c'}$

For #14 bars ....$\ell_{db} = 0.85\ f_y/\sqrt{f_c'}$

For #18 bars ....$\ell_{db} = 0.125\ f_y/\sqrt{f_c'}$

Table 4-1—Summary of Modification Factors for Development Length in Tension

Item	Parameters	Value of Modification Factor	Corresponding Code Section
1	Bar spacing, cover, and enclosing transverse reinforcement	1.0, 2.0, or 1.4	12.2.3.1 to 12.2.3.3
2	Bar spacing	0.8	12.2.3.4
3	Enclosing transverse reinforcement	0.75	12.2.3.5
4	Top bar effect*	1.3	12.2.4.1
5	Lightweight aggregate concrete	1.3 or $6.7\sqrt{f_c'}/f_{ct} \not< 1$	12.2.4.2
6	Epoxy coated bars*	1.5 or 1.2	12.2.4.3
7	Excess reinforcement	A_s required / A_s provided	12.2.5

*The product of items 4 (top bar effect) and 6 (epoxy coated bars) need not exceed 1.7.

where $\sqrt{f_c'}$ must not be taken greater than 100 psi.

Basic development lengths ℓ_{db} for Grade 60 bars in tension are tabulated in Table 4-2. Tabulated values are for bars embedded in normal weight concrete with specified compressive strengths ranging between 3000 and 10,000 psi.

Table 4-2—"Basic" Tension Development Length ℓ_{db} (inches) for Grade 60 Bars*

Bar Size	f_c' (Normal Weight Concrete), psi					
	3000	4000	5000	6000	8000	10,000
#3	4.8	4.2	3.7	3.4	3.0	2.6
#4	8.8	7.6	6.8	6.2	5.4	4.8
#5	13.6	11.8	10.5	9.6	8.3	7.4
#6	19.3	16.7	14.9	13.6	11.8	10.6
#7	26.3	22.8	20.4	18.6	16.1	14.4
#8	34.6	30.0	26.8	24.5	21.2	19.0
#9	43.8	37.9	33.9	31.0	26.8	24.0
#10	55.6	48.2	43.1	39.3	34.1	30.5
#11	68.4	59.2	52.9	48.3	41.9	37.4
#14	93.1	80.6	72.1	65.8	57.0	51.0
#18	136.9	118.6	106.1	96.8	83.9	75.0

*Development length, ℓ_d (including all applicable modification factors), must not be less than 12 in.

After establishing the basic tension development length, consideration must be given to the applicable modification factor or factors in Sections 12.2.3 through 12.2.5. The modification factors are multipliers for the basic development lengths (ℓ_{db}) to account for the following conditions.

12.2.3 Modification Factors for Bar Spacing and Cover, and Transverse Reinforcement

To account for clear bar spacing, amount of bar cover, and enclosing transverse reinforcement, appropriate modification factors are given in Sections 12.2.3.1 through 12.2.3.5. A minimum development length is also specified in Section 12.2.3.6 for the product of ℓ_{db} and the above modification factors for bar spacing and cover, and enclosing transverse reinforcement.

12.2.3.1 Modification Factor...1.0—Use of a modification factor of 1.0 is permitted if the bars being developed satisfy any one of the following four conditions:

(a) The bars are in beams or columns and satisfy all the following three conditions which are illustrated in Fig. 4-1.

1. Bars to be developed have a cover equal to or greater than the minimum cover specified for cast-in-place concrete (Section 7.7.1).

2. Bars are enclosed in transverse reinforcement satisfying the minimum tie requirements for columns (Section 7.10.5) or the minimum shear reinforcement requirements for beams (Sections 11.5.4 and 11.5.5.3) along the development length.

3. Bars have a clear spacing equal or greater than three times the diameter of the bar to be developed ($3d_b$).

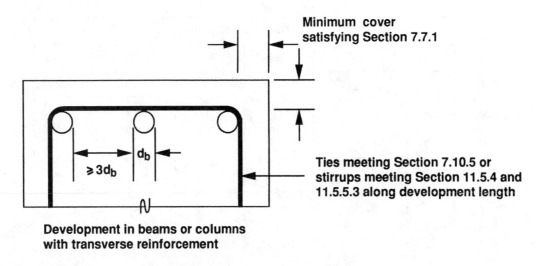

Development in beams or columns with transverse reinforcement

Fig. 4-1 Modification factor = 1 per Section 12.2.3.1(a)

(b) The bars are in beams or columns and satisfy both of the following conditions which are illustrated in Fig. 4-2.

1. Bars to be developed have a cover equal to or greater than the minimum cover specified for cast-in-place concrete (Section 7.7.1), and

2. Bars are enclosed along the development length in transverse reinforcement satisfying the equation

$$A_{tr} \geq \frac{d_b s\, N}{40}$$

where A_{tr} is the total cross-sectional area (in.^{2}) of stirrups or ties within a spacing s(in.) and perpendicular to plane of bars being developed or spliced; N is the number of bars, in a layer, being developed or spliced; and d_b is the nominal diameter (in.) of the largest bar being developed in the layer.

(c) The bars are in the inner layer of slab or wall reinforcement and have a clear spacing $\geq 3d_b$, as illustrated in Fig. 4-3, or

(d) The bars have a cover $\geq 2d_b$ and a clear spaceing $\geq 3d_b$, as illustrated in Fig. 4-4.

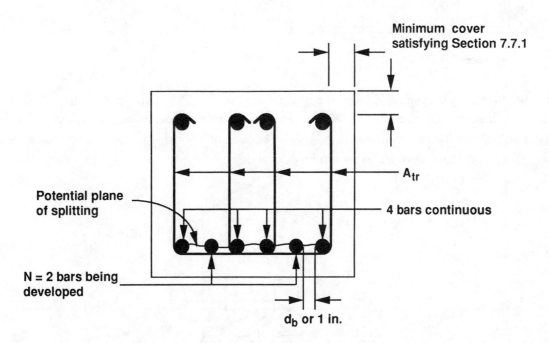

Confining reinforcement for closely spaced bars

Fig. 4-2   Modification factor = 1 per Section 12.2.3.1(b)

4-5

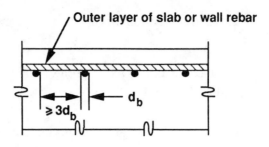

Fig. 4-3   Modification factor = 1 per Section 12.2.3.1(c)

Table 4-3—Transverse Reinforcement $A_{tr}/s = Nd_b/40$

Bar Size	Number of Bars Being Developed					
	N = 2		N = 3		N = 4	
	A_{tr}/s (in.^{2}/in.)	Stirrups or ties	A_{tr}/s (in.^{2}/in.)	Stirrups or ties	A_{tr}/s (in.^{2}/in.)	Stirrups or ties
#5	0.031	#4@13	0.047	#4@8.5	0.063	#4@6
#6	0.038	#4@10.5	0.056	#4@7	0.075	#4@5
#7	0.044	#4@9	0.066	#4@6	0.088	#5@7
#8	0.050	#4@8	0.075	#4@5	0.100	#5@6
#9	0.056	#4@7	0.085	#5@7	0.113	#5@5.5
#10	0.064	#4@6	0.095	#5@6.5	0.127	#5@5
#11	0.071	#4@5.5	0.106	#5@6	0.141	#5@4.5

Minimum concrete covers specified for precast concrete (Section 7.7.2) are smaller than those specified for cast-in-place concrete (Section 7.7.1). If the lesser covers permitted for precast concrete are used, these members do not qualify for conditions (a) and (b). Amount of enclosing transverse reinforcement required to qualify for above condition (b)2 is shown in Table 4-3.

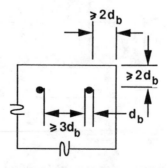

Cover greater than or equal to $2d_b$

Fig. 4-4   Modification factor = 1 per Section 12.2.3.1(d)

12.2.3.2   Modification Factor...2.0—Where clear spacing between bars being developed and bar cover are at or close to Code minimums (Section 7.6 and 7.7), splitting of concrete as shown in Commentary Fig. 12.2.3 can occur, reducing the transfer of force from bar to concrete. For such a condition, an increased length of bar embedment must be provided to develop the yield strength of the bar. For bars with clear spacing of two bar diameters ($2d_b$) or less, and with cover of one bar diameter (d_b) or less, the basic development length ℓ_{db} must be doubled as illustrated in Fig. 4-5.

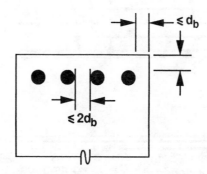

Small cover or closely spaced bars

Fig. 4-5   Modification factor = 2 per Section 12.2.3.2

12.2.3.3   Modification Factor...1.4—If bars to be developed do not qualify for the modification factors 1.0 (adequate cover, bar spacing and/or enclosing transverse reinforcement) or 2.0 (small cover or small clear bar spacing) discussed above, a modification factor of 1.4 must be used.

Note the phrase..."bar(s) being developed"; bar spacing to be used for the bar (or bars) being developed includes consideration of the location within the span where adjacent bars are developed. "Effective" bar spacing (to preclude a splitting failure) may be greater than actual bar spacing. If adjacent bars are all being developed at the same location within the span, development of the bars will be affected by the actual clear spacing between the bar, and development length must be based on the actual clear spacing between bars. If, however, an adjacent bar has already been developed at another location within the span, the "effective" spacing for the bar being developed is greater than the actual spacing to the adjacent bar. Effective bar spacing criteria for bars being developed is illustrated in Fig. 4-6. For development of Bars x, effective bar spacing "s" may be used, since Bars x are developed in length AB while Bars y are already developed in length BC. If adjacent Bars x and Bars y were being terminated at the same location within the span, then the effective bar spacing would be the actual spacing between the bars.

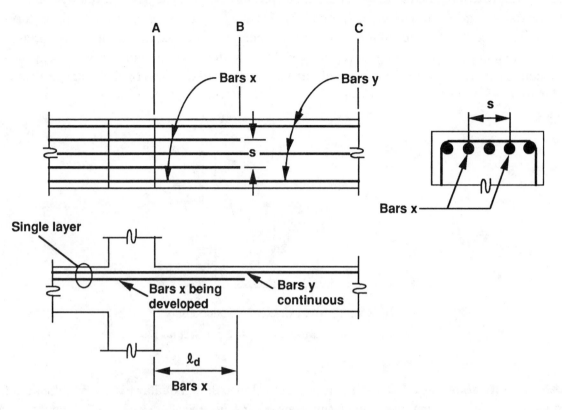

Fig. 4-6   Spacing Criteria for Bars Being Developed

12.2.3.4   Widely Spaced Bars...0.8—For #11 and smaller bars spaced sufficiently far apart to preclude splitting of the concrete through the plane of the bars, a modification factor of 0.8 is permitted in addition to those required by Sections 12.2.3.1 through 12.2.3.3. Clear spacing between bars being developed must not be less than $5d_b$ and cover to side or edge bar, measured in plane of bars, must not be less than $2.5d_b$. This is illustrated in Fig. 4-7.

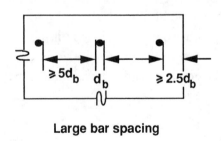

Large bar spacing

Fig. 4-7   Modification factor = 0.8 per Section 12.2.3.4

12.2.3.5   Closely Spaced Spiral or Ties...0.75—For closely spaced spirals or circular ties (circular bar arrangement), an additional modification factor of 0.75 may be applied to those of Sections 12.2.3.1 through 12.2.3.4.

12.2.3.6   Minimum Development Length—To guard against a pullout failure, a minimum development length ℓ_d is specified for the product of ℓ_{db} and the modification factors for bar spacing, cover, and enclosing transverse reinforcement (factors from Sections 12.2.3.1 through 12.2.3.5). Even though restraint to splitting may be provided by sufficient bar spacing and cover, tests indicate that a pullout failure is still possible; thus, the lower limit on development length. Minimum basic development lengths, corresponding to the pullout failure equation $0.03\,d_b f_y/\sqrt{f_c'}$ are given in Table 4-4 for grade 60 bars and concrete compressive strengths of 3000, 4000, 5000, 6000, and 10,000 psi. In no event is the development length ℓ_d permitted to be less than 12 in. (Section 12.2.1).

Table 4-4—Minimum Basic Development Length $\ell_{db} = 0.03 d_b f_y/\sqrt{f_c'}$ (inches) for Grade 60 Bars

Bar Size	f_c' (Normal Weight Concrete), psi					
	3000	4000	5000	6000	8000	10,000
#3	12.3	10.7	9.5	8.7	7.5	6.8
#4	16.4	14.2	12.7	11.6	10.1	9.0
#5	20.5	17.8	15.9	14.5	12.6	11.3
#6	24.6	21.3	19.1	17.4	15.1	13.5
#7	28.8	24.9	22.3	20.3	17.6	15.8
#8	32.9	28.5	25.5	23.3	20.1	18.0
#9	37.1	32.1	28.7	26.2	22.7	20.3
#10	41.7	36.1	32.3	29.5	25.6	22.9
#11	46.3	40.1	35.9	32.8	28.4	25.4
#14	55.6	48.2	48.2	39.3	34.1	30.5
#18	74.2	64.2	57.5	52.4	45.4	40.6

12.2.4   Modification Factors for Top Bar Effect, Type of Aggregate, and Epoxy Coating

Basic development length ℓ_{db} as modified by Section 12.2.3 must also be multiplied by applicable factors for top reinforcement, lightweight aggregate concrete, and epoxy-coated reinforcement as follows:

12.2.4.1   Top Reinforcement...1.3—This modification reflects the condition that top reinforcement may have reduced anchorage bond due to settlement of fresh concrete below the reinforcement, excess water used in the mix for workability, and entrapped air during mixing and placing. Water and air entrapped below the reinforcing bar decrease the bond of concrete to the reinforcement. The ℓ_d must be multiplied by a factor of 1.3 to account for the so-called "top bar effect". Top reinforcement is defined as horizontal reinforcement where more than 12 in. of fresh concrete is cast in the member below the development length or splice.

12.2.4.2 Lightweight Aggregate Concrete...1.3—The modification factor for lightweight concrete is based on the generally lower splitting tensile strength of lightweight aggregate concretes. For the '89 Code, the factor for lightweight concrete has been made the same for all types of lightweight aggregates. A lower factor may be used when the splitting tensile strength f_{ct} is specified. This factor is $6.7\sqrt{f'_c}/f_{ct}$, but must not be less than 1.0. The value of f_{ct} is a function of the concrete mix design and lightweight aggregate characteristics.

12.2.4.3 Epoxy-Coated Reinforcement...1.2 or 1.5—To account for reduced anchorage of epoxy-coated bars due to reduced adhesion between a coated bar and the concrete, the development length of coated bars must be increased, depending on bar spacing and cover, by a factor of 1.2 or 1.5. Bar spacing and cover requirements to determine the modification factor for epoxy-coated bars are as follows:

(a)   Clear Spacing $\geq 6d_b$ and
          Cover $\geq 3d_b$....$\ell_d = 1.2\,\ell_{db}$*

**Table 4-5—Multiples of Bar Diameter d_b for Bar Spacing
and Cover Check (Sections 12.2.3 and 12.2.4)**

Bar	d_b	$2d_b$	$2.5d_b$	$3d_b$	$5d_b$	$6d_b$
#3	0.375	0.75	0.94	1.13	1.89	2.25
#4	0.500	1.00	1.25	1.50	2.50	3.00
#5	0.625	1.25	1.56	1.88	3.13	3.75
#6	0.750	1.50	1.88	2.25	3.75	4.50
#7	0.875	1.75	2.19	2.63	4.38	5.25
#8	1.000	2.00	2.50	3.00	5.00	6.00
#9	1.128	2.25	2.81	3.38	5.63	6.75
#10	1.270	2.54	3.18	3.81	6.35	7.62
#11	1.410	2.82	3.53	4.23	7.06	8.46
#14	1.693	3.39	—	5.08	—	10.16
#18	2.257	4.51	—	6.77	—	13.54

(b)   Clear Spacing $< 6d_b$ or
          Cover $< 3d_b$....$\ell_d = 1.5\,\ell_{db}$*⁺

For reference, multiples of bar diameter d_d for bar spacing and cover requirements of Section 12.2.3 and 12.2.4 are given in Table 4-5.

---

*In addition to other modifiers of Section 12.2.3.

+For top bars, 1.3(1.5) = 1.95. However, Section 12.2.4.3 permits the use of a multiplier of 1.7 for the combined effects of top reinforcement (12.2.4.1) and epoxy coating (12.2.4.3).

12.2.5 Excess Reinforcement...(A_s required)/(A_s provided)

Development length may be reduced when excess reinforcement is provided in a flexural member. Use of a reduced development length implies that the bar will not be expected to develop full yield strength at factored design loads. Note that this reduction does not apply when the full f_y development is required, as for tension lap splices in Section 12.15.1, development of positive moment reinforcement at supports in Section 12.11.2, and for development of shrinkage and temperature reinforcement according to Section 7.12.2.3. Note also that this reduction in development length is not permitted for reinforcement in structures located in regions of high seismic risk (see Section 21.2.1.4).

SUMMARY

Development length ℓ_d for deformed bars and deformed wire in tension must be computed as the product of the basic development length ℓ_{db} of Section 12.2.2 and the applicable modification factors of Sections 12.2.3 through 12.2.5, but ℓ_d must not be less than 12 in.:

$$\ell_d = \ell_{db} \times (\text{applicable modification factors}) \geq 12 \text{ in.}$$

The following step-by-step procedure illustrates proper application of the various modification factors to determine development length ℓ_d.

	Calculations	Code Reference
(1)	Determine basic development length ℓ_{db}	12.2.2
(2)	Multiply ℓ_{db} by applicable factor for bar spacing, cover, and enclosing transverse reinforcement	12.2.3.1-12.2.3.3
(3)	Multiply by factor for wide bar spacing if applicable.	12.2.3.4
(4)	Multiply by factor for closely spaced spirals or ties, if applicable	12.2.3.5
(5)	Check minimum development length	12.2.3.6
(6)	Multiply by factor if top reinforcement	12.2.4.1
(7)	Multiply by factor if lightweight concrete	12.2.4.2
(8)	Multiply by factor if epoxy-coated reinforcement after checking that the product of multipliers for top bar and epoxy coating does not exceed 1.7	12.2.4.3
(9)	Multiply by ratio if excess reinforcement	12.2.4.4
(10)	Check minimum development length $\geq$ 12 in.	12.2.1

Examples 4-1 and 4-2 illustrate proper application of the various modification factors to determine tension development length.

OBSERVATIONS

Substantial increase in development lengths will result in several cases over that required by previous code requirements. The longer development lengths required ($\ell_d = 2 \ell_{db}$) for bars spaced at or near code minimum (Section 7.6) may create construction problems requiring the designer to alter his design in some way (i.e., use smaller bars, increase cover and/or spacing, or use enclosing transverse reinforcement or confining steel, etc.) to permit the use of shorter development lengths.

A comparison between minimum beam widths required by the '83 Code and those required by the '89 Code, for equivalent development lengths ($\ell_d = \ell_{db}$) is presented in Table 4-6. For the '83 Code, there is no multiplier ($\ell_d = \ell_{db}$) for clear bar spacing at Code minimums of d_b or 1 in. (Section 7.6.1). For the '89 Code, clear bar spacing must be increased to $3d_b$ to permit $\ell_d = \ell_{db}$. In essence, to use traditional beam widths with bar spacing at or near code minimums, development lengths by the '89 Code will be double those required by the '83 Code; otherwise, beam widths will need to be significantly increased as shown in Table 4-6 to permit $\ell_d = \ell_{db}$.

The longer required development lengths by the '89 Code may also make bar cutoff impractical and uneconomical. Also, where stirrups are not require for shear, it may be beneficial to provide stirrups to reduce bar development length.

Table 4-6—Minimum Beam Width (inches)

| Bar Size | Number of Bars in Single Layer | | | | | |
| | 3 | | 4 | | 5 | |
	'83 Code*	'89 Code**	'83 Code	'89 Code	'83 Code	'89 Code
#5	8.5	10.3	10.1	12.8	11.8	15.3
#6	8.8	11.3	10.5	14.3	12.3	17.3
#7	9.0	12.3	10.9	15.8	12.8	19.3
#8	9.3	13.3	11.3	17.3	13.3	21.3
#9	9.8	14.3	12.0	18.8	14.3	23.3
#10	10.3	15.4	12.9	20.5	15.4	25.6
#11	10.9	16.5	13.7	22.2	16.5	27.8

*To satisfy larger of one-bar diameter (d_b) or 1 in. clear spacing
**To satisfy three-bar diameters ($3d_b$) clear spacing

Assumptions made in computing minimum beam widths of Table 4-6:

1. Side Cover 1.5 in. each side

2. #3 Stirrups

3. Distance from centroid of bar nearest side face of beam to inside face of #3 stirrup taken as $2d_b = 0.75$ in. to satisfy minimum bend diameter for #3 stirrup (Section 7.2.2).

4. Distribution of reinforcement for crack control does not govern (Section 10.6).

12.3 DEVELOPMENT OF DEFORMED BARS IN COMPRESSION

Shorter development lengths are required for bars in compression since the weakening effect of flexural tension cracks in the concrete is not present. The basic compression development length is $\ell_{db} = 0.02d_bf_y/\sqrt{f'_c}$ but not less than $0.0003d_bf_y$ or 8 in. The basic development length may be reduced where

Table 4-7—"Basic" Compression Development Length ℓ_{db} (inches) for Grade 60 Bars

Bar Size	f'_c (Normal Weight Concrete), psi		
	3000	4000	≥ 4444*
#3	8.2	7.1**	6.8**
#4	11.0	9.5	9.0
#5	13.7	11.9	11.3
#6	16.4	14.2	13.5
#7	19.2	16.6	15.8
#8	21.9	19.0	18.0
#9	24.7	21.4	20.3
#10	27.8	24.1	22.9
#11	30.9	26.8	25.4
#14	37.1	32.1	30.5
#18	49.4	42.8	40.6

*For $f'_c \geq 4444$ psi, minimum basic development length $0.0003d_bf_y$ governs; for Grade 60, $\ell_d = 18d_b$.

**Development length ℓ_d (including applicable modification factors) must not be less than 8 in.

excess bar area is provided and where "confining" ties or spirals are provided around the bars (Section 12.3.3). Note that the tie and spiral requirements to permit the 25 percent reduction in development length are somewhat more restrictive than those required for "regular" column ties in Section 7.10.5 and less restrictive than those required for spirals in Section 7.10.4. For reference, the basic compression development lengths for Grade 60 bars are given in Table 4-7.

12.4 DEVELOPMENT OF BUNDLED BARS

Increased development length for individual bars within a bundle, whether in tension or compression, is required when 3 or 4 bars are bundled together. The additional length is needed because the grouping makes it more difficult to mobilize resistance to slippage from the "core" between the bars. The modification factor is 1.2 for a 3-bar bundle, and 1.33 for a 4-bar bundle. Note Section 7.6.6.4 relating to cut-off points of individual bars within a bundle, and Section 12.14.2.2 relating to lap splices of bundled bars.

Where the modification factors of Section 12.2.3 and 12.2.4 are based on bar diameter db, a unit of bundled bars must be treated as a single bar of a diameter derived from the total equivalent area. See Part 3, Table 3-5.

12.5 DEVELOPMENT OF STANDARD HOOKS IN TENSION

The current provisions for hooked bar development were first introduced in the 1983 ACI Code, and represented a major departure from the hooked-bar anchorage provisions of earlier codes in that they uncoupled hooked-bar anchorages from straight bar development and gave total hooked-bar embedment length directly. The current provisions not only simplify calculations for hook anchorage lengths but also result in a required embedment length considerably less, especially for the larger bar sizes, than that required by earlier codes. Provisions are given in Section 12.5 for determining the development length of deformed bars with standard end hooks. End hooks can only be considered effective in developing bars in tension, and not in compression; see Section 12.1 and 12.5.5. Only "standard" end hooks (Section 7.1) are considered; anchorage capacity of end hooks with larger end diameters cannot be determined by the provisions of Section 12.5.

Application of the hook development provisions is essentially the same as calculating development lengths of straight bars. The first step is to calculate the basic development length of the hooked bar, ℓ_{hb}. The basic development length is then multiplied by the applicable modification factor or factors to determine the development length of the hook, $\ell_{dh} = \ell_{hb} \times$ applicable modification factors. Development length ℓ_{dh} is measured from the critical section to the outside end of the standard hook, i.e., the straight embedment length between the critical section and the start of the hook, plus the radius of bend of the hook, plus one-bar diameter. For reference, Fig. 4-8 shows ℓ_{dh} and the standard hook details (Section 7.1) for all standard bar

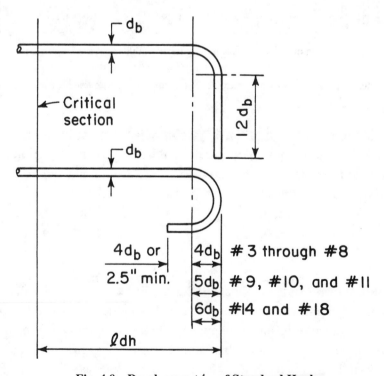

Fig. 4-8   Development ℓ_{dh} of Standard Hooks

sizes. For 180-degree hooks normal to exposed surfaces, the embedment length should provide for a minimum distance of 2 in. beyond the tail of the hook[4.2]. This distance is denoted by an asterisk (*) in Fig. 4-8.

12.5.2 Basic Development Length ℓ_{hb}

The basic development length, ℓ_{hb}, for standard hooks in tension is given in Section 12.5.2 for grade 60 bars as:

Table 4-8—Basic Development Length ℓ_{hb} of Standard Hooks for Grade 60 Bars

Bar Size	f_c' (Normal Weight Concrete), psi					
	3000	4000	5000	6000	8000	10,000
#3	8.2	7.1	6.4	5.8	5.0	4.5
#4	11.0	9.5	8.5	7.7	6.7	6.0
#5	13.7	11.9	10.6	9.7	8.4	7.5
#6	16.4	14.2	12.7	11.6	10.1	9.0
#7	19.2	16.6	14.8	13.6	11.7	10.5
#8	21.9	19.0	17.0	15.5	13.4	12.0
#9	24.7	21.4	19.1	17.5	15.1	13.5
#10	27.8	24.1	21.6	19.7	17.0	15.2
#11	30.9	26.8	23.9	21.8	18.9	16.9
#14	37.1	32.1	28.7	26.2	22.7	20.3
#18	49.4	42.8	38.3	35.0	30.3	27.1

$$\ell_{hb} = 1200 \, d_b / \sqrt{f_c'}$$

Table 4-8 lists the basic development length of hooked bars embedded in normal weight concrete with specified compressive strengths of 3000, 4000, 5000, 6000, 8000, and 10,000 psi.

12.5.3 Modification Factors

The ℓ_{hb} modification factors listed in Section 12.5.3 account for:

- Bar yield strength other than Grade 60;
- Favorable confinement conditions provided by increased cover

- Favorable confinement provided by transverse ties or stirrups to resist splitting of the concrete,
- More reinforcement provided than required by analysis;
- Lightweight aggregate concrete.

After multiplying the basic development length ℓ_{hb} by the applicable modification factor or factors, the resulting development length ℓ_{dh} must not be less than $8d_b$ nor less than 6 in.

The side cover (normal to plane of hook), and the cover on bar extension beyond 90- degree hook referred to in Section 12.5.3.2 are illustrated in Fig. 4-9.

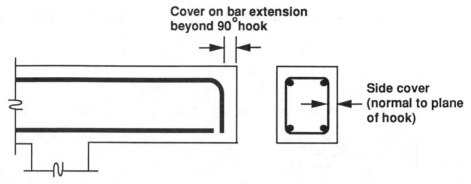

Fig. 4-9   Concrete Covers Referenced in Section 12.5.3.2

12.5.4   Standard Hook at Discontinuous Ends

Section 12.5.4 is a special provision for hooked bars terminating at discontinuous ends of members, such as at the ends of simply-supported beams, at free ends of cantilevers, and at ends of members framing into a joint where the member does not extend beyond the joint. If the full strength of a hooked bar must be developed, and both side cover and top (or bottom) cover over the hook are less than 2½ in., Section 12.5.4 requires the hook to be enclosed within ties or stirrup-ties for the full development length, ℓ_{dh}. Spacing of the ties or stirrup-ties must not exceed $3d_b$, where d_b is the diameter of the hooked bar. In addition,the modification factor of 0.8 for confinement provided by ties or stirrups (Section 12.5.3.3) does not apply to the special condition covered by Section 12.5.4. At discontinuous ends of slabs with concrete confinement provided by the slab continuous on both sides normal to the plane of the hook, the provisions of Section 12.5.4 do not apply.

12.6   MECHANICAL ANCHORAGE

Section 12.6 permits the use of mechanical devices for development of reinforcement, provided their adequacy without damaging the concrete has been confirmed by tests. Section 12.6.3 reflects the concept that development of reinforcement may consist of a combination of mechanical anchorage plus additional embedment length of the reinforcement. For example, when a mechanical device cannot develop the design strength of a bar, an additional embedment length of the bar must be provided between the mechanical device and the critical section.

12.7   DEVELOPMENT OF WELDED DEFORMED WIRE FABRIC IN TENSION

For welded deformed wire fabric, development length is measured from the critical section to the end of the wire. The provisions of Section 12.7.2 for calculating the basic development length ℓ_{db} are based on the condition that at least one cross wire is located within the development length, and the cross wire is no less than 2 in. distant from the critical section. The basic development length is given as $\ell_{db} = 0.03 d_b (f_y - 20{,}000)/\sqrt{f_c'}$, but not less than $0.20 A_w f_y/(s_w \sqrt{f_c'})$.

The required development length is established by multiplying the basic development length ℓ_{db} by the applicable modification factor or factors of Sections 12.2.3 through 12.2.5. The modification factors for bar "or wire" spacing and cover (Section 12.2.3), introduced in the '89 Code, have limited application for wire fabric with customary spacings used for slab and wall reinforcement. Deformed wire reinforcement is available in sizes up to D31 (nominal diameter = 0.628 in.). For wire fabric, two conditions for wire spacing and cover need to be considered; both include the wide wire spacing factor of Section 12.2.3.4:

(1) Clear spacing $\geq 5 d_w$ and
    cover $\geq 2 d_w \dots \ell_d = 0.8(1.0 \, \ell_{db}) = 0.8 \, \ell_{db}$

(2) Clear spacing $\geq 5 d_w$ and
  $2 d_w >$ cover $\geq d_w \dots \ell_d = 0.8(1.4 \, \ell_{db}) = 1.12 \, \ell_{db}$

With wire size available in diameters up to ⅝ in. (D31), the $5 d_w$ clear spacing is easily satisfied for customary spacings of wire fabric used for slab or wall reinforcement. For the D31 wire size, $2 d_w$ cover = 2(0.628) = 1.26 in. Minimum cover permitted by Section 7.7.1 for interior exposure is ¾ in., less than the $2 d_w$ cover to qualify for $\ell_d = 0.8 \, \ell_{db}$. For D11 wire ($d_w$ = 0.374 in.) and smaller, the $2 d_w$ cover (2 × 0.374 = 0.75 in.) is satisfied, permitting $\ell_d = 0.8 \, \ell_{db}$. Since most commonly used wire sizes are less than D11, the $\ell_d = 0.8 \, \ell_{db}$ will apply for development of most deformed wire fabric used for slab and wall reinforcement.

In addition to the wire spacing and cover modification factor ($\ell_d = 0.8 \, \ell_{db}$), the top reinforcement factor of 1.3 and the lightweight concrete factor of 1.3 must be applied where applicable. Note that the top reinforcement factor applies only when 12 in. of fresh concrete is cast below the reinforcement being developed. The excess reinforcement ratio may also be applied where applicable. The resulting development length ℓ_d cannot be less than 8 in., except in computation of lap splice lengths (Section 12.18) and development of web reinforcement (Section 12.13). Fig. 4-10 shows the development length requirements for welded deformed wire fabric.

If no cross wires are located within the development length, the basic development length for the fabric must be based on the provisions for deformed wire in Section 12.2.

12.8 DEVELOPMENT OF WELDED PLAIN WIRE FABRIC IN TENSION

For welded plain wire fabric, the development length is measured from the point of critical section to the outermost cross wire. Full development of plain fabric ($A_w f_y$) is achieved by embedment of at least two cross wires beyond the critical section, with the closer cross wire located not less than 2 in. from the critical section. Section 12.8 further requires that the length of embedment from critical section to outermost cross wire must not be less than $\ell_{db} = 0.27 \, A_w f_y/(s_w \sqrt{f_c'})$, nor less than 6 in. For lightweight aggregate concrete, the basic development length ℓ_{db} must be modified by the factor in Section 12.2.4.2. If more reinforcement is provided than that required by analysis, the basic development length ℓ_{db} may be reduced by the ratio of (A_s required)/

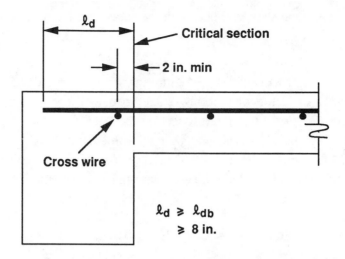

Fig. 4-10 Development of Welded Deformed Wire Fabric

A_s provided). The 6 in. minimum development length does not apply to computation of lap splice lengths (Section 12.19). Fig. 4-11 shows the development length requirements for plain wire fabric.

For fabrics made with smaller wires, embedment of two cross wires, 2 in. or more beyond the point of critical section, is usually adequate to develop the full yield strength of the anchored wire. Fabrics made with larger (closely spaced) wires will require a longer embedment based on the ℓ_{db} basic development length.

For example, check fabric 6 × 6-W4xW4 with $f_c' = 3000$ psi and $f_y = 60,000$ psi.

$$\ell_{db} = 0.27 \times (A_w/s_w) \times (f_y/\sqrt{f_c'})$$
$$= 0.27 \times (0.04/6) \times (60,000/\sqrt{3000}) = 1.97 \text{ in.}$$
$$< 6 \text{ in.}$$
$$< (1 \text{ space} + 2 \text{ in.}) \text{ See Fig. 4-12}$$

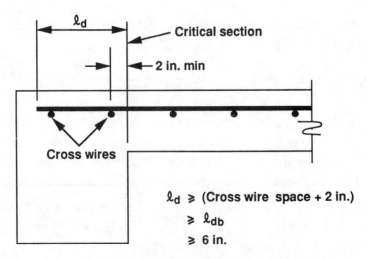

Fig. 4-11 Development of Welded Plain Wire Fabric

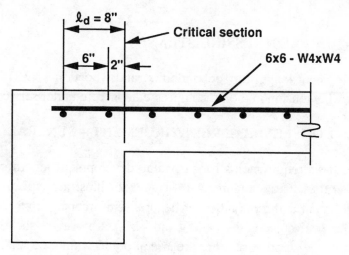

Two cross wire embedment plus 2 in. is satisfactory.

Fig. 4-12 Development of 6x6-W4xW4 Fabric

Check fabric 6 × 6-W20xW20:

$$\ell_{bd} = 0.27 \times (0.20/6) \times (60{,}000/\sqrt{3000}) = 9.9 \text{ in.}$$
$$> 6 \text{ in.}$$
$$> (1 \text{ space} + 2 \text{ in.})$$

As shown in Fig. 4-13, an additional 2 in. beyond the two cross wires + 2 in. embedment is required to fully develop the W20 fabric. If the longitudinal spacing is reduced to 4 in. (4x6-W20xW20), a minimum ℓ_{db} of 15 in. is required for full development...e.g., 3 cross wires + 3 in. embedment.

References 4.1 and 4.2 provide design aid data in the use of welded wire fabric, including development length tables for both deformed and plain welded wire fabric.

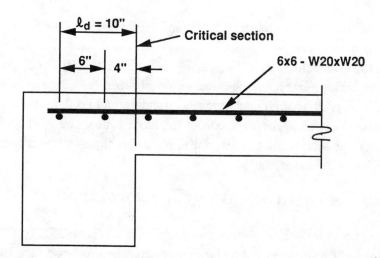

Note: If end support is not wide enough for straight embedment, the ℓ_d length may be bent down (hooked) into support.

Fig. 4-13 Development of 6x6-W20xW20 Fabric

12.9 DEVELOPMENT OF PRESTRESSING STRAND

Note that double development length for "debonded" strands (Section 12.9.3) is required only when the member is designed allowing tension in the precompressed concrete under service load conditions.

12.10 DEVELOPMENT OF FLEXURAL REINFORCEMENT—GENERAL

Section 12.10 gives the basic requirements for providing development length of reinforcement from the points of maximum or critical stress. Figures 4-14(a) and (b) illustrate typical critical sections and Code requirements for development and termination of flexural reinforcement in a continuous beam. Points of maximum positive and negative moments ($+M_u$ and $-M_u$) are critical sections, from which adequate anchorage l_d must be provided. Critical sections are also at points within the span where adjacent reinforcement is terminated; continuing bars must have adequate anchorage l_d from the theoretical cut-off points of terminated bars (Section 12.10.4). Note also that terminated bars must be extended beyond the theoretical cut-off points in accordance with Section 12.10.3. This extension requirement is to guard against possible shifting of the moment diagram due to load variation, settlement of supports, and other unforeseen changes in the moment conditions. In determining development or embedment lengths l_d, effect of bar spacing and cover, and enclosing stirrups within the development length, along with other modifying conditions, must be considered in accordance with Section 12.2.

Sections 12.10.1 and 12.10.5 concern the option of anchoring tension reinforcement in a compression zone. Research has confirmed the need for restrictions on terminating bars in a tension zone. When flexural bars are cut off in a tension zone, flexural cracks tend to open early. If the shear stress in the area of bar cut-off and tensile stress in the remaining bars at the cut-off location are near the permissible limits, diagonal tension cracking tends to develop from the flexural cracks. One of the three alternatives of Section 12.10.5 must be satisfied to reduce the possible occurrence of diagonal tension cracking near bar cut-offs in a tension zone. Section 12.10.5.2 requires excess stirrup area over that required for shear and torsion. Requirements of Section 12.10.5 are not intended to apply to tension splices.

Section 12.10.6 is for end anchorage of tension bars in special flexural members such as brackets, members of variable depth, and others where bar stress, f_s, does not decrease linearly in proportion to a decreasing moment. In Fig. 4-15, the l_d into the support is probably less critical than the required development length. In such a case, safety depends primarily on the outer end anchorage provided. A welded cross bar of equal diameter should provide an effective end anchorage. A standard end hook in the vertical plane may not be effective because an essentially plain concrete corner might exist near the load and could cause localized failure. Where brackets are wide and loads are not applied too close to the corners, U-shaped bars in a horizontal plane provide effective end hooks.

12.11 DEVELOPMENT OF POSITIVE MOMENT REINFORCEMENT

To further guard against possible shifting of moments due to various causes, Section 12.11.1 requires specific amounts of positive moment reinforcement to be extended along the same face of the member into the support, and for beams to be embedded into the support at least 6 in. The specified amounts are one-third for simple members and one-fourth for continuous members. In Fig. 4-14(b), for example, the area of Bars "B" would have to be at least one-fourth of the area of reinforcement required at the point of maximum $+M_u$.

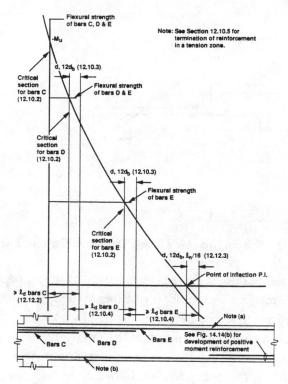

Note: Portion of total negative reinforcement (-A_s) must be continuous along full length of perimeter beams (Section 7.13).

(a) Negative Moment Reinforcement

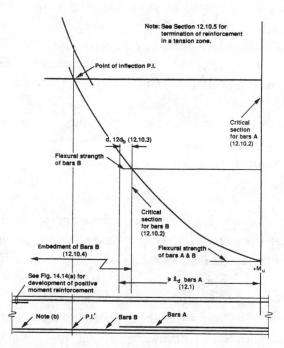

Note: (b) Portion of total positive reinforcement (+A_s) must be continuous along full length of perimeter beams and beams without closed stirrups (Section 7.13).

(b) Positive Moment Reinforcement

Fig. 4-14 Development of Positive and Negative Moment Reinforcement

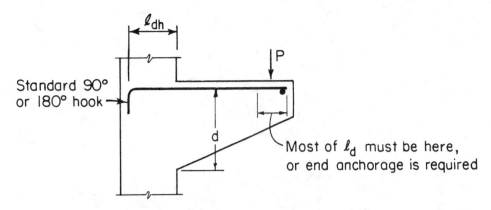

Fig. 4-15   Special Member Largely Dependent on End Anchorage

Section 12.11.2 is intended to assure ductility in the structure under severe overload, as might be experienced in a strong wind or earthquake. In a lateral load resisting system, full anchorage of the reinforcement extended into the support provides for possible stress reversal under such overload. Anchorage must be provided to develop the full yield strength in tension at the face of the support. The provision will require such members to have bottom bars lapped at interior supports or hooked at exterior supports. The full anchorage requirement does not apply to any excess reinforcement provided at the support.

Section 12.11.3 limits bar sizes for the positive moment reinforcement at simple supports and at points of inflection. In effect, this places a design restraint on flexural bond stress in areas of small moment and large shear. Such a condition could exist in a heavily loaded beam of short span, thus requiring large size bars to be developed within a short distance. Bars should be limited to a diameter such that the development length ℓ_d computed for f_y according to Section 12.2 does not excess $(M_n/V_u) + \ell_a$. The limit on bar size at simple supports is waived if the bars have standard end hooks or mechanical anchorages terminating beyond the centerline of the support. Mechanical anchorages must be equivalent to standard hooks.

The length (M_n/V_u) corresponds to the development length of the maximum size bar permitted by the previously used flexural bond equation. The length (M_n/V_u) may be increased 30% when the ends of the bars are confined by a compressive reaction, such as provided by a column below, but not when a beam frames into a girder.

For the simply-supported beam shown in Fig. 4-16, the maximum permissible ℓ_d for Bars "a" is 1.3 $M_n/V_u + \ell_a$. This has the effect of limiting the size of bar to satisfy flexural bond. Even though the total embedment length from the critical section for Bars "a" is greater than 1.3 $M_n/V_u + \ell_a$, the size of Bars "a" must be limited so that $\ell_d \leq$ 1.3 $M_n/V_u + \ell_a$. Note that M_n is the nominal flexural strength of the cross section (without the φ factor). As noted previously, larger bar sizes can be accommodated by providing a standard hook or mechanical anchorage at the end of the bar within the support. At a point of inflection (see Fig. 4-17), the positive moment reinforcement must have a development length ℓ_d, as computed by Section 12.2, not to exceed the value of $(M_n/V_u) + \ell_a$, with ℓ_a not greater than d or 12 d_b, whichever is greater. For example, a #11 bar requires a basic tension development length of $0.04A_b f_y/\sqrt{f_c'} = 59$ in. for Grade 60 and 4000 psi normal weight concrete. For a short span beam, the #11 bar may be too large to satisfy the flexural bond requirement.

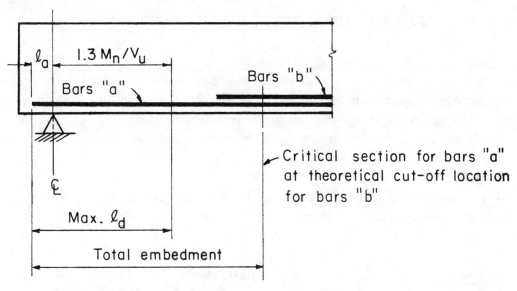

Fig. 4-16   Development Length Requirements at Simple Support (straight bars)

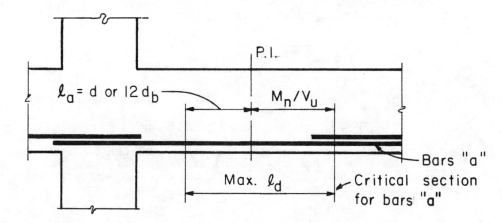

Fig. 4-17   Concept for Determining Maximum Size of Bars "a" at Point of Inflection

12.12   DEVELOPMENT OF NEGATIVE MOMENT REINFORCEMENT

The requirements in Section 12.12.3 guard against possible shifting of the moment diagram at points of inflection. At least one-third of the negative moment reinforcement provided at a support must be extended a specified embedment length beyond a point of inflection. The embedment length must be the effective depth of the member d, $12d_b$, or $\frac{1}{16}$ the clear span, whichever is greater...as shown in Figs. 4-14 and 4-18. The area of Bars "E" in Fig. 4-14(a) must be at least one-third the area of reinforcement provided for $-M_u$ at the face of the support. Anchorage of top reinforcement in tension beyond interior support of continuous members usually becomes part of the adjacent span top reinforcement... as shown in Fig. 4-18.

Standard end hooks are an effective means of developing top bars in tension at exterior supports as shown in Fig. 4-19. The Code requirements for development of standard hooks are discussed in Section 12.5.

12.13 DEVELOPMENT OF WEB REINFORCEMENT

Stirrups must be properly anchored so that the full tensile force in the stirrup can be developed at or near mid-depth of the member. To function properly, stirrups must be extended as close to the compression and tension surfaces of the member as cover requirements and proximity of other reinforcement permit (Section 12.13.1). It is equally important for stirrups to be anchored as close to the compression face of the member as possible because flexural tension cracks initiate at the tension face and extend towards the compression zone as member strength is approached.

The ACI Code anchorage details for stirrups evolved over many editions of the Code and are based primarily on past experience and performance in laboratory tests.

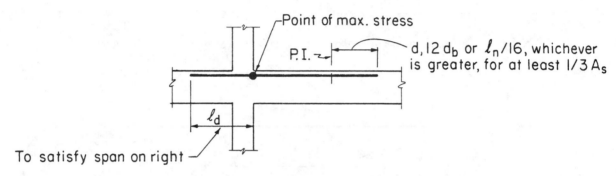

Fig. 4-18   Anchorage into Adjacent Beam

(Usually such anchorage becomes part of adjacent beam reinforcement)

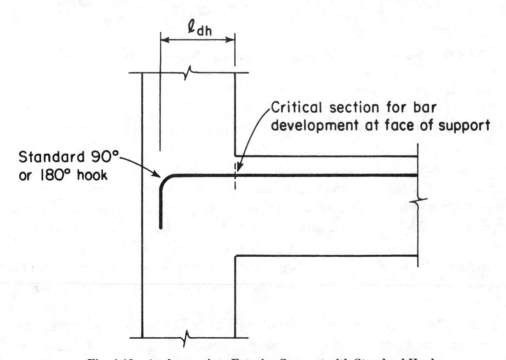

Fig. 4-19   Anchorage into Exterior Support with Standard Hook

Designers and fabricators have had considerable confusion interpreting the d/2 and mid-depth anchorage requirements of the '83 Code. This prompted the ACI Committee 318 to review the web reinforcement anchorage details. For the '89 Code, U-stirrup anchorage details for deformed bars and deformed wire have been significantly simplified; for the commonly used stirrup bar sizes, the designer need only provide a standard hook at the ends of stirrups for anchorage. For #5 bar and smaller, stirrup anchorage is now provided, simply by a standard stirrup hook (90° bend plus $6d_b$ extension at free end of bar)* around a longitudinal bar (Section 12.13.2.1). The same anchorage detail is permitted for the larger stirrup bar sizes, #6, #7, and #8, in Grade 40. Note that for the larger bar sizes, the 90° hook detail requires a $12d_b$ extension at the free end of the bar. Fig. 4-20 illustrates the new anchorage requirement for U-stirrups fabricated from deformed bars and deformed wire.

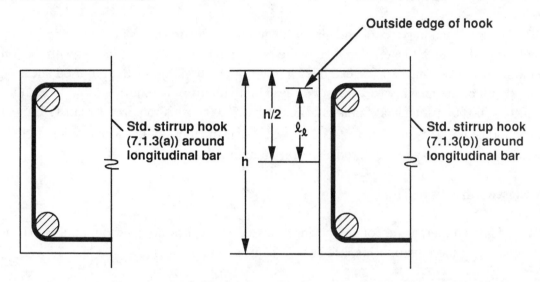

Fig. 4-20 Anchorage Details for U-Stirrups (Deformed Bars and Deformed Wire)

For the larger stirrup bar sizes (#6, #7, or #8) in Grade 60, in addition to a standard stirrup hook, an embedment of $0.014d_bf_y/\sqrt{f_c'}$ between midheight of member and outside end of hook is required. The available embedment length, denoted ℓ_ℓ, must be checked to ensure adequate anchorage at the higher bar force (Section 12.13.2.2). The embedment length required is tabulated in Fig. 4-20. Minimum depth of member required to accommodate the #6, #7, or #8 stirrups fabricated in Grade 60 is also tabulated in Fig. 4-20. For practical size of beams where the loads are of such magnitude to require #6, #7, or #8 bar sizes for shear reinforcement, the embedment length required should be easily satisfied, and the designer needonly be concerned with providing a standard stirrup hook around a longitudinal bar for proper stirrup end anchorage.

Note that the straight bar stirrup anchorage detail of the 1983 Code is no longer permitted. That anchorage detail was difficult to hold in place during concrete placement and the lack of an end hook anchorage made the stirrup effectiveness questionable at high shear cracking.

---

*For structures located in regions of high seismic risk, stirrups required to be hoops must be anchored with a 135-deg bend plus $6d_b$ (but not less than 3 in.) extension of free end of bar. See Sections 21.1 and 21.7.2.2.

Provisions covering the use of welded smooth wire fabric as simple U-stirrups are shown in Fig. 4-21. Stirrup anchorage detail for single leg stirrups formed with welded plain or deformed wire fabric is shown in Fig. 4-22. Anchorage of the single leg is provided primarily by the longitudinal wires. Use of welded wire fabric for shear reinforcement has become commonplace in the precast, prestressed concrete industry.

Note that Section 12.13.3 requires that each bend in the continuous portion of U-stirrups must enclose a longitudinal bar; a requirement usually satisfied for simple U-stirrups but requiring special attention in bar detailing when multiple U-stirrups are used.

12.13.4   Anchorage for Bent-Up Bars

Section 12.13.4 gives anchorage requirements for longitudinal (flexural) bars bent up to resist shear. If the bent-up bars are extended into a tension region, the bent-up bars must be continuous with the longitudinal reinforcement. If the bent-up bars are extended into a compression region, the required anchorage length beyond mid-depth of the member ($d/2$) must be based on that part of f_y required to satisy Eq. (11-19). For example, if $f_y = 60,000$ psi and calculations indicate that 30,000 psi is required to satisfy Eq. (11-19), the required anchorage length $\ell'_d = (30,000/60,000)\,\ell_d$, where ℓ_d is the tension development length for full f_y per Section 12.2. Fig. 4-23 shows the required anchorage length ℓ'_d.

12.13.5   Closed Stirrups or Ties

Section 12.13.5 gives requirements for lap splicing double U-stirrups or ties (without hooks) to form a closed stirrup. Legs are considered properly spliced when the laps are $1.3\,\ell_d$ as shown in Fig. 4-24.

Embedment Length ℓ (in.) for Grade 60 Bars

Bar Size	Concrete Compressive Strength, f'_c psi					
	3000	4000	5000	6000	8000	10,000
#6	11.5	10.0	8.9	8.1	7.0	6.3
#7	13.4	11.6	10.4	9.5	8.2	7.4
#8	15.3	13.3	11.9	10.8	9.4	8.4

Minimum depth of member (in.) to accommodate #6, #7, and #8 in Grade 60

Min. Cover and Exposure	Bar Size	3000	4000	5000	6000	8000	10,000
1½ in.	#6	26	23	21	20	17	16
Interior	#7	30	27	24	22	20	18
Exposure	#8	34	30	27	25	22	20
2 in.	#6	27	24	22	21	18	17
Exterior	#7	31	28	25	23	21	19
Exposure	#8	35	31	28	26	23	21

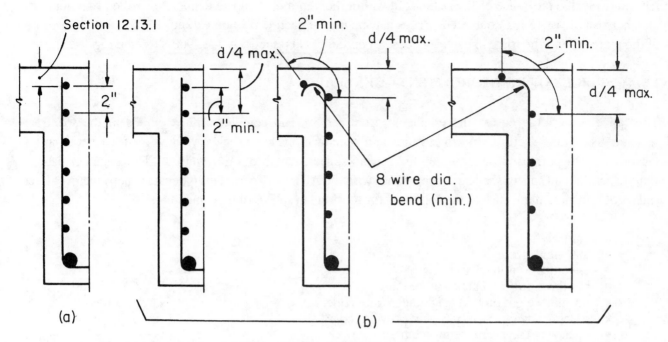

Fig. 4-21   Anchorage Details for Welded Plain Wire Fabric U-Stirrups (Section 12.13.2.3)

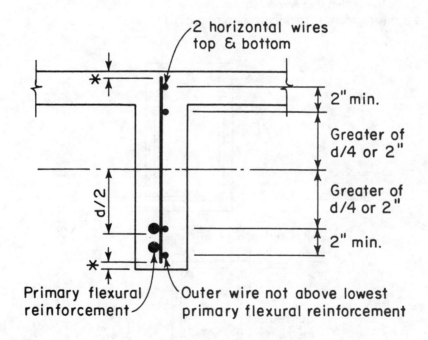

Fig. 4-22   Anchorage Details for Welded Wire Fabric Single Leg Stirrups (Section 12.13.2.4)

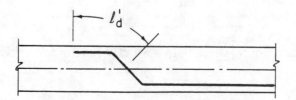

Fig. 4-23   Anchorage for Bent-Up Bars

4-27

Alternatively, if a lap splice of $1.3\ell_d$ cannot fit within the depth of shallow members, provided that depth of members is at least 18 in., double U-stirrups may be used if each U portion extends the full available depth of the member and the force in each leg does not exceed 9000 pounds; $A_b f_y \le 9000$ lbs (see Fig. 4-25).

12.14 SPLICES OF REINFORCEMENT—GENERAL

The splice provisions require the engineer to show clear and complete splice details in the Contract Documents. The structural drawings, notes and specifications should clearly show or describe all splice locations, types permitted or required, and for lap splices, length of lap required. The engineer cannot simply state that all splices shall be in accordance with the ACI-318 Code. This is because many factors affect splices of reinforcement, such as the following for tension lap splices of deformed bars:

- bar size
- bar yield strength
- concrete compressive strength
- bar location (top bars or other bars)
- normal weight or lightweight aggregate concrete
- spacing and cover of bars being developed
- enclosing transverse reinforcement
- epoxy coating
- number of bars spliced at one location
- excess reinforcement (provided versus required)

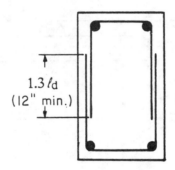

Fig. 4-24   Overlapping U-Stirrups to Form Closed Unit

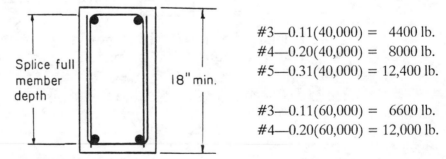

#3—0.11(40,000) =	4400 lb.
#4—0.20(40,000) =	8000 lb.
#5—0.31(40,000) =	12,400 lb.
#3—0.11(60,000) =	6600 lb.
#4—0.20(60,000) =	12,000 lb.

If stirrups are designed for the full yield strength f_y, #3 and #4 stirrups of Grade 40 and only #3 of Grade 60 satisfy the 9000 lb. limitation.

Fig. 4-25   Lap Splice Alternative

It is virtually impossible for a reinforcing bar detailer to know what splices are required at a given location in a structure, unless the engineer explicitly illustrates or defines the splice requirements. Section 12.14.1 states: "Splices of reinforcement shall be made only as required or permitted on the design drawings, or in specifications, or as authorized by the Engineer."

Two industry publications are suggested as design reference material for proper splicing of reinforcement. Reference 4.1 provides design aid data in the use of welded wire fabric,including development length and splice length tables for both deformed and plain wire fabric. Reference 4.2 provides accepted practices in splicing reinforcement; use of lap splices, welded splices and splice devices are described, including simplified design data for lap splice lengths.

12.14.2   Lap Splices

Lap splices are not permitted for bars larger than #11, either in tension or compression, except:

- #14 and #18 bars in compression only may be lap spliced to #11 and smaller bars (Section 12.16.2), and

- #14 and #18 in compression only may be lap spliced to smaller size footing dowels (Section 15.8.2.3).

Section 12.14.2.2 gives the provisions for lap splicing of bars in a bundle (tension or compression). The lap lengths required for individual bars within a bundle must be increased by 20 percent and 33 percent for 3- and 4-bar bundles, respectively. Overlapping of individual bar splices within a bundle is not permitted. Two bundles must not be lap-spliced as individual bars.

Bars in flexural members may be spliced by noncontact lap splices. To prevent a possible unreinforced section in a spaced (noncontact) lap splice, Section 12.14.2.3 limits the maximum distance between bars in a splice to one-fifth the lap length, or 6 in. whichever is less. Contact lap splices are preferred for the practical reason that when the bars are wired together, they are more easily secured against displacement during concrete placement.

12.14.3   Welded Splices and Mechanical Connections

Section 12.14.3 permits the use of welded splices or other mechanical connections. In a full welded splice, the bars must be butted and the splice must develop in tension at least 125 percent of the specified yield strength of the bar. Likewise, a full mechanical connection must develop, in tension or compression, at least 125 percent of the specified yield strength of the bar. The Code permits the use of welded splices or mechanical connections having less than 125 percent of the specified yield strength of the bar in regions of low computed stress as specified in Section 12.15.4.

Section 12.14.3.2 requires all welding of reinforcement to conform to *Structural Welding Code-Reinforcing Steel* (AWS D1.4). Section 3.5.2 requires that the reinforcement to be welded must be indicated on the drawings, and the welding procedure to be used must be specified. To carry out these Code requirements properly, the engineer should be familiar with provisions in AWS D1.4 and the ASTM specifications for reinforcing bars.

Since the standard rebar specifications ASTM A615, A616 and A617 specifically state that "weldability of the steel is not part of this specification," there are no limits on the chemical elements that affect weldability of the steels. A key item in AWS D1.4 is carbon equivalent (C.E.). The minimum preheat and interpass temperatures specified in AWS D1.4 are based on C.E. and bar size. Thus, as indicated in Section 3.5.2 and in Commentary Section 3.5.2, when welding is required, the ASTM A615, A616 and A617 rebar specifications must be supplemented to require a report of the chemical composition to assure that the welding procedure specified is compatible with the chemistry of the bars.

ASTM A706 reinforcing bars are intended for welding. The A706 specification contains restrictions on chemical composition, including carbon, and C.E. is limited to 0.55 percent. The chemical composition and C.E. must be reported. By limiting C.E. to 0.55 percent, little or no preheat is required by AWS D1.4. Thus, the engineer does not need to supplement the A706 specification when the bars are to be welded. However, before specifying ASTM A706 reinforcing bars, local availability should be investigated.

Reference 4.2 contains a detailed discussion of welded splices. Included in the discussion are requirements for other important items such as field inspection, supervision, and quality control.

Note that careful review of AWS D1.4 reveals that the document essentially covers the welding of reinforcing bars only. For welding of wire to wire, and of wire or welded wire fabric to reinforcing bars or structural steels, such welding should conform to applicable provisions of AWS D1.4 and to supplementary requirements specified by the engineer. Also, the engineer should be aware that there is a potential loss of yield strength and ductility of low carbon cold-drawn wire if wire is welded by a process other than controlled resistance welding used in the manufacture of welded wire fabric.

In the discussion of Section 7.5 in Part 3, it was noted that welding of crossing bars (tack welding) is not permitted for assembly of reinforcement unless authorized by the engineer. An example of tack welding would be a column cage where the ties are secured to the longitudinal bars by small arc welds. Such welding can cause a metallurgical notch in the longitudinal bars, which may affect the strength of the bars. Tack welding seems to be particularly detrimental to ductility (impact resistance) and fatigue resistance. Reference 4.2 recommends, "Never permit field welding of crossing bars ("tack" welding, "spot" welding, etc.). Tie wire will do the job without harm to the bars."

12.15 SPLICES OF DEFORMED BARS AND DEFORMED WIRE IN TENSION

Tension lap splices of deformed bars and deformed wire are designated as Class A and B with the length of lap being a multiple of the tension development length l_{bd}. For the '89 Code, the Class C tension lap splice has been eliminated in consideration of the multitude of new modification factors for l_{bd} to account for closely spaced bars and bars with minimal cover. The two-level splice classification (Class A & B) has been retained to encourage designers to splice bars at points of minimum stress and to stagger lap splices along the length of the bars to improve behavior of critical details.

The development length l_d (Section 12.2) used in the calculation of lap length must be that for the full f_y because the splice classifications already reflect any excess reinforcement at the splice location (factor of Section 12.2.5 for excess A_s must not be used). To account for clear spacing, amount of cover, and enclosing transverse reinforcement of bars being spliced, the modification factors of Section 12.2.3 must be applied. The factors of Section 12.2.4 to reflect influence of casting position (top bar effect), type of aggregate, and epoxy coating must also be applied where applicable. The minimum length of lap is 12 in.

For lap splices of slab and wall reinforcement, effective clear spacing of bars being spliced at the same location is taken as the clear spacing between the spliced bars, less one-bar diameter. This clear spacing criterion is illustrated in Fig. 4-26(a). Spacing for noncontact lap splices (spacing between lapped bars not greater than (1/5) lap length nor 6 in) should be considered the same as for contact lap splices. For lap splices of column and beam bars, effective clear spacing between bars being spliced will depend on the orientation of the lapped bars; see Fig. 4-26(b) and (c), respectively.

The designer must specify the class of tension lap splice to be used. The class of splice depends on the magnitude of tensile stress in the reinforcement and the percentage of total reinforcement to be lap spliced within any given splice length as shown in Table 4-9. If the area of tensile reinforcement provided at the splice location is more than twice that required for strength (low tensile stress) and ½ or less of the total steel area is lap spliced within the required splice length, a Class A splice may be used. Both splice conditions must be satisfied, otherwise, a Class B splice must be used. In other words, if the area of reinforcement provided at the splice location is less than twice that required for strength (high tensile stress) and/or more than ½ of the total area is to be spliced within the lap length, a Class B splice must be used.

Welded splices or mechanical connections conforming to Section 12.14.3 may be used in lieu of lap splices. Section 12.15.4 allows a reduction in the requirements of Section 12.14.3.3 for welded splices or 12.14.3.4 for mechanical connections if certain conditions are met.

Splices in tension tie members are required to be made with a full welded splice or full mechanical connection, with a 30 in. stagger between adjacent bar splices (welded splices or mechanical connections). See Commentary Section 12.15.5 definition of "tension-tie member".

Table 4-9—Lap Splice Conditions (at splice location)

CLASS A...1.0l_d	CLASS B...1.3l_d
(A_s provided) $\geq$ 2 (A_s required) and Percent A_s Splice $\leq$ 50	All other Conditions

12.16 SPLICES OF DEFORMED BARS IN COMPRESSION

Since bond behavior of reinforcing bars in compression is not complicated by the potential problem of transverse tension cracking in the concrete, compression lap splices do not require such strict provisions as those specified for tension lap splices. Tests have shown that the strength of compression lap splices depends primarily on end bearing of the bars on the concrete, without a proportional increase in strength

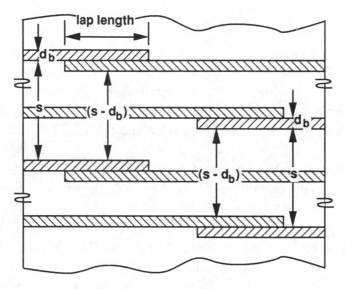

(a)   Wall and Slab Reinforcement

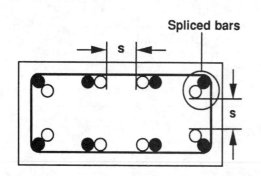

(b)   Column With Offset Corner Bars

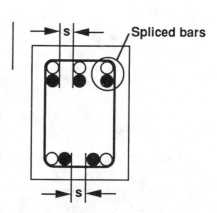

(c)   Beam Bar Splices

Fig. 4-26   Effective Clear Spacing of Spliced Bars

even when the lap length is doubled. Thus, the Code requires significant longer lap length for bars with a yield strength greater than 60,000 psi.

12.16.1 Compression Lap Splices

For the '89 Code, calculation of compression lap splices has been simplified by removing the redundant calculation for development length in compression. For compression lap splices, Section 12.16.1 requires the minimum lap length to be simply, $0.0005d_bf_y$, but not less than 12 in. For reinforcing bars with a yield strength greater than 60,000 psi, a minimum lap length of $(0.0009f_y - 24)d_b$ but not less than 12 in. is specified. Lap splice lengths must be increased by one-third for concrete with a specified compressive strength less than 3000 psi.

Lap splice lengths for columns may be reduced by a factor of 0.83 or 0.75 when the splice is enclosed throughout its length by minimum specified ties (Section 12.17.2.4), or by minimum specified spiral (Section 12.17.2.5), respectively. The 12 in. minimum lap length also applies to these permitted reductions.

With the "basic" lap length for compression lap splices a function of bar diameter d_b and bar yield strength f_y, and three modification factors for ties and spirals and for lower concrete strength, it is convenient to establish compression lap splices simply as a multiple of bar diameter.

For Grade 60 bars . . . . $30d_b$
   enclosed within ties . . . . $25d_b$
   enclosed within spirals . . . . $22.5d_b$

For Grade 75 bars . . . . $43.5d_b$
   enclosed within ties . . . . $36d_b$
   enclosed within spirals . . . . $33d_b$

but not less than 12 in. For f_c' less than 3000 psi, multiply by a factor of 1.33. Compression lap splice tables for the standard bar sizes can be readily developed using the above values.

As noted in the discussion of Section 12.14.2, #14 and #18 bars may be lap spliced, in compression only, to #11 and smaller bars or to smaller size footing dowels. Section 12.16.2 requires that when bars of a different size are lap spliced in compression, the length of lap must be the compression development length of the larger bar, or the compression lap splice length of the smaller bar, whichever is the longer length.

12.16.4 End-Bearing Splices

Section 12.16.4 specifies the requirements for end-bearing compression splices. End-bearing splices are only permitted in members containing closed ties, closed stirrups or spirals. Commentary Section 12.16.4.1 cautions the engineer in the use of end-bearing splices for bars inclined from the vertical. End-bearing splices for compression bars have been used almost exclusively in columns and the intent is to limit use to essentially vertical bars because of the field difficulty of getting adequate end bearing on horizontal bars or bars significantly inclined from the vertical. Welded splices or mechanical connections are also permitted for compression splices and must meet the requirements of Section 12.14.3.3 or 12.14.3.4, respectively.

12.17   SPECIAL SPLICE REQUIREMENTS FOR COLUMNS

For the '89 Code, the special splice requirements for columns have been significantly simplified. The new column splice requirements simplify the amount of calculations that are required compared to the previous provisions by assuming that a compression lap splice (Section 12.17.2.1) has a tensile capacity of at least one-fourth f_y.

The column splice provisions are based on the concept of providing some tensile resistance at all column splice locations even if analysis indicates compression only at a splice location. In essence, Section 12.17 establishes the required tensile strength of spliced longitudinal bars in columns. Lap splices, butt-welded splices, mechanical connections or end-bearing splices may be used.

12.17.2   Lap Splices in Columns

Lap splices are permitted in column bars required for compression or tension. Type of lap splice to be used will depend on the bar stress at the splice location, compression or tension, and magnitude if tension, due to all factored load combinations considered in the design of the column. Type of lap splice to be used will be governed by the load combination producing the greatest amount of tension in the bars being spliced. The design requirements for lap splices in column bars can be illustrated by a typical column load-moment strength interaction as shown in Fig. 4-27.

Bar stress at various locations along the strength interaction curve define segments of the strength curve where the different types of lap splices may be used. For factored load combinations along the strength curve, bar stress can be readily calculated to determine type of lap splice required. However, a design dilemma exists for load combinations that do not fall exactly on the strength curve (below the strength curve) as there is no simple exact method to calculate bar stress for this condition.

A seemingly rational approach is to consider factored load combinations below the strength curve as producing bar stress of the same type, compression or tension, and of the same approximate magnitude as that produced along the segment of the strength curve intersected by radial lines (lines of equal eccentricity) through the load combination point. This assumption becomes more exact as the factored load combinations being investigated fall nearer to the actual strength interaction curve of the column. Using this approach, zones of "bar stress" can be established as shown in Fig. 4-27.

For factored load combinations in Zone 1, all column bars are considered to be in compression. For load combinations in Zone 2, bar stress on the tension face of the column is considered to vary from zero to $0.5f_y$ in tension. For load combinations in Zone 3, bar stress on the tension face is considered to be greater than $0.5f_y$ in tension. Type of lap splice to be used will then depend on which zone, or zones, all factored load combinations considered in the design of the column are located. The designer need only locate the factored load combinations on the load-moment strength diagram for the column and bars selected in the design to determine type of lap splice required. Use of load-moment design charts in this manner will greatly facilitate

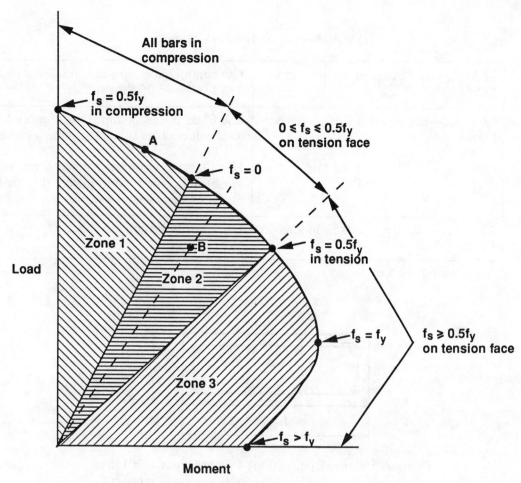

Fig. 4-27 Special Splice Requirements for Columns

the design of column bar splices. For example, if factored gravity load combination governed design of the column, say Point A in Fig. 4-27, all bars in compression, but a load combination including wind, say Point B in Fig. 4-27, produces some tension in the bars, the lap splice must be designed for a Zone 2 condition (bar stress is tensile but does not exceed $0.5f_y$ in tension). As a design convenience, column design charts in Design Aid SP 17A, referenced at the end of Part 11, indicate bar stress locations along the design strength interaction diagrams.

The design requirements for lap splices in columns are summarized in Table 4-10. Note that the compression lap splice permitted when all bars are in compression (Section 12.17.2.1) considers a compression lap length adequate as a minimum tensile strength requirement. See Example 4.8 for design application of the lap splice requirements for columns.

Sections 12.17.2.4 and 12.17.2.5 provide reduction factors for the compression lap splice when the splice is enclosed throughout its length by ties (0.83 reduction) or by a spiral (0.75 reduction). Spirals must meet the requirements of Sections 7.10.4 and 10.9.3. When ties are used to reduce the lap splice length, the ties must have a minimum effective area of $0.0015hs$. The tie legs in both directions must provide the minimum effective area to permit the 0.83 reduction factor. See Fig. 4-28.

4-35

Table 4-10—Lap Splices in Columns

12.17.2.1—Bar Stress in compression (Zone 1)*	Use compression lap splice modified by factor of 0.83 for ties or 0.75 for spirals.
12.17.2.2—Bar Stress ≤ 0.5f_y in tension (Zone 2)*	Use Class B tension lap splice if more than ½ of total column bars spliced at same location. Use Class A tension lap splice if not more than ½ of total column bars spliced at same location. Stagger alternate splices l_d.
12.17.2.3—Bar Stress > 0.5f_y in tension (Zone 3)*	Use Class B tension lap splice.

*For Zones 1, 2, and 3, see Fig. 4-27

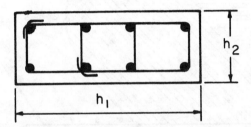

(perpendicular to h_1 direction) 4 tie bar areas 0.0015h_1s
(perpendicular to h_2 direction) 2 tie bar areas 0.0015h_2s

Fig. 4-28   Application of Section 12.17.2.4

12.17.3   Welded Splices or Mechanical Connections in Columns

Welded splices or mechanical connectors are permitted in column bars where bar stress is either compressive or tensile for all factored load combinations (Zones 1,2, and 3 in Fig. 4-27). "Full" welded or "full" mechanical connections must be used; that is, the welded splice or mechanical connection must develop at least 125 percent of the bar yield strength, 1.25A_bf_y. Use of welded splices or mechanical connectors of lesser strength is not permitted for splicing.

12.17.4   End Bearing Splices in Columns

End bearing splices are permitted for column bars stressed in compression for all factored load combinations (Zone 1 in Fig. 4-27). Even though there is no calculated tension, a minimum tensile strength of the continuing (unspliced) bars must be maintained when end bearing splices are used. Continuing bars on each face of the column must provide a tensile strength of $A_sf_y/4$, where A_s is the total area of bars on the face of the column. Thus, not more than ¾ of the bars can be spliced on each face of the column at any one location. End bearing splices must be staggered or additional bars must be added at the splice location if more than ¾ of the bars are to be spliced.

12.18   SPLICES OF WELDED DEFORMED WIRE FABRIC IN TENSION

For tension lap splices of deformed wire fabric, the Code requires a minimum lap length of 1.3 l_d, but not less than 8 in. Lap length is measured between the ends of each fabric sheet. The development length l_d is the value calculated by the provision in Section 12.7. The Code also requires that the overlap measured between the outermost cross wires be at least 2 in. Fig. 4-29 shows the lap length requirements.

If there are no cross wires within the splice length, the provisions in Section 12.15 for deformed wire must be used to determine the length of the lap.

12.19   SPLICES OF WELDED PLAIN WIRE FABRIC IN TENSION

The minimum length of lap for tension lap splices of plain wire fabric is dependent upon the ratio of the area of reinforcement provided to that required by analysis. Lap length is measured between the outermost cross wires of each fabric sheet. The required lap lengths are shown in Fig. 4-30.

CLOSING REMARKS

One additional comment concerning splicing of temperature and shrinkage reinforcement at the exposed surfaces of walls or slabs...one must assume all temperature and shrinkage reinforcement to be stressed to the full specified yield strength f_y. The purpose of this reinforcement is to prevent excess cracking. At some point in the member, it is likely that cracking will occur, thus fully stressing the temperature and shrinkage reinforcement. Therefore, all splices in temperature and shrinkage reinforcement must be assumed to be those required for development of yield tensile strength. A Class B tension lap splice must be provided for this steel.

REFERENCES

4.1   *Welded Wire Fabric Manual of Standard Practice*, 3rd Edition, Wire Reinforcement Institute, McLean, VA, 1979.

4.2   *Reinforcement Anchorages and Splices*, 2nd Edition, Concrete Reinforcing Steel Institute, Schaumburg, IL, 1984.

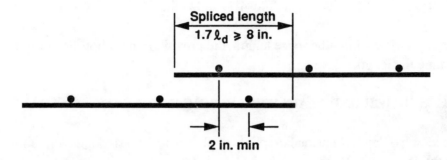

Fig. 4-29   Lap Splice Length for Deformed Wire Fabric

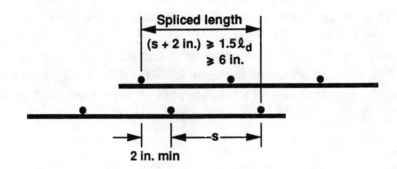

(a) Lap Splice for (A_s provided) < 2 (A_s required)

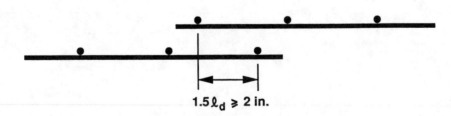

(b) Lap Splice for (A_s provided) ≥ 2 (A_s required)

Fig. 4-30   Lap Splice Length for Plain Wire Fabric

Example 4.1—Development of Bars in Tension

A beam at the perimeter of a structure has 7#9 top bars over the support. Structural integrity provisions require that at least one-sixth of the tension reinforcement required for negative moment at the support be made continuous (7.13.2.2). The continuity bars are to be spliced with a Class A splice at midspan. Determine the required length of a Class A lap splice.

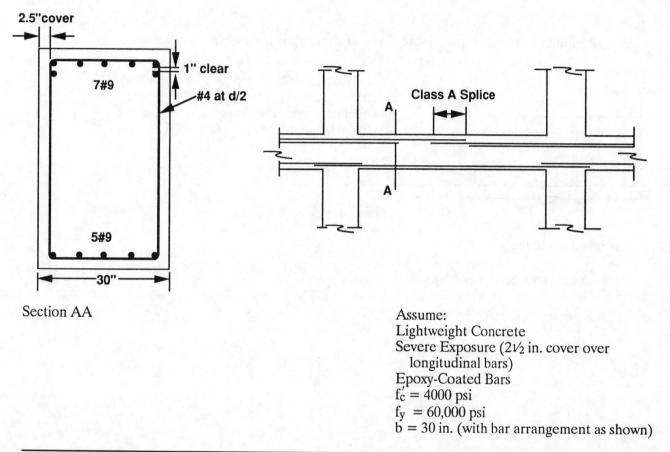

Section AA

Assume:
Lightweight Concrete
Severe Exposure (2½ in. cover over
   longitudinal bars)
Epoxy-Coated Bars
$f_c' = 4000$ psi
$f_y = 60,000$ psi
b = 30 in. (with bar arrangement as shown)

Calculations and Discussion	Code Reference
Minimum number of bars to be made continuous is 7/6 bars. Two corner bars will be spliced at midspan.	7.13.2.2
Class A lap splice requires a 1.0 ℓ_d length of bar lap	12.15.1

To illustrate proper application of the various modification factors to determine development length ℓ_d, a step-by-step procedure will be followed.

Example 4.1—Continued

	Code
Calculations and Discussion	**Reference**

(1) Basic development length $\ell_{db} = 0.04 A_b f_y / \sqrt{4000}$ 12.2.2

$$= 0.04(1.00)(60,000)/\sqrt{4000}$$

$$= 37.9 \text{ in.} \qquad \text{(Table 4-2)}$$

(2) Modification factor for bar spacing and cover, and enclosing transverse reinforcement   12.2.3

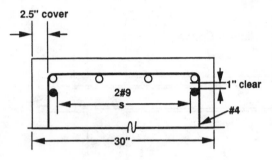

Note: Development of top 2#9 bars is not affected by the close spacing of the 2#9 bars along the beam sides since they have been terminated away from length of 2#9 bars being developed.

Section at Splice

Clear spacing between spliced bars

$$= [30 - 2(\text{cover}) - 2(\#4 \text{ stirrup}) - 2 \,(\#9 \text{ bar})]$$

$$= [30 - 2 \,(2.5) - 2 \,(0.5) - 2(1.128)]$$

$$= 21.7 \text{ in.}$$

$$= 19 \, d_b$$

Side cover $= 2.5 + 0.5 = 3.0$ in. $= 2.66 d_b$

Referring to Section 12.2.3.1(a)

Clear spacing $> 3d_b$ 12.2.3.1(a)

cover $>$ Sectionn7.7.1

stirrups $>$ Section 11.5.4 and 11.5.5.

$\ell_d = 1.0 \, \ell_{db} = 37.9$ in.

Example 4.1—Continued

Calculations and Discussion	Code Reference

(3) Modification factor for widely spaced bars   12.2.3.4

Clear spacing > $5d_b$

   Side cover > $2.5d_b$

   $\ell_d = 0.8(37.9) = 30.3$ in.

(4) Check minimum development length   12.2.3.6

$0.03d_bf_y/\sqrt{f_c'} = 0.03(1.128)(60,000)/\sqrt{4000}$

   $= 32.1$ in. $>$ 30.3 in. Use $\ell_d = 32.1$ in.   (Table 4-8)

(5) Modification for top reinforcement = 1.3   12.2.4.1

(6) Modification for lightweight concrete = 1.3   12.2.4.2

(7) Modification for epoxy-coated bars   12.2.4.3

Clear spacing = $19d_b > 6d_b$

Cover = $2.5 + 0.5 = 3.0$ in. = $2.66d_b < 3d_b$

$\ell_d = 1.5\,\ell_{db}$

However, product for top bars and epoxy-coating = 1.3(1.5)

$= 1.97 > 1.7$ 12.2.4

Use 1.7 factor for combined effect of top bars and epoxy coating.

$\ell_d = 32.1(1.3)(1.7) = 70.9$ in.

(8) Modification for excess reinforcement   12.2.5

Not permitted for tension lap splices..."where ℓ_d is the tensile development length for the specified yield strength f_y."   12.15.1

(9) Check minimum development length   12.2.1

Example 4.1—Continued

l_d = 70.9 in. > 12. in.

(10) Required lap splice for 2#9 top bars:

Class A splice = $1.0 l_d$ = 70.9 in. = 5.9 ft, say 6 ft.

The critical development effect of lightweight concrete and epoxy-coating, and top bar effect, more than doubles the required splice length.

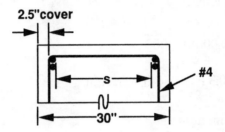

Example 4.2—Development of Bars in Tension

Calculate required tension development length for the #8 bars (alternate short bars) in the "sand-lightweight" one-way slab shown below. Use $f_c' = 4000$ psi and $f_y = 60,000$ psi

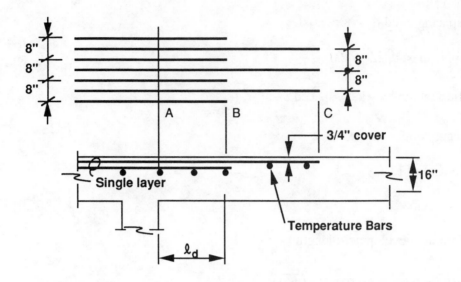

Calculations and Discussion	Code Reference

(1) Basic development length $\ell_{db} = 0.04A_b f_y/\sqrt{f_c'}$ 12.2.2

$$= 0.40(0.79)(60,000)/\sqrt{4000}$$

$$= 30.0 \text{ in.} \qquad \text{(Table 4-2)}$$

(2) Modification factor for bar spacing and cover   12.2.3

Note: Spacing for short bars may be considered the same as for long bars for required development length since short bars are developed in length AB, while long bars are already developed in length BC.

Effective clear spacing for bar development $= 8 - d_b = 8 - 1.0 = 7.0$ in. $= 7d_b$

Example 4.2—Continued

Calculations and Discussion	Code Reference

Cover = 0.75 in. = $0.75d_b$

Clear spacing > $2d_b$

Cover < $d_b..l_d = 2.0\,l_{db}$ 12.2.3.2

(3) Modification factor for widely spaced bars   12.2.3.4

Clear spacing >$5d_b$

$l_d = 0.8(2.0\,l_{db}) = 1.6\,l_{db}$

$= 1.6(30) = 48$ in.

(4) Check minimum development length   12.2.3.6

$0.03d_b f_y/\sqrt{f_c'} = 0.03(1.0)(60{,}000)/\sqrt{4000}$

$= 28.5$ in. < 48 in.   (Table 4-4)

(5) Modification for top reinforcement   12.2.4.1

(Greater than 12 in. of fresh concrete cast below the bars)

$l_d = 1.3(48) = 62.4$ in.

(6) Modification for "sand-lightweight" concrete   12.2.4.2

$l_d = 1.3(62.4) = 81.1$ in. 12 in. minimum   12.2.1

l_d(required) = 81 in.

Required development length based on '83 Code:
l_{db} = 30 in.
factor for top bars = 1.4
factor for sand-lightweight = 1.18
factor for wide bar spacing = 0.8
Required l_d = (30)(1.4)(1.18)(0.8) = 40 in. << 81 in.

Example 4.3—Development of Bars in Tension

Calculate required development length for the inner 2#8 bars in the beam shown below. The 2#8 outer bars are to be made continuous along the full length of the beam. Use $f_c' = 4000$ psi (normal weight concrete) and $f_y = 60,000$ psi

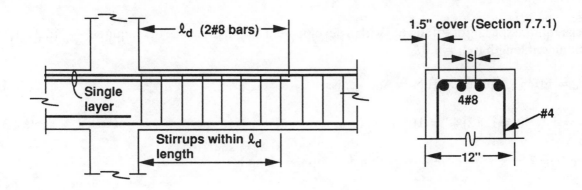

	Code
Calculations and Discussion	**Reference**

(1) Basic development length $\ell_{db} = 30.0$ in.   (Table 4-2)

(2) Modification factor for bar spacing and cover

Clear spacing $= [12 - 2(\text{cover}) - 2(\#4\ \text{stirrups}) - 4(\#8\ \text{bars})]/3$

$\qquad = [12 - 2(1.5) - 2(0.50) - 4(1.00)]/3$

$\qquad = 1.33$ in.

$\qquad = 1.33d_b$

Clear spacing $< 2d_b \ldots \ell_d = 2.0\ell_{db}$ 12.2.3.2

$\ell_d = 2(30) = 60$ in. > 28.5 in.   (Table 4-4)

(3) Modification factor for top reinforcement

$\ell_d = 1.3(0) = 78$ in. > 12 in.   12.2.1

Example 4.3—Continued

Calculations and Discussion	Code Reference

For the closely spaced 2#8 inner bars being developed, required development length from face of support, $\ell_d = 78$ in.

(4) Check required spacing of #4 stirrups within development length to reduce the required development length to $\ell_d = 1.0 \ell_{db}$ 12.2.3.1(b)

For $\ell_d = 1.0 \ell_{db}$, required stirrups = $A_{tr} = d_b Ns/40$ Eq. (12-1)

$A_{tr}/s = d_b N/40 = 1.0(2)/40 = 0.05$ in.^{2}/ in.   (Table 4-3)

where N = 2 bars being developed within development length

For #4 stirrups (2 legs), $A_{tr} = 2(0.20) = 0.40$ in.^{2}

    $s = 0.40/0.05 = 8.0$ in.   (Table 4-3)

#4 stirrups @ 8 in. will permit $\ell_d = 1.0 \ell_{db}$:

    $\ell_d = 1.3(30) = 39$ in. > 12 in.

      Use $\ell_d = 39$ in.

With the #4 stirrups spaced at 8 in. within the development length $\ell_d = 39$ in., the required development length from face of support for the 2#8 terminated bars is reduced by one-half. Note that the #4 stirrups must enclose the 2#8 bars being developed (closed stirrups). Note also that the 8 in. stirrup spacing is required over development length; the required spacing for shear at the support must also be checked, with the lesser spacing used in the final design.

Example 4.4—Development of Flexural Reinforcement

Determine lengths of top and bottom bars for the exterior span of the continuous beam shown below. Concrete is normal weight and bars are Grade 60. Total uniformly distributed factored gravity load on the beam is w_u = 6.0 kips/ft (including weight of beam).

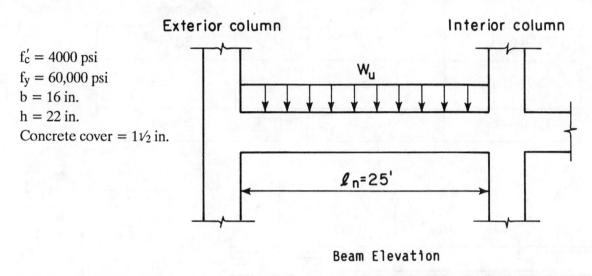

f'_c = 4000 psi
f_y = 60,000 psi
b = 16 in.
h = 22 in.
Concrete cover = $1\frac{1}{2}$ in.

Beam Elevation

	Calculations and Discussion	**Code Reference**

1. Preliminary design for moment and shear reinforcement

   (a) Use approximate analysis for moment and shear values

   8.3.3

Location	Factored moments & shears
Interior face of exterior support	$-M_u = w_u \ell_n^2/16 = 6(25^2)/16 + 234.4$ ft−kips
End span positive	$+M_u = w_u \ell_n^2/14 = 6(25^2)/14 = 267.9$ ft−kips
Exterior face of first interior support	$-M_u = w_u \ell_n^2/10 = 6(25^{2)})/10 = 375.0$ ft−kips
Exterior face of first interior support	$V_u = 1.15 w_u \ell_n /2 = 1.15(6)(25)/2 = 86.3$ kips

   (b) Determine required flexural reinforcement using procedures of Part 9. With 1.5 in. cover, #4 bar stirrups, and #9 or #10 flexural bars; d ≈ 19.4 in.

   (c) Determine required shear reinforcement

Example 4.4—Continued

M_u	A_s required	Bars	A_s provided
–234.4 ft–kips	2.93 in.^{2}	4#8	3.16 in.^{2}
+267.9 ft–kips	3.40 in.^{2}	2#8 2#9	3.58 in.^{2}
–375.0 ft–kips	5.01 in.^{2}	4#10	5.06 in.^{2}

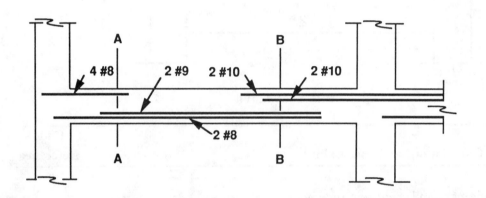

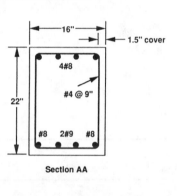

Section AA

Section BB

V_u at "d" distance from face of support:

$V_u = 86.3 – 6(19.4/12) = 76.6$ kips

$\varphi V_c = \varphi 2\sqrt{f_c'}\, b_w d = 0.85 \times 2\sqrt{4000} \times 16 \times 19.4/1000 = 33.4$ kips

with $s_{max} = d/2 = 19.4/2 = 9.7$ in., try #4 U-stirrups @ 9 in. spacing

$\varphi V_s = \varphi A_v f_y d/s = 0.85(0.40)(60)(19.4)/9 = 44.0$ kips

$\varphi V_n = \varphi V_c + \varphi V_s = 33.4 + 44.0 = 77.4$ kips $>$ 76.6 kips   O.K.

Distance from support where stirrups not required:

$V_u < \phi V_c /2 < 33.4/2 < 16.7$ kips

$V_u = 86.3 – 6x = 16.7$ kips

$x = 11.6$ ft $\approx$ 1/2 span

Use #4 U-stirrups @ 9 in. (entire span)

Example 4.4—Continued

	Code
Calculations and Discussion	Reference

2. Bar lengths for bottom reinforcement

   (a) Required number of bars to be extended into supports.

   One-fourth of ($+A_s$) must be extended at least 6 in. into the supports. One #9 bar could be full span length with the other #9 and the 2#8 bars cut off within the span. With a longitudinal bar required at each corner of the stirrups (Section 12.13.3), at least 2 bars should be extended full length:

   Consider extending the 2#8 bars full span length (plus 6 in. into the supports) and cut off the 2#9 bars within the span.

   (b) Determine cut-off locations for the 2#9 bars and check other development requirements.

   Shear and moment diagrams for loading condition causing maximum factored positive moment are shown below.

12.11.1

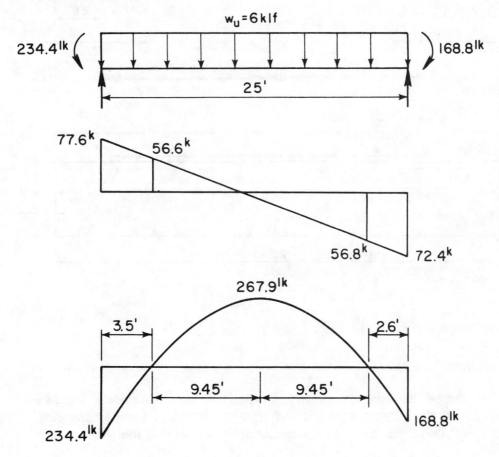

Example 4.4—Continued

The positive moment portion of the M_u diagram is shown below at a larger scale, including the design moment strengths φM_n for the total positive A_s (2#8 and 2#9) and for (2#8 bars) separately; and the necessary dimensions. For 2#8 and 2#9, φM_n = 280.7 ft–kips. For 2#8, φM_n = 131.8 ft-kips

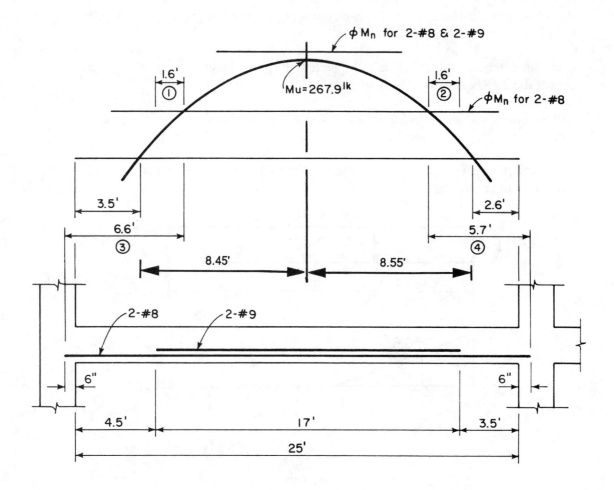

As shown, the 2#8 bars extend full span length plus 6 in. into the supports. The 2#9 bars are cut off tentatively at 4.5 ft and 3.5 ft from the exterior and interior supports, respectively. These tentative cut-off locations are determined as follows:

Example 4.4—Continued

Calculations and Discussion	Code Reference

dimensions (1) and (2) must be the larger of d or $12d_b$:

12.10.3

d = 19.4 in. (1.6 ft) (governs)

[Fig. 4-13(b)]

$12d_b$ = 12 (1.128) = 13.5 in.

dimensions (3) and (4) must be equal to or larger than l_d:

12.10.4
[Fig. 4-12(b)]

Within the development length l_d, only 2#8 bars are being developed (2#9 bars are already developed in length 8.45 ft)

For #8 bars, l_{db} = 30 in.

(Table 4-2)

Clear spacing of 2#8 outer bars

[16 – 2(1.5) – 2(0.5) – 2(1) = 10 in.] = $10d_b$

> $3d_b$*

Cover (1.5 + 0.5 = 2.0 in.) = $2d_b$..l_b = 1.0l_{db}

12.2.3.1(d)

l_d = 30 in > 28.5 in.

12.2.3.6
(Table 4-4)

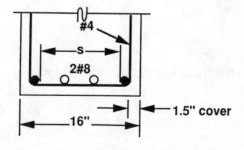

*spacing > $5d_b$, but side cover < 2.5 d_b;
 additional 0.8 factor does not apply.

Example 4.4—Continued

Calculations and Discussion	Code Reference

dimension (3) 6.6 ft > 2.5 ft   O.K.
dimension (4) 5.7 ft > 2.5 ft   O.K.

For illustration, calculate required development length ℓ_d for 2#9 bars. 2#8 bars already developed in length 2.5 ft from bar end. For #9 bars, $\ell_{db} = 37.9$ in.   (Table 4-2)

Clear spacing between 2#9 bars

$[16 - 2(1.5) - 2(0.5) - 2(1.0) - 2(1.128)]/3 = 2.58$ in. $= 2.29\, d_b$

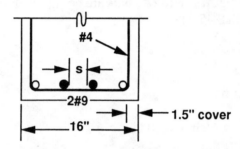

Clear spacing > $2d_b$

$\qquad$ Cover > d_b .. $\ell_d = 1.4\ell_{db}$ 12.2.3.3

$\ell_d = 1.4(37.9) = 53.1$ in. (4.43 ft) > 32.1 in.   (Table 4-4)

8.45 ft >> 4.43   O.K.

For #8 bars, check development requirements at points of inflection (PI):   12.11.3

$\ell_d \leq \dfrac{M_n}{V_u} + \ell_a$ Eq. (12-2)

For 2#8 bars, $M_n = 131.8/0.9 = 146.4$ ft–kips at left PI, $V_u = 56.6$ kips

ℓ_a = larger of $12d_b = 12(1.0) = 12$ in.

$\qquad$ or d = 19.4 in. (governs)

$\ell_d \leq \dfrac{146.4 \times 12}{56.6} + 19.4 = 50.4$ in.

For #8 bars, $\ell_d = 30$ in. < 50.4   O.K.

Example 4.4—Continued

Calculations and Discussion	Code Reference

at right PI, $V_u = 56.8$ kips; by inspection, the #8 bars are O.K.

With both tentative cut-off points located in a zone of flexural tension, one of the three conditions of Section 12.10.5 must be satisfied.

At left cut-off point (4.5 ft from support):

$V_u = 77.6 - 4.5 \times 6 = 50.6$ kips

$V_n = 77.4$ kips (#4 U-stirrups @ 9 in.)

$2/3(77.4) = 51.6$ kips > 50.6 O.K.   12.10.5.1

For illustrative purposes, determine if the condition of Section 12.10.5.3 is also satisfied:

$M_u = 54.1$ ft–kips at 4.5 ft from support

A_s required $= 0.63$ in.^{2}

for 2#8 bars, A_s provided $= 2.00$ in.^{2}

$2.00 > 2(0.63) = 1.26$ in.^{2} O.K.   12.10.5.3

$3/4(77.4) = 58.1$ kips > 50.6 O.K.   12.10.5.3

Therefore, Section 12.10.5.3 is also satisfied at cut-off location.

At right cut-off point (3.5 ft from support):

$V_u = 72.4 - 3.5 \times 6 = 51.4$ kips

$2/3(\varphi V_n) = 51.6$ kips > 51.4 O.K.   12.10.5.1

Summary: The tentative cut-off locations for the bottom reinforcement meet all code development requirements. The 2#9 bars x 17ft would have to be placed unsymmetrically within the span. To assure proper placing of the #9 bars, it would be prudent to specify a 19 ft length for symmetrical bar placement within the span, i.e., 3.5 ft from each support. The ends of the cut off bars would then be at or close to the points of inflection...thus, eliminating the need to satisfy the conditions of Section 12.10.5 when bars are terminated in a tension zone. The recommended bar arrangement is shown at the end of the example.

Example 4.4—Continued

Calculations and Discussion

3.   Bar lengths for top reinforcement

Shear and moment diagrams for loading condition causing maximum factored negative moments are shown below.

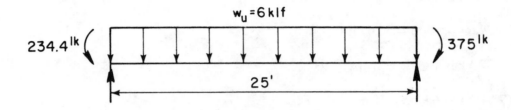

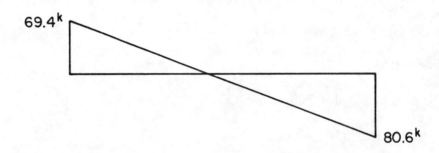

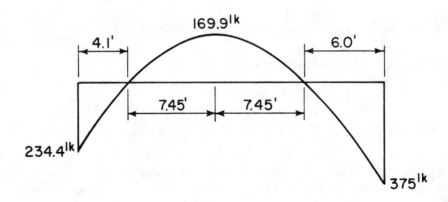

Example 4.4—Continued

| **Calculations and Discussion** | **Code Reference** |

The negative moment portions of the M_u diagram are shown below at a larger scale, including the design moment strengths φM_n for the total negative A_s at each support (4#8 and 4#10) and for 2#10 bars at the interior support; and the necessary dimensions. For 4#8, $\varphi M_n = 251.1$ ft–kips. For 4#10, $\varphi M_n = 379.5$ ft–kips. For 2#10, $\varphi M_n = 194.3$ ft–kips.

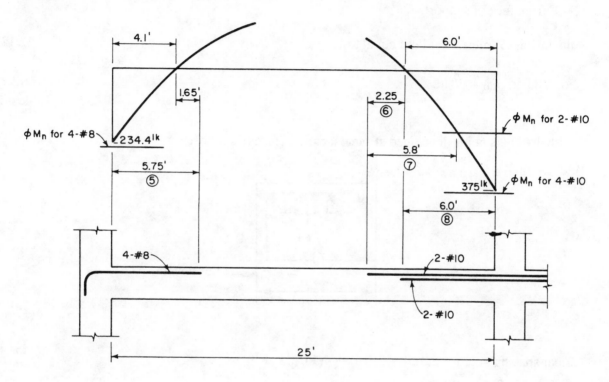

4. Development requirements for 4#8 bars

   (a) Required number of bars to be extended.

   One-third of $(-A_s)$ provided at supports must be extended beyond the point of inflection a distance equal to the greater of d, $12d_b$, or $\ell_n/16$.   12.12.3 [Eq.4-12(a)]

   d = 19.4 in. or 1.6 ft (governs)

   $12d_b = 12(1.0) = 12.0$ in.

Example 4.4—Continued

Calculations and Discussion	Code Reference

$l_n/16 = 25 \times 12/16 = 18.75$ in.

Since the inflection point is located only 4.1 ft from the support, total length of the #8 bars will be relatively short even with the required 1.6 ft extension beyond the point of inflection. Check required development length l_d for a cut-off location at 5'-9" from face of support.

dimension (5) must be at least equal to l_d — 12.12.2

For #8 bars, $l_{db} = 30$ in. — 12.2.2 (Table 4-2)

With 4#8 bars being developed at same location (face of support):

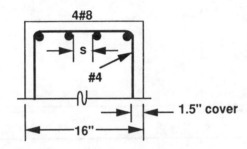

Clear spacing $= [16 - 2(1.5) - 2(0.5) - 4(1.0)]/3$

$= 2.67$ in. $= 2.67\ d_b$

Clear spacing $>2d_b$ — 12.2.3.3

Cover $> d_b..l_d = 1.4 l_{db}$

$l_d = 1.4(30) = 42$ in. > 28.5 in. — 12.2.3.6 (Table 4-4)

Including top bar effect, $l_d = 1.3(42) = 54.6$ in.

For #8 top bars, $l_d = 54.6$ in. (4.55 ft) < 5.75 ft.　　　O.K.

(b) Anchorage into exterior column.

Example 4.4—Continued

Calculations and Discussion	Code Reference

The #8 bars can be anchored into the column with a standard end hook. From Table 4-8, ℓ_{hb} = 19.0 in. For a 90° hook with side cover ≥ 2½ in. and end cover ≥ 2 in., a modification factor of 0.7 applies (12.5.3.2). Therefore, the required total embedment ℓ_{dh} is 0.7(19.0) = 14 in. for a #8 hooked bar. Overall depth of column required would be 16 in. The required ℓ_{dh} for the hook could be reduced by 1 in. (a refinement) if excess reinforcement is considered.

$$\frac{(A_s \text{ required})}{(A_s \text{ provided})} = \frac{2.93}{3.16} = 0.93$$

12.5.3.4

ℓ_{dh} = 14 × 0.93 = 13 in.

5. Development requirements for 4#10 bars

    (a) Required extension for one-third of (–A_s)

    12.12.3
    [Fig. 4-12(a)]

    d = 19.4 in. (1.6 ft)   (governs)

    $12d_b$ = 12(1.27) = 15.24 in.

    $\ell_n/16$ = 18.75 in.

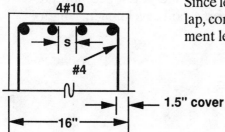

4#10

s

#4

1.5" cover

16"

Since lengths of 4#10 bars being developed will overlap, consider all bars being developed within development length ℓ_d.

For #10 bars, ℓ_{db} = 48.2 in.

12.2.3.2
(Table 4-2)

Clear spacing = [16 – 2(1.5) – 2(0.5) – 4(1.27)]/3

= 2.31 in. = 1.82 d_b

Example 4.4—Continued

Calculation and Discussion	Code Reference

Cover $= 1.5 + 0.5 = 2.0 = 1.57 \, d_b$

Clear spacing $< 2 d_b$ 12.2.3.2

    Cover $< d_b .. l_{db} = 2.0 \, l_{db}$ 12.2.3.6
    (Table 4-4)

Including top bar effect, $l_d = 1.3(96.4) = 125.3$ in.   12.2.4.1

dimension (6) $= 6.0$ ft $<< l_d = 125.3$ in (10.4 ft)   N.G.

Could cut off total negative reinforcement (4#10 bars) at 10.5' from support

Check required stirrups within development length l_d to permit $l_d = 1.0 \, l_{db}$ (Required $l_d = 125.3$ in./2 $= 5.2$ ft)   12.2.3.1(b)

For N $= 4$#10 bars being developed:

$A_{tr}/s = d_b N/40 = 1.27(4)/40 = 0.127$ in.^{2}/in.   (Table 4-3)

    #5 @ 5 in. required

Alternatively, check required stirrups if 2#10 outer bars are developed beyond development length for 2#10 inner bars (2#10 inner bars cut off development length beyond cut-off of 2#10 outer bars).

Using N $= 2$:

$A_{tr}/s = 1.27(2)/40 = 0.064$ in.^{2}/in.   (Table 4-3)

#4 @ 6 in. required

For this development condition:

Cover $>$ Section 7.7.1   12.2.3.1(b)

Stirrups $> A_{tr}$.........$l_d = 1.0 \, l_{db}$

$l_d = 48.2$ in. > 36.1 in.   12.2.3.6
(Table 4-4)

Example 4.4—Continued

Calculations and Discussion	**Code Reference**

$l_d = 1.3(48.2) = 62.7$ in. (5.2 ft) — 12.2.4.1

2#10 outer bars must extend l_d distance beyond cut-off point of 2#10 inner bars. For 2#10 outer bars:

Clear spacing > $3d_b$*

    Cover > Section 7.7.1

    Stirrups > Section 11.5.4....$l_d = 1.0\, l_{db}$

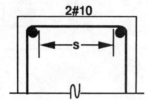

2#10

*Clear spacing > $5d_b$, but side cover < $2.5d_b$; therefore, additional 0.8 factor does not apply.

$l_d = 1.3(48.2) = 62.7$ in. (5.2 ft) — 12.2.4.1

2#10 outer bar cut-off = 5.2 + 5.2 = 10.4 ft from support.

Alternatives for termination of 4 #10 bars:

(1) Cut off 4#10 bars at 10'-6" from support, or

(2) Cut off 2#10 bars at 5'-3" from support and provide #4 @ 6 in. stirrups along 5'-3" development length; and cut off 2#10 bars at 10'-6" from support. Note: #4 @ 9 in. stirrups required for bar development beyond first cut-off location; also, the #4 @ 6 in. stirrups must be closed stirrups (Section 7.11.3).

Another option is to select 5#9 bars ($A_s = 5.00$ in.^{2}) for the negative reinforcement at interior support, and cut off alternate 2#9 bars (2 short #9 bars and 3 long #9 bars). With the 5 bar arrangement and alternate bar cut-off, the effective clear spacing for bar development will be increased, possibly eliminating the need for the extra stirrups along the short bar development length.

6. Summary: Selected bar lengths for the top and bottom reinforcement are shown below.

Example 4.4—Continued

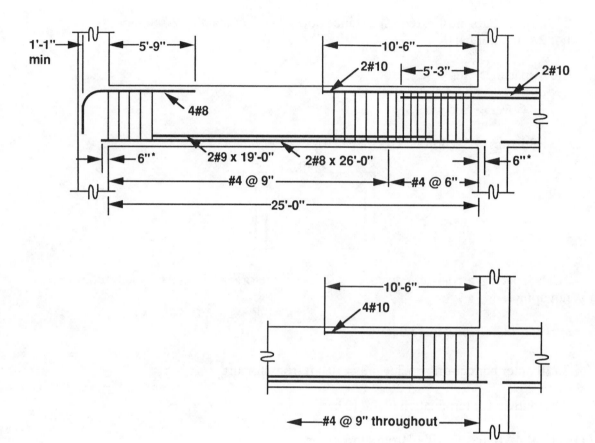

7. Supplementary Requirements

   (a) If the beam were part of a primary lateral load resisting system, the 2#8 bottom bars   12.11.2
       extending into the supports would have to be anchored to develop the bar yield
       strength at the face of supports. At the exterior column, anchorage can be provided
       by a standard end hook. Minimum width of support (overall column depth) required

Example 4.4—Continued

	Code
Calculations and Discussion	**Reference**

for anchorage of the #8 bar with a standard hook is a function of the basic development length, ℓ_{hb} from Table 4-8, and the appropriate modification factors (12.5.3.2 through 12.5.3.5):

- 0.7(19 in.) = 13.3 in. for 90° hook with 2 in. end cover on hook   12.5.3.2

- 19.0 in. (0.7)(0.8) = 10.7 in. for 90° hook with 2 in. end cover and hook enclosed within ties or stirrup-ties spaced not greater than $3d_b$ 12.5.3.3

At the interior column, the 2#8 bars could be extended ℓ_d distance beyond the face of support into the adjacent space or lap spliced with extended bars from the adjacent span. Consider a Class A lap splice adequate to satisfy the intent of Section 12.11.2. For the 2#8 bars, ℓ_{db} = 30 in.

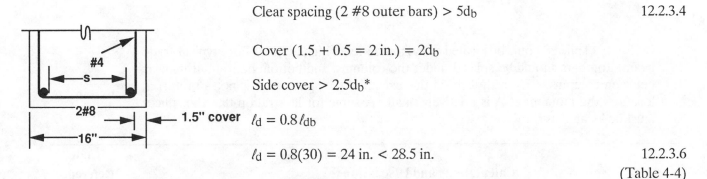

Clear spacing (2 #8 outer bars) > $5d_b$ 12.2.3.4

Cover (1.5 + 0.5 = 2 in.) = $2d_b$

Side cover > $2.5d_b$*

$\ell_d = 0.8\ell_{db}$

$\ell_d = 0.8(30) = 24$ in. < 28.5 in.   12.2.3.6
(Table 4-4)

Class A splice = $1.0\ell_d$ = 28.5 in. (say 29 in.)   12.15.1

*For bar anchorage in beam column joints, the beam bars are usually placed (fitted) inside the column vertical bars with side cover greater than $2.5d_b$.

(b) If closed stirrups (Section 7.11.3) are **not used** along the length of the beam where stirrups are required for shear, at least $\frac{1}{4}$ (+A_s) = $\frac{1}{4}$(3.58) = 0.90 in.^{2} must be made continuous or spliced with a Class A tension lap splice at the interior column support, and anchored by a standard end hook at the exterior column support. Consider 2#8 bars (A_s = 1.58 in.^{2}) as adequate to satisfy the intent of Section 7.13.   7.13.2.3

(c) If the beam were a perimeter member of a structure, $\frac{1}{4}$ (+A_s) = $\frac{1}{4}$(3.58) = 0.90 in.^{2} must be made continuous or spliced with a Class A tension splice at the interior column support, and anchored with a standard end hook at the exterior column support. This reinforcement should be enclosed in closed stirrups throughout the member but need not extend into the supports (7.13.2.2). With the bar arrangement selected, 2#8 (A_s = 1.58 in.^{2}) would need to be extended into the supports.

Example 4.5—Lap Splices in Tension

Design the tension lap splices for the grade beam shown below.

$f_c' = 4000$ psi
$f_y = 60,000$ psi
$b = 16$ in.
$h = 30$ in.
Bar cover = 3 in.
4#9 bars top and bottom (continuous)
#4 stirrups @ 14 in. (entire span)
$+M_u$ @ B = 340 ft–kips
$-M_u$ @ A = 120 ft–kips

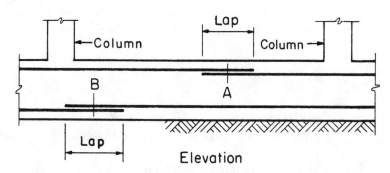

Elevation

Preferably, splices should be located away from zones of high tension. For a typical grade beam, top bars should be spliced under the columns, and bottom bars about midway between columns. Even though, in this example, the splice at A is not a preferred location, the moment at A is relatively small. Assume for illustration that the splices must be located as shown.

Calculations and Discussion	Code Reference

1. Determine required lap splice for bottom bars at B.

   (a) Required tension lap splice .... 12.15.1 (Table 4-9)

   A_s required ($+M_u$ @ B = 340 ft–kips) = 3.11 in.^{2}

   A_s provided (4#9 bars) = 4.00 in.^{2}

   A_s provided /A_s required = 4.00/3.11 = 1.29 < 2

   Class B splice required....$1.3 \ell_d$

   Note: Even if lap splices were staggered (% A_s spliced = 50), a Class B splice must be used with (A_s provided /A_s required) < 2

   (b) Basic development length

Example 4.5—Continued

Calculations and Discussion	Code Reference

For #9 bars, l_{db} = 37.9 in.

(Table 4-2)

(c) Factor for bar spacing and cover

With all bars spliced at same location (% A_s spliced = 100)

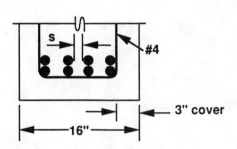

Clear spacing between spliced bars

$[16 - 2(3.0) - 2(0.5) - 4(1.128)]/3$
$= 1.50$ in. $= 1.33 \, d_b$

Cover $= 3.0 + 0.5 = 3.5$ in. $= 3.1 \, d_b$

Clear spacing $< 2d_b .. l_d = 2.0 \, l_{db}$

12.2.3.2

$l_d = 2(37.9) = 75.8$ in. > 32.1 in.

12.2.3.6
(Table 4-4)

Class B Splice $= 1.3 \, l_d = 1.3(75.8) = 98.5$ in. (8.2 ft)

Stagger alternate lap splices a lap length to increase clear spacing between spliced bars.

Clear spacing $= 2(1.50) + 1.128 = 4.13$ in. $= 3.66d_b$

Clear spacing $> 3d_b$

12.2.3.2(b)

Cover $> 2d_b ... l_d = 1.0 \, l_{db}$

$l_d = 37.9$ in. > 32.1 in.

Class B Splice $= 1.3(37.9) = 49.3$ in. (4.1 ft)

Use 4'-3" lap splice @ B and stagger alternate lap splices

2. Determine tension lap splice for top bars at A.

(a) Required tension lap splice

12.15.1
(Table 4-9)

Example 4.5—Continued

	Code
Calculations and Discussion	**Reference**

A_s required ($+M_u$ @ A = 120 ft–kips) = 1.05 in.^{2}

A_s provided /A_s required = 4.00/1.05 = 3.81 > 2

If alternate lap splices are staggered at least a lap length (% A_s spliced = 50):

Class A splice may be used...1.0 ℓ_d

If all bars are lap spliced at the same location (within req'd lap length):

Class B splice must be used...1.3 ℓ_d

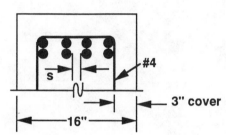

Including top bar effect (12.2.4.1), Class B splice = 1.3(1.3 ×75.8) = 128 in. (10.67 ft)

Stagger alternate lap splices and use Class A splice:

Class A splice = 1.0(1.3 × 37.9) = 49.3 in. (4.1 ft)

Use 4'-3" lap splice @ A and stagger alternate lap splices

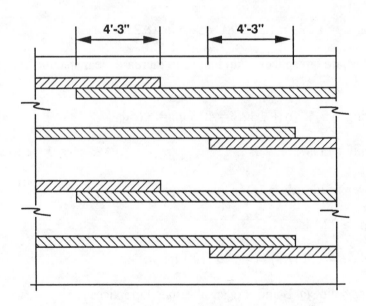

**Alternate lap splice stagger arrangement
(Note: Bar laps are positioned vertically).**

4-64

Example 4.6—Lap Splices in Compression

The following two examples illustrate typical calculations for compression lap splices in tied and spirally reinforced columns.

	Code
Calculations and Discussion	**Reference**

1. Design a compression lap splice for the tied column shown below. Assume all bars in compression for factored load combinations considered in design (Zone 1 in Fig. 4-26). See also Table 4-10.

    b = 16 in.
    h = 16 in.
    f'_c = 4000 psi
    f_y = 60,000 psi
    8#9 bars

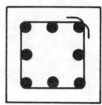

    (a) Determine lap splice length: 12.16.1

    For f_y = 60,000 psi

    Length of lap = $0.0005f_yd_b$, but not less than 12 in.

$$= 0.0005(60,000)d_b = 30d_b$$

$$= 30(1.128) = 34 \text{ in.}$$

    If f'_c <3000 psi, calculate lap length as above and increase by factor of 1.33.   12.16.1

    (b) Determine column tie requirements to allow 0.83 reduced lap length: 12.17.2.4

    Required column ties, #3 @ 16 in. o.c.   7.10.5

    Required spacing of #3 ties for reduced lap length:

    effective area of ties ≥ 0.0015hs

    $(2 \times 0.11) = 0.0015 \times 16s$

    s = 9.17 in.

    Spacing of the #3 ties must be reduced to 9 in. o.c. throughout the lap splice length to allow a lap length of 0.83 $(30d_b) = 25d_b = 28$ in.

Example 4.6—Continued

Calculations and Discussion	Code Reference

2. Determine compression lap splice for spiral column shown.

$f'_c = 4000$ psi
$f_y = 60,000$ psi
8#9 bars
#3 spirals

(a) Determine lap splice length   12.16.1

For bars enclosed within spirals, "basic" lap splice length may be multiplied by a factor   12.17.2.5
of 0.75.

For $f_y = 60,000$ psi:

lap = 0.75(34) = 26 in.

End bearing, welded, or mechanical connections may also be used.   12.16.3
                                                                                    12.16.4

Example 4.7—Lap Splices in Columns

Design the lap splice for the tied column detail shown below.

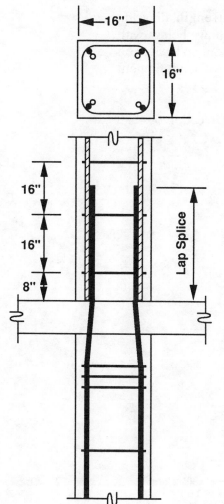

- Continuing bars from column above (4#8 bars)
- Offset bars from column below (4#8 bars)

$f_c' = 4000$ psi (normal weight)
$f_y = 60,000$ psi
b = h = 16 in.
4#8 bars (above and below floor level)
#3 ties @ 16 in.

lap splice to be designed for the following factored load combinations:

Gravity Loads

(1) Eq. (9-1) $\Rightarrow$ $P_u = 465$ kips
$\qquad\qquad\qquad M_u = 20$ ft−kips

Gravity + Wind Loads

(2) Eq. (9-2) $\Rightarrow$ $P_u = 360$ kips
$\qquad\qquad\qquad M_u = 120$ ft−kips

(3) Eq. (9-3) $\Rightarrow$ $P_u = 310$ kips
$\qquad\qquad\qquad M_u = 100$ ft−kips

	Code
Calculations and Discussion	**Reference**

1. Determine type of lap splice required.   12.17.2

Type of lap splice to be used depends on the bar stress at the splice location due to all factored load combinations considered in the design of the column. For design purposes, type of lap splice will be based on which zone, or zones, of bar stress all factored load combinations are located on the column load-moment strength diagram. See discussion

Example 4.7—Continued

Calculations and Discussion

for code Section 12.17.2, and Fig. 4-27. The load-moment strength diagram (columndesign chart) for the 16 in. x 16 in. column with 4#8 bars is shown below, with the three factored load combinations considered in the design of the column located on the interaction strength diagram. See Example 6.4 for calculation of load-moment strength values at $f_s = 0$ and $f_s = 0.5f_y$.

Note: that load combination (2) governed the design of the column (selection of 4 #8 bars).

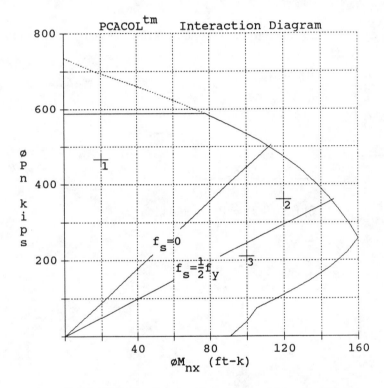

PCACOL On-Screen Results for 16 in. × 16 in.

For load combination (1), all bars in compression (Zone 1), a compression lap splice could be used. For load combination (2), bar stress not greater than $0.5f_y$ (Zone 2), a Class B tension lap splice is required or, a Class A splice may be used if alternate lap splices are staggered. For load combination (3), bar stress greater than $0.5f_y$ (Zone 3), a Class B splice must be used.

(Table 4-9)

Lap splice required for the 4#8 bars must be based on the load combination producing the greatest amount of tension in the bars; for this example, load combination (3) governs the type of lap splice to be used.

Class B splice required...$l_d = 1.3 l_{db}$

12.15.1

Example 4.7—Continued

Calculations and Discussion	Code Reference

2. Determine lap splice length

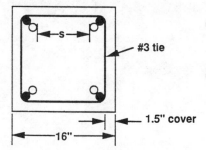

#3 tie

1.5" cover

16"

(a) Basic development length for #8 bars, $\ell_{db} = 30$ in. (Table 4-2)

(b) Factor for bar spacing and cover
Clear spacing = 8 in. (scaled) = $8d_b$

Referring to 12.2.3.1(a)

Clear spacing > $3d_b$

    Cover > Section 7.7.1

      Ties > Section 7.10.5.. $\ell_d = 1.0\,\ell_{db}$

$\ell_d = 30$ in. > 28.5 in. (Table 4-4)

Class B splice = 1.3(30) = 39 in.

Use 39 in. lap splice for the 4#8 bars at the floor level indicated.

Design Methods and Strength Requirements

8.1 DESIGN METHODS

Two philosophies of design for reinforced concrete have long been prevalent. Working Stress Design was the principal method used from the early 1900's until the early 1960's. Since publication of the 1963 edition of the ACI Code, there has been a rapid transition to Ultimate Strength Design, largely because of its more rational approach. Ultimate strength design (referred to in the Code as the Strength Design Method) is conceptually more realistic in its approach to structural safety.

The 1956 ACI Code (ACI 318-56) was the first code edition which officially recognized and permitted the ultimate strength method of design. Recommendations for the design of reinforced concrete structures by ultimate strength theories were included in an appendix.

The 1963 ACI Code (ACI 318-63) treated the working stress and the ultimate strength methods on an equal basis. However, a major portion of the working stress method was modified to reflect ultimate strength behavior. The working stress provisions of the 1963 Code, relating to bond, shear and diagonal tension, and combined axial compression and bending, had their basis in ultimate strength.

The 1971 ACI Code (ACI 318-71) was based entirely on the strength approach for proportioning reinforced concrete members, except for a small section devoted to what was called an Alternate Design Method. Even in that section, the service load capacities (except for flexure) were given as various percentages of the ultimate strength capacities of other parts of the Code. The transition to ultimate strength theories for reinforced concrete design was essentially complete in the 1971 ACI Code, with ultimate strength design definitely established as being preferred.

In the 1977 ACI Code (ACI 318-77) the Alternate Design Method (ADM) was relegated to an appendix, just as the Strength Design Method was introduced by way of an appendix in ACI 318-56. The appendix location served to separate and clarify the two methods of design, with the main body of the Code devoted exclusively to the Strength Design Method. The Alternate Design Method was retained in Appendix B of ACI 318-83, and has been retained in Appendix A of ACI 318-89. Since an appendix location is sometimes not considered to be an official part of a legal document (unless specifically adopted), specific reference is made in the main body of the code (Section 8.1.2) to make Appendix A a legal part of the Code. Regardless of whether the Strength Design Method of the Code or the Alternate Design Method of Appendix A is used in proportioning for strength, the general serviceability requirements of the Code, such as the provisions for deflection control and crack control, must always be satisfied.

8.1.1 Strength Design Method

The Strength Design Method requires that the computed nominal strengths reduced by specified strength reduction factors, i.e., design strengths, equal or exceed the service load effects (internal forces and moments) increased by specified load factors, i.e. required strengths.

Since the distinction between "design strength" and "required strength" is crucial to an understanding of the Strength Design Method, the definitions and notations used with the Strength Design Method are summarized below:

Definitions

Service load—load specified by general building code (without load factors)

Factored load—load multiplied by appropriate load factor, used to proportion members by the Strength Design Method

Required strength—strength of a member or cross section required to resist factored loads or related internal moments and forces in such combinations as are stipulated

Nominal strength—strength of a member or cross section calculated in accordance with provisions and assumptions of the Strength Design Method before application of any strength reduction factor

Design strength—nominal strength multiplied by a strength reduction factor

Notation

Required Strength:

M_u	=	factored moment (required flexural strength)
P_u	=	factored axial load (required axial load strength) at given eccentricity
V_u	=	factored shear force (required shear strength)
T_u	=	factored torsional moment (required torsional strength)

Nominal Strength:

M_n	=	nominal moment strength
M_b	=	nominal moment strength at balanced strain conditions
P_n	=	nominal axial load strength at given eccentricity
P_o	=	nominal axial load strength at zero eccentricity

P_b = nominal axial load strength at balanced strain conditions
V_n = nominal shear strength
V_c = nominal shear strength provided by concrete
V_s = nominal shear strength provided by shear reinforcement
T_n = nominal torsional moment strength
T_c = nominal torsional moment strength provided by concrete
T_s = nominal torsional moment strength provided by torsion reinforcement

Design Strength:

φM_n = design moment strength
φP_n = design axial load strength at given eccentricity
φV_n = design shear strength = $\varphi(V_c + V_s)$
φT_n = design torsional moment strength = $\varphi(T_c + T_s)$

The following discussion is essentially reproduced from the Commentary on Chapter 2 of ACI 318-89:

A number of definitions for loads are given as the Code contains requirements that must be met at various load levels. The terms "dead load" and "live load" refer to the unfactored loads (service loads) specified or defined by a general building code. Service loads (loads without load factors) are to be used where specified in the Code to proportion or investigate members for adequate serviceability. Loads used to proportion a member for adequate strength are defined as "factored loads." Factored loads are service loads multiplied by the appropriate load factors specified for required strength. The term "design loads" is not used in ACI 318 to avoid confusion with the design load terminology used in general building codes to denote service loads, or posted loads in buildings. The factored load terminology used in ACI 318 clearly indicates whether load factors are applied to a particular load, moment, or shear value as used in the Code provisions.

The required axial load, moment, and shear strengths used to proportion members are referred to as factored axial loads, factored moments, and factored shears. The factored load effects are calculated from the applied factored loads and forces in such load combinations as are stipulated in Section 9.2.

The subscript "u" is used to denote the required strengths: required axial load strength (P_u), required moment strength (M_u), required shear strength (V_u), and required torsional moment strength (T_u) calculated from the applied factored loads and forces.

Strength of a member or cross section calculated using standard assumptions and strength equations, and nominal (specified) values of material strengths and dimensions, is referred to as "nominal strength" (see Part 6). The subscript "n" is used to denote the nominal strengths: nominal axial load strength (P_n), nominal moment strength (M_n), nominal shear strength (V_n), and nominal torsional moment strength (T_n).

"Design strength" or usable strength of a member or cross section is the nominal strength reduced by the strength reduction factor φ, and may be denoted as φP_n, φM_n, φV_n, or φT_n.

9.1 STRENGTH AND SERVICEABILITY—GENERAL

9.1.1 Strength Requirements

The basic criterion for strength design may be expressed as follows:

$$\text{Design Strength} \geq \text{Required Strength}$$

or,   [Load Factor] [Service Load Effects] ≤ [Strength Reduction Factor] [Nominal Strength]

All members and all sections of members must be proportioned to meet the above criterion under the most critical load combination and under all possible actions (flexure, axial load, shear, etc.):

$$P_u \leq \varphi P_n$$
$$M_u \leq \varphi M_n$$
$$V_u \leq \varphi V_n$$
$$T_u \leq \varphi T_n$$

The above criterion provides for the margin of structural safety in two ways:

(1) The required strength is computed in terms of factored loads or the related internal moments and forces. Factored loads are defined in Section 2.1 as service loads multiplied by the appropriate load factors. The loads to be used are described in Section 8.2. Thus, for example, the required flexural strength for dead and live loads is:

$$M_u = 1.4\,M_d + 1.7\,M_\ell$$

where M_d and M_ℓ are the moments due to service dead and live loads, respectively.

(2)   The design strength is computed by multiplying the nominal strength with the appropriate strength reduction factor. The nominal strength is computed by the Code procedures assuming that the member or the section will have the exact dimensions and material properties assumed in the computations. Thus, for example, the design moment strength for a singly reinforced cross section is:

$$\varphi M_n = \varphi[A_s f_y\,(d-a/2)]$$

For the example section without compression reinforcement and subjected to flexure, the basic criterion for strength design reduces to:

$$1.4 M_d + 1.7 M_\ell \leq \varphi[A_s f_y\,(d-a/2)]$$

Similarly, for shear in a beam, the basic criterion for strength design can be stated as:

$$V_u \leq \varphi V_n = \varphi(V_c + V_s)$$

$$1.4\,V_d + 1.7\,V_\ell \leq \varphi[\,2\,\sqrt{f_c'}\,b_w d + \frac{A_v f_y d}{s}\,]$$

The following reasons for requiring load and strength reduction factors in structural design are given by MacGregor in Ref. 5.1:

1. The strength of materials or elements may be less than expected. The following factors contribute:

(A) Material strengths may differ from those assumed in design because of:

- Variability in material strengths—the compression strength of concrete as well as the yield strength and ultimate tensile strength of reinforcement are variable.

- Effect of speed of testing—the strengths of both concrete and steel are affected by the rate of loading.

- In situ strength vs. specimen strength—the strength of concrete in a structure is somewhat different from the strength of the same concrete in a control specimen.

- Effect of variability of shrinkage stresses or residual stresses—the variability of the residual stresses due to shrinkage may affect the cracking load of a member, and is significant where cracking is the critical limit state. Similarly, the transfer of compression loading from concrete to steel due to creep and shrinkage in columns may lead to premature yielding of the compression steel, possibly resulting in instability failures of slender columns with small amounts of reinforcement.

(B) Members may vary from those assumed, due to fabrication errors. The following are significant:

- Rolling tolerances in reinforcing bars.

- Geometric errors in cross section and errors in placement of reinforcement.

(C) Simplified assumptions and equations, such as use of the rectangular stress block and the maximum usable strain of concrete equal to 0.003, introduce both systematic and random errors.

(D) The use of discrete bar sizes leads to variations in the actual capacity of members.

2. Overloads may occur.

(A) Magnitudes of loads may vary from those assumed. Dead loads may vary because of:

- Variations in member sizes.

- Variations in material density.

- Structural and nonstructural alterations.

Live loads vary considerably from time to time and from building to building.

(B) Uncertainties exist in the calculation of load effects—the assumptions of stiffnesses, span lengths, etc., and the inaccuracies involved in modeling three-dimensional structures for structural analysis lead to differences between the stresses which actually occur in a building and those estimated in the designer's analysis.

3. Consequences of failure may be severe. A number of factors ought to be considered:

(A) The type of failure, warning of failure, and existence of alternative load paths.

(B) Potential loss of life.

(C) Costs to society in lost time, lost revenue, or indirect loss of life or property due to failure.

(D) The importance of the structural element in the structure.

(E) Cost of replacing the structure.

By way of background to the numerical values of load factors and strength reduction factors specified in the Code, it may be worthwhile reproducing the following paragraph from Ref. 5.1:

> "The ACI ... design requirements ... are based on an underlying assumption that if the probability of understrength members is roughly 1 in 100 and the probability of overload is roughly 1 in 1000, the probability of overload on an understrength structure is about 1 in 100,000. Load factors were derived to achieve this probability of overload. Based on values of concrete and steel strength corresponding to probability of 1 in 100 of understrength, the strengths of a number of typical sections were computed. The ratio of the strength based on these values to the strength based on nominal strengths of a number of typical sections were arbitrarily adjusted to allow for the consequences of failure and the mode of failure of a particular type of member, and for a number of other sources of variation in strength."

An Appendix to Ref. 5.1 traces the history of development of the current ACI load and strength reduction factors.

9.1.2 Serviceability Requirements

The provision of adequate strength does not necessarily ensure acceptable behavior at service load levels. Therefore, the Code includes additional requirements designed to provide satisfactory service load performance.

There is not always a clear separation between the provisions for strength and those for serviceability. For actions other than flexure, the detailing provisions in conjunction with the strength requirements are meant to ensure adequate performance at service loads.

For flexural action, there are special serviceability requirements concerning deflection, distribution of reinforcement, and permissible stresses in prestressed concrete. A consideration of service load deflections is particularly important in view of the extended use of high-strength materials and sophisticated methods of design which result in increasingly slender reinforced concrete members.

9.2   REQUIRED STRENGTH

As previously stated, the required strength U is expressed in terms of factored loads, or their related internal moments and forces. Factored loads are the loads specified in the general building code, multiplied by appropriate load factors.

While considering gravity loads (dead and live), a designer using the Code moment coefficients (same coefficients for dead and live loads—Section 8.3.3) has three choices: (1) multiplying the loads by the appropriate load factors, adding them into the total factored load, and then computing the forces and moments due to the total load, (2) computing the effects of factored dead and live loads separately, and then superimposing the effects, or (3) computing the effects of unfactored dead and live loads separately, multiplying the effects by the appropriate load factors, and then superimposing them. Under the principle of superposition, all three procedures yield the same answer. To a designer doing a more exact analysis using different coefficients for dead and live loads (pattern loading for live loads), choice (1) does not exist. While considering gravity as well as lateral loads, load effects (due to factored or unfactored loads), of course, have to be computed separately before any superposition can be made.

Section 9.2 prescribes load factors for specific combinations of loads. A list of these combinations is given in Table 5-1. The numerical value of the load factor assigned to each type of load is influenced by the degree of accuracy with which the load can usually be assessed, and by the variation which may be expected in the load during the lifetime of a structure. Hence, dead loads, because they can usually be more accurately determined and are less variable, are assigned a lower load factor than live loads. For weight and pressure of liquids with well-defined densities and controllable maximum heights, a reduced load factor of 1.4 is permitted (Section 9.2.5) recognizing the lesser probability of overloading with such liquid loads. A higher load factor of 1.7 is required (Section 9.2.4) where there is considerable uncertainty of pressures such as earth and groundwater pressures and ponding of water.

While most usual combinations of loads are included, it should not be assumed that all cases are covered. In assigning factors to combinations of loads, some consideration is given to the likelihood of simultaneous occurrence.

In determining the required strength for combinations of loads, due regard must be given to the proper sign (positive or negative), since one type of loading may produce effects which are opposite in sense to those produced by another type. Equation (9-3) for wind and earthquake loads is specifically included for the case

Table 5-1—Required Strength for Combinations of Loads

9.2.1 — <u>Dead & Live Loads</u>

$$U = 1.4D + 1.7L \qquad (9\text{-}1)$$

9.2.2 — <u>Dead, Live & Wind Loads</u>

$$U = 1.4D + 1.7L$$
$$\text{or} \quad U = 0.75 (1.4D + 1.7L + 1.7W) \qquad (9\text{-}2)$$
$$= 1.05D + 1.275L + 1.275W$$
$$\text{or} \quad U = 0.9D + 1.3W \qquad (9\text{-}3)$$

9.2.3 — <u>Dead, Live & Earthquake Loads</u>

$$U = 1.4D + 1.7L$$
$$\text{or} \quad U = 0.75 (1.4D + 1.7L + 1.87E)$$
$$= 1.05D + 1.275L + 1.402E$$
$$\text{or} \quad U = 0.9D + 1.43E$$

9.2.4 — <u>Dead & Live Loads Plus Earth and Groundwater Pressure*</u>

$$U = 1.4D + 1.7L$$
$$\text{or} \quad U = 1.4D + 1.7L + 1.7H \qquad (9\text{-}4)$$
or (D reducing H)
$$U = 0.9D + 1.7L + 1.7H$$
or (L reducing H)
$$U = 1.4D + 1.7H$$
or (D & L reducing H)
$$U = 0.9D + 1.7H$$

9.2.5 — <u>Dead & Live Load Plus Liquid Pressure**</u>

$$U = 1.4D + 1.7L$$
$$\text{or} \quad U = 1.4D + 1.7L + 1.4F$$
or (D reducing F)
$$U = 0.9D + 1.7L + 1.4F$$
or (L reducing F)
$$U = 1.4D + 1.4F$$
or (D & L reducing F)
$$U = 0.9D + 1.4F$$

9.2.6 — <u>Impact</u>

In all equations substitue (L + Impact) for (L) when impact must be considered.

9.2.7 — <u>Dead & Live Loads Plus Differential Settlement, Creep, Shrinkage or Temperature Change</u>

$$U = 1.4D + 1.7L$$
$$\text{or} \quad U = 0.75 (1.4D + 1.4T + 1.7L) \qquad (9\text{-}5)$$
$$= 1.05D + 1.05T + 1.275L$$
$$\text{or} \quad U = 1.4D + 1.4T \qquad (9\text{-}6)$$

---

*Weight and pressure of soil and water in soil. (Groundwater pressure is to be considered part of earth pressure with a 1.7 load factor.)
**Weight and pressure of liquids with well-defined densities and controllable maximum heights.

where E or W produces effects opposite in sense to those caused by D and L. Typical cases are uplift on the windward columns and the reversal of moments in beams due to W or E.

Consideration must be given to various combinations of loads in determining the most critical design combination. This is of particular importance when strength is dependent on more than one load effect, such as strength under combined moment and axial load, or the shear strength of members carrying axial load.

9.3 DESIGN STRENGTH

9.3.1 Nominal Strength vs. Design Strength

The design strength provided by a member, its connections to other members, and its cross sections, in terms of flexure, axial load, shear, and torsion, is equal to the nominal strength calculated in accordance with the provisions and assumptions stipulated in the Code, multiplied by a strength reduction factor φ which is less than unity. The rules for computing the nominal strength are based generally on conservatively chosen limiting states of stress, strain, cracking or crushing, and conform to research data for each type of structural action. An understanding of all aspects of the strengths computed for various actions can only be obtained by reviewing the background to the Code provisions.

The purpose of the strength reduction factor φ is (1) to define a design strength level that is somewhat lower than would be expected if all dimensions and material properties were those used in computations, (2) to reflect the degree of ductility, toughness, and reliability of the member under the load effects being considered, and (3) to reflect the importance of the member. For example, a lower φ is used for columns than for beams because columns generally have less ductility, are more sensitive to variations in concrete strength, and carry larger loaded areas than beams. Furthermore, spiral columns are assigned a higher φ than tied columns because the former have greater toughness or ductility.

9.3.2 Strength Reduction Factors

The φ factors prescribed for the different types of action are listed in Table 5-2.

For members subject to flexure and axial load, the design strengths are determined by multiplying both P_n and M_n by the appropriate single value of φ. For members subject to flexure with axial tension, the value of φ given in Section 9.3.2.2(a) is used. For members subject to flexure with axial compression, the value of φ given in Section 9.3.2.2(b) is used for both P_n and M_n.

For members subject to relatively small axial loads (and flexure) it is reasonable to permit an increase in the φ factor from that required for compression members, so that when the axial load reduces to zero and the member is subjected to pure flexure, the strength reduction factor equals 0.90. This is also justified in view of the fact that failure under flexure and small axial loads is initiated by yielding of the tension reinforcement and takes place in an increasingly more ductile manner as the axial load decreases. At the same time, the variability of the strength also decreases. Thus, a varying φ factor is permitted in the Code for members

Table 5-2—Strength Reduction Factors

Action	φ
Flexure, without axial load	0.90
Axial tension, and axial tension with flexure	0.90
Axial compression, and axial compression with flexure:	
Members with spiral reinforcement conforming to Section 10.9.3	0.75*
Other reinforced members	0.70*
Shear and torsion	0.85
Bearing on concrete	0.70**

*May be increased linearly to 0.90 as φP_n decreases from $0.10 f_c' A_g$ or φP_b, whichever is smaller, to zero.

**Does not apply to post-tensioning anchorage bearing plates. See Section 18.13.

subjected to bending and small axial loads. The φ may be increased from that for compression members to the 0.90 for flexure, as the axial load decreases from a specified value to zero.

The value of the axial load strength φP_n below which an increase in φ can be made is $0.10\,f_c'\,A_g$ or φP_b, whichever is less. For sections with symmetrical reinforcement and with $f_y \leq 60,000$ psi, in which the distance γh (distance between A_s and A_s') is not less than $0.7h$, φP_b will always be greater than $0.10\,f_c'\,A_g$. Thus, for such sections, the computation of P_b is not required.

9.3.3 Development Lengths for Reinforcement

Development lengths for reinforcement, as specified in Chapter 12, do not require a strength reduction modification. Likewise, φ factors are not required for splice lengths, since these are expressed in multiples of development lengths.

9.4 DESIGN STRENGTH FOR REINFORCEMENT

An upper limit of 80,000 psi is placed on the yield strength of reinforcing steels other than prestressing tendons. A steel strength above 80,000 psi is not recommended because the yield strain of 80,000 psi steel is about equal to the maximum usable strain of concrete in compression. Currently there is no ASTM specification for Grade 80 reinforcement. However, #11, #14, and #18 deformed reinforcing bars with a yield strength of 75,000 psi (Grade 75) are included in the 1987 edition of ASTM A615 Specification.

In accordance with Section 3.5.3.2, use of reinforcing bars with a specified yield strength f_y exceeding 60,000 psi requires that f_y be measured at a strain of 0.35 percent. The A615-87 Specification for Grade 75 bars includes the same requirement. The 0.35 percent strain requirement also applies to welded wire fabric with wire having a specified yield strength greater than 60,000 psi. Higher-yield-strength wire is available and a value of f_y greater than 60,000 psi can be used in design, provided compliance with the 0.35 percent strain requirement is certified.

There are limitations on the yield strength of reinforcement in other sections of the Code:

(1) Sections 11.5.2, 11.6.7.4, and 11.7.6: The maximum f_y that may be used in design for shear and torsion reinforcement is 60,000 psi.

(2) Sections 19.3.2 and 21.2.5.1: The maximum specified f_y is 60,000 psi in shells, folded plates and structures governed by the special seismic provisions of Chapter 21.

(3) Appendix A: The useful f_y is controlled by permissible stresses in the Alternate Design Method.

In addition, the deflection provisions of Section 9.5 and the limitations on distribution of flexural reinforcement of Section 10.6 will become increasingly critical as f_y increases.

8.1.2 Alternate Design Method

An alternate method of design employing load factors and strength reduction factors equal to unity (i.e., service load effects and allowable service load stresses) is permitted for nonprestressed members. The method is outlined in Appendix A.

The Alternate Design Method requires that a structural member (in flexure) be so proportioned that the stresses resulting from the action of service loads (without load factors) and computed by the straight line theory for flexure do not exceed permissible service load stresses. The permissible stresses are limited to values well within the elastic range of the materials, so that the linear relationship between stress and strain is applicable.

The method is similar to the working stress design method of previous ACI Codes. For members subjected to flexure without axial load, the method is identical to that given in the 1963 Code. Differences in procedure occur in all other cases, including the design of columns, and the design for shear, anchorage length, and splices. In view of the simplifications permitted, the Alternate Design Method will generally result in designs that are more conservative than those based on the Strength Design Method.

Although prestressed members may not be designed for strength under the provisions of Appendix A, Chapter 18 permits linear stress-strain assumptions in the computation of service load stresses and of transfer stresses for use in serviceability control.

It should be noted that all relevant provisions of the Code, except those permitting moment redistribution, apply also to members designed by the Alternate Design Method. These include control of deflections and distribution of flexural reinforcement, as well as the provisions related to slenderness effects in compression members.

REFERENCE

5.1 MacGregor, J.G., "Safety and Limit States Design for Reinforced Concrete," *Canadian Journal of Civil Engineering,* Vol. 3, No. 4, December 1976, pages 484-513.

General Principles of Strength Design

UPDATE FOR THE '89 CODE

The only revisions to the general principles of strength design are editorial clarifications to the procedure for calculating the bearing strength of concrete, in Section 10.15. These changes were made in the 1986 Supplement to ACI 318-83.

GENERAL CONSIDERATIONS

Historically, ultimate strength was the earliest method used in design, since the ultimate load could be measured by test without a knowledge of the magnitude or distribution of internal stresses. Since the early 1900's, experimental and analytical investigations have been conducted to develop ultimate strength design theories that would predict the ultimate load measured by test. Some of the early theories that resulted from the experimental and analytical investigations are reviewed in Fig. 6-1.

Structural concrete and reinforcing steel both behave inelastically as ultimate strength is approached. In theories dealing with the ultimate strength of reinforced concrete, the inelastic behavior of both materials must be considered and must be expressed in mathematical terms. For reinforcing steel with a distinct yield point, the inelastic behavior may be expressed by a bilinear stress-strain relationship (Fig. 6-2). For concrete, the inelastic stress distribution is more difficult to measure experimentally and to express in mathematical terms.

Studies of inelastic concrete stress distribution have resulted in numerous proposed stress distributions as outlined in Fig. 6-1. The development of our present ultimate strength design procedures has its basis in these early experimental and analytical studies. Ultimate strength of reinforced concrete in American design specifications is based primarily on the 1912 and 1932 theories (Fig. 6-1).

10.2 DESIGN ASSUMPTIONS

10.2.1—Design Assumptions Based on Satisfaction of Equilibrium of Forces and Compatibility of Strains.

Computation of the strength of a member or cross section by the Strength Design Method requires that two basic conditions be satisfied: (1) static equilibrium, and (2) compatibility of strains.

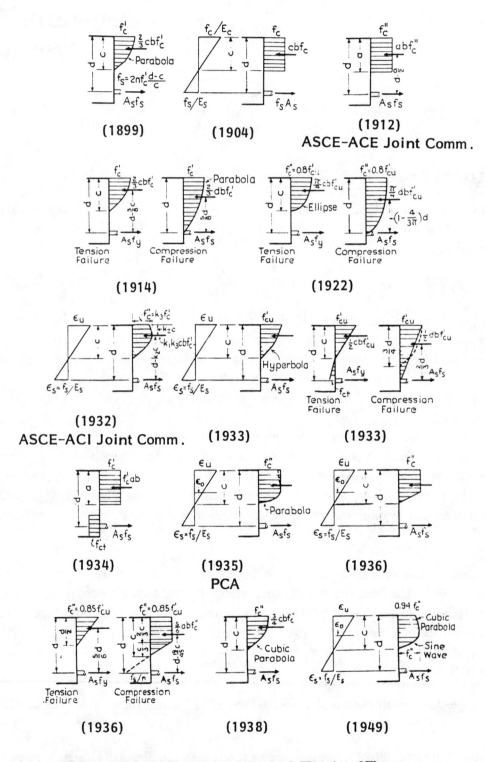

Fig. 6-1   Development of Ultimate Strength Theories of Flexure

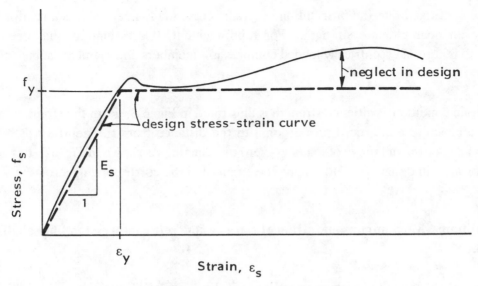

Fig. 6-2   Stress-Strain Relationship for Reinforcement

Equilibrium between the compressive and tensile forces acting on the cross section at "ultimate" strength must be satisfied.   Compatibility between the strains in the concrete and the reinforcement at "ultimate" conditions must also be established within the design assumptions permitted by the Code (Section 10.2).

The term "ultimate" is used frequently in reference to the Strength Design Method; however, it should be realized that the "nominal" strength computed under the provisions of the Code may not necessarily be the actual ultimate value.   Within the design assumptions permitted, certain properties of the materials are neglected and other conservative limits are established for practical design.   These contribute to a possible lower "ultimate strength" than that obtained by test.   The computed nominal strength should be considered a Code-defined strength only.   Accordingly, the term "ultimate" is not used when defining the computed strength of a member.   The term "nominal" strength is used instead.

Furthermore, in discussing the strength method of design for reinforced concrete structures, attention must be called to the difference between loads on the structure as a whole and load effects on the cross sections of individual members.   Elastic methods of structural analysis are used first to compute service load effects on the individual members due to the action of service loads on the entire structure.   Only then are the load factors applied to the service load effects acting on the individual cross sections. Inelastic (or limit) methods of structural analysis, in which design load effects on the individual members are determined directly from the ultimate loads acting on the whole structure, are not considered.   Section 8.4, however, does permit a limited redistribution of negative moments in continuous members.   The provisions of Section 8.4 recognize the inelastic behavior of concrete structures and constitute a move toward "limit design."   This subject is presented in Part 9.

The computed "nominal strength" of a member must satisfy the design assumptions given in Section 10.2.

10.2.2—Assumption: Strain in reinforcement and concrete shall be assumed directly proportional to the distance from the neutral axis.

In other words, plane sections normal to the axis of bending are assumed to remain plane after bending.

Many tests have confirmed that the distribution of strain is essentially linear across a reinforced concrete cross section, even near ultimate strength. For reinforcement, this assumption has been verified by numerous tests to failure of eccentrically loaded compression members and members subjected to bending only.

The assumed strain condition at ultimate strength is illustrated in Fig. 6-3. Both the strain in the reinforcement and in the concrete are directly proportional to the distance from the neutral axis. Actually this assumption is valid over the full range of loading—zero to ultimate. As shown in Fig. 6-3, this assumption is of primary importance in design for determining the strain (and the corresponding stress) in the reinforcement.

10.2.3—Assumption: Maximum usable strain at extreme concrete compression fiber shall be assumed equal to $\varepsilon_u = 0.003$.

The maximum concrete compressive strain at crushing of the concrete has been measured in many tests of both plain and reinforced concrete members. The test results from a series of reinforced concrete beam and column specimens are shown in Fig. 6-4.

The maximum concrete compressive strain varies from 0.003 to as high as 0.008, as shown in Fig. 6-4; however, the maximum strain for practical cases is 0.003 to 0.004 (see stress-strain curves in Fig. 6-5).

Though the maximum strain decreases somewhat with increasing compressive strength of concrete, the 0.003 value allowed for design is reasonably conservative. The codes of some countries specify a value of 0.0035 for design, which makes little difference in the computed strength of a member.

10.2.4—Assumption: Stress in reinforcement below the yield strength f_y shall be taken as E_s times the steel strain ($f_s = E_s\varepsilon_s$). For strains greater than f_y/E_s, stress in reinforcement shall be considered independent of strain and equal to f_y.

For deformed reinforcement, it is reasonably accurate to assume that the stress in reinforcement is proportional to strain below the yield strength. For practical design, the increase in strength due to the effect of strain hardening of the reinforcement is neglected for strength computations. The actual vs. the design (bilinear) stress-strain relationship is shown in Fig. 6-2.

In strength computations, the force developed in tensile or compressive reinforcement is a function of the reinforcement strain,

when $\varepsilon_s \leq \varepsilon_y$ (yield strain $= f_y/E_s$)

$$A_s f_s = A_s E_s \varepsilon_s$$

when $\varepsilon_s \geq \varepsilon_y$

$$A_s f_s = A_s f_y$$

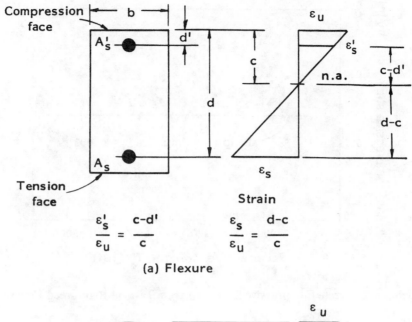

$$\frac{\varepsilon'_s}{\varepsilon_u} = \frac{c-d'}{c} \qquad \frac{\varepsilon_s}{\varepsilon_u} = \frac{d-c}{c}$$

(a) Flexure

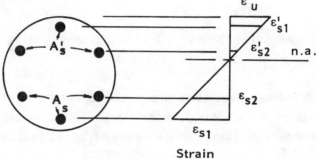

(b) Flexure and axial load

Fig. 6-3   Assumed Strain Variation

where ε_s is the value from the strain diagram at the location of the reinforcement (see Fig. 6-3). For design, the modulus of elasticity of steel reinforcement, E_s, may be taken as 29,000,000 psi (Section 8.5.2).

10.2.5—Assumption: Tensile strength of concrete shall be neglected in flexural calculations of reinforced concrete.

The tensile strength of concrete in flexure, known as the modulus of rupture, is a more variable property than the compressive strength, and is about 10% to 15% of the compressive strength. The generally accepted value for design is $7.5\sqrt{f'_c}$ for normal-weight concrete. This tensile strength in flexure is neglected in strength design. For practical percentages of reinforcement, the resulting computed strengths are in good agreement with test results. For very small percentages of reinforcement, neglect of the tensile strength of concrete is conservative.

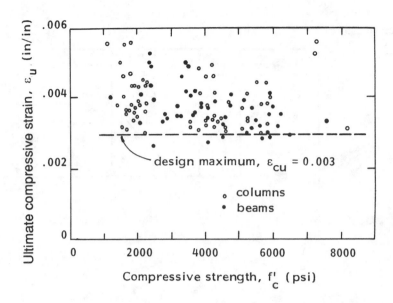

Fig. 6-4   Maximum Concrete Compressive Strain, ε_u, from Tests of Reinforced Concrete Members

It should be realized, however, that the strength of concrete in tension is important in cracking and deflection (serviceability) considerations.

10.2.6—Assumption: Relationship between concrete compressive stress distribution and concrete strain may be assumed to be rectangular, trapezoidal, parabolic, or any other shape that results in prediction of strength in substantial agreement with results of comprehensive tests.

This assumption recognizes the inelastic stress distribution in concrete at high stresses. As maximum stress is approached, the stress-strain relationship of concrete is not a straight line (stress is not proportional to strain). The general stress-strain behavior is shown in Fig. 6-5. The shape of the curves is primarily a function of concrete strength and consists of a rising curve from zero stress to a maximum at a compressive strain between 0.0015 and 0.002, followed by a descending curve to an ultimate strain (corresponding to crushing of the concrete) varying from 0.003 to as high as 0.008. As discussed under Assumption 10.2.3, the Code sets the maximum usable strain at 0.003 for design. The curves show that the stress-strain behavior for concrete is nonlinear at stress levels above about $0.5f_c'$.

The actual distribution of concrete compressive stress in a practical case is complex and usually not known. However, research has shown that the important properties of the concrete stress distribution can be approximated closely using any one of several different assumptions as to the form of stress distribution. Many stress distributions have been proposed (see Fig. 6-1). The three most common are the parabolic, the trapezoidal, and the rectangular. All yield reasonable results. At the theoretical strength of a member in flexure, the compressive stress distribution should conform closely to the actual variation of stress, as shown in Fig. 6-6. The maximum stress is indicated by k_3f_c', the average stress is indicated by $k_1k_3f_c'$, and the depth of the centroid of the approximate parabolic distribution from the extreme compression fiber by k_2c, where c is the neutral axis depth.

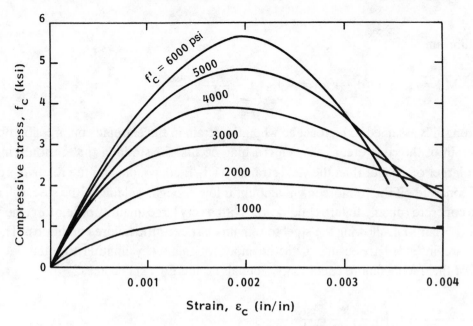

Fig. 6-5　Typical Stress-Strain Curves for Concrete

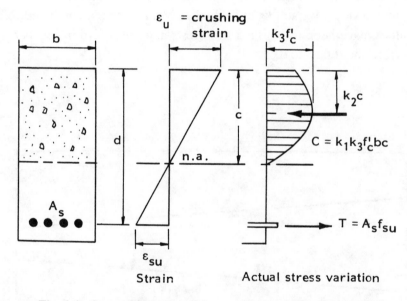

Fig. 6-6　Stress-Strain Conditions at Nominal Strength in Flexure

For the stress conditions at ultimate, the nominal moment strength, M_n, may be computed by equilibrium of forces and moments.

For force equilibrium:

$$C = T \qquad \text{or,} \qquad k_1 k_3 f_c' bc = A_s f_{su}$$

so that

$$c = \frac{A_s f_{su}}{k_1 k_3 f_c' b}$$

For moment equilibrium:

$$M_n = (C \text{ or } T)(d - k_2c) = A_sf_{su}\left(d - \frac{k_2}{k_1k_3}\frac{A_sf_{su}}{f_c'\,b}\right) \tag{1}$$

The maximum strength is assumed to be reached when the strain in the extreme compression fiber is equal to the crushing strain of the concrete, ε_u. When crushing occurs, the strain in the tension reinforcement, ε_{su}, may be either larger or smaller than the yield strain, $\varepsilon_y = f_y/E_s$, depending on the relative proportion of reinforcement to concrete. If the reinforcement amount is low enough, yielding of the steel will occur prior to crushing of the concrete (ductile failure condition). With a very large quantity of reinforcement, crushing of the concrete will occur first, allowing the steel to remain elastic (brittle failure condition). The Code has provisions which are intended to ensure a ductile mode of failure by limiting the amount of tension reinforcement. For the ductile failure condition, f_{su} equals f_y, and Eq. (1) becomes:

$$M_n = A_sf_y\left(d - \frac{k_2}{k_1k_3}\frac{A_sf_y}{f_c'\,b}\right) \tag{2}$$

If the quantity $k_2/(k_1k_3)$ is known, the moment strength can be computed directly from Eq. (2). It is not necessary to know the values of k_1, k_2, and k_3 individually. Values for the combined term, as well as the individual k_1 and k_2 values, have been established from tests and are shown in Fig. 6-7. As shown in the figure, $k_2/(k_1k_3)$ varies from about 0.55 to 0.63.

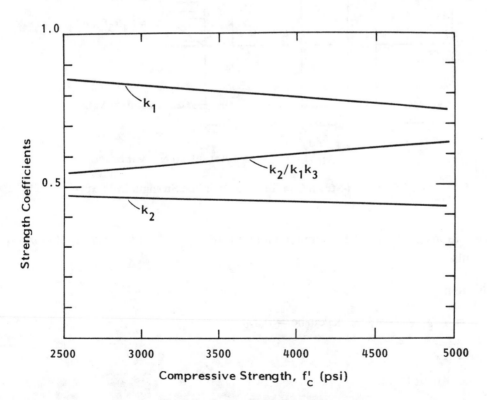

Fig. 6-7 Stress-Block Parameters

The Portland Cement Association has adopted the parabolic stress-strain relationship shown in Fig. 6-8 for much of its experimental and analytical research work. Most PCA-published strength design aids and computer programs are based on the parabolic stress variation shown. Such "more exact" stress distributions have their greatest application with electronic computers and are not recommended for longhand calculations.

10.2.7—Assumption: Requirements of Section 10.2.6 may be considered satisfied by an equivalent rectangular concrete stress distribution defined as follows: A concrete stress of 0.85 f_c' shall be assumed uniformly distributed over an equivalent compression zone bounded by edges of the cross section and a straight line located parallel to the neutral axis at a distance a = $\beta_1 c$ from the fiber of maximum compressive strain. Distance c from the fiber of maximum compressive strain to the neutral axis shall be measured in a direction perpendicular to that axis. Fraction β_1 shall be taken as 0.85 for strengths f_c' up to 4000 psi and shall be reduced continuously at a rate of 0.05 for each 1000 psi of strength in excess of 4000 psi, but β_1 shall not be taken less than 0.65.

Computation of the flexural strength based on the approximate parabolic stress distribution of Fig. 6-6 may be done using Eq. (2) with given values of $k_2/(k_1 k_3)$. However, for practical design purposes, a method based on simple static equilibrium is desirable. The Code allows the use of a rectangular compressive stress block to replace the more exact stress distributions of Fig. 6-6 (or Fig. 6-8). In this equivalent rectangular stress block, as shown in Fig. 6-9, a uniform stress of 0.85 f_c' is used over a depth a = $\beta_1 c$, determined so that a/2 = $k_2 c$. A β_1 of 0.85 for concrete with $f_c' \leq$ 4000 psi and 0.05 less for each 1000 psi of f_c' in excess of 4000 psi was determined experimentally to agree with test data. For high-strength concretes, above 8000 psi, a lower limit

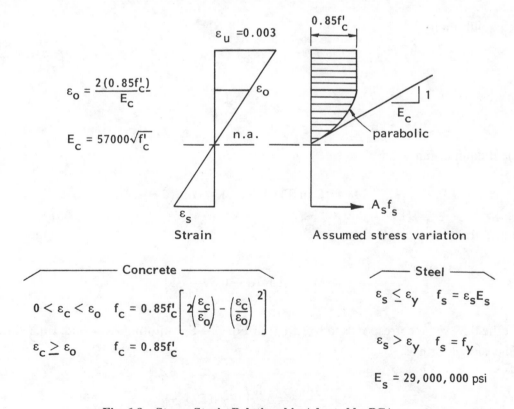

$$\varepsilon_o = \frac{2(0.85 f_c')}{E_c}$$

$$E_c = 57000\sqrt{f_c'}$$

— Concrete —

$$0 < \varepsilon_c < \varepsilon_o \quad f_c = 0.85 f_c'\left[2\left(\frac{\varepsilon_c}{\varepsilon_o}\right) - \left(\frac{\varepsilon_c}{\varepsilon_o}\right)^2\right]$$

$$\varepsilon_c \geq \varepsilon_o \quad f_c = 0.85 f_c'$$

— Steel —

$$\varepsilon_s \leq \varepsilon_y \quad f_s = \varepsilon_s E_s$$

$$\varepsilon_s > \varepsilon_y \quad f_s = f_y$$

$$E_s = 29,000,000 \text{ psi}$$

Fig. 6-8   Stress-Strain Relationship Adopted by PCA

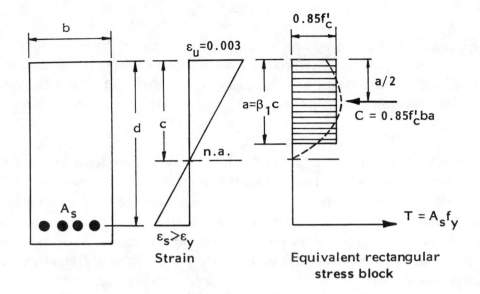

Fig. 6-9   Equivalent Rectangular Concrete Stress Distribution

of 0.65 is placed on the β_1 factor. Variation in β_1 vs. concrete strength f'_c is shown in Fig. 6-10. Effect of the limit of 0.65 on β_1 for high-strength concretes is illustrated in Example 6.1.

Using the equivalent rectangular stress distribution, and assuming that the reinforcement yields prior to crushing of the concrete ($\varepsilon_s > \varepsilon_y$), the nominal moment strength M_n may be computed by equilibrium of forces and moments.

For force equilibrium:

$$C = T \qquad \text{or,} \qquad 0.85f'_c ba = A_s f_y$$

so that

$$a = \frac{A_s f_y}{0.85 f'_c b}$$

For moment equilibrium:

$$M_n = (C \text{ or } T)\,(d - a/2) = A_s f_y\,(d - a/2)$$

substituting a from force equilibrium,

$$M_n = A_s f_y \left(d - 0.59\,\frac{A_s f_y}{f'_c b} \right) \tag{3}$$

Note that the 0.59 value corresponds to $k_2/(k_1 k_3)$ of Eq. (2). Substituting $A_s = \rho bd$, Eq. (3) may be written in nondimensional form as:

$$\frac{M_n}{bd^2 f'_c} = \rho\,\frac{f_y}{f'_c}\left(1 - 0.59\rho\,\frac{f_y}{f'_c} \right) \tag{4}$$

6-10

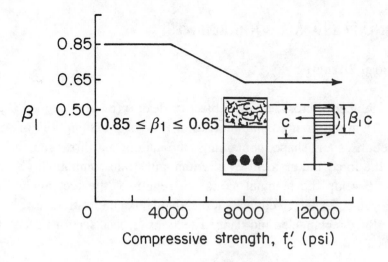

Fig. 6-10   Strength Factor β_1

As shown in Fig. 6-11, Eq. (4) is "in substantial agreement with the results of comprehensive tests." It must, however, be realized that the rectangular stress block does not represent the actual stress distribution in the compression zone at ultimate, but does provide essentially the same strength results as those obtained in tests. Computation of moment strength using equivalent rectangular stress distribution and static equilibrium is illustrated in Example 6.2 .

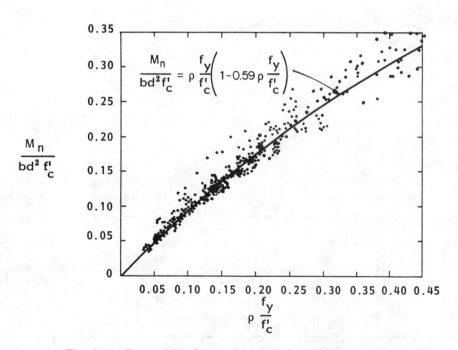

Fig. 6-11   Tests of 364 Beams Controlled by Tension ($\varepsilon_s > \varepsilon_y$)

6-11

10.3 GENERAL PRINCIPLES AND REQUIREMENTS

10.3.1 Nominal Flexural Strength

Nominal strength of a member or cross section subject to flexure (or to combined flexure and axial load) must be based on equilibrium and strain compatibility using the design assumptions of Section 10.2. Nominal strength of a cross section of any shape, containing any amount and arrangement of reinforcement, is computed by applying the force and moment equilibrium and strain compatibility conditions in a manner similar to that used to develop the nominal moment strength of the rectangular section with tension reinforcement only in Fig. 6-9. Using the equivalent rectangular concrete stress distribution, expressions for nominal moment strength of rectangular and flanged sections (typical sections used in concrete construction) are summarized as follows:

(a) Rectangular section with tension reinforcement only (See Fig. 6-9):

   Expressions are given on page 6-10.

(b) Flanged section with tension reinforcement only:

   When the compression flange thickness is equal to or greater than the depth of the equivalent rectangular stress block a, moment strength M_n is calculated by Eq. (3), as for a rectangular section with width equal to the flange width. When compression flange thickness is less than a, moment strength M_n is (Fig. 6-12):

$$M_n = (A_s - A_{sf}) f_y (d - a/2) + A_{sf} f_y (d - h_f/2) \tag{5}$$

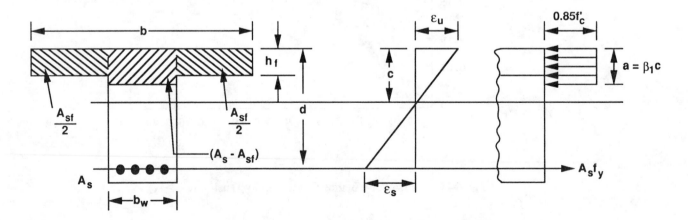

Fig. 6-12 Strain and Equivalent Stress Distribution for Flanged Section

Where A_{sf} = area of reinforcement required to equilibrate compressive strength of overhanging flanges = $0.85f'_c (b - b_w) h_f/f_y$

$$a = (A_s - A_{sf}) f_y/0.85f'_c b_w$$

b = width of effective flange (Section 8.10)

b_w = width of web

h_f = thickness of flange

(c)　Rectangular section with compression reinforcement (see Fig. 6-3):

If
$$\frac{(A_s - A'_s)}{bd} \geq 0.85 \frac{\beta_1 f'_c d'}{f_y d} \left(\frac{87,000}{87,000 - f_y}\right) \qquad (6)$$

Moment strength M_n is:

$$M_n = (A_s - A'_s) f_y(d - a/2) + A'_s f_y (d - d') \qquad (7)$$

where $a = (A_s - A'_s) f_y/0.85f'_c b$

and d' = distance from extreme compression fiber to centroid of compression reinforcement.

When the value of $(A_s - A'_s)/bd$ is less than that given above, stress in the compression reinforcement is less than f_y(i.e. $\varepsilon'_s < \varepsilon_y$) at ultimate, and a general analysis must be made based on stress and strain compatibility (see Fig. 6-3). Alternatively, the contribution of compression reinforcement may be neglected and the moment strength calculated by Eq. (3), as for a rectangular section with tension reinforcement only.

(d)　For other cross sections, the nominal moment strength M_n is calculated by a general analysis based on equilibrium and strain compatibility using the design assumptions of Section 10.2.

(e)　Nominal flexural strength M_n of a cross section of a composite flexural member consisting of cast-in-place and precast concrete is computed in a manner similar as that for a regular reinforced concrete section. Since the "ultimate" strength is unrelated to the sequence of loading, no distinction is made between shored and unshored members in strength computations (Section 17.2.4).

10.3.2 Balanced Strain Condition

A balanced strain condition exists at a cross section when the maximum strain at the extreme compression fiber just reaches $\varepsilon_u = 0.003$ simultaneously with the first yield strain of $\varepsilon_s = \varepsilon_y = f_y/E_s$ in the tension reinforcement. This balanced strain condition is shown in Fig. 6-13.

The required reinforcement ratio, ρ_b, to produce a balanced strain condition in a rectangular section with tension reinforcement only may be obtained by applying equilibrium and strain compatibility conditions. Referring to Fig. 6-13, for the linear strain condition:

$$\frac{c_b}{d} = \frac{\varepsilon_u}{\varepsilon_u + \varepsilon_y}$$

$$= \frac{0.003}{0.003 + f_y/29,000,000} = \frac{87,000}{87,000 + f_y}$$

For force equilibrium:

$$C_b = T_b \quad \text{or,} \quad 0.85f_c' \, ba_b = A_{sb}f_y \quad \text{or,} \quad 0.85f_c'b \, (\beta_1 c_b) = \rho_b bd f_y$$

thus,

$$\rho_b = \frac{0.85\,\beta_1 f_c'}{f_y} \times \frac{c_b}{d}$$

$$= \frac{0.85\,\beta_1 f_c'}{f_y} \times \frac{87,000}{87,000 + f_y} \qquad \text{Eq. (8-1)}$$

Values of ρ_b for various concrete and reinforcement strengths are listed in Table 6-1.

The balanced reinforcement ratio ρ_b for flanged sections and rectangular sections with compression reinforcement may be obtained by applying the equilibrium and compatibility conditions in a similar manner:

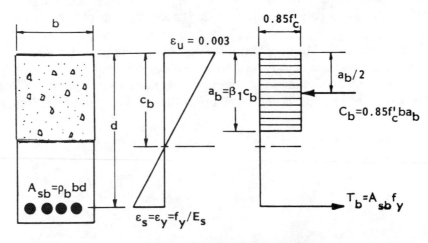

Fig. 6-13 Balanced Strain Condition in Flexure

Table 6-1—Balanced Ratio of Reinforcement ρ_b for Rectangular Sections with Tension Reinforcement Only

f_y	$f_c' = 3000$ $\beta_1 = 0.85$	$f_c' = 4000$ $\beta_1 = 0.85$	$f_c' = 5000$ $\beta_1 = 0.80$	$f_c' = 6000$ $\beta_1 = 0.75$	$f_c' = 8000$ $\beta_1 = 0.65$	$f_c' = 10,000$ $\beta_1 = 0.65$
40,000	0.0371	0.0495	0.0582	0.0655	0.0757	0.0946
60,000	0.0214	0.0285	0.0335	0.0377	0.0436	0.0545
75,000	0.0155	0.0207	0.0243	0.0274	0.0316	0.0396

For a flanged section with tension reinforcement only:

$$\rho_b = \frac{b_w}{b} \, (\overline{\rho_b} + \rho_f) \tag{8}$$

where

$$\rho_f = \frac{A_{sf}}{b_w d} \quad \text{and} \quad A_{sf} = 0.85 \frac{f_c'}{f_y} (b - b_w) h_f$$

For a rectangular section with compression reinforcement:

$$\rho_b = \overline{\rho_b} + \rho' \frac{f_{sb}'}{f_y} \tag{9}$$

where $f_{sb}' = $ stress in compression reinforcement at balanced strain condition

$$= 87,000 - \frac{d'}{d} (87,000 + f_y) \le f_y$$

and $\overline{\rho_b} = $ balanced reinforcement ratio for a rectangular section with tension reinforcement only.

10.3.3   Maximum Reinforcement Ratio

The flexural strength of a member is ultimately reached when the strain in the extreme compression fiber reaches the ultimate (crushing) strain of the concrete, ε_u. At that stage, the strain in the tension reinforcement could just reach the strain at first yield ($\varepsilon_s = \varepsilon_y = f_y/E_s$), be less than the yield strain, or exceed the yield strain. Which steel strain condition exists at ultimate concrete strain depends on the relative proportion of reinforcement to concrete. If the steel amount is low enough, the strain in the tension steel will greatly exceed the yield strain ($\varepsilon_s \gg \varepsilon_y$) when the concrete strain reaches ε_u, with large deflection and ample warning of impending failure (ductile failure condition). With a larger quantity of steel, the strain in the tension steel may not reach the yield strain ($\varepsilon_s < \varepsilon_y$) when the concrete strain reaches ε_u, which would mean small deflection and little warning of impending failure (brittle failure condition). For design it is desirable to restrict the ultimate strength condition so that a ductile failure mode would be expected.

The Code has provisions that are intended to ensure a ductile mode of failure by limiting the amount of tension reinforcement to 75% of the amount that will cause the strain in the tension steel to just reach yield strain at crushing strain of the concrete. This strain condition is defined as the "balanced condition," and the amount of reinforcement required to produce a balanced condition at ultimate strength is defined as the "balanced reinforcement ratio ρ_b."

The maximum amount of reinforcement permitted in a rectangular section with tension reinforcement only is

$$\rho_{max} = 0.75\,\overline{\rho_b} = 0.75\left[0.85\,\beta_1\frac{f_c'}{f_y} \times \frac{87,000}{87,000 + f_y}\right] \tag{10}$$

The maximum amount of reinforcement permitted in a flanged section with tension reinforcement only is

$$\rho_{max} = 0.75\left[\frac{b_w}{b}\,(\overline{\rho_b} + \rho_f)\right] \tag{11}$$

The maximum amount of reinforcement permitted in a rectangular section with compression reinforcement is

$$\rho_{max} = 0.75\,\overline{\rho_b} + \rho'\,\frac{f_{sb}'}{f_y} \tag{12}$$

Note that with compression reinforcement, the portion of ρ_b contributed by the compression reinforcement ($\rho'f_{sb}'/f_y$) need not be reduced by the 0.75 factor. For ductile behavior of beams with compression reinforcement, only that portion of the total tension steel balanced by compression in the concrete ($\overline{\rho_b}$) need be limited (see Example 6.3).

It should be realized that the limit on the amount of tension reinforcement for flexural members is a Code-defined limitation for ductile behavior. Tests have shown that beams reinforced with the computed amount of balanced reinforcement actually behave in a ductile manner with gradually increasing deflections and cracking up to failure. Sudden compression failures do not occur until the amount of reinforcement is considerably higher than the computed balanced amount.

One reason for the above is the limit on the ultimate concrete strain assumed at $\varepsilon_u = 0.003$ for design. The actual maximum strain based on physical testing may be much higher than this value. The 0.003 value serves as a lower bound on limiting strain. Note discussion under Section 10.2.3. Unless unusual amounts of ductility are required, the $0.75\rho_b$ limitation will provide ample ductile behavior for most designs.

10.3.5 Maximum Axial Load Strength

The strength of a member in pure compression (zero eccentricity) is computed simply as:

$$P_O = 0.85f_c' A_g + f_y A_{st}$$

where A_{st} is the total area of reinforcement and A_g is the gross area of the concrete section. Refinement in concrete area can be considered by subtracting the area of concrete displaced by the steel:

$$P_O = 0.85f_c' (A_g - A_{st}) + f_y A_{st} \qquad (13)$$

Pure compression strength P_O represents a hypothetical loading condition. Prior to the 1977 ACI Code, all compression members were required to be designed for a minimum eccentricity of 0.05h for spirally reinforced members or 0.10h for tied reinforced members (h = overall thickness of member). The specified minimum eccentricities were originally intended to serve as a means of reducing the axial design load strength of a section in pure compression, (1) to account for accidental eccentricities, not considered in the analysis, that may exist in a compression member, and (2) to recognize that concrete strength is less than f_c' at sustained high loads.

Since the primary purpose of the minimum eccentricity requirement was to limit the axial load strength for design of compression members with small or zero computed end moments, the 1977 Code was revised to accomplish this directly by limiting the axial load strength to 85% and 80% of the axial load strength at zero eccentricity, P_O, depending on whether spiral or tie reinforcement is used.

For spirally reinforced members,

$$P_{n(max)} = 0.85P_O = 0.85 \left[0.85f_c' (A_g - A_{st}) + f_y A_{st} \right] \qquad (14)$$

For tied reinforced members,

$$P_{n(max)} = 0.80 P_O = 0.80 \left[0.85 f_c' (A_g - A_{st}) + f_y A_{st} \right] \qquad (15)$$

The maximum axial load strength, $P_{n(max)}$, is illustrated in Fig. 6-14. In essence, design within the cross-hatched portion of the load-moment interaction diagram is not permitted. The 85% and 80% values approximate the axial load strengths at e/h ratios of 0.05 and 0.10 specified in the 1971 Code for spirally reinforced and tied reinforced members, respectively (see Example 6.5). The designer should note that Commentary Section 10.3.5 states that "Design aids and computer programs based on the minimum eccentricity requirement of the 1963 and 1971 ACI Building Codes may be considered equally applicable for usage."

The current provisions for maximum axial load strength also eliminate the concerns expressed by engineers about the excessively high minimum design moments required for large column sections, and the often asked question as to whether the minimum moments were required to be transferred to other interconnecting members (beams, footings, etc.).

Note that a minimum moment (minimum eccentricity requirement) for slender compression members is given in Section 10.11.5.4. If factored column moments are very small or zero, the design of slender columns must be based on a minimum moment of $P_u(0.6 + 0.003h)$.

10.3.6 Nominal Strength for Combined Flexure and Axial Load

The strength of a member or cross section subject to combined flexure and axial load, M_n and P_n, must satisfy the same two conditions as required for a member subject to flexure only, (1) static equilibrium and (2) compatibility of strains. Equilibrium between the compressive and tensile forces includes the axial load P_n acting on the cross section. The general condition of the stress and the strain in concrete and steel at nominal strength of a member under combined flexure and axial compression is shown in Fig. 6-15. The tensile or compressive force developed in the reinforcement is determined from the strain condition at the location of the reinforcement.

Referring to Fig. 6-15,

For A_s': $\quad C_s = A_s' f_s' = A_s' (E_s \varepsilon_s')$ when $\varepsilon_s' < \varepsilon_y$ (yield strain)

or $\quad C_s = A_s' f_y$ $\quad$ when $\varepsilon_s' \geq \varepsilon_y$

For A_s: $\quad T = A_s f_s = A_s (E_s \varepsilon_s)$ when $\varepsilon_s < \varepsilon_y$

or $\quad T = A_s f_y$ $\quad$ when $\varepsilon_s \geq \varepsilon_y$

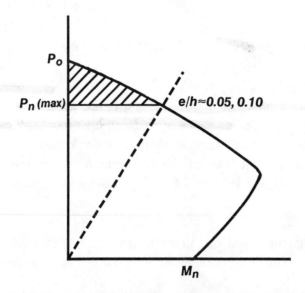

Fig. 6-14   Maximum Axial Load Strength (Section 10.3.5)

6-18

The combined load-moment strength (P_n and M_n) may be computed by equilibrium of forces and moments.

For force equilibrium:

$$P_n = C_c + C_s - T \tag{16}$$

where
$$C_c = 0.85 \, f'_c \, ba$$

For moment equilibrium about the mid-depth of the section:

$$M_n = P_n e = C_c \left(\frac{h}{2} - \frac{a}{2}\right) + C_s \left(\frac{h}{2} - d'\right) + T\left(d - \frac{h}{2}\right) \tag{17}$$

For a known strain condition, the corresponding load-moment strength, P_n and M_n, can be computed directly. Assume the strain in the tension steel, A_s, is at first yield ($\varepsilon_s = \varepsilon_y$). This strain condition (simultaneous strain of 0.003 in the extreme compression fiber and first yield strain ε_y in the tension steel) defines the "balanced" load-moment strength, P_b and M_b, for the cross section.

For the linear strain condition:

$$\frac{c_b}{d} = \frac{\varepsilon_u}{\varepsilon_u + \varepsilon_y} = \frac{0.003}{0.003 + f_y/29,000,000} = \frac{87,000}{87,000 + f_y}$$

so that
$$a_b = \beta_1 c_b = \left(\frac{87,000}{87,000 + f_y}\right)\beta_1 d$$

Also
$$\frac{c_b}{c_b - d'} = \frac{\varepsilon_u}{\varepsilon'_s}$$

so that
$$\varepsilon'_s = 0.003\left(1 - \frac{d'}{c_b}\right) = 0.003\left[1 - \frac{d'}{d}\left(\frac{87,000 + f_y}{87,000}\right)\right]$$

and
$$f'_{sb} = E_s \varepsilon'_s = 87,000\left[1 - \frac{d'}{d}\left(\frac{87,000 + f_y}{87,000}\right)\right] \quad \text{but not greater than } f_y$$

For force equilibrium:

$$P_b = 0.85 f'_c \, ba_b + A'_s \, f'_{sb} - A_s f_y \tag{18}$$

For moment equilibrium:

$$M_b = P_b e_b = 0.85 \, f'_c \, ba_b \left(\frac{h}{2} - \frac{b}{2}\right) + A'_s \, f'_{sb}\left(\frac{h}{2} - d'\right) + A_s f_y\left(\frac{d}{2} - h\right) \tag{19}$$

6-19

The "balanced" load-moment strength defines only one of many load-moment combinations possible over the full range of the load-moment interaction relationship of a cross section subject to combined flexure and axial load. The general form of a strength interaction diagram is shown in Fig. 6-16. The load-moment combination may be such that compression exists over most or all of the section, so that the compressive strain in the concrete reaches 0.003 before the tension steel yields ($\varepsilon_s < \varepsilon_y$), (compression-controlled segment); or the load combination may be such that tension exists over a large portion of the section, so that

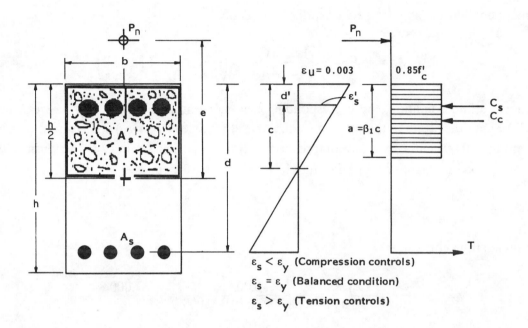

$\varepsilon_s < \varepsilon_y$ (Compression controls)

$\varepsilon_s = \varepsilon_y$ (Balanced condition)

$\varepsilon_s > \varepsilon_y$ (Tension controls)

Fig. 6-15   Strain and Equivalent Stress Distribution for Section Subject to Combined Flexure and Axial Load

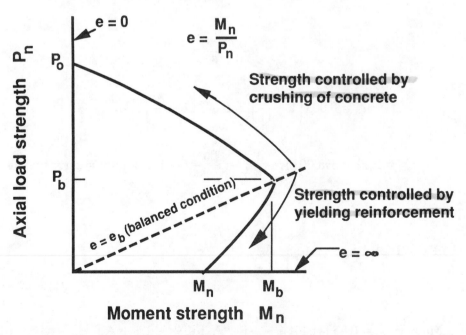

Fig. 6-16   Axial Load-Moment Interaction Diagram

6-20

the strain in the tension steel is greater than the yield strain ($\varepsilon_s > \varepsilon_y$) when the compressive strain in the concrete reaches 0.003 (tension-controlled segment). The "balanced" strain condition ($\varepsilon_s = \varepsilon_y$) divides these two segments of the strength curve.

The linear strain variation for the full range of the load-moment interaction relationship is illustrated in Fig. 6-17.

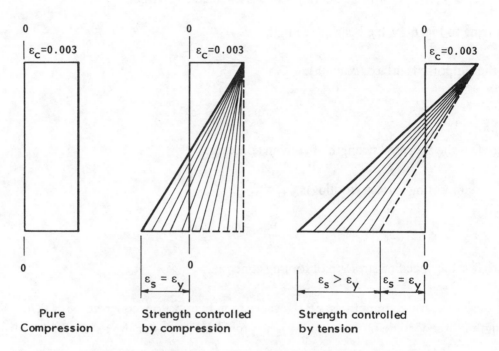

Fig. 6-17   Strain Variation for Full Range of Load-Moment Interaction

Under pure compression, the strain is uniform over the entire cross section and equal to 0.003. With increasing load eccentricity (moment), the compressive strain at the "tension face" gradually decreases to zero, then becomes tensile, with the tensile strain in the steel most distant from the neutral axis reaching the yield strain ($\varepsilon_s = \varepsilon_y$) at the balanced strain condition. For this range of strain variations, the strength of the section is controlled by compression ($\varepsilon_s = -0.003$ to ε_y). Beyond the balanced strain condition, the steel strain gradually increases ($\varepsilon_s \gg \varepsilon_y$) up to the state of pure flexure corresponding to an infinite load eccentricity ($e = \infty$). For this range of strain variations, strength is controlled by tension ($\varepsilon_s > \varepsilon_y$). With increasing eccentricity, more and more tension exists over the cross section. Each of the many possible strain conditions illustrated in Fig. 6-17 describes a point, P_n and M_n, on the load-moment curve. Calculation of P_n and M_n for two different strain conditions along the load-moment strength curve is illustrated in Example 6.4.

10.15   BEARING STRENGTH ON CONCRETE

Code-defined bearing strength of concrete is expressed in terms of an average bearing stress of 0.85 f_c' over a bearing area (loaded area) A_1. When the supporting concrete area is wider than the loaded area **on all sides**, the surrounding concrete acts to confine the loaded area, resulting in an increase in the bearing strength of the supporting concrete. With confining concrete, the bearing stress may be increased by the factor $\sqrt{A_2/A_1}$, but not greater than 2, where $\sqrt{A_2/A_1}$ is a measure of the confining effect of the surrounding concrete. Evaluation of the stress increase factor $\sqrt{A_2/A_1}$ is illustrated in Fig. 6-18.

For the usual case of a supporting concrete area considerably greater than the loaded area, $\sqrt{A_2/A_1} > 2$, the permissible bearing stress is doubled to $2(0.85\,f_c')$.

Referring to Fig. 6-19, the bearing strength is

For the supported surface (column):

$$P_{nb} = 0.85\,f_c'\,A_1$$

where f_c' is the specified strength of column concrete.

For the supporting surface (footing):

$$P_{nb} = 2(0.85\,f_c'\,A_1)$$

where f_c' is the specified strength of footing concrete.

The design bearing strength is φP_{nb}, where, for bearing on concrete, $\varphi = 0.70$. Where bearing strength is exceeded, reinforcement must be provided to transfer the excess load.

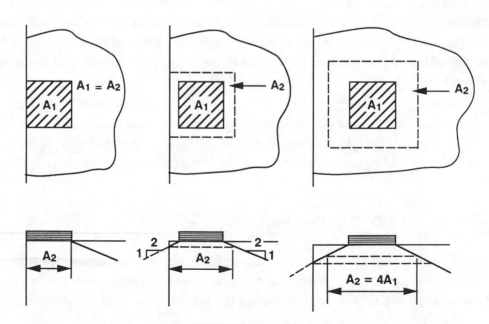

Fig. 6-18   Measure of Confinement $\sqrt{A_2/A_1} \leq 2$ Provided by Surrounding Concrete

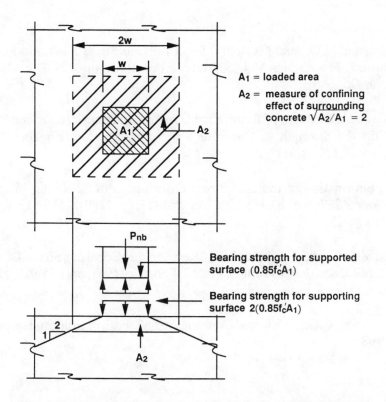

Fig. 6-19   Nominal Bearing Strength of Concrete (Section 10.15)

SUMMARY

The nominal strength of a member or cross section subject to flexure (M_n), or to combined flexure and axial load (M_n, P_n), must satisfy both equilibrium and strain compatibility using the design assumptions of Section 10.2. For design, the nominal strength must be multiplied (reduced) by the strength reduction factor φ to obtain the design strength or "usable" strength of the member or cross section (φM_n, φP_n). See Part 5 for discussion on Section 9.1.1 - Strength Requirements. In review, the basic criterion for strength design is

$$\text{Required Strength} \leq \text{Design Strength}$$

$$M_u \leq \varphi M_n$$

$$P_u \leq \varphi P_{n,} \text{ etc.}$$

All members and all member cross sections must be designed to satisfy this basic criterion. Design for flexure, and design for flexure and axial load are presented in Part 10 and Part 11, respectively.

REFERENCES

6.1 Hognestad, E., Hanson, N.W., and McHenry, D., "Concrete Stress Distribution in Ultimate Strength Design," *ACI Journal, Proceedings* Vol. 52, December 1955, pp 455-479; also *PCA Development Department Bulletin D6.*

6.2 Hognestad, E., "Ultimate Strength of Reinforced Concrete in American Design Practice," Proceedings of a Symposium on the Strength of Concrete Structures, London, England, May 1955; also *PCA Development Department Bulletin D12.*

6.3 Hognestad, E., "Confirmation of Inelastic Stress Distribution in Concrete," *Journal of the Structural Division, Proceedings ASCE,* Vol. 83, No. ST2, March 1957; pp. 1189-1–1189-17 also PCA Development Department Bulletin D15.

6.4 Mattock, A.H., Kriz, L.B., and Hognestad, E., "Rectangular Concrete Stress Distribution in Ultimate Strength Design," *ACI Journal, Proceedings,* Vol. 57, February 1961, pp. 875-928; also *PCA Development Department Bulletin D49.*

6.5 Wang, C.K., and Salmon, C.G., *Reinforced Concrete Design,* Fourth Edition, Harper & Row Publishers, New York, N.Y. 1985.

Example 6.1—Strength Factor β_1 for High-Strength Concretes, $f'_c \geq 8000$ psi

Plot the load-moment strength interaction diagram for a 20-in.x20-in. column section with four #18 bars ($\rho_g = 4\%$), $f'_c = 12,000$ psi and $f_y = 60,000$ psi.

Calculations and Discussion	Code Reference

The interaction diagram is plotted using two β_1 strength factors as follows:

(1) $\beta_1 = 0.85 - 0.05\left(\dfrac{f'_c - 4000}{1000}\right)$

   for $f'_c = 12,000, \beta_1 = 0.45$

(2) $\beta_1 = 0.85 - 0.05\left(\dfrac{f'_c - 4000}{1000}\right)$ but not less than 0.65   10.2.7.3

   for $f'_c = 12,000, \beta_1 = 0.65$

The lower limit of 0.65 for β_1 primarily affects members subject to axial load plus bending within the intermediate range of load-moment interaction strength, where strength is controlled by compression.

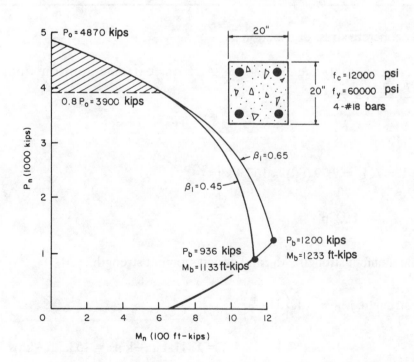

6-25

Example 6.2—Moment Strength Using Equivalent Rectangular Stress Distribution

For the beam section shown, calculate moment strength based on static equilibrium using the equivalent rectangular stress distribution shown in Fig. 6-9. Assume f'_c = 4000 psi and f_y = 60,000 psi.

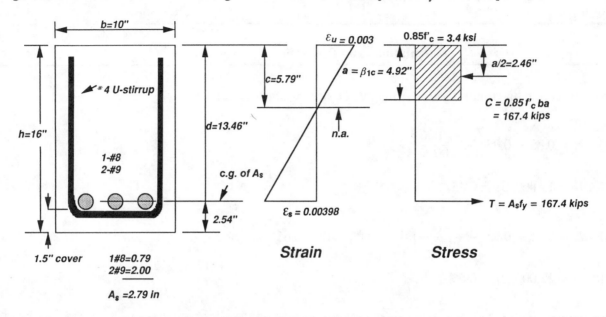

	Code
Calculations and Discussion	**Reference**

1. Define rectangular concrete stress distribution — 10.2.7

   With mixed bar sizes, distance of c.g. of A_s from extreme tension fiber
   $$= \frac{2.0\,(2.56) + 0.79\,(2.5)}{2.79} = 2.54 \text{ in.}$$

   d = 16 − 2.54 = 13.46 in. — 10.0

   Assuming $\varepsilon_s > \varepsilon_y$, T = $A_s f_y$ = 2.79 (60) = 167.4 kips — 10.2.4

   $$a = \frac{A_s f_y}{0.85\,f'_c\,b} = \frac{167.4}{3.4(10)} = 4.92 \text{ in.}$$

2. Determine nominal moment strength, M_n, and design moment strength, φM_n.

   Nominal moment strength, $M_n = A_s f_y \left(d - \dfrac{a}{2} \right)$

   $$= 167.4\,(13.46 - 2.46) = 1841.4 \text{ in.}-\text{kips} = 153.5 \text{ ft}-\text{kips}$$

   Design moment strength = φM_n = 0.9(153.3) = 138 ft−kips — 9.3.2.1

Example 6.2—continued

3. Check code limit on reinforcement.   10.3.3

$$\rho = \frac{A_s}{bd} = \frac{2.79}{10(13.46)} = 0.0207$$

$$< \rho_{max} = 0.75\rho_b = 0.75(0.0285) = 0.0214 \quad \text{(Table 6−1)} \quad \text{O.K.}$$

This confirms that $\varepsilon_s > \varepsilon_y$ at nominal strength.

Example 6.3—Design of Beams with Compression Reinforcement

A beam cross section is limited to the size shown. Determine required area of reinforcement for a factored moment M_u = 580 ft–kips. f'_c = 4000 psi, f_y = 60,000 psi.

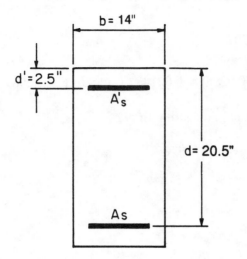

	Code
Calculations and Discussion	**Reference**

1. Check if doubly reinforced section is needed.

   ρ_b = 0.0285 from Table 6−1

   ρ_{max} = 0.75ρ_b = 0.0214

   $A_{s(max)}$ = 0.0214×14×20.5 = 6.14 in.^{2}

   Corresponding $a = \dfrac{A_{s(max)}f_y}{0.85f'_cb} = \dfrac{6.14 \times 60}{0.85 \times 4 \times 14}$ = 7.74 in.

   $\varphi M_{n(max)} = \varphi\, A_{s(max)}f_y\left(d - \dfrac{a}{2}\right)$ = 0.9 × 6.14 × 60 (20.5 − 3.87)

   $\qquad$ = 5514 in.−kips = 459.5 ft−kips $\;<\;$ M_u = 580 ft−kips

   Therefore, a doubly reinforced section is needed.

Example 6.3—continued

Calculations and Discussion

2. Design doubly reinforced section.

$M_u - \varphi M_{n(max)} = 580 - 459.5 = 120.5 \text{ ft} - \text{kips}$

Assuming that $f'_s = f_y$ at the ultimate stage

$A'_s = \dfrac{M_u - \varphi M_{n(max)}}{\varphi f_y (d - d')}$ (see Eq. 7)

$= \dfrac{120.5 \times 12}{0.9 \times 60 (20.5 - 2.5)} = 1.49 \text{ in.}^2$

Provide 2 #8 bars — $A'_s = 1.58 \text{ in.}^2$

$A_s = A_{s(max)} + A'_s = 6.14 + 1.49 = 7.63 \text{ in.}^2$

Provide 6 #7 bars — $A_s = 7.62 \text{ in.}^2$

Note that the above would require two layers of tension reinforcement in a 14 in. wide beam.

3. Check yielding of compression reinforcement.

$\rho - \rho' = \dfrac{A_s - A'_s}{bd} = \dfrac{7.62 - 1.58}{14 \times 20.5} = 0.021$

$> \dfrac{0.85\beta_1 f'_c d'}{f_y d} \left(\dfrac{87000}{87000 - f_y} \right) = \dfrac{0.85 \times 0.85 \times 4 \times 2.5}{60 \times 20.5} \left(\dfrac{87}{27} \right) = 0.019$

Therefore, the compression reinforcement yields at nominal strength, as assumed.

For the column section shown, calculate load-moment strength, P_n and M_n, for two strain conditions; (1) bar stress near tension face of member equal to zero ($f_s = 0$), and (2) bar stress near tension face of member equal to $0.5f_y$ ($f_s = 0.5f_y$). Use $f'_c = 4000$ psi and $f_y = 60,000$ psi.

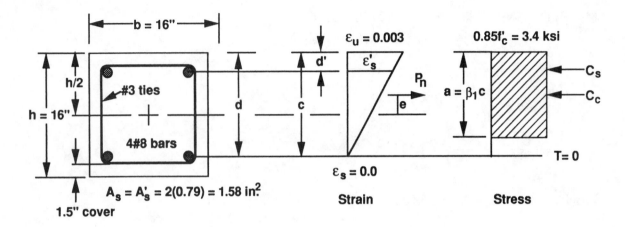

Calculations and Discussion	Code Reference

1. Load-moment strength, P_n and M_n, for strain condition $\varepsilon_s = 0$,

    $f_s = 0$, near tension face of member.

    (a) Define stress distribution and determine force values   10.2.7

    $d' = $ Cover + #3 tie dia. $+ \dfrac{d_b}{2} = 1.5 + 0.375 + 0.5 = 2.38$ in.

    $d = 16 - 2.38 = 13.62$ in.

    For strain condition $\varepsilon_s = 0$, $c = d = 13.62$ in.   10.2.7.2

    $a = \beta_1 c = 0.85(13.62) = 11.58$ in.   10.2.7.1

    where $\beta_1 = 0.85$ for $f'_c = 4000$ psi   10.2.7.3

    $C_c = 0.85\, f'_c\, ba = 3.4(16)(11.58) = 630.0$ kips   10.2.7

    $\varepsilon_y = \dfrac{f_y}{E_s} = \dfrac{60}{29,000} = 0.00207$ 10.2.4

Example 6.4—Continued

$$\varepsilon_s' = \varepsilon_u \left(1 - \frac{d'}{c}\right) = 0.003\left(1 - \frac{2.38}{13.62}\right) = 0.0025$$ 10.2.2

$$> \varepsilon_y = 0.00207$$

$$C_s = A_s' f_y = 1.58(60) = 94.8 \text{ kips}$$

(b) Determine P_n and M_n from static equilibrium.

$$P_n = C_c + C_s = 630.0 + 94.8 = 724.8 \text{ kips} \qquad \text{Eq. (16)}$$

$$M_n = P_n e = C_c\left(\frac{h}{2} - \frac{a}{2}\right) + C_s\left(\frac{h}{2} - d'\right) \qquad \text{Eq. (17)}$$

$$= 630(8.0 - 5.79) + 94.8(8.0 - 2.38) = 1925.1 \text{ in.} - \text{kips} = 160.4 \text{ ft} - \text{kips}$$

$$e = \frac{M_n}{P_n} = \frac{1925.1}{724.8} = 2.66 \text{ in.}$$

For strain condition $\varepsilon_s = 0$,

Design axial load strength, $\varphi P_n = 0.7(724.8) = 507.4 \text{ kips}$ 9.3.2.2

Design moment strength, $\varphi M_n = 0.7(160.4) = 112.3 \text{ ft–kips}$

2. Load-moment strength, P_n and M_n, for strain condition

$\varepsilon_s = 0.5\varepsilon_y$, $f_s = 0.5f_y$, near tension face of member.

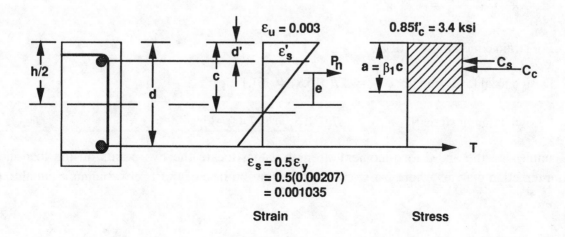

Strain   Stress

Example 6.4—Continued

	Code
Calculations and Discussion	**Reference**

(a) Define stress distribution and determine force values.   10.2.7

$d' = 2.38$ in., $d = 13.62$ in.

$$c = \frac{d}{1 + \dfrac{\varepsilon_s}{\varepsilon_u}} = \frac{13.62}{1 + \dfrac{0.001035}{0.003}} = 10.13 \text{ in.}$$

$$\varepsilon'_s = \varepsilon_u \left(1 - \frac{d'}{c}\right) = 0.003\left(1 - \frac{2.38}{10.13}\right) = 0.0023 \quad > \quad \varepsilon_y = 0.00207$$

$a = \beta_1 c = 0.85(10.13) = 8.61$ in.   10.2.7.1

$C_c = 0.85\, f'_c\, ba = 3.4(16)(8.61) = 468.4$ kips   10.2.7

$C_s = A'_s f_y = 1.58(60) = 94.8$ kips

$T = A_s f_s = A_s(0.5 f_y) = 1.58(30) = 47.4$ kips

(b) Determine P_n and M_n from static equilibrium.

$P_n = C_c + C_s - T = 468.4 + 94.8 - 47.4 = 515.8$ kips   Eq. (16)

$$M_n = P_n e = C_c\left(\frac{h}{2} - \frac{a}{2}\right) + C_s\left(\frac{h}{2} - d'\right) + T\left(d - \frac{h}{2}\right)$$ Eq. (17)

$$= 468.4(8.0 - 4.31) + 94.8(8.0 - 2.38) + 47.4(13.62 - 8.0)$$

$$= 2527.6 \text{ in.–kips} = 210.6 \text{ ft–kips}$$

$$e = \frac{M_n}{P_n} = \frac{2527.6}{515.8} = 4.90 \text{ in.}$$

For strain condition $\varepsilon_s = 0.5\varepsilon_y$,

Design axial load strength, $\varphi P_n = 0.7(515.8) = 361.1$ kips   9.3.2.2

Design moment strength, $\varphi M_n = 0.7(210.6) = 147.4$ ft–kips

Note Example 4.8; the above load-moment strength values locate the two points on the load-moment strength interaction diagram where bar stress near the tension face of the 16x16 column is equal to 0 and $0.5 f_y$.

Example 6.5—Maximum Axial Load Strength vs. Minimum Eccentricity

For the tied reinforced concrete column section shown, compare the nominal axial load strength P_n equal to 0.80 P_0 with P_n at 0.1h eccentricity. f'_c = 5000 psi, f_y = 60,000 psi.

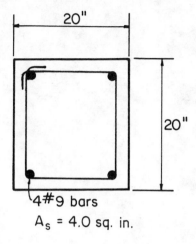

20"

20"

4#9 bars
A_s = 4.0 sq. in.

Calculations and Discussion	**Code Reference**

Prior to ACI 318-77, columns were required to be designed for a minimum eccentricity of 0.1h (tied) or 0.05h (spiral). This required tedious computation to find the axial load strength at these minimum eccentricities. With the 1977 ACI Code, the minimum eccentricity provision was replaced with a maximum axial load strength: 0.80P_0 (tied) or 0.85 P_0 (spiral). The 80% and 85% values were chosen to approximate the axial load strengths at e/h ratios of 0.1 and 0.05, respectively.

1. In accordance with the minimum eccentricity criterion:

   At e/h = 0.10: P_n = 1543 kips (computer solution)

2. In accordance with maximum axial load strength criterion: 10.3.5.2

   $P_{n(max)}$ = 0.80 P_0 = 0.80 [0.85f'_c (A_g − A_{st}) + f_yA_{st}]   Eq. (10-2)

   = 0.80 [0.85 × 5 (400 − 4) + 60 × 4.0] = 1538 kips

Depending on material strengths, size, and amount of reinforcement, the comparison will vary slightly. Both solutions are considered equally acceptable.

7

Distribution of Flexural Reinforcement

UPDATE FOR THE '89 CODE

For ACI 318-89, the requirements for longitudinal crack control reinforcement (new name: skin reinforcement) in deep flexural members (Section 10.6.7) has been modified to reflect new research resulting from reported cases where wide cracks had developed on the side faces of deep beams between the main tension reinforcement and the neutral axis. Tests have confirmed that, although crack widths at the main reinforcement level are adequately controlled by present design requirements (Section 10.6.4), cracks tend to widen as they extend towards the neutral axis. Tests also confirmed that the amount and spacing of the side face crack control reinforcement of the '83 Code were inadequate to control crack widths in the upper regions of the flexural tension zones of the deeper beams; thus Section 10.6.7 was revised for the '89 Code. See discussion on Section 10.6.7.

GENERAL CONSIDERATIONS

Provisions of Section 10.6 require proper distribution of tension reinforcement in beams and one-way slabs to control flexural cracking. Structures built in the past using Working Stress Design and reinforcement with a yield strength of 40,000 psi or less had low tensile stresses in the reinforcement at service loads. Laboratory investigations have shown that cracking is generally in proportion to the steel tensile stress. Thus, with low tensile stresses in the reinforcement at service loads, these structures exhibited few flexural cracking problems.

With the advent of high-strength steels with yield stresses of 60,000 and 75,000 psi, and even higher, and the use of Strength Design with steel reinforcement stressed to higher proportions of the yield strength, control of flexural cracking by proper reinforcing details has assumed more importance. For example, if a beam were designed using Working Stress Design and a steel yield strength of 40,000 psi, stress in the reinforcement at service loads would be about 20,000 psi. Using Strength Design and a steel yield strength of 60,000 psi, stress at service loads could be as high as 36,000 psi. If flexural cracking is indeed proportional to steel tensile stress, then it is quite evident that criteria for crack control must be included in the design process.

From the above, it is apparent that the effective moment of inertia of a section is more dependent upon the cracked section when using Strength Design concepts. This, of course, is the reason for the emphasis on the cracked section in the Code provisions for calculating deflections in Section 9.5.

Early investigations of crack width in beams and members subject to axial tension indicated that crack width was proportional to steel stress and bar diameter, but was inversely proportional to reinforcement percentage. More recent research using modern deformed bars has confirmed that crack width is proportional to steel stress. However, other variables were also found to affect crack width; these reflect steel detailing and include thickness of concrete cover and the area of concrete in the zone of maximum tension surrounding each individual reinforcing bar. It should be kept in mind that there are large variations in crack widths, even in careful laboratory-controlled work. For this reason, only a simple crack control expression is presented in the Code; it is designed to give reasonable reinforcing details that are in accord with laboratory work and practical experience.

10.6 BEAMS AND ONE-WAY SLABS

10.6.4 Distribution of Tension Reinforcement

The Code requires that when the yield strength of the reinforcement exceeds 40,000 psi, detailing of the flexural tension reinforcement must be such that the quantity z given by

$$z = f_s \sqrt[3]{d_c A}$$
Eq. (10-4)

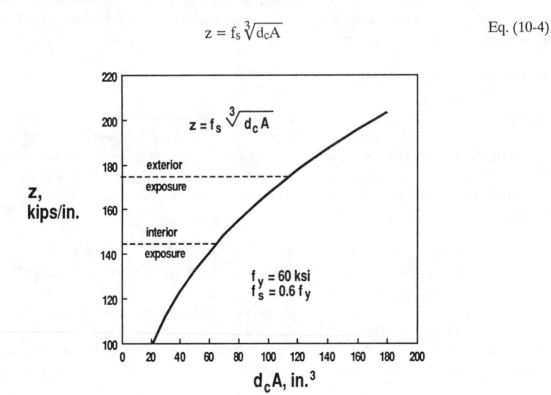

Fig. 7-1 Design Curve for Z, Quantity Limiting Distribution of Flexural Reinforcement

does not exceed certain specified limits. These limits are 175 for interior exposure and 145 for exterior exposure. Equation (10-4) is written in a form emphasizing reinforcing details rather than crack width itself. The equation will provide a distribution of the flexural reinforcement that should ensure reasonable control of flexural cracking—a large number of small diameter bars at close spacing (see Examples 7.1 and 7.2). A plot of z vs. d_cA is given in Fig. 7-1 which may serve as a design aid.

In Eq. (10-4), f_s (ksi) is the calculated stress in reinforcement at service loads; it may be taken as 60 percent of the specified yield strength, f_y. d_c (in.) is thickness of concrete cover measured from extreme tension fiber to center of bar or wire located closest to it. A (in.^{2}) is defined as the effective tension area of concrete surrounding the flexural tension reinforcement and having the same centroid as the reinforcement, divided by the number of bars or wires. When the flexural reinforcement consists of different bar or wire sizes, the number of bars or wires is computed as the total area of reinforcement divided by the area of the largest bar

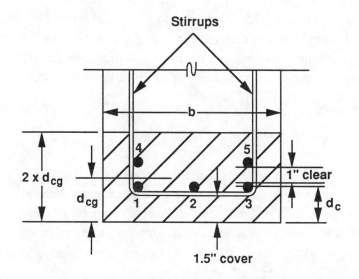

Examples

1. Bars 1, 2, and 3 are #10 (area = 1.27 in.^{2})
    Bars 4 and 5 are #9 (area = 1.00 in.^{2})
    Equivalent no. of #10 bars
    = (3 x 1.27 + 2 x 1.00)/1.27 = 4.57
    A = $2bd_{cg}$/4.57

2. Bars 1, 3, 4, and 5 are #8 (area = 0.79 in.^{2})
    Bar 2 is #9 (area = 1.00 in.^{2})
    Equivalent no. of #9 bars
    = (4 x 0.79 + 1.00)/1.00 = 4.16
    A = $2bd_{cg}$/4.16

Fig. 7-2 Effective Tension Area A, When Different Bar Sizes Are Used

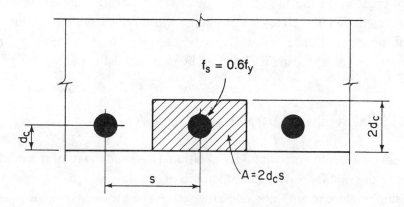

Fig. 7-3. Criteria Used to Develop Tables 7-1 and 7-2

or wire used (Fig. 7-2). This definition is satisfactory for all reinforcement details except bundled bars. For guidance in determining the number of equivalent bars to use in calculating A for bundled bars, consult Ref. 7.1.

In the original crack width expression from which Eq. (10-4) was derived, a factor β was included as one of the parameters, β being the ratio of the distances of the neutral axis from the extreme tension fiber and from the centroid of the reinforcement. To simplify practical design of beams, an approximate value of 1.2 for β was used in Code Eq. (10-4). Since derivation of the original expression, additional tests have indicated that the crack width expression is also applicable to one-way slabs, with a value of β around 1.35. Accordingly, the Commentary to the Code suggests that the maximum value of z for one-way slabs be reduced by the ratio of 1.2/1.35, which yields z = 156 for interior exposure and z = 129 for exterior exposure. Similar adjustments are prudent for other cases where the value of β exceeds 1.2.

Tables 7-1 and 7-2 are meant to serve as design aids, and show maximum bar spacings allowed by the Code for Grade 60 reinforcement (see Fig. 7-1). The tables are based on a service load stress $f_s = 0.6f_y$, as permitted by Section 10.6.4. Computations for f_s would yield about $0.56f_y$ for a dead-to-live load ratio of 0.5, and $0.60 f_y$ for a dead-to-live load ratio of 2. The tables apply to reinforcement in a single layer. Tables 7-1 and 7-2 show that normal spacing of reinforcement generally will satisfy crack control requirements for 1½ in. or less cover. However, for 2 in. or more cover, the maximum spacing of bars will often be limited by these requirements.

10.6.5   Corrosive Environments

Data are not available regarding crack width beyond which a danger of corrosion exists. Exposure tests indicate that concrete quality, adequate compaction, and ample cover may be of greater importance for corrosion protection than crack width at the concrete surface. The limiting z values of Section 10.6.4 were chosen primarily to give reasonable reinforcing details in terms of practical experience with existing structures. The Code requirements do not apply to structures subject to very aggressive exposure or designed to be watertight; special precautions are required and must be investigated for such cases.

10.6.6   Distribution of Tension Reinforcement in Flanges of T-Beams

For control of flexural cracking in the flanges of T-beams, the flexural tension reinforcement must be distributed in accordance with Section 10.6.4 over a flange width not exceeding the effective flange width (Section 8.10) or $^1\!/_{10}$ of the span. If the effective flange width is greater than $^1\!/_{10}$ the span, some additional longitudinal reinforcement, as illustrated in Fig. 7-4, must be provided in the outer portions of the flange (see Example 7.3).

10.6.7   Crack Control Reinforcement in Deep Flexural Members

In deep flexural members (web depth exceeding 3 ft), additional longitudinal reinforcement for crack control must be distributed along the side faces over the full depth of the flexural tension zone. For the '89 Code, the requirements for "skin reinforcement" have been revised, as the previous code provisions were found to be inadequate in some cases. The new side face crack control requirements provide an increase in the

Table 7-1—Maximum Bar Spacing in Beams for Crack Control*
(Grade 60 Reinforcement)

Bar Size	Outside Exposure $z = 145$			Inside Exposure $z = 175$		
	Cover—in.			Cover—in.		
	1½	2	3	1½	2	3
# 4	10.7	6.5	3.1	18.8	11.3	5.4
# 5	9.9	6.1	3.0	17.5	10.7	5.2
# 6	9.3	5.8	2.9	16.3	10.2	5.0
# 7	8.7	5.5	2.8	15.3	9.7	4.9
# 8	8.2	5.2	2.7	14.4	9.2	4.7
# 9	7.7	5.0	2.6	13.5	8.7	4.5
#10	7.2	4.7	2.5	12.6	8.3	4.3
#11	6.7	4.5	2.4**	11.8	7.8	4.2

* Values in inches, $f_s = 0.6 f_y = 36$ ksi, single layer of reinforcement.
** Spacing less than permitted by Section 7.6.1

Table 7-2—Maximum Bar Spacing in One-Way Slabs for Crack Control*
(Grade 60 Reinforcement)

Bar Size	Outside Exposure $z = 129$				Inside Exposure $z = 156$			
	Cover—in.				Cover—in.			
	¾	1	1½	2	¾	1	1½	2
# 4	—	14.7	7.5	4.5	—	—	13.3	8.0
# 5	—	13.4	7.0	4.3	—	—	12.4	7.6
# 6	—	12.2	6.5	4.1	—	—	11.6	7.2
# 7	16.3	11.1	6.1	3.9	—	—	10.8	6.8
# 8	14.7	10.2	5.8	3.7	—	—	10.2	6.5
# 9	13.3	9.4	5.4	3.5	—	16.6	9.6	6.2
#10	12.0	8.6	5.0	3.3	—	15.2	8.9	5.9
#11	10.9	7.9	4.7	3.1	—	14.0	8.4	5.6

* Values in inches, $f_s = 0.6f_y = 36$ ksi, single layer of reinforcement. Spacing should not exceed 3 times slab thickness nor 18 in. (Section 7.6.5). No value indicates spacing greater than 18 in.

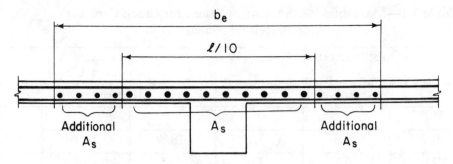

Fig. 7-4 Negative Moment Reinforcement for Flanged Floor Beams

amount of skin reinforcement for increasing depth of web, with maximum spacing also limited as a function of web depth. For most cases, more crack control reinforcement at closer spacing will be required by the new provisions. Note that Section 10.6, including revised Section 10.6.7, does not apply to prestressed concrete members, as the behavior of a prestressed member is considerably different from that of a nonprestressed member. Experience and judgment must be used for proper distribution of reinforcement in prestressed members.

The required skin reinforcement (see Fig. 7-5) must be uniformly distributed along both side faces of the member within the flexural tension zone, considered to extend over a distance d/2 nearest the main tension reinforcement. The area required **per side** is calculated as $A_{sk} = 0.012(d - 30)$ in.^{2}/ft. depth, with a spacing

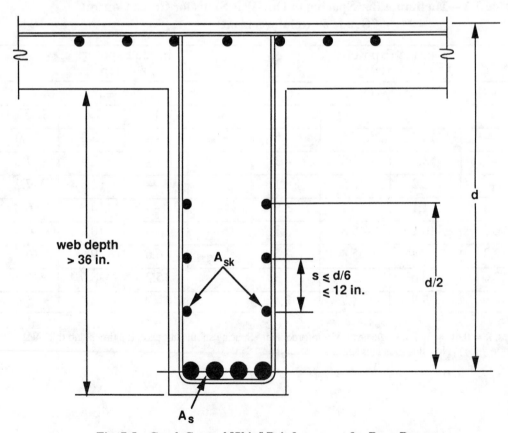

Fig. 7-5 Crack Control "Skin" Reinforcement for Deep Beams

not greater than d/6 or 12 in. The total area (in.^{2}) of skin reinforcement provided on both faces (d/2 distance) need not be greater than one-half the total area of the main tensile reinforcement(A_s).

13.4 TWO-WAY SLABS

Control of flexural cracking in two-way slabs, including flat plates and flat slabs, is usually not a problem, and is not specifically covered in the Code. However, Code Section 13.4.2 restricts spacing of slab reinforcement at critical moment sections to 2 times the slab thickness, and the area of reinforcement in each direction for two-way slab systems must not be less than that required for shrinkage and temperature (Section 13.4.2). These limitations are intended in part to control cracking. Also, the minimum thickness requirements for two-way construction for deflection control (Section 9.5.3) indirectly serve as a control on excessive cracking.

REFERENCE

7.1   Lutz, L.A., "Crack Control Factor for Bundled Bars and for Bars of Different Sizes," *ACI Journal, Proceedings* Vol. 71, No. 1, January 1974, pp. 9-10.

Example 7.1—Distribution of Reinforcement for Effective Crack Control

Assume a 16 in. wide beam with A_s (required) = 3.00 in.^{2}, and f_y = 60,000 psi. Select various bar arrangements to satisfy Eq. (10-4) for control of flexural cracking (use of a larger number of smaller bars at closer spacing is usually advantageous).

	Code
Calculations and Discussion	**Reference**

(a) Select 2 #11 bars—A_s = 3.12 in.^{2}

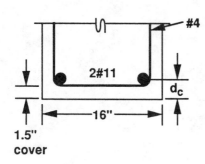

$$z = f_s \sqrt[3]{d_c A}$$ Eq. (10-4)

$$d_c = 1.5 + 0.5 + \frac{1.41}{2} = 2.71 \text{ in.}$$

$$A = \frac{2(2.71)\,16}{2} = 43.4 \text{ in.}^2/\text{bar}$$

Use $f_s = 0.6 f_y = 0.6(60) = 36$ ksi

$$z = 36 \sqrt[3]{2.71\,(43.4)} = 176 \text{ kips/in.}$$

(From Fig. 7-1, z = 176 kips/in. for $d_c A$ = 2.71×43.4 = 117.6 in.^{3})

(b) Select 4 #8 bars—A_s = 3.16 in.^{2}

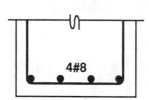

$$d_c = 2.0 + \frac{1.00}{2} = 2.50 \text{ in.}$$

$$A = 2(2.5)\frac{16}{4} = 20.0 \text{ in.}^2/\text{bar}$$

$$z = 36 \sqrt[3]{2.50\,(20.0)} = 133 \text{ kips/in.}$$

(From Fig. 7-1, z = 133 kips/in. for $d_c A$ = 2.5 × 20 = 50 in.^{3})

(c) Select 7 #6 bars—A_s = 3.08 in.^{2}

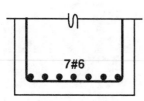

$$d_c = 2.0 + \frac{0.75}{2} = 2.38 \text{ in.}$$

$$A = \frac{2(2.38)16}{7} = 10.9 \text{ in.}^2 / \text{bar}$$

$$z = 36 \sqrt[3]{2.38(10.9)} = 107 \text{ kips/in.}$$

Example 7.1—Continued

	Code
Calculations and Discussion	**Reference**

(d)   Alternative arrangement for 7 #6 bars

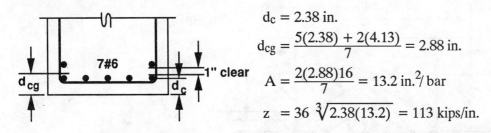

$d_c = 2.38$ in.

$d_{cg} = \dfrac{5(2.38) + 2(4.13)}{7} = 2.88$ in.

$A = \dfrac{2(2.88)16}{7} = 13.2$ in.^{2}/ bar

$z = 36 \sqrt[3]{2.38(13.2)} = 113$ kips/in.

As illustrated, selection of a larger number of smaller diameter bars at closer spacing, and with bars located as close as possible to the tension face of the beam (single layer), significantly reduces the quantity z. Placement considerations would favor the single layer 4 #8 bar selection.

Example 7.2—Distribution of Flexural Reinforcement

Check distribution of reinforcement for beam section shown, using Eq. (10-4). Use limit of z = 145 kips/in. for exterior exposure, and Grade 60 reinforcement.

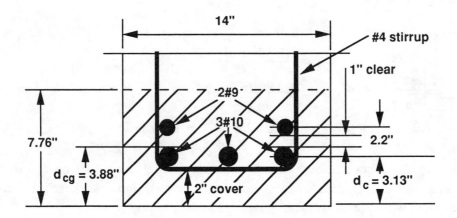

Calculations and Discussion	Code Reference

1. Reinforcement centroid

$$d_{cg} = \frac{3(1.27)\,3.13 + 2\,(1.0)\,5.33}{3\,(1.27) + 2\,(1.0)} = 3.88 \text{ in.}$$

2. Effective tension area of concrete

$$= 2 \times 3.88 \times 14 = 108.6 \text{ in.}^2$$

3. Equivalent number of #10 bars

$$= \frac{3\,(1.27) + 2\,(1.0)}{1.27} = 4.57$$

4. $A = \dfrac{\text{effective area}}{\text{no. of bars}} = \dfrac{108.6}{4.57} = 23.8 \text{ in.}^2/\text{bar}$

5. $z = f_s \sqrt[3]{d_c A}$

Eq.(10-4)

$= 0.6\,(60)\,\sqrt[3]{3.13 \times 23.8} = 152 \text{ kips/in.} > 145 \text{ kips/in.}$ N.G.

10.6.4

7-10

Example 7.2—Continued

Calculations and Discussion

6. Try 4 #10 (bottom row) and 2 #6 (top row)

$$d_{cg} = \frac{4\,(1.27)\,3.13 + 2\,(0.44)\,5.13}{4\,(1.27) + 2\,(0.44)} = 3.43 \text{ in.}$$

$$A = \frac{2 \times 3.43 \times 14}{\dfrac{4\,(1.27) + 2\,(0.44)}{1.27}} = 20.45 \text{ in.}^2$$

$$z = 0.6\,(60)\,\sqrt[3]{3.13 \times 20.45} = 144 \text{ kips/in.} < 145 \text{ kips/in.} \qquad \text{O.K.}$$

Select reinforcement for T-section shown below.

Exposure: Outside $f_c' = 4000$ psi
Span: 50 ft continuous $f_y = 60,000$ psi
Service load moments:

Positive Moment	Negative Moment
$M_d = +265$ ft-kips	$M_d = -290$ ft-kips
$M_\ell = +690$ ft-kips	$M_\ell = -760$ ft-kips

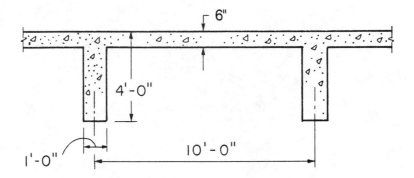

Calculations and Discussion	**Code Reference**

A.   Distribution of positive moment reinforcement

1. A_s required $= 7.76$ in.^{2}
     Try 5 #11 bars, $A_s = 7.80$ in.^{2}

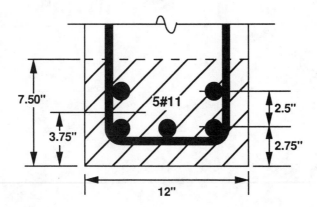

Example 7.3—Continued

	Code Reference
Calculations and Discussion	

2. $d_{cg} = \dfrac{3\,(1.56)\,2.75 + 2\,(1.56)\,5.25}{5\,(1.56)} = 3.75$ in.

Effective concrete area $= 2 \times 3.75 \times 12 = 90$ in.^{2}

$A = \dfrac{90}{5} = 18$ in.^{2}/ bar

10.0

Stress in reinforcement at service load

10.6.4

$f_s = \dfrac{+M}{jdA_s} = \dfrac{(265 + 690)\,12}{0.98 \times 44.3 \times 7.80} = 33.9$ ksi

$z = f_s \sqrt[3]{2.75 \times 18} = 125$ kips/in. < 145 kips/in.   O.K.

Eq.(10-4)

B. Distribution of negative moment reinforcement

1. A_s required $= 10.0$ in.^{2}

Effective flange width $= 12 + 2 \times 8 \times 6 = 108$ in.

8.10.2

Effective width for tension reinforcement $= \frac{1}{10} \times 50 \times 12 = 60$ in.

10.6.6

Try 10 #9 bars @ ≈ 6.5 in., $A_s = 10.0$ in.^{2}

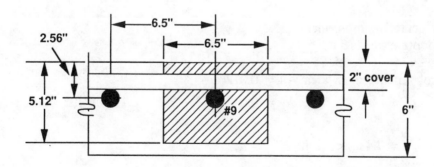

Example 7.3—Continued

Calculations and Discussion	Code Reference

2. $d_c = 2 + \dfrac{1.127}{2} = 2.56$ in.

   $A = 2d_c s = 2 \times 2.56 \times 6.5 = 33.3$ in.^{2}/bar

   In lieu of computations, use $f_s = 0.6 f_y$ as permitted in Section 10.6.4   10.6.4

   $z = f_s \sqrt[3]{d_c A} = 0.6\,(60)\ \sqrt[3]{2.56 \times 33.3} = 158$ kips/in. > 145 kips/in.   N.G.

   Check maximum bar spacing by Table 7-1 for outside exposure (2 in. cover).

   For #9 bar, $s_{max} = 5.0$ in. < 6.5 in.   N.G.

3. Try 13 #8 bars @ 5 in., $A_s = 10.27$ in.^{2}

   From Table 7-1, $s_{max} = 5.2$ in.   O.K.

4. Longitudinal reinforcement in slab outside 60-in. width.   10.6.6

   For crack control outside the 60-in. width, use shrinkage and temperature reinforcement according to Section 7.12.   7.12

   For Grade 60 reinforcement, $A_s = 0.0018 \times 12 \times 6 = 0.130$ in.^{2}/ft

   Use #4 bars @ 18 in., $A_s = 0.133$ in.^{2}/ft

C. Side face crack control reinforcement   10.6.7
          (web depth exceeds 3 ft)

   Crack control reinforcement area $A_{sk} = 0.012(d - 30)$
   $$= 0.012(44.3 - 30) = 0.172 \text{ in.}^2/\text{ft}$$

   $s_{max} = \dfrac{d}{6} = \dfrac{44.3}{6} = 7.4$ in. < 12 in.

   Use #3 bars @ 6 in., providing $0.11 \times \dfrac{12}{6} = 0.22$ in.^{2}/ft

   Use 4 # 3 bars evenly spaced along each side face of beam, extending 25.5 in. ($> d/2$) beyond the centroid of tension reinforcement

Example 7.3—Continued

D.   Detail section as shown below.

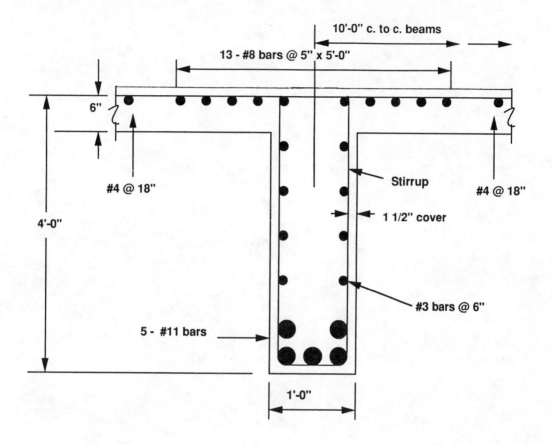

8

Deflections

UPDATE FOR THE '89 CODE

For ACI 318-89, presentation of the minimum thickness requirements for two-way slab systems (Section 9.5.3) has been simplified. Minimum thickness for two-way slabs without beams is presented directly as a function of span length only in a new Table 9.5(c). The engineer need only refer to Table 9.5(c) for minimum thickness of flat plates and flat slabs. Furthermore, the minimum thickness Eqs. (9-11) and (9-12) for two-way slabs with beams have been simplified by eliminating the parameter β_s related to discontinuous edges. See discussion on Section 9.5.3.

GENERAL CONSIDERATIONS

The ACI Code provisions for control of deflections are concerned only with deflections that occur at service load levels under static conditions and may not apply to loads with strong dynamic characteristics such as those due to earthquakes, transient winds, and vibration of machinery. Because of the variability of concrete structural deformations, it is essential in most cases to use relatively simple procedures, so that designers will not place undue reliance on computed or predicted deflection results. In-depth treatments of the subject of deflection control may be found in Refs. 8.1 and 8.2.

9.5   CONTROL OF DEFLECTIONS

Two methods are given in the Code for controlling deflections of one-way and two-way flexural members. Deflections may be controlled by means of minimum thickness [Table 9.5(a) for one-way systems, and Table 9.5(c) and Eqs. (9-11), (9-12), (9-13) for two-way systems] or directly by limiting computed deflections [Table 9.5(b)].

9.5.2   Minimum Thickness for Beams and One-Way Slabs (Nonprestressed)

Deflections of beams and one-way slabs supporting loads commonly experienced in buildings will normally be satisfactory when the minimum thickness from Table 9.5(a) (essentially reproduced in Table 8-1) are met or exceeded.

Table 8-1—Minimum Thickness for Nonprestressed Beams and One-Way Slabs
(Grade 60 Reinforcement and Normal Weight Concrete)

Member	Minimum Thickness, h			
	Simply Supported	One End Continuous	Both Ends Continuous	Cantilever
One-Way Slabs	$\ell/20$	$\ell/24$	$\ell/28$	$\ell/10$
Beams	$\ell/16$	$\ell/18.5$	$\ell/21$	$\ell/8$

(1) For Grade 40 reinforcement, multiply values by 0.80

(2) For structural lightweight concrete, multiply values by $(1.65 - 0.005w_c)$ but not less than 1.09, where w_c is the unit weight in lb per cu ft

The designer should especially note that the minimum thickness requirement is intended only for members **not** supporting or attached to partitions or other construction likely to be damaged by large deflections. For all other members, deflections need to be computed.

For shored composite members, the minimum thicknesses of Table 9.5(a) apply as for monolithic structural members. For unshored construction, if the thickness of a nonprestressed precast member meets the minimum thickness requirements, deflections need not be computed. Section 9.5.5 also states that, if the thickness of an unshored nonprestressed composite member meets the minimum thickness requirements, deflections occurring after the member becomes composite need not be computed, but the long-term deflection of the precast member should be investigated for the magnitude and duration of load prior to beginning of effective composite action.

9.5.3 Minimum Thickness for Two-Way Slab Systems (Nonprestressed)

Deflections of two-way slab systems with and without beams, drop panels, and column capitals need not be computed when the minimum thickness requirements of Section 9.5.3 are met. The minimum thickness requirements include the effects of panel location (interior or exterior); panel shape; span ratios; beams on panel edges; supporting columns and capitals; drop panels; and the yield strength of the reinforcing steel. For the '89 Code, required minimum thickness for flat plates and flat slabs (slabs without beams between interior columns) is presented directly as a function of span length in a new Table 9.5(c). The table values are based on Eq. (9-13) which represents the upper limit of slab thicknesses given by the three limiting equations, Eqs. (9-11), (9-12), and (9-13), for minimum thickness of two-way slab systems. The flat plate and flat slab minimum thickness values are summarized in Table 8-2 and illustrated in Fig. 8.1

For two-way beam-supported slabs (slabs with beams on all four edges of a panel), selection of minimum thickness is still based on Eqs. (9-11), (9-12), and (9-13). For the '89 Code, Eqs. (9-11) and (9-12) have been simplified by eliminating the term related to discontinuous edges. Studies have shown that the discontinuous edge parameter β_s has negligible effect on slab thickness and that the modified Eqs. (9-11) and (9-12) lead to

Table 8-2—Minimum Thickness for Nonprestressed Two-Way Slab Systems without Interior Beams between Supports (Grade 60 Reinforcement)

Two-Way Slab System	Minimum Thickness, h		
	Exterior Panels		Interior Panels
	Without Edge Beams	With Edge Beams	
Flat Plate (without drop panels)	$\ell_n/30$	$\ell_n/33$	$\ell_n/33$
Flat Slab (with drop panels)	$\ell_n/33$	$\ell_n/36$	$\ell_n/36$

(1) For Grade 40 reinforcement, multiply values by 0.90

(2) Drop Panels: length $\geq \ell/3$ and depth $\geq 1.25h$

(3) Edge Beams: $\alpha \geq 0.8$. For corner panels, refer to the edge beam with the lesser relative stiffness. α = ratio of flexural stiffness of beam section to flexural stiffness of a width of slab bounded laterally by centerlines of adjacent panels, if any, on each side of the beam.

virtually the same results as the corresponding original equations. The minimum thickness equations provide for a transition, Eq. (9-11), from slabs on stiff beams, Eq. (9-12), to slabs without beams, Eq. (9-13). Fig. 8-2 may be used to simplify minimum thickness calculations for two-way slabs. It may be noted in Fig. 8-2 that the difference between the controlling minimum thicknesses for square panels and rectangular panels having a 2-to-1 panel side ratio is not very large.

9.5.4   Minimum Thickness for Prestressed Members

Typical span-depth ratios for general use in design of prestressed members are given in the PCI Design Handbook[8.3], and summarized in Ref. 8.2 from several sources.

9.5.2.6, 9.5.3.6   Maximum Permissible Computed Deflections—The allowable computed deflections specified in Table 9.5(b) apply to both one-way and two-way nonprestressed and prestressed members.

Where excessive deflections may cause damage to nonstructural or structural elements, only that part of the deflection occurring after the construction of the elements needs to be considered. The most stringent deflection limit of $\ell/480$ in Table 9.5(b) is an example of such a case.

Where excessive deflections may result in either esthetic or functional problems, such as objectionable visual sagging, ponding of water, vibration, and improper operation of machinery or sliding doors, the total deflection should be considered. Such examples are not included in Table 9.5(b) and must be dealt with by the designer on a case-by-case basis.

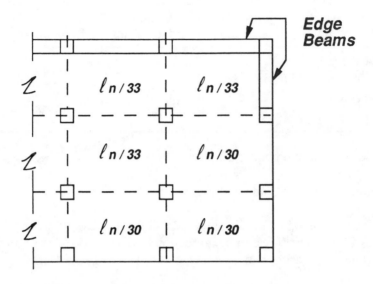

(a) Flat Plates (without drop panels)

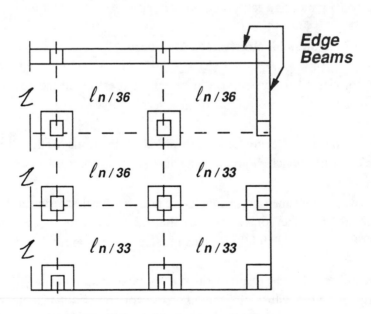

(b) Flat Slabs (with drop panels)

Fig. 8-1   Minimum Thickness of Slabs without Interior Beams (Grade 60 Reinforcement)

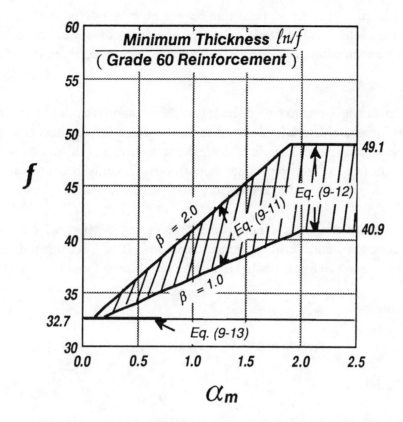

Fig. 8-2   Minimum Thickness for Two-Way Beam-Supported Slabs

9.5.2.3, 9.5.2.4   Initial or Short-Term Deflection of Beams and One-Way Slabs (Nonprestressed)—The effective moment of inertia for cantilevers, simple beams and continuous beams between inflection points is given by:

$$I_e = (M_{cr}/M_a)^3 I_g + [1 - (M_{cr}/M_a)^3] I_{cr} \le I_g \qquad \text{Eq. (9-7)}$$

where
$$M_{cr} = f_r I_g / y_t \qquad \text{Eq. (9-8)}$$

For normal weight concrete,
$$f_r = 7.5 \sqrt{f_c'} \qquad \text{Eq. (9-9)}$$

For lightweight concrete, f_r is modified according to Section 9.5.2.3. See Example 8.2.

Values of I_g and I_{cr} may be computed using the equations in Fig. 8.3. M_a in Eq. (9-7) is the maximum service load moment (unfactored) at the stage for which deflections are being considered. For different load combinations, the deflection should be computed in each case using Eq. (9-7) for the load combination being considered, such as dead load or dead plus live load. The incremental deflection, such as for live load, is then computed as the difference between these values.

8-5

The effective moment of inertia I_e provides a transition between the well-defined upper and lower bounds of I_g and I_{cr}, as a function of the level of cracking as represented by M_{cr}/M_a. The equation empirically accounts for the effect of tension stiffening—undamaged tensile concrete between cracks and in regions of low tensile stress.

For prismatic members (including T-beams with different cracked sections in positive and negative moment regions), I_e (and thus M_a) may be determined at the support section for cantilevers and at the midspan section for simple and continuous spans. The use of the midspan section properties for continuous prismatic members is considered satisfactory in approximate calculations primarily because the midspan rigidity (including the effect of cracking) has the dominant effect on deflections.

Alternatively, for continuous prismatic and nonprismatic members, Section 9.5.2.4 suggests using the average I_e at the critical positive and negative moment sections. The '83 Commentary Section 9.5.2.4 suggests the following average values for somewhat improved results over the other two methods:

Beams with one end continuous:

$$\text{Avg. } I_e = 0.85 \, I_m + 0.15(I_{cont.end}) \tag{1}$$

Beams with both ends continuous:

$$\text{Avg. } I_e = 0.70 \, I_m + 0.15(I_{e1} + I_{e2}) \tag{2}$$

where I_m refers to I_e at the midspan section, and I_{e1} and I_{e2} refer to I_e at the respective beam ends. Moment envelopes should be used in computing both positive and negative values of I_e. Moment envelopes based on the approximate moment coefficients of Section 8.3.3 are accurate enough. See Example 8.2. For a single heavy concentrated load, only the midspan I_e should be used.

The initial or short-term deflection (a_i) may be computed using the following elastic equation given in '83 Commentary Section 9.5.2.4 for cantilevers, and simple and continuous beams. For continuous beams, the midspan deflection may normally be used as an approximation of the maximum deflection.

$$a_i = K \, (5/48) \, M_a \, \ell^2 / E_c \, I_e \tag{3}$$

where M_a is the support moment for cantilevers and the midspan moment (when K is so defined) for simple and continuous beams, and ℓ is the span length as defined in Section 8.7. For uniformly distributed loading w, the theoretical values of the deflection coefficient K are as follows:

Cantilevers (the deflection due to rotation at the supports must be determined in addition)	$K = 12/5 = 2.40$
Simple Beams	$K = 1.00$
Continuous Beams where M_o is the simple span moment at midspan and M_a is the net midspan moment.	$K = 1.20 - 0.20 \, M_o/M_a$

Continuous Beams
 Fixed-Hinged Beams, Midspan Deflection $K = 0.80$
 Fixed-Hinged Beams, Maximum Deflection $K = 0.738$
  using Maximum Moment
 Fixed-Fixed Beams $K = 0.60$
For other types of loading, K values are given in Ref. 8.2.

Since deflections are logically computed for a given continuous span based on the same loading pattern as for maximum positive moment, Eq. (3) is thought to be the most convenient form of a deflection equation. In addition, when using Eq. (3) with only the midspan I_e, the negative moments are not required in the deflection calculation for continuous beams. The use of the ACI approximate moment coefficients for maximum positive moment are considered to be satisfactory in most cases for computing deflections.

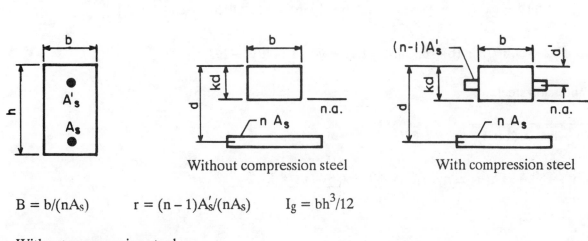

Without compression steel   With compression steel

$B = b/(nA_s)$ $r = (n-1)A_s'/(nA_s)$ $I_g = bh^3/12$

Without compression steel

$$kd = (\sqrt{2dB + 1} - 1)/B$$

$$I_{cr} = bk^3d^3/3 + nA_s(d - kd)^2$$

With compression steel

$$kd = [\sqrt{2dB(1 + rd'/d) + (1 + r)^2} - (1 + r)]/B$$

$$I_{cr} = bk^3d^3/3 + nA_s(d - kd)^2 + (n-1)A_s'(kd - d')^2$$

(a) Rectangular Sections

Fig. 8.3 Moment of Inertia of Gross Section, I_g, and Cracked Transformed Section, I_{cr}

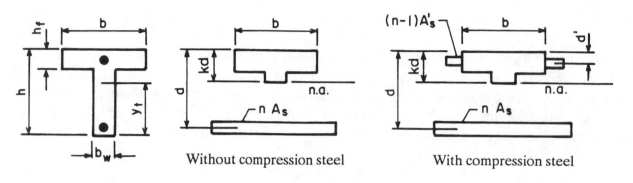

Without compression steel   With compression steel

$$C = b_w/(nA_s), \quad f = h_f(b - b_w)/(nA_s), \quad y_t = h - 1/2[(b - b_w)h_f^2 + b_w h^2]/[(b - b_w)h_f + b_w h]$$

$$I_g = (b - b_w)h_f^3/12 + b_w h^3/12 + (b - b_w)h_f(h - h_f/2 - y_t)^2 + b_w h(y_t - h/2)^2$$

Without compression steel

$$kd = [\sqrt{C(2d + h_f f) + (1 + f)^2} - (1 + f)]/c$$

$$I_{cr} = (b - b_w)h_f^3/12 + b_w k^3 d^3/3 + (b - b_w)h_f(kd - h_f/2)^2 + nA_s(d - kd)^2$$

With compression steel

$$kd = [\sqrt{C(2d + h_f f + 2rd') + (f + r + 1)^2} - (f + r + 1)]/c$$

$$I_{cr} = (b - b_w)h_f^3/12 + b_w k^3 d^3/3 + (b - b_w)h_f(kd - h_f/2)^2 + nA_s(d - kd)^2 + (n - 1)A_s'(kd - d')^2$$

(b) Flanged Sections

Fig. 8.3 (contd.)   **Moment of Inertia of Gross Section, I_g, and Cracked Transformed Section, I_{cr}**

9.5.2.5 Long-Term Deflection of Beams and One-Way Slabs (Nonprestressed)—According to Section 9.5.2.5, additional long-term deflections due to the combined effects of shrinkage and creep due to sustained loads may be estimated by:

$$\lambda = \frac{\xi}{1 + 50\rho'} \qquad \text{Eq. (9-10)}$$

and

$$a_{(cp + sh)} = \lambda(a_i)_{sus} \qquad (4)$$

where ρ' is determined at the support section for cantilevers and the midspan section for simple and continuous spans, and $\xi = 2.0$ for sustained loading of 5 years or longer duration, 1.4 for 12 months, 1.2 for 6 months, and 1.0 for 3 months duration. The multiplier λ for additional long-term deflection was first introduced in the 1983 Code. The ξ multiplier for creep and shrinkage effects is considered to better represent the effect of compression reinforcement by more appropriately relating it to the compression reinforcement ratio ρ', rather than to the ratio of compression to tension reinforcement A_s'/A_s as used in earlier code editions.

Alternatively, creep and shrinkage deflections may be considered separately using the following expressions from Refs. 8.2, 8.4, and 8.5.

$$a_{cp} = \lambda_{cp}(a_i)_{sus} \qquad (5)$$

$$a_{sh} = K_{sh}\,\varphi_{sh}\,\ell^2 \qquad (6)$$

where

$$\lambda_{cp} = k_r C_t = \frac{0.85 C_t}{1 + 50\rho'}$$

and

$$\varphi_{sh} = A_{sh}\,\varepsilon_{sh}/h$$

For average conditions, ultimate values for C_t and ε_{sh} may be taken as $C_u = 1.6$ and $(\varepsilon_{sh})_u = 400 \times 10^{-6}$. A_{sh} may be taken from Fig. 8-4. Assuming equal positive and negative shrinkage curvatures with an inflection point at the quarter-point of continuous spans (generally satisfactory for deflection computation), the following values for the shrinkage deflection coefficient K_{sh} may be used:

Cantilevers	$K_{sh} = 1/2 = 0.50$
Simple Spans	$K_{sh} = 1/8 = 0.125$
Spans with One End Continuous — Multi-span Beams	$K_{sh} = 0.09$
Spans with One End Continuous — Two-Span Beams	$K_{sh} = 0.084$
Spans With Both Ends Continuous —	$K_{sh} = 0.065$

The compression reinforcement ratio ρ', and the reinforcement percentages ρ and ρ' used in determining A_{sh} from Fig. 8-4, refer to the support section of cantilevers and the midspan section of simple and continuous beams. For T-beams use $\rho = 100(\rho + \rho_w)/2$ and a similar calculation for any compression steel ρ' in determining A_{sh}, where $\rho_w = A_s/b_w d$. See Example 8.2

As to the choice of computing creep and shrinkage deflections by the combined ACI Eq. (9-10) or separately by Eqs. (5),(6), the combined calculation is simpler but provides only a rough approximation, since shrinkage deflections are only indirectly related to the loading (primarily by means of the steel content). One case in which the separate calculation of creep and shrinkage deflections may be preferable is when part of the live load is considered as a sustained load.

All procedures and properties for computing creep and shrinkage deflections apply equally to normal weight and lightweight concrete.

9.5.5   Deflection of Composite Flexural Members (Nonprestressed)

The ultimate (in time) deflection of unshored and shored composite flexural members may be computed by Eqs. (7) to (10). These equations are derived in detail in Refs. 8.2 and 8.7. Subscripts 1 and 2 are used to refer to the slab (or effect of the slab, such as under slab dead load) and the precast beam, respectively.

The following procedures are for composite beams using both unshored and shored construction. Examples 8.3 and 8.4 demonstrate the beneficial effect of shoring in reducing deflections.

9.5.5.2   Unshored Composite Members—

$$
a_u = \underset{(1)}{(a_i)_2} + \underset{(2)}{0.77\,k_r\,(a_i)_2} + \underset{(3)}{0.83\,k_r\,(a_i)_2\frac{I_2}{I_c}} + \underset{(4)}{0.36\,a_{sh}} + \underset{(5)}{0.64\,a_{sh}\frac{I_2}{I_c}} + \underset{(6)}{(a_i)_1} + \underset{(7)}{1.22\,k_r\,(a_i)_1\frac{I_2}{I_c}}
$$

$$
+ \underset{(8)}{a_{ds}} + \underset{(9)}{(a_i)_\ell} + \underset{(10)}{(a_{cp})_\ell} \tag{7}
$$

With $k_r = 0.85$ (no compression steel in the precast beam) and a_{ds} assumed to be equal to $0.50\,(a_i)_1$, Eq. (7) reduces to Eq. (8).

$$
a_u = \underset{(1+2+3)}{\left(1.65 + 0.71\frac{I_2}{I_c}\right)(a_i)_2} + \underset{(4+5)}{\left(0.36 + 0.64\frac{I_2}{I_c}\right)a_{sh}} + \underset{(6+7+8)}{\left(1.50 + 1.04\frac{I_2}{I_c}\right)(a_i)_1} + \underset{(9)}{(a_i)_\ell} + \underset{(10)}{(a_{cp})_\ell} \tag{8}
$$

In Eqs. (7) and (8), the parts of the total creep and shrinkage occurring before and after slab casting are based on the assumption of a precast beam age of 20 days when its dead load is applied and of 2 months when the composite slab is cast.

Term (1) is the initial or short-term dead load deflection of the precast beam, using Eq.(3), with $M_a = M_2 =$ midspan moment due to the precast beam dead load. For computing $(I_e)_2$ in Eq. (9-7), M_a refers to the precast beam dead load, and M_{cr}, I_g, and I_{cr} to the precast beam section at midspan.

Term (2) is the dead load creep deflection of the precast beam up to the time of slab casting, using Eq. (5), with $C_t = 0.48$ (for 20 days to 2 months) $\times 1.60 = 0.77$, and with ρ' referring to the compression steel in the precast beam at midspan when computing k_r.

Term (3) is the creep deflection of the composite beam following slab casting, due to the precast beam dead load, using Eq. (5), with $C_t = 1.60 - 0.77 = 0.83$. ρ' is the same as in Term (2). The ratio I_2/I_c modifies the initial stress (strain) and accounts for the effect of the composite section in restraining additional creep curvature (strain) after the composite section becomes effective. As a simple approximation, $I_2/I_c = [(I_2/I_c)_g + (I_2/I_c)_{cr}]/2$ may be used.

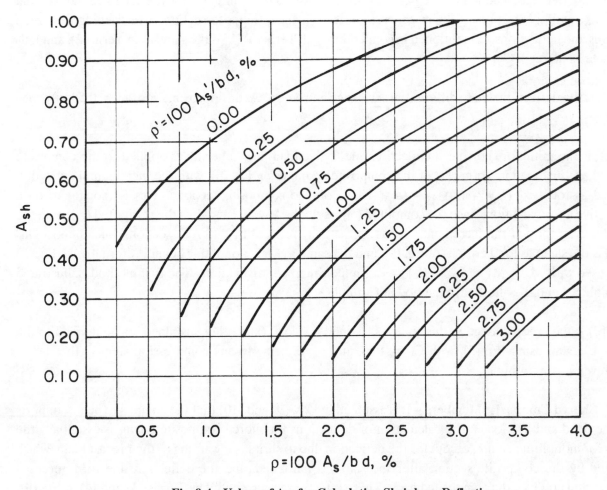

Fig. 8-4 Values of A_{sh} for Calculating Shrinkage Deflection

8-11

Term (4) is the deflection due to shrinkage warping of the precast beam up to the time of slab casting, using Eq. (6), with $\varepsilon_{sh} = 0.36 (\varepsilon_{sh})_u$ at age 2 months for steam cured concrete (assumed to be the usual case for precast beams). $(\varepsilon_{sh})_u = 400 \times 10^{-6}$ in./in.

Term (5) is the shrinkage deflection of the composite beam following slab casting, due to the shrinkage of the precast beam concrete, using Eq.(6), with $\varepsilon_{sh} = 0.64(\varepsilon_{sh})_u$. This term does not include the effect of differential shrinkage and creep, which is given by Term (8). I_2/I_c is the same as in Term (3).

Term (6) is the initial or short-term deflection of the precast beam under slab dead load, using Eq. (3), with the incremental deflection computed as follows: $(a_i)_1 = (a_i)_{1+2} - (a_i)_2$, where $(a_i)_2$ is the same as in Term (1). For computing $(I_e)_{1+2}$ and $(a_i)_{1+2}$ in Eqs. (9-7) and (3), $M_a = M_1 + M_2$ due to the precast beam plus slab dead load at midspan, and M_{cr}, I_g, and I_{cr} refer to the precast beam section at midspan. When partitions, roofing, etc., are placed at the same time as the slab, or soon thereafter, their dead load should be included in M_1 and M_a.

Term (7) is the creep deflection of the composite beam due to slab dead load using Eq. (5), with $C_t = 0.76$ (loading age correction factor at age 2 months) $\times 1.60 = 1.22$. In this term, the initial strains, curvatures and deflections under slab dead load were based on the precast section only. Hence the creep curvatures and deflections refer to the precast beam concrete, although the composite section is restraining the creep curvatures and deflections, as mentioned in connection with Term (3). k_r is the same as in Term (2), and I_2/I_c is the same as in Term (3).

Term (8) is the deflection due to differential shrinkage and creep. As an approximation, $a_{ds} = 0.50(a_i)_1$, may be used.

Term (9) is the initial or short-term deflection due to live load [and other loads applied to the composite beam and not included in Term (6)] of the composite beam, using Eq. (4), with the incremental deflection estimated as follows: $(a_i)_\ell = (a_i)_{d+\ell} - (a_i)_d$, based on the composite section. This is thought to be a conservative approximation, since the computed $(a_i)_d$ is on the low side and thus the computed $(a_i)_\ell$ is on the high side, even though the incremental loads are actually resisted by different sections (members). This method is the same as for Term (5) of Eq. (9), and the same as for a monolithic beam. Alternatively, Eq. (3) may be used with $M_a = M_1$ and $I_e = (I_c)_{cr}$ as a simple rough approximation. The first method is illustrated in Example 8.4 and the alternative method in Example 8.3.

Term (10) is the creep deflection due to any sustained live load (and other sustained loads) applied to the composite beam, using Eq. (5), with $C_u = 1.60$, and ρ' referring to any compression steel in the slab at midspan when computing k_r.

9.5.5.1 Shored Composite Members—It is assumed in Eqs. (9) and (10) that the composite beam supports all of the dead and live load. The calculation of deflections for shored composite beams is essentially the same as for monolithic beams, except for the deflection due to shrinkage warping of the precast beam which is resisted by the composite section after the slab has hardened, and the deflection due to differential shrinkage and creep of the composite beam. These effects are represented by Terms (3) and (4) in Eq. (9).

$$\overset{\text{(1)} \quad\quad\quad \text{(2)} \quad\quad\quad\quad \text{(3)} \quad\quad \text{(4)} \quad \text{(5)} \quad\quad \text{(6)}}{a_u = (a_i)_{1+2} + 1.80\,k_r\,(a_i)_{1+2} + a_{sh}\frac{I_2}{I_c} + a_{ds} + (a_i)_\ell + (a_{cp})_\ell} \tag{9}$$

When $k_r = 0.85$ (neglecting any effect of slab compression steel) and a_{ds} is assumed to be equal to $(a_i)_{1+2}$, Eq. (9) reduces to Eq. (10).

$$\overset{\text{(1 + 2 + 4)} \quad\quad\quad \text{(3)} \quad\quad \text{(5)} \quad\quad \text{(6)}}{a_u = 3.53\,(a_i)_{1+2} + a_{sh}\frac{I_2}{I_c} + (a_i)_\ell + (a_{cp})_\ell} \tag{10}$$

Term (1) is the initial or short-term deflection of the composite beam due to slab plus precast beam dead load (plus partitions, roofing, etc.), using Eq. (3), with $M_a = M_1 + M_2$ = midspan moment due to slab plus precast beam dead load. For computing $(I_e)_{1+2}$ in Eq. (1), M_a refers to the moment $M_1 + M_2$, and M_{cr}, I_g, and I_{cr} to the composite beam section at midspan.

Term (2) is the creep deflection of the composite beam due to the dead load in Term (1), using Eq. (5). C_u = 1.80 (based on the shores being removed at about 10 days of age for a moist-cured slab), and ρ' refers to any compression steel in the slab at midspan when computing k_r.

Term (3) is the shrinkage deflection of the composite beam after the shores are removed, due to the shrinkage of the precast beam concrete, but not including the effect of differential shrinkage and creep which is given by Term (4). Equation (6) may be used to compute a_{sh}. Assuming the slab is cast at a precast (steam-cured) beam concrete age of 2 months and that shores are removed about 10 days later, $(\varepsilon_{sh})_u = (1 - 0.37)(400 \times 10^{-6}) = 252 \times 10^{-6}$ in./in.

Term (4) is the deflection due to differential shrinkage and creep. As an approximation, $a_{ds} = (a_i)_{1+2}$ may be used.

Term (5) is the initial or short-term live load deflection of the composite beam, using Eq. (3). The calculation of the incremental live load deflection follows the same procedure as that for a monolithic beam. This is the same as the first method described in connection with Term (9) of Eq. (7).

Term (6) is the creep deflection due to any sustained live load, using Eq. (5). This is the same as Term (10) of Eq. (7).

These procedures suggest using midspan values only, which may normally be satisfactory for both simple composite beams and those with a continuous slab as well. See Ref. 8.7 for an example of a continuous slab in composite construction.

9.5.3.6 Deflection of Nonprestressed Two-Way Slab Systems—Initial or Short-Term Deflection:

An approximate procedure[8.6] that is compatible with the Direct Design and Equivalent Frame Methods of Code Chapter 13 will be used to compute the initial or short-term deflection of two-way slab systems. The method is essentially the same for flat plates, flat slabs, and two-way beam-supported slabs, once the appropriate stiffnesses are computed. In this procedure, the midpanel deflection is computed as the sum of the midspan column strip deflection in one direction, such as a_{cx}, and the midspan middle strip deflection in the orthogonal direction, such as a_{my}, as shown in Fig. 8.5

Under vertical loads, the midspan deflection of an equivalent frame can be considered as the sum of three parts: that of a panel assumed to be fixed at both ends of its span, plus that due to the rotation at each of the two support lines.

Midspan fixed-end deflection of the equivalent frame under uniform loading is given by Eq. (11).

$$\text{Fixed } a_{frame} = w \ell_2 \ell^4 / 384 \, E_c \, I_{frame} \tag{11}$$

where $w\ell_2$ is the uniformly distributed load on the frame, and ℓ is the span center-to-center of columns. To include the effect of different positive and negative moment region I values [primarily when using drop panels and/or I_e in Eq. (9-7)], an average may be used, as given by Eqs. (1) and (2).

Calculation of the midspan fixed-end deflection of the column and middle strips is then based on the M/EI ratio of the strips to the frame.

$$\text{Fixed } a_{c,m} = (LDF)_{c,m} \, (\text{Fixed } a_{frame}) \, \frac{(EI)_{frame}}{(EI)_{c,m}} \tag{12}$$

where $(LDF)_{c,m} = M_{c,m}/M_{frame}$. Typical values of the lateral distribution factor, LDF, are shown in Table 8-3.

If the ends of the columns at the floors above and below are assumed to be fixed (usual case in Equivalent Frame analysis), or ideally pinned, the rotation of the column at the floor in question is equal to the net applied moment divided by the stiffness of the equivalent column. This is given by Eq. (13) for the frame, column strip, and middle strip.

$$\text{End } \theta_c = \text{End } \theta_m = \text{End } \theta_{frame} = \text{End } \theta = (M_{net})_{frame}/K_{ec} \tag{13}$$

where K_{ec} is the gross-section flexural stiffness of the equivalent column. See Equivalent Columns: Commentary Section 13.7.4.

In the Direct Design Method, moments are based on the clear span and should theoretically be adjusted to obtain moments and rotations at the column centerlines. However, the use in Eq. (3) of moments at the faces of the columns should cause little error. Particularly in the case of flat plates and flat slabs, the span center-to-center of columns is thought to be more appropriate in deflection calculations than the clear span.

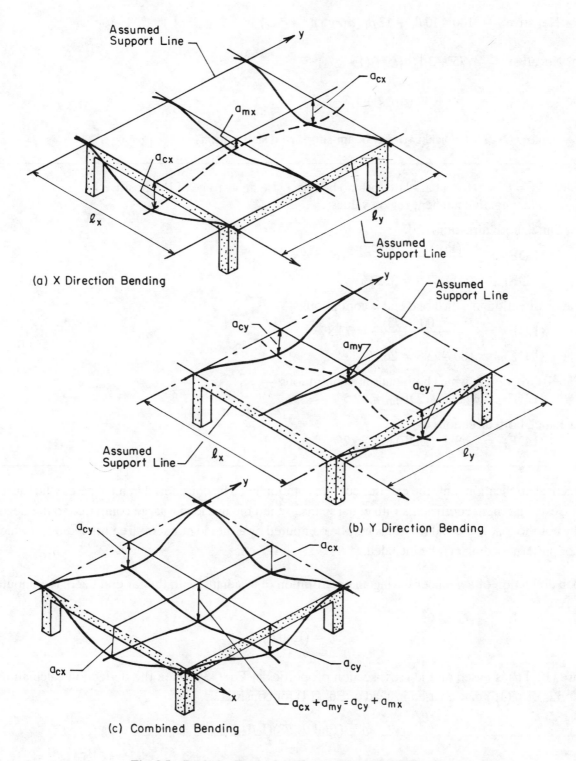

(a) X Direction Bending

(b) Y Direction Bending

(c) Combined Bending

**Fig. 8.5   Basis for Equivalent Frame Method of Deflection
Analysis of Two-Way Slab Systems, with or without Beams**

Table 8-3—Lateral Distribution Factors (LDF) for Column and Middle Strips

Section 13.6.4—the column strip percentages of total slab strip moments are:

Exterior Negative — $100 - 10\beta_t + 12\beta_t (\alpha_1 \ell_2/\ell_1)(1 - \ell_2/\ell_1)$

Interior Negative — $75 + 30 (\alpha_1 \ell_2/\ell_1)(1 - \ell_2/\ell_1)$

Positive — $60 + 30 (\alpha_1 \ell_2/\ell_1)(1.5 - \ell_2/\ell_1)$

except when $\alpha_1 \ell_2/\ell_1 > 1$ (typical two-way beam-supported slab system), use $\alpha_1 \ell_2/\ell_1 = 1$

Example: $\alpha_1 = \beta_t = 0$ for a flat plate with no beams. Use an average of the positive and negative moment region values:

Interior Panel, both directions

$$(LDF)_c = \frac{60 \text{ (Pos)} + 75 \text{ (Neg)}}{2} = 67.5\%$$

$$(LDF)_m = 100 - 67.5 = 32.5\%$$

Side Panel, direction in which one end is continuous —

$$(LDF)_c = \frac{60 + (100 + 75)/2}{2} = 73.8\%$$

$$(LDF)_m = 100 - 73.8 = 26.2\%$$

Side Panel, direction in which both ends are continuous —
$(LDF)_c = 67.5\%$, $(LDF)_m = 32.5\%$

Corner Panel, both directions —
$(LDF)_c = 73.8\%$, $(LDF)_m = 26.2\%$

For practical application, only the exterior column rotation need be considered in most cases when using the Direct Design moment coefficients with equal spans. When the live load is large compared to the dead load (usually not the case), the end rotations may be computed by a simple moment-area procedure in which the effect of pattern loading may be included.

Midspan deflection of a member having an end rotation of θ radians, with the far end fixed, is computed by Eq. (14).

$$a_\theta = (\text{End } \theta) \, \ell/8 \tag{14}$$

Because (End θ) is based on the gross-section properties in Eq. (13), when the deflection calculations are based on I_e, Eq. (15) may be used instead of Eq. (14) for consistency.

$$a_\theta = (\text{End } \theta)(\ell/8)(I_g/I_e)_{frame} \tag{15}$$

The combined midspan deflection of a column or middle strip is the sum of the three parts, as computed by Eq. (16).

$$a_{c,m} = \text{Fixed } a_{c,m} + (a_{\theta 1})_{c,m} + (a_{\theta 2})_{c,m} \tag{16}$$

where $a_{\theta 1}$ and $a_{\theta 2}$ refer to the midspan deflections due to rotations at both ends. As shown in Fig. 8-5, the total midpanel deflection is given by Eq (17).

$$a = a_{cx} + a_{my} = a_{cy} + a_{mx} \tag{17}$$

For other than square symmetrical panels, Eq. (18) may be used.

$$a = [(a_{cx} + a_{my}) + (a_{cy} + a_{mx})]/2 \tag{18}$$

<u>Effective Moment of Inertia:</u> The effective moment of inertia by Eq. (9-7) is recommended for computing deflections of partially cracked two-way construction. An average I_e of the positive and negative regions in accordance with Eqs. (1) and (2) may also be used. The following typical cracking locations have been found empirically, and the corresponding values of Ie have been shown to apply in most cases.[8.2]

Slabs without beams (flat plates, flat slabs)
  All dead load deflections — I_g
  Dead-plus-live load deflections:
    For the column strips in both directions — I_e
    For the middle strips in both directions — I_g
These conditions are demonstrated in Example 8.5.
Slab with beams (two-way beam-supported slabs)
  All dead load deflections — I_g
  Dead-plus-live load deflections:
    For the column strips in both directions — I_g
    For the middle strips in both directions — I_e

The I_e of the equivalent frame in each direction is taken as the sum of the column and middle strip I_e values.

<u>Long-Term Deflection:</u>   Since the available data on long-term deflections of two-way construction is too limited to justify more elaborate procedures, the same procedures as those used for one-way members are recommended. ACI Eq. (9-10), with $\xi = 2.5$ for sustained loading of five years or longer duration may be used.

9.5.4 Deflection of Noncomposite Prestressed Members

The ultimate (in time) camber and deflection of prestresed members may be computed by Eq. (19). The a_u expression is based on a procedure described in Ref. 8.2. The procedure includes the use of I_e for partially prestressed members, as presented in the PCI Design Handbook[8.3], as a suggested method of satisfying Code Section 18.4.2(c) for deflection analysis when the computed tensile stress exceeds the modulus of rupture, but does not exceed $12\sqrt{f_c'}$. For more information on cracked prestressed beam deflections and on composite prestressed beam deflections, see Refs. 8.2 and 8.7.

$$
a_u = \overset{(1)}{-a_{po}} + \overset{(2)}{a_o} - \overset{(3)}{\left[-\frac{\Delta P_u}{P_o} + (k_r\, C_u)\left(1 - \frac{\Delta P_u}{2P_o} \right) \right]} + \overset{(4)}{a_{po}} + \overset{(5)}{(k_r\, C_u)\, a_o} + a_s
$$

$$
\overset{(6)}{+ (\beta_s\, k_r\, C_u)\, a_s} + \overset{(7)}{a_\ell} + \overset{(8)}{(a_{cp})_\ell} \tag{19}
$$

Term (1) is the initial camber due to the initial prestressing moment after elastic loss, P_{oe}. For example, $a_{po} = P_{oe}\ell^2/8E_{ci}\,I_g$ for a straight tendon.

Term (2) is the initial deflection due to self-weight of the beam. $a_o = 5M_o\ell^2/48\, E_{ci}\,I_g$ for a simple beam, where M_o = midspan self-weight moment.

Term (3) is the creep (time-dependent) camber of the beam due to the prestressing moment. This term includes the effects of creep and loss of prestress; that is, the creep effect under variable stress. Average values of the prestress loss ratio after transfer (excluding elastic loss), $\Delta P_u/P_o$, are about 0.18, 0.21, and 0.23 for normal, sand, and all-lightweight concretes, respectively. An average value of $C_u = 2.0$ might be reasonable for the creep factor due to ultimate prestress force and self-weight. The k_r factor takes into account the effect of any nonprestressed tension steel in reducing time-dependent camber, using Eq. (20). It is also used in the PCI Design Handbook[8.3] in a slightly different form.

$$
k_r = 1/[1 + (A_s/A_{ps})], \text{ for } A_s/A_{ps} < 2 \tag{20}
$$

When $k_r = 1$ and $\Delta P_u = P_o - P_e$, Terms (1) + (3) can be combined as:

$$
-a_{po} - \left[-a_{po} + a_{pe} + C_u\left(\frac{a_{po} + a_{pe}}{2} \right) \right] = -a_{pe} - C_u\left(\frac{a_{po} + a_{pe}}{2} \right)
$$

Term (4) is the creep deflection due to self-weight of the beam. Use the same value of C_u as in Term (3). Since creep due to prestress and self-weight takes place under the combined stresses caused by them, the effect of any nonprestressed tension steel in reducing the creep deformation is included in both the camber Term (3) and the deflection Term (4).

Term (5) is the initial deflection of the beam under a superimposed dead load. $a_s = 5M_s\ell^2/48E_cI_g$ for a simple beam, where M_s = midspan moment due to superimposed dead load (uniformly distributed).

Term (6) is the creep deflection of the beam caused by a superimposed dead load. k_r is the same as in Terms (3) and (4), and is included in this deflection term for the same reason as in Term (4). An average value of

$C_u = 1.6$ is recommended, as in Eq. (7) for nonprestressed members. β_s is the creep correction factor for the age of the beam concrete when the superimposed dead load is applied (same values apply for normal as well as lightweight concrete): $\beta_s = 0.85$ for age 3 weeks, 0.83 for age 1 month, 0.76 for age 2 months, 0.74 for age 3 months, and 0.71 for age 4 months.

Term (7) is the initial live load deflection of the beam. $a_l = 5M_l l^2/48E_c I_e$ for a simple beam under uniformly distributed live load, where M_l = midspan live load moment. For uncracked members, $I_e = I_g$. For partially cracked noncomposite and composite members, see Refs. 8.2 and 8.3. See also Example 8.7 for a partially cracked case.

Term (8) is the live load creep deflection of the beam. This deflection increment may be computed as $(a_{cp})_l = (M_s/M_l)C_u a_l$, where M_s is the sustained portion of the live load moment and $C_u = 1.6$ as in Term (6).

DESIGN EXAMPLES

In the examples that follow, all 4 of the allowable deflection categories in Table 9.5(b) are satisfied. Only h_{min} is checked for the two-way slab in Example 8.6, as required.

REFERENCES

8.1   *Deflections of Concrete Structures,* Special Publication SP 43, American Concrete Institute, Detroit, MI, 1974.

8.2   Branson, D.E., *Deformation of Concrete Structures,* McGraw-Hill Book Co., New York, N.Y., 1977.

8.3   *PCI Design Handbook—Precast and Prestressed Concrete,* 3rd Ed., Prestresed Concrete Institute, Chicago, IL, 1985.

8.4   *Designing for Creep and Shrinkage in Concrete Structures,* Special Publication SP 76, American Concrete Institute, Detroit, MI, 1982.

8.5   *Designing for Effects of Creep, Shrinkage, and Temperature in Concrete Structures,* Special Publication SP 27, American Concrete Institute, Detroit, MI, 1971.

8.6   Nilson, A. H., and Walters, D. B., "Deflection of Two-Way Floor Systems by the Equivalent Frame Method," *ACI Journal, Proceedings* Vol. 72, No. 5, May 1975, pp. 210-218.

8.7   Branson, D. E., "Reinforced Concrete Composite Flexural Members," Chapter 4, pp. 97-174, and "Prestressed Concrete Composite Flexural Members," Chapter 5, pp. 148-210, *Handbook of Composite Construction Engineering,* G.M. Sabanis, Editor, Van Nostrand Reinhold Co., New York, N.Y., 1979.

Example 8.1—Simple-Span Nonprestressed Rectangular Beam

Required: Analysis of short-term deflections, and long-term deflections at ages 3 months and 5 years (ultimate value)

Data: $f_c' = 3000$ psi (normal weight concrete)
$f_y = 40,000$ psi
$A_s = 3\#7 = 1.80$ in.^{2}
$\rho = A_s/bd = 0.0077$
$A_s' = 3\#4 = 0.60$ in.^{2}
$\rho' = A_s'/bd = 0.0026$
(A_s' not required for strength)
Superimposed dead load (not including beam weight) = 120 lb/ft
Live load = 300 lb/ft (50% sustained)
Span = 25 ft

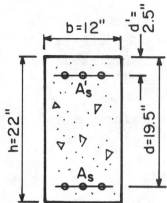

	Code
Calculations and Discussion	**Reference**

1. Minimum thickness, for members not supporting or attached to partitions or other construction likely to be damaged by large deflections:

   $h_{min} = (\ell/16)(0.80$ for $f_y) = (300/16)(0.80)$ Table 9.5(a)

   $= 15.0$ in. $< h = 22$ in.   O.K.

2. Moments:

   $w_d = 0.120 + (12)(22)(0.150)/144 = 0.395$ kip/ft

   $M_d = w_d \ell^2/8 = (0.395)(25)^2/8 = 30.9$ ft-kips

   $M_\ell = w_\ell \ell^2/8 = (0.300)(25)^2/8 = 23.4$ ft-kips

   $M_{d+\ell} = 54.3$ ft-kips

   $M_{sus} = M_d + 0.50M_\ell = 30.9 + (0.50)(23.4) = 42.6$ ft-kips

3. Modulus of rupture, modulus of elasticity, modular ratio:

   $f_r = 7.5\sqrt{f_c'} = 7.5\sqrt{3000} = 411$ psi   Eq. (9-9)

   $E_c = 33\sqrt{w_c^3 f_c'} = 33\sqrt{(150)^3(3000)} = 3.32 \times 10^6$ psi   8.5.1

   $n_s = E_c/E_s = 29/3.32 = 8.7$

Example 8.1—Continued

Calculations and Discussion

Code Reference

4. Gross and cracked section moments of inertia, using Fig. 8.3:

$I_g = b h^3/12 = (12)(22)^3/12 = 10,650 \text{ in.}^4$

$B = b/(n A_s) = 12/(8.7)(1.80) = 0.766 \text{ 1/in.}$

$r = (n - 1)A_s'/(n A_s) = (7.7)(0.60)/(8.7)(1.80) = 0.295$

$kd = [\sqrt{2dB(1 + rd'/d) + (1 + r)^2} - (1 + r)]/B$

$$= \left[\sqrt{(2)(19.5)(0.766)(1 + \left(\frac{0.295 \times 2.5}{19.5}\right) + 1.295^2} - 1.295\right]/0.766 = 5.77 \text{ in.}$$

$I_{cr} = b k^3 d^3/3 + nA_s(d - kd)^2 + (n - 1)A_s'(kd - d')^2$

$= (12)(5.77)^3/3 + (8.7)(1.80)(19.5 - 5.77)^2 + (7.7)(0.60)(5.77 - 2.5)^2$

$= 3770 \text{ in.}^4$

$I_g/I_{cr} = 2.8$

5. Effective moments of inertia, using Eq. (9-7):

$M_{cr} = f_r I_g/y_t = (411)(10,650)/(11)(12,000) = 33.2 \text{ ft-kips}$ Eq. (9-8)

$M_{cr}/M_d = 33.2/30.9 > 1.$ Hence $(I_e)_d = I_g = 10,650 \text{ in.}^4$

$(M_{cr}/M_{sus})^3 = (33.2/42.6)^3 = 0.473$

$(I_e)_{sus} = (M_{cr}/M_a)^3 I_g + [1 - (M_{cr}/M_a)^3] I_{cr} \leq I_g$ Eq. (9-7)

$= (0.473)(10,650) + (1 - 0.473)(3770)$

$= 7025 \text{ in.}^4$

$(M_{cr}/M_{d+l})^3 = (33.2/54.3)^3 = 0.229$

$(I_e)_{d+l} = (0.229)(10,650) + (1 - 0.229)(3770)$

$= 5345 \text{ in.}^4$

Example 8.1—Continued

	Code
Calculations and Discussion	**Reference**

6. Initial or short-time deflections, using Eq. (3):

$$(a_i)_d = \frac{K(5/48)M_d \ell^2}{E_c (I_e)_d} = \frac{(1)(5/48)(30.9)(25)^2(12)^3}{(3320)(10,650)}$$

$$= 0.098 \text{ in.}$$

$$(a_i)_{sus} = \frac{K(5/48) M_{sus} \ell^2}{E_c(I_e)_{sus}} = \frac{(1)(5/48)(42.6)(25)^2(12)^3}{(3320)(7025)}$$

$$= 0.205 \text{ in.}$$

$$(a_i)_{d+\ell} = \frac{K(5/48) M_{d+\ell} \ell^2}{E_c(I_e)_{d+\ell}} = \frac{(1)(5/48)(54.3)(25)^2(12)^3}{(3320)(5345)}$$

$$= 0.344 \text{ in.}$$

$(a_i)_\ell = (a_i)_{d+\ell} - (a_i)_d = 0.344 - 0.098 = 0.246 \text{ in.}$

versus the following allowable deflections from Table 9.5(b):

Flat roofs not supporting and not attached to nonstructural elements likely to be damaged by large deflections—
$(a_i)_\ell \leq \ell/180 = 300/180 = 1.67 \text{ in.}$ O.K.

Floors not supporting and not attached to nonstructural elements likely to be damaged by large deflections—
$(a_i)_\ell \leq \ell/360 = 300/360 = 0.83 \text{ in.}$ O.K.

7. Additional long-term deflections at ages 3 mos. and 5 yrs. (ult. value):

Combined creep and shrinkage deflections, using Eqs. (9-10) and (4):

$$\lambda = \frac{\xi}{1 + 50\rho'} = \frac{2.0(\text{ult.value})}{1 + (50)(0.0026)} = 1.77$$

$a_{(cp+sh)} = \lambda(a_i)_{sus} = (1.77)(0.205) = 0.363 \text{ in.}$

$a_{(cp+sh)} + (a_i)_\ell = 0.363 + 0.246 = 0.61 \text{ in.}$—ult. value

versus 0.43 in. at age 1 year using $\xi = 1.0$ instead of 2.0 in Eq. (5)

Example 8.1—Continued

Calculations and Discussion

Code Reference

Separate creep and shrinkage deflections, using Eqs. (5) and (6):

$$\lambda_{cp} = \frac{0.85 C_u}{1 + 50 \rho'} = \frac{(0.85)(1.60)}{1 + (50)(0.0026)} = 1.20$$

$$a_{cp} = \lambda_{cp} (a_i)_{sus} = (1.20)(0.205) = 0.246 \text{ in.}$$

$$A_{sh} \text{ (from Fig. 7-2)} = 0.455$$

$$\varphi_{sh} = a_{sh} (\varepsilon_{sh})_u/h = (0.455)(400 \times 10^{-6})/22$$

$$= 8.27 \times 10^{-6} \text{ 1/in.}$$

$$a_{sh} = K_{sh} \varphi_{sh} \ell^2 = (1/8)(8.27 \times 10^{-6})(25)^2(12)^2 = 0.093 \text{ in.}$$

$$a_{cp} + a_{sh} + (a_i)_\ell = 0.246 + 0.093 + 0.246 = 0.59 \text{ in.—ult. value}$$

versus 0.44 in. at age 3 months using $C_t = (0.56)(1.60)$
= 0.90 and $\varepsilon_{sh} = (0.60)(400 \times 10^{-6}) = 240 \times 10^{-6}$.

These computed ultimate deflections (long-term deflections due to sustained loads plus immediate deflections due to live loads) of 0.59 in. and 0.61 in. are compared with the allowable deflections in Table 9.5(b) as follows:

Roof or floor construction supporting or attached to nonstructural elements likely to be damaged by large deflections (very stringent limitation)—

$$a_{cp} + a_{sh} + (a_i)_\ell \le 1/480 = 300/480 = 0.63 \text{ in.} \qquad \text{O.K. by both methods}$$

Roof or floor construction supporting or attached to nonstructural elements <u>not likely</u> to be damaged by large deflections—

$$a_{cp} + a_{sh} + (a_i)_\ell \le \ell/240 = 300/240 = 1.25 \text{ in.} \qquad \text{O.K. by both methods}$$

Example 8.2—Continuous Nonprestressed T-Beam

Required: Analysis of short-term and ultimate long-term deflections of end-span of multi-span beam shown below.

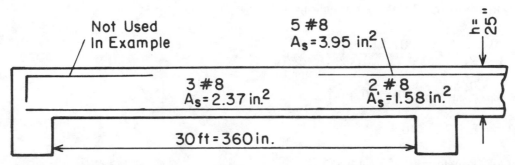

Not Used In Example

5 #8 $A_s = 3.95$ in.^{2}

$h = 25"$

3 #8 $A_s = 2.37$ in.^{2}

2 #8 $A'_s = 1.58$ in.^{2}

30 ft = 360 in.

Beam Spacing = 10', b = 360/4 = 90" or 120" or 16(5) + 12 = 92". Use 90"

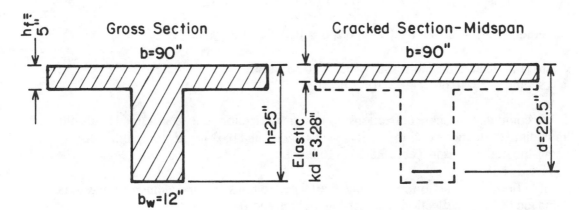

Gross Section

b = 90"

$h_f = 5"$

$h = 25"$

$b_w = 12"$

Cracked Section–Midspan

b = 90"

Elastic kd = 3.28"

d = 22.5"

$n A_s = (11.3)(2.37) = 26.8$ in.^{2}
$\rho = 2.37/(90)(22.5) = 0.00117$
$\rho_w = 2.37/(12)(22.5) = 0.00878$
$\rho' = 0$

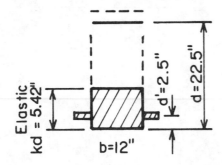

Cracked Section–Interior Support

Elastic kd = 5.42"

$d' = 2.5"$

d = 22.5"

b = 12"

Example 8.2—Continued

Data: $f_c' = 4000$ psi (sand-lightweight concrete)
        $f_y = 50{,}000$ psi
        $w_c = 120$ pcf
        Superimposed Dead Load (not including beam weight) $= 20$ psf
        Live Load $= 100$ psf (30% sustained)

Beam will be assumed to be continuous at one end only for h_{min} in Table 9.5(a), for avg. I_e in Eq. (1), and for K_{sh} in Eq. (6), since the exterior end is supported by a spandrel beam. The end span might be assumed to be continuous at both ends when supported by an exterior column.

(A_s' is not required for strength)

Calculations and Discussion	Code Reference

1. Minimum thickness, for members not supporting or attached to partitions or other construction likely to be damaged by large deflections:

   $h_{min} = (\ell/18.5)(0.90 \text{ for } f_y)(1.05 \text{ for } w_c)$ Table 9.5(a)

         $= (360/18.5)(0.90)(1.05) = 18.4$ in. $<$ h $= 25$ in.   O.K.

2. Loads and moments:

   $w_d = (20 \times 10) + (120)(12 \times 20 + 120 \times 5)/144 = 900$ lb/ft

   $w_\ell = (100 \times 10) = 1000$ lb/ft

   In lieu of a moment analysis, the ACI approximate moment coefficients may be used as follows: Pos. $M = w\ell_n^2/14$ for positive I_e and maximum deflection, an Neg. $M = w\ell_n^2/10$ for negative I_e.   8.3.3

   Pos. $M_d = w_d \ell_n^2/14 = (0.900)(30)^2/14 = 57.9$ ft-kips

   Pos. $M_\ell = (1.000)(30)^2/14 = 64.3$ ft-kips

   Pos. $M_{d+\ell} = 122.2$ ft-kips

   Pos. $M_{sus} = M_d + 0.30\, M_\ell = 57.9 + (0.30)(64.3)$

          $= 77.2$ ft$-$kips

   Neg. $M_d = w_d \ell_n^2/10 = (0.900)(30)^2/10 = 81.0$ ft-kips

Example 8.2—Continued

Calculations and Discussion	**Code Reference**

Neg. $M_l = (1.000)(30)^2/10 = 90.0$ ft-kips

Neg. $M_{d+l} = 171.0$ ft-kips

Neg. $M_{sus} = M_d + 0.30\,M_l = 81.0 + (0.30)(90.0)$

$\qquad = 108.0$ ft-kips

3. Modulus of rupture, modulus of elasticity, modular ratio:

$f_r = (0.85)(7.5)\,\sqrt{f_c'} = 6.38\,\sqrt{4000} = 404$ psi   Eq. (9-9)

$E_c = 33\sqrt{w_c^3\,f_c'} = 33\sqrt{(120)^3(4000)} = 2.74 \times 10^6$ psi   9.5.2.3(b)

$n = E_s/E_c = 29/2.74 = 10.6$ 8.5.1

4. Gross and cracked section moments of inertia, using Fig. 8.3:

Positive moment section

$y_t = h - (1/2)[(b - b_w)h_f^2 + b_w\,h^2]\,/\,[(b - b_w)h_f + b_w\,h]$

$\quad = 25 - (1/2)[(78)(5)^2 + (12)(25)^2]\,/\,[(78)(5) + (12)(25)]$

$\quad = 18.15$ in.

$I_g = (b - b_w)h_f^3/12 + b_w h^3/12 + (b - b_w)h_f\,(h - h_f/2 - y_t)^2 + b_w h\,(y_t - h/2)^2$

$\quad = (78)(5)^3/12 + (12)(25)^3/12 + (78)(5)(25 - 2.5 - 18.15)^2$

$\qquad + (12)(25)(18.15 - 12.5)^2 = 33{,}390$ in.^{4}

$B = b/(n\,A_s) = 90/(10.6)(2.37) = 3.58$ l/in.

$kd = (\sqrt{2\,d\,B + 1} - 1)\,/B = [\sqrt{(2)(22.5)(3.58) + 1} - 1]\,/3.58$

$\quad = 3.28$ in. $< h_f = 5$ in.

Hence, treat as a rectangular compression area.

$I_{cr} = b\,k^3 d^3/3 + n\,A_s(d - kd)^2 = (90)(3.28)^3/3 + (10.6)(2.37)(22.5 - 3.28)2$
$\quad = 10{,}340$ in.^{4} $I_g/I_{cr} = 3.2$

Negative moment section

$I_g = 12 \times 25^3/12 = 15{,}625$ in.^{4}

Example 8.2—Continued

Calculations and Discussion	Code Reference

I_{cr} = 3820 in.[4] (similar to Example 8.1, for b = 12 in., d = 22.5 in., d' = 2.5 in., A_s = 3.95 in.[2], A_s' = 1.58 in.[2]).

5. Effective moments of inertia, using Eqs. (9-7) and (1):

Positive moment section

M_{cr} = f_r I_g/y_t = (404)(33,390)/(18.15)(12,000) = 61.9 ft-kips   Eq. (9-8)

M_{cr}/M_d = 61.9/57.9 > 1. Hence $(I_e)_d$ = I_g = 33,390 in.[4]

$(M_{cr}/M_{sus})^3$ = $(61.9/77.2)^3$ = 0.515

$(I_e)_{sus}$ = $(M_{cr}/M_a)^3$ I_g + $[1 - (M_{cr}/M_a)^3]$ I_{cr} ≤ I_g Eq. (9-7)

       = (0.515)(33,390) + (1 − 0.515)(10,950) = 22,510 in.[4]

$(M_{cr}/M_{d+l})^3$ = $(61.9/122.2)^3$ = 0.130

$(I_e)_{d+l}$ = (0.130)(33,390) + (1 − 0.130)(10,950) = 13,870 in.[4]

Negative moment section

M_{cr} = (404)(15,625)/(12.5)(12,000) = 42.1 ft-kips   Eq. (9-8)

$(M_{cr}/M_d)^3$ = $(42.1/81.0)^3$ = 0.14

$(I_e)_d$ = (0.14)(15,625) + (1 − 0.14)(3820) = 5475 in.[4]   Eq. (9-7)

$(M_{cr}/M_{sus})^3$ = $(42.1/108.0)^3$ = 0.06

$(I_e)_{sus}$ = (0.06)(15,625) + (1 − 0.06)(3820) = 4530 in.[4]   Eq. (9.7)

$(M_{cr}/M_{d+l})^3$ = $(42.1/171.0)^3$ = 0.015

$(I_e)_{d+l}$ = (0.015)(15,625) + (1 − 0.015)(3820) = 4,000 in.[4]   Eq. (9-7)

Average values

Avg. $(I_e)_d$ = $0.85I_m$ + 0.15 ($I_{cont.\ end}$) = (0.85)(33,390) + (0.15)(5475) = 29,205 in.[4]

Example 8.2—Continued

	Code
Calculations and Discussion	**Reference**

Avg. $(I_e)_{sus} = 0.85\ I_m + 0.15\ (I_{cont.\ end}) = (0.85)(22,510) + (0.15)(4530) = 19,815\ \text{in.}^4$

Avg. $(I_e)_{d+l} = (0.85)(13,870) + (0.15)(4000) = 12,390\ \text{in.}^4$

6. Initial or short-time deflections, using Eq. (3), with midspan I_e and with avg. I_e: 9.5.2.4

$K = 1.20 - 0.20\ M_o/M_a = 1.20 - (0.20)(w\ell_n^2/8)/(w\ell_n^2/14) = 0.850$

$(a_i)_d = \dfrac{K\ (5/48)\ M_d\ \ell^2}{E_c\ (I_e)_d} = \dfrac{(0.85)(5/48)(57.9)(30)^2(12)^3}{(2740)(33,390)} = 0.087\text{in.}$

    $= 0.100\ \text{in., using avg. } I_e \text{ Eq. (1) of } 29,205\ \text{in.}^4$

$(a_i)_{sus} = \dfrac{K(5/48)\ M_{sus}\ \ell^2}{E_c(I_e)_{sus}} = \dfrac{(0.85)(5/48)(77.2)(30)^2(12)^3}{(2740)(22,510)} = 0.172\ \text{in.}$

    $= 0.195\ \text{in., using avg. } I_e \text{ Eq. (1) of } 19,815\ \text{in.}^4$

$(a_i)_{d+l} = \dfrac{K\ (5/48)\ M_{d+l}\ \ell^2}{E_c(I_e)_{d+l}} = \dfrac{(0.85)(5/48)(122.2)(30)^2(12)^3}{(2740)(13,870)} = 0.442\ \text{in.}$

    $= 0.495\ \text{in., using avg. } I_e \text{ from Eq. (1) of } 12,390\ \text{in.}^4$

$(a_i)_l = (a_i)_{d+l} - (a_i)_d = 0.442 - 0.087 = 0.355\ \text{in.}$

    $= 0.395\ \text{in., using avg. } I_e \text{ from Eq. (1)}$

versus the following allowable deflections from Table 9.5(b):

Flat roofs not supporting and not attached to nonstructural elements likely to be damaged by large deflections—$(a_i)_l \le \ell/180 = 2.00\ \text{in.}$ O.K.

Floors not supporting and not attached to nonstructural elements likely to be damaged by large deflections—$(a_i)_l \le \ell/360 = 360/360 = 1.00\ \text{in.}$ O.K.

7. Ultimate long-term deflections:

Combined creep and shrinkage deflections, using Eqs. (9-10) and (4):

Example 8.2—Continued

Calculations and Discussion	**Code Reference**

$$\lambda = \frac{\xi}{1 + 50\rho'} = \frac{2.0 \text{ (ult.value)}}{1 + 0} = 2.0$$

9.5.2.5
Eq. (9-10)

$a_{(cp + sh)} = \lambda\,(a_i)_{sus} = (2.0)(0.172) = 0.344$ in.

$a_{(cp + sh)} + (a_i)_\ell = 0.344 + 0.355 = 0.699$ in.

$\qquad\qquad = 0.785$ in., using avg. I_e Eq. (1)

Separate creep and shrinkage deflections, using Eqs. (5) and (6):

$$\lambda_{cp} = \frac{0.85\,C_u}{1 + 50\rho'} = \frac{(0.85)(1.60)}{1 + 0} = 1.36$$

$a_{cp} = \lambda_{cp}\,(a_i)_{sus} = (1.36)(0.172) = 0.234$ in.

$\qquad = 0.265$ in., using avg. I_e Eq. (1)

$$\rho = 100(\rho + \rho_w)/2 = 100\left(\frac{2.37}{90 \times 22.5} + \frac{2.37}{12 \times 22.5}\right)$$

$\qquad = 100(0.00117 + 0.00878)/2 = 0.498\%$

A_{sh} (from Fig. 8-4) $= 0.555$

$\varphi_{sh} = A_{sh}\,(\varepsilon_{sh})_u/h = (0.555)(400 \times 10^{-6})/25 = 8.88 \times 10^{-6}$ 1/in.

$a_{sh} = K_{sh}\,\varphi_{sh}\,\ell^2 = (0.090)(8.88 \times 10^{-6})(30)^2(12)^2 = 0.104$ in.

$a_{cp} + a_{sh} + (a_i)_\ell = 0.234 + 0.104 + 0.355 = 0.693$ in.

$\qquad\qquad = 0.764$ in., using avg. I_e from Eq. (1)

These computed ultimate deflections of 0.69 in. to 0.76 in. are compared with the allowable deflections in Table 9.5(b) as follows:

Roof or floor construction supporting or attached to nonstructural elements likely to be damaged by large deflections (very stringent limitation)—

$a_{cp} + a_{sh} + (a_i)_\ell \leq \ell/480 = 360/480 = 0.75$ in.

Example 8.2—Continued

All results O.K., 0.76 in. being just marginally greater than 0.75 in.

Roof or floor construction supporting or attached to nonstructural elements not likely to be damaged by large deflections—

$a_{cp} + a_{sh} + (a_i)_\ell \leq \ell/240 = 360/240 = 1.50$ in.   All results O.K.

Example 8.3—Unshored Nonprestresed Composite Beam

Required: Analysis of short-term and ultimate long-term deflections.

Data: Normal weight concrete
  Slab $f_c' = 3000$ psi
  Precast beam $f_c' = 4000$ psi
  $f_y = 40{,}000$ psi
  $A_s = 3\#9 = 3.00$ in.^{2}
  Superimposed Dead Load (not including
    beam and slab weight) = 10 psf
  Live Load = 75 psf (20% sustained)
  Simple span = 26 ft = 312 in., spacing = 8 ft = 96 in.
  $b_e = 312/4 = 78.0$ in. or spacing = 96.0 in.
    or $16(4) + 12 = 76.0$ in.

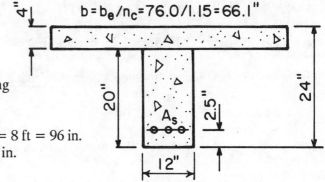

Calculations and Discussion	Code Reference

1. Minimum thickness for members not supporting or attached to partitions or other construction likely to be damaged by large deflections:

   $h_{min} = (\ell/16)(0.80 \text{ for } f_y) = (312/16)(0.80) = 15.6$ in. $<$ $h = 20$ in. or 24 in.   Table 9.5(a)

2. Loads and moments:

   $w_1 = (10 \text{ psf})(8) + (150 \text{ pcf})(96)(4)/144 = 480$ lb/ft

   $w_2 = (150 \text{ pcf})(12)(20)/144 = 250$ lb/ft

   $w_\ell = (75 \text{ psf})(8) = 600$ lb/ft

   $M_1 = w_1 \ell^2/8 = (0.480)(26)^2/8 = 40.6$ ft-kips

   $M_2 = w_2 \ell^2/8 = (0.250)(26)^2/8 = 21.1$ ft-kips

   $M_\ell = w_\ell \ell^2/8 = (0.600)(26)^2/8 = 50.7$ ft-kips

3. Modulus of rupture, modulus of elasticity, modular ratio:

   $(E_c)_1 = 33\sqrt{w_c^3 f_c'} = 33\sqrt{(150)^3(3000)} = 3.32 \times 10^6$ psi   8.5.1

   $(f_r)_2 = 7.5\sqrt{f_c'} = 7.5\sqrt{4000} = 474$ psi   Eq. (9-9)

Example 8.3—Continued

Calculations and Discussion	Code Reference

$(E_c)_2 = 33\sqrt{(150)^3(4000)} = 3.83 \times 10^6$ psi

$\qquad$ 8.5.1

$n_c = (E_c)_2/(E_c)_1 = 3.83/3.32 = 1.15$, $n = E_s/(E_c)_2 = 29/3.83 = 7.56$

4. Gross and cracked section moments of inertia, using Fig. 8.3

Precast Section

$I_g = (12)(20)^3/12 = 8000$ in.^{4} $\quad$ B = $b/(nA_s) = 12/(7.56)(3.00) = 0.529$ 1/in.

$kd = (\sqrt{2dB + 1} - 1)/B = [\sqrt{(2)(17.5)(0.529) + 1} - 1]/0.529 = 6.46$ in.

$I_{cr} = b\,k^3d^3/3 + n\,A_s(d - kd)^2 = (12)(6.46)^3/3 + (7.56)(3.00)(17.5 - 6.46)^2 = 3840$ in.^{4}

Composite Section

$y_t = h - (1/2)[(b - b_w)h_f^2 + b_w h^2]/[(b - b_w)h_f + b_w h]$

$\quad = 24 - (1/2)[(54.1)(4)^2 + (12)(24)^2]/[(54.1)(4) + (12)(24)] = 16.29$ in.

$I_g = (b - b_w)h_f^3/12 + b_w h^3/12 + (b - b_w)h_f(h - h_f/2 - y_t)^2 + b_w h(y_t - h/2)^2$

$\quad = (54.1)(4)^3/12 + (12)(24)^3/12 + (54.1)(4)(24 - 2 - 16.29)^2$

$\quad + (12)(24)(16.29 - 12)^2 = 26{,}470$ in.^{4}

B = $b/(n\,A_s) = 66.1/(7.56)(3.00) = 2.914$

d = $(\sqrt{2\,d\,B + 1} - 1)/B = [\sqrt{(2)(21.5)(2.914) + 1} - 1]/2.914$

$\quad = 3.51$ in. $< h_f = 4$ in. Hence, treat as a rectangular compression area.

$I_{cr} = b\,k^3d^3/3 + n\,A_s(d - kd)^2 = (66.1)(3.51)^3/3 + (7.56)(3.00)(21.5 - 3.51)^2$

$\quad = 8295$ in.^{4}

$I_2/I_c = [(I_2/I_c)_g + (I_2/I_c)_{cr}]/2 = [(8000/26{,}470) + (3840/8295)]/2 = 0.383$

5. Effective moments of inertia, using Eq. (9-7):

For Term (1), Eq. (7)—Precast Section,

Example 8.3—Continued

	Code
Calculations and Discussion	**Reference**

$M_{cr} = f_r I_g/y_t = (474)(8000)/(10)(12,000) = 31.6$ ft-kips $\qquad$ Eq. (9-8)

$M_{cr}/M_2 = 31.6/21.1 > 1$. Hence $(I_e)_2 = I_g = 8000$ in.^{4}

For Term (6), Eq. (7)—Precast Section,

$[M_{cr}/(M_1 + M_2)]^3 = [31.6/(40.6 + 21.1)]^3 = 0.134$

$(I_e)_{1+2} = (M_{cr}/M_a)^3 I_g + [1 - (M_{cr}/M_a)^3] I_{cr} \leq I_g$ $\qquad$ Eq. (9-7)

$\qquad = (0.134)(8000) + (1 - 0.134)(3840) = 4400$ in.^{4}

6. Deflection, using Eqs. (9) and (10):

Term (1)—$(a_i)_2 = \dfrac{K(5/48)\,M_2\,\ell^2}{(E_c)_2(I_e)_2} = \dfrac{(1)(5/48)(21.1)(26)^2(12)^3}{(3830)(8000)} = 0.084$ in.

Term (2)—$k_r = 0.85$ (no compression steel in precast beam).

$\qquad 0.77\,k_r\,(a_i)_2 = (0.77)(0.085)(0.084) = 0.055$ in.

Term (3)—$0.83\,k_r\,(a_i)_2\,\dfrac{I_2}{I_c} = (0.83)(0.85)(0.084)(0.383) = 0.023$ in.

Term (4)—$K_{sh} = 1/8$. Precast Section: $\rho = (100)(3.00)/(12)(17.5) = 1.43\%$

$\qquad$ From Fig. 8-4, $A_{sh} = 0.789$

$\qquad \varphi_{sh} = A_{sh}\,(\varepsilon_{sh})_u/h = (0.789)(400 \times 10^{-6})/20 = 15.78 \times 10^{-6}$ 1/in.

$\qquad a_{sh} = K_{sh}\,\varphi_{sh}\,\ell^2 = (1/8)(15.78 \times 10^{-6})(26)^2(12)^2 = 0.192$ in.

$\qquad 0.36\,a_{sh} = (0.36)(0.192) = 0.069$ in.

Term (5)—$0.64\,a_{sh}\,\dfrac{I_2}{I_c} = (0.64)(0.192)(0.383) = 0.047$ in.

Term (6)—$(a_i)_1 = \dfrac{K(5/48)(M_1 + M_2)\,\ell^2}{(E_c)_2(I_e)_{1+2}} - (a_i)_2$

$\qquad = \dfrac{(1)(5/48)(40.6 + 21.1)(26)^2(12)^3}{(3830)(4400)} - 0.088 = 0.358$ in.

Example 8.3—Continued

Term (7)—$1.22\, k_r\, (a_i)_1\, \dfrac{I_2}{I_c} = (1.22)(0.85)(0.358)(0.383) = 0.142$ in.

Term (8)—$a_{ds} = 0.50(a_i)_1 = (0.50)(0.358) = 0.179$ in.　　(rough estimate)

Term (9)—Using the alternative method,

$$(a_i)_\ell = \frac{K\,(5/48)\,M_\ell \ell^2}{(E_c)_2\,(I_c)_{cr}} = \frac{(1)(5/48)(50.7)(26)^2(12)^3}{(3830)(8295)} = 0.194 \text{ in.}$$

Term (10)—$k_r = 0.85$ (neglecting the effect of any compression steel in slab)

$(a_{cp})_\ell = k_r\, C_u\, [0.20\, (a_i)\,\ell]$

$\qquad = (0.85)(1.60)(0.20 \times 0.194) = 0.053$ in.

In Eq. (7), $a_u = 0.084 + 0.055 + 0.023 + 0.069 + 0.047 + 0.358 + 0.142 + 0.179$
$\qquad + 0.194 + 0.053 = 1.20$ in.

Checking Eq. (8) (same solution),

$$a_u = \left(1.65 + 0.71\,\frac{I_2}{I_c}\right)(a_i)_2 + \left(0.36 + 0.64\,\frac{I_2}{I_c}\right) a_{sh}$$

$$\qquad + \left(1.50 + 1.04\,\frac{I_2}{I_c}\right)(a_i)_1 + (a_i)_\ell + (a_{cp})_\ell$$

$$\qquad = (1.65 + 0.71 \times 0.383)(0.084) + (0.36 + 0.64 \times 0.383)(0.192)$$

$$\qquad + (1.50 + 1.04 \times 0.383)(0.358) + 0.194 + 0.053 = 1.20 \text{ in.}$$

(same as above)

Assuming nonstructural elements are installed after the composite slab has hardened,

$a_{cp} + a_{sh} + (a_i)_\ell$

= Terms (3) + (5) + (7) + (8) + (9) + (10)

= $0.023 + 0.047 + 0.142 + 0.179 + 0.194 + 0.053 = 0.64$ in.

Comparisons with the allowable deflections in Table 9.5(b) are shown at the end of Example 8.4.

Example 8.4—Shored Nonprestressed Composite Beam

Data: Same as in Example 8.3, except that shored construction is used.

Required: Analysis of short-term and ultimate long-term deflections, to show the beneficial effect of shoring in reducing deflections.

Calculations and Discussion	Code Reference

1. Effective moments of inertia, using Eq. (9-7):

$$M_{cr} = f_r I_g/y_t = (474)(26,470)/(16.29)(12,000) = 64.2 \text{ ft-kips}$$ Eq. (9-8)

$$M_{cr}/(M_1 + M_2) = [64.2/(40.6+21.1)] = 1.04 \; > \; 1.$$

Hence $(I_e)_{1+2} = I_g = 26,470 \text{ in.}^4$

In Term (5), Eq. (9)—Composite Section,

$$[M_{cr}/(M_1 + M_2 + M_l)]^3 = [64.2/(40.6 + 21.1 + 50.7)]^3 = 0.186$$

$$(I_e)_{d+l} = (M_{cr}/M_a)^3 I_g + [1 - (M_{cr}/M_a)^3] I_{cr} \le I_g$$ Eq. (9-7)

$$= (0.186)(26,470) + (1 - 0.186)(8295) = 11,675 \text{ in.}^4$$

versus the alternative method of Example 8.3 where $I_e = (I_c)_{cr} = 8295 \text{ in.}^4$ was used with the live load moment directly.

2. Deflections, using Eqs. (9) and (10):

$$\text{Term (1)}—(a_i)_{1+2} = \frac{K \, (5/48) \, (M_1 + M_2) \, l^2}{(E_c)_2(I_e)_{1+2}}$$

$$= \frac{(1)(5/48)(40.6+21.1)(26)^2(12)^3}{(3830)(26,470)} = 0.074 \text{ in.}$$

Term (2)—$k_r = 0.85$ (neglecting the effect of any compression steel in slab).

$$1.80 \, k_r \, (a_i)_{1+2} = (1.80)(0.85)(0.074) = 0.113 \text{ in.}$$

Term (3)—From Term (4) of Example 8.3, and using $(\varepsilon_{sh})_u = 252 \times 10^{-6} \text{ in./in.}$,

$$a_{sh} \frac{I_2}{I_c} = (252/400)(0.192)(0.383) = 0.046 \text{ in.}$$

Example 8.4—Continued

Term (4)—$a_{ds} = (a_i)_{1+2} = 0.074$ in. (rough estimate)

Term (5)—$(a_i)_\ell = \dfrac{K\,(5/48)\,(M_1 + M_2 + M_\ell)\,\ell^2}{(E_c)_2\,(I_e)_{d+\ell}} - (a_i)_{1+2}$

$$= \frac{(1)(5/48)(40.6 + 21.1 + 50.7)(26)^2(12)^3}{(3830)(11{,}675)} - 0.074 = 0.232 \text{ in.}$$

Term (6)—$k_r = 0.85$ (neglecting the effect of any compression steel in slab),

$(a_{cp})_\ell = k_r C_u[0.20(a_i)_\ell] = (0.85)(1.60)(0.20 \times 0.232) = 0.063$ in.

In Eq. (9), $a_u = 0.074 + 0.113 + 0.046 + 0.074 + 0.232 + 0.063 = 0.60$ in.

versus 1.20 in. in Example 8.3 with unshored construction.

This shows the beneficial effect of shoring in reducing the total deflection.

Checking by Eq. (10) (same solution),

$$a_u = 3.53\,(a_i)_{1+2} + a_{sh}\frac{I_2}{I_c} + (a_i)_\ell + (a_{cp})_\ell$$

$$= (3.53)(0.074) + 0.046 + 0.232 + 0.063 = 0.60 \text{ in.}$$
(same as above)

Assuming that nonstructural elements are installed after shores are removed,

$a_{cp} + a_{sh} + (a_i)_\ell = a_u - (a_i)_{1+2} = 0.60 - 0.07 = 0.53$ in.

The computed deflections of $(a_i)_\ell = 0.19$ in. in Example 8.3 and 0.23 in. in Example 8.4; and $a_{cp} + a_{sh} + (a_i)_\ell = 0.64$ in. in Example 8.3 and 0.53 in. in Example 8.4 are compared with the allowable deflections in Table 9.5(b) as follows:

Flat roofs not supporting and not attached to nonstructural elements likely to be damaged by large deflections—

$(a_i)_\ell \leq \ell/180 = 312/180 = 1.73$ in.   O.K.

Floors not supporting and not attached to nonstructural elements likely to be damaged by large deflections—

$(a_i)_\ell \leq \ell/360 = 312/360 = 0.87$ in.   O.K.

Example 8.4—Continued

Calculations and Discussion	Code Reference

Roof or floor construction supporting or attached to nonstructural elements likely to be damaged by large deflections (very stringent limitation)—

$$a_{cp} + a_{sh} + (a_i)_\ell \leq \ell/480 = 312/480 = 0.65 \text{ in.} \qquad \text{O.K.}$$

Roof or floor construction supporting or attached to nonstructural elements not likely to be damaged by large deflections—

$$a_{cp} + a_{sh} + (a_i)_\ell \leq \ell/240 = 312/240 = 1.30 \text{ in.} \qquad \text{O.K.}$$

All computed deflections are found to be satisfactory in all four categories.

Example 8.5—Slab System without Beams (Flat Plate)

Required: Analysis of short-term and ultimate long-term deflections of a corner panel.

Data: Flat plate with no edge beams, designed by direct design method
Slab $f_c' = 3000$ psi, Column $f_c' = 5000$ psi, (Normal Weight Concrete)
$f_y = 40,000$ psi
Square panels—15 ft × 15 ft center-to-center of columns
Square columns—14 in. × 14 in., Clear span, $\ell_n = 15 - 1.17 = 13.83$ ft.
Story height = 10 ft, Slab thickness, h = 6 in.
Column Strip -- Pos. $A_s = 4\#5 = 1.24$ in.2, Neg. $A_s = 6\#5 = 1.86$ in.^{2}
   Pos. d = 6.0 – 0.75 – 0.31 (1#5) = 4.94 in.
   Neg. d = 6.0 – 0.75 – 0.63 (2#5) = 4.62 in.
Middle Strip reinforcement and d values are not required for deflection
   computations, since the slab remains uncracked in the middle strips.
Superimposed Dead Load = 10 psf
Live Load = 50 psf. Check for 0% and 40% Sustained Live Load

Calculations and Discussion	Code Reference

1. Minimum thickness, using Table 8-2:

   From Table 8-2, and using the factor 0.90 for fy = 40 ksi

   Interior panel $h_{min} = (0.90) \ell_n/33 = (0.90) (13.83 \times 12)/33 = (0.90) (5.03) = 4.53$ in.

   Exterior panel $h_{min} = (0.90) \ell_n/30 = (0.90) (13.83 \times 12)/30 = (0.90) (5.53) = 4.98$ in.

   Since the actual slab thickness is 6 in., deflection calculations are not required; however, as an illustration, deflections will be checked for a corner panel, to make sure that all allowable deflections per Table 9.5(b) are satisfied.

   9.5.3.2

2. Comment on trial design with regard to deflections:

   Based on the minimum thickness limitations versus the actual slab thickness, it appears likely that computed deflections will meet most or all of the Code deflection limitations. It turns out that all are met.

3. Modulus of rupture, modulus of elasticity, modular ratio:

   $f_r = 7.5\sqrt{f_c'} = 7.5\sqrt{3000} = 411$ psi

   Eq. (9-9)

   $E_{cs} = 33\sqrt{w_c^3 f_c'} = 33\sqrt{(150)^3(3000)} = 3.32 \times 10^6$ psi

   8.5.1

   $E_{cc} = 33\sqrt{(150)^3(5000)} = 4.29 \times 10^6$ psi, n = E_s/E_{cs} = 29/3.32 = 8.73

Example 8.5—Continued

Calculations and Discussion

Code Reference

4. Service load moments and cracking moment:

$w_d = 10 + (150)(6.0)/12 = 85.0$ psf

$(M_o)_d = w_d \ell_2 \ell_n^2/8 = (85.0)(15)(13.83)^2/8000 = 30.48$ ft-kips

$(M_o)_{d+l} = w_{d+l} \ell_2 \ell_n^2/8 = (85.0 + 50.0)(15)(13.83)^2/8000 = 48.41$ ft-kips

See the following table for the half-column strip and half-middle strip moments:

$(M_{cr})_{c/2} = (M_{cr})_{m/2} = f_r I_g/y_t = (411)(15 \times 12)(6.0)^3/(4)(12)(3.0)(12,000)$

$= 9.25$ ft-kips > All $(M_d)_{c/2}$, $(M_d)_{m/2}$, and $(M_{d+l})_{m/2}$ on Lines 5, 8, and 9

of the following table.

Hence, $I_e = I_g$ for all dead load and all middle strip dead-plus-live load deflections. The $(I_e)_{d+l}$ calculations for the half-column strips are shown in the following table.

$d(I_e)_{d+l} = (0.391)(810) + (1 - 0.391)(208) = 443$ in.[4] ... Eq. (1)

where $I_g = (15/4)(12)(6.0)^3/12 = 810$ in.[4]

Example 8.5—Continued

Calculations and Discussion

Table 8-4—Slab Strip Moments (ft-kips) and Moments of Inertia (in.^{4})^a

		Ext. Frame $1/\alpha_{ec} = 0.898$		Int. Frame $1/\alpha_{ec} = 1.029$	
		Pos.	Int. Neg.	Pos.	Int. Neg.
1.	Moment Ratios^b	0.482	0.697	0.492	0.701
2.	Panel M_d = (Line 1)$(M_O)_d$	14.69	21.24	15.00	21.37
3.	Panel M_{d+l} = (Line 1)$(M_O)_{d+l}$	23.33	33.74	23.82	33.94
4.	(LDF)$_c$—From Table 8-3 with $\alpha_1 = \beta_t = 0$	0.60	0.75	0.60	0.75
5.	$(M_d)_{c/2}$ = (Line 2)(Line 4)/2	4.41	7.97	4.50	8.01
6.	$(M_{d+l})_{c/2}$ = (Line 3) (Line 4)/2	7.00	12.65	7.15	12.73
7.	$(M_l)_{c/2}$ = Line 6 - Line 5	2.59	4.68	2.65	4.72
8.	$(M_d)_{m/2}$ = (Line 2)/2 - Line 5	2.94	2.65	3.00	22.68
9.	$(M_{d+l})_{m/2}$ = (Line 3)/2 - Line 6	4.67	4.22	4.76	4.24
10.	$[(M_{cr})_{c/2}/(\text{Line } 6)]^3$	>1	0.391	>1	0.384
For Half-Column Strips					
11.	No. of #5 Bars	4	6	4	6
12.	A_s = (Line 11)(0.31) in.^{2}	1.24	1.86	1.24	1.86
13.	I_{cr}^c in.^{4}	—	208	—	208
14.	$(I_e)_{d+l}^d$ = (Line 10)I_g + [1 – (Line 10)I_{cr}], in.^{4}	810	443	810	439

^{a}Ext. Neg. Moments and Ext. Neg. I_e are not required in Eq. (1).

^{b}Using the Modified Stiffness Method of Commentary Section 13.6.3.3:

$$+M = 0.63 - \frac{0.28}{1+1/\alpha_{ec}}, \quad -M = 0.75 - \frac{0.10}{1+1/\alpha_{ec}}$$

^{c}Referring to Fig. 8-3, B = b/nA_s = (15)(12)/(4)(8.7)(1.86) = 2.78 1/in.

kd = $(\sqrt{2dB + 1} - 1)/B$ = $(\sqrt{2 \times 4.62 \times 2.78 + 1} - 1)/2.78$ = 1.50 in.

I_{cr} = $bk^3d^3/3 + nA_s(d - kd)^2$ = $\frac{(15)(12)}{4} \frac{(1.50)^3}{3}$ + $(8.7)(1.86)(4.62 - 1.50)^2$

= 208 in.^{4}

$^d(I_e)_{d + l}$ = (0.391)(810) + (1–0.391)(208) = 443 in.^{4} Eq. (1)   where I_g = (15/4)(12)(6.0)^3/12 = 810 in.^{4}

Example 8.5 —Continued

	Code
Calculations and Discussion	**Reference**

5. Flexural stiffness (K_{ec} and α_{ec}) of an exterior equivalent column:

$K_b = 0$ (no beams)

Commentary 13.7.4

$I_s = (I_g)_{frame} = \ell_2\, h^3/12 = (15 \times 12)(6.0)^3/12 = 3240$ in.^{4}

$K_s = 4\, E_{cs}\, I_s/\ell_1 = 4\, E_{cs}\,(3240)/(15)(12) = 72.0\, E_{cs}$

For Ext. Frame, $K_s = 72.0\, E_{cs}/2 = 36.0\, E_{cs}$

$K_c = 4\, E_{cc}\, I_c/\ell_c = 4\, E_{cc}(14)^4/(12)(10)(12) = 106.7\, E_{cc}$

$\Sigma K_c = 2\, K_c = (2)(106.7\, E_{cc}) = 213.4\, E_{cc}$

$C = (1 - 0.63\, x/y)(x^3 y/3) = (1 - 0.63 \times 6.0/14)\,(6.0^3 \times 14/3) = 735.8$ in.^{4}

$$K_t = \frac{\Sigma 9\, E_{cs}\, C}{\ell_2(1 - c_2/\ell_2)^3} = \frac{(2)(9)E_{cs}(735.8)}{(15)(12)\left(1 - \dfrac{14}{15 \times 12}\right)^3} = 93.9\, E_{cs}$$

For Ext. Frame, $K_t = 93.9\, E_{cs}/2 = 47.0\, E_{cs}$, $E_{cc} = (4.29/3.32)E_{cs} = 1.292\, E_{cs}$

$$K_{ec} = \frac{1}{1/\Sigma K_c + 1/K_t} = \frac{E_{cs}}{[1/(213.4)(1.292)]+(1/93.9)} = 70.0\, E_{cs}$$

For Ext. Frame, $K_{ec} = \dfrac{E_{cs}}{[1/(213.4)(1.292)] + (1/47.0)} = 40.1\, E_{cs}$

To use in Eq. (13), avg. $K_{ec} = (70.0 + 40.1)(3.32 \times 10^6)/(2)(12,000) = 15,230$ ft-kips

$\alpha_{ec} = K_{ec}/\Sigma(K_s + K_b) = 70.0\, E_{cs}/72.0\, E_{cs} = 0.972$

$1/\alpha_{ec} = 1.029$

For Ext. Frame, $\alpha_{ec} = 40.1\, E_{cs}/36.0\, E_{cs} = 1.114$

$1/\alpha_{ec} = 0.898$

To compute M_{net} for use in Eq. (13), avg. $\alpha_{ec} = 1.043$

$1/\text{avg. } \alpha_{ec} = 0.959$

6. Summary of effective moments of inertia:

Example 8.5—Continued

	Code
Calculations and Discussion	**Reference**

Middle Strips—$(I_e)_d = (I_e)_{d+l} = I_g = (7.5)(12)(6.0)^3/12 = 1620$ in.^{4}

Column Strips—$(I_e)_d = I_g = 1620$ in.^{4}

  —Using the half-column strip values of $(I_e)_{d+l}$ from Line 14 of the preceding Table,

   avg. $(I_e)_{d+l} = 0.85\, I_m + 0.15\, (I_{cont.end})$ Eq. (1)

   $= (0.85)(810 + 810) + (0.15)(443 + 439) = 1510$ in.^{4}

Equivalent Frame—$(I_e)_d = (I_g)_c + (I_g)_m = (I_g)_{frame} = (2)(1,620) = 3240$ in.^{4}

  $(I_e)_{d+l} = (I_e)_c + (I_g)_m = 1510 + 1620 = 3130$ in.^{4}

7. Deflections, using Eqs. (11) to (17):

Fixed $a_{frame} = w\, l_2\, l^4/384\, E_{cs}\, I_{frame}$ Eq. (11)

$$(\text{Fixed } a_{frame})_{d,d+l} = \frac{(85.0 \text{ or } 135.0)(15)^5(12)^3}{(384)(3.32 \times 10^6)(3240 \text{ or } 3130)} = 0.027 \text{ in.}, 0.044 \text{ in.}$$

Fixed $a_{c,m} = (LDF)_{c,m}(\text{Fixed } a_{frame})(I_{frame}/I_{c,m})$ Eq. (12)

Pos. and Neg. avg. $(LDF)_c = 0.738$, $(LDF)_m = 0.262$ (corner panel) Table 8.3

$(\text{Fixed } a_c)_d = (0.738)(0.027)(2) = 0.040$ in.

$(\text{Fixed } a_c)_{d+l} = (0.738)(0.044)(3130/1510) = 0.067$ in.

$(\text{Fixed } a_c)_l = 0.067 - 0.040 = 0.027$ in.

$(\text{Fixed } a_m)_d = (0.262)(0.027)(2) = 0.014$ in.

$(\text{Fixed } a_m)_{d+l} = (0.262)(0.044)(3130/1620) = 0.022$ in.

$(\text{Fixed } a_m)_l = 0.022 - 0.014 = 0.008$ in.

$(M_{net})_{d+l} = 0.65\,(M_o)_{d+l}/(1 + 1/\text{avg. } \alpha_{ec}) = (0.65)(48.41)/(1 + 0.959) = 16.06$ ft-kips Commentary 13.6.3.3

$(M_{net})_d = (30.48/48.41)(16.06) = 10.11$ ft-kips

Example 8.5—Continued

For both column and middle strips,

End θ_d = $(M_{net})_d$/avg. K_{ec} = 10.11/15,230 = 0.000664 rad   Eq. (13)

End θ_{d+l} = 16.06/15,230 = 0.001055 rad

a_θ = (End θ)(l/8)(I_g/I_e)$_{frame}$ Eq. (15)

$(a_\theta)_d$ = (0.000664)(15)(12)(1)/8 = 0.015 in.

$(a_\theta)_{d+l}$ = (0.001055)(15)(12)(3240/3130)/8 = 0.025 in.

$(a_\theta)_l$ = 0.025 – 0.015 = 0.010 in.

$a_{c,m}$ = Fixed $a_{c,m}$ + $(a_{\theta 1})_{c,m}$ + $(a_{\theta 2})_{c,m}$ Eq. (16)

$(a_c)_d$ = 0.040 + 0.015 + 0 = 0.055 in.

$(a_m)_d$ = 0.014 + 0.015 + 0 = 0.029 in.

$(a_c)_l$ = 0.027 + 0.010 + 0 = 0.037 in.

$(a_m)_l$ = 0.008 + 0.010 + 0 = 0.018 in.

a = a_{cx} + a_{my} = midpanel deflection of corner panel   Eq. (17)

$(a_i)_d$ = 0.055 + 0.029 = 0.084 in.

$(a_i)_l$ = 0.037 + 0.018 = 0.055 in.

Using Eqs. (9-10) with ξ = 2.5 (ult. value),

$a_{(cp+sh)}$ = λ $(a_i)_{sus}$ = $\dfrac{2.5}{1 + (50)(\rho' = 0)}$ $[(a_i)_d + 0.40(a_i)_l]$

   + (2.5)[0.084 + (0.40)(0.055)] = 0.265 in.   (with 40% sustained LL)

   = (2.5)(0.084) = 0.210 in. (with 0% sustained LL)

$a_{(cp + sh)}$ + $(a_i)_l$ = 0.265 + 0.055 = 0.342 in. (with 40% sustained LL)

   = 0.210 + 0.55 = 0.287 in.   (with 0% sustained LL)

Example 8.5—Continued

Calculations and Discussion	Code Reference

These computed deflections are compared with the code allowable deflections in Table 9.5(b) as follows:

Flat roofs not supporting and not attached to nonstructural elements likely to be damaged by large deflections—

$(a_i)\ell \le (\ell_n$ or $\ell)/180 = (13.83$ or $15)(12)/180 = 0.92$ in. or 1.00 in., versus 0.06 in. O.K.

Floors not supporting and not attached to nonstructural elements likely to be damaged by large deflections—

$(a_i)\ell \le (\ell_n$ or $\ell)/360 = 0.46$ in. or 0.50 in., versus 0.06 in. O.K.

Roof or floor construction supporting or attached to nonstructural elements likely to be damaged by large deflections—

$a_{(cp + sh)} + (a_i)\ell \le (\ell_n$ or $\ell)/480 = 0.35$ in. or 0.38 in., versus 0.29 in. and 0.34 in. O.K.

Roof or floor construction supporting or attached to nonstructural elements not likely to be damaged by large deflections—

$a_{(cp + sh)} + (a_i)\ell \le (\ell_n$ or $\ell)/240 = 0.69$ in. or 0.75 in., versus 0.29 in. and 0.34 in.   O.K.

All computed deflections are found to be satisfactory in all four categories.

Example 8.6—Two-Way Beam-Supported Slab System

Required: Minimum thickness for deflection control

Data:

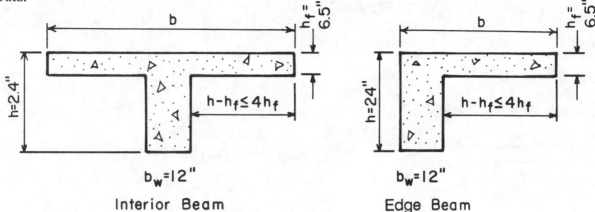

Interior Beam   Edge Beam

f_y = 60,000 psi, Slab thickness h_f = 6.5 in.
Square panels— 22 ft x 22 ft center-to-center of columns
All beams—b_w = 12 in. and h = 24 in. ℓ_n = 22 – 1 = 21 ft
It is noted that f_c' and the loading are not required in this analysis.

Calculations and Discussion	**Code Reference**

1. Effective width b and section properties, using Fig.8-3:

Interior Beam

$I_s = (22)(12)(6.5)^3/12 = 6040$ in.^{4}

$h - h_f = 24 - 6.5 = 17.5$ in. $\leq 4\,h_f = (4)(6.5) = 26$ in. O.K.

Hence, b = 12 + (2)(17.5) = 47 in.

$y_t = h - (1/2)[(b - b_w)\,h_f^2 + b_w\,h^2]/[(b - b_w)h_f + b_w\,h]$

$\quad = 24 - (1/2)[(35)(6.5)^2 + (12)(24)^2]/[(35)(6.5) + (12)(24)]$

$\quad = 15.86$ in.

$I_b = (b - b_w)\,h_f^3/12 + b_w\,h^3/12 + (b - b_w)h_f(h - h_f/2 - y_t)^2 + b_w\,h(y_t - h/2)^2$

$\quad = (35)(6.5)^3/12 + (12)(24)^3/12 + (35)(6.5)(24 - 3.25 - 15.86)^2$

Example 8.6—Continued

	Code
Calculations and Discussion	**Reference**

$$+ (12)(24)(15.86 - 12)^2 = 24{,}360 \text{ in.}^4$$

$\alpha = E_{cb} I_b/E_{cs} I_s = I_b/I_s = 24{,}360/6040 = 4.03$

Edge Beam

$I_s = (11)(12)(6.5)^3/12 = 3020 \text{ in.}^4$

$b = 12 + (24 - 6.5) = 29.5 \text{ in.}$

$y_t = 24 - (1/2)[(17.5)(6.5)^2 + (12)(24)^2]/[(17.5)(6.5) + (12)(24)] = 14.48 \text{ in.}$

$I_b = (17.5)(6.5)^3/12 + (12)(24)^3/12 + (17.5)(6.5)(24 - 3.25 - 14.48)^2$

$$+ (12)(24)(14.48 - 12)^2 = 20{,}470 \text{ in.}^4$$

$\alpha = I_b/I_s = 20{,}470/3020 = 6.78$

α_m and β values

α_m (average value of α for all beams on the edges of a panel):

Interior panel—$\alpha_m = 4.03$

Side panel—$\alpha_m = [(3)(4.03) + 6.78]/4 = 4.72$

Corner panel—$\alpha_m = [(2)(4.03) + (2)(6.78)]/4 = 5.41$

For square panels, β = ratio of clear spans in the two directions = 1

2. Minimum thickness:

9.5.3.3

Interior panel

$$h_{min} = \frac{\ell_n(0.8 + f_y/200{,}000)}{36 + 5\beta\,[\alpha_m - 0.12(1 + 1/\beta)]}$$

Eq. (9-11)

$$= \frac{(21 \times 12)(0.8 + 60{,}000/200{,}000)}{36 + 5(1)[4.03 - 0.12(1 + 1)]} = 5.04 \text{ in.}$$

Also $h_{min} = 4.75$ in. (side panel), and $h_{min} = 4.48$ in. (corner panel) but not less than

8-46

Example 8.6—Continued

Calculations and Discussion	**Code Reference**

$$h_{min} = \frac{\ell_n(0.8 + f_y/200{,}000)}{36 + 9(\beta)}$$ Eq. (9-12)

$$= \frac{(21 \times 12)(0.8 + 60{,}000/200{,}000)}{36 + 9(1)} = 6.16 \text{ in.}$$

(all panels)

and need not be more than

$$h_{min} = \frac{\ell_n(0.8 + f_y/200{,}000)}{36}$$ Eq. (9-13)

$$= \frac{(21 \times 12)(0.8 + 60{,}000/200{,}000)}{36} = 7.70 \text{ in.}$$

(all panels)

Hence, the slab thickness of 6.5 in. > 6.16 in. is satisfactory for all panels, and deflections need not be checked.

From Fig. 8-2, since all α_m > 2.0, Eq. (9-12) controls for minimum slab thickness. For Grade 60 reinforcement with $\beta = 1.0$, $h_{min} = \ell_n/40.9 = (21 \times 12)/40.9 = 6.16$ in.

Example 8.7—Simple-Span Prestressed Single T-Beam

Required: Analysis of short-term and ultimate long-term camber and deflection.

Data: 8ST36 (Design details from PCI Handbook)
Span = 80 ft, beam is partially cracked
f'_{ci} = 3500 psi, f'_c = 5000 psi (normal weight concrete)
f_{pu} = 270,000 psi
14–½ in. dia. depressed (1 Pt.) strands
4 –½ in. dia. nonprestressed strands
(Assume same centroid when computing I_{cr})
P_i = (0.7)(14)(0.153)(270) = 404.8 kips
P_o = (0.90)(404.8) = 364 kips
P_e = (0.78)(404.8) = 316 kips
e_e = 11.15 in., e_c = 22.51 in.
y_t = 26.01 in., A_g = 570 in.^{2}, I_g = 68,920 in.^{4}
Self weight w_o = 594 lb/ft
Superimposed DL w_s = (8)(10 psf) = 80 lb/ft is applied at age 2 mos
   (β_s = 0.76 in Term 6 of Eq. (19))
Live load w_ℓ = (8)(51 psf) = 408 lb/ft
Capacity is governed by flexural strength

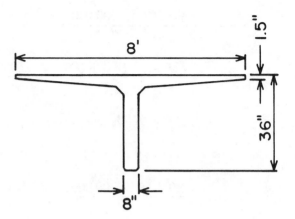

	Code
Calculations and Discussion	**Reference**

1. Span-depth ratios (using PCI Handbook):
   Typical span-depth ratios for single T beams are 25 to 35 for floors and 35 to 40 for roofs, versus (80)(12)/36 = 27, which indicates a relatively deep beam. It turns out that all allowable deflections in Table 9.5(b) are satisfied.

2. Moments for computing deflections:

   $M_o = w_o \ell^2/8$ = (0.594)(80)^2/8 = 475 ft-kips

       (× 0.96 = 456 ft-kips at 0.4ℓ for computing stresses

       and I_e—tendons depressed at one point)

   $M_s = w_s \ell^2/8$ = (0.080)(80)^2/8 = 64 ft-kips (61 ft-kips at 0.4ℓ)

   $M_\ell = w_\ell \ell^2/8$ = (0.408)(80)^2/8 = 326 ft-kips (313 ft-kips at 0.4ℓ)

3. Modulus of rupture, modulus of elasticity:

   $f_r = 7.5\sqrt{f'_c} = 7.5\sqrt{5000}$ = 530 psi   Eq. (9-9)

   $E_{ci} = 33\sqrt{w^3 f'_c} = 33\sqrt{(150)^3(3500)} = 3.59 \times 10^6$ psi   8.5.1

Example 8.7—Continued

	Code
Calculations and Discussion	**Reference**

$E_c = 33\sqrt{(150)^3(5000)} = 4.29 \times 10^6$ psi

$n = E_p/E_c$

$\quad = 27/4.29 = 6.3$

4. Camber and deflection, using Eq. (19):

Term (1)—$a_{po} = \dfrac{P_o(e_c - e_e)\,\ell^2}{12\,E_{ci}\,I_g} + \dfrac{P_o\,e_e\,\ell^2}{8\,E_{ci}\,I_g}$

$\quad = \dfrac{(364)(22.51 - 11.15)(80)^2(12)^2}{(12)(3590)(68,920)} + \dfrac{(364)(11.15)(80)^2(12)^2}{(8)(3590)(68,920)}$

$\quad = 3.17$ in.

Term (2)—$a_o = \dfrac{5\,M_o\,\ell^2}{48\,E_{ci}\,I_g} = \dfrac{(5)(475)(80)^2(12)^3}{(48)(3590)(68,590)} = 2.21$ in.

Term (3)—$k_r = 1/[1 + (A_s/A_{ps})] = 1/[1 + (4/14)] = 0.78$

$\quad \left[\dfrac{\Delta P_u}{P_o} + (k_r C_u)\left(1 - \dfrac{P_u}{2P_o}\right)\right] a_{po}$

$\quad = [-0.18 + (0.78 \times 2.0)(1 - 0.09)](3.17) = 4.32$ in.

Term (4)—$(k_r\,C_u)\,a_o = (0.78)(2.0)(2.21) = 3.45$ in.

Term (5)—$a_s = \dfrac{5\,M_s\,\ell^2}{48\,E_c\,I_g} = \dfrac{(5)(64)(80)^2(12)^3}{(48)(4290)(68,920)} = 0.25$ in.

Term (6)—$(\beta_s\,k_r\,C_u)\,a_s = (0.76)(0.78)(1.6)(0.25) = 0.24$ in.

Term (7)—Determination of I_e at $0.4\,\ell$ for tendons depressed at one point

Method 1, Ref. 8.2, for $(M_\ell)_{cr}/M_\ell$:

$\quad (M_\ell)_{cr} = \dfrac{P_e\,I_g}{A_g\,y_t} + P_e\,e - M_{o+s} + \dfrac{f_r\,I_g}{y_t} = \dfrac{(316)(68,920)}{(570)(26.01)(12)}$

$\quad\quad + \dfrac{(316)(20.24)}{12} - 517 + \dfrac{(0.530)(68,920)}{(26.01)(12)} = 255$ ft−kips

Example 8.7—Continued

Code
Reference

Calculations and Discussion

$(M_l)_{cr}/M_l = 255/313 = 0.815$ at $0.4l$

Method 2, Ref. 8.2, for $(M_l)_{cr}/M_l$:

$$f_{pe} = \frac{P_e}{A_g} + \frac{P_e\,e}{S} = \frac{316}{570} + \frac{(316)(20.25)}{2650} = 2.968 \text{ ksi}$$

where $S = I_g/y_t = 68{,}920/26.01 = 2650 \text{ in.}^3$

$$f_{o+s} = \frac{M_{o+s}}{S} = \frac{(517)(12)}{2650} = 2.341 \text{ ksi}$$

$$f_{to} = -f_{pe} + \frac{M_{to}}{S} = -2.968 + \frac{(456 + 61 + 313)(12)}{2650} = 0.790 \text{ ksi}$$

versus $f_r = 0.530$ ksi and $12\sqrt{5000} = 0.849$ ksi.

Hence use I_e.

$$\frac{(M_l)_{cr}}{M_l} = \frac{f_{pe} - f_{o+s} + f_r}{f_{pe} - f_{o+s} + f_{to}} = \frac{2.968 - 2.341 + 0.530}{2.968 - 2.341 + 0.790} = 0.817 \text{ at } 0.4l$$

PCI Handbook Method for $(M_l)_{cr}/M_l$:

$f_{tl} = f_{to}$ (above) $= 0.790$ ksi

$$f_l = \frac{M_l}{S} = \frac{(313)(12)}{2650} = 1.417 \text{ ksi}$$

$$\frac{(M_l)_{cr}}{M_l} = 1 - \frac{f_{tl} - f_r}{f_l} = 1 - \frac{790 - 530}{1417} = 0.817 \text{ at } 0.4l$$

All 3 methods yield the same results.

Neglecting the taper (compression flange area = 96 in. by 1.5 in.),

$(96)(1.5)(x - 0.75) = (6.3)(18 \times 0.153)(30.23 - x)$

or, $x = 3.92$ in.

$$I_{cr} = \frac{(96)(1.5)^3}{12} + (96)(1.5)(3.92 - 0.75)^2 + (6.3)(18 \times 0.153)(30.23 - 3.92)2$$
$$= 13{,}480 \text{ in.}^4 \text{ at } 0.4l$$

Example 8.7—Continued

PCI Handbook Method for I_{cr}:

$$I_{cr} = nA_{st}d^2(1 - \sqrt{\rho}) = (6.3)(18 \times 0.153)(30.23)^2(1 - \sqrt{0.000949})$$

$$= 15{,}370 \text{ in.}^4 \text{ at } 0.4\ell \text{ (use } I_{cr} = 15{,}370 \text{ in.}^4)$$

where $\rho = (18)(0.153)/(96)(30.23) = 0.000949$

$0.817^3 = 0.545$

$$(I_e)_\ell = (M_{cr}/M_\ell)^3 I_g + [1 - (M_{cr}/M_\ell)^3] I_{cr}$$

Eq. (9-7)

$$= (0.545)(68{,}920) + (1 - 0.545)(15{,}370) = 44{,}550 \text{ in.}^4 \text{ at } 0.4\ell$$

It is noted that for other prestress profiles (straight tendons, 2 point depressed, draped, etc.) the stresses and I_e are computed at midspan instead of 0.4ℓ. Check using PCI Handbook Chart, Fig. 4.10.15, p.4-75 (3rd Edition)

$f_e = f_{t\ell} - f_r = 790 - 530 = 260 \text{ psi}$

$f_e/f_\ell = 260/1417 = 0.18$, $I_{cr}/I_g = 15{,}370/68{,}920 = 0.22$

From chart, $I_e/I_g = 0.66$, $I_e = (0.66)(68{,}920) = 45{,}490 \text{ in.}^4$, comparable to the previously computed value of 44,550 in.^{4}

Live load deflection at midspan,

$$a_\ell = \frac{5 \, M_\ell \ell^2}{48 \, E_c \, (I_e)_\ell} = \frac{(5)(362)(80)^2(12)^3}{(48)(4290)(44{,}550)} = 2.18 \text{ in.} \downarrow$$

Combined results and comparisons with Code Limitations

(1)　　(2)　　(3)　　(4)　　(5)　　(6)　　(7)

$a_u = -3.17 + 2.21 - 4.32 + 3.45 + 0.25 + 0.24 + 2.18 = 0.84 \text{ in.} \downarrow$　Eq. (19)

Initial Camber $= a_{po} - a_o = 3.17 - 2.21 = 0.96 \text{ in.} \uparrow$ versus

1.6 in. at erection in PCI Handbook

Calculations and Discussion

Residual Camber = $a_\ell - a_u$ = 2.18 – 0.84 = 1.34 in. ↑ versus 1.1 in.

Time-Dependent plus Superimposed Dead Load and Live Load Deflection

> = – 4.32 + 3.45 + 0.25 + 0.24 + 2.18 = 1.80 in. or

> = $a_u - (a_o - a_{po})$ = 0.84 – (– 0.96) = 1.80 in. ↓

These computed deflections are compared with the allowable deflections in Table 9.5(b) as follows:

$\ell/180$ = (80)(12)/180 = 5.33 in. versus a_ℓ = 2.18 in.   O.K.

$\ell/360$ = (80)(12)/360 = 2.67 in. versus a_ℓ = 2.18 in.   O.K.

$\ell/480$ = (80)(12)/480 = 2.00 in. versus Time-Dep. etc. = 1.80 in.   O.K.

Moment Redistribution

8.4 REDISTRIBUTION OF NEGATIVE MOMENTS IN CONTINUOUS NONPRESTRESSED FLEXURAL MEMBERS

Section 8.4 permits a redistribution of negative moments in continuous flexural members if reinforcement percentages do not exceed a specified amount. This provision recognizes the inelastic behavior of concrete structures and constitutes a move toward "limit design."

A maximum 10 percent adjustment of negative moments was first permitted in the 1963 ACI Code. Experience with the use of that provision was satisfactory, and showed the provision to be conservative. The 1971 Code increased the maximum adjustment percentage to that shown in Fig. 9-1. The increase was justified by additional knowledge of ultimate and service load behavior obtained from tests and analytical studies. The 1989 Code retains the same adjustment percentage criteria.

Application of Section 8.4 will permit, in many cases, substantial reduction in total reinforcement required without reducing safety, and allow relief of reinforcement congestion in negative moment regions.

According to Section 8.9, continuous members must be designed to resist more than one configuration of live loads. An elastic analysis is performed for each loading configuration, and an envelope moment value is obtained for the design of each section. Thus, for any of the loading conditions considered, certain sections in a given span will reach the ultimate moment while others will have reserve capacity. Tests have shown that a structure can continue to carry additional loads if the sections that already have reached their moment capacities can continue to rotate as "plastic hinges." This allows the moments at other sections to increase (moment redistribution) until a collapse mechanism forms.

Recognition of this additional load capacity beyond the intended original design suggests the possibility of redesign with resulting savings in material. Section 8.4 allows a redesign by decreasing or increasing the elastic negative moments for each loading condition (with the corresponding changes in positive moment required by statics). These moment changes may be such as to reduce both the maximum positive and negative moments in the final moment envelope. Also, to insure proper rotation capacity, the percentage of steel in the sections that may be required to act as plastic hinges must conform to Fig. 9-1 after redesign. Example 9.1 illustrates this requirement.

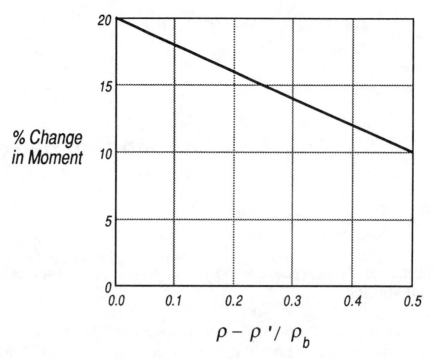

$$\rho - \rho \,'/\, \rho_b$$

Fig. 9-1   Permissible Moment Redistribution

In certain cases, the primary benefit to be derived from Section 8.4 will be simply a reduction of negative moment at the supports, to avoid reinforcement congestion or reduce concrete dimensions. In this case, the steel percentage must still conform to Fig. 9-1. Example 9.2 illustrates this situation.

Limits of applicability of Section 8.4 may be summarized as follows:

(1)   Provisions apply to continuous flexural nonprestressed members. Moment redistribution for prestressed members is addressed in Section 18.10.

(2)   Provisions do not apply to members designed by the Alternate Design Method of Appendix A or to slab systems designed by the Direct Design Method, (Section 13.6.1.7).

(3)   Bending moments must be determined by analytical methods, such as moment distribution, slope deflection, etc. Approximate methods cannot be used to determine the original bending moments.

(4)   Reinforcement ratio at a cross section where moment is to be adjusted must not exceed one-half of the balanced steel ratio, ρ_b, as defined by Eq. (8-1), and as given in Table 6-1.

(5)   Maximum allowable percentage increase or decrease of negative moment is given by the expression, $20\left(1 - \dfrac{\rho - \rho'}{\rho_b}\right)$.

(6)   Adjustment of negative moments is made for each loading configuration considered. Members are then proportioned for the maximum adjusted moments resulting from all loading conditions.

(7) Adjusted negative support moments for any span require adjustment of positive moments in the same span. A decrease of a negative support moment requires a corresponding increase in the positive moment for the same span.

(8) Static equilibrium must be maintained at all joints before and after moment redistribution.

(9) In the case of unequal negative moments on the two sides of a fixed support (i.e., where adjacent spans are unequal), the difference between these two moments is taken into the support. Should either or both of these negative moments be adjusted, the resulting difference between the adjusted moments is taken into the support.

(10) Moment redistribution may be carried out for as many cycles as deemed practical, provided that, after each cycle of redistribution, a new allowable percentage increase or decrease in negative moment is calculated, based on the final steel ratios provided for the adjusted support moments from the previous cycle.

(11) After the design is completed and the reinforcement selected, the actual steel ratios provided must comply with Fig. 9-1 for the percent moment redistribution taken, to ensure that the requirements of Section 8.4 are met.

Example 9.1— Moment Redistribution

Determine required reinforcement for the one-way joist floor shown, using moment redistribution to reduce total reinforcement required.

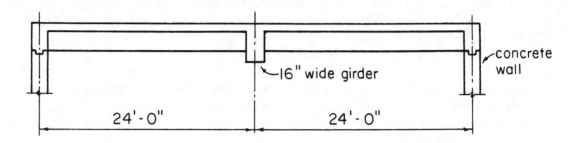

5-in. wide joist @ 25 in. o.c. (10-in. deep forms + 2½-in. slab)

$f_c' = 4000$ psi

$f_y = 60,000$ psi

D = 80 psf

L = 100 psf

For simplicity, continuity at concrete walls is not considered.

Calculations and Discussion	Code Reference

1. Determine factored loads and balanced reinforcement ratio ρ_b

   $\rho_b = 0.0285$ (from Table 6-1) 8.4

   $w_d = 1.4 \times 0.08 \times 25/12 = 0.234$ kip/ft

   $w_\ell = 1.7 \times 0.10 \times 25/12 = \underline{0.354}$ kip/ft 9.2

   Total $w_u = 0.588$ kip/ft per joist

2. Obtain moment diagrams by elastic analysis
   (moments shown in ft–kips)

Example 9.1—Continued

Calculations and Discussion	Code Reference

2. Obtain moment diagrams by elastic analysis (moments shown in ft. kips)

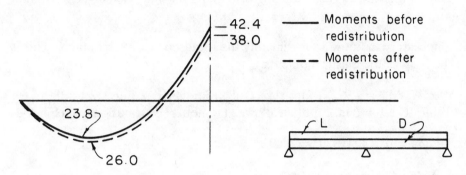

Diag. 1 LOAD PATTERN I

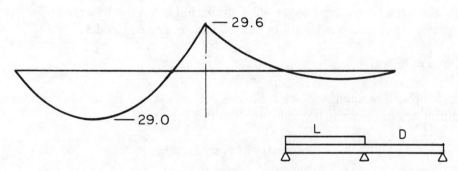

Diag. 2 LOAD PATTERNS II & III (Reverse of II)

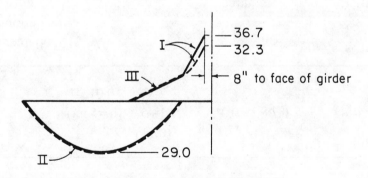

Diag. 3 FACTORED MOMENT ENVELOPE

Example 9.1—Continued

3. Redistribution of negative moments.

Intent is to decrease the negative moment in load pattern I (Diag. 1) to obtain a new moment envelope with smaller maximum negative moments and unchanged maximum positive moments.

To begin, it is necessary to know approximately the required steel percentage. This is obtained on the basis of the elastic moments.

From load pattern I: $M_u = -36.7$ ft-kips at face of girder which requires: $\rho_b = 0.014$ for $d = 11.5$ in. (Table 10-2, b = 5 in.). Obtain allowed moment redistribution from Fig. 9-1.

Neglecting ρ' : $\rho/\rho_b = 0.0140/0.0285 = 0.492$

Percent permissible adjustment = 10.2%

By decreasing the negative moment $M_u = -42.4$ ft-kips in Diag. 1 by 10%, redistributed moment diagrams are obtained as shown by dashed lines in Diags. 1 and 3.

4. Design factored moments.

From the redistributed moment envelope, factored moments and required reinforcement are determined as shown in the following Table.

Summary of Final Design

Section	Load Pattern		Required		Provided		Redistribution, percent	Permitted Max. ρ (Fig. 9-1)
	I	II	A_s, in.^{2}	ρ	A_s, in.^{2}	ρ		
Support Moment* ft-kips	−32.3	—	0.70	0.0122	2#4 0.0124 1#5 (b=5 in.) (0.71)		10.2	0.014
Midspan Moment ft-kips	—	29.0	0.57	0.0021	1#4 0.0023** 1#6 (b=25 in.) (0.64)		—	—

* calculated at face of support

** check $\rho_{min} = \dfrac{200}{60,000} = 0.0033 < \dfrac{A_s}{b_w d} = \dfrac{0.64}{5 \times 11.5} = 0.0111$ provided   10.5.1

Example 9.1—Continued

Calculations and Discussion

Since the provided steel ratios ρ are smaller than the maximum permitted by Section 8.4, the design is O.K.

Final Note:
Moment redistribution has permitted a saving of 10.2% in the negative reinforcement. Since the 29 ft-kips positive factored moment remained unchanged after redistribution, a **total steel saving** was obtained without reduction in safety.

Example 9.2—Moment Redistribution

Determine required reinforcement for the spandrel beam at an intermediate floor level as shown, using moment redistribution to reduce total reinforcement required.

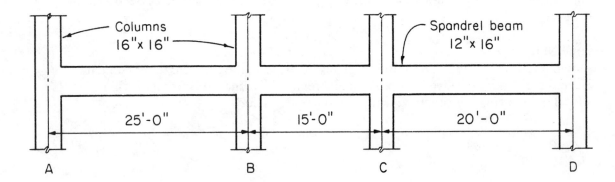

All columns = 16 in. x 16 in.
Story height = 10 ft
Spandrel beam = 12 in. wide x 16 in. deep
$f_c' = 4000$ psi
$f_y = 60,000$ psi
D on beam = 1000 lb/ft
L on beam = 424 lb/ft

Calculations and Discussion	Code Reference
1. Determine factored loads	9.2
$\quad$ U $\;= 1.4D + 1.7L$	Eq. (9-1)
$\quad$ $w_d = 1.4 \times 1 \quad = 1.4$ kip/ft	
$\quad$ $w_l = 1.7 \times 0.424 = \underline{0.72}$ kip/ft	
$\qquad$ Total $w_u = 2.12$ kip/ft	
2. Determine bending moment diagrams for the five load patterns shown in Diags. 1 to 5.	8.9.2

(Maximum negative moments at supports and positive mid-span moments were determined by computer analysis for each of the five loading configurations. Respective bending moment diagrams were then determined graphically).

Example 9.2—Continued

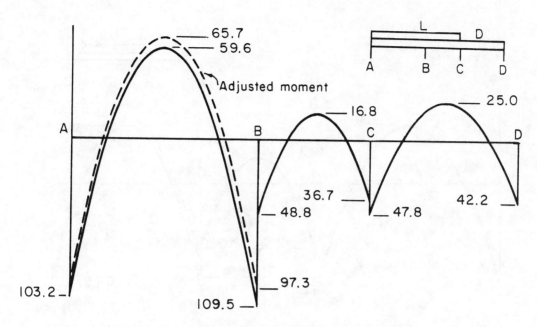

Diag. 1 LOAD PATTERN I (moments in ft. kips)

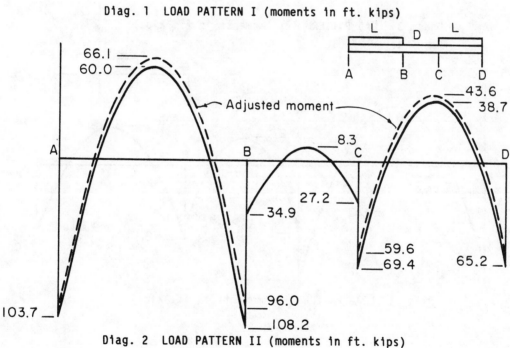

Diag. 2 LOAD PATTERN II (moments in ft. kips)

Example 9.2—Continued

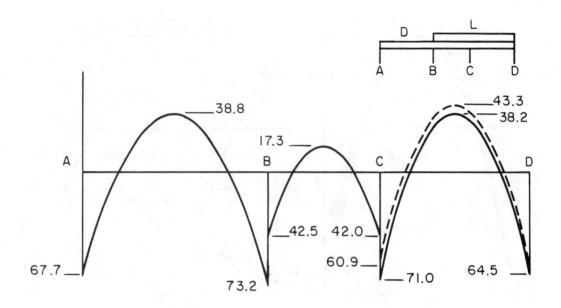

Diag. 3   LOAD PATTERN III (moments in ft. kips)

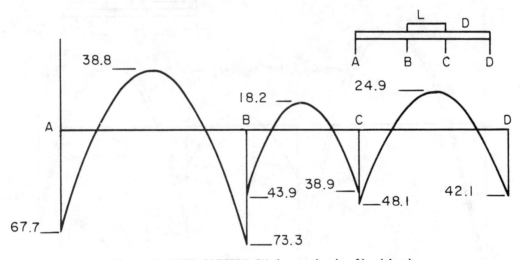

Diag. 4   LOAD PATTERN IV (moments in ft. kips)

Example 9.2—Continued

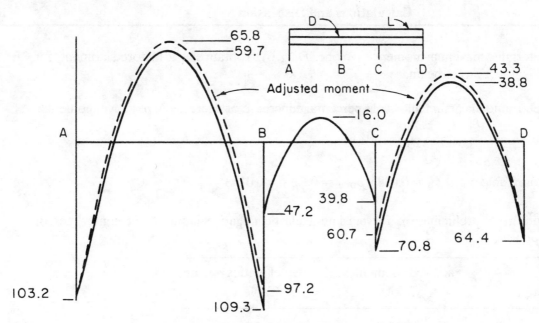

Diag. 5 LOAD PATTERN V (moments in ft. kips)

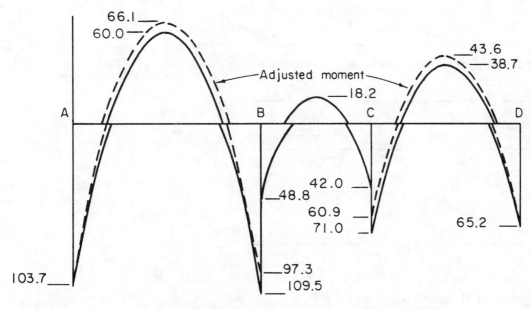

Diag. 6 MAXIMUM MOMENT ENVELOPES FOR PATTERN LOADING
(moments in ft. kips)

Example 9.2—Continued

	Calculations and Discussion	Code Reference

3. Determine maximum moment envelope (Diag. 6) to obtain elastic factored moments for critical sections of beam.

4. Determine maximum allowable percentage increase or decrease in negative moments:

   $d = 13.5$ in.; cover $= 2$ in.   7.7.1

   From Table 6-1: $\rho_b = 0.0285, \rho_{max} = 0.5\,\rho_b = 0.0143$ 8.4.3

   Determine preliminary ρ for maximum elastic negative moments at centerlines of supports:

Support	Factored Moment M_u ft-kips	Steel Ratios ρ from Table 10-2
A	−103.7	0.0118
B	−109.5	0.0125
C	− 71.0	0.0078
D	− 65.2	0.0071

   Obtain percent allowable increase or decrease in negative moments from Fig. 9-1:

Support	$\dfrac{\rho - \rho'}{\rho_b}$ *	Percent Allowable Adjustment
A	0.414	11.7
B	0.439	11.2
C	0.274	14.5
D	0.249	15.0

   *$\rho' = 0$

5. Adjustment of moments.

   Note: Adjustment of negative moments, either increase or decrease, is a decision to be made by the engineer. In this example, it was decided to reduce the negative moments at supports B and C and accept the increase in the corresponding positive moments, and not to adjust the negative moments at the exterior supports, A and D.

   Referring to Diag. 1 through Diag. 5, the following adjustment in moments is made.

Example 9.2—Continued

	Code
Calculations and Discussion	**Reference**

Load Pattern I—Diag. 1

Adjust $M_B = 109.5$ ft-kips   (adjustment = 11.2%)

Reduction to $M_B = -109.5 \times 0.1120 = -12.2$ ft-kips

Adjusted $M_B = -109.5 - (-12.2) = -97.3$ ft-kips

Increase in positive moment in span A-B

$M_A = -103.2$ ft-kips

Adjusted $M_B = -97.3$ ft-kips

Mid-span ordinate on line M_A to $M_B = \dfrac{-103.2 + (-97.3)}{2} = -100.3$ ft-kips

Moment due to uniform load $= w_u \ell^2/8 = 2.12 \times 25^2/8 = 166.0$ ft-kips

Adjusted positive moment at mid-span $= 166.0 + (-100.3) = +65.7$ ft-kips

The adjusted moment is approximately equal to the maximum positive moment. Similar calculations are made to determine the adjusted support and mid-span moments for other loading cases if deemed beneficial.

Results of the additional calculations made for this design indicating the adjusted moments for the various load cases are shown in the following table.

Maximum Moments for one Cycle of Redistribution (moments in ft-kips)

	Load Pattern I		Load Pattern II		Load Pattern III		Load Pattern V	
Location	M_u	M_{adj}	M_u	M_{adj}	M_u	M_{adj}	M_u	M_{adj}
A	−103.2	—	−103.7	—	−67.7	—	−103.2	—
Mid-span A-B	+59.6	+65.7	+60.0	+66.1	+38.8	—	+59.7	+65.8
B	−109.5	−97.3	−108.2	−96.0	−73.2	—	−109.3	−97.2
Mid-span B-C	+16.8	—	+8.3	—	+17.3	—	+16.0	—
C	−47.8	—	−69.4	−59.6	−71.0	−60.9	−70.8	−60.7
Mid-span C-D	+25.0	—	+38.7	+43.6	+38.2	+43.3	+38.8	+43.3
D	−42.2	—	−65.2	—	−64.5	—	−64.4	—

No adjustment is made to the elastic support moments obtained from load pattern IV. The negative support moments are less than the maximum adjusted moments from the other loading patterns and, therefore, would not control the maximum adjusted moment envelope.

Example 9.2—Continued

	Code
Calculations and Discussion	**Reference**

6. After the adjusted moments have been determined analytically, the adjusted bending moment diagrams for each loading pattern can be determined. For the solution of this problem, the adjusted moment curves were determined graphically and are indicated by the dashed lines in Diags. 1 to 5.

7. An adjusted maximum moment envelope can now be obtained from the adjusted moment curves as shown in Diag. 6 by dashed lines.

   Since the provided steel ratios ρ are smaller than or equal to the maximum ratios permitted by Section 8.4, the design is O.K.

8. Final steel ratios ρ can now be obtained on the basis of the adjusted moments. A final check of these ratios must be made for compliance with Fig. 9-1.

   From the redistributed moment envelopes of Diag. 6, the design factored moments and required reinforcements are obtained as shown in the following table.

Summary of Final Design

Location	Moment, ft-kips	Load Case	Required As, in.^{2}	ρ	Provided As, in.^{2}	ρ	Redistribution, percent	Permitted Max. ρ (Fig. 9-1)
Mid-Span A-B	+66.1	II	1.16	0.0072	1#5 2#6 (1.19)	0.0073**	—	—
Support B*	−97.3	I	1.76	0.0108	1#6 1#7 1#8 (1.83)	0.0112	11.2	0.0125
Support C*	-60.9	III	1.06	0.0066	2#5 1#6 (1.06)	0.0065	14.5	0.008
Mid-Span C-D	+43.6	II	0.75	0.0046	3#5 (0.93)	0.0057**	—	—

*centerline moments. May use those at face of support

**check $\rho_{min} = \dfrac{200}{f_y} = \dfrac{200}{60,000} = 0.0033$ 10.5.1

Example 9.2—Continued

	Code
Calculations and Discussion	**Reference**

9. Second cycle of moment redistribution.

Should the engineer decide additional adjustment of the moments is warranted, further cycles of redistribution may be made. The procedure to be followed is similar to that used for the first cycle, except that the allowable percentage redistributions are obtained by using the final steel ratios from the previous cycle. To illustrate this, the percentage allowable adjustments for a second cycle redistribution are calculated for supports B and C, and shown in the following Table.

Second Cycle of Redistribution

Support	Factored Moment M_u , ft-kips	Steel Ratio from 1st Cycle	$\rho/\rho b$ ($\rho b = 0.0285$)	Allowable Redistribution (Fig. 9-1)
B	−97.3	0.0112	0.386	12.3%
C	−60.9	0.0065	0.228	15.4%

To continue with the second cycle of redistribution, the percentage adjustments calculated are then applied to the **original** elastic moment curves for each of the loading patterns and a second adjusted maximum moment envelope obtained. The moments from this second adjusted maximum moment envelope are then used to determine the required reinforcement. The **actual** steel ratio provided must comply with Fig. 9-1 for the percent moment redistribution taken. This is an important final step to assure that the design meets the requirements of Section 8.4.

The second cycle distribution at support B results in an adjusted moment of -109.5 (1-0.125) = -95.8 ft-kips. This is only a reduction of 1.5% from the first cycle redistribution.

For this design example, the design moments were not reduced to the face of the column because the intent was to illustrate moment redistribution alone. In Example 9.1, both moment redistribution and moments at the face of the support were used in design.

Design for Flexure

GENERAL CONSIDERATIONS

For design or investigation of members subject to flexure (beams and slabs), the nominal strength of the member cross section (M_n) must be reduced by the strength reduction factor ϕ for flexure ($\phi = 0.90$) to obtain the design strength (ϕM_n) of the section. The design strength must be equal to or greater than the required strength ($\phi M_n \geq M_u$).

All members subject to flexure must be designed to satisfy this basic criterion. The serviceability requirements of deflection control and crack control (distribution of reinforcement) must also be satisfied.

Six design examples are presented in Part 10, each illustrating proper application of the various code provisions that govern design of members subject to flexure only. The design examples are prefaced by a step-by-step procedure for design of rectangular sections with tension reinforcement only.

DESIGN OF RECTANGULAR SECTIONS WITH TENSION REINFORCEMENT ONLY[10.1]

In the design of rectangular sections with tension reinforcement only (Fig. 10-1), the two conditions of equilibrium are:

$$C = T \tag{1}$$

and

$$M_n = (C \text{ or } T)\left(d - \frac{a}{2}\right) \tag{2}$$

When the reinforcement ratio $\rho = A_s/bd$ is preset, from Eq. (1):

$$0.85f'_c\, ba = A_s f_y = \rho bd f_y \text{ or, } a = \frac{A_s f_y}{0.85f'_c\, b} = \frac{\rho d f_y}{0.85f'_c}$$

from Eq. (2):

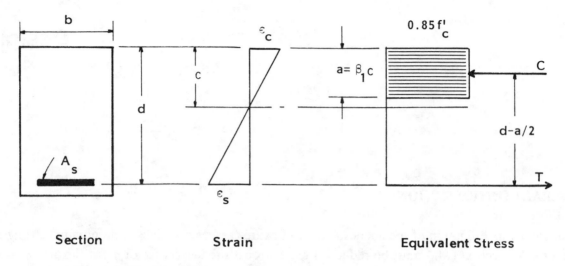

Section　　　　　Strain　　　　　Equivalent Stress

Fig. 10-1　Strain and Equivalent Stress Distribution in Rectangular Section

$$M_n = \rho b d f_y \left[d - 0.5 \frac{\rho d}{0.85} \frac{f_y}{f_c'} \right]$$

A nominal strength coefficient of resistance R_n is obtained when both sides are divided by bd^2

$$R_n = \frac{M_n}{bd^2} = \rho f_y \left(1 - 0.5 \frac{\rho f_y}{0.85 f_c'} \right) \tag{3}$$

When b and d are preset, ρ is obtained by solving the quadratic equation for R_n:

$$\rho = \frac{0.85 f_c'}{f_y} \left(1 - \sqrt{1 - \frac{2R_n}{0.85 f_c'}} \right) \tag{4}$$

The relationship between ρ and R_n for Grade 60 reinforcement and various values of f_c' is shown in Fig. 10-2.

Using Eqs. (3) and (4), a design procedure for rectangular sections with tension reinforcement only is outlined as follows:

Step 1　Select an approximate value of tension reinforcement ratio ρ equal to or less than $0.75\rho_b$ (Section 10.3.3), but greater than the minimun (Section 10.5.1), where the balanced reinforcement ratio ρ_b is given by:

$$\rho_b = \frac{0.85\,\beta_1\,f_c'}{f_y}\left(\frac{87{,}000}{87{,}000+f_y}\right), \text{ and } \beta = 0.85 \text{ for } f_c' \le 4000 \text{ psi}$$

$$= 0.85 - 0.05\left(\frac{f_c'-4000}{1000}\right) \text{ for } 4000 \text{ psi} < f_c' < 8000 \text{ psi}$$

$$= 0.65 \text{ for } f_c' \ge 8000 \text{ psi}$$

Values of ρ_b and $0.75\rho_b$ are given in Table 6-1.

Step 2   With ρ preset $\left(\frac{200}{f_y} \le \rho \le 0.75\,\rho_b\right)$ compute bd^2 required.

$$bd^2(\text{required}) = \frac{M_u}{\varphi R_n}$$

where $R_n = \rho f_y\left(1 - 0.5\,\frac{\rho f_y}{0.85 f_c'}\right)$

$\varphi = 0.90$ for flexure

M_u = applied factored moment (required flexural strength)

Step 3   Size and member so that the value of bd^2 provided is approximately equal to the value of bd^2 required.

Step 4   Compute a revised value of ρ by one of the following methods:

(1) By Eq. (4) (exact method) where $R_n = M_u/\varphi(bd^2 \text{ provided})$

(2) By strength curves such as shown in Fig. 10-2. Values of ρ are given in terms of $R_n = M_u/\varphi bd^2$ for Grade 60 reinforcement.

(3) By moment strength tables such as shown in Table 10-1. Values of $\omega = \rho f_y/f_c'$ are given in terms of moment strength $M_u/\varphi f_c' bd^2$.

(4) By approximate proportion

$$\rho \approx (\text{original }\rho)\,\frac{(\text{revised } R_n)}{(\text{original } R_n)}$$

Note from Fig. 10-2, that the relationship between R_n and ρ is approximately linear.

Step 5   Compute A_s required
A_s = (revised ρ) (bd provided)

When b and d are preset, the A_s required is computed directly from
$A_s = \rho$(bd provided)

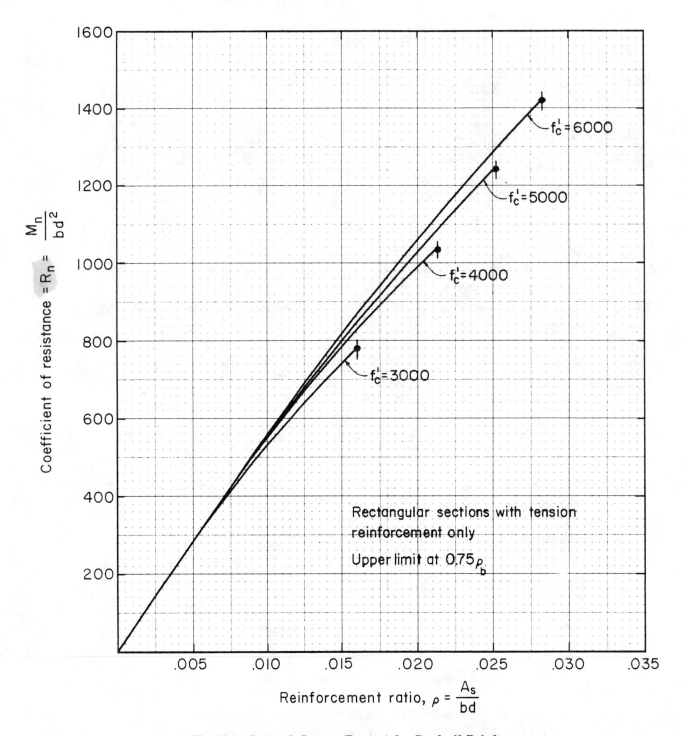

Fig. 10-2   Strength Curves (R_n vs ρ) for Grade 60 Reinforcement

where ρ is computed using one of the methods outlined in Step 4.

REFERENCE

10.1 Wang, C.K. and Salmon, C.G., *Reinforced Concrete Design,* Fourth Edition, Harper & Row Publishers, New York, N.Y., 1985.

Table 10-1—Flexural Strength $M_u/\varphi\, f'_c\, bd^2$ or $M_n/f'_c\, bd^2$
of Rectangular Sections with Tension Reinforcement only

ω	.000	.001	.002	.003	.004	.005	.006	.007	.008	.009
0.0	0	.0010	.0020	.0030	.0040	.0050	.0060	.0070	.0080	.0090
0.01	.0099	.0109	.0119	.0129	.0139	.0149	.0159	.0168	.0178	.0188
0.02	.0197	.0207	.0217	.0226	.0236	.0246	.0256	.0266	.0275	.0285
0.03	.0295	.0304	.0314	.0324	.0333	.0343	.0352	.0362	.0372	.0381
0.04	.0391	.0400	.0410	.0420	.0429	.0438	.0448	.0457	.0467	.0476
0.05	.0485	.0495	.0504	.0513	.0523	.0532	.0541	.0551	.0560	.0569
0.06	.0579	.0588	.0597	.0607	.0616	.0625	.0634	.0643	.0653	.0662
0.07	.0671	.0680	.0689	.0699	.0708	.0717	.0726	.0735	.0744	.0753
0.08	.0762	.0771	.0780	.0789	.0798	.0807	.0816	.0825	.0834	.0843
0.09	.0852	.0861	.0870	.0879	.0888	.0897	.0906	.0915	.0923	.0932
0.10	.0941	.0950	.0959	.0967	.0976	.0985	.0994	.1002	.1011	.1020
0.11	.1029	.1037	.1046	.1055	.1063	.1072	.1081	.1089	.1098	.1106
0.12	.1115	.1124	.1133	.1141	.1149	.1158	.1166	.1175	.1183	.1192
0.13	.1200	.1209	.1217	.1226	.1234	.1243	.1251	.1259	.1268	.1276
0.14	.1284	.1293	.1301	.1309	.1318	.1326	.1334	.1342	.1351	.1359
0.15	.1367	.1375	.1384	.1392	.1400	.1408	.1416	.1425	.1433	.1441
0.16	.1449	.1457	.1465	.1473	.1481	.1489	.1497	.1506	.1514	.1522
0.17	.1529	.1537	.1545	.1553	.1561	.1569	.1577	.1585	.1593	.1601
0.18	.1609	.1617	.1624	.1632	.1640	.1648	.1656	.1664	.1671	.1679
0.19	.1687	.1695	.1703	.1710	.1718	.1726	.1733	.1741	.1749	.1756
0.20	.1764	.1772	.1779	.1787	.1794	.1802	.1810	.1817	.1825	.1832
0.21	.1840	.1847	.1855	.1862	.1870	.1877	.1885	.1892	.1900	.1907
0.22	.1914	.1922	.1929	.1937	.1944	.1951	.1959	.1966	.1973	.1981
0.23	.1988	.1995	.2002	.2010	.2017	.2024	.2031	.2039	.2046	.2053
0.24	.2060	.2067	.2075	.2082	.2089	.2096	.2103	.2110	.2117	.2124
0.25	.2131	.2138	.2145	.2152	.2159	.2166	.2173	.2180	.2187	.2194
0.26	.2201	.2208	.2215	.2222	.2229	.2236	.2243	.2249	.2256	.2263
0.27	.2270	.2277	.2284	.2290	.2297	.2304	.2311	.2317	.2324	.2331
0.28	.2337	.2344	.2351	.2357	.2364	.2371	.2377	.2384	.2391	.2397
0.29	.2404	.2410	.2417	.2423	.2430	.2437	.2443	.2450	.2456	.2463
0.30	.2469	.2475	.2482	.2488	.2495	.2501	.2508	.2514	.2520	.2527
0.31	.2533	.2539	.2546	.2552	.2558	.2565	.2571	.2577	.2583	.2590
0.32	.2596	.2602	.2608	.2614	.2621	.2627	.2633	.2639	.2645	.2651
0.33	.2657	.2664	.2670	.2676	.2682	.2688	.2694	.2700	.2706	.2712
0.34	.2718	.2724	.2730	.2736	.2742	.2748	.2754	.2760	.2766	.2771
0.35	.2777	.2783	.2789	.2795	.2801	.2807	.2812	.2818	.2824	.2830
0.36	.2835	.2841	.2847	.2853	.2858	.2864	.2870	.2875	.2881	.2887
0.37	.2892	.2898	.2904	.2909	.2915	.2920	.2926	.2931	.2937	.2943
0.38	.2948	.2954	.2959	.2965	.2970	.2975	.2981	.2986	.2992	.2997
0.39	.3003	.3008	.3013	.3019	.3024	.3029	.3035	.3040	.3045	.3051

$M_n/f'_c\, bd^2 = A_s f_y\, (d-a/2)\, f'_c\, bd^2 = \omega(1-0.59\omega)$, where $\omega = \rho f_y/f'_c$ and $a = A_s f_y/0.85 f'_c\, b$.

Design: Using factored moment M_u enter table with $M_u/\varphi f'_c\, bd^2$; find ω and compute steel percentage ρ from $\rho = \omega f'_c/f_y$

Investigation: Enter table with ω from $\omega = \rho f_y/f'_c$; find value of $M_n/f'_c\, bd^2$ and solve for nominal strength, M_n.

Example 10.1—Design of Rectangular Beam with Tension Reinforcement Only

Select a rectangular beam size and required reinforcement A_s to carry service load moments of: $M_d = 55$ ft-kips and $M_l = 36$ ft-kips. Select reinforcement to control flexural cracking for exterior exposure.

$f_c' = 4000$ psi

$f_y = 60,000$ psi

$z = 145$ kips/in. (exterior exposure)

Calculations and Discussion	Code Reference

1. To illustrate a complete design procedure for rectangular sections with tension reinforcement only, a minimum beam depth will be computed using the maximum reinforcement permitted for flexural members, $0.75\rho_b$. The design procedure will follow the method outlined on the preceding pages. → 10.3.3

   Step 1. Determine maximum reinforcement ratio for material strengths $f_c' = 4000$ psi and $f_y = 60,000$ psi.

   $\rho_b = 0.0285$, from Table 6-1

   $\rho_{max} = 0.75\rho_b = 0.75(0.0285) = 0.0214$ → 10.3.3

   Step 2. Compute bd^2 required.

   Required moment strength:

   $M_u = 1.4 \times 55 + 1.7 \times 36 = 138$ ft-kips → Eq. (9-1)

   $$R_n = \rho f_y \left(1 - 0.5\frac{\rho f_y}{0.85 f_c'}\right)$$

   $$= 0.0214 \times 60,000 \left(1 - \frac{0.5 \times 0.0214 \times 60,000}{0.85 \times 4000}\right) = 1042 \text{ psi}$$

   $$bd^2 \text{ (required)} = \frac{M_u}{\varphi R_n} = \frac{138 \times 12 \times 1000}{0.90 \times 1042} = 1766 \text{ in.}^3$$

   Step 3. Size member so that

   bd^2 required $\leq bd^2$ provided

   Set $b = 10$ in. (column width)

Example 10.1—Continued

	Code
Calculations and Discussion	**Reference**

$d = \sqrt{1766/10} = 13.3$ in.

Minimum beam depth $\approx 13.3 + 2.5 = 15.8$ in.

For moment strength, a 10 in. x 16 in. beam size is adequate. However, deflection is an essential consideration in designing beams by the Strength Design Method. Control of deflection is discussed in Part 8.

Step 4. Using the 16 in. beam depth, compute a revised value of ρ. For illustration, ρ will be computed by all four methods outlined earlier.

$d = 16 - 2.5 = 13.5$ in.

(1) by Eq. (4) (exact method):

$$R_n = \frac{M_u}{\varphi(bd^2 \text{ provided})} = \frac{138 \times 12 \times 1000}{0.90(10 \times 13.5^2)} = 1010 \text{ psi}$$

$$\rho = \frac{0.85f_c'}{f_y}\left(1 - \sqrt{1 - \frac{2R_n}{0.85f_c'}}\right)$$

$$= \frac{0.85 \times 4}{60}\left(1 - \sqrt{1 - \frac{2 \times 1010}{0.85 \times 4000}}\right) = 0.0206$$

(2) by strength curves such as that shown in Fig. 10-2:

for $R_n = 1010$ psi, $\rho \approx 0.0205$

(3) by strength tables such as Table 10-1:

$$\text{for } \frac{M_u}{\varphi f_c' bd^2} = \frac{138 \times 12 \times 1000}{0.90 \times 4000 \times 10 \times 13.5^2} = 0.252$$

$\omega = 0.308$

$\rho = \omega f_c'/f_y = 0.308 \times 4/60 = 0.0205$

(4) by approximate proportion:

$$\rho \approx (\text{original } \rho) \frac{(\text{revised } R_n)}{(\text{original } R_n)}$$

Example 10.1—Continued

Calculations and Discussion

$$\rho \approx 0.0214 \times \frac{1010}{1042} = 0.0207$$

Step 5.   Compute A_s required.

$$A_s = (\text{revised } \rho)(\text{bd provided})$$

$$= 0.0206 \times 10 \times 13.5 = 2.78 \text{ in.}^2$$

2.   A review of the correctness of the computations can be made by considering simple statics.

$$T = \rho b d f_y = A_s f_y = 2.78 \times 60 = 166.8 \text{ kips}$$

$$a = \frac{C \text{ or } T}{0.85 f_c' b} = \frac{166.8}{0.85 \times 4 \times 10} = 4.91 \text{ in.}$$

Design moment strength:

$$\varphi M_n = \varphi \left[A_s f_y \left(d - \frac{a}{2} \right) \right] = 0.9 \left[166.8 \left(13.5 - \frac{4.91}{2} \right) \right]$$

$$= 1658.1 \text{ in.-kips} = 138.2 \text{ ft-kips}$$

Required moment strength $M_u \leq$ design moment strength, ϕM_n

138 ft-kips $\leq$ 138.2 ft-kips   O.K.

3.   Select reinforcement to satisfy distribution of flexural reinforcement requirements of Section 10.6. Use $z = 145$ kips/in. for exterior exposure.

10.6

A_s required $= 2.78 \text{ in.}^2$

For illustrative purposes, select 2#9 and 1#8 bars. $A_s = 2.79 \text{ in.}^2$ For practical design and detailing, one bar size for total A_s is preferable. See Example 6.2 for moment strength calculations with A_s provided $= 2.79 \text{ in.}^2$

$$z = f_s \sqrt{d_c A}$$

Eq. (10-4)

With mixed bar sizes

$$d_{cg} = \frac{0.79 d_c + 2.0 (d_c + 1/16)}{2.79} = 2.50 \text{ in.}$$

Example 10.1—Continued

	Code
Calculations and Discussion	**Reference**

where d_c = distance from extreme tension fiber to center of bar located closest thereto (#8 bar)

Solving, d_c = 2.46 in.

Effective tension area = $2(2.50)10 = 50.0$ in.^{2}

Equivalent number of #9 bars = $2.79/1.00 = 2.79$

A = Effective area / no. bars = $50/2.79 = 17.92$ in.^{2}/bar   10.0

Use $f_s = 0.6f_y = 0.6(60) = 36$ ksi   10.6.4

$z = 36 \sqrt[3]{2.46\,(17.92)} = 127$ kips/in. $<$ 145   O.K.

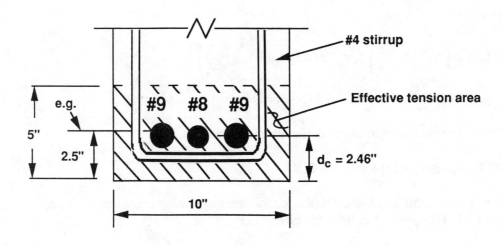

Example 10.1—Continued

	Code
Calculations and Discussion	**Reference**

4. Check beam width

$b \geq 2(\text{cover}) + 2(\text{stirrup dia.}) + \Sigma d_b + 2(\text{min. bar spacing})$ 7.6.1

$= 2(1.5) + 2(0.50) + 3.26 + 2(1.128) = 9.52 \text{ in.} \quad < \quad 10 \text{ in.} \qquad \text{O.K.}$

Example 10.2—Design of Rectangular Beam with Compression Reinforcement

A beam cross section is limited to the size shown. Determine required area of reinforcement for a factored moment M_u = 900 ft-kips.

f_c' = 4000 psi

f_y = 60,000 psi

z = 145 kips/in. (exterior exposure)

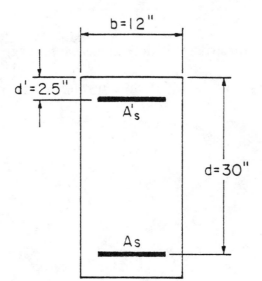

	Code
Calculations and Discussion	**Reference**

1. Check design for tension reinforcement only.

   Compute required tension reinforcement using strength Table 10-1:

   $$\frac{M_u}{\varphi f_c' bd^2} = \frac{900 \times 12 \times 1000}{0.90 \times 4000 \times 12 \times 30^2} = 0.2777$$

   From Table 10-2, ω = 0.35

   Ratio of tension reinforcement required:

   $$\rho = \omega f_c' / f_y = 0.35 \times 4/60 = 0.0233$$

   With tension reinforcement only:

   $$\rho_{max} = 0.75\rho_b$$ 10.3.3

   From Table 6-1, with f_c' = 4000 psi and f_y = 60,000 psi

   $$\rho_{max} = 0.0214$$

Example 10.2—Continued

	Code
Calculations and Discussion	**Reference**

$0.0233 > 0.0214$

Therefore, compression reinforcement is required.

2. Compute reinforcement required, A_s and A_s'

Maximum ω permitted for singly reinforced beam (tension reinforcement only):

$\omega \leq 0.75 \rho_b f_y / f_c' \leq 0.0214 \times 60/4 = 0.321$

From Table 10-2, with $\omega = 0.321$

$M_n / f_c' \, bd^2 = 0.2602$

Maximum design moment strength carried by concrete:

$(\varphi M_n)_{max} = 0.9(0.2602 \times 4 \times 12 \times 30^2/12)$

$= 843 \text{ ft−kips}$

Required moment strength to be carried by compression reinforcement:

$M_u' = 900 - 843 = 57 \text{ ft−kips}$

Assume compression reinforcement yields at the ultimate stage, $(f_s' = f_y)$

$\rho' = \dfrac{A_s'}{bd} = \dfrac{M_u'}{\varphi f_y (d - d') bd}$

$\dfrac{57 \times 12 \times 1000}{0.9 \times 60,000(30 - 2.5)12 \times 30} = 0.00128$

$\rho = 0.75 \rho_b + \rho' = 0.0124 + 0.00128 = 0.0227$

Note that for members with compression reinforcement, the portion of ρ_b contributed by compression reinforcement need not be reduced by the 0.75 factor.   10.3.3

$A_s' = \rho' bd = 0.00128 \times 12 \times 30 = 0.46 \text{ in.}^2$

$A_s = \rho bd = 0.0227 \times 12 \times 30 = 8.17 \text{ in.}^2$

Check yielding of compression reinforcement:

Example 10.2—Continued

$$\frac{A_s - A_s'}{bd} \geq \frac{0.85\,\beta_1\,f_c'\,d'}{f_y d}\left(\frac{87,000}{87,000 - f_y}\right)$$

$$0.0227 - 0.00128 \geq \frac{0.85 \times 0.85 \times 4 \times 2.5}{60 \times 30}\left(\frac{87,000}{87,000 - 60,000}\right)$$

$$0.0214 \geq 0.0129$$

Therefore, compression reinforcement yields at the ultimate stage, as assumed,   O.K.

3. A review of the correctness of the computations may be made by using Eq. (7) developed in Part 6. When the compression reinforcement yields:

$$\varphi M_n = \varphi\left[(A_s - A_s')f_y\left(d - \frac{a}{2}\right) + A_s'f_y\,(d - d')\right]$$

$$= 0.9\left[7.71 \times 60\left(30 - \frac{11.34}{2}\right) + 0.46 \times 60(30 - 2.5)\right]/12$$

$$= 901\,\text{ft}-\text{kips} > M_u = 900\,\text{ft}-\text{kips}\qquad\text{O.K.}$$

where $a = \dfrac{(A_s - A_s')f_y}{0.85f_c'\,b} = \dfrac{7.71 \times 60}{0.85 \times 4 \times 12} = 11.34$ in.

4. Select reinforcement to satisfy control of flexural cracking criteria of Section 10.6 for exterior exposure.

Compression reinforcement:

Select 2#5 bars ($A_s' = 0.62$ in.^{2} > 0.46 in.^{2})

Tension reinforcement:

Select 8#9 bars ($A_s = 8.00$ in.$^2 \approx 8.17$ in.^{2})
(2% less than required, should be sufficient)

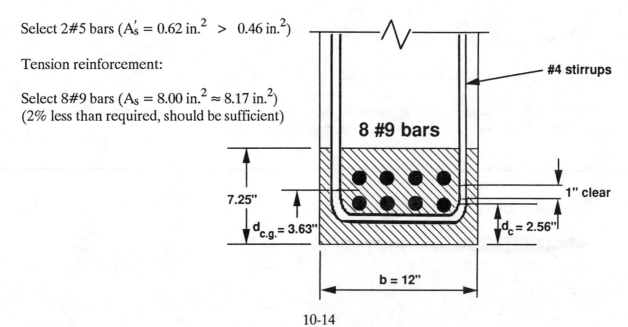

Example 10.2—Continued

Calculations and Discussion	Code Reference

$z = f_s \sqrt[3]{d_c A}$ Eq. (10-4)

d_c = cover + ½ bar dia. + stirrup dia.   10.0

    = 1.5 + 0.56 + 0.5 = 2.56 in.

d_{cg} = 1.5 + 0.5 + 1.128 + 0.5 = 3.63 in.

$A = 3.63 \times 2 \times 12/8 = 10.9$ in.^{2}/bar   10.0

Use $f_s = 0.6 f_y = 36$ ksi   10.6.4

$z = 36\sqrt[3]{2.56 \times 10.9} = 109$ kips/in.   <   145 kips/in.   O.K.

5. Check beam width:

   $b = 2 \times$ cover $+ 4 \times 1.128 + 3 \times 1.128$ 7.6.1

     $= 2 \times 2 + 4.51 + 3.38 = 11.9$ <   12 in. provided,   O.K.   7.7.1

6. Stirrups or ties are required throughout distance where compression reinforcement is required for strength.   7.11.1

   Max spacing $\leq 16 \times 0.625 = 10$ in.   7.10.5.2

         $48 \times 0.500 = 24$ in.

         least dimension of member = 12 in.

   Use $s_{max} = 10$ in. for #4 stirrups

Example 10.3—Design of One-Way Solid Slab

Determine required thickness and reinforcement for a one-way slab continuous over two or more equal spans. Clear span, ℓ_n = 18 ft.

f_c' = 4000 psi
f_y = 60,000 psi
Service loads: w_ℓ = 50 psf, w_d = 75 psf (assume 6-in. slab)

Calculations and Discussion	Code Reference

1. Compute required moment strengths using approximate moment analysis permitted by Section 8.3.3. Design will be based on end span.

   Factored load: Eq. (9-1)
   w_u = 1.4 × 75 + 1.7 × 50 = 190 psf

   Positive moment:
   $+M_u = w_u \ell_n^2 / 14$ 8.3.3
          = 0.190 × 18² /14 = 4.40 ft−kips/ft of width

   Negative moment at exterior face of first interior support:
   $-M_u = w_u \ell_n^2 / 10$ 8.3.3
          = 0.190 × 18² /10 = 6.16 ft−kips/ft of width

2. Compute required slab thickness.   10.3.3
   Choose a reinforcement percentage ρ equal to about $0.375\rho_b$, or one-half the maximum permitted, to have reasonable deflection control.

   From Table 6-1, for f_c' = 4000 psi and f_y = 60,000 psi

   ρ_b = 0.0285

   Set ρ = 0.375(0.0285) = 0.0107

   Design procedure will follow method outlined earlier:

   $$R_n = \rho f_y \left(1 - 0.5 \frac{\rho f_y}{0.85 f_c'} \right)$$

   $$= 0.0107 \times 60,000 \left(1 - \frac{0.5 \times 0.0107 \times 60,000}{0.85 \times 4000} \right)$$

   $$= 581 \text{ psi}$$

Example 10.3—Continued

Required d = $\sqrt{\dfrac{M_u}{\varphi R_n b}}$ = $\sqrt{\dfrac{6.16 \times 12,000}{0.90 \times 581 \times 12}}$ = 3.43 in.

9.3.2.1

Assume #5 bars,
required h = 3.43 + 0.31 + 0.75 = 4.49 in.

The above design indicates a slab thickness of 4½ in. is adequate. However, Table 8-1 indicates a minimum thickness of $\ell/28 > 7\frac{3}{4}$ in., unless deflections are computed. Also note that Table 8-1 is applicable only to "members not supporting or attached to partitions or other construction likely to be damaged by large deflections." Otherwise deflections must be computed.

For purposes of illustration, the required reinforcement will be computed for h = 4½ in., d = 3.43 in.

3. Compute required reinforcement.

    $-A_s$ (required) = ρbd = 0.0107 × 12 × 3.43 = 0.440 in.^{2}/ft

    Use #5 @ 8 in., A_s = 0.47 in.^{2}/ft
    or #6 @ 12 in., A_s = 0.44 in.^{2}/ft

    For positive moment, using strength Table 10-1:

    $\dfrac{M_u}{\varphi f_c' bd^2} = \dfrac{4.40 \times 12,000}{0.9 \times 4000 \times 12 \times 3.43^2} = 0.1039$

    From Table 10-2, ω = 0.111

    $\rho = \omega f_c' / f_y$ = 0.111 × 4/60 = 0.0074

    $+A_s$ (required) = ρbd = 0.0074 × 12 × 3.43 = 0.305 in.^{2}/ft

    Use #4 @ 8 in., A_s = 0.30 in.^{2}/ft
    or #5 @ 12 in., A_s = 0.31 in.^{2}/ft

Example 10.4—Design of Flanged Section with Tension Reinforcement Only

Select reinforcement for the T section shown, to carry service dead and live load moments of $M_d = 72$ ft-kips and $M_l = 88$ ft-kips.

$f'_c = 4000$ psi
$f_y = 60,000$ psi
$z = 145$ kips/in. (exterior exposure)

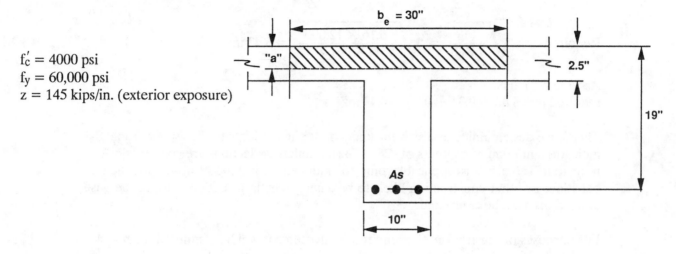

	Code
Calculations and Discussion	**Reference**

1. Determine required flexural strength.

   $M_u = 1.4\,M_d + 1.7\,M_l$ Eq. (9-1)
        $= 1.4 \times 72 + 1.7 \times 88$
        $= 250$ ft$-$kips

2. Using Table 10-1, determine depth of equivalent stress block a, as for a rectangular section.

   For $\dfrac{M_u}{\varphi f'_c\, bd^2} = \dfrac{250 \times 12}{0.9 \times 4 \times 30 \times 19^2} = 0.077$ 9.3.2.1

   From Table 10-1, $\omega = \rho f_y / f'_c = 0.081$

   $a = \dfrac{A_s f_y}{0.85 f'_c\, b} = \dfrac{\rho d f_y}{0.85 f'_c} = 1.18\,\omega d$

       $= 1.18 \times 0.081 \times 19 = 1.82$ in. $<$ 2.5 in.

   With a less than the flange thickness, determine reinforcement as for a rectangular section. See Example 10.5 for a greater than flange depth.

3. Compute A_s required

   $A_s f_y = 0.85 f'_c\, ba$

Example 10.4—Continued

Calculations and Discussion	**Code Reference**

$$A_s = \frac{0.85 \times 4 \times 30 \times 1.82}{60} = 3.09 \text{ in.}^2$$

Try 2#11 bars, $A_s = 3.12$ in.^{2}

4. Check minimum required reinforcement   10.5

$\rho_{min} = 200/f_y = 200/60,000 = 0.0033 <$ Eq. (10-3)
$A_s/b_w d = 3.12/(10 \times 19) = 0.0164$ O.K.   10.5.1

5. Check distribution of reinforcement for exterior exposure ($z = 145$ kips/in.).   10.6

$$z = f_s \sqrt[3]{d_c A}$$ Eq. (10-4)

d_c = cover + ½ bar diameter
     = 2 + 0.71 = 2.71 in.

$A = 2d_c b_w/$no. of bars
     $= 2 \times 2.71 \times 10/2 = 27.1$ in.^{2}/bar

$z = 0.6 \times 60 \sqrt[3]{2.71 \times 27.1} = 150.8$ kips/in. $>$ 145 kips/in.   N.G.   10.6.4

Since the limiting value of z is exceeded for exterior exposure, unacceptable tensile cracking is indicated. Smaller bar sizes must be used.

Try 3#9 bars, $A_s = 3.00$ in.^{2}
(3% less than required, assumed sufficient)

$d_c = 2 + 0.56 = 2.56$ in.

$A = 2 \times 2.56 \times 10/3 = 17.1$ in.^{2}/bar

$z = 0.6 \times 60 \sqrt[3]{2.56 \times 17.1} = 127$ kips/in. $<$ 145 kips/in.   O.K.

6. Check minimum web width.

$b_w \geq 2 \times$ cover $+ 3 \times 1.128 + 2 \times 1.128$ 7.6.1
     $= 2 \times 2 + 5.64 = 9.64 <$ 10 in. provided,   O.K.   7.7.1

Example 10.5—Design of Flanged Section with Tension Reinforcement Only

Select reinforcement for the T section shown, to carry a factored moment of M_u = 400 ft-kips.

f_c' = 4000 psi

f_y = 60,000 psi

z = 145 kips/in. (exterior exposure)

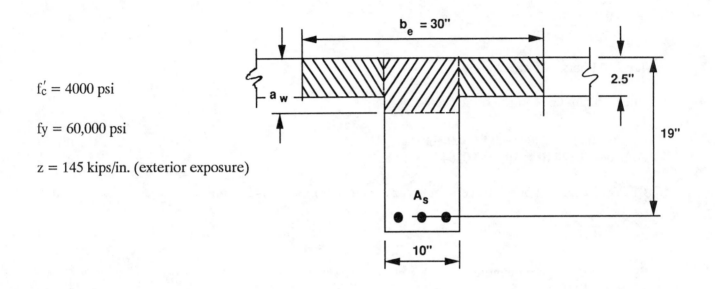

	Code
Calculations and Discussion	**Reference**

1. Using Table 10-1, determine depth of equivalent stress block a, as for a rectangular section.

   For $\dfrac{M_u}{\varphi f_c'\, bd^2} = \dfrac{400 \times 12}{0.9 \times 4 \times 30 \times 19^2} = 0.123$ 9.3.2.1

   From Table 10-1, $\omega = \rho f_y/f_c' = 0.133$

   a = 1.18 ωd

      = 1.18 × 0.133 × 19 = 2.98 in. > 2.5 in.

   Since the value of a as a rectangular section exceeds the flange thickness, the equivalent stress block is not rectangular, and the design must be based on a T section. See Example 10.4 when a less than flange depth.

2. Compute required reinforcement A_{sf} and moment strength φM_{nf} corresponding to the overhanging beam flange in compression.

   Compressive strength of flange:

   $C_f = 0.85 f_c'\, (b - b_w) h_f$

Example 10.5—Continued

$$= 0.85 \times 4(30 - 10)2.5 = 170 \text{ kips}$$

Required A_{sf} to equilibrate C_f:

$$A_{sf} = C_f/f_y = 170/60 = 2.83 \text{ in.}^2$$

Design moment stength of flange:

$$\varphi M_{nf} = \varphi[A_{sf}f_y(d - 0.5h_f)]$$

$$= 0.9[2.83 \times 60(19 - 1.25)]/12 = 226 \text{ ft-kips}$$

Required moment strength to be carried by beam web:

$$M_{uw} = M_u - \varphi M_{nf} = 400 - 226 = 174 \text{ ft-kips}$$

3. Using Table 10-1, compute reinforcement A_{sw} required to develop moment strength to be carried by web.

$$\text{For } \frac{M_{uw}}{\varphi f_c' \, bd^2} = \frac{174 \times 12}{0.9 \times 4 \times 10 \times 19^2} = 0.161$$

from Table 10-1, $\omega_w = 0.180$

$$a_w = 1.18 \, \omega d = 1.18 \times 0.180 \times 19 = 4.04 \text{ in.}$$

$$A_{sw} = \frac{0.85 f_c' \, b_w a_w}{f_y} = \frac{0.85 \times 4 \times 10 \times 4.04}{60} = 2.29 \text{ in.}^2$$

Alternatively, A_{sw} can be computed directly from:

$$A_{sw} = \frac{\omega f_c' \, b_w d}{f_y} = \frac{0.180 \times 4 \times 10 \times 19}{60} = 2.28 \text{ in.}^2$$

4. Total reinforcement required to carry factored moment $M_u = 400$ ft-kips

$$A_s = A_{sf} + A_{sw} = 2.83 + 2.29 = 5.12 \text{ in.}^2$$

5. A review of the correctness of the computations may be made using Eq. (6) developed in Part 6.

$$\phi M_n = \phi[(A_s - A_{sf}) \, f_y \, (d - a_w/2) + A_{sf} \, f_y \, (d - h_f/2)] \qquad \text{Eq. (6), Part 6}$$

Example 10.5—Continued

$$= 0.9[(5.12 - 2.83)60(19 - 4.04/2) + 2.83 \times 60(19 - 2.5/2)]/12$$

$$= 401 \text{ ft-kips} \ > \ M_u = 400 \text{ ft-kips} \qquad \text{O.K.}$$

where $A_{sf} = 0.85 f_c' (b - b_w) h_f / f_y$

$$= 0.85 \times 4(30 - 10)2.5/60 = 2.83 \text{ in.}^2$$

$$a_w = (A_s - A_{sf}) f_y / 0.85 f_c' b_w$$

$$= (5.12 - 2.83)60/0.85 \times 4 \times 10 = 4.04 \text{ in.}$$

6. Check maximum tension reinforcement permitted according to Section 10.3.3. See Part   10.3.3
   6, Eqs. (8) and (11).

   For flanged section with tension reinforcement only:

   $$\rho_{max} = \left[0.75 \frac{b_w}{b} (\bar{\rho}_b + \rho_f)\right] \qquad \text{Eq. (11), Part 6}$$

   $$\rho_f = 0.85 \frac{f_c'}{f_y} (b - b_w) h_f / b_w d \qquad \text{Eq. (8), Part 6}$$

   $$= 0.85 \frac{4}{60} (30 - 10)2.5/(10 \times 19) = 0.0149$$

   From Table 10-1, $\bar{\rho}_b = 0.0285$

   $$\rho_{max} = 0.75 \left[\frac{10}{30} (0.0285 + 0.0149)\right] = 0.0109$$

   $$A_{s(max)} = 0.0109 \times 30 \times 19 = 6.21 \text{ in.}^2 \ > \ 5.12 \text{ in.}^2 \qquad \text{O.K.}$$

7. Select reinforcement to satisfy crack control criteria for exterior exposure ($z = 145$ 10.6
   kips/in.)

   Try 5#9 bars, $A_s = 5.00 \text{ in.}^2$
   (2.34% less than required, assumed sufficient)

Example 10.5—Continued

	Code
Calculations and Discussion	**Reference**

For exterior exposure:

$d_c = 2 + 1.128/2 = 2.56$ in.   10.0

$d_{cg} = [3(2.56) + 2(4.69)]/5 = 3.41$ in.

$A = 2(3.41)10/5 = 13.64$ in.^{2}/bar   10.0

$z = f_s\sqrt[3]{d_c A} = 0.6 \times 60\sqrt[3]{2.56 \times 13.64}$ Eq. (10-4)

$\qquad\qquad = 117.7$ kips/in. $<$ 145 kips/in.   O.K.

8. Check required web width   7.6.1

Required $b_w = 2$ (cover) $+ 3d_b + 2d_b{}^*$ 7.7.1

$\qquad\qquad 2 \times 2 + 5 \times 1.188 = 9.64$ in. $<$ 10 in.   O.K.

*Clear distance between bars not less than d_b or 1 in.

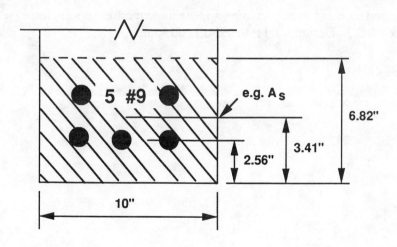

Example 10.6—Design of One-Way Joist

Compute required depth and reinforcement for a one-way joist system continuous over two or more equal spans. Assume the joists are 10 in. deep and 5 in. wide @ 35 in. centers, as shown below. The slab is 3 in. thick. Clear span, l_n = 20 ft, f_c' = 4000 psi and f_y = 60,000 psi.

Service loads: w_l = 100 psf, w_d = 58 psf

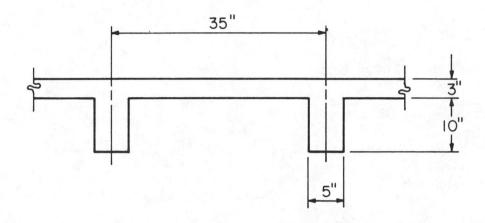

	Code
Calculations and Discussion	**Reference**

1. Compute required moment strengths using approximate moment coefficients permitted by Code Section 8.3.3. Design will be based on end span.

   Factored load:

   w_u = 1.4 × 58 + 1.7 × 100 = 251 psf   Eq. (9-1)

   $\quad$ = 251 × 35/12 = 732 lb/ft of joist span

   Positive moment:

   $+M_u = w_u l_n^2 /14$ 8.3.3

   $\quad = 0.732 \times 20^2/14 = 20.9$ ft$-$kips

   Negative moment at exterior face of first interior support:

   $-M_u = w_u l_n^2 /10$ 8.3.3

   $\quad = 0.732 \times 20^2/10 = 29.3$ ft$-$kips

Example 10.6—Continued

2. Compute required joist depth.
   For reasonable deflection control, choose a reinforcement percentage ρ about one-half
   the maximum permitted. From Table 6-1, for $f_c' = 4000$ and $f_y = 60,000$ psi,

   $\rho_{max} = 0.75\rho_b = 0.0214$

   $0.5(0.0214) = 0.0107$ Set $\rho = 0.012$

   Note: The design meets the minimum thickness requirement of Table 9.5(a). $h_{min} = \ell/18.5 > 12 \times 20/18.5 = 12.9$ in., $h = 13$ in.

   Using Table 10-1 as a design aid:
   $\omega = \rho f_y / f_c' = 0.012 \times 60/4 = 0.18$

   From Table 10-1, $M_u / \varphi f_c' bd^2 = 0.1609$

   For negative moment section, $M_u = 29.3$ ft-kips

   $$d = \sqrt{\frac{M_u}{\varphi f_c' b(0.1609)}} = \sqrt{\frac{29.3 \times 12}{0.9 \times 4 \times 5(0.1609)}} = 11.0 \text{ in.}$$

   (10 + 3 joist is O.K.)

3. Compute required reinforcement.

   $-A_s \text{ (required)} = \rho bd = 0.0120 \times 5 \times 11.0 = 0.66 \text{ in.}^2$

   Use 1#6 bar $= 0.44 \text{ in.}^2$
   Plus 1#5 bar $= \underline{0.31 \text{ in.}^2}$
                   0.75 in.^2

   For positive moment, $M_u = 20.9$ ft-kips

   Using Table 10-1: $\dfrac{M_u}{\varphi f_c' bd^2} = \dfrac{20.9 \times 12}{0.9 \times 4 \times 35 \times 11.75^2} = 0.0144$

   $\omega = 0.0145$

   $\rho = \omega f_c' / f_y = 0.0145 \times 4/60 = 0.00097$

   $+A_s \text{ (required)} = 0.00097 \times 35 \times 11.75 = 0.398 \text{ in.}^2$

Example 10.6—Continued

Calculations and Discussion	Code Reference

Use 1#6 bar = 0.44 in.^{2}

4. Shear strength needs to be checked at supports.

11

Design for Flexure and Axial Load

GENERAL CONSIDERATIONS

Design or investigation of a short compression member is based primarily on the strength of its cross section. Strength of a cross section under combined flexure and axial load must satisfy both force equilibrium and strain compatibility (See Part 6). The combined nominal axial load and moment strength (P_n, M_n) is then multiplied by the appropriate strength reduction factor φ to obtain the design strength $(\varphi P_n, \varphi M_n)$ of the section. The design strength must be equal to or greater than the required strength:

$$(\varphi P_n, \varphi M_n) \geq (P_u, M_u)$$

All members subject to combined flexure and axial load must be designed to satisfy this basic criterion. Note that the required strength (P_u, M_u) represents the structural effects of the various combinations of loads and forces to which a structure may be subjected; see Part 5 for discussion on Section 9.2—Required Strength.

A "strength interaction diagram" can be generated by plotting the design axial load strength φP_n against the corresponding design moment strength about an axis φM_n; this diagram defines the "usable" strength of a section. A typical design load-moment strength interaction diagram is shown in Fig. 11-1, illustrating the various segments of the strength curve permitted for design. The "flat-top" segment of the design strength curve defines the limiting axial load strength $P_{n(max)}$; see Part 5 discussion on Section 10.3.5. The value of φ may be increased linearly from the value for compression members ($\varphi = 0.75$ or 0.70) to the value for flexure ($\varphi = 0.90$) as the design axial load strength φP_n decreases from $0.10\, f_c' A_g$ or φP_b, whichever is smaller, to zero. Note that the φ factor increase in Fig. 11-1 is indicated at $0.10 f_c' A_g$. For $f_y \leq 60,000$ psi and symmetrical reinforcement with distance between tension and compression reinforcement not less than 0.7 times the total member depth, the $0.10 f_c' A_g$ value will be less than P_b. For this usual design condition, the expressions for φ factor increase are:

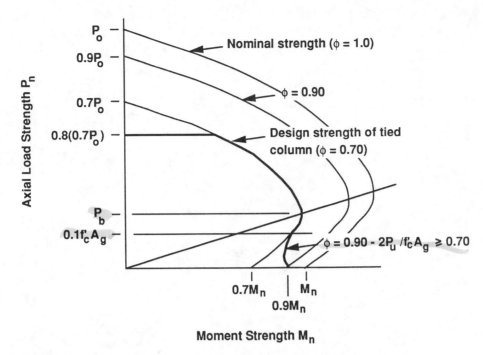

Fig. 11-1   Design Load-Moment Strength Diagram (Tied Column $\varphi = 0.70$)

For members with ties as lateral reinforcement

$$\varphi = 0.90 - 0.20 \frac{P_u}{0.10 f_c' A_g} \geq 0.70$$

For members with spirals as lateral reinforcement

$$\varphi = 0.90 - 0.15 \frac{P_u}{0.10 f_c' A_g} \geq 0.75$$

See also Part 5 for discussion on the φ factor increase.

Column design by manual calculation can be a tedious and time-consuming process requiring trial designs, especially if column slenderness is a design consideration, since the amount of strength reduction for member slenderness is directly dependent on the cross-sectional properties, including reinforcement, of the column selected. Without the use of design aids, the trial-and-error process must continue until a trial column section and selected reinforcement satisfy the required strength including strength reduction due to slenderness effects. It will quickly become apparent to the designer that a manual calculation process for column design can be "overwhelming," especially if several trials are required.

For practical design of the many columns (or column stacks) that comprise a building, a designer must resort to design aids in the form of tables, charts, or computer programs.

Examples 11.1 through 11.6 consider design for combined flexure and axial load. The examples illustrate use of the selected design aids that follow. The design examples treat uniaxial bending only for square, rectangular, and circular sections; for circular sections, biaxial bending is also considered. For a complete treatise on biaxial loading, including design examples, see Part 12. Note also that the Design Examples of Part 11 consider design of "short" columns only; it is assumed that slenderness effects may be neglected for all selected design conditions. Consideration of slenderness in column design is treated in Part 13.

It should be further noted that no design aid can include all possible variables and their variations, including bar size, number and arrangement; mixed bar sizes; grade of reinforcement; concrete strength; and shape and dimensions of column sections. Computer programs possibly treat the widest range of variations and variables; even then, computer solutions need to be supplemented by manual calculations using basic force equilibrium and strain compatibility, as illustrated in Example 6.4.

Some design aids use the rectangular concrete stress "block;" others use the parabolic or some other more realistic concrete stress distribution; this is permitted by Section 10.2.6. The user can expect some slight differences in designs given by design aids that use different concrete stress distributions. However, all these designs should be considered equally valid within the realm of practicality.

SELECTED DESIGN AIDS

Strength Design of Reinforced Concrete Columns, **Publication EB009, Portland Cement Association, Skokie, IL, 1978.** Provides tables of column strength in terms of load in kips versus moment in ft-kips, for a concrete strength of 5000 psi and Grade 60 reinforcement.

Column strength by this design aid is based on the parabolic stress-strain relationship shown in Fig. 6-8. The strength reduction factor φ is not included in the tabulated values which represent nominal strengths (P_n, M_n). When designing with this aid, the factored load and moment (required strength) values must be divided by the φ factor to enter the strength tables directly:

$$\frac{P_u}{\varphi} \leq P_n \text{ (tabulated axial load strength)}$$

$$\frac{M_u}{\varphi} \leq M_n \text{ (tabulated moment strength)}$$

For members with small or zero computed end moments, a design based on the maximum axial load strength equal to 0.85 or 0.80 times the axial load strength at zero eccentricity is required. The design data for this condition can be found under the "special conditions" P_o. To enter the strength tables directly,

For spirally reinforced members:

$$\frac{P_u}{(0.85)(0.75)} \leq P_o \text{ (tabulated axial load strength)}$$

For tied reinforced members:

$$\frac{P_u}{(0.80)(0.70)} \leq P_o \text{ (tabulated axial load strength)}$$

Design Handbook, Volume 2, Columns, **Publication SP-17A (85), American Concrete Institute, Detroit, MI, 1985.** Provides comprehensive design aids (tables and graphs) for use in the design and analysis of reinforced concrete columns by the Strength Design Method. Design examples illustrating the use of these design aids are included.

Design Aid SP17A contains a series of nondimensional load-moment strength interaction diagrams for rectangular and circular column sections with varying reinforcement ratios ρ_g (0.01 → 0.08) and different reinforcement patterns. The nondimensional format of this design aid makes it ideally suited for any size column section, including unusually large rectangular or circular sections which are beyond the scope of load capacity tables for specific column sizes. The strength reduction factor φ is already included in the strength interaction curves. φ factors of 0.75 and 0.70 are used for spirally reinforced and tied reinforced columns, respectively. For both types of laterally reinforced columns, the φ factor increase permitted by Section 9.3.2.2 is considered. For circular columns with circular ties, in lieu of spirals, design data can be extracted simply by multiplying the required strength P_u and M_u by the factor (0.70/0.75 = 0.93), then entering the strength diagrams with the modified required strength values.

CRSI Handbook—1984, **Concrete Reinforcing Steel Institute, Schaumburg, IL, 1984.** Provides designs for square, rectangular, and round column sections. The tabulated data are presented in "column load capacity tables" for a wide range of column sections with quantities of reinforcement given in terms of a range of bar sizes and number of bars for each section. Concrete strengths of 4, 5, 6, and 8 ksi are considered in combination with Grade 60 reinforcement. The designer need only enter the appropriate table with the required factored load (P_u) and load eccentricity ($e = M_u/P_u$) to obtain the required reinforcement pattern (bar size and number of bars). The strength reduction factor φ is already included in the load capacity tables. The single φ factor for tied reinforced columns ($\varphi = 0.70$) is used for all column sections, including round sections. The round section capacities apply for circular sections with circular ties, not spirals. Load capacities for round columns with spirals in lieu of ties can be extracted simply by multiplying the round column capacities by the factor (0.75/0.70 = 1.07).

PCACOL— Graphical Program for Investigation and Design of Reinforced Concrete Columns, **Portland Cement Association, Skokie, IL, 1989.** Provides the design and permits the investigation of reinforced concrete column sections. This microcomputer software program designs reinforced concrete columns to resist a given combination of loads, or investigates the adequacy of a given column section to resist a similar set of loadings. Each load case consists of an axial force combined with uniaxial or biaxial bending. The program is also capable of taking column slenderness into account. Column strength by this design aid is based on either the parabolic stress-strain relationship shown in Fig. 6-8 or the equivalent rectangular stress "block" shown in Fig. 6-9, either of which may be specified by the user.

The program investigates or designs reinforced concrete column sections graphically. With the irregular section module, IRRCOL, PCACOL can investigate irregularly shaped column sections that may contain openings. The on-screen graphics show both an interaction diagram and a scaled column cross section with the vertical reinforcement in place.

Example 11.1—Design for Pure Compression

Design a concentrically loaded square column with ties providing lateral reinforcement. Service dead and live loads are 320 and 190 kips, respectively. The column has an unsupported height of 8.5 ft and is braced against sidesway. Use $f_c' = 4000$ psi and $f_y = 60,000$ psi.

Calculations and Discussion	Code Reference

1. Determine required strength.

   $P_u = 1.4D + 1.7L = 1.4(320) + 1.7(190) = 771$ kips

   Eq. (9-1)

2. Check column slenderness. Assume 18 in. square column.

   $k = 1.0$ for braced compression member

   10.11.2.1

   $r = 0.3 \times 18 = 5.4$ in.

   10.11.3

   $kl_u/r = 1.0 \times 8.5 \times 12/5.4 = 18.9 < 34 - 12(M_1/M_2) = 22$

   10.11.4.1

   Therefore, slenderness may be neglected.

3. <u>Using Design Aid EB9</u>

   (a) Required $P_u/\varphi = 771/0.70 = 1101$ kips

   9.3.2.2(b)

   Adjust required P_n for concrete strength lower than tabulated values.
   $P_n = 1101 \left(\dfrac{5000}{4000}\right) = 1376$ kips

   (b) For columns with zero computed end moments, design is based on the limiting axial load strength of $0.80 P_o$. To enter the tables directly,

   10.3.5

   Required $P_o = \dfrac{1376}{0.80} = 1720$ kips

   From strength table for 18 in. x 18 in. column size, (see Table 11-1), read PCNT = 1.93 for $P_o = 1725$ kips.

   Adjust PCNT for 4000 psi concrete.

   Required $\rho = 1.93 \left(\dfrac{4000}{5000}\right) = 1.54\%$

   Use 4 #10 bars ($\rho = 1.57\%$)

Example 11.1—Continued

4. Select lateral reinforcement.

   Use #3 ties with #10 longitudinal bars.

   7.10.5.1

   Spacing not greater than: 16×1.27 = 20.3 in.
   48×0.375 = 18 in.
   Column size = 18 in.

   Use #3 ties @ 18 in.

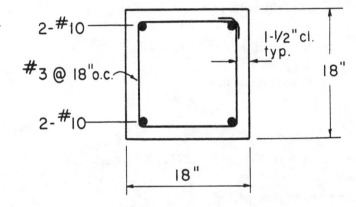

   2-#10

   #3 @ 18" o.c.

   2-#10

   1-1/2" cl. typ.

   18"

   18"

Selected Column

5. Using Design Aid SP17A

   (a) When the column section is preset, selection of reinforcement by Design Aid SP17A is simple and direct. Select reinforcement for 18 in. x 18 in. column.

   Compute $\dfrac{P_u}{A_g} = \dfrac{771}{18 \times 18} = 2.38$

   Estimate $\gamma = \dfrac{h-5}{h} = \dfrac{18-5}{18} = 0.72$

   where 2(cover + tie diameter + ½ bar diameter) = 5 in.

   (b) For rectangular section with bars along end faces, $f_c' = 4$ ksi, $f_y = 60$ ksi, and $\gamma = 0.72$, use Column Chart 7.11.3 for E4-60.75 Columns (reproduced in Fig. 11-2). Read $\rho_g = 0.015$.

   $A_{st} = \rho_g A_g = 0.015 (324) = 4.86$ in.^{2}

   Use 4 #10 Bars ($A_s = 5.08$ in.^{2})

Example 11.1—Continued

Table 11-1—Design Aid Reproduced from Page 12 of EB9

PCA—LOAD AND MOMENT STRENGTH TABLES FOR CONCRETE COLUMNS

```
         ┌──────────┐  ← 1 1/2" clear
 No      │  No.A2   │              COLUMN SIZE =18X18 INCHES
 A1      │          │  ← Axis of bending       FPC=5,000  FY=60,000
         └──────────┘
```

BARS				SPECIAL CONDITIONS		AXIAL LOAD STRENGTH P_n (KIPS)																
NO A1	NO A2	SIZE NO	PCNT	0. PO	BALANCE PB	0 MB	40	80	120	160	260	360	460	560	660	760	860	960	1060	1160	1260	1360	1560
						MOMENT STRENGTH M_n (FT·KIPS)																
8	0	6	1.09	1573	540	355	133	156	179	202	223	272	312	341	353	342	327	307	282	249	208		
6	0	7	1.11	1578	538	356	135	159	181	203	225	273	313	342	354	343	328	309	283	250	210		
4	0	9	1.23	1600	533	366	148	171	193	215	236	283	323	353	363	351	336	316	291	259	219		
10	0	6	1.36	1622	538	383	163	187	209	231	252	300	341	370	380	366	350	329	303	271	232		
6	0	8	1.46	1641	533	391	174	196	218	240	261	309	348	378	387	373	356	335	310	278	240	194	
8	0	7	1.48	1645	535	395	176	199	221	243	264	312	352	382	391	376	359	338	312	281	242	197	
4	0	10	1.57	1660	528	398	183	206	227	249	269	316	355	386	393	378	361	341	316	285	247	202	
6	0	9	1.85	1711	528	429	215	237	259	280	300	347	387	417	424	406	387	365	339	309	272	229	
4	0	11	1.93	1725	523	432	220	242	263	284	304	350	389	420	425	408	389	367	342	312	276	234	
8	0	8	1.93	1729	530	441	226	249	270	292	312	359	399	429	436	417	397	375	348	318	282	239	
6	0	10	2.35	1802	523	477	266	288	309	330	350	396	435	467	469	449	427	403	377	347	312	272	226
8	0	9	2.47	1823	524	492	281	302	324	344	364	411	451	481	484	462	440	415	388	358	323	284	238
4	0	14	2.78	1879	512	510	305	326	346	366	385	430	468	499	500	477	454	430	404	375	342	304	261
6	0	11	2.89	1899	516	528	321	342	362	383	402	447	487	518	518	494	470	445	418	388	354	316	273
4	0	18	4.94	2269	468	689	506	525	543	562	579	620	657	687	671	648	619	590	561	530	499	465	430	349
0	0	0	6.00	2461	439	769	599	617	635	652	669	709	745	765	744	723	699	668	637	606	574	541	506	432
0	0	0	7.00	2641	410	840	683	701	718	735	752	790	825	829	808	787	766	740	708	676	644	611	577	505
0	0	0	8.00	2822	377	908	765	782	799	816	832	869	903	890	869	847	826	804	779	746	713	680	646	575
6	2	6	1.09	1573	544	326	134	157	179	200	221	263	292	314	325	318	307	291	268	237	199		
8	4	5	1.15	1584	547	326	141	164	186	207	225	264	297	315	326	319	308	293	271	241	203		
6	2	7	1.48	1645	541	356	177	199	220	241	260	297	324	345	354	344	332	315	293	264	229		
10	6	5	1.53	1654	547	353	183	205	224	242	260	297	327	344	352	343	331	315	294	267	232	190	
8	4	6	1.63	1671	543	361	194	215	236	253	269	306	335	352	360	350	337	321	300	273	239	198	
6	2	8	1.95	1729	537	391	226	248	268	288	306	336	361	381	388	375	361	343	321	295	262	223	
10	6	6	2.17	1769	543	398	249	267	285	302	319	351	378	391	396	383	369	352	332	307	277	241	
8	4	7	2.22	1778	539	403	256	275	291	306	321	357	381	395	401	387	373	356	335	310	280	243	
6	2	9	2.47	1823	532	429	280	300	320	338	353	378	401	420	425	409	393	374	353	327	297	262	220
8	4	8	2.93	1906	533	453	322	337	353	367	382	416	434	446	448	432	416	397	377	353	325	293	255
6	2	10	3.14	1943	526	471	346	365	384	396	406	429	452	470	471	453	435	415	393	368	340	308	271
8	4	9	3.70	2046	527	506	389	404	419	433	447	474	490	501	500	482	463	444	422	399	373	343	310
6	2	11	3.85	2073	520	528	415	432	442	451	460	483	504	522	520	500	480	458	436	412	385	354	321	240
4	4	6	1.09	1573	547	304	134	157	178	197	212	249	278	294	304	300	292	279	258	230	193		
4	2	7	1.11	1578	543	317	136	159	181	202	221	259	285	305	311	301	286	264	235	197		
4	6	6	1.36	1622	549	312	164	182	199	217	233	265	291	304	312	308	300	288	269	244	211		
4	4	8	1.46	1641	540	340	174	196	217	237	255	287	311	330	331	320	305	284	257	222		
4	4	7	1.48	1645	544	326	177	198	215	230	245	279	303	317	325	319	311	297	278	253	220		
4	4	9	1.85	1711	536	366	215	236	256	274	292	316	338	356	364	353	341	326	306	280	249	210	
4	4	8	1.95	1729	540	352	224	239	254	268	282	314	333	345	350	342	332	319	301	278	249	213	
4	2	10	2.35	1802	531	398	265	285	304	320	329	351	372	390	394	382	368	352	333	309	280	245	204
4	2	9	2.47	1823	535	380	267	281	294	308	321	350	363	373	378	368	356	343	326	305	279	247	208
4	2	11	2.89	1899	526	432	317	336	350	359	367	388	408	425	427	413	397	380	361	338	312	280	243
10	2	5	1.15	1584	545	342	141	164	187	208	229	275	306	329	341	332	319	301	278	246	207		
8	2	6	1.36	1622	543	355	164	186	208	229	250	291	320	343	353	343	330	312	289	259	222		
10	2	6	1.63	1671	541	383	194	216	238	258	278	320	349	372	381	368	353	334	310	281	246	203	
8	2	7	1.85	1711	538	395	217	239	260	280	299	336	363	384	392	378	363	344	321	293	260	219	
8	2	8	2.44	1817	533	441	278	299	319	339	357	387	412	432	437	420	402	382	360	333	302	265	222
8	2	9	3.09	1934	528	492	344	364	384	402	414	440	465	484	486	466	446	425	402	375	346	312	274
10	4	5	1.34	1619	546	347	162	185	207	228	245	285	317	336	346	337	324	308	286	256	220		
6	4	6	1.36	1622	545	332	164	186	207	225	241	278	306	323	332	325	314	299	279	251	216		
6	4	7	1.85	1711	541	365	216	237	253	268	283	318	342	356	363	353	341	326	306	281	250	212	
10	4	6	1.90	1720	541	390	223	245	264	281	297	335	364	381	387	375	360	343	321	294	261	222	
6	4	8	2.44	1817	537	402	273	288	303	318	332	363	383	396	399	387	373	358	338	315	287	253	213
6	4	9	3.09	1934	531	443	328	342	356	370	384	412	427	438	439	424	409	392	373	351	326	296	260
6	6	5	1.15	1584	550	312	141	164	183	201	219	257	285	302	312	307	299	285	264	236	200		
8	6	5	1.34	1619	548	332	162	184	203	221	239	277	306	323	332	325	314	300	279	251	216		
6	6	6	1.63	1671	547	341	192	210	228	245	262	293	320	333	340	333	323	309	290	265	233	193	
8	6	6	1.90	1720	545	369	220	238	256	273	290	322	349	362	368	358	346	330	311	286	255	217	

11-7

Example 11.1—Continued

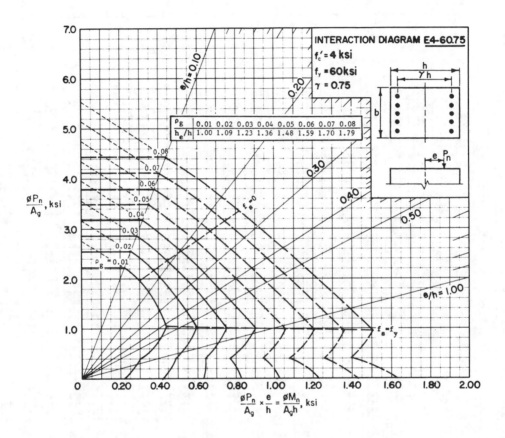

**Fig. 11-2   Chart Reproduced from Page 96 of SP17A
(Courtesy of American Concrete Institute)**

Note that axial load strength for pure compression is limited by Section 10.3.5, defined by the flat-top portion of the strength interaction diagram. A check of the above solution can be easily made.

$$\varphi P_{n(max)} = 0.80\,\varphi\,[0.85 f_c' (A_g - A_{st}) + f_y A_{st}]$$

Eq. (10-2)

$$= 0.80(0.70)[0.85 \times 4(324 - 5.08) + 60(5.08)] = 778 \text{ kips}$$

$P_u = 771 \text{ kips} < 778 \text{ kips}$ O.K.

Example 11.1—Continued

6. <u>Using CRSI Handbook</u>

   (a) P_u = 771 kips

   (b) For Grade 60 bars with f_c' = 4000 psi, and 18 in. square section, use Load Capacity Table—SQUARE TIED COLUMNS 18" x 18", page 3-21 (reproduced in Table 11-2).

   (c) Select 4#10 bars @ 0.80(φP_o) = 778 kips. For the 18 in. x 18 in. column, other bar selections can easily be determined by the CRSI load tables; alternatively, 8#7 bars @ 769 kips ≈ 771 kips. Note that with the CRSI Handbook, bar selection is simple and direct, allowing for alternative bar patterns when selecting bars for a column stack. Note also that CRSI load tables already include the φ factor; enter the load tables directly with the factored load P_u.

   (d) For illustration, note that a 16 in. x 16 in. square column with 4#14 bars @ 773 kips (Table 11-2) would be an alternative solution for this design. A recheck of column slenderness indicates that slenderness may also be neglected in design for the 16 in. x 16 in. column size.

   Alternatively, use 16 in. x 16 in. column with 4#14 bars

Example 11.1—Continued

Table 11-2—Design Aid Reproduced from Page 3-21 of CRSI Handbook
(Courtesy of Concrete Reinforcing Steel Institute)

Bars Grade 60		SQUARE TIED COLUMNS 16″ × 16″ Short columns; no sidesway [1] Bars symmetrical in 4 faces													0.10f′$_c$A$_g$ = 102 kips [2]			
Concrete f′$_c$ = 4,000 psi		φP$_n$ (kips)—Ultimate Usable Capacity													For φP$_n$ at [2]			
															OT [3]	Balance		102 k [2]	O [2]
Bars	p %	0.80 × φP$_o$ (k)	e 0.80 × φP$_o$ (in.)	M_n/P_n = e (in.) (φ = 0.70)										e (in.)	e (in.)	φP$_b$ (k)	e (in.)	φM$_n$ (k-ft.)
				2″	3″	4″	6″	8″	12″	16″	20″	24″	28″					
4-#8	1.23	587	1.59	553	476	408	307	235	126	0	0	0	0	2.63	7.46	257	13.78	91
4-#9	1.56	614	1.58	578	499	430	327	261	150	0	0	0	0	2.74	8.23	255	15.54	112
4-#10	1.98	648	1.56	609	527	456	351	283	177	118	0	0	0	2.87	9.24	252	17.80	138
4-#11	2.44	685	1.51	640	554	481	372	300	199	135	0	0	0	3.01	10.38	241	19.85	163
4-#14	3.52	773	1.46	717	623	543	425	346	246	179	135	108	0	3.22	13.25	226	25.15	225
4-#18	6.25	994	1.33	907	789	691	547	444	318	248	203	170	142	3.54	23.05	178	37.76	367
8-#6	1.37	599	1.50	554	472	400	296	220	128	0	0	0	0	2.43	6.95	261	14.36	101
8-#7	1.87	639	1.46	589	503	429	322	253	153	107	0	0	0	2.53	7.84	259	16.67	133
8-#8	2.47	688	1.42	630	538	461	350	280	179	127	0	0	0	2.63	8.91	256	19.35	170
8-#9	3.13	741	1.38	674	578	496	379	306	207	147	114	0	0	2.72	10.09	253	22.10	209
8-#10	3.97	809	1.34	732	627	540	415	337	240	172	134	109	0	2.82	11.66	248	25.58	255
8-#11	4.87	883	1.27	789	675	581	448	364	259	193	150	123	104	2.91	13.68	231	28.53	288
8-#14	7.03	1058	1.19	931	795	685	531	435	310	240	194	159	135	3.04	18.99	205	36.47	376
12-#10	5.95	970	1.26	865	739	636	492	399	289	222	175	144	123	2.89	15.01	238	33.40	344
12-#11	7.31	1081	1.20	950	809	696	538	436	314	243	196	162	138	2.96	18.53	213	37.28	393
		SQUARE TIED COLUMNS 18″ × 18″													0.10f′$_c$A$_g$ = 129 kips			
4-#9	1.23	744	1.81	726	637	557	430	344	199	0	0	0	0	2.90	8.40	330	15.76	131
4-#10	1.57	778	1.80	758	667	586	457	369	233	152	0	0	0	3.03	9.30	328	17.88	162
4-#11	1.93	815	1.77	791	697	613	481	391	262	175	0	0	0	3.19	10.25	322	19.90	193
4-#14	2.78	902	1.72	872	771	681	541	445	326	230	171	136	0	3.44	12.56	314	24.96	267
4-#18	4.94	1124	1.60	1072	951	844	680	565	419	329	267	215	179	3.86	19.29	279	37.14	442
8-#6	1.09	728	1.73	702	610	526	395	298	172	0	0	0	0	2.59	7.21	338	14.64	118
8-#7	1.48	769	1.69	738	643	557	424	336	203	140	0	0	0	2.70	8.01	336	17.10	156
8-#8	1.95	817	1.66	781	681	593	456	366	236	165	0	0	0	2.81	8.96	334	19.63	200
8-#9	2.47	870	1.62	828	723	631	489	396	270	191	147	0	0	2.91	10.01	331	22.37	247
8-#10	3.14	939	1.58	888	776	679	531	432	308	222	171	139	0	3.03	11.36	327	25.66	306
8-#11	3.85	1012	1.52	950	829	725	568	465	338	249	193	157	133	3.15	12.84	319	28.63	361
8-#14	5.56	1187	1.43	1102	961	841	664	544	401	315	247	203	171	3.32	16.52	306	36.33	470
12-#10	4.70	1100	1.51	1030	898	786	618	504	368	282	223	183	156	3.13	14.13	321	33.43	416
12-#11	5.78	1210	1.45	1122	976	853	671	549	401	316	251	206	175	3.23	16.39	309	37.50	485
16-#10	6.27	1261	1.46	1175	1023	896	706	579	425	335	270	224	191	3.23	16.82	321	40.78	527

11-10

Example 11.1—Continued

7. Using PCACOL Program

Using the design option of the PCACOL program, design the 18 in. x 18 in. column section with a range of 4#8 to 6#10 bars for a required axial load strength of $P_u = 771$ kips.

The "on-screen" graphic results of PCACOL are reproduced in Fig. 11-3. For the design option, both an interaction diagram and a scaled column cross section showing vertical reinforcement are displayed. The factored load combination $P_u = 771$ kips is shown as point "1" relative to the strength interaction curve for the column section and reinforcement provided. Clear cover to longitudinal bars (cc = 1.88 in.) and clear spacing between longitudinal bars (spacing = 11.71 in.) are included in the displayed results.

Example 11.1—Continued

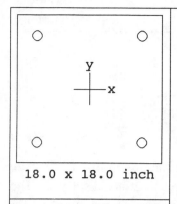

18.0 x 18.0 inch

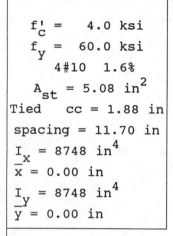

f'_c = 4.0 ksi

f_y = 60.0 ksi

4#10 1.6%

A_{st} = 5.08 in^2

Tied   cc = 1.88 in

spacing = 11.70 in

I_x = 8748 in^4

$\bar{x}$ = 0.00 in

I_y = 8748 in^4

$\bar{y}$ = 0.00 in

© 1989 PCA

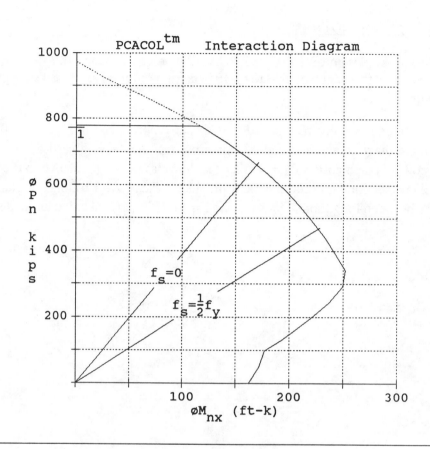

PCACOL^{tm} Interaction Diagram

ϕP_n kips (y-axis)

ϕM_{nx} (ft-k) (x-axis)

$f_s = 0$

$f_s = \frac{1}{2}f_y$

Licensed To: Portland Cement Association, Skokie, IL

Project: Example 11.1   File name: C:\PCACOL\DATA\EX11_1.COL

Column Id: Design for Compression   Material Properties:

Engineer: E_c = 3834 ksi ϵ_u = 0.003 in/in

Date: Time: f_c = 3.40 ksi E_s = 29000 ksi

Code: ACI 318-89 β_1 = 0.85

Version: 2.10   Stress Profile : Block

                                     Reduction: ϕ_c = 0.70 ϕ_b = 0.90

Slenderness not considered x-axis

**Fig. 11-3   PCACOL On-Screen Graphic Results
for 18 in. x18 in. Column with 4#10 Bars**

Example 11.2—Design for Small Axial Load

Design a square reinforced concrete column with ties providing lateral reinforcement for service dead and live loads of 20 and 15 kips respectively. Service dead and live load moments at each end have values of 90 and 70 ft-kips respectively. The column has an unsupported height of 12 ft and is bent in double curvature. Use $f'_c = 5000$ psi and $f_y = 60,000$ psi. Column size is set at 18 in. minimum by architectural considerations and is braced against sidesway.

Calculations and Discussion	Code Reference

1. Determine required strength.

   $P_u = 1.4\,(20) + 1.7\,(15) = 53.5$ kips   Eq. (9-1)

   $M_u = 1.4\,(90) + 1.7\,(70) = 245$ ft$-$kips

2. Check column slenderness. **Assume** 18 in. x 18 in. column section.

   k = 1.0 for biaxial compression member   10.11.2.1

   $r = 0.3\,(18) = 5.4$ in.   10.11.3

   $kl_u/r = 1.0 \times 12 \times 12/5.4 = 26.7 \; < \; 34 - 12\left(\dfrac{-245}{245}\right) = 46$ 10.11.4.1

   Therefore, slenderness may be neglected.

3. Since the column section is large and required strength $P_u = 53.5$ kips is relatively small, check whether the strength reduction factor φ may be increased ($\varphi = 0.70 \rightarrow 0.90$):

   $\dfrac{h - d' - d_s}{h} = \dfrac{18 - 2.5 - 2.5}{18} = 0.72 > 0.70$ O.K.

   $0.10\,f'_c\,A_g = 0.10 \times 5\,(18 \times 18) = 162$ kips > 53.5 kips   O.K.

   For members with ties providing lateral reinforcement:

   $\varphi = 0.90 - 0.2\,\dfrac{P_u}{0.10\,f'_c\,A_g} \geq 0.70$

   $= 0.9 - 0.2\,\dfrac{53.5}{162} = 0.83$

Example 11.2—Continued

	Code
Calculations and Discussion	**Reference**

4. <u>Using Design Aid EB9</u>

   (a) Required $P_n = \dfrac{P_u}{\varphi} = \dfrac{53.5}{0.83} = 64.5$ kips

   $$M_n = \dfrac{M_u}{\varphi} = \dfrac{245}{0.83} = 295 \text{ ft} - \text{kips}$$

   (b) From strength table for 18 in. x 18 in. column size, $f_c' = 5000$ psi and $f_y = 60,000$ psi (see Table 11-3), <u>select 6#10 bars</u> (PCNT = 2.35) to provide a load moment strength of $P_n = 65$ kips and $M_n = 301$ ft-kips.

5. Select lateral reinforcement.

   Use #3 ties with #10 longitudinal bars.   7.10.5.1

   Spacing not greater than: $16 \times 1.27 \quad = 20.3$ in   7.10.5.2
   $\qquad\qquad\qquad\qquad\quad\; 48 \times 0.375 \; = 18$ in.
   $\qquad\qquad\qquad\qquad\quad\; \text{Column size} = 18$ in.

   Use #3 ties @ 18 in.

6. <u>Using Design Aid SP17A</u>

   (a) Select reinforcement for 18 in. x 18 in. column section ($A_g = 324$ in.^{2})

   Compute $\dfrac{P_u}{A_g} = \dfrac{53.5}{324} = 0.17$

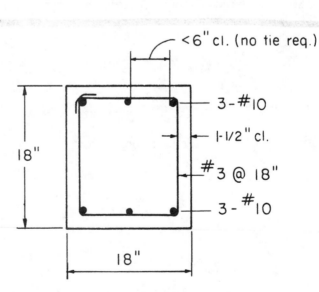

Example 11.2—Continued

Table 11-3—Design Aid Reproduced from Page 12 of EB9

PCA–LOAD AND MOMENT STRENGTH TABLES FOR CONCRETE COLUMNS

1 1/2" clear — Axis of bending
No A1, No A2

COLUMN SIZE =18X18 INCHES
FPC=5,000 FY=60,000

NO A1	NO A2	SIZE NO	PCNT	PO	PB	MB	0	40	80	120	160	260	360	460	560	660	760	860	960	1060	1160	1260	1360	1560
					BALANCE		MOMENT STRENGTH Mn (FT-KIPS)																	
8	0	6	1.09	1573	540	355	133	156	179	202	223	272	312	341	353	342	327	307	282	249	208			
6	0	7	1.11	1578	538	356	135	159	181	203	225	273	313	342	342	343	328	309	283	250	210			
4	0	9	1.23	1600	533	366	148	171	193	215	236	283	323	353	363	351	336	316	291	259	219			
10	0	6	1.36	1622	538	383	163	187	209	231	252	300	341	370	380	366	350	329	303	271	232			
6	0	8	1.46	1641	533	391	174	196	218	240	261	309	348	378	387	373	356	335	310	278	240	194		
8	0	7	1.48	1645	535	395	176	199	221	243	264	312	352	382	391	376	359	338	312	281	242	197		
4	0	10	1.57	1660	528	398	183	206	227	249	269	314	355	386	393	378	361	341	316	285	247	202		
6	0	9	1.85	1711	528	429	215	237	259	280	300	347	387	417	424	406	387	365	339	309	272	229		
4	0	11	1.93	1725	523	432	220	242	263	284	304	350	389	420	425	408	389	367	342	312	276	234		
8	0	8	1.95	1729	530	441	226	249	270	292	312	359	399	429	436	417	397	375	348	318	282	239		
6	0	10	2.35	1802	523	477	266	288	309	330	350	396	435	467	469	449	427	403	377	347	312	272	226	
8	0	9	2.47	1823	524	492	281	302	324	344	364	411	451	481	484	462	440	415	388	358	323	284	238	
4	0	14	2.78	1879	512	510	305	326	346	366	385	430	468	499	500	477	454	430	404	375	342	304	261	
6	0	11	2.89	1899	516	528	321	342	362	383	402	447	487	518	518	494	470	445	418	388	354	316	273	
4	0	18	4.94	2269	468	689	506	525	543	562	579	620	657	687	671	648	619	590	561	530	499	465	430	349
0	0	0	6.00	2461	439	769	599	617	635	652	669	709	745	765	744	723	699	668	637	606	574	541	506	432
0	0	0	7.00	2641	410	840	683	701	718	735	752	790	825	829	808	787	766	740	708	676	644	611	577	505
0	0	0	8.00	2822	377	908	765	782	799	816	832	869	903	890	869	847	826	804	779	746	713	680	646	575
6	2	6	1.09	1573	544	326	134	157	179	200	221	263	292	314	325	318	307	291	268	237	199			
8	4	5	1.15	1584	547	326	141	164	186	207	225	264	297	315	326	319	308	293	271	241	203			
6	2	7	1.48	1645	541	356	177	199	220	241	260	297	324	345	354	344	332	315	293	264	229			
10	6	5	1.53	1654	547	353	183	205	224	242	260	297	327	344	352	343	331	315	294	267	232	190		
8	4	6	1.63	1671	543	361	194	215	236	253	269	306	335	352	360	350	337	321	300	273	239	198		
6	2	8	1.95	1729	537	391	226	248	268	288	306	336	361	381	388	375	361	343	321	295	262	223		
10	6	6	2.17	1769	543	398	249	267	285	302	319	351	378	391	396	383	369	352	332	307	277	241		
8	4	7	2.22	1778	539	403	256	275	291	306	321	357	381	395	401	387	373	356	335	310	280	243		
6	2	9	2.47	1823	532	429	280	300	320	338	353	378	401	420	425	409	393	374	353	327	297	262	220	
8	4	8	2.93	1906	533	453	322	337	353	367	382	416	434	446	448	432	416	397	377	353	325	293	255	
6	2	10	3.14	1943	526	477	346	365	384	396	406	429	452	470	471	453	435	415	393	368	340	308	271	
8	4	9	3.70	2046	527	506	389	404	419	433	447	474	490	501	500	482	463	444	422	399	373	343	310	
6	2	11	3.85	2073	520	528	415	432	442	451	460	483	504	522	520	480	458	436	412	385	354	321		240
4	4	6	1.09	1573	547	304	134	157	178	197	212	249	278	294	304	300	292	279	258	230	193			
4	2	7	1.11	1578	543	317	136	159	181	202	221	259	285	305	316	311	301	286	264	235	197			
4	6	6	1.36	1622	549	312	164	182	199	217	233	265	291	304	312	308	300	288	269	244	211			
4	2	8	1.46	1641	540	340	174	196	217	237	255	287	311	330	339	331	320	305	284	257	222			
4	4	7	1.48	1645	544	326	177	198	215	230	245	279	303	317	325	319	311	297	278	253	220			
4	2	9	1.85	1711	536	366	215	236	256	274	292	316	338	356	364	353	341	326	306	280	249	210		
4	4	8	1.95	1729	540	352	224	239	254	268	282	314	333	345	350	342	332	319	301	278	249	213		
4	2	10	2.35	1802	531	398	265	285	304	320	329	351	372	390	394	382	368	352	333	309	280	245	204	
4	4	9	2.47	1823	535	380	267	281	294	308	321	350	363	374	378	368	356	343	326	305	279	247	208	
4	2	11	2.89	1899	526	432	317	336	350	359	367	388	408	425	427	413	397	380	361	338	312	280	243	
10	2	5	1.15	1584	545	342	141	164	187	208	229	275	306	329	341	332	319	301	278	246	207			
8	2	6	1.36	1622	543	355	164	186	208	229	250	291	320	343	353	343	330	312	289	259	222			
10	2	6	1.63	1671	541	383	194	216	238	258	278	320	349	372	381	368	353	334	310	281	246	203		
8	2	7	1.85	1711	538	395	217	239	260	280	299	336	363	384	392	378	363	344	321	293	260	219		
8	2	8	2.44	1817	533	441	278	299	319	339	357	386	412	432	437	420	402	382	360	333	302	265	222	
8	2	9	3.09	1934	528	492	344	364	384	402	414	440	465	484	486	466	446	425	402	375	346	312	274	
10	4	5	1.34	1619	546	347	162	185	207	228	245	285	317	336	346	337	324	308	286	256	220			
6	4	6	1.36	1622	545	332	164	186	207	225	241	278	306	323	332	325	314	299	279	251	216			
6	4	7	1.85	1711	541	365	216	237	253	268	283	318	342	356	363	353	341	326	306	281	250	212		
10	4	6	1.90	1720	541	390	223	245	264	281	297	335	364	381	387	375	360	343	321	294	261	222		
6	4	8	2.44	1817	537	402	273	288	303	318	332	365	383	396	399	387	373	358	338	315	287	253	213	
6	4	9	3.09	1934	531	443	328	342	356	370	384	412	427	438	439	424	409	392	373	351	326	296	260	
6	6	5	1.15	1584	550	312	141	164	183	201	219	257	285	302	312	307	299	285	264	236	200			
8	6	5	1.34	1619	548	332	162	184	203	221	239	277	306	323	332	325	314	300	279	251	216			
6	6	6	1.63	1671	547	341	192	210	228	245	262	293	320	333	340	333	323	309	290	265	233	193		
8	6	6	1.90	1720	545	369	220	238	256	273	290	322	349	362	368	358	346	330	311	286	255	217		

Example 11.2—Continued

$$\frac{M_u}{A_g h} = \frac{245 \times 12}{(324)18} = 0.50$$

$$\text{Estimate } \gamma = \frac{h-5}{h} = \frac{18-5}{18} = 0.72$$

(b) For rectangular section with bars along end faces, $f'_c = 5$ ksi, $f_y = 60$ ksi, and $\gamma = 0.72$, use Column Chart 7.12.3 for E5-60.75 Columns (reproduced in Fig. 11-4).

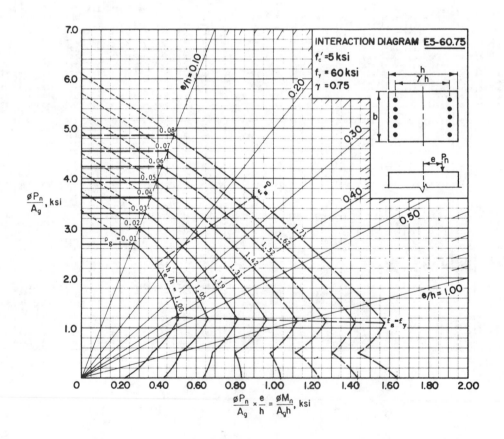

**Fig. 11-4   Chart Reproduced from Page 98 of SP17A
(Courtesy of American Concrete Institute)**

11-16

Example 11.2—Continued

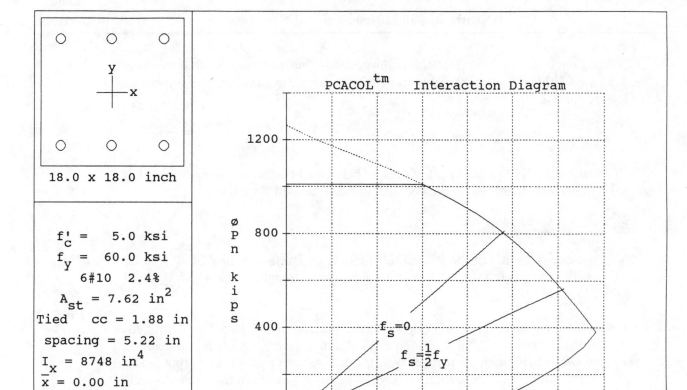

18.0 x 18.0 inch

f'_c =   5.0 ksi
f_y = 60.0 ksi
    6#10 2.4%
A_{st} = 7.62 in^2
Tied   cc = 1.88 in
 spacing = 5.22 in
I_x = 8748 in^4
$\bar{x}$ = 0.00 in
I_y = 8748 in^4
$\bar{y}$ = 0.00 in

© 1989 PCA

PCACOL^{tm} Interaction Diagram

ϕP_n kips

$f_s=0$

$f_s=\frac{1}{2}f_y$

ϕM_{nx} (ft-k)

Licensed To: Portland Cement Association, Skokie, IL

Project: Example 11.2

Column Id: Design for Small Axial

Engineer:

Date: Time:

Code: ACI 318-89

Version: 2.10

File name: C:\PCACOL\DATA\EX11_2.COL

Material Properties:

E_c =   4287 ksi ϵ_u = 0.003 in/in

f_c = 4.25 ksi E_s = 29000 ksi

β_1 = 0.80

Stress Profile : Block

Reduction: ϕ_c = 0.70 ϕ_b = 0.90

Slenderness not considered x-axis

**Fig. 11-5   PCACOL On-Screen Graphic Results
for 18 in. x 18 in. Column with 6#10 Bars**

Example 11.2—Continued

Read $\rho_g = 0.022$. Note that the required load-moment strength combination is located within the segment of the strength interaction curve where φ factor increase is permitted.

$A_{st} = \rho_g A_g = 0.022\,(324) = 7.13$ in.^{2}

Use 6 #10 Bars ($A_g = 7.62$ in.^{2}). Note that bars must be oriented as shown in sketch on p. 11-14, with axis of bending parallel to side faces with 3#10 bars.

7. Using PCACOL Program

Using the design option of the PCACOL program, design the 18 in. x 18 in. column section with a range of 4#8 to 8#10 bars for a required strength of $P_u = 53.5$ kips and $M_u = 245$ ft-kips.

The "on-screen" graphic results of PCACOL are reproduced in Fig. 11-5. Both a strength interaction diagram and a scaled column cross section showing vertical reinforcement, with the 6#10 bars oriented parallel with the x-axis, are displayed as output. The factored load-moment combination is shown as point "1" relative to the strength interaction curve for the column section and reinforcement provided. Note that the required strength is in the region of the strength interaction curve where the strength reduction factor φ may be increased ($\varphi = 0.83$).

Example 11.3—Design for General Loading (Rectangular Section)

Design a rectangular reinforced concrete column with ties providing lateral reinforcement for service dead and live loads of 350 and 240 kips respectively. Service dead and live load moments at the top about the strong axis have values of 100 and 80 ft-kips respectively. Moments are negligible about the weak axis. Assume moments at the bottom of the column are half those at the top. The column has an unsupported height of 7 ft 6 in. is bent in double curvature about the strong axis and single curvature about the weak axis. Use $f_c' = 5000$ psi and $f_y = 60,000$ psi Architectural considerations limit the width of the column to 14 in. and the column is braced against sidesway.

Calculations and Discussion	Code Reference

1. Determine required strength.

   $P_u = 1.4\,(350) + 1.7\,(240) = 898$ kips   Eq. (9-1)

   $M_u = 1.4\,(100) + 1.7\,(80) = 276$ ft-kips

2. Check column slenderness. Assume 14 in. x 22 in. column size.

   (a) Slenderness about weak axis (14 in. width):

   k = 1.0 for braced compression member   10.11.2.1

   $r = 0.3 \times 14 = 4.2$ in.   10.11.3

   $kl_u/r = (1.0 \times 7.5 \times 12)/4.2 = 21.4$

   With negligible moments about weak axis, assume $M_1/M_2 = 1.0$ for slenderness consideration.

   $kl_u/r < 34 - 12(M_1/M_2) = 22$ 10.11.4.1

   Therefore, slenderness may be neglected about weak axis.

   (b) Slenderness about strong axis (22 in. width):

   k = 1.0   10.11.2.1

   $r = 0.3 \times 22 = 6.6$ 10.11.3

   $kl_u/r = (1.0 \times 7.5 \times 12)/6.6 = 13.6 \; < 34 - 12\left(-\dfrac{1}{2}\right) = 40$ 10.11.4.1

Example 11.3—Continued

Therefore, slenderness may be neglected about strong axis.

3. Using Design Aid EB9

(a) Required $P_n = \dfrac{P_u}{\varphi} = \dfrac{898}{0.70} = 1283$ kips

$M_n = \dfrac{M_u}{\varphi} = \dfrac{276}{0.70} = 394$ ft-kips

9.3.2.2(b)

(b) Since strength tables in EB9 are given for square sections only, systematic interpolation is required. Select the smallest square column capable of supplying the required strength, then a series of larger rectangular sections to suit the required loading. From strength tables, 18 in. x 18 in. is the smallest tabulated size to support $P_n = 1282$ kips. Adjust for smaller width as follows:

Column Size	Ratio of Actual to Square Width	Adjusted Required Nominal Strength	
		$\dfrac{P_n}{\text{Width ratio}}$	$\dfrac{M_n}{\text{Width ratio}}$
18 x 18	14/18 = 0.78	1645	506
20 x 20	14/20 = 0.70	1835	564
22 x 22	14/22 = 0.64	2005	617
24 x 24	14/24 = 0.58	2215	681

(c) An 18 in. x 18 in. section (see Table 11-3 of Example 11.2) is not adequate to support the adjusted required nominal strength $P_n = 1645^k$.

A 20 in. x 20 in. section (see Table 11-4) with 6% reinforcement has combined load-moment strengths of 1835 kips and 642 ft-kips which exceed the adjusted nominal strength requirements (1835 kips and 564 ft-kips). For a 14 in. x 20 in. section, the required area of reinforcement, $A_{st} = 0.06 (14 \times 20) = 16.8$ in.^{2}

A 22 in. x 22 in. section (see Table 11-5) with 4% reinforcement has combined load-moment strengths of 2005 kips and 689 ft-kips, which exceed the adjusted strength requirements (2005 kips and 617 ft-kips). For a 14 in. x 22 in. section, $A_{st} = 0.04 (14 \times 22) = 12.3$ in.^{2}

Example 11.3—Continued

	Code
Calculations and Discussion	Reference

A 24 in. x 24 in. section (see Table 11-6) with 2.34% reinforcement has combined load-moment strengths of 2215 kips and 675 ft-kips (say O.K.). For a 14 in. x 24 in. section, $A_{st} = 0.0234 \, (14 \times 24) = 7.86$ in.^{2}

Assume no architectural limitation on the long dimension. To allow room for lap splicing of the bars, select 14 in. x 24 in. section with 4#14 bars ($A_{st} = 9.0$ in.^{2}).

4. Select lateral reinforcement

Use #4 ties with #14 longitudinal bars   7.10.5.1

Spacing not greater than: 16×1.693 = 27 in.   7.10.5.2
                              48×0.50 = 24 in.
                              Column Size = 14 in.

Use #4 ties @ 14 in.

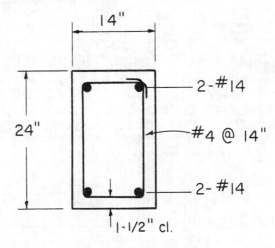

Selected Design

5. Check limiting axial load strength. For the 14 in. x 24 in. column with 4#14 bars:

$$0.8 \, \varphi P_0 = 0.8 \, \varphi \, [0.85 \, f_c' \, (A_g - A_{st}) + f_y \, A_{st}]$$ Eq. (10-2)

$$= 0.8 \, (0.70) \, [0.85 \times 5 \, (336 - 9) + 60 \times 9]$$

$$= 1081 \text{ kips}$$

$P_u = 898$ kips $<$ 1081 kips

Therefore, the column selected is adequate for required strength about both axes.

11-21

Example 11.3—Continued

Table 11-4—Design Aid Reproduced from Page 13 of EB9

PCA—LOAD AND MOMENT STRENGTH TABLES FOR CONCRETE COLUMNS

No A1 — No A2 — ↕ 1 1/2" clear ← Axis of bending   COLUMN SIZE = 20X20 INCHES   FPC=5,000 FY=60,000

NO A1	NO A2	SIZE NO	PCNT	PO (O*)	PB (BALANCE)	MB	0	40	140	240	340	440	540	640	740	840	940	1040	1140	1240	1340	1540	1740	1940
							MOMENT STRENGTH Mn (FT-KIPS)																	
4	0	9	1.00	1923	670	475	170	196	260	318	371	415	449	471	468	456	440	419	392	359	319			
10	0	6	1.10	1945	677	495	187	213	278	337	390	435	468	487	473	455	433	406	373	333				
6	0	8	1.19	1964	671	504	199	225	289	347	399	444	478	499	495	480	462	440	413	380	341			
8	0	7	1.20	1968	673	508	202	228	292	351	403	448	481	503	499	484	466	443	416	383	344			
4	0	10	1.27	1983	665	513	211	236	299	357	408	452	487	509	502	487	469	447	420	388	349			
10	0	7	1.50	2034	671	553	248	274	337	396	448	493	526	548	541	522	502	478	450	417	379	283		
6	0	9	1.50	2034	666	548	247	272	334	392	444	489	523	544	536	518	498	475	447	415	377	282		
4	0	11	1.56	2048	660	552	253	279	340	397	448	492	527	549	539	521	501	478	450	419	381	288		
8	0	11	1.58	2052	668	562	260	286	348	406	458	503	537	558	549	530	509	485	458	425	387	293		
6	0	10	1.90	2125	660	604	307	332	393	450	501	545	580	601	588	566	544	519	491	458	422	332		
10	0	8	1.98	2140	664	620	320	346	407	465	517	562	596	617	604	581	557	531	503	470	433	343		
8	0	9	2.00	2146	662	621	323	348	409	466	518	563	597	618	604	582	558	532	503	471	434	345		
4	0	14	2.25	2202	649	644	353	377	436	491	541	585	621	643	623	600	576	550	522	491	455	370		
6	0	11	2.34	2222	653	664	370	395	454	510	561	605	641	662	643	618	593	566	537	505	469	384		
8	0	10	2.54	2266	654	696	402	427	486	543	594	638	673	694	674	648	621	593	563	530	494	410	306	
6	0	14	3.38	2453	639	802	516	540	597	651	701	744	782	801	771	742	712	681	649	616	581	501	406	
4	0	18	4.00	2592	623	866	589	612	667	718	765	808	844	861	829	798	767	736	704	671	636	560	473	370
0	0	0	5.00	2815	606	985	718	739	792	842	888	929	965	977	943	909	875	841	807	773	738	663	581	487
0	0	0	6.00	3038	577	1094	841	862	913	962	1007	1047	1083	1079	1053	1017	981	946	910	875	838	764	684	597
0	0	0	7.00	3261	545	1198	961	982	1031	1078	1122	1162	1197	1174	1149	1123	1086	1049	1012	976	939	863	788	701
0	0	0	8.00	3484	509	1297	1078	1098	1146	1191	1234	1273	1289	1264	1238	1212	1187	1151	1113	1075	1038	961	883	801
6	2	7	1.20	1968	680	464	202	228	289	344	386	417	440	458	459	448	435	416	393	363	326			
10	6	5	1.24	1977	687	460	210	235	292	343	385	418	441	455	456	446	433	416	394	365	330			
8	4	6	1.32	1994	683	470	222	247	307	353	395	430	450	465	465	454	441	423	401	372	337			
12	8	5	1.55	2046	688	491	257	280	335	382	423	459	475	487	485	474	460	442	420	393	360	272		
6	2	8	1.58	2052	676	504	260	285	344	397	432	461	483	500	497	484	468	449	425	397	363	276		
10	6	6	1.76	2092	684	512	289	311	363	411	447	476	496	508	508	493	478	459	437	411	379	297		
8	4	7	1.80	2101	679	518	294	318	370	414	454	484	501	514	511	497	482	463	441	414	382	300		
6	2	9	2.00	2146	672	548	322	346	403	452	481	509	529	544	538	522	505	485	462	434	402	321		
8	4	8	2.37	2229	675	576	377	398	442	484	522	546	561	572	565	549	531	511	489	463	433	359		
10	6	7	2.40	2235	680	576	377	397	446	489	522	550	563	573	566	550	533	513	491	466	436	363		
6	2	10	2.54	2266	666	604	400	423	477	514	542	568	587	601	591	573	553	532	508	481	451	377		
8	4	9	3.00	2369	669	638	460	477	519	559	596	613	625	635	625	605	586	564	541	516	487	419	333	
6	2	11	3.12	2396	660	664	481	503	552	579	605	630	648	661	647	626	605	582	557	531	501	432	345	
10	6	8	3.16	2405	675	651	477	496	543	577	608	631	641	648	638	619	599	578	555	530	502	436	352	
8	4	10	3.81	2550	662	717	559	575	615	654	682	707	707	715	700	678	656	633	609	584	556	493	416	324
6	2	14	4.50	2703	646	802	666	677	702	726	750	773	790	801	778	753	728	702	676	649	620	556	483	395
4	6	6	1.10	1945	690	413	188	212	264	311	348	378	397	409	412	408	399	385	365	339	305			
4	2	9	1.19	1964	679	446	200	225	285	338	374	402	424	441	442	434	422	405	383	354	319			
4	4	7	1.20	1968	684	429	203	227	282	325	364	394	411	425	427	421	411	396	376	349	315			
4	6	7	1.50	2034	688	442	245	265	313	356	387	414	429	439	440	433	424	410	392	368	337			
4	2	9	1.50	2034	676	475	247	272	329	380	408	434	455	471	470	459	446	429	407	380	347	260		
4	4	8	1.58	2052	681	459	259	282	327	367	405	429	444	456	455	447	436	421	402	377	347	265		
4	2	10	1.90	2125	672	513	306	330	385	425	450	475	494	509	505	492	477	460	438	412	382	303		
4	6	8	1.98	2140	685	476	304	323	368	403	431	455	465	473	472	464	453	439	422	400	373	302		
4	4	9	2.00	2146	678	492	318	335	375	413	448	466	479	489	486	476	464	449	430	407	379	306		
4	2	11	2.34	2222	667	552	367	390	442	470	494	517	535	549	549	527	511	492	471	446	417	345		
4	4	10	2.54	2266	673	533	379	395	433	470	498	512	522	531	526	514	500	485	466	445	419	354		
4	2	14	3.38	2453	656	644	507	527	549	571	593	614	631	642	629	610	591	571	549	525	499	436	357	
12	2	5	1.08	1942	683	471	185	212	275	333	383	418	445	465	466	454	439	419	393	361	323			
8	2	6	1.10	1945	682	462	187	214	276	333	379	412	438	456	457	447	432	413	389	358	320			
10	2	6	1.32	1994	680	495	222	248	310	366	412	446	471	490	488	475	458	438	413	383	346			
8	2	7	1.50	2034	677	508	249	274	335	389	430	462	485	503	500	486	470	450	426	396	361	270		
10	2	7	1.80	2101	675	553	295	320	380	434	475	507	530	548	542	525	506	484	459	430	395	308		
8	2	8	1.98	2140	673	562	320	345	403	456	490	520	542	558	551	534	515	494	469	441	407	324		
10	2	8	2.37	2229	669	620	380	404	462	514	548	579	600	617	606	585	564	540	514	485	452	372		
8	2	9	2.50	2257	667	621	397	421	477	523	554	582	602	618	607	587	566	543	518	489	457	381		
8	2	10	3.17	2408	661	696	493	516	570	605	633	661	679	694	678	655	631	606	580	552	521	449	360	

11-22

Example 11.3—Continued

Table 11-5—Design Aid Reproduced from Page 14 of EB9

PCA—LOAD AND MOMENT STRENGTH TABLES FOR CONCRETE COLUMNS

No A1 — No. A2 — ⟵ Axis of bending — 1 1/2" clear

COLUMN SIZE = 22X22 INCHES
FPC = 5,000 FY = 60,000

NO A1	NO A2	SIZE NO	PCNT	PO	PB	MB	0	100	200	300	400	500	600	700	800	900	1000	1200	1400	1600	1800	2000	2200	2400
4	0	10	1.05	2340	818	648	238	310	379	444	502	552	594	624	645	637	622	583	525	444	338			
12	0	6	1.09	2351	828	664	249	323	393	459	518	569	609	639	660	654	637	596	536	455	348			
10	0	7	1.24	2391	824	692	280	353	423	488	547	598	638	668	689	680	662	618	555	477	373			
6	0	9	1.24	2391	819	688	278	350	420	484	542	593	634	664	685	675	657	614	555	475	371			
4	0	11	1.29	2405	812	693	287	358	426	490	547	598	640	670	691	679	661	618	559	481	377			
8	0	8	1.31	2409	821	703	293	365	435	499	558	609	650	680	700	690	671	626	567	487	383			
12	0	7	1.49	2458	822	742	332	405	474	539	598	649	689	719	739	727	706	658	596	517	415			
6	0	10	1.57	2482	812	752	347	417	486	549	607	658	700	730	750	750	713	665	604	526	427			
10	0	8	1.63	2497	817	769	362	433	502	566	625	676	717	746	767	751	729	679	617	538	439			
8	0	9	1.65	2503	815	770	364	435	504	568	626	677	719	748	768	752	730	680	618	540	441			
4	0	14	1.86	2559	800	798	400	469	535	597	654	704	748	778	798	775	752	701	641	564	470			
6	0	11	1.93	2579	806	819	419	488	555	618	676	727	769	799	818	797	772	719	657	580	485	370		
10	0	9	2.07	2614	810	853		520	588	652	710	762	803	832	851	830	803	747	683	605	510	396		
8	0	10	2.10	2623	807	856	455	524	592	655	712	764	806	835	855	831	805	749	685	608	514	400		
8	0	11	2.58	2753	799	946	550	618	684	747	804	856	898	927	946	916	886	824	757	680	589	481		
6	0	14	2.79	2810	791	977	586	653	718	779	836	887	931	960	974	944	913	850	783	706	617	513		
4	0	18	3.31	2949	774	1053	673	737	799	858	913	962	1006	1039	1045	1012	980	914	846	770	685	588	474	
0	0	0	4.00	3136	762	1172	798	861	921	979	1032	1081	1124	1160	1158	1122	1087	1016	944	868	784	692	585	467
0	0	0	5.00	3406	745	1338	973	1034	1092	1148	1200	1248	1291	1328	1316	1277	1239	1163	1086	1007	924	834	737	627
0	0	0	6.00	3676	729	1499	1142	1201	1257	1312	1362	1409	1452	1489	1469	1429	1388	1307	1227	1145	1061	974	880	779
0	0	0	7.00	3946	697	1645	1306	1363	1418	1471	1521	1567	1609	1644	1615	1576	1534	1450	1366	1283	1197	1110	1019	923
0	0	0	8.00	4216	659	1784	1465	1521	1575	1626	1675	1720	1761	1771	1741	1712	1676	1590	1504	1419	1333	1245	1154	1061
10	6	5	1.02	2334	843	587	236	308	370	427	477	515	547	568	583	584	575	545	494	422				
8	4	6	1.09	2351	839	598	250	321	387	439	487	529	558	578	594	594	584	554	504	430	331			
12	8	5	1.28	2403	844	622	291	357	418	472	519	556	585	605	618	617	606	575	526	456	359			
6	2	8	1.31	2409	830	638	294	364	429	488	533	566	594	616	634	630	617	583	532	460	363			
10	6	6	1.45	2449	840	647	326	392	450	504	548	584	613	630	643	640	627	594	546	478	386			
8	4	7	1.46	2458	835	654	332	401	459	509	554	594	617	636	650	646	633	599	550	482	390			
14	10	5	1.54	2472	845	657	342	407	465	517	562	597	623	643	654	651	638	605	557	488	397			
6	2	9	1.65	2503	826	688	364	433	496	553	590	622	647	668	684	677	662	624	573	504	414			
12	8	6	1.82	2548	841	696	398	459	517	565	609	640	665	682	692	687	673	637	589	524	437			
8	4	8	1.96	2586	831	719	428	492	541	589	633	668	687	703	716	708	691	653	604	540	456			
10	6	7	1.98	2592	837	719	432	490	545	596	634	666	691	705	716	709	693	655	607	543	461			
6	2	10	2.10	2623	821	752	453	519	580	630	661	691	714	734	749	738	719	677	626	560	476	372		
12	8	7	2.48	2726	838	785	523	581	631	676	714	740	763	775	783	774	765	714	665	604	526	428		
8	4	9	2.48	2726	826	789	530	582	630	675	718	746	762	776	787	776	756	714	664	602	525	429		
6	2	11	2.58	2753	816	819	547	610	668	705	736	764	785	803	817	802	780	735	682	618	539	442		
10	6	8	2.61	2762	833	804	547	602	655	698	733	762	781	792	801	791	771	729	679	619	543	450		
8	4	10	3.15	2907	820	880	647	694	740	784	825	843	856	868	878	862	840	793	741	681	608	522	419	
10	6	9	3.31	2949	828	896	670	723	771	806	839	866	878	887	895	880	858	811	759	700	630	546	443	
6	2	14	3.72	3060	803	977	761	819	849	877	905	931	949	964	977	952	926	872	816	753	681	598	500	
8	4	11	3.87	3101	814	975	766	812	856	898	934	944	956	966	974	954	929	878	824	764	695	615	522	
4	8	6	1.09	2351	848	547	249	310	368	415	457	489	515	532	543	546	542	521	490	413				
4	6	7	1.24	2391	844	568	279	339	393	444	481	513	539	553	564	566	560	537	496	432	340			
4	2	9	1.24	2391	831	605	279	348	412	469	507	538	563	585	601	600	590	560	512	442	346			
4	4	8	1.31	2409	838	587	293	359	411	457	499	535	554	571	583	583	576	550	507	442	351			
4	8	7	1.49	2458	848	584	321	379	429	472	511	536	560	573	581	582	575	553	515	456	370			
4	2	10	1.57	2482	827	648	347	414	475	527	557	586	609	624	644	640	627	595	548	482	393			
4	6	8	1.63	2497	843	607	351	405	456	499	533	562	582	594	603	603	595	570	531	473	392			
4	4	9	1.65	2503	835	624	362	419	465	509	550	579	595	610	621	619	609	582	540	480	397			
4	2	11	1.93	2579	822	693	417	482	540	580	609	636	657	675	690	682	667	632	586	523	441			
4	6	9	2.07	2614	841	649	423	475	524	557	588	615	629	646	643	633	607	569	516	444				
4	4	10	2.10	2623	831	672	444	489	532	574	613	633	647	659	669	664	652	622	582	526	431			
4	2	11	2.58	2753	827	722	517	560	602	642	676	689	701	711	720	712	698	666	626	573	505	418		
4	2	14	2.79	2810	813	798	579	637	674	700	725	749	768	783	796	782	763	722	674	616	544	455		
14	2	5	1.02	2334	836	626	236	310	379	443	499	541	575	602	621	619	606	569	514	435	329			
10	2	6	1.09	2351	834	627	249	322	391	453	508	546	578	603	622	620	607	571	517	440	337			
8	2	7	1.24	2391	832	642	280	352	419	480	530	566	595	620	638	634	621	585	532	457	357			
10	2	7	1.49	2458	829	692	333	404	470	531	581	617	646	670	688	681	665	624	569	496	399			
8	2	8	1.63	2497	827	703	362	431	496	555	603	633	660	683	700	692	675	635	581	510	417			
10	2	8	1.96	2586	824	769	429	498	563	622	666	700	727	749	766	754	733	687	631	560	471			
8	2	9	2.07	2614	822	770	449	517	580	637	673	705	730	751	768	755	735	690	636	567	480	373		
10	2	9	2.48	2726	818	853	534	601	663	720	757	789	814	833	851	833	809	758	700	630	546	443		
8	2	10	2.62	2765	816	856	560	625	686	733	766	796	819	839	854	836	812	762	706	639	558	459		
8	2	11	3.22	2927	809	946	676	739	797	831	863	892	913	931	949	922	895	841	782	715	637	546	439	

Example 11.3—Continued

Table 11-6—Design Aid Reproduced from Page 15 of EB9

PCA—LOAD AND MOMENT STRENGTH TABLES FOR CONCRETE COLUMNS

No A1 / No A2 — 1 1/2" clear — ← Axis of bending

COLUMN SIZE =24X24 INCHES
FPC=5,000 FY=60,000

NO A1	NO A2	SIZE NO	PCNT	O₀ PO	BALANCE PB	MB	0	100	200	300	400	500	600	700	900	1100	1300	1500	1700	1900	2100	2300	2500	2700
							\<— MOMENT STRENGTH Mₙ (FT-KIPS)																
10	0	7	1.04	2782	993	854	312	393	472	547	616	678	733	776	837	837	799	746	676	584	469			
6	0	9	1.04	2782	987	850	310	391	469	543	612	674	729	773	834	832	795	743	673	582	467			
14	0	6	1.07	2791	996	864	320	402	482	556	626	688	742	786	847	847	807	754	683	591	476			
4	0	11	1.08	2796	980	855	320	399	477	550	618	680	735	779	841	837	799	748	679	589	474			
8	0	8	1.10	2800	989	867	326	407	486	560	629	691	746	789	850	849	809	756	686	595	480			
12	0	7	1.25	2849	990	910	370	451	530	604	673	736	790	833	894	889	845	790	719	628	516			
6	0	10	1.32	2873	980	921	387	466	544	616	685	747	802	846	906	898	854	798	728	639	529			
10	0	10	1.37	2888	986	940	403	483	561	635	703	766	821	864	925	916	870	813	742	652	542			
8	0	9	1.39	2894	983	942	406	486	563	637	705	768	823	866	927	917	871	814	743	654	545			
4	0	14	1.56	2950	968	974	447	525	600	672	739	800	855	901	961	945	896	840	770	683	577			
6	0	11	1.63	2970	973	997	468	546	622	694	762	824	879	924	984	967	917	858	787	699	593	468		
12	0	8	1.65	2977	983	1013	479	559	636	709	778	841	896	939	999	984	932	871	798	710	603	477		
10	0	9	1.74	3005	973	1034	502	581	658	731	799	862	918	960	1020	1003	949	887	815	727	621	496		
8	0	10	1.76	3014	975	1037	508	586	662	735	803	865	921	964	1024	1005	951	890	818	730	625	501		
8	0	11	2.17	3144	967	1139	615	692	767	839	906	968	1024	1068	1127	1099	1038	972	897	811	709	590		
6	0	14	2.34	3201	958	1175	656	732	806	877	943	1004	1060	1106	1164	1131	1068	1001	926	841	742	626		
4	0	18	2.78	3340	940	1263	756	829	900	969	1033	1093	1147	1196	1256	1210	1143	1073	998	914	815	710	585	
0	0	0	4.00	3732	914	1542	1051	1121	1189	1254	1316	1374	1428	1477	1539	1468	1389	1309	1228	1142	1050	950	840	715
0	0	0	5.00	4054	895	1762	1282	1350	1416	1479	1540	1597	1650	1698	1759	1673	1587	1502	1415	1327	1235	1138	1034	921
0	0	0	6.00	4375	876	1976	1507	1573	1636	1698	1757	1813	1865	1913	1964	1873	1782	1692	1602	1511	1418	1322	1220	1114
0	0	0	7.00	4696	857	2184	1724	1789	1851	1911	1968	2023	2074	2121	2163	2068	1974	1881	1788	1694	1599	1503	1403	1299
0	0	0	8.00	5017	826	2378	1936	1999	2059	2118	2174	2227	2278	2324	2352	2259	2163	2067	1971	1875	1779	1681	1582	1480
12	8	5	1.08	2794	1015	776	324	401	471	537	594	644	685	718	762	769	743	703	644	561	453			
6	2	8	1.10	2800	1000	793	327	406	482	551	614	662	697	728	777	783	755	711	650	566	458			
10	6	8	1.22	2840	1011	803	364	441	508	571	630	677	716	750	790	794	766	725	666	586	483			
8	4	7	1.25	2849	1006	811	370	448	521	580	635	685	727	756	797	801	772	730	670	590	487			
14	10	5	1.29	2863	1016	815	383	456	526	588	644	691	731	761	803	806	778	736	677	597	495			
6	2	9	1.39	2894	996	850	407	484	557	625	685	726	759	788	835	835	802	757	696	616	514			
12	8	6	1.53	2939	1013	858	447	517	584	645	697	743	779	809	847	846	814	772	714	638	540			
8	4	8	1.65	2977	1002	884	478	552	618	672	724	772	812	835	871	869	834	790	732	656	561			
10	6	7	1.67	2983	1008	884	484	554	617	676	732	772	808	840	872	870	836	792	735	661	567			
6	2	10	1.76	3014	991	921	507	582	653	718	771	807	837	864	907	902	863	815	755	679	584	468		
14	10	6	1.83	3037	1015	913	525	595	658	715	767	807	843	869	904	900	864	819	762	688	595	482		
12	8	7	2.08	3117	1010	958	588	655	719	772	821	862	892	919	949	942	903	857	800	730	640	531		
8	4	9	2.08	3117	998	963	593	664	719	771	821	867	900	920	952	944	904	857	799	727	639	531		
6	2	11	2.17	3144	986	997	612	685	753	816	858	892	919	944	985	974	929	879	818	744	655	547		
10	6	8	2.19	3153	1005	979	618	681	742	800	847	883	916	943	969	961	920	873	816	746	660	554		
8	4	10	2.65	3298	993	1064	738	793	845	895	944	986	1010	1027	1055	1041	994	943	885	816	734	636	522	
12	8	8	2.74	3329	1007	1076	751	815	868	918	964	996	1023	1047	1069	1056	1009	959	902	834	753	655	540	
10	6	9	2.78	3340	1001	1083	758	819	877	930	969	1002	1033	1053	1075	1061	1014	963	905	838	757	663	548	
6	2	13	3.13	3451	974	1175	856	923	986	1023	1054	1085	1109	1130	1165	1141	1087	1030	967	897	815	722	612	
8	4	11	3.25	3492	987	1172	875	927	977	1026	1072	1110	1125	1140	1164	1143	1091	1037	977	909	831	742	637	
4	6	7	1.04	2782	1016	716	312	386	448	506	561	601	638	670	703	713	697	664	612	535	432			
4	2	9	1.04	2782	1000	757	311	390	464	531	592	633	665	695	742	750	726	687	628	544	439			
4	10	6	1.07	2791	1024	707	316	386	450	505	557	597	634	660	696	705	691	661	609	533	431			
4	4	8	1.10	2800	1009	737	327	403	472	525	576	622	663	687	724	732	713	678	624	546	444			
4	8	7	1.25	2849	1020	734	364	429	492	545	592	635	665	692	724	731	714	684	635	562	464			
4	2	10	1.32	2873	997	805	387	463	535	600	648	689	719	747	790	794	767	726	669	591	491			
4	6	8	1.37	2888	1015	759	399	461	520	576	623	660	693	721	748	754	735	702	653	583	490			
4	4	9	1.39	2894	1006	779	405	477	535	586	634	679	714	734	767	771	749	714	662	590	495			
4	2	11	1.63	2970	993	855	467	541	610	672	716	748	776	801	842	841	810	768	712	638	544			
4	8	8	1.65	2977	1021	783	456	519	574	620	665	699	726	751	775	777	758	727	682	617	529			
4	4	10	1.74	3005	1014	806	482	541	598	651	688	722	753	775	797	799	776	743	697	633	549			
4	4	10	1.76	3014	1003	832	502	563	613	661	707	749	776	793	822	822	795	759	710	643	557			
4	2	11	2.17	3144	1000	888	597	645	693	739	784	823	840	855	879	875	845	808	760	697	618	520		
4	2	14	2.34	3201	984	974	651	719	782	823	853	882	906	927	962	952	913	867	812	744	662	562		
8	2	7	1.04	2782	1001	798	312	393	469	540	605	658	696	730	781	787	757	712	649	562	452			
14	2	6	1.22	2840	1000	864	364	445	521	594	659	718	758	794	847	848	810	760	694	607	498			
10	2	7	1.25	2849	999	854	370	450	526	597	662	715	753	787	838	839	803	755	690	605	498			
8	2	8	1.37	2888	997	867	403	482	556	625	688	736	771	803	851	851	815	767	704	622	518			
10	2	8	1.65	2977	993	940	479	557	631	700	763	811	846	877	925	919	879	824	759	678	578			
8	2	9	1.74	3005	992	942	502	578	651	718	779	820	853	882	928	921	879	828	765	686	589	471		
10	2	8	2.08	3117	988	1034	597	673	745	812	873	914	946	975	1021	1007	958	901	836	756	661	547		
8	2	10	2.20	3156	986	1037	626	701	771	836	888	925	955	982	1025	1011	962	907	843	767	675	565		
8	2	11	2.71	3318	979	1139	757	830	897	960	1000	1036	1063	1088	1127	1107	1052	993	928	852	765	663	544	

Example 11.3—Continued

Calculations and Discussion

Code Reference

Note: Design for rectangular columns by Design Aid EB9 is an approximate trial procedure. A more exact and direct procedure is available using the other selected design aids as follows.

6. Using Design Aid SP17A

(a) When the column section is preset, selection of reinforcement by Design Aid SP17A is simple and direct. Select reinforcement for 14 in. x 24 in. column section (A_g = 336 in.^{2}).

When designing with SP17A, the factored load and moment (required strength values) are used directly; the strength reduction factor φ is already included in the load-moment strength curves.

Compute $\quad \dfrac{P_u}{A_g} = \dfrac{898}{336} = 2.67$

Compute $\quad \dfrac{M_u}{A_g h} = \dfrac{276 \times 12}{336 \times 24} = 0.41$

Estimate $\quad \gamma \approx \dfrac{h - 5.5}{h} = \dfrac{24 - 5.5}{24} = 0.77$

where 2 (cover + tie diameter + ½ bar diameter) ≈ 5.5 in.

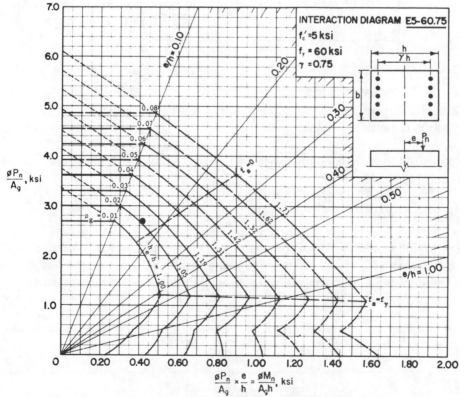

Fig. 11-6   Chart Reproduced from Page 98 of SP17A
(Courtesy of American Concrete Institute)

Example 11.3—Continued

(b) For rectangular sections with bars along end faces, $f_c' = 5$ ksi, $f_y = 60$ ksi, and $\gamma = 0.77$, use column chart 7.12.3 for E5 - 60.75 columns (reproduced in Fig. 11-6). Read $\rho_g = 0.02$.

$$A_{st} = \rho_g A_g = 0.02 \, (336) = 6.72 \text{ in.}^2$$

As seen above, the approximate interpolation procedure required for rectangular sections by Design Aid EB9 yields somewhat conservative results for this design.

7. Using CRSI Handbook

(a) $P_u = 898$ kips, $M_u = 276$ ft-kips
    $e = M_u/P_u = (276 \times 12)/898 = 3.7$ in.

(b) For Grade 60 bars with $f_c' = 5000$ psi, and 14 in. x 24 in. rectangular section, use Load Capacity Table—RECTANGULAR TIED COLUMNS 14" x 24", Page 3-118 (reproduced in Table 11-7). With e = 3.7 in., interpolate for solution:

Bars	$\rho\%$	φP_n
4#10	1.51	852
4#11	1.86	884
4#14	2.68	963
6#10	2.27	928

Note that the EB9 "approximate" solution required $\rho = 2.34\%$, and the SP17A graphical solution required $\rho = 2.0\%$.

Alternatively, use 14 in. x 24 in. column with 6#10 bars. Note: 3#10 bars must be located on each 14 in. side.

8. Using PCACOL Program

Using the design option of the PCACOL program, design the column for a required strength, $P_u = 898$ kips and $M_u = 276$ ft-kips. The section width is 14 in. and the depth may vary from 18 in. to 24 in., the bars will range from 4#8 to 8#14. The maximum reinforcement ratio will be set at 3%.

Example 11.3—Continued

The graphical results of PCACOL are reproduced in Fig. 11-7. 6#10 bars were chosen by the program. Both a strength interaction diagram and a scaled column section showing the vertical reinforcement are displayed. The required factored load-moment combination is shown as point "1" relative to the strength curve provided. The alternative 6#10 bar arrangement from the CRSI solution gave a closer agreement between the strength required and the strength provided.

Example 11.3—Continued

Table 11-7—Design Aid Reproduced from Page 3-118 of CRSI Handbook
(Courtesy of Concrete Reinforcing Steel Institute)

BARS Grade 60			RECTANGULAR TIED COLUMNS Short columns (1)—no sidesway / Bending on major or minor axis MI. MA. 14" × 20"													0.10f'cAg = 140 kips (2) For φPn at (2)			
Concrete f'c = 5,000 psi		0.80 φPo			φPn (kips) ULTIMATE USABLE CAPACITY, φ = 0.70 (2) Eccentricity, e = φMn/φPn										OT (3)	Balance		140 k (2)	O (2)
(4) Bars	P %	φPn (kips)	AXIS	e (in.)	2"	3"	4"	6"	8"	12"	16"	20"	24"	28"	e (in.)	e (in.)	φPb (in.)	e (in.)	φMn (k-ft.)
8-#8	2.26	864	MA	1.81	842	737	641	486	383	250	180	0	0	0	3.14	8.98	346	19.53	223
4L-2S			MI	1.35	774	649	547	406	317	181	0	0	0	0	2.76	8.17	311	14.42	149
8-#9	2.86	916	MA	1.76	889	777	676	516	410	277	202	158	0	0	3.20	9.84	343	22.16	265
4L-2S			MI	1.33	819	689	584	439	345	215	146	0	0	0	2.86	9.31	302	16.52	183
8-#10	3.63	983	MA	1.72	948	828	721	554	443	309	227	179	147	0	3.26	10.96	340	25.20	318
4L-2S			MI	1.30	876	740	630	478	378	255	176	0	0	0	2.96	10.85	290	19.23	225
8-#11	4.46	1056	MA	1.66	1009	879	764	589	473	337	251	197	162	0	3.33	12.24	331	27.69	368
4L-2S			MI	1.25	933	787	671	507	400	281	201	153	0	0	3.05	12.87	264	21.53	264
8-#14	6.43	1228	MA	1.58	1158	1005	873	677	547	394	305	244	201	171	3.41	15.30	320	34.12	489
4L-2S			MI	1.19	1074	908	778	590	467	329	254	203	164	0	3.18	18.64	221	27.70	361
10-#10	4.54	1063	MA	1.75	1030	906	796	625	509	368	278	220	182	154	3.50	13.37	336	30.68	399
4L-3S			MI	1.19	918	764	644	484	380	258	184	143	0	0	2.76	11.03	286	20.42	271
10-#11	5.57	1153	MA	1.70	1109	973	854	673	548	399	310	246	203	173	3.60	15.24	326	34.26	465
4L-3S			MI	1.13	984	817	688	513	402	280	206	160	0	0	2.82	13.32	255	22.64	306
10-#10	4.54	1063	MA	1.64	1015	882	763	583	467	329	245	194	162	0	3.19	11.44	345	27.51	368
5L-2S			MI	1.28	945	802	686	526	421	297	212	161	0	0	3.05	12.74	282	22.55	276
10-#11	5.57	1153	MA	1.59	1091	944	815	623	501	356	269	214	178	152	3.25	12.87	335	30.41	420
5L-2S			MI	1.23	1017	861	737	565	448	316	242	185	149	0	3.14	15.58	250	25.46	324
12-#10	5.44	1142	MA	1.68	1097	962	840	655	533	385	294	234	194	166	3.43	13.82	342	33.16	449
5L-3S			MI	1.18	988	828	701	533	423	297	217	168	0	0	2.86	12.95	277	23.62	317
12-#11	6.69	1251	MA	1.62	1191	1040	908	708	578	418	327	261	217	186	3.50	15.83	331	36.96	516
5L-3S			MI	1.12	1068	892	756	572	451	316	243	189	155	0	2.92	16.16	241	26.42	360
RECTANGULAR TIED COLUMNS 14" × 24" 0.10f'cAg = 168 kips																			
4-#9	1.19	924	MA	2.47	924	878	795	647	530	361	232	0	0	0	3.98	10.69	419	19.33	185
2E			MI	1.37	827	684	563	396	259	0	0	0	0	0	2.61	6.27	379	10.33	102
4-#10	1.51	958	MA	2.47	958	911	827	678	561	405	273	192	0	0	4.14	11.73	416	21.76	231
2E			MI	1.36	856	711	589	420	295	0	0	0	0	0	2.68	6.85	372	11.46	124
4-#11	1.86	994	MA	2.45	994	945	858	707	589	433	309	222	169	0	4.33	12.84	410	24.13	277
2E			MI	1.33	884	734	610	435	324	177	0	0	0	0	2.77	7.50	354	12.46	147
4-#14	2.68	1081	MA	2.44	1081	1027	936	778	656	492	389	295	228	185	4.66	15.49	404	30.08	388
2E			MI	1.29	955	796	667	482	373	228	0	0	0	0	2.89	9.14	330	15.06	198
4-#18	4.76	1299	MA	2.37	1299	1229	1125	947	809	621	503	423	357	298	5.27	22.32	388	44.64	658
2E			MI	1.19	1128	943	790	574	449	312	233	178	0	0	3.06	14.74	258	21.09	316
6-#7	1.07	912	MA	2.47	912	866	784	636	519	345	217	0	0	0	3.89	10.30	422	18.52	168
2L-3S			MI	1.31	801	650	523	340	223	0	0	0	0	0	2.39	5.45	389	9.81	94
6-#8	1.41	948	MA	2.48	948	902	818	670	553	394	263	183	0	0	4.06	11.40	419	21.13	217
2L-3S			MI	1.29	828	673	546	371	250	0	0	0	0	0	2.43	5.86	383	10.76	119
6-#9	1.79	987	MA	2.48	987	940	855	705	588	433	307	221	0	0	4.24	12.63	416	24.01	271
2L-3S			MI	1.26	857	698	568	396	275	0	0	0	0	0	2.47	6.32	376	11.78	145
6-#10	2.27	1037	MA	2.48	1037	989	902	749	630	471	357	267	205	0	4.45	14.21	412	27.68	339
2L-3S			MI	1.23	895	729	596	421	306	184	0	0	0	0	2.50	6.93	367	12.99	178
6-#11	2.79	1092	MA	2.46	1092	1040	949	791	669	504	402	307	240	194	4.67	15.88	406	31.24	407
2L-3S			MI	1.18	931	757	620	436	329	201	0	0	0	0	2.56	7.70	346	13.99	208

Example 11.3—Continued

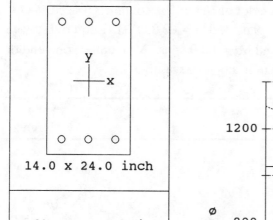

14.0 x 24.0 inch

f'_c = 5.0 ksi

f_y = 60.0 ksi

6#10 2.3%

A_{st} = 7.62 in^2

Tied   cc = 1.88 in

spacing = 3.21 in

I_x = 16128 in^4

$\bar{x}$ = 0.00 in

I_y = 5488 in^4

$\bar{y}$ = 0.00 in

© 1989 PCA

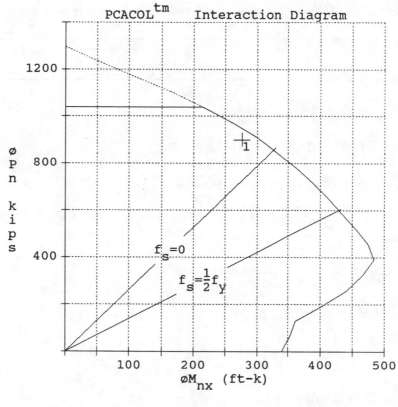

PCACOL^{tm} Interaction Diagram

ϕP_n kips

$f'_s = 0$

$f_s = \frac{1}{2} f_y$

ϕM_{nx} (ft-k)

Licensed To: Portland Cement Association, Skokie, IL

Project: Example 11.3

Column Id: Design for Small Axial

Engineer:

Date: Time:

Code: ACI 318-89

Version: 2.10

Slenderness not considered x-axis

File name: C:\PCACOL\DATA\EX11_3.COL

Material Properties:

E_c = 4287 ksi ϵ_u = 0.003 in/in

f_c = 4.25 ksi E_s = 29000 ksi

β_1 = 0.80

Stress Profile : Block

Reduction: ϕ_c = 0.70 ϕ_b = 0.90

**Fig. 11-7   PCACOL On-Screen Graphic Results
for 14 in. x 24 in. Column with 6#10 Bars**

Example 11.4—Design for General Loading (Circular Section)

Design a circular spirally reinforced column with a minimum diameter of 16 in. The column is braced against sidesway and has an unsupported height of 8 ft 6 in. Use $f'_c = 5000$ psi and $f_y = 60,000$ psi. Required service load strengths: $P_d = 80$ kips, $P_\ell = 60$ kips, $M_d = 100$ ft-kips, and $M_\ell = 80$ ft-kips. Moments at one end of the column are half those of the other end and the column is bent in single curvature.

Calculations and Discussion	Code Reference

1. Determine required strength.

    $P_u = 1.4\,(80) + 1.7\,(60) = 214$ kips

    Eq. (9-1)

    $M_u = 1.4\,(100) + 1.7\,(80) = 276$ ft-kips

2. Check if slenderness need be considered in the design. Assume 16 in. column size for slenderness evaluation.

    $k = 1.0$ for braced compression member

    10.11.2.1

    $r = 0.25\,(16) = 4$ in.

    10.11.3

    $kl_u/r = 1.0 \times 8.5 \times 12/4 = 25.5 \; < 34 - 12\left(\dfrac{1}{2}\right) = 28$

    10.11.4.1

    Therefore, slenderness need not be considered.

3. <u>Using Design Aid EB9</u>

    (a) Required $Pn = \dfrac{P_u}{\varphi} = \dfrac{214}{0.75} = 285$ kips

    .3.2.2(b)

    $Mn = \dfrac{M_u}{\varphi} = \dfrac{276}{0.75} = 368$ ft$-$kips

    (b) A review of the strength tables indicates that the smallest column section that will support the required load-moment strength is a 20 in. diameter section with approximately 3 percent reinforcement (See Table 11-8).

Example 11.4—Continued

	Calculations and Discussion	**Code Reference**

Possible Reinforcement Selections

Bars	PCNT	P_n	M_n
12#8	3.02	285	374
10#9	3.18	285	382
8#10	3.23	285	386
6#11	2.98	285	368

Use 20 in. diameter column with 6#11 bars.

4. Select Spiral Reinforcement.

$$\rho_s = 0.45 \left(\frac{A_g}{A_c} - 1\right) \frac{f_c'}{f_y}$$

Eq. (10-5)

$$A_g = 20^2 \times \frac{\pi}{4} = 314 \text{ in.}^2, \ A_c = (20-3)^2 \left(\frac{\pi}{4}\right) = 227 \text{ in.}^2$$

$$\rho_s = 0.45 \left(\frac{314}{227} - 1\right) \frac{5}{60} = 0.0144$$

$$\rho_s = \frac{\text{Volume of spiral in one hoop}}{\text{Volume of core in one pitch}} = \frac{A_s \pi (D_c - D_s)}{A_c s}$$

where s = spiral pitch, D_c = core diameter, D_s = spiral diameter, and A_s = spiral area

Minimum spiral diameter = $\frac{3}{8}$ in.

7.10.4.2

Solving for maximum spiral pitch,

$$s = \frac{0.11 \pi (17 - 0.375)}{227 (0.0144)} = 1.8 \text{ in.}$$

Clear spacing limitations: s ≥ 1 in.

7.10.4.3

  ≤ 3 in.

  > 4/3 (maximum aggregate size)

3.3.3

Example 11.4—Continued

Table 11-8—Design Aid Reproduced from Page 39 of EB9

PCA–LOAD AND MOMENT STRENGTH TABLES FOR CONCRETE COLUMNS

1 1/2" clear

COLUMN SIZE =20X20 INCHES
FPC=5,000   FY=60,000

NO TOT	SIZE NO	PCNT	PO	PB	MB	0	20	120	220	320	420	520	620	720	820	920	1020	1120	1220	1320	1420	1520	1620
			P_O	P_B	M_B																		
11	5	1.09	1525	493	278	131	140	186	225	252	271	279	277	271	258	240	214	181	141				
12	5	1.18	1543	511	285	139	149	196	232	261	277	285	282	276	263	245	220	188	149				
13	5	1.28	1560	499	291	149	159	204	240	266	285	291	288	280	268	251	226	195	156	113			
14	5	1.38	1577	511	296	159	169	212	248	274	290	296	293	286	274	256	232	201	163	121			
15	5	1.48	1594	502	302	170	178	221	255	281	297	302	298	291	279	262	238	207	170	128			
16	5	1.58	1612	512	308	178	187	230	263	288	303	308	304	296	284	267	243	213	177	136			
17	5	1.68	1629	504	314	187	197	238	271	295	310	314	309	301	289	272	249	219	184	143			
18	5	1.78	1646	512	320	197	207	246	278	302	316	320	314	306	294	277	254	226	191	150			
19	5	1.87	1664	506	326	206	215	255	286	309	322	326	320	312	299	282	260	232	198	157			
8	6	1.12	1531	506	279	132	142	191	226	254	273	279	278	272	261	242	216	183	143				
9	6	1.26	1556	481	288	147	158	201	237	266	281	288	286	279	268	249	224	192	153	111			
10	6	1.40	1580	508	296	162	171	213	250	273	291	296	293	287	274	256	232	201	163	122			
11	6	1.54	1605	490	305	173	183	226	259	284	299	305	301	293	281	263	240	210	174	133			
12	6	1.68	1630	510	313	186	196	238	269	294	309	313	308	300	288	271	248	219	184	143			
13	6	1.82	1654	496	322	201	210	248	282	303	318	322	316	307	295	278	256	228	194	153			
14	6	1.96	1679	510	330	213	221	260	291	314	326	329	323	314	302	286	264	237	203	163	121		
15	6	2.10	1703	499	339	225	233	272	301	323	337	338	331	322	310	293	272	245	212	173	132		
16	6	2.24	1728	511	346	238	247	282	313	332	344	346	339	329	317	301	280	253	221	183	143		
17	6	2.38	1752	502	355	251	259	294	322	343	354	354	346	337	324	308	287	261	230	193	153		
6	7	1.15	1536	502	279	135	144	188	228	259	272	279	277	271	260	242	218	186	147				
7	7	1.34	1569	460	290	152	161	208	245	267	285	290	288	282	270	253	230	198	160	117			
8	7	1.53	1603	504	302	172	182	226	256	282	298	302	298	292	280	264	240	209	172	131			
9	7	1.72	1636	476	314	192	202	238	271	296	309	313	309	302	290	273	250	221	185	145			
10	7	1.91	1670	506	325	207	215	254	287	307	323	325	320	312	299	282	260	232	198	159			
11	7	2.10	1703	486	337	223	232	273	299	322	333	336	330	321	309	292	270	244	211	173	133		
12	7	2.29	1737	508	348	242	251	285	314	334	346	347	341	331	318	302	281	255	224	187	146		
13	7	2.48	1770	492	361	258	265	300	328	346	358	359	351	340	328	312	291	267	237	200	159		
14	7	2.67	1803	509	371	273	281	317	341	361	370	370	361	350	338	322	302	278	249	213	173	132	
15	7	2.86	1837	496	383	291	298	329	355	372	382	382	371	360	347	332	313	289	260	226	187	146	
6	8	1.51	1599	499	299	167	176	218	256	283	293	298	295	288	277	260	238	209	173	132			
7	8	1.76	1643	453	315	191	200	244	275	294	311	313	309	302	290	274	253	225	190	148			
8	8	2.01	1688	502	329	217	227	262	290	314	326	328	323	315	304	288	267	239	205	165	124		
9	8	2.26	1732	470	345	239	245	280	311	331	341	343	337	328	317	301	279	253	221	183	142		
10	8	2.51	1776	503	359	257	265	302	328	346	359	358	351	342	329	313	292	267	236	201	161		
11	8	2.77	1820	481	375	280	288	321	345	366	372	365	355	342	325	305	281	252	218	180	140		
12	8	3.02	1864	506	389	302	308	338	365	379	390	387	379	368	354	338	318	295	268	235	198	157	
13	8	3.27	1908	488	405	320	327	359	380	396	404	402	392	380	367	351	332	309	283	252	215	175	
14	8	3.52	1952	507	418	341	348	376	398	413	420	417	406	394	380	364	345	324	298	268	232	192	152
6	9	1.91	1670	495	320	201	210	250	287	308	316	319	314	306	295	280	259	232	200	162			
7	9	2.23	1725	445	342	232	241	283	306	324	338	338	332	323	312	297	277	253	221	182	140		
8	9	2.55	1781	499	358	266	272	300	326	349	357	357	349	340	329	314	296	270	239	203	162		
9	9	2.86	1837	463	379	285	292	324	354	367	376	375	367	357	346	331	311	286	257	223	184	144	
10	9	3.18	1893	500	395	311	318	353	372	388	397	393	385	375	363	346	326	303	276	244	207	168	
11	9	3.50	1948	474	416	341	347	371	394	410	414	412	403	392	378	361	342	320	294	264	229	191	152
12	9	3.82	2004	502	432	361	367	395	418	428	436	431	421	408	394	377	359	337	312	284	251	214	173
6	10	2.43	1760	491	348	243	252	290	325	338	345	346	339	330	319	304	285	261	232	197	157		
7	10	2.83	1831	434	375	284	293	326	350	360	374	369	361	352	340	325	307	285	259	224	183	142	
8	10	3.23	1902	495	395	316	321	347	372	393	395	392	383	373	361	347	329	308	280	247	210	169	
9	10	3.64	1972	453	422	343	349	379	404	413	420	416	406	395	382	368	350	328	301	271	236	198	158
10	10	4.04	2043	495	441	377	384	409	425	439	444	439	428	417	404	389	370	348	323	295	262	226	187
6	11	2.98	1857	486	377	287	295	332	363	370	375	374	365	355	343	329	311	289	263	232	197	156	
7	11	3.48	1944	422	411	338	346	366	383	398	410	402	392	382	370	355	338	318	294	265	229	188	146
8	11	3.97	2031	489	433	366	371	396	419	435	434	430	420	408	396	381	364	345	321	292	258	220	180
9	11	4.47	2118	442	467	402	408	437	452	460	466	458	447	434	421	407	391	371	346	319	288	253	215
6	14	4.30	2088	472	444	387	394	428	437	441	443	438	427	419	402	387	371	352	331	306	277	245	210
7	14	5.01	2213	392	493	440	443	457	472	485	489	477	465	453	439	425	408	390	370	347	322	292	294

Example 11.4—Continued

Calculations and Discussion

**Code
Reference**

Use #3 Spiral @ 1¾ in. pitch

5. <u>Using Design Aid SP17A</u>

(a) Check required reinforcement for 20 in. diameter column (A_g = 314 in.²).

Compute $\dfrac{P_u}{A_g} = \dfrac{214}{314} = 0.68$

Compute $\dfrac{M_u}{A_g h} = \dfrac{276 \times 12}{314 \times 20} = 0.53$

Estimate $\gamma = \dfrac{h - 5}{h} = \dfrac{20 - 5}{20} = 0.75$

(b) For circular section, $f_c' = 5$ ksi, $f_y = 60$ ksi, and $\gamma = 0.75$, use Column Chart 7.23.3 for C5-60.75 columns (see Fig. 11-8). Read $\rho_g = 0.03$.

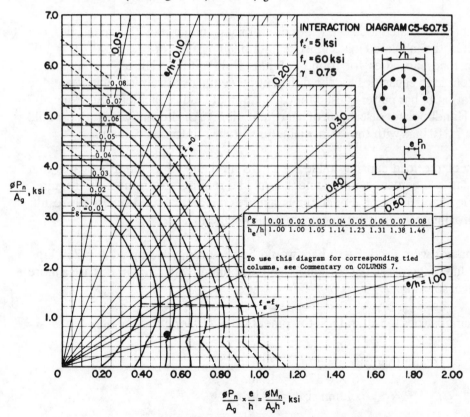

**Fig. 11-8 Chart Reproduced from Page 115 of SP17A
(Courtesy of American Concrete Institute)**

Example 11.4—Continued

Calculations and Discussion	Code Reference

6. Using CRSI Handbook

   (a) P_u = 214 kips, M_u = 276 ft-kips

       e = M_u/P_u = (276 × 12)/214 = 15.5 in.

   (b) For Grade 60 bars with f_c' = 5000 psi, and 20 in. diameter circular section, use Load Capacity Table—ROUND TIED COLUMNS 20" DIA., Page 4-37 (reproduced in Table 11-9). With e = 15.5 in., interpolate for solution.

Bars	$\rho\%$	φP_n	$1.07(\varphi P_n)$
5#14	3.58	224	240
6#11	2.98	202	214
7#10	2.83	202	214
8#10	3.23	219	234

   Note that the tabulated load capacities are for round columns with circular ties, not spirals; thus, the tabulated loads must be multiplied by a factor of (0.75/0.70) = 1.07 to obtain load capacities for round columns with spirals. The above seemingly inconsistent load capacities for the different percentages of reinforcement are due to orientation of bar patterns used for calculation of load capacities. (See Page 4-7 of CRSI Handbook for discussion of bar orientation.) An alternative solution would be to use 7#10 bars with the 20 in. diameter spirally reinforced column.

   A review of the load capacity table for an 18 in. diameter column indicates that approximately 6 percent reinforcement would be required to support the required factored load and moment. For columns, a more economical solution usually results with approximately 3 percent maximum reinforcement.

   (c) If circular ties were used, 8#10 bars would be required, yielding φP_n = 219 kips. Usually, due to increased cost of spirals, round columns with circular ties are a more economical solution.

       Use #3 circular ties for lateral reinforcement with #10 longitudinal bars   7.10.5.1

       Spacing not greater than: 16 × 1.37   = 20.3 in.
                          48 × 0.375 = 18 in.
                   column size = 20 in.

       Use #3 circular ties @ 18 in.

Example 11.4—Continued

7. Using PCACOL Program

Using the investigative option of the PCACOL program, investigate the 20 in. spirally reinforced circular column with 6#11 bars for a required strength P_u = 214 kips and M_u = 276 ft-kips

The graphic results of PCACOL are reproduced in Fig. 11-9. The required factored load-moment combination is shown as point "1" relative to the strength curve provided.

Example 11.4—Continued

Table 11-9—Design Aid Reproduced from Page 4-37 of CRSI Handbook
(Courtesy of Concrete Reinforcing Steel Institute)

Bars Grade 60				ROUND TIED COLUMNS 20″ DIA. Short columns; no sidesway [1]										$0.10 f'_c A_g = 157$ kips [2] For ϕP_n at [2]				
Concrete $f'_c = 5{,}000$ psi				ϕP_n (kips)—Ultimate Usable Capacity										OT [3]	Balance		157 k [2]	$(\phi = 0.90)$ O [2]
Bars	P %	0.80 ϕP_o (k)	e 0.80 ϕP_o (in.)	2″	3″	4″	6″	8″	$M_n/P_n = e$ (in.) 12″	$(\phi = 0.70)$ 16″	20″	24″	28″	e (in.)	e (in.)	ϕP_b (k)	e (in.)	ϕM_n (k-ft.)
4-#8	1.01	846	1.65	803	682	571	399	276	0	0	0	0	0	2.25	6.44	371	11.94	114
4-#9	1.27	872	1.63	825	701	589	418	299	173	0	0	0	0	2.30	6.81	369	12.94	141
4-#10	1.62	906	1.60	853	725	611	440	325	193	0	0	0	0	2.36	7.27	367	14.17	174
4-#11	1.99	942	1.56	881	749	632	459	349	211	0	0	0	0	2.44	7.80	360	15.31	206
4-#14	2.86	1029	1.48	950	808	685	506	394	252	179	0	0	0	2.54	9.80	353	18.01	264
4-#18	5.09	1247	1.34	1123	953	810	613	484	340	247	193	158	0	2.70	12.20	335	24.23	376
5-#8	1.26	871	1.67	830	706	593	421	308	175	0	0	0	0	2.45	7.08	357	12.97	132
5-#9	1.59	904	1.65	859	732	617	443	333	199	0	0	0	0	2.50	7.58	354	14.29	161
5-#10	2.02	946	1.62	895	764	646	470	360	227	0	0	0	0	2.56	8.24	351	15.95	197
5-#11	2.48	991	1.58	933	794	672	493	381	247	175	0	0	0	2.65	9.00	342	17.52	231
5-#14	3.58	1099	1.52	1025	872	740	551	431	293	215	169	0	0	2.75	10.78	329	21.29	304
5-#18	6.37	1372	1.39	1252	1061	903	680	542	379	291	231	191	163	2.91	16.13	289	29.14	461
6-#8	1.51	896	1.64	849	725	613	443	334	192	0	0	0	0	2.76	8.07	331	13.94	156
6-#9	1.91	935	1.61	882	754	640	469	358	217	0	0	0	0	2.82	8.82	326	15.45	192
6-#10	2.43	985	1.57	925	792	675	501	387	247	172	0	0	0	2.89	9.81	319	17.31	236
6-#11	2.98	1040	1.53	969	830	708	529	410	273	192	0	0	0	2.98	11.01	304	19.02	271
6-#14	4.30	1169	1.46	1077	922	790	600	469	324	240	186	0	0	3.10	14.01	281	23.28	348
6-#18	7.64	1497	1.31	1345	1150	988	755	600	421	323	263	221	188	3.29	25.02	213	33.18	521
7-#8	1.76	920	1.64	872	745	631	457	346	213	0	0	0	0	2.44	7.63	364	15.11	176
7-#9	2.23	966	1.61	912	780	663	486	374	238	169	0	0	0	2.50	8.30	361	16.94	214
7-#10	2.83	1025	1.58	963	825	703	521	405	269	193	0	0	0	2.57	9.16	358	19.23	261
7-#11	3.48	1088	1.53	1015	868	741	551	431	293	214	167	0	0	2.66	10.18	347	21.21	301
7-#14	5.01	1239	1.46	1142	977	836	628	497	348	262	207	171	0	2.78	12.56	334	25.99	396
8-#8	2.01	945	1.62	894	767	651	474	362	227	0	0	0	0	2.37	7.75	374	16.02	199
8-#9	2.55	997	1.59	940	807	688	506	392	259	181	0	0	0	2.44	8.46	372	18.00	243
8-#10	3.23	1065	1.55	998	858	734	546	426	288	210	161	0	0	2.51	9.37	370	20.50	292
8-#11	3.97	1137	1.50	1058	909	776	580	455	312	234	182	0	0	2.60	10.43	359	22.85	334
8-#14	5.73	1309	1.42	1204	1035	885	665	529	372	282	227	189	161	2.71	12.91	348	28.64	436
9-#10	3.64	1104	1.54	1033	887	758	569	448	307	226	176	0	0	2.60	10.15	362	22.13	318
9-#11	4.47	1186	1.49	1100	943	806	606	479	333	249	197	161	0	2.69	11.44	349	24.60	368
9-#14	6.45	1380	1.41	1264	1082	927	704	562	396	304	244	202	173	2.80	14.48	335	30.73	485
10-#10	4.04	1144	1.53	1070	918	785	590	467	324	240	190	0	0	2.72	11.26	346	23.73	346
10-#11	4.97	1235	1.48	1146	980	837	631	503	354	265	210	174	0	2.81	12.85	333	26.56	399
10-#14	7.16	1450	1.40	1329	1135	971	738	593	422	326	261	216	185	2.92	16.78	312	32.90	526
11-#10	4.45	1184	1.51	1103	947	813	614	487	340	254	200	165	0	2.63	11.16	365	25.20	375
11-#11	5.46	1283	1.46	1185	1015	872	660	525	370	281	224	184	0	2.71	12.73	351	27.97	433
12-#10	4.85	1223	1.52	1139	977	838	637	509	356	267	211	174	0	2.65	11.44	373	26.51	400
12-#11	5.96	1332	1.46	1228	1051	902	687	550	387	296	236	195	166	2.74	13.08	359	29.57	463
13-#10	5.26	1263	1.49	1172	1009	867	659	526	371	282	223	184	0	2.68	12.19	366	27.92	429
13-#11	6.46	1381	1.43	1269	1090	936	714	570	404	311	249	206	176	2.77	13.88	356	31.35	494
14-#10	5.66	1303	1.48	1207	1038	894	683	545	385	293	235	194	165	2.77	13.14	356	29.33	455

Example 11.4—Continued

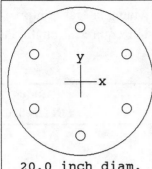

20.0 inch diam.

f'_c = 5.0 ksi

f_y = 60.0 ksi

6#11 3.0%

A_{st} = 9.36 in^2

Spiral cc = 2.00 in

spacing = 5.89 in

I_x = 7854 in^4

$\bar{x}$ = 0.00 in

I_y = 7854 in^4

$\bar{y}$ = 0.00 in

© 1989 PCA

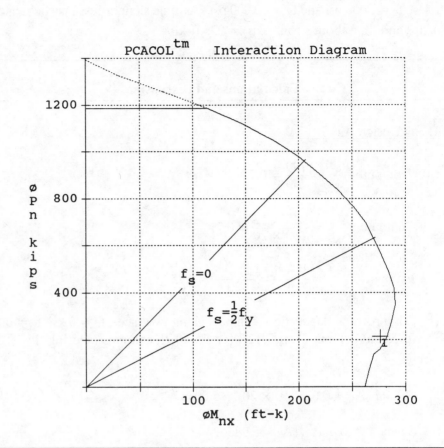

Licensed To: Portland Cement Association, Skokie, IL

Project: Example 11.4

Column Id: Design for Gen'l Loads

Engineer:

Date: Time:

Code: ACI 318-89

Version: 2.10

Slenderness not considered x-axis

File name: C:\PCACOL\DATA\EX11_4.COL

Material Properties:

E_c = 4287 ksi ϵ_u = 0.003 in/in

f_c = 4.25 ksi E_s = 29000 ksi

β_1 = 0.80

Stress Profile : Block

Reduction: ϕ_c = 0.75 ϕ_b = 0.90

**Fig. 11-9 PCACOL On-Screen Graphic Results
for 20-in. Circular Column with 6#11 Bars**

Example 11.5—Design for Reinforcement with Column Size Preset

Select reinforcement for a 10 in. x 20 in. reinforced concrete column with ties providing lateral reinforcement. Use $f_c' = 5000$ psi and $f_y = 60,000$ psi. Assume slenderness may be neglected. Required strength: $P_u = 190$ kips and M_u (about strong axis) = 235 ft-kips.

Calculations and Discussion	Code Reference

1. <u>Using Design Aid SP17A</u>

    (a) Compute $\dfrac{P_u}{A_g} = \dfrac{190}{200} = 0.95$

    Compute $\dfrac{M_u}{A_g h} = \dfrac{235 \times 12}{200 \times 20} = 0.71$

    Estimate $\gamma = \dfrac{h-5}{h} = \dfrac{20-5}{20} = 0.75$

    (b) For a rectangular section with bars along two faces, $f_c' = 5$ ksi, $f_y = 60$ ksi, and $\gamma = 0.75$, use column chart 7.12.3 for E5-60.75 columns (see Fig. 11-10). Read $\rho_g = 0.025$.

    $A_{st} = \rho_g A_g = 0.025 \,(10 \times 20) = 5.00$ in.^{2}

    Select 4 #10 bars   ($A_s = 5.08$ in.^{2})

2. Select lateral reinforcement.

    Use #3 ties with #10 longitudinal bars.   7.10.5.1

    Spacing not greater than: $16 \times 1.27 \quad = 20.3$ in.   7.10.5.2
    $48 \times 0.375 \ = 18$ in.
    Column size $= 10$ in.

    Use #3 ties @ 10 in.

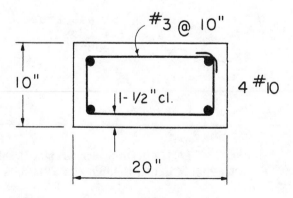

Example 11.5—Continued

3. Using CRSI Handbook

(a) $P_u = 190$ kips, $M_u = 235$ ft-kips

$e = \dfrac{M_u}{P_u} = \dfrac{235 \times 12}{190} = 14.8$ in.

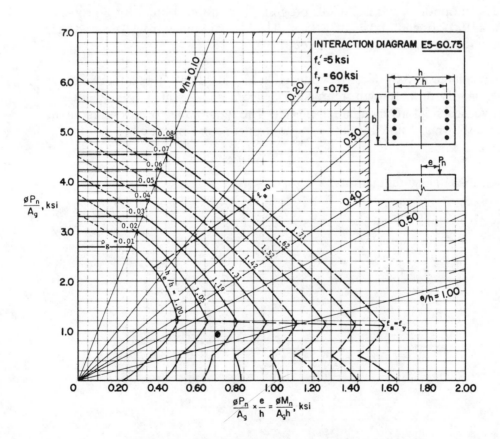

**Fig. 11-10   Chart Reproduced from Page 98 of SP17A
(Courtesy of American Concrete Institute)**

Example 11.5—Continued

(b) For Grade 60 bars with $f_c' = 5000$ psi, and 10 in. x 20 in. rectangular section, use Load Capacity Table—RECTANGULAR TIED COLUMNS 10" x 20", Page 3-108 (reproduced in Table 11-10). With e = 14.8 in., interpolate for solution:

Bars	$\rho\%$	φP_n
4#10	2.54	196
6#8	2.37	190

Note that the SP17A graphical solution required $\rho = 2.5\%$ (4#10 bars). Design aids using tabulated values can be considered to provide a slightly higher degree of accuracy by avoiding the estimating error of a graphical solution.

Alternatively, use 6#8 bars. Note: 3#8 bars must be located on each 10 in. side.

4. Using PCACOL Program

Using the design option of the PCACOL program, design the column for a required strength, $P_u = 190$ kips and $M_u = 235$ ft-kips. The bars will range from 4#8 to 8#11. The maximum reinforcement ratio will be set at 3%.

The graphic results of PCACOL are reproduced in Fig. 11-11. 4#10 bars were chosen by the program. Both a strength interaction diagram and a scaled column section showing the vertical reinforcement are displayed. The required factored load-moment combination is shown as point "1" relative to the strength curve provided.

Example 11.5—Continued

Table 11-10—Design Aid Reproduced from Page 3-108 of CRSI Handbook
(Courtesy of Concrete Reinforcing Steel Institute)

					RECTANGULAR TIED COLUMNS											0.10f'ₐAₐ = 100 kips (2)			

| BARS Grade 60 | | | Short columns (1)—no sidesway Bending on major or minor axis MI, MA. 10″ × 20″ | | | | | | | | | | | | | For φPₙ at (2) | | | |

| Concrete f'꜀ = 5,000 psi | | 0.80 φPₒ | | | φPₙ (kips) ULTIMATE USABLE CAPACITY, φ = 0.70 (2) | | | | | | | | | | | OT (3) | Balance | | 100 k (2) | O (2) |

(4) Bars	P %	φPₙ (kips)	A X I S	e (in.)	2″	3″	4″	6″	8″	12″	16″	20″	24″	28″	e (in.)	e (in.)	φPᵦ (in.)	e (in.)	φMₙ (k-ft.)
4-#7	1.20	551	MA	2.04	551	491	434	338	270	161	0	0	0	0	3.40	8.94	245	15.88	91
2E			MI	.95	414	303	228	128	0	0	0	0	0	0	2.01	4.59	198	6.99	43
4-#8	1.58	574	MA	2.03	574	514	455	359	290	190	122	0	0	0	3.55	9.95	243	18.18	117
2E			MI	.93	430	316	241	148	0	0	0	0	0	0	2.05	5.11	189	7.79	53
4-#9	2.00	601	MA	2.03	601	538	479	381	311	218	146	104	0	0	3.70	11.06	241	20.70	145
2E			MI	.91	447	329	253	168	111	0	0	0	0	0	2.08	5.75	178	8.64	63
4-#10	2.54	634	MA	2.02	634	568	507	407	336	246	175	128	0	0	3.87	12.50	239	23.90	181
2E			MI	.88	467	345	267	181	128	0	0	0	0	0	2.11	6.68	164	9.64	76
4-#11	3.12	671	MA	1.99	670	599	536	432	358	265	200	148	116	0	4.04	14.01	234	26.94	216
2E			MI	.84	478	350	270	184	139	0	0	0	0	0	2.13	8.19	136	10.47	87
4-#14	4.50	757	MA	1.94	752	675	606	494	413	309	248	198	157	130	4.35	17.72	228	34.68	302
2E			MI	.79	516	377	292	200	152	102	0	0	0	0	2.14	13.92	89	12.36	113
6-#6	1.32	558	MA	2.04	558	499	442	346	278	172	107	0	0	0	3.43	9.26	246	16.71	100
2L-3S			MI	.90	406	291	216	125	0	0	0	0	0	0	1.90	4.28	201	7.08	47
6-#7	1.80	588	MA	2.04	588	527	469	372	303	207	136	0	0	0	3.61	10.54	244	19.66	133
2L-3S			MI	.87	422	304	228	140	0	0	0	0	0	0	1.91	4.76	191	7.84	59
6-#8	2.37	624	MA	2.03	624	560	500	401	330	242	168	122	0	0	3.79	12.07	241	23.14	171
2L-3S			MI	.83	441	318	241	156	109	0	0	0	0	0	1.92	5.41	179	8.65	73
6-#6	1.32	558	MA	1.90	552	485	423	320	250	147	100	0	0	0	3.12	8.01	250	16.08	100
3L-2S			MI	.95	421	311	236	137	0	0	0	0	0	0	2.00	4.73	200	7.31	46
6-#7	1.80	588	MA	1.87	579	510	445	342	271	170	118	0	0	0	3.21	8.82	249	18.44	133
3L-2S			MI	.92	442	328	253	162	106	0	0	0	0	0	2.05	5.39	189	8.34	59
6-#8	2.37	624	MA	1.83	611	538	472	365	293	195	138	106	0	0	3.30	9.79	247	21.06	170
3L-2S			MI	.90	466	348	270	184	126	0	0	0	0	0	2.09	6.28	176	9.50	73
6-#9	3.00	663	MA	1.80	647	570	500	390	316	219	158	122	0	0	3.38	10.85	246	23.90	210
3L-2S			MI	.87	492	367	286	197	145	0	0	0	0	0	2.12	7.42	161	10.75	87
6-#10	3.81	714	MA	1.75	692	610	536	421	343	248	181	141	115	0	3.47	12.23	244	27.25	260
3L-2S			MI	.84	522	389	305	211	161	102	0	0	0	0	2.14	9.28	140	12.22	105
6-#11	4.68	768	MA	1.70	739	650	572	449	368	267	202	158	129	109	3.57	13.75	239	30.35	298
3L-2S			MI	.80	538	397	310	214	163	111	0	0	0	0	2.15	13.17	101	13.36	120
6-#14	6.75	897	MA	1.62	854	750	659	522	428	316	249	200	164	139	3.71	17.40	232	38.38	385
3L-2S			MI	.73	596	438	342	237	181	123	0	0	0	0	2.14	46.69	32	14.92	156
8-#8	3.16	673	MA	1.85	662	586	518	410	335	242	174	135	108	0	3.54	11.87	245	25.97	223
3E			MI	.82	478	350	270	184	133	0	0	0	0	0	1.96	6.67	166	10.18	92
8-#8	3.16	673	MA	1.76	652	571	497	381	304	208	152	120	0	0	3.20	10.25	247	23.69	206
4L-2S			MI	.87	502	378	296	205	154	0	0	0	0	0	2.12	7.62	164	11.21	92
8-#9	4.00	726	MA	1.71	699	610	532	411	330	233	172	136	112	0	3.26	11.46	245	26.85	247
4L-2S			MI	.84	537	403	317	221	169	108	0	0	0	0	2.15	9.50	144	12.83	111
8-#10	5.08	793	MA	1.66	758	661	577	448	361	259	195	155	129	110	3.32	13.04	242	30.73	298
4L-2S			MI	.81	577	433	341	239	183	125	0	0	0	0	2.17	12.92	117	14.76	134
8-#11	6.24	865	MA	1.60	819	713	620	482	390	281	217	173	144	122	3.39	14.88	234	34.12	346
4L-2S			MI	.76	599	444	349	243	186	127	0	0	0	0	2.16	23.31	67	15.39	152
10-#10	6.35	872	MA	1.58	825	716	619	477	386	276	210	168	140	120	3.24	13.67	247	33.71	338
5L-2S			MI	.78	632	475	377	265	204	140	106	0	0	0	2.20	18.42	93	17.11	162
10-#11	7.80	963	MA	1.51	902	778	671	517	419	301	234	187	155	133	3.28	15.71	238	37.42	388
5L-2S			MI	.73	659	490	387	270	208	142	108	0	0	0	2.18	55.38	32	17.37	183

Example 11.5—Continued

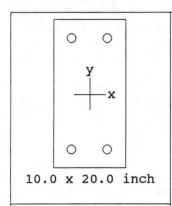

10.0 x 20.0 inch

f'_c =   5.0 ksi

f_y =   60.0 ksi

4#10   2.5%

A_{st} = 5.08 in^2

Tied   cc = 1.88 in

spacing = 3.70 in

I_x = 6667 in^4

$\bar{x}$ = 0.00 in

I_y = 1667 in^4

$\bar{y}$ = 0.00 in

© 1989 PCA

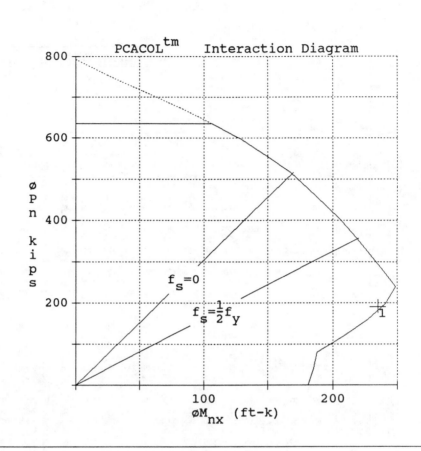

Project: Example 11.5

Column Id: Design of Rect Column

Engineer:

Date: Time:

Code: ACI 318-89

Version: 2.10

File name: C:\PCACOL\DATA\EX11_5.COL

Material Properties:

E_c =   4287 ksi ϵ_u = 0.003 in/in

f_c = 4.25 ksi E_s = 29000 ksi

β_1 = 0.80

Stress Profile : Block

Reduction: ϕ_c = 0.70 ϕ_b = 0.90

Slenderness not considered x-axis

**Fig. 11-11   PCACOL On-Screen Graphic Results
for 10 in. x 20 in. Column with 4#10 Bars**

Example 11.6—Design for Biaxial Loading (Circular Section)

Design a circular spirally reinforced column for the service load conditions given below. The column has an unsupported height of 7 ft and is braced against sidesway. Use $f_c' = 5000$ psi and $f_y = 60,000$ psi. Column size is limited to 16 in. diameter by architectural considerations.

	Dead	Live
Axial Loads (kips)	80	73
EW Moments (ft-kips)	40	30
NS Moments (ft-kips)	52	42

Calculations and Discussion	Code Reference

1. Determine required strength.

$P_u = 1.4 (80) + 1.7 (73) = 236$ kips

M_u (EW) $= 1.4 (40) + 1.7 (30) = 107$ ft-kips

M_u (NS) $= 1.4 (52) + 1.7 (42) = 144$ ft-kips

Eq. (9-1)

Since this is a circular column, moments about both axes can be combined into a resultant moment, and the column can be treated as one subject to uniaxial loading.

M_u (resultant) $= \sqrt{107^2 + 144^2} = 179$ ft-kips

2. Check if slenderness need be considered in the design. Assume 16 in. column size for slenderness evaluation.

$k = 1.0$ for braced compression member

10.11.2.1

$r = 0.25 (16) = 4$ in.

10.11.3

$kl_u/r = 1.0 \times 7 \times 12/4 = 21 < 34 - 12 = 22$ (ratio of end moments assumed equal to 1.0)

10.11.4.1

Therefore, slenderness need not be considered.

3. Using Design Aid EB9

Required $\quad P_n = \dfrac{P_u}{\varphi} = \dfrac{236}{0.75} = 315$ kips

9.3.2.2(b)

11-43

Example 11.6—Continued

$$M_n = \frac{M_u}{\varphi} = \frac{179}{0.75} = 239 \text{ ft-kips}$$

From strength table for 16-in. diameter column size (see Table 11-11), 7#11 bars (PCNT = 5.43) are required, with a strength provided of P_n = 315 kips and M_n = 239 ft-kips.

4. Select spiral reinforcement as illustrated in Example 11.4

5. <u>Using Design Aid SP17A</u>

   (a) Check required reinforcement for 16 in. diameter column (A_g = 201 in.^{2})

   Compute $\dfrac{P_u}{A_g} = \dfrac{236}{201} = 1.17$

   $$\frac{M_u}{A_g h} = \frac{179 \times 12}{(201)\,16} = 0.67$$

   Estimate $\quad \gamma = \dfrac{h-5}{h} = \dfrac{16-5}{16} = 0.69$

   (b) For circular section, f_c' = 5 ksi, f_y = 60 ksi, and γ = 0.69, use Column Chart 7.23.3 for C5-60.75 Columns (reproduced in Fig. 11-12). Read ρ_g = 0.042.

   For this combination of material strengths (5 ksi & 60 ksi), only column charts for γ of 0.60 and 0.75 are included in SP17A. For C5-60.60 Columns (γ = 0.60), ρ_g = 0.063. Use ρ_g by interpolation = 0.053 $\approx$ 0.054 by EB9 table.

6. <u>Using CRSI Handbook</u>

   (a) P_u = 236 kips, M_u = 179 ft-kips

   $$e = \frac{M_u}{P_u} = \frac{179 \times 12}{236} = 9.1 \text{ in.}$$

   (b) For Grade 60 bars with f_c' = 5000 psi, and 16 in. diameter circular section, use Load Capacity Table—ROUND TIED COLUMN 16" DIA., Page 4-35 (reproduced in Table 11-12). With e = 9.1 in., interpolate for solution.

Example 11.6—Continued

Table 11-11—Design Aid Reproduced from Page 37 of EB9

PCA—LOAD AND MOMENT STRENGTH TABLES FOR CONCRETE COLUMNS

1 1/2" clear

COLUMN SIZE = 16X16 INCHES
FPC=5,000 FY=60,000

BARS			SPECIAL CONDITIONS			AXIAL LOAD STRENGTH P_n (KIPS)																	
NO TOT	SIZE NO	PCNT	0. PO	BALANCE PB	MB	0	20	40	60	80	100	120	160	200	240	340	440	540	640	740	840	940	1040
						MOMENT STRENGTH M_n (FT-KIPS)																	
7	5	1.08	975	283	137	64	72	79	87	94	101	107	117	124	131	138	136	128	111	85			
8	5	1.23	993	308	142	72	80	87	95	102	107	112	121	130	137	142	140	132	115	89	56		
9	5	1.39	1010	291	146	80	88	95	100	106	112	117	128	135	141	146	144	135	119	94	62		
10	5	1.54	1027	307	150	87	94	100	106	113	119	124	133	140	146	150	147	139	123	99	68		
11	5	1.70	1045	295	155	93	100	107	113	120	125	129	138	145	150	154	151	142	127	104	74		
12	5	1.85	1062	307	159	100	107	114	120	125	130	135	144	150	155	158	155	146	131	109	80		
13	5	2.00	1079	297	163	108	114	120	125	131	136	141	149	155	160	163	159	150	135	114	85		
14	5	2.16	1096	306	168	114	120	126	132	137	142	146	154	161	164	167	162	153	139	119	90		
6	6	1.31	1002	304	143	75	82	89	96	102	108	114	126	133	138	143	141	132	117	92	60		
7	6	1.53	1026	277	149	85	92	99	106	113	119	124	131	138	144	149	146	138	123	100	68		
8	6	1.75	1051	304	155	96	103	111	116	121	125	130	139	146	152	155	151	143	129	106	75		
9	6	1.97	1075	285	161	106	112	117	123	128	133	138	148	153	157	160	156	148	134	112	83		
10	6	2.19	1100	302	167	114	120	126	132	138	143	147	153	160	165	166	162	154	140	119	91		
11	6	2.41	1124	289	174	123	130	136	141	145	149	153	161	168	171	172	167	159	145	125	98	67	
12	6	2.63	1149	302	179	133	138	143	148	153	157	162	169	173	177	178	172	164	150	131	106	75	
6	7	1.79	1055	300	155	95	102	108	114	121	127	132	143	147	151	155	151	143	129	107	78		
7	7	2.09	1089	268	164	109	116	123	129	136	139	142	149	155	160	162	158	150	136	117	88		
8	7	2.39	1122	299	171	124	129	134	138	143	147	151	159	166	169	170	165	157	144	125	97	65	
9	7	2.69	1156	277	180	133	138	144	149	154	159	163	170	174	177	178	172	164	152	133	107	76	
10	7	2.98	1189	297	187	145	151	156	162	166	169	172	178	183	187	185	179	171	159	140	116	86	
11	7	3.28	1222	280	196	158	163	166	170	174	178	181	188	192	194	193	187	178	166	148	125	97	
6	8	2.36	1119	295	169	118	124	130	136	142	148	153	160	163	166	168	163	155	142	123	97	65	
7	8	2.75	1163	258	181	137	143	150	153	156	159	162	168	174	179	178	172	164	152	134	110	78	
8	8	3.14	1207	294	190	151	155	159	163	167	171	175	182	188	189	188	182	173	161	145	122	91	
9	8	3.54	1251	266	202	164	168	173	178	183	187	190	193	197	200	198	191	182	171	155	133	105	73
10	8	3.93	1295	290	211	180	185	189	192	194	197	200	205	210	211	208	200	192	181	164	144	117	87
6	9	2.98	1189	288	184	142	147	153	159	165	170	174	177	179	182	182	176	167	156	139	116	88	
7	9	3.48	1245	245	199	166	170	172	175	178	181	183	189	194	198	194	188	179	168	152	132	104	72
8	9	3.98	1301	286	210	178	182	186	189	193	197	200	207	209	210	207	200	191	179	165	146	119	88
9	9	4.48	1356	253	225	196	200	205	209	212	214	216	218	221	224	219	211	202	191	178	158	134	105
6	10	3.79	1279	279	203	171	177	182	187	193	194	195	197	199	201	200	193	184	172	158	138	113	85
7	10	4.42	1350	229	221	194	197	199	202	204	207	209	214	218	221	215	208	198	187	173	156	134	105
6	11	4.66	1376	268	222	201	207	212	214	215	216	217	218	220	221	218	210	201	190	176	159	138	112
7	11	5.43	1463	209	245	222	224	226	229	231	233	235	240	244	243	236	228	219	208	195	179	160	138

Example 11.6—Continued

Bars	$\rho\%$	φP_n	$1.07(\varphi P_n)*$
5 #14	5.60	229	245
6 #11	4.66	216	231
7 #10	4.42	217	232
7 #11	5.43	232	248
8 #10	5.05	231	247

* Correction factor for spirals. See Example 11.4.

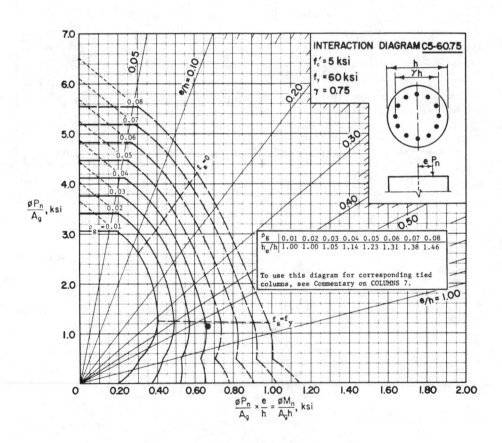

Fig. 11-12   Chart Reproduced from Page 120 of SP17A
(Courtesy of American Concrete Institute)

Example 11.6—Continued

Calculations and Discussion

The above table indicates the EB9 solution to be adequate.

Note: The slight difference between the designs of EB9 and the CRSI Handbook is possibly due to interpolation approximations and possibly due to different concrete stress distributions; EB9 uses a parabolic stress distribution, while CRSI uses the rectangular stress block. Both designs should, however, be considered equally valid.

7. Using PCACOL Program

Using the design option of the PCACOL program, design the column for a required strength $P_u = 236$ kips and $M_u = 179$ ft-kips. The bars will range from 4#8 to 8#11.

The graphic results of PCACOL are reproduced in Fig. 11-13. 8#10 bars were chosen by the program. Both a strength interaction diagram and a scaled column section showing the vertical reinforcement are displayed. The required strength combination is shown as point "1" relative to the strength curve provided.

Example 11.6—Continued

Table 11-12—Design Aid Reproduced from Page 4-35 of CRSI Handbook
(Courtesy of Concrete Reinforcing Steel Institute)

Bars Grade 60				ROUND TIED COLUMNS 16" DIA. Short columns; no sidesway [1]											$0.10f'_cA_g = 100$ kips [2] For ϕP_n at [2]				
Concrete $f'_c = 5{,}000$ psi				ϕP_n (kips)—Ultimate Usable Capacity											OT [3]	Balance	100 k [2]	($\phi = 0.90$) O [2]	
Bars	P %	**0.80** ϕP_o (k)	e 0.80 ϕP_o (in.)	$M_n/P_n = e$ (in.) ($\phi = 0.70$)											e (in.)	e (in.)	ϕP_b (k)	e (in.)	ϕM_n (k-ft.)
				2"	3"	4"	6"	8"	12"	16"	20"	24"	28"					
4-#7	1.19	553	1.29	481	386	308	200	134	0	0	0	0	0	1.95	5.47	226	9.88	66
4-#8	1.57	577	1.26	499	401	323	219	149	0	0	0	0	0	1.99	5.88	224	10.90	84
4-#9	1.99	603	1.23	518	418	338	234	165	0	0	0	0	0	2.03	6.34	222	11.94	103
4-#10	2.53	637	1.19	542	438	356	251	183	112	0	0	0	0	2.07	6.95	218	13.21	121
4-#11	3.10	673	1.14	566	457	372	263	198	122	0	0	0	0	2.14	7.71	207	14.27	134
4-#14	4.48	759	1.06	626	505	414	297	228	148	0	0	0	0	2.20	9.52	193	17.06	167
4-#18	7.96	978	.93	771	620	510	369	285	196	149	118	0	0	2.28	16.10	148	23.22	236
5-#6	1.09	547	1.33	480	385	306	199	133	0	0	0	0	0	2.05	5.56	219	9.64	59
5-#7	1.49	572	1.31	500	402	322	218	153	0	0	0	0	0	2.09	6.06	216	10.84	76
5-#8	1.96	602	1.28	523	422	340	235	170	102	0	0	0	0	2.14	6.65	212	12.20	95
5-#9	2.49	634	1.25	548	443	359	252	187	117	0	0	0	0	2.18	7.34	207	13.67	116
5-#10	3.16	677	1.21	579	469	383	271	207	132	0	0	0	0	2.22	8.26	201	15.41	138
5-#11	3.88	722	1.17	610	493	402	285	218	144	105	0	0	0	2.28	9.47	186	16.84	157
5-#14	5.60	830	1.10	687	554	455	326	251	172	127	101	0	0	2.34	12.64	164	20.21	203
6-#6	1.31	561	1.31	490	396	318	212	143	0	0	0	0	0	2.28	6.27	202	10.30	69
6-#7	1.79	591	1.27	513	416	337	230	164	0	0	0	0	0	2.33	7.00	197	11.63	90
6-#8	2.36	626	1.24	541	440	359	249	188	112	0	0	0	0	2.38	7.93	190	13.19	114
6-#9	2.98	666	1.20	571	466	382	268	204	128	0	0	0	0	2.44	9.03	182	14.77	134
6-#10	3.79	716	1.16	609	498	411	291	223	148	106	0	0	0	2.49	10.60	171	16.75	157
6-#11	4.66	771	1.11	646	528	433	308	237	162	118	0	0	0	2.55	12.86	151	18.52	177
6-#14	6.71	900	1.03	740	605	495	356	276	190	144	116	0	0	2.62	19.87	117	22.94	227
7-#5	1.08	546	1.33	479	386	306	199	133	0	0	0	0	0	1.97	5.43	225	9.65	59
7-#6	1.53	575	1.30	502	405	325	220	155	0	0	0	0	0	2.03	5.96	222	11.04	78
7-#7	2.09	609	1.27	529	429	347	241	177	108	0	0	0	0	2.08	6.62	219	12.70	101
7-#8	2.75	651	1.23	561	457	371	262	198	125	0	0	0	0	2.14	7.41	215	14.54	126
7-#9	3.48	697	1.19	596	486	397	283	217	141	102	0	0	0	2.18	8.34	209	16.38	149
7-#10	4.42	756	1.15	641	523	429	309	239	160	118	0	0	0	2.24	9.60	202	18.57	178
7-#11	5.43	819	1.10	684	556	456	329	254	173	129	102	0	0	2.29	11.35	184	20.44	204
7-#14	7.83	970	1.03	794	642	528	384	298	206	157	125	104	0	2.36	16.09	156	24.98	260
8-#8	3.14	676	1.21	583	474	386	274	209	136	0	0	0	0	2.08	7.59	221	15.54	140
8-#9	3.98	728	1.18	623	507	415	298	230	153	112	0	0	0	2.13	8.55	216	17.65	165
8-#10	5.05	796	1.13	675	549	451	327	254	172	129	102	0	0	2.18	9.86	210	20.36	195
8-#11	6.21	868	1.08	726	586	481	350	272	187	140	112	0	0	2.23	11.68	192	22.37	222
9-#10	5.68	835	1.12	704	574	475	345	269	185	138	109	0	0	2.24	11.06	200	21.80	217
9-#11	6.98	917	1.07	759	617	509	370	288	199	151	121	0	0	2.29	13.48	178	23.93	245
10-#10	6.32	875	1.11	735	599	495	364	284	196	147	117	0	0	2.33	12.85	184	23.25	234

Example 11.6—Continued

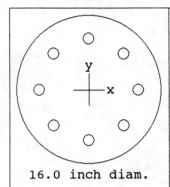

16.0 inch diam.

f'_c = 5.0 ksi

f_y = 60.0 ksi

   8#10 5.1%

A_{st} = 10.16 in^2

Spiral cc = 1.88 in

 spacing = 2.93 in

I_x = 3217 in^4

$\bar{x}$ = 0.00 in

I_y = 3217 in^4

$\bar{y}$ = 0.00 in

© 1989 PCA

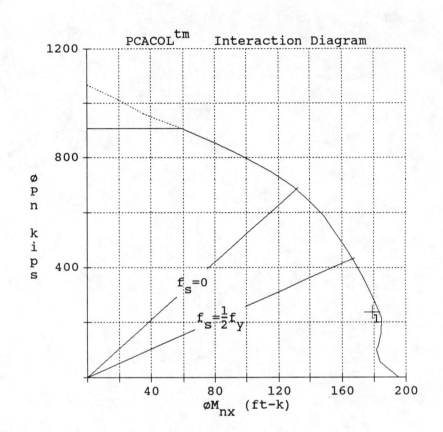

Licensed To: Portland Cement Association, Skokie, IL

Project: Example 11.6

Column Id: Design of Sprl Column

Engineer:

Date: Time:

Code: ACI 318-89

Version: 2.10

Slenderness not considered x-axis

File name: C:\PCACOL\DATA\EX11_6.COL

Material Properties:

E_c = 4287 ksi ϵ_u = 0.003 in/in

f_c = 4.25 ksi E_s = 29000 ksi

β_1 = 0.80

Stress Profile : Block

Reduction: ϕ_c = 0.75 ϕ_b = 0.90

**Fig. 11-13   PCACOL On-Screen Graphic Results
for 16 in. Circular Column with 8#10 Bars**

11-49

Design for
Biaxial Loading

GENERAL CONSIDERATIONS

Biaxial bending of columns occurs when the column loading causes bending simultaneously about both principal axes; the commonly encountered case of such loading occurs in corner columns. Design for biaxial bending and axial load is mentioned in Commentary Section 10.3.6. Also, for considering column slenderness, Section 10.11.7 addresses moment magnifiers for biaxial loading. Commentary Section 10.3.6 states that "corner and other columns exposed to known moments about each axis simultaneously should be designed for biaxial bending and axial load." For the '89 Code, Commentary Section 10.3.6 has been expanded to address two satisfactory methods for combined biaxial bending and axial load design: the Reciprocal Load Method and the Load Contour Method. Both methods, and an extension of the Load Contour Method (PCA Load Contour Method), are presented in the following text.

As with uniaxial bending and axial load design, use of design aids is not just convenient, but necessary for a practical design solution involving biaxial bending. Examples 12.1 and 12.2 illustrate the use of design aids SP17A, CRSI Handbook and the PCACOL Program for biaxial bending design. Note that circular column sections possess essential polar symmetry; so that biaxial moments can simply be combined into $M_u = \sqrt{M_{ux}^2 + M_{uy}^2}$, with the circular column designed for uniaxial bending and axial load (P_u and M_u). See Example 11.6.

BIAXIAL INTERACTION STRENGTH

A uniaxial interaction diagram defines the load-moment strength along a single plane of a section under an axial load P and a uniaxial moment M. The biaxial bending resistance of an axially loaded column can be represented schematically (see Fig. 12-1) as a surface formed by a series of uniaxial interaction curves drawn radially from the P axis. Data for these intermediate curves are obtained by varying the angle of the neutral axis (for assumed strain configurations) with respect to the major axes (see Fig. 12-2).

The difficulty associated with the determination of the strength of reinforced columns subject to combined axial load and biaxial bending is primarily an arithmetic one. The bending resistance of an axially loaded

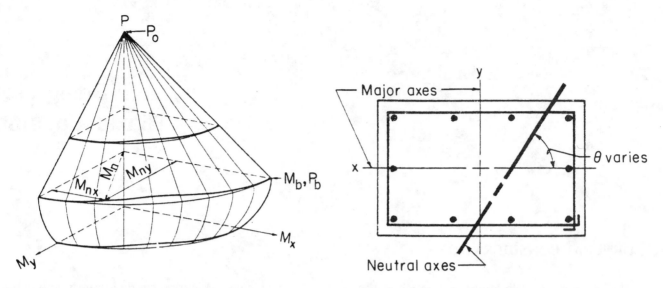

| Fig. 12-1   Biaxial Interaction Surface | Fig. 12-2   Neutral Axis at an Angle to Major Axes |

column about a particular skewed axis is determined through iteration of simple, but lengthy calculations. These extensive calculations are compounded when optimization of the reinforcement or cross section is sought.

For uniaxial bending, it is customary to utilize design aids in the form of interaction curves or tables. However, for biaxial bending, because of the voluminous nature of the data and the difficulty in multiple interpolations, the development of interaction curves or tables for the various ratios of bending moments about each axis is impractical. Instead, several approaches (based on acceptable approximations) have been developed that relate the response of a column in biaxial bending to its uniaxial resistance about each major axis.[12.1]

Failure Surfaces

The nominal strength of a section under biaxial bending and compression is a function of three variables P_n, M_{nx} and M_{ny} which may be expressed in terms of an axial load acting at eccentricities $e_x = \dfrac{M_{ny}}{P_n}$ and $e_y = \dfrac{M_{nx}}{P_n}$ as shown in Fig. 12-3. A failure surface may be described as a surface produced by plotting the failure load P_n as a function of its eccentricities e_x and e_y or of its associated bending moments M_{ny} and M_{nx}.

Three types of failure surfaces have been defined.[12.2, 12.3, 12.4] The basic surface S_1 is defined by a function which is dependent upon the variables P_n, e_x and e_y, as shown in Fig. 12-4. A reciprocal surface can be derived from S_1 in which the reciprocal of the nominal axial load P_n is employed to produce the surface $S_2 \left(\dfrac{1}{P_n}, e_x, e_y \right)$ as illustrated in Fig. 12-5. The third type of failure surface, shown in Fig. 12-6, is obtained by

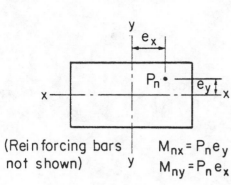

Fig. 12-3   Notation for Biaxial Loading

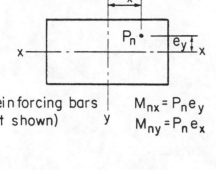

Fig. 12-4   Failure Surface S_1

relating the nominal axial load P_n to moments M_{nx} and M_{ny} to produce surface S_3 (P_n, M_{nx}, M_{ny}). Failure surface S_3 is the three dimensional extension of the uniaxial interaction diagram previously described.

A number of investigators have made approximations for both the S_2 and S_3 failure surfaces for use in design and analysis.[12.4 – 12.10] An explanation of these methods used in current practice, along with design examples, is given below.

A.   Bresler Reciprocal Load Method

This method approximates the ordinate $\frac{1}{P_n}$ on the surface S_2 $\left(\frac{1}{P_n}, e_x, e_y\right)$ by a corresponding ordinate $\frac{1}{P_n'}$ on the plane S_2' $\left(\frac{1}{P_n'}, e_x, e_y\right)$, which is defined by the characteristic points A, B and C, as indicated in Fig. 12-7. For any particular cross section, the value P_o (corresponding to point C) is the load strength under pure axial compression; P_{ox} (corresponding to point B) and P_{oy} (corresponding to point A) are the load strengths under uniaxial eccentricities e_y and e_x, respectively. Each point on the true surface is approximated by a different plane; therefore, the entire surface is approximated using an infinite number of planes.

The general expression for axial load strength for any values of e_x and e_y is as follows:[12-4]

$$\frac{1}{P_n} \approx \frac{1}{P_n'} = \frac{1}{P_{ox}} + \frac{1}{P_{oy}} - \frac{1}{P_o}$$

Rearranging variables yields:

$$P_n \approx \frac{1}{\dfrac{1}{P_{ox}} + \dfrac{1}{P_{oy}} - \dfrac{1}{P_o}} \tag{1}$$

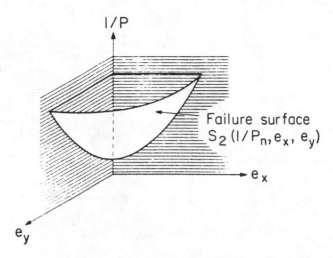

Fig. 12-5   Reciprocal Failure Surface S₂

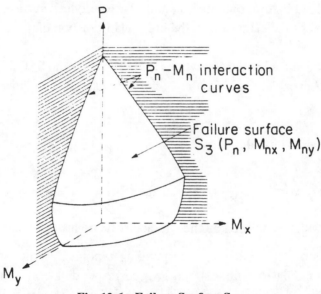

Fig. 12-6   Failure Surface S₃

This equation is simple in form and the variables are easily determined. Axial load strengths P_o, P_{ox}, and P_{oy} are determined using any of the methods presented in Part 11. Experimental results have shown the above equation to be reasonably accurate when flexure does not govern design. The equation should only be used when:

$$P_n \geq 0.1\, f_c'\, A_g \qquad (2)$$

B. Bresler Load Contour Method

In this method, the surface S_3 (P_n, M_{nx}, M_{ny}) is approximated by a family of curves corresponding to constant values of P_n. These curves, as illustrated in Fig. 12-8, may be regarded as "load contours."

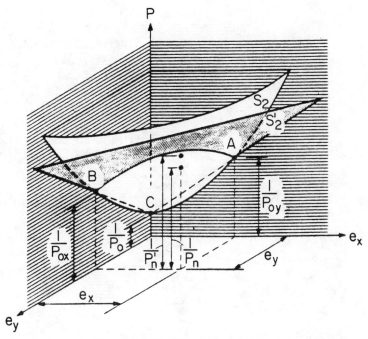

Fig. 12-7   Reciprocal Load Method

The general expression for these curves can be approximated[12.4] by a nondimensional interaction equation of the form

$$\left(\frac{M_{nx}}{M_{nox}}\right)^{\alpha} + \left(\frac{M_{ny}}{M_{noy}}\right)^{\beta} = 1.0 \tag{3}$$

In the above equation, M_{nx} and M_{ny} are the nominal biaxial moment strengths in the direction of the x and y axes respectively. M_{nx} and M_{ny} are the vectorial equivalent of the nominal uniaxial moment strength M_n. M_{nox} and M_{noy} are the nominal uniaxial moment strengths, with bending considered in the direction of the x and y axes separately. The values of the exponents α and β are a function of the amount, distribution and location of reinforcement, the dimensions of the column, and the strength and elastic properties of the steel and concrete. Bresler[12.4] indicates that it is reasonably accurate to assume that $\alpha = \beta$; therefore, Eq. (3) becomes

$$\left(\frac{M_{nx}}{M_{nox}}\right)^{\alpha} + \left(\frac{M_{ny}}{M_{noy}}\right)^{\alpha} = 1.0 \tag{4}$$

which is shown graphically in Fig. 12-9.

When using Eq. (4) or Fig. 12-9, it is still necessary to determine the α value for the cross section being designed. Bresler indicated that, typically, α varied from 1.15 to 1.55, with a value of 1.5 being reasonably accurate for most square and rectangular sections with uniformly distributed reinforcement.

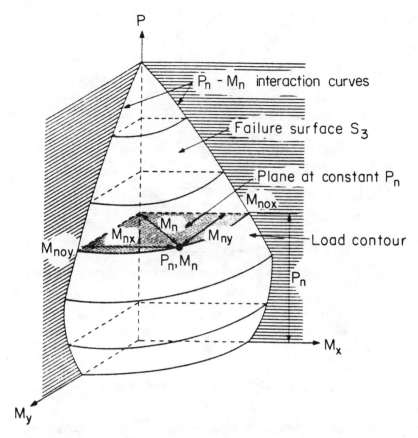

Fig. 12-8   Load Contours for Constant P_n on Failure Surface S_3

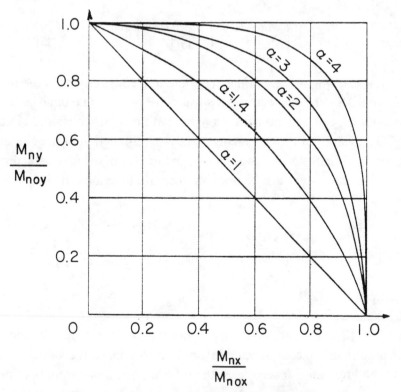

Fig. 12-9   Interaction Curves according to Eq. (4)

With α set at unity, the interaction equation becomes linear, as shown in Fig. 12-9, and will always yield conservative results:

$$\frac{M_{nx}}{M_{nox}} + \frac{M_{ny}}{M_{noy}} = 1.0 \tag{5}$$

The use of Eq. (5) becomes overly conservative for high axial loads or low percentages of reinforcement. It should only be used when

$$P_n < 0.1 \, f_c' \, A_g \tag{6}$$

C. PCA Load Contour Method

The PCA approach described below was developed as an extension of the Bresler Load Contour Method. The Bresler interaction equation [Eq. (4)] was chosen as the most viable method in terms of accuracy, condensation of design aids and simplification potential.

A typical Bresler load contour is shown in Fig. 12-10. In the PCA method,[12.9] a point B is defined such that the nominal biaxial moment strengths M_{nx} and M_{ny} at this point are in the same ratio as the uniaxial moment strengths M_{nox} and M_{noy}; therefore, at point B

$$\frac{M_{nx}}{M_{ny}} = \frac{M_{nox}}{M_{noy}} \tag{7}$$

When the load contour of Fig. 12-10 is nondimensionalized, it takes the form shown in Fig. 12-11, and the point B will have x and y coordinates of β. When the bending resistance is plotted in terms of the dimensionless parameters $\frac{P_n}{P_o}$, $\frac{M_{nx}}{M_{nox}}$, and $\frac{M_{ny}}{M_{noy}}$ (the latter two designated as the relative moments), the generated failure surface $S_4 \left(\frac{P_n}{P_o}, \frac{M_{nx}}{M_{nox}}, \frac{M_{ny}}{M_{noy}} \right)$ assumes the typical shape shown in Fig. 12-12. The advantage of expressing the behavior in relative terms is that the contours of the surface (Fig. 12-11)—i.e., the

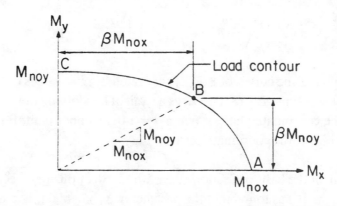

Fig. 12-10 Load Contour of Failure Surface S_3 along Plane of Constant P_n

12-7

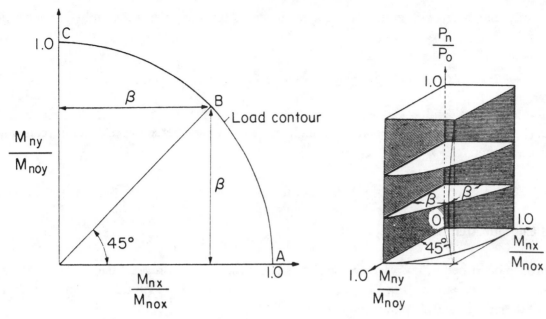

Fig. 12-11   Nondimensional Load Contour
at Constant P_n

Fig. 12-12   Failure Surface S_4
$$\left(\frac{P_n}{P_o}, \frac{M_{nx}}{M_{nox}}, \frac{M_{ny}}{M_{noy}}\right)$$

intersection formed by planes of constant P_n/P_o and the surface—can be considered for design purposes to be symmetrical about the vertical plane bisecting the two coordinate planes. Even for sections that are rectangular or have unequal reinforcement on the two adjacent faces, this approximation yields values sufficiently accurate for design.

The relationship between α from Eq. (4) and β is obtained by substituting the coordinates of point B from Fig. 12-10 into Eq. (4), and solving for α in terms of β. This yields:

$$\alpha = \frac{\log 0.5}{\log \beta}$$

Thus Eq. (4) may be written as:

$$\left(\frac{M_{nx}}{M_{nox}}\right)^{\left(\frac{\log 0.5}{\log \beta}\right)} + \left(\frac{M_{ny}}{M_{noy}}\right)^{\left(\frac{\log 0.5}{\log \beta}\right)} = 1.0 \qquad (8)$$

For design convenience, a plot of the curves of Eq. (8) generated by nine values of β are given in Fig. 12-13. Note that when $\beta = 0.5$, its lower limit, Eq. (8) describes a straight line joining the points at which the relative moments equal 1.0 along the coordinate planes. When $\beta = 1.0$, its upper limit, Eq. (8) describes two lines, each of which is parallel to one of the coordinate planes.

Values of β were computed on the basis of Section 10.2 of the Code, utilizing a rectangular stress block and basic principles of equilibrium. It was found that the parameters γ, b/h, and f_c' had minor effect on β values.

The maximum difference in β amounted to about 5% for a given value of P_n/P_o ranging from 0.1 to 0.9. The bulk of the values, especially those in the most frequently used range of P_n/P_o, did not differ by more than 3%. In view of these small differences, only envelopes of the lowest β values were developed for two values of f_y and different bar arrangements. These are shown in Figs. 12-14 to 12-17.

As can be seen from an inspection of Figs. 12-14 to 12-17, β is dependent primarily on the ratio P_n/P_o and to a lesser, though still significant extent, on the bar arrangement, the reinforcement index ω and the strength of the reinforcement.

Figure 12-13, in combination with Figs. 12-14 to 12-17, furnish a convenient and direct means of determining the biaxial moment strength of a given cross section subject to an axial load, since the values of P_o, M_{nox}, and M_{noy} can be readily obtained by methods described here and in Part 11.

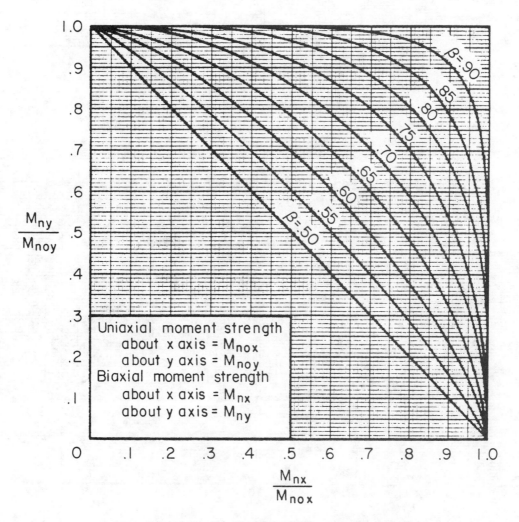

Fig. 12-13   Biaxial Moment Strength Relationship

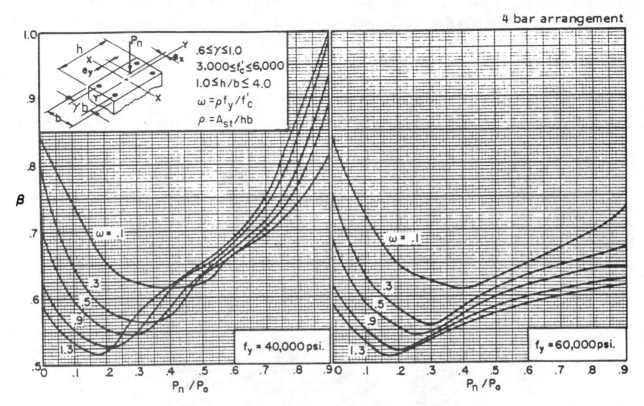

Fig. 12-14 Biaxial Bending Design Constants

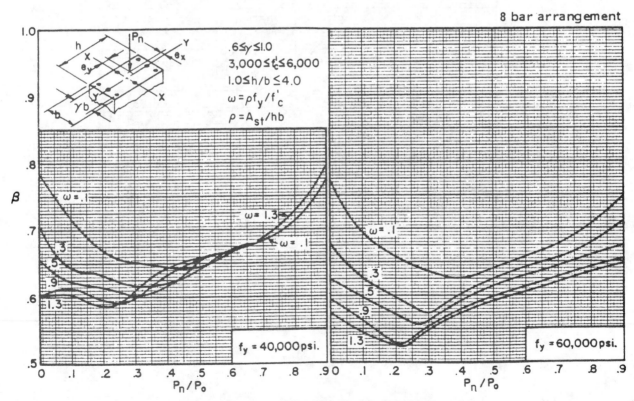

Fig. 12-15 Biaxial Bending Design Constants

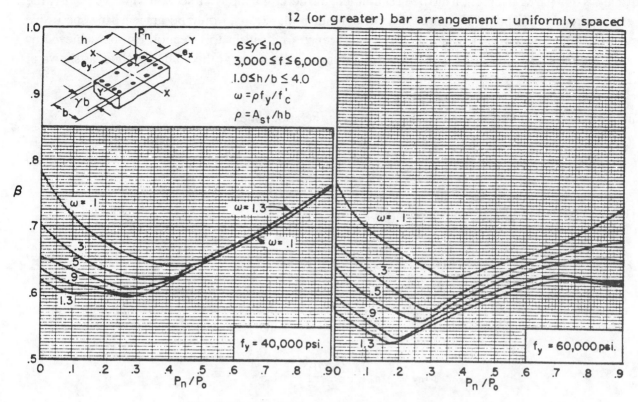

Fig. 12-16 Biaxial Bending Design Constants

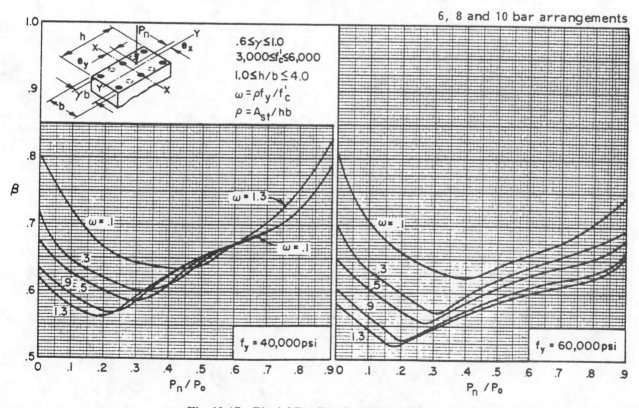

Fig. 12-17 Biaxial Bending Design Constants

While analysis of a particular section has been simplified, the determination of a section which will satisfy the strength requirements imposed by a load eccentric about both axes can only be achieved by successive analyses of assumed sections. Rapid and easy convergence to a satisfactory section can be achieved by approximating the curves in Fig. 12-13 by two straight lines intersecting at the 45 degree line, as shown in Fig. 12-18.

By simple geometry, it can be shown that the equation of the upper lines is:

$$\frac{M_{ny}}{M_{noy}} + \frac{M_{nx}}{M_{nox}}\left(\frac{1-\beta}{\beta}\right) = 1 \quad \text{for} \quad \frac{M_{ny}}{M_{nx}} > \frac{M_{noy}}{M_{nox}} \tag{9}$$

which can be restated for design convenience as follows:

$$M_{ny} + M_{nx}\left(\frac{M_{noy}}{M_{nox}}\right)\left(\frac{1-\beta}{\beta}\right) = M_{noy} \tag{10}$$

For rectangular sections with reinforcement equally distributed on all faces, Eq. (10) can be approximated by:

$$M_{ny} + M_{nx}\frac{b}{h}\left(\frac{1-\beta}{\beta}\right) \approx M_{noy} \tag{11}$$

The equation of the lower line of Fig. 12-18 is:

$$\frac{M_{ny}}{M_{noy}}\left(\frac{1-\beta}{\beta}\right) + \frac{M_{nx}}{M_{nox}} = 1 \quad \text{for} \quad \frac{M_{ny}}{M_{nx}} < \frac{M_{noy}}{M_{nox}} \tag{12}$$

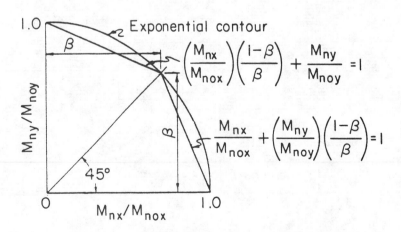

Fig. 12-18 Bilinear Approximation of Nondimensionalized Load Contour

12-12

or,

$$M_{ny} \left(\frac{M_{nox}}{M_{noy}} \right) \left(\frac{1 - \beta}{\beta} \right) + M_{nx} = M_{nox} \tag{13}$$

or, for rectangular sections with reinforcement equally distributed on all faces,

$$M_{ny} \frac{h}{b} \left(\frac{1 - \beta}{\beta} \right) + M_{nx} \approx M_{nox} \tag{14}$$

In design Eqs. (11) and (14), the ratio b/h or h/b must be chosen and the value of β must be assumed. For lightly loaded columns, β will generally vary from 0.55 to about 0.70. Hence, a value of 0.65 for β is generally a good initial choice in a biaxial bending analysis.

Manual Design Procedure

To aid the engineer in designing columns for biaxial bending, a procedure for manual design is outlined below:

1. Choose the value of β at 0.65 or use Figs. 12-14 to 12-17 to make an estimate.

2. If M_{ny}/M_{nx} is greater than b/h, use Eq. (11) to calculate an approximate equivalent uniaxial moment strength M_{noy}. If M_{ny}/M_{nx} is less than b/h, use Eq. (14) to calculate an approximate equivalent uniaxial moment strength M_{nox}.

3. Design the section using any of the methods presented in Part 11 (i.e., uniaxial bending with axial load) to provide an axial load strength P_n and an equivalent uniaxial moment strength M_{noy} or M_{nox}.

4. Verify the section chosen by any one of the following three methods:

    A. Bresler Reciprocal Load Method:

    $$P_n \leq \frac{1}{\dfrac{1}{P_{ox}} + \dfrac{1}{P_{oy}} - \dfrac{1}{P_o}} \tag{1}$$

    B. Bresler Load Contour Method:

    $$\frac{M_{nx}}{M_{nox}} + \frac{M_{ny}}{M_{noy}} \leq 1.0 \tag{5}$$

    C. PCA Load Contour Method: Use Eq. (8) or,

    $$\frac{M_{ny}}{M_{noy}} + \frac{M_{nx}}{M_{nox}} \left(\frac{1 - \beta}{\beta} \right) \leq 1.0 \left[\text{for } \frac{M_{ny}}{M_{nx}} > \frac{M_{noy}}{M_{nox}} \right] \tag{9}$$

    $$\frac{M_{ny}}{M_{noy}} \left(\frac{1 - \beta}{\beta} \right) + \frac{M_{nx}}{M_{nox}} \leq 1.0 \left[\text{for } \frac{M_{ny}}{M_{nx}} < \frac{M_{noy}}{M_{nox}} \right] \tag{12}$$

REFERENCES

12.1 Wang, C. K., and Salmon, C. G., *Reinforced Concrete Design,* 4th Edition, Harper & Row Publishers, New York, 1985.

12.2 Pannell, F. N., "The Design of Biaxially Loaded Columns by Ultimate Load Methods," *Magazine of Concrete Research,* London, July 1960, pp. 103-104.

12.3 Pannell, F. N., "Failure Surfaces for Members in Compression and Biaxial Bending," *ACI Journal, Proceedings* Vol. 60, January 1963, pp. 129-140.

12.4 Bresler, Boris, "Design Criteria for Reinforced Columns under Axial Load and Biaxial Bending," *ACI Journal, Proceedings* Vol. 57, November 1960, pp. 481-490, discussion pp. 1621-1638.

12.5 Furlong, Richard W., "Ultimate Strength of Square Columns under Biaxially Eccentric Loads," *ACI Journal, Proceedings* Vol. 58, March 1961, pp. 1129-1140.

12.6 Meek, J. L., "Ultimate Strength of Columns with Biaxially Eccentric Loads," *ACI Journal, Proceedings* Vol. 60, August 1963, pp. 1053-1064.

12.7 Aas-Jakobsen, A., "Biaxial Eccentricities in Ultimate Load Design," *ACI Journal, Proceedings* Vol. 61, March 1964, pp. 293-315.

12.8 Ramamurthy, L. N., "Investigation of the Ultimate Strength of Square and Rectangular Columns under Biaxially Eccentric Loads," Symposium on Reinforced Concrete Columns, American Concrete Institute, Detroit, 1966, pp. 263-298.

12.9 *Capacity of Reinforced Rectangular Columns Subject to Biaxial Bending,* Publication EB011D, Portland Cement Association, Skokie, IL, 1966.

12.10 *Biaxial and Uniaxial Capacity of Rectangular Columns,* Publication EB031D, Portland Cement Association, Skokie, IL, 1967.

Example 12.1—Design of a Square Column for Biaxial Loading

Determine the required square column size and reinforcement for the factored load and moments given. Assume the reinforcement is equally distributed on all faces.

$$P_u = 1200 \text{ kips}, \quad M_{ux} = 1800 \text{ ft-kips}, \quad M_{uy} = 750 \text{ ft-kips}$$
$$f_c' = 4000 \text{ psi}, \quad f_y = 60,000 \text{ ksi}$$

Calculations and Discussion	Code Reference

1. Determine required nominal strengths. For tied reinforced column, $\varphi = 0.70$ 9.3.2.2(b)

$$P_n = \frac{P_u}{\varphi} = \frac{1200}{0.7} = 1714 \text{ kips}$$

$$M_{nx} = \frac{M_{ux}}{\varphi} = \frac{1800}{0.7} = 2571 \text{ ft-kips}$$

$$M_{ny} = \frac{M_{uy}}{\varphi} = \frac{750}{0.7} = 1071 \text{ ft-kips}$$

2. Assume $\beta = 0.65$

3. Determine an equivalent uniaxial moment strength M_{nox} or M_{noy}.

$$\frac{M_{ny}}{M_{nx}} = \frac{1071}{2571} = 0.42 \text{ is less than } \frac{b}{h} = 1.0 \text{ (square column)}$$

Therefore, using Eq. (14)

$$M_{nox} \approx M_{ny} \frac{h}{b} \frac{1-\beta}{\beta} + M_{nx} = 1071 \, (1.0) \frac{1-0.65}{0.65} + 2571 = 3148 \text{ ft-kips}$$

4. Assuming a 36 in. square column, determine reinforcement required to provide an axial load strength $P_n = 1714$ kips and an equivalent uniaxial moment strength $M_{nox} = 3148$ ft–kips.

5. <u>Using Design Aid SP17A</u> (See Part 10, Page 10-6)

   For a square section with bars uniformly distributed on all sides, $f_c' = 4000$ psi, $f_y = 60,000$ psi, and $\gamma \approx 0.9$, use column chart 7.4.4 for R4-60.90 columns (See Fig. 12-19). Note that the strength reduction factor φ is already included in the load-moment strength curves.

Example 12.1—Continued

Compute $\dfrac{\varphi P_n}{A_g} = \dfrac{0.7\,(1714)}{1296} = 0.93$

Compute $\dfrac{\varphi M_{nox}}{A_g h} = \dfrac{0.7\,(3148)\,12}{(1296)\,36} = 0.57$

From Fig. 12-19, read $\rho_g \approx 0.022$

$A_{st} = \rho_g A_g = 0.022\,(1296) = 28.51 \text{ in.}^2$

Select 24 #10 bars ($A_s = 30.40 \text{ in.}^2$), ρ_g (actual) $= 0.023$

6. Selected section will now be checked for biaxial loading strength by each of the three methods presented in the discussion.

   A. <u>Bresler Reciprocal Load Method</u>

   Check $P_n \geq 0.1\,f_c'A_g$ (Eq. 2)

   $1714 \text{ kips} > 0.1\,(4)\,(1296) = 518 \text{ kips}$

   To employ this method, P_o, P_{ox}, and P_{oy} must be determined.

   $P_o = 0.85\,f_c'\,(A_g - A_{st}) + A_{st}\,f_y$

        $= 0.85\,(4)\,(1296 - 30.4) + 30.4\,(60) = 6127 \text{ kips}$

   Knowing ρ_g (actual) and required nominal load-moment strengths (P_n, M_{nx}, and M_{ny}), P_{ox} and P_{oy} can be determined.

   <u>x-axis</u>

   $P_n = 1714 \text{ kips}, M_{nx} = 2571 \text{ ft-kips}$

   $e_y = \dfrac{M_{nx}}{P_n} = \dfrac{(2571)\,(12)}{1714} = 18 \text{ in.}, \dfrac{e_y}{h} = \dfrac{18}{36} = 0.5$

   From Fig. 12-19, with $e_y/h = 0.5$ and $\rho_g = 0.023$, read $\varphi P_{ox}/A_g = 1.16$

   Therefore, $P_{ox} = \dfrac{1.16\,(1296)}{0.70} = 2148 \text{ kips}$

Example 12.1—Continued

COLUMNS 7.4.4—Load-moment strength interaction diagram for R4-60.90 columns

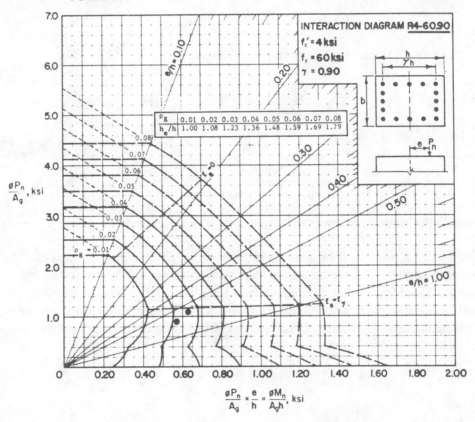

Fig. 12-19   Chart Reproduced from Page 82 of SP17A
(Courtesy of American Concrete Institute)

Example 12.1—Continued

	Code
Calculations and Discussion	**Reference**

<u>y-axis</u>

$P_n = 1714$ kips, $M_{ny} = 1071$ ft$-$kips

$$e_x = \frac{M_{ny}}{P_n} = \frac{1071\,(12)}{1714} = 7.5 \text{ in.,} \quad \frac{e_x}{h} = \frac{7.5}{36} = 0.21$$

From Fig. 12-19, with $e_x/h = 0.21$ and $\rho_g = 0.023$, read $\varphi P_{oy}/A_g = 2.05$

Therefore, $P_{oy} = \dfrac{2.05\,(1296)}{0.70} = 3795$ kips

Using the above values, Eq. (1) can now be evaluated.

$$P_n = 1714 \text{ kips} \ \le\ \frac{1}{\dfrac{1}{P_{ox}} + \dfrac{1}{P_{oy}} - \dfrac{1}{P_o}} \qquad \text{(Eq. 1)}$$

$$= \frac{1}{\dfrac{1}{2148} + \dfrac{1}{3795} - \dfrac{1}{6127}} = 1767 \text{ kips,} \qquad \text{O.K.}$$

B. Bresler Load Contour Method

Due to a lack of available data, a conservative α value of 1.0 is chosen; i.e., Eq. (5). Although $P_u > 0.1\,f_c'\,A_g$, the necessary calculations will be carried out for example purposes. Since the section is symmetrical, M_{nox} is equal to M_{noy}. Knowing ρ_g (actual) and the required axial load strength P_u, M_{nox} is determined as the nominal moment strength with bending considered in the direction of the x axis only.

$$P_n = 1714 \text{ kips,} \quad \frac{\varphi P_n}{A_g} = \frac{0.7\,(1714)}{1296} = 0.93$$

From Fig. 12-19, with $\dfrac{\varphi P_n}{A_g} = 0.93$ and $\rho_g = 0.023$, read $\dfrac{\varphi M_{nox}}{A_g h} = 0.59$

Therefore, $M_{nox} = \dfrac{0.59\,(1296)\,36}{12(0.70)} = 3277$ ft$-$kips

Using the above value, Eq. (5) can now be evaluated.

$$\frac{M_{nx}}{M_{nox}} + \frac{M_{ny}}{M_{noy}} = \frac{2571}{3277} + \frac{1071}{3277} = 0.785 + 0.327 = 1.11 > 1.0 \qquad \text{N.G.}$$

Example 12.1—Continued

Due to the inherent conservatism when using $\alpha = 1.0$, the section is inadequate when checked by this method.

C. PCA Load Contour Method

To employ this method, P_O, M_{nox}, M_{noy} and the true value of β must first be found.

$$P_O = 0.85\, f_c'\, (A_g - A_{st}) + A_{st}\, f_y$$

$$= 0.85\,(4)\,(1296 - 30.4) + 30.4\,(60) = 6127 \text{ kips}$$

Since the section is symmetrical, M_{nox} and M_{noy} are equal. Knowing ρ_g (actual) and the required axial load strength P_n, M_{nox} is determined as follows:

$$P_n = 1714 \text{ kips}, \quad \frac{\varphi P_n}{A_g} = \frac{0.7\,(1714)}{1296} = 0.93$$

From Fig. 12-19, with $\dfrac{\varphi P_n}{A_g} = 0.93$ and $\rho_g = 0.023$, read $\dfrac{\varphi M_{nox}}{A_g h} = 0.59$

Therefore, $\quad M_{nox} = \dfrac{0.50\,(1296)\,36}{12(0.70)} = 3277 \text{ ft-kips}$

Having found P_O and using ρ_g (actual), the true β value is determined as follows:

$$\frac{P_n}{P_O} = \frac{1714}{6127} = 0.28, \quad \omega = \frac{\rho_g\, f_y}{f_c'} = \frac{0.023\,(60)}{4} = 0.345$$

From Fig. 12-16, read $\beta = 0.573$

Using the above values, Eq. (8) can now be evaluated.

$$\left(\frac{M_{nx}}{M_{nox}}\right)^{\left(\frac{\log 0.5}{\log \beta}\right)} + \left(\frac{M_{ny}}{M_{noy}}\right)^{\left(\frac{\log 0.5}{\log \beta}\right)} \le 1.0$$

$\log 0.5 \;\; = -0.3$
$\log \beta \;\;\; = \log 0.573 = -0.242$
$\dfrac{\log 0.5}{\log \beta} = 1.24$

$$\left(\frac{2571}{3277}\right)^{1.24} + \left(\frac{1071}{3277}\right)^{1.24} = 0.740 + 0.250 = 0.990 < 1.0 \qquad \text{O.K.}$$

Example 12.1—Continued

	Code
Calculations and Discussion	**Reference**

The section can also be checked using the bilinear approximation.

Since $\dfrac{M_{ny}}{M_{nx}} < \dfrac{M_{noy}}{M_{nox}}$, Eq. (12) should be used.

$$\frac{M_{ny}}{M_{noy}}\left(\frac{1-\beta}{\beta}\right) + \frac{M_{nx}}{M_{nox}}$$

$$= \frac{1071}{3277}\left(\frac{1-0.573}{0.573}\right) + \frac{2571}{3277} = 0.244 + 0.785 = 1.029 \approx 1.0 \qquad \text{O.K.}$$

7. Using CRSI Handbook

Designing for biaxial loading is a simple and direct procedure using the load capacity tables in the CRSI handbook. The first three steps in the manual design procedure outlined on page 12-13 are, in essence, a procedure to establish a first trial design, with the fourth step, a check of the trial design for biaxial bending capacity. The Bresler Reciprocal Load Method is ideally suited for use with the CRSI Handbook as the tabulated strengths are given in terms of loads directly. Note that most column designs will be above the limiting P_n of 0.10 f_c' A_g, above which the reciprocal load equation will give reasonably accurate results. Once the trial design section is selected, the actual load-moment values (P_u, M_{ux}, M_{uy}) are used with Eq. (1) to check the trial section for biaxial loading capacity. Expressing Eq. (1) in terms of design load strengths for direct substitution of the CRSI load capacities, and using CRSI notation:

$$\varphi P_{nxy} = \frac{1}{\dfrac{1}{\varphi P_{nx}} + \dfrac{1}{\varphi P_{ny}} - \dfrac{1}{\varphi P_o}}$$

where φP_{nx}, φP_{ny}, and φP_o are taken directly from the load tables—the first two values at the required eccentricities e_x and e_y.

(a) Select a first trial section using the factored load $P_u = 1200$ kips and an equivalent uniaxial moment $M_u = \varphi M_{nox}$ (p. 12-15) = 3148 (0.70) = 2204 ft-kips. Note that the φ factor is already included in the CRSI load capacity tables.
$$e = \frac{M_u}{P_u} = \frac{2204 \times 12}{1200} = 22 \text{ in.}$$

(b) For Grade 60 bars with $f_c' = 4000$ psi, and 36 in. square column, use Load Capacity Table—SQUARE TIED COLUMNS 36 in. x 36 in., Page 3-26 (reproduced in Table 12-1). With e = 22 in., interpolate for solution:

Example 12.1—Continued

Bars	$\rho\%$	φP_n (kips)
8#18	2.47	1289
12#14	2.08	1155

Note that the CRSI load capacity table for the 36 in. square column section does not include load data for 24#10 bars.

Using the CRSI load capacity tables, a check of other section size that may provide the required strength within a reasonable percentage of reinforcement ($\pm 3\%$) can easily be made.

Section	Bars	$\rho\%$	φP_n (kips)
34 in. x 34 in.	8#18	2.77	1146
	16#14	3.11	1190
32 in. x 32 in.	12#18	4.69	1226
	20#14	4.39	1186

(c) Check trial design, 36 in. x 36 in. square column with 12#14 bars, for biaxial loading capacity.

$P_u = 1200$ kips
$M_{ux} = 1800$ ft-kips
$e_x = \dfrac{M_{ux}}{P_u} = \dfrac{1800 \times 12}{1200} = 18$ in.
$M_{uy} = 750$ ft-kips
$e_y = \dfrac{M_{uy}}{P_u} = \dfrac{750 \times 12}{1200} = 7.5$ in.

From load capacity table, by interpolation:

@ $e_x = 18$ in., $\varphi P_{nx} = 1418$ kips
@ $e_y = 7.5$ in., $\varphi P_{ny} = 2564$ kips
$\varphi P_o = \dfrac{3323}{0.8} = 4154$ kips

Example 12.1—Continued

Table 12-1   Design Aid Reproduced from Page 3-26 of CRSI Handbook
(Courtesy of Concrete Reinforcing Steel Institute)

Bars Grade 60	**SQUARE TIED COLUMNS 36″ × 36″** Short columns; no sidesway [1] Bars symmetrical in 4 faces														0.10$f'_c A_g$ = 518 kips [2]				
Concrete f'_c = 4,000 psi	ϕP_n (kips)—Ultimate Usable Capacity														For ϕP_n at [2]				
															OT [3]	Balance		518 k [2]	O [2]
Bars	P %	0.80 × ϕP_o (k)	e 0.80 × ϕP_o (in.)	2″	3″	4″	6″	8″	12″	16″	20″	24″	28″	e (in.)	e (in.)	ϕP_b (k)	e (in.)	ϕM_n (k-ft.)
4-#18	1.23	2975	3.78	2975	2975	2935	2594	2286	1794	1453	1135	883	701	5.30	16.81	1398	33.88	1115
8-#14	1.39	3038	3.54	3038	3038	2951	2585	2257	1741	1380	1059	843	696	4.82	15.43	1435	35.35	1259
8-#18	2.47	3482	3.45	3482	3482	3364	2963	2609	2059	1680	1417	1160	975	5.44	19.90	1425	49.16	2157
12-#10	1.18	2950	3.57	2950	2950	2870	2507	2180	1661	1273	976	764	612	4.53	14.19	1454	31.55	1085
12-#11	1.44	3061	3.54	3061	3061	2972	2600	2267	1743	1375	1066	860	698	4.74	15.26	1446	34.96	1311
12-#14	2.08	3323	3.50	3323	3323	3219	2825	2477	1934	1565	1270	1039	875	5.09	17.75	1442	43.09	1845
12-#18	3.70	3989	3.38	3989	3989	3837	3381	2984	2373	1950	1651	1431	1221	5.77	23.97	1432	62.83	3055
16-#10	1.57	3111	3.55	3111	3111	3023	2651	2313	1784	1420	1112	893	732	4.72	15.52	1465	36.75	1425
16-#11	1.93	3259	3.51	3259	3259	3159	2774	2427	1886	1522	1219	998	828	4.95	16.88	1458	41.20	1711
16-#14	2.78	3608	3.44	3608	3608	3486	3072	2701	2126	1733	1461	1212	1030	5.35	20.04	1459	51.49	2362
16-#18	4.94	4496	3.29	4496	4496	4307	3811	3370	2690	2222	1888	1645	1453	6.09	27.88	1459	76.11	3937
20-#10	1.96	3273	3.50	3273	3273	3172	2789	2446	1905	1537	1236	1010	840	4.90	16.92	1470	41.55	1737
20-#11	2.41	3456	3.45	3456	3456	3341	2941	2587	2027	1645	1359	1125	947	5.15	18.59	1464	46.93	2080
20-#14	3.47	3894	3.37	3894	3894	3747	3309	2924	2319	1901	1607	1373	1175	5.58	22.47	1466	59.80	2898
20-#18	6.17	5003	3.22	5003	5003	4768	4222	3752	3010	2497	2128	1853	1640	6.35	32.12	1467	89.66	4728

M_n/P_n = e (in.)   (ϕ = 0.70)

Example 12.1—Continued

Calculations and Discussion

Code
Reference

$$\varphi P_{ny} = \cfrac{1}{\cfrac{1}{1418} + \cfrac{1}{2564} - \cfrac{1}{4154}} = 1170 \text{ kips}$$

$P_u = 1200 \text{ kips} \approx 1170 \text{ kips}$ say O.K.

Other trial sections can be easily checked in a similar manner.

8. Using PCACOL Program

Using the investigative option of the PCACOL program, check the 36 in. square column with 24#10 bars for the biaxial loading combination, $P_u = 1200$ kips, $M_{ux} = 1800$ ft-kips, and $M_{uy} = 750$ ft-kips.

For biaxial bending strength, PCACOL computes the M_x and M_y moment capacities of the section bending about the centroidal axis rotated $360°$ in $10°$ increments. This method produces exact results. The three-dimensional biaxial bending interaction diagram (see Fig. 12-8) is represented by horizontal slices (load contours) at constant axial load values. The PCACOL results show a biaxial bending diagram looking down the axial load axis at the required axial load. The on-screen graphic results of PCACOL are reproduced in Fig. 12-20. The interaction diagram displayed is a horizontal slice at the constant $P_u = \varphi P_n = 1200$ kips. The factored biaxial moments $M_{ux} = \varphi M_{nx} = 1800$ ft-kips and $M_{uy} = \varphi M_{uy} = 750$ ft-kips are shown as point "1" relative to the moment strength interaction curve for the 36 in. x 36 in. column section with 24#10 bars. Clear cover to longitudinal bars (cc = 1.88 in.) and clear spacing between longitudinal bars (spacing = 3.85 in.) are included in the displayed results.

Example 12.1—Continued

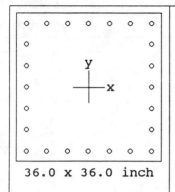

36.0 x 36.0 inch

f'_c = 4.0 ksi

f_y = 60.0 ksi

24#10 2.4%

A_{st} = 30.48 in^2

Tied cc = 1.88 in

spacing = 3.89 in

I_x = 139968 in^4

$\bar{x}$ = 0.00 in

I_y = 139968 in^4

$\bar{y}$ = 0.00 in

© 1989 PCA

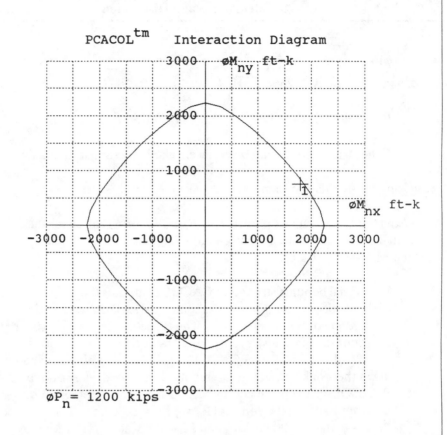

PCACOL^{tm} Interaction Diagram

ϕM_{ny} ft-k

ϕM_{nx} ft-k

ϕP_n = 1200 kips

Project: Example 12.1

Column Id: Square Biaxial Column

Engineer:

Date: Time:

Code: ACI 318-89

Version: 2.10

File name: C:\PCACOL\DATA\EX12_1.COL

Material Properties:

E_c = 3834 ksi ϵ_u = 0.003 in/in

f_c = 3.40 ksi E_s = 29000 ksi

β_1 = 0.85

Stress Profile : Block

Reduction: ϕ_c = 0.70 ϕ_b = 0.90

Slenderness not considered x-axis

Slenderness not considered y-axis

Fig. 12-20   PCACOL On-Screen Graphic Results for
36x36 Column with 24 #10 Bars

Example 12.2—Design of a Rectangular Column for Biaxial Loading

Determine the required rectangular column size and reinforcement for the load and moments given. Assume h/b = 1.5, and that the reinforcement is equally distributed on all faces.

P_u = 1700 kips
M_{ux} = 2900 ft-kips
M_{uy} = 1200 ft-kips
f_c' = 4000 psi
f_y = 60,000 ksi

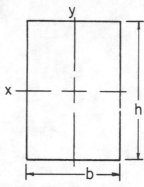

h/b = 1.5

	Code
Calculations and Discussion	Reference

1. Determine required nominal strengths. For tied reinforced column, φ = 0.70   9.3.2.2(b)

$$P_n = \frac{P_u}{\varphi} = \frac{1700}{0.7} = 2429 \text{ kips}$$
$$M_{nx} = \frac{M_{ux}}{\varphi} = \frac{2900}{0.7} = 4143 \text{ ft}-\text{kips}$$
$$M_{ny} = \frac{M_{uy}}{\varphi} = \frac{1200}{0.7} = 1714 \text{ ft}-\text{kips}$$

2. Assume β = 0.65

3. Determine an equivalent uniaxial moment strength M_{nox} or M_{noy}.

$$\frac{M_{ny}}{M_{nx}} = \frac{1714}{4143} = 0.41 \text{ is less than } \frac{b}{h} = 0.67$$

Therefore, using Eq. (14),

$$M_{nox} = M_{ny}\frac{h}{b}\left(\frac{1-\beta}{\beta}\right) + M_{nx}$$

$$= 1714\,(1.5)\left(\frac{1-0.65}{0.65}\right) + 4143 = 5527 \text{ ft}-\text{kips}$$

4. Assuming b = 32 in. and h = 48 in., determine reinforcement required to provide an axial load strength P_n = 2429 kips and an equivalent uniaxial moment strength M_{nox} = 5527 ft kips.

5 <u>Using Design Aid SP17A</u> (See Part 11, page 11-4)

For rectangular section with bars uniformly distributed on all sides, f_c' = 4000 psi, f_y = 60,000 psi, and $\gamma \approx 0.90$, use column chart 7.4.4 for R4 - 60.90 columns (see Fig. 12-19).

Note that the strength reduction factor φ is already included in the strength curves.

Example 12.2—Continued

Compute $\dfrac{\varphi P_n}{A_g} = \dfrac{0.7\,(2429)}{32 \times 48} = 1.11$

Compute $\dfrac{\varphi M_{nox}}{A_g\,h} = \dfrac{0.7\,(5527)\,12}{(32 \times 48)\,48} = 0.63$

From Fig. 12-19, read $\rho_g \approx 0.027$

$A_{st} = \rho_g\,A_g = 0.027\,(32 \times 48) = 41.47$ in.^{2}

Select 30 #11 bars ($A_s = 46.8$ in.^{2}), ρ_g (provided) $= 46.8/32 \times 48 = 0.030$

6. Selected section will now be checked for biaxial loading strength by the first and third methods previously described. The Bresler load contour method will not be used due to the conservatism involved in the procedure, as noted in Example 12.1.

   A. Bresler Reciprocal Load Method

   Check $P_n \geq 0.1\,f'_c\,A_g$ (Eq. 2)

   2429 kips $\geq 0.1\,(4)\,(32 \times 48) = 614$ kips

   To employ this method, P_o, P_{ox}, and P_{oy} must be determined.

   $$P_o = 0.85\,f'_c\,(A_g - A_{st}) + A_{st}\,f_y$$

   $$= 0.85\,(4)\,(1536 - 46.8) + 46.8\,(60) = 7871 \text{ kips}$$

   Knowing ρ_g and the required nominal load-moment strengths (P_n, M_{nx}, and M_{ny}), P_{ox} and P_{oy} can be determined.

   <u>x-axis</u>

   $P_n = 2429$ kips, $M_{nx} = 4143$ ft-kips

   $e_y = \dfrac{M_{nx}}{P_n} = \dfrac{4143\,(12)}{2429} = 20.47$ in., $e_y/h = \dfrac{20.47}{48} = 0.43$

   From Fig. 12-19, with $e_y/h = 0.43$ and $\rho_g = 0.030$, read $\varphi P_{ox}/A_g = 1.50$

   Therefore, $P_{ox} = \dfrac{1.50\,(32 \times 48)}{0.70} = 3291$ kips

Example 12.2—Continued

Calculations and Discussion

Code
Reference

y-axis

P_n = 2429 kips, M_{ny} = 1714 ft-kips

$e_x = \dfrac{M_{ny}}{P_n} = \dfrac{1714\,(12)}{2429} = 8.47$ in., $\dfrac{e_x}{b} = \dfrac{8.47}{32} = 0.26$

From Fig. 12-19, with $e_x/b = 0.26$ and $\rho_g = 0.030$, read $\varphi P_{oy}/A_g = 2.05$

Therefore, $P_{oy} = \dfrac{2.05\,(32 \times 48)}{0.70} = 4498$ kips

Using the above values, Eq. (1) can now be evaluated.

$P_n = 2429$ kips $\leq \dfrac{1}{\dfrac{1}{P_{ox}} + \dfrac{1}{P_{oy}} - \dfrac{1}{P_o}}$

Eq. (1)

$= \dfrac{1}{\dfrac{1}{3291} + \dfrac{1}{4498} - \dfrac{1}{7871}} = 2505$ kips,   O.K.

B. PCA Load Contour Method

To employ this method, P_o, M_{nox}, M_{noy}, and the true value of β must be found.

$P_o = 0.85\,f_c'\,(A_g - A_{st}) + A_{st}\,f_y$

$= 0.85(4)(1536 - 46.8) + 46.8(60) = 7871$ kips

Knowing ρ_g and the required axial load strength P_n, M_{nox} and M_{noy} are determined as follows:

$P_n = 2429$ kips, $\dfrac{\varphi P_n}{A_g} = \dfrac{0.7\,(2429)}{32 \times 48} = 1.11$

From Fig. 12-19, with $\dfrac{\varphi P_n}{A_g} = 1.11$ and $\rho_g = 0.030$, read $\dfrac{\varphi M_n}{A_g\,h} = 0.68$

Therefore, $M_{nox} = \dfrac{0.68\,A_g\,h}{\varphi} = \dfrac{0.68\,(32 \times 48)\,48}{12(0.70)} = 5968$ ft-kips

Example 12.2—Continued

Calculations and Discussion

$$M_{noy} = \frac{0.68\, A_g\, b}{\varphi} = \frac{0.68\,(32 \times 48)\,32}{12(0.70)} = 3979 \text{ ft−kips}$$

Having found P_o and using ρ_g, the true value of β is determined as follows:

$$\frac{P_n}{P_o} = \frac{2429}{7871} = 0.309$$

$$\omega = \frac{\rho_g\, f_y}{f_c'} = \frac{0.030\,(60)}{4} = 0.45$$

From Fig. 12-16, read $\beta = 0.57$

Using the above values, Eq. (8) can now be evaluated.

$$\left(\frac{M_{nx}}{M_{nox}}\right)^{\left(\frac{\log 0.5}{\log \beta}\right)} + \left(\frac{M_{ny}}{M_{noy}}\right)^{\left(\frac{\log 0.5}{\log \beta}\right)} \le 1.0$$

$$\log 0.5 = -0.3$$
$$\log \beta = \log 0.57 = -0.244$$
$$\frac{\log 0.5}{\log \beta} = 1.23$$

$$\left(\frac{4143}{5968}\right)^{1.23} + \left(\frac{1714}{3979}\right)^{1.23} = 0.638 + 0.355 = 0.993 \ < \ 1.0 \qquad \text{O.K.}$$

The section can also be checked using the bilinear approximation.
Since $\dfrac{M_{ny}}{M_{nx}} < \dfrac{M_{noy}}{M_{nox}}$, Eq. (12) should be used.

$$\left(\frac{M_{ny}}{M_{noy}}\right)\frac{1-\beta}{\beta} + \frac{M_{nx}}{M_{nox}} = \frac{1714}{3979}\left(\frac{1-0.57}{0.57}\right) + \frac{4143}{5968} = 0.325 + 0.694 \qquad \text{Eq. (12)}$$

$$= 1.019 \approx 1.0 \qquad \text{O.K.}$$

7. Using CRSI Handbook

(a) Required
$P_u = 1700$ kips
$M_u = 5527\,(0.70) = 3869$ ft−kips   (p. 12−36)
$e = \dfrac{M_u}{P_u} = \dfrac{3869 \times 12}{1700} = 27.3$ in.

Example 12.2—Continued

	Code
Calculations and Discussion	Reference

Size of section and amount of reinforcement required to support the above factored load and moment is beyond the scope of the CRSI Load Capacity Tables. For such unusually large sections, use of nondimensional design aids such as SP17A, or use of a computer program solution is required.

8. <u>Using PCACOL Program</u>

Using the investigative option of the PCACOL program, investigate the 32 in. x 48 in. column section with 30#11 bars for the biaxial loading combination, P_u = 1700 kips, M_{ux} = 2900 ft−kips, and M_{uy} = 1200 ft−kips.

The on-screen graphic results of PCACOL are reproduced in Fig. 12-21. For biaxial bending, the interaction diagram displayed is a horizontal slice at the constant $P_u = \varphi P_n$ = 1700 kips. The factored biaxial moments, $M_{ux} = \varphi M_{nx}$ = 2900 ft-kips and $M_{uy} = \varphi M_{ny}$ = 1200 ft-kips are shown as point "1" relative to the biaxial moment strength interaction curve for the 32 in. x 48 in. column section reinforced with 30#11 bars.

Example 12.2—Continued

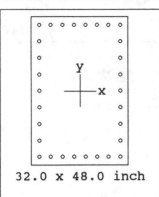

32.0 x 48.0 inch

f'_c = 4.0 ksi
f_y = 60.0 ksi
   30#11 3.0%
A_{st} = 46.80 in^2
Tied cc = 2.00 in
 spacing = 2.39 in
I_x = 294912 in^4
$\overline{x}$ = 0.00 in
I_y = 131072 in^4
$\overline{y}$ = 0.00 in

© 1989 PCA

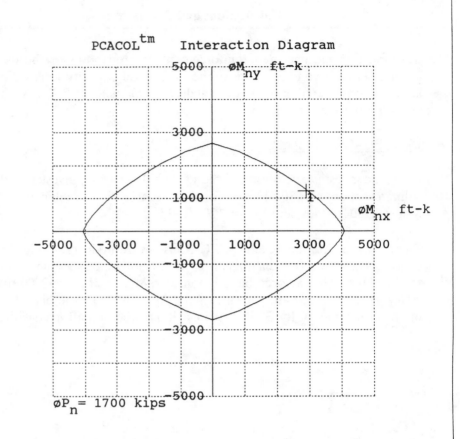

PCACOL^{tm} Interaction Diagram

$\emptyset P_n$ = 1700 kips

Project: Example 12.2   File name: C:\PCACOL\DATA\EX12_2.COL

Column Id: Rect. Biaxial Column   Material Properties:

Engineer: E_c = 3834 ksi $\in_u$ = 0.003 in/in

Date: Time: f_c = 3.40 ksi E_s = 29000 ksi

Code: ACI 318-89 β_1 = 0.85

Version: 2.10   Stress Profile : Block

                                 Reduction: $\emptyset_c$ = 0.70 $\emptyset_b$ = 0.90

Slenderness not considered x-axis

Slenderness not considered y-axis

**Fig. 12-21 PCACOL On-Screen Graphic Results
for 32 in. x 48 in. Column with 30#11 Bars**

13

Design for Slenderness Effects

UPDATE FOR THE '89 CODE

The moment magnifier method for consideration of column slenderness effects was first introduced in the 1971 edition of the ACI Code. The design provisions remained essentially the same in the 1977 Code edition. With the publication of ACI 318-83, application of the moment magnification procedure was clarified. Equation (10-6) was revised to express the column secondary moments as the sum of (1) the magnification of non-sway moments (gravity load effects) plus (2) the magnification of sway moments (lateral load effects). Application of the revised Eq. (10-6) to unbraced frames generally results in substantially lessened column design moments as compared with values given by the previous code edition, since gravity load moments are magnified by a braced frame magnification factor, with resulting economy in both column and floor system designs. For an in-depth discussion addressing Eq. (10-6), the reader is referred to Commentary Section 10.11.5.1.

For ACI 318-89, the definition of the creep effect factor β_d, used in Eqs. (10-10) and (10-11) to reduce the effective value of the stiffness parameter EI, has been modified to make it a function of a sustained load ratio rather than a sustained moment ratio. The previous Code definition of β_d, as a ratio of dead to total moments, caused confusion and misinterpretation as to the selection of a proper moment ratio, particularly when calculated moments were small or zero. In the case of braced columns, the ratio of sustained-to-total vertical loads is a more important consideration in evaluating the effect of creep in reducing the effective member stiffness, whereas in unbraced columns, the ratio of the sustained-to-total lateral load effects is the more important consideration. See discussion on Section 10.11.5.2.

GENERAL CONSIDERATIONS

Design of columns consists essentially of selecting an adequate column cross section with reinforcement to support a required combination of factored axial load P_u and factored (primary) moment M_u, including consideration of column slenderness (secondary moments) in determining the design strength of the selected column. Column slenderness is expressed in terms of its slenderness ratio $k\ell_u/r$, where k is an effective length factor (dependent on rotational and lateral restraints at the ends of the column), ℓ_n is the unsupported column length, and r is the radius of gyration of the column cross section. A column is said to be slender if its cross sectional dimensions are small in comparison to its length.

For design purposes, the term "short column" is used to denote a column that has a strength equal to that computed for its cross section, using the forces and moments obtained from an analysis for combined bending and axial load. A "slender column" is defined as a column whose strength is reduced by second order deformations (secondary moments). By these definitions, a column with a given slenderness ratio may be considered a short column for design under one condition of restraints, and a long column under another combination of restraints. With the use of higher strength concrete and reinforcement, and with more accurate design methods, it is possible to design much smaller cross sections than earlier, resulting in slenderer members. The need for reliable and rational design procedures for slender columns thus becomes a more important consideration in column design.

A short column may fail due to a combination of moment and axial load that exceeds the strength of the cross section. This type of a failure is known as "material failure." As an illustration, consider the column shown in Fig. 13-1. Due to loading, the column has a deflection Δ which will cause an additional (secondary) moment in the column. In the free body diagram, it can be seen that the maximum moment in the column occurs at section A-A, equal to the applied moment plus the moment due to member deflection, that is $M = P(e + \Delta)$. Failure of a short column can occur at any point along the strength interaction curve, depending on the combination of applied moment and axial load. As discussed above, some deflection will occur and a "material failure" will result when a particular combination of load P and moment $M = P(e + \Delta)$ intersects the strength interaction curve. If the column is very slender, it may reach a deflection due to axial load P and a moment Pe, such that, deflections will increase indefinitely with increase in load P. This type of failure is known as a "stability failure," as shown on the strength interaction curve.

The basic concept on the behavior of straight, concentrically loaded, slender columns was originally developed by Euler more than 200 years ago. It states that a member will fail by buckling at the critical load $P_c = \pi^2 EI/(\ell_e)^2$, where EI is the flexural stiffness of the member cross section, and ℓ_e is the effective length, equal to $k\ell_u$. For a "stocky" short column, the value of the buckling load will exceed the direct crushing

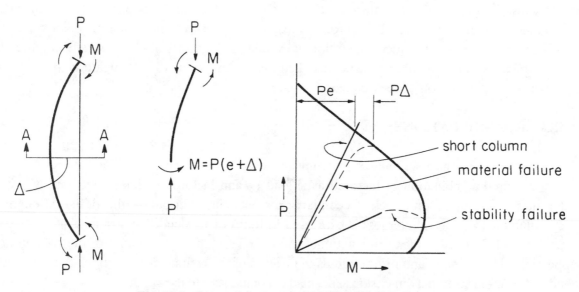

Fig. 13-1 Strength Interaction for Slender Columns

13-2

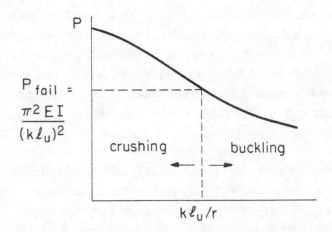

Fig. 13-2   Failure Load as a Function of Column Slenderness

strength (corresponding to material failure). In slenderer members (with larger $k\ell_u/r$ values), failure may occur by buckling (stability failure), with the buckling load decreasing with increasing member slenderness (see Fig. 13-2).

As shown above, it is possible to depict slenderness effects and amplified moments on a typical strength interaction curve. Hence, a "family" of strength interaction diagrams for slender columns with varying slenderness ratios can be developed, as shown in Fig. 13-3. The strength interaction diagram for $k\ell_u/r = 0$ corresponds to the combinations of moment and axial load where strength is not affected by member slenderness (short column strength).

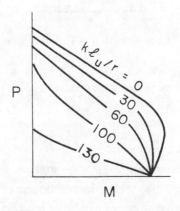

Fig. 13-3   Strength Interaction Diagrams for Slender Columns

13-3

CONSIDERATION OF SLENDERNESS EFFECTS

Three ranges of slenderness ratios are defined in Section 10.11.4. Slenderness limits are prescribed for both braced and unbraced frames, including design methods permitted for each slenderness range. Lower-bound slenderness limits are given, below which secondary moments may be disregarded and only axial load and primary moment need be considered to select a column cross section and reinforcement (short column design). It should be noted that for ordinary beam-column sizes and story heights of concrete framing systems, effects of slenderness may be neglected for more than 90 percent of columns in braced frames and around 40 percent of columns in unbraced frames. For moderate slenderness ratios, an approximate analysis of slenderness effects based on a moment magnifier (Section 10.11) is permitted. For columns with high slenderness ratios, a more exact second-order analysis is required (Section 10.10.1), taking into account the influence of axial loads and variable moment of inertia on column stiffness and forces, and the effects of load duration. No upper limits for column slenderness are prescribed. The slenderness ratio limits as prescribed in Section 10.11.4, and design methods permitted for consideration of column slenderness, are summarized in Fig. 13-4. See also discussion on Section 10.11.4.

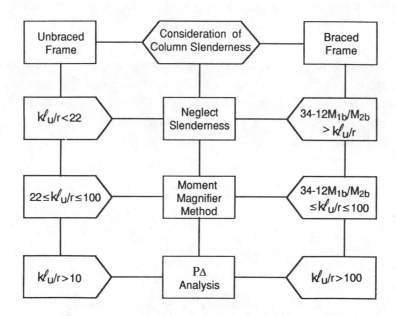

Fig. 13-4 Consideration of Column Slenderness

10.10 SLENDERNESS EFFECTS IN COMPRESSION MEMBERS

10.10.1 Second-Order Frame Analysis

The Code encourages the use of a second-order frame analysis or $P-\Delta$ analysis for consideration of slenderness effects in compression members, taking into account effects of sway deflections on the axial loads and moments in the frame. Extensive studies have been made over the past two or three decades, and it is now feasible for a designer to use second-order analyses in the design of reinforced concrete buildings. With increasing use of computers, and the greater availability of more sophisticated computer programs for column design, more exact analyses are becoming not only possible but also practical. Generally, the moments from a second-order analysis provide a better approximation of the real moments than those obtained from an analysis by Section 10.11. For sway frames or lightly braced frames, economies can be achieved by the use of second-order analyses. Procedures for carrying out a second-order analysis are given in Commentary Refs. 10.21, 10.22, and 10.23. The reader is referred to Commentary Section 10.10.1, which discusses minimum requirements for an adequate second-order analysis under Section 10.10.1.

If, however, such more exact analyses are not feasible or practical for a particular design, Section 10.10.2 permits an approximate moment-magnifier method to account for column slenderness in design. Note, however, that for all compression members with a column slenderness ratio ($k\ell_u/r$) greater than 100 (See Fig. 13-4), a more-exact analysis as defined in Section 10.10.1 must be used for consideration of slenderness effects.

10.11 APPROXIMATE EVALUATION OF SLENDERNESS EFFECTS

The approximate moment magnifier method prescribed in Section 10.11 is similar in concept to the method used for structural steel design; a moment magnification factor δ is used to magnify the primary moments to account for increased moments due to member curvature and lateral drift. The moment magnifier δ is a function of the ratio of the applied axial load to the critical or buckling load of the column, the ratio of the applied moments at the ends of the column, and the deflected shape of the column.

10.11.1-10.11.2 Unsupported and Effective Lengths of Compression Members

Unsupported length ℓ_u of a column is defined in Section 10.11.1. The unsupported length is to be taken as the clear distance between lateral supports as shown in Fig. 13-5. Note that the length ℓ_u may be different for buckling about each of the principal axes of the column cross section.

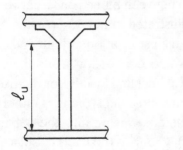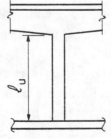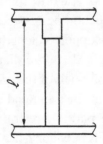

Fig. 13-5 Unsupported Length (ℓ_u)

The basic Euler equation for critical buckling load can be expressed as $P_c = \pi^2 EI/(\ell_e)^2$, where ℓ_e is the effective length $k\ell_u$. In most cases, the designer can determine the effective length $k\ell_u$ of a column by analyzing the moments acting on the column by means of a free body diagram. The basic equations for the

13-5

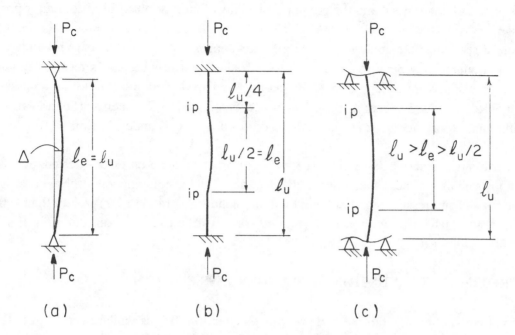

Fig. 13-6   Effective Length l_e (Members Braced Against Sidesway)

design of slender columns were derived for hinged ends, and thus, must be modified to account for the effects of end restraint. Effective column length kl_u, as contrasted to actual unbraced length l_u, is the term used in estimating slender column strength, and considers end restraint as well as bracing against sidesway.

At the critical load defined by the Euler equation, an originally straight member buckles into a half sine wave as shown in Fig. 13-6(a). In this configuration, an additional moment PΔ acts at every section, where Δ is the lateral deflection at the specific location under consideration along the length of the member. This deflection continues to increase until the bending stress caused by the increasing moment (PΔ), plus the original compression stress caused by the applied loading, exceeds the compressive strength of concrete and the member fails. The effective length l_e ($=$kl_u) is the length between pinned ends, between zero moments or between inflection points. For the pin-ended condition illustrated in Fig. 13-6(a), the effective length is equal to the unsupported length l_u. If the member is fixed against rotation at both ends, as shown in Fig. 13-6(b), it will buckle in the shape shown; inflection points will occur as shown, and the effective length l_e ($=$kl_u) will be one-half of the unsupported length. The critical buckling load P_c for the fixed-end condition is four times that for a pin-end condition. Rarely are columns in actual structures either hinged or fixed, rather they are partially restrained against rotation by members framing into the column, and thus the effective length is between $l_u/2$ and l_u as shown in Fig. 13-6(c) as long as the lateral displacement of one end of the column with respect to the other end is prevented. The actual value of effective length will depend on the rigidity of the members framing into the top and bottom ends of the column.

A column that is fixed at one end and entirely free at the other end (cantilever) will buckle as shown in Fig. 13-7(a). The upper end will deflect laterally relative to the lower end; this is known as sidesway. The deflected shape of such a member is similar to one-half of the sinusoidal deflected shape of the pin-ended

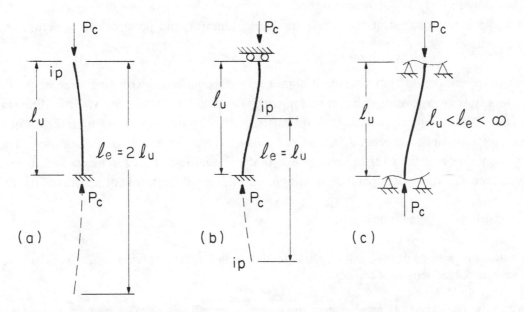

Fig. 13-7   Effective Length ℓ_e (Members Not Braced Against Sidesway)

member illustrated in Fig. 13-6(a), with an effective length equal to twice the actual length. If the column is fixed against rotation at both ends but one end can move laterally with respect to the other, it will buckle as shown in Fig. 13-7(b). The effective length ℓ_e will be equal to the actual length ℓ_u, with an inflection point (ip) occurring as shown. The buckling load of the column in Fig. 13-7(b), where sidesway is not prevented, is one-quarter that of the column in Fig. 13-6(b), where sidesway is prevented. Again, rarely are the ends of columns either completely hinged or completely fixed, but rather they are partially restrained against rotation by members framing into the ends of the columns, and thus the effective length will vary between ℓ_u and ∞, as shown in Fig. 13-7(c). If restraining members (beams or slab) are very rigid as compared to the

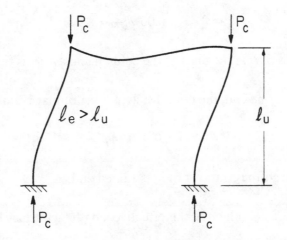

Fig. 13-8   Rigid Frame   (Not Braced Against Sidesway)

column, the buckling in Fig. 13-7(b) is approached. If, however, the restraining members are quite flexible, a hinged condition is approached at both ends and the column(s), and possibly the structure as a whole approaches instability.

In typical reinforced concrete structures, the designer rarely is concerned with single members, but rather with rigid framing systems consisting of beam-column and slab-column assemblies. The buckling behavior of a frame that is not braced against sidesway can be illustrated by the simple portal frame shown in Fig. 13-8. Without lateral restraint at the upper end, the entire (unbraced) frame is free to move sideways. The bottom end may be pin-ended or partially restrained against rotation as indicated. It can be seen that the effective length l_e may exceed $2 l_u$, and will depend on the degree of rotational restraint at the ends of the columns.

In summary, the following comments can be made:

1. For columns braced against sidesway, the effective length l_e falls between $l_u/2$ and l_u, where l_u is the actual unsupported length of the column.

2. For columns not braced against sidesway, the effective length l_e is always longer than the actual length of the column l_u, and may be $2 l_u$ and higher. A value of l_e or $k l_u$ less than 1.2 for columns not braced against sidesway normally would not be realistic.

3. Use of the alignment charts shown in Figs. 13-9 and 13-10, also given in Commentary Fig. 10.11.12 allows graphical determination of the effective length factors for braced and unbraced frames, respectively. If both ends of a braced member have minimal rotational stiffness, or approach $\psi = \infty$, then $k = 1.0$. If both ends have or approach full fixity, $\psi = 0$, then $k = 0.5$. If both ends of an unbraced member have minimal rational stiffness, or approach $\psi = \infty$, then $k = \infty$. If both ends have or approach full fixity, $\psi = 0$, then $k = 1.0$.

An alternative method for computing the effective length factors for braced and unbraced members is given in Commentary Section 10.11.2. For braced columns, an upper bound to the effective length factor may be taken as the smaller of the values given by the following two expressions:

$$k = 0.7 + 0.05 \, (\psi_A + \psi_B) \leq 1.0$$

$$k = 0.85 + 0.05 \, \psi_{(min)} \leq 1.0$$

where ψ_A and ψ_B are the values of ψ at the ends of the column and $\psi_{(min)}$ is the smaller of the two values.

For unbraced columns hinged at one end, the effective length factor may be taken as:

$$k = 2.0 + 0.34\psi$$

where ψ is the column-to-beam stiffness ratio at the restrained end.

For unbraced columns restrained at both ends, the effective length factor may be taken as:

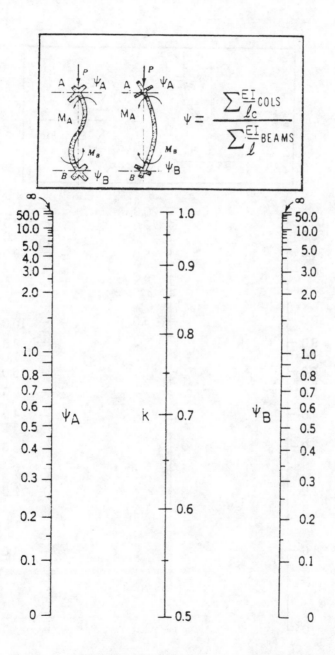

Fig. 13-9 Effective Length Factor (k) for Braced Columns

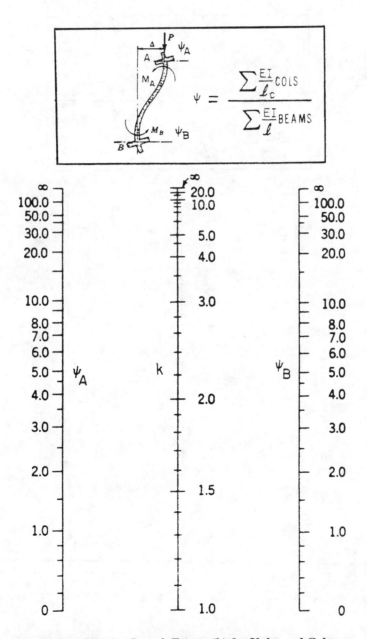

Fig. 13-10   Effective Length Factor (k) for Unbraced Columns

$$k = \frac{20-\Psi_m}{20} \sqrt{1+\Psi_m} \text{ for } \Psi_m < 2$$

or
$$k = 0.9 \sqrt{1 + \Psi_m} \text{ for } \Psi_m > = 2$$

where ψ_m is the average of the ψ values at the two ends of the column.

In determining the effective length factor k from Figs. 13-9 and 13-10, or from the Commentary equations, the rigidity (EI) of the beams (or slabs) may be calculated on the basis of the moment of inertia of the cracked transformed section and the rigidity of the columns using Eq. (10-10) with $\beta_d = 0$. Alternately, using a value of $0.5I_g$ for the beams (to account for effects of cracking and reinforcement on relative stiffness) and I_g for the columns, when computing ψ, will usually result in reasonable member sizes for columns with $k \ell_u/r$ less than about 60.

In an actual structure, however, there is rarely a completely braced or a completely unbraced condition. For the purposes of applying Sections 10.11.2.1 and 10.11.2.2, a column braced against sidesway is a member within a story in which horizontal displacements do not significantly affect the moments in the structure. When the stability index for a story, defined as $Q = (\Sigma P_u \Delta_u / H_u h_s)$, is not greater than 0.04, the $P\Delta$ secondary moments should not exceed 5 percent of the primary moments and the structure can be considered as braced. Note that H_u is the total factored shear force acting within the story and Δ_u is the elastically-computed (first-order) lateral deflection due to H_u (neglecting P-Δ effects) at the top of the story, relative to the bottom of the story. In many cases, the engineer will be able to determine by inspection whether a story can be considered as braced or unbraced.

Alternatively, a more approximate procedure can be used to determine if a story is braced or unbraced. A column may be assumed braced if located in a story in which the bracing elements (shearwalls, or other lateral bracing elements) have a total stiffness, resisting lateral movement of the story, at least six (6) times the sum of the stiffnesses of all columns within the story. With this amount of lateral stiffness, lateral deflections of the story will not be large enough to affect the column strength significantly. What constitutes adequate bracing in a given case (layout and arrangement of the structural framing members) must be left to the judgment of the engineer.

10.11.3 Radius of Gyration

Radius of gyration may be taken as 0.3 of the overall dimension of a rectangular section and 0.25 of the diameter of a circular section, as shown in Fig. 13-11. For other shapes, the radius of gyration must be computed.

10.11.4 Consideration of Slenderness Effects

For columns braced against sidesway, effects of slenderness may be neglected when $k\ell_u/r$ is less than $34 - 12$ M_{1b}/M_{2b}, where M_{2b} is the larger end moment on the braced column, obtained by elastic frame analysis, and M_{1b} is the smaller end moment. The ratio M_{1b}/M_{2b} is positive if the column is bent in single curvature,

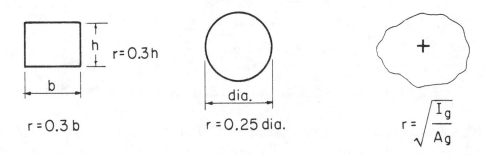

Fig. 13-11   Radius of Gyration (r)

negative if bent in double curvature. Note that M_{1b} and M_{2b} are factored end moments. For columns not braced against sidesway, effects of slenderness may be neglected when $k\ell_u/r$ is less than 22. The moment magnifier method may be used for columns with slenderness ratios exceeding these lower limits.

The upper slenderness limit for columns that may be designed by the approximate moment magnifier method is $k\ell_u/r$ equal to 100. When $k\ell_u/r$ is greater than 100, an analysis as defined in Section 10.10.1 must be used, taking into account the influence of axial loads and variable moment of inertia on member stiffness and fixed-end moments, the effect of deflections on the moments and forces, and the effects of duration of loading (sustained load effects). Consideration of column slenderness is summarized in Fig. 13-4.

The lower slenderness ratio limits will allow a large number of columns to be exempt from slenderness consideration. Considering the slenderness ratio $k\ell_u/r$ in terms of ℓ_u/h for rectangular columns, the effects of slenderness may be neglected in design when ℓ_u/h is less than 10 for columns braced against sidesway and with zero restraint at both ends. This lower limit increases to 18 for a braced column in double curvature with equal end moments and a column-to-beam stiffness ratio equal to one at each end. In the case of most braced frames, it is sufficiently accurate to use estimated values of the effective length factor k for preliminary slenderness evaluation. For columns with minimal or zero restraint at both ends, k of 1.0 should be used. For stocky columns restrained by flat slab floors, k ranges from about 0.95 to 1.0 and can be conservatively estimated as 1.0. For columns in beam-column frames, k ranges from about 0.75 to 0.9, and can be conservatively estimated as 0.90. If the initial computation of slenderness ratio based on estimated values of k indicates that effects of slenderness must be considered in the design, a more accurate value of k should be calculated and slenderness re-evaluated. For an unbraced column with a column-to-beam stiffness ratio equal to one at both ends, effects of slenderness may be neglected when ℓ_u/h is less than 5. This value reduces to 3 if the beam stiffness is reduced to one-fifth of the column stiffness at each end of the column. Thus, beam stiffnesses at the top and bottom of a column of a high-rise structure where sidesway is not prevented by structural walls or other means will have a significant effect on the degree of slenderness of the column.

The upper limit on the slenderness ratio of $k\ell_u/r$ equal to 100 corresponds to an ℓ_u/h equal to 30 for a column braced against sidesway with zero restraint at both ends. This ℓ_u/h limit increases to 39 for a column-to-beam stiffness ratio equal to one at each end.

10.11.5 Moment Magnification

The approximate slender column design equations of Section 10.11.5 are based on the concept of a moment-magnifier δ which amplifies the column moments to account for the effect of the axial load on the column moments. The column is then designed for the axial load and the magnified moment. In application, δ is a function of the ratio of the axial load on the column to the critical "buckling" load of the column (P_c), and the ratio of end moments and the deflected shape of the column (C_m). Code provisions prescribe different methods of calculating δ for braced and unbraced conditions. For a braced column, the magnifier δ is based on an effective length factor of 1.0 or less. For an unbraced column, the magnifier δ is based on an effective length factor greater than 1.0. C_m factors for braced columns range from 0.4 to 1.0; for unbraced columns, a value of 1.0 is used. Thus, critical loads P_c are higher and the magnification factors are smaller for braced columns.

In ACI 318-83, Eq. (10-6) was revised to express the column secondary moments as the sum of (1) the magnification of essentially nonsway moments (gravity load effects) plus (2) the magnification of sway moments (lateral load effects):

$$M_c = \delta_b M_{2b} + \delta_s M_{2s} \qquad \text{Eq. (10-6)}$$

The first term is the magnified gravity load moment, where δ_b is a braced frame magnifier:

$$\delta_b = \frac{C_m}{1 - P_u/\varphi P_{cb}} \geq 1.0 \qquad \text{Eq. (10-7)}$$

$$P_{cb} = \frac{\pi^2 EI}{(k_b \ell_u)^2} \qquad \text{Eq. (10-9)}$$

and
$$C_m = 0.6 + 0.4(M_{1b}/M_{2b}) \geq 0.4 \qquad \text{Eq. (10-12)}$$

The critical load P_{cb} is computed for a braced condition using an effective length factor k_b of 1.0 or less according to Section 10.11.2.1.

The second term of Eq. (10-6) is the magnified lateral load moment, where δ_s is a sway magnifier:

$$\delta_s = \frac{1}{1 - \Sigma P_u/\varphi\Sigma P_{cs}} \geq 1.0 \qquad \text{Eq. (10-8)}$$

$$P_{cs} = \frac{\pi^2 EI}{(k_s \ell_u)^2} \qquad \text{Eq. (10-9)}$$

The critical load P_{cs} is computed for an unbraced condition using an effective length factor k_s greater than 1.0 according to Section 10.11.2.2. ΣP_u and ΣP_{cs} are the sums of factored axial loads and critical loads, respectively, for all columns within a story.

When Eq. (10-6) is used for design of the columns of a moment resisting frame (beam or slab and column framing) that is not braced by structural walls or other bracing elements, both moment magnifiers must be evaluated. M_{2b} is the larger end moment due to gravity loads (dead and live loads). M_{2s} is the larger end moment due to lateral loads (wind or earthquake forces). Both moments, M_{2b} and M_{2s}, are computed using a conventional (first-order) frame analysis. The member stiffnesses used in the analysis for lateral load effects should allow, at least approximately, for cracking of the flexural members (beams and slabs).

When Eq. (10-6) is used for design of the columns of a moment resisting frame that is effectively braced against sidesway by shearwalls or other bracing elements, the δ_s term is taken equal to 1.0, since the drift of a braced frame is small under lateral loads. When shearwalls are tall and slender (flexible), the δ_s term may need to be evaluated, as the moment resisting frame may not be effectively braced by the shearwalls. Designer's judgment combined with appropriate analysis are essential elements in determining the effectiveness of such "braced" framing systems.

To illustrate proper application of Eq. (10-6) for slender column design, the following summary of design equations, and how they are applied in combination with Eq. (10-6), may be helpful. Examples 13.1 and 13.2 illustrate application of Eq. (10-6) for the design of columns in a braced and an unbraced frame, respectively.

For the general case of an unbraced frame resisting gravity loads (dead and live loads) plus wind loads:

(1) The factored load combinations to be considered are:

   (a) Gravity loads
   $P_u = 1.4D + 1.7L$ Eq. (9-1)
   $M_{1b} = 1.4D + 1.7L$ (smaller end moment)
   $M_{2b} = 1.4D + 1.7L$ (larger end moment)

   (b) Gravity plus wind loads
   $P_u = 0.75(1.4D + 1.7L + 1.7W)$ Eq. (9-2)
   $M_{2b} = 0.75(1.4D + 1.7L)$
   $M_{2s} = 0.75(1.7W)$

   or $P_u = 0.9D + 1.3W$ Eq. (9-3)
   $M_{2b} = 0.9D$
   $M_{2s} = 1.3W$

(2) For the gravity load combination:

   $$M_c = \delta_b M_{2b}$$ Eq. (10-6)

   where $M_{2b} = 1.4D + 1.7L$
   but not less than $P_u(0.6 + 0.03h)$ (Section 10.11.5.4)

$$\delta_b = \frac{C_m}{1 - P_u / \varphi P_{cb}} \geq 1.0 \qquad\qquad\text{Eq. (10-7)}$$

with $C_m = 0.6 + 0.4(M_{1b}/M_{2b}) \geq 0.4$ Eq. (10-12)

and $P_{cb} = \dfrac{\pi^2 EI}{(k_b \ell_u)^2}$ Eq. (10-9)

k_b is evaluated as for a braced condition according to Section 10.11.2.1. In the calculation of column stiffness EI by Eq. (10-10) or (10-11), $\beta_d = 1.4D/(1.4D + 1.7L)$, where D and L are dead and live load, respectively.

(3) For gravity plus wind load combinations:

$$M_c = \delta_b M_{2b} + \delta_s M_{2s} \qquad\qquad\text{Eq. (10-6)}$$

where $M_{2b} = 0.75(1.4D + 1.7L)$

$$\delta_b = \frac{C_m}{1 - P_u / \varphi P_{cb}} \geq 1.0 \qquad\qquad\text{Eq. (10-7)}$$

with $P_{cb} = \dfrac{\pi^2 EI}{(k_b \ell_u)^2}$ Eq (10-9)

C_m, P_{cb}, k_b, and β_d are the same as for the gravity load combination

$M_{2s} = 0.75(1.7W)$
but not less than $P_u(0.6 + 0.03h)$ 10.11.5.5

$$\delta_s = \frac{1.0}{1 - \Sigma P_u / \varphi \Sigma P_{cs}} \qquad\qquad\text{Eq. (10-8)}$$

with $P_{cs} = \dfrac{\pi^2 EI}{(k_s \ell_u)^2}$ Eq. (10-9)

k_s is evaluated as for an unbraced condition according to Section 10.11.2.2. ΣP_u and ΣP_{cs} are the sums of factored axial loads and critical loads, respectively, for all columns within the story under consideration. Note that in the calculation of column stiffness EI by Eq. (10-10) or (10-11), if the sway moment M_{2s} is entirely caused by wind (nonsustained load), $\beta_d = 0$.

The same general procedure is applied for the load combination given by Eq. (9-3), using the appropriate P_u, M_{2b}, and M_{2s} values.

(4) Final selection of column size and reinforcement is based on the most severe of the three load combinations considered, Eq. (9-1), Eq. (9-2) or Eq. (9-3).

(5) For a frame effectively braced by shearwalls, where frame-shearwall interaction is considered in the lateral load analysis, design of the frame columns is somewhat simplified by taking δ_s equal to 1.0 in the gravity plus wind load combination. For this condition, the M_{2s} column moments are directly additive to the magnified gravity load moments. Thus, Eq. (10-6) reduces to:

$$M_c = \delta_b M_{2b} + M_{2s} \qquad \text{Eq. (10-6)}$$

(6) For a braced frame with columns resisting gravity loads only (lateral load effects assumed in the analysis to be resisted by shearwalls), design of the columns reduces to Step (2) only. Equation (10-6) reduces to:

$$M_c = \delta_b M_{2b} \qquad \text{Eq. (10-6)}$$

Since the moment magnifier method of Section 10.11 is an approximate procedure, some judgment must be used in its application. The application of gravity loads to an unbraced frame in an unsymmetrical loading pattern, or to an unsymmetrical unbraced frame, will result in some calculated sidesway of the frame. Unless the lateral deflection due to gravity loading is appreciable ($\Delta/\ell_u >$ 1/1500), the minor effect of this sway component can be neglected and the corresponding moments can be considered as nonsway moments, magnified by the braced frame magnifier. The definitions of M_{2b} and M_{2s} both contain the terminology "appreciable sway." For use in the approximate method, a deflection of $\Delta = \ell_u/1500$ is a reasonable upper limit above which sway deflections become "appreciable." In computing Δ for this purpose, only the nonlateral loads should be considered.

In the case of columns subjected to transverse loading between supports, it is possible that maximum moment will occur at a section away from the ends of a member. If this occurs, the value of the largest calculated moment occurring anywhere along the member should be used for the value of M_{2b} in Eq. (10-6). In accordance with Section 10.11.5.3, C_m must be taken as 1.0 for this case.

In Eq. (10-12), the ratio of the end moments indicates the shape of the deformed column. The positive algebraic sign should be used for moments if each moment induces compression on the same face of the column. For example, if end moments are equal, C_m is equal to unity, but if one end of a column is restrained, and a carryover moment is taken as minus one-half the top moment, C_m is equal to 0.4 (see Fig. 13-12).

10.11.5.2 Column Stiffness EI—In defining the critical column load, the difficult problem is the choice of a stiffness parameter EI in Eq. (10-9) which reasonably approximates the stiffness variations due to cracking, creep, and the nonlinearity of the concrete stress-strain curve. When more exact values are not available, EI defined by Eqs. (10-10) or (10-11) may be used:

$$EI = [(E_c I_g /5) + E_s I_{se}] /(1 + \beta_d) \qquad \text{Eq. (10-10)}$$

$$EI = (E_c I_g /2.5)/(1 + \beta_d) \qquad \text{Eq. (10-11)}$$

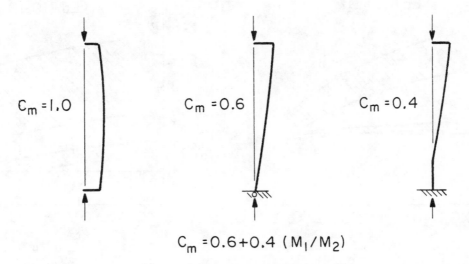

$$C_m = 0.6 + 0.4 \, (M_1/M_2)$$

Fig. 13-12   Moment Factor (C_m)

The Code equations for EI approximate the lower limits of EI for practical cross sections and thus are conservative for secondary moment calculations. The approximate nature of the EI equations is shown in Fig. 13-13 where they are compared with values derived from load-moment-curvature diagrams for the case of no sustained load ($\beta_d = 0$).

Equation (10-10) represents the lower limit of the practical range of stiffness values. This is especially true for heavily reinforced columns. Equation (10-11) is simpler to use but greatly underestimates the effect of reinforcement in heavily reinforced columns (see Fig. 13-13).

Both EI equations were derived for small e/h values and high P_u/P_o values, where the effect of axial load is most pronounced. P_o is the nominal axial load strength at zero eccentricity.

For reinforced concrete columns subject to sustained loads, creep of concrete transfers some of the load from the concrete to the steel, thus increasing steel stresses. For lightly reinforced columns, this load transfer may cause the compression steel to yield prematurely, resulting in a loss in the effective value of EI. This is taken into account by dividing the EI term by $(1 + \beta_d)$, where β_d is the creep effect factor used in Eqs. (10-10) and (10-11) to reduce the effective value of EI. For ACI 318-89, the definition of the β_d factor has been modified to reflect a sustained load ratio rather than a sustained moment ratio.

For moment magnification due to gravity load effects by Eq. (10-7), stiffness parameter EI should be based on:

$$\beta_d = \frac{\text{factored dead load}}{\text{total factored load}} = \frac{1.4D}{1.4D + 1.7L}$$

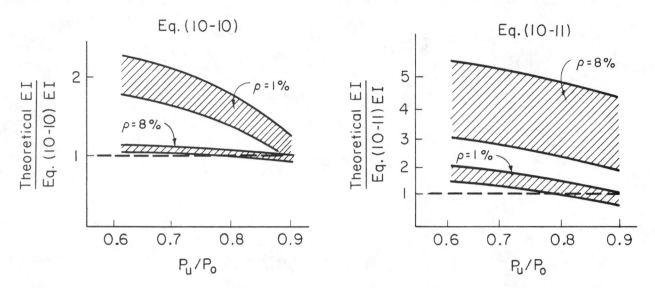

Fig. 13-13   Comparison of Equations for EI with EI Values from Moment-Curvature Diagrams

For moment magnification due to sustained lateral load effects by Eq. (10-8), stiffness parameter EI should be based on:

$$\beta_d = \frac{\text{factored sustained lateral load}}{\text{total factored lateral load}}$$

Note that sustained lateral load represents an unusual load condition, and applies only when the lateral load is of long duration, or when unsymmetrical gravity loading or frame geometry causes lateral shear forces (sustained dead loads) to act on frame columns. Note also that if the sway moments M_{2s} are due to nonsustained lateral loads, as in the usual cases of wind and earthquake loading, $\beta_d = 0$ for calculation of moment magnification by Eq. (10-8).

For composite columns in which a steel pipe or structural steel shape makes up a large percentage of the total column cross section, load transfer due to creep is not significant. Accordingly, in Eq. (10-14), only the EI of the concrete portion is reduced by $(1 + \beta_d)$ to account for sustained load effects.

$$EI = (E_c I_g /5)/(1 + \beta_d) + E_s I_t \qquad\qquad \text{Eq. (10-14)}$$

where I_t = moment of inertia of structural steel shape, pipe or tubing about centroidal axis of composite member cross section.

10.11.5.4   Minimum Moment Magnification—Column slenderness is accounted for by magnifying the column end moments. If the computed column moments M_{2b} are small or zero, design of a braced slender column must be based on a minimum moment, $M_{2b} = P_u(0.6 + 0.03h)$. The minimum moment is to be taken about each column axis separately, not about both axes simultaneously. When design is based on the minimum moment, the moment correction factor C_m is, however, evaluated using the computed column

13-18

moments M_{1b} and M_{2b} in Eq. (10-12). If computations show that there is no moment at both ends of a column, due to greater relative flexibility of the restraining members at the column ends, the ratio M_{1b}/M_{2b} is taken equal to 1.0. If the computed column moments due to sidesway, M_{2s}, are small or zero, design of an unbraced slender column must be based on a minimum moment, $M_{2s} = P_u (0.6 + 0.03h)$.

10.11.6   Moment Magnification for Flexural Members

The strength of a laterally unbraced frame is governed by the stability of the columns and by the degree of end restraint provided by the beams in the frame. If plastic hinges form in the restraining beams, the structure approaches a mechanism and its axial load capacity is drastically reduced. Section 10.11.6 requires that the designer make certain that the restraining flexural members (beams or slabs) have the capacity to resist the magnified column moments. The ability of the moment magnifier method to provide a good approximation of the actual magnified moments at the member ends in an unbraced frame is a significant improvement over the reduction factor method for long columns prescribed in earlier ACI Codes to account for member slenderness in design.

10.11.7   Moment Magnifier δ for Biaxial Bending

When biaxial bending occurs in a column, the computed moments about each of the principal axes must be magnified. The magnification factors δ_b and δ_s are computed considering the buckling load P_c about each axis separately, based on the appropriate effective lengths and the related stiffness ratios of columns to beams in each direction. Thus, different buckling capacities about the two axes are reflected in different magnification factors. The moments about each of the two axes are magnified separately, and the cross section is then proportioned for an axial load P_u and magnified biaxial moments.

**Example 13.1—Slenderness Effects by Approximate Design Method (Section 10.11)—
Columns Braced Against Sidesway**

Design columns for the first story of a 10-story office building. Clear height of first story is 18 ft. Clear height above first story and below first story (basement) is 11 ft. Building extends over 7 x 3 bays. Assume lateral load effects of wind are resisted by shearwalls,   with columns resisting gravity load effects of dead and live loads only. Assume shearwall bracing is adequate to classify the building frame as braced against sidesway for column design. f_y = 60,000 psi, f_c' = 5000 psi

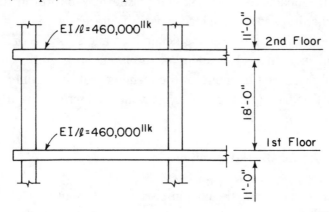

		Code
	Calculations and Discussion	**Reference**

1. Load data from frame analysis

	Interior Columns	Exterior Columns
Service Loads (kips)	D = 363 L = 128	D = 282 L = 64
Service Moments (ft-kips)	Top: D = 16.5 　　　 L = 13.5	Top: D = 35 　　　 L = 20
	Btm: D = 43 　　　 L = 34	Btm: D = 58 　　　 L = 30

	Code
Calculations and Discussion	**Reference**

2. Factored gravity loads.

   Interior columns

   $U\ = 1.4D+1.7L$
   $P_u\ = 1.4(363)+1.7(128)=726$ kips
   $M_{1b} = 1.4(16.5)+1.7(13.5)=46$ ft$-$kips
   $M_{2b} = 1.4(43)+1.7(34)=118$ ft$-$kips

   Exterior columns

   $U\ = 1.4D+1.7L$
   $P_u\ = 1.4(282)+1.7(64)=504$ kips
   $M_{1b} = 1.4(35)+1.7(20)=83$ ft$-$kips
   $M_{2b} = 1.4(58)+1.7(30)=132$ ft$-$kips

3. Preliminary selection of column size and reinforcement.

   Base preliminary selection on load combinations excluding slenderness effects (if any).

   Interior columns

   $$\text{required } P_n = \frac{P_u}{\varphi} = \frac{726}{0.7} = 1037 \text{ kips}$$ 9.3.2.2

   $$M_n = \frac{M_u}{\varphi} = \frac{118}{0.7} = 169 \text{ ft}-\text{kips}$$

   Exterior columns

   $$\text{required } P_n = \frac{P_u}{\varphi} = \frac{504}{0.7} = 720 \text{ kips}$$ 9.3.2.2

   $$M_n = \frac{M_u}{\varphi} = \frac{132}{0.7} = 189 \text{ ft}-\text{kips}$$

   Using Design Aid EB9, try 16 in. x16 in. column size (see Table 13-1 reproduced from EB9).

Example 13.1—Continued

Table 13-1 Reproduced from Page 11 of EB9

PCA—LOAD AND MOMENT STRENGTH TABLES FOR CONCRETE COLUMNS

COLUMN SIZE =16X16 INCHES
FPC=5,000 FY=60,000

NO A1	NO A2	SIZE NO	PCNT	O* PO	BALANCE PB	MB	0	40	80	120	160	200	240	280	320	420	520	620	720	820	920	1020	1120	1220
							MOMENT STRENGTH Mn (FT-KIPS)																	
6	0	6	1.03	1235	419	241	88	108	128	147	165	181	196	209	221	241	233	220	202	175	140			
10	0	5	1.21	1261	420	255	102	122	142	161	178	195	210	223	235	255	245	231	212	186	151			
4	0	8	1.23	1264	415	253	103	123	142	160	177	194	208	222	233	252	243	230	211	186	151			
8	0	6	1.38	1284	418	265	114	134	153	172	189	206	220	234	246	265	254	239	220	194	161			
6	0	7	1.41	1289	416	266	116	136	155	173	191	207	221	235	247	266	254	240	221	195	162			
4	0	9	1.56	1311	411	274	127	146	164	182	199	215	230	243	255	273	261	246	227	202	170			
6	0	8	1.85	1352	411	296	148	167	186	204	221	237	251	265	276	294	280	263	244	219	188	149		
8	0	7	1.87	1356	413	299	151	170	189	207	224	240	255	268	280	298	283	266	246	221	190	151		
4	0	10	1.98	1371	405	301	156	175	193	210	227	243	257	270	282	299	284	268	248	224	194	156		
6	0	9	2.34	1422	406	328	183	202	220	237	254	269	283	297	308	325	308	290	269	244	215	179		
4	0	11	2.44	1436	398	328	187	205	223	240	256	271	285	298	310	325	309	291	270	246	218	182		
6	0	10	2.98	1513	400	368	226	244	262	279	295	310	324	337	349	364	344	323	301	276	248	215	175	
4	0	14	3.52	1590	377	389	258	275	291	308	323	338	351	364	375	383	367	346	323	299	272	241	204	
0	0	0	4.00	1659	367	415	288	305	321	337	352	366	380	392	403	407	392	370	347	322	296	266	231	192
0	0	0	5.00	1802	344	466	349	365	381	396	411	424	437	449	460	454	438	421	396	370	344	315	284	248
0	0	0	6.00	1944	319	515	408	424	439	453	467	480	493	505	515	498	481	464	444	418	391	363	333	300
0	0	0	7.00	2087	292	562	465	480	494	508	522	535	547	558	557	539	522	505	487	465	438	410	380	349
0	0	0	8.00	2230	262	606	520	534	548	562	575	588	599	602	595	578	560	543	525	508	484	456	427	396
6	2	6	1.38	1284	421	241	115	134	153	170	186	198	208	217	225	241	234	223	207	184	153			
8	4	5	1.45	1295	422	241	121	140	158	173	186	200	212	221	228	241	234	224	209	186	156			
6	2	7	1.87	1356	417	266	151	169	187	204	216	226	235	244	252	266	256	243	227	206	178	141		
10	4	5	1.94	1365	422	264	156	172	188	203	216	227	237	247	253	263	254	243	228	207	180	145		
8	4	6	2.06	1382	418	271	165	182	196	210	223	235	246	253	259	270	260	248	233	212	186	152		
6	2	8	2.47	1440	412	296	193	210	227	240	249	258	266	275	282	295	282	268	251	231	205	173		
8	4	7	2.81	1489	419	306	214	227	241	253	266	278	284	291	297	305	292	278	262	243	219	190	153	
6	2	9	3.13	1534	407	328	237	253	266	275	283	292	300	307	315	326	311	295	278	258	234	205	170	
6	2	10	3.97	1654	399	368	292	302	310	318	326	334	341	349	356	365	348	330	312	292	269	243	213	176
4	2	6	1.03	1235	422	216	89	109	128	145	162	174	184	193	201	216	213	205	189	165	132			
4	6	5	1.21	1261	426	211	103	121	136	151	165	175	185	195	201	211	209	203	190	169	139			
4	2	6	1.38	1284	418	222	115	133	149	162	175	187	198	204	210	222	218	211	197	176	147			
4	2	7	1.41	1289	419	233	117	136	154	170	185	194	203	211	219	233	227	218	203	181	151			
4	2	8	1.85	1352	415	253	149	167	184	199	208	216	224	232	239	253	245	234	219	199	172	137		
4	2	7	1.87	1356	418	240	150	164	177	189	201	212	220	225	231	240	234	226	213	195	169			
4	2	9	2.34	1422	411	274	182	199	215	224	232	240	247	255	261	273	263	252	237	218	193	161		
4	2	10	2.98	1513	405	301	224	239	247	254	262	269	276	282	289	299	287	274	259	241	218	191	156	
8	2	5	1.21	1261	423	237	102	122	142	160	176	191	202	212	221	237	231	220	203	179	146			
10	2	5	1.45	1295	421	255	121	140	159	177	194	208	219	229	238	255	246	233	216	192	160			
8	2	6	1.72	1333	419	265	140	159	177	195	211	221	232	241	250	265	255	242	225	202	173			
8	2	7	2.34	1422	414	299	185	203	220	237	248	258	267	276	285	298	285	270	252	230	203	170		
6	4	5	1.21	1261	424	224	103	122	141	155	169	182	194	204	211	224	220	211	196	173	142			
10	4	5	1.70	1330	421	259	139	158	175	190	203	217	229	238	246	259	250	238	222	199	170			
6	4	6	1.72	1333	420	246	140	158	172	186	199	211	222	229	235	246	239	229	215	194	167			
6	4	7	2.34	1422	415	273	182	196	209	221	233	245	252	258	264	273	263	252	237	219	194	163		
6	6	5	1.45	1295	424	229	121	138	153	169	182	193	203	212	218	228	224	216	202	182	153			
8	6	5	1.70	1330	423	246	139	155	170	186	199	210	220	229	236	246	239	229	215	194	167			

Example 13.1—Continued

Try 8#9 bars for both interior and exterior columns. $A_s = 8.00$ in.^{2} $\rho = 3.13\%$

Check $P_{n(max)} = 0.80\, P_o = 0.80(1534) = 1227$ kips > 1037 kips   O.K.

For $P_n = 1037$ kips, $M_n \approx 199$ ft$-$kips > 169 ft$-$kips   O.K.
  $P_n = 720$ kips, $M_n \approx 278$ ft$-$kips > 189 ft$-$kips   O.K.

4. Check if slenderness need be considered for 16 in. x 16 in. column with 8#9 bars.

(a) Effective length factors (k)

   k for braced condition must be taken as 1.0, unless an analysis shows that a lower
   value may be used. Determine k factors from Fig. 13-9. With $k\ell_u/r < 60$, use $0.5EI_g$
   for beams (to account for effects of cracking and reinforcement), and EI_g for
   columns to evaluate k factors.

   $$I_g = \frac{16^4}{12} = 5461 \text{ in.}^4$$
   $$E_c = 57000\sqrt{5000} = 4.03 \times 10^3 \text{ ksi}$$

   For 18 ft long columns $\dfrac{EI_g}{\ell_c} = \dfrac{4.03 \times 10^3 \times 5461}{18 \times 12} = 102 \times 10^3 \text{ in.}-\text{kips}$

   For 11 ft long columns $\dfrac{EI_g}{\ell_c} = \dfrac{4.03 \times 10^3 \times 5461}{11 \times 12} = 167 \times 10^3 \text{ in.}-\text{kips}$

   Interior columns

   $$\psi_A = \psi_B = \frac{\Sigma EI/\ell_c}{\Sigma EI/\ell} = \frac{102 + 167}{2(0.5 \times 460)} = 0.58$$

   From Fig. 13-9, $k_b = 0.71$

   $$\frac{k_b \ell_u}{r} = \frac{0.71(18 \times 12)}{0.3 \times 16} = 32$$

Example 13.1—Continued

	Code
Calculations and Discussion	**Reference**

To neglect slenderness for braced condition, $k\ell_u/r$ must be less than (see Fig. 13-4):

$$34 - 12\left(\frac{M_{1b}}{M_{2b}}\right)$$

10.11.4.1

Assuming member bent in single curvature:

$$34 - 12\left(\frac{46}{118}\right) = 29.3 < 32$$

Assuming member bent in double curvature:

$$34 + 12\left(\frac{46}{118}\right) = 38.7 > 32$$

Column slenderness need not be considered for the interior columns if bent in double curvature, the more common condition for conventional cast-in-place construction. However, to illustrate design procedure including slenderness effects for braced columns, assume single curvature bending.

Exterior columns

$$\psi_A = \psi_B = \frac{102 + 167}{0.5 \times 460} = 1.17$$

From Fig. 13-9, $k_b = 0.79$

$$\frac{k_b\ell_u}{r} = \frac{0.79\,(18 \times 12)}{0.3 \times 16} = 35.6$$

Single curvature bending:

$$34 - 12\left(\frac{83}{132}\right) = 26.5 < 35.6$$

Double curvature bending:

$$34 + 12\left(\frac{83}{132}\right) = 41.5 > 35.6$$

Slenderness also need not be considered for the exterior columns if bent in double curvature. To illustrate design procedure, assume single curvature bending and evaluate slenderness effects.

Example 13.1—Continued

(b) Critical load $P_{cb} = \dfrac{\pi^2 EI}{k_b \ell_u^2}$

Eq. (10-9)

Compute EI from Eq. (10-10)

$I_{se} = 6.0\,(5.5)^2 = 181.5\ \text{in.}^4$

$E_s = 29 \times 10^3\ \text{ksi}$

$\beta_d = \dfrac{\text{factored dead load}}{\text{total factored load}} = \dfrac{1.4D}{1.4D + 1.7L}$

10.0

$\beta_d\ (\text{interior}) = \dfrac{1.4(363)}{726} = 0.70$

$\beta_d\ (\text{exterior}) = \dfrac{1.4(282)}{504} = 0.78$

$EI = \dfrac{\dfrac{E_c I_g}{5} + E_s I_{se}}{1 + \beta_d}$

Eq. (10-10)

$EI_{(int)} = \dfrac{\dfrac{4.03 \times 10^3 \times 5461}{5} + 29 \times 10^3 \times 181.5}{1 + 0.70} = 5.7 \times 10^6$

$EI_{(ext)} = \dfrac{\dfrac{4.03 \times 10^3 \times 5461}{5} + 29 \times 10^3 \times 181.5}{1 + 0.78} = 5.4 \times 10^6$

$P_{cb(int)} = \dfrac{\pi^2\,(5.7 \times 10^6)}{(0.71 \times 18 \times 12)^2} = 2392\ \text{kips}$

$P_{cb(ext)} = \dfrac{\pi^2\,(5.4 \times 10^6)}{(0.79 \times 18 \times 12)^2} = 1830\ \text{kips}$

5. Final design including slenderness effects.

(a) Interior Columns

$C_m = 0.6 + 0.4\left(\dfrac{M_{1b}}{M_{2b}}\right) = 0.6 + 0.4\left(\dfrac{46}{118}\right) = 0.76$

Eq. (10-12)

Example 13.1—Continued

	Code
Calculations and Discussion	**Reference**

$$\delta_b = \frac{C_m}{1 - \dfrac{P_u}{\varphi P_{cb}}} = \frac{0.76}{1 - \dfrac{726}{0.7 \times 2392}} = 1.34$$

Eq. (10-7)

Minimum moment for slenderness effects

10.11.5.4

$$M_{2b} \geq P_u(0.6 + 0.03h)$$

$$\geq \frac{726(0.6 + 0.03 \times 16)}{12} = 65 \text{ ft−kips} < 118 \text{ ft−kips}$$

$$M_c = \delta_b M_{2b} = 1.34(118) = 158 \text{ ft−kips}$$

Eq. (10-6)

$$\text{required } P_n = \frac{P_u}{\varphi} = \frac{726}{0.7} = 1037 \text{ ft−kips}$$

$$M_n = \frac{M_u}{\varphi} = \frac{158}{0.7} = 226 \text{ ft−kips}$$

Review of Table 13-1 indicates that the 8#9 bars are not adequate for the interior columns ($M_n = 199 < 226$).

16x16—8#9 bars			
P_n (kips)	1020	1037	1120
M_n (ft-kips)	205	199	170

The next bar arrangement (8#10 bars) is more than adequate.

(b) Exterior Columns

$$C_m = 0.6 + 0.4\left(\frac{83}{132}\right) = 0.85$$

Eq. (10-12)

$$\delta_b = \frac{0.85}{1 - \dfrac{504}{0.7 \times 1830}} = 1.40$$

Eq. (10-7)

Example 13.1—Continued

Calculations and Discussion	**Code Reference**

Minimum moment for slenderness effects

10.11.5.4

$$M_{2b} \geq \frac{504(0.6 + 0.03 \times 16)}{12} = 45 \text{ ft}-\text{kips} < 132 \text{ ft}-\text{kips}$$

$$M_c = 1.40(132) = 185 \text{ ft}-\text{kips}$$

Eq. (10-6)

$$\text{rerquired} \quad P_n = \frac{P_u}{\varphi} = \frac{504}{0.7} = 720 \text{ kips}$$

$$M_n = \frac{M_u}{\varphi} = \frac{185}{0.7} = 264 \text{ ft}-\text{kips}$$

Review of Table 13-2 indicates that the 8#9 bars are adequate for the exterior columns (M_n = 278 ft-kips > 264 ft-kips).

Use 16 in. x 16 in. section with 8#10 bars for interior columns and 8#9 bars for exterior columns of the first story.

6. Using the investigative option of <u>PCACOL Program</u>, compare design results for the interior and exterior 16 in. x 16 in. column size with 8#10 bars for interior columns and 8#9 bars for exterior columns.

The on-screen graphic results for the interior and exterior columns are reproduced in Fig. 13-14 and Fig. 13-15, respectively. The required gravity load combinations are shown as point "1" relative to the strength interaction curves.

Example 13.1—Continued

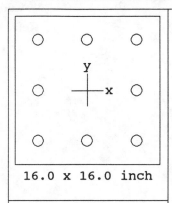

16.0 x 16.0 inch

f'_c = 5.0 ksi

f_y = 60.0 ksi

8#10   4.0%

A_{st} = 10.16 in²

Tied   cc = 1.94 in

spacing = 4.15 in

I_x = 5461 in⁴

$\overline{x}$ = 0.00 in

I_y = 5461 in⁴

$\overline{y}$ = 0.00 in

© 1989 PCA

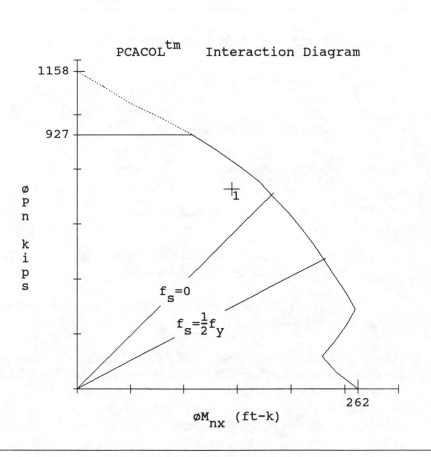

PCACOL^{tm} Interaction Diagram

1158

927

$f_s=0$

$f_s=\frac{1}{2}f_y$

ϕP_n kips

ϕM_{nx} (ft-k)

262

Licensed To: Portland Cement Association, Skokie, IL

Project: Example 13.1 Interior   File name: C:\PCACOL\DATA\EX13_1I.COL

Column Id: Slenderness Braced   Material Properties:

Engineer: E_c = 4030 ksi ϵ_u = 0.003 in/in

Date: Time: f_c = 4.25 ksi E_s = 29000 ksi

Code: ACI 318-89 β_1 = 0.80

Version: 2.10   Stress Profile : Block

                                        Reduction: ϕ_c = 0.70 ϕ_b = 0.90

Slenderness considered x-axis k_b = 0.71

**Fig. 13-14   PCACOL On-Screen Results for 16 in. x 16 in. Column
with 8#10 Bars (Interior Columns)**

Example 13.1—Continued

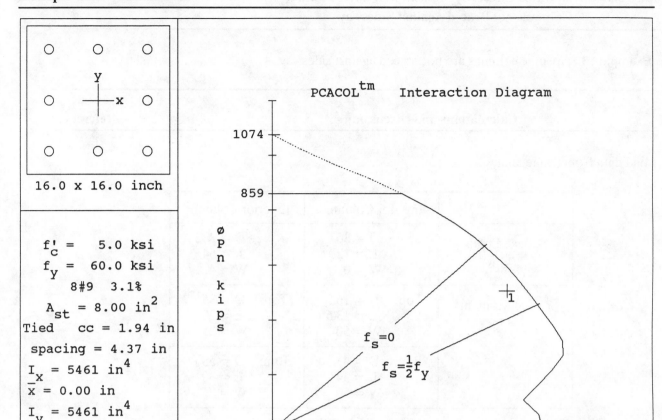

16.0 x 16.0 inch

$f_c' = 5.0$ ksi

$f_y = 60.0$ ksi

8#9 3.1%

$A_{st} = 8.00$ in^2

Tied cc = 1.94 in

spacing = 4.37 in

$I_x = 5461$ in^4

$\overline{x} = 0.00$ in

$I_y = 5461$ in^4

$\overline{y} = 0.00$ in

© 1989 PCA

Licensed To: Portland Cement Association, Skokie, IL

Project: Example 13.1 Exterior   File name: C:\PCACOL\DATA\EX13_1E.COL

Column Id: Slenderness Braced   Material Properties:

Engineer: $E_c = 4030$ ksi $\epsilon_u = 0.003$ in/in

Date: Time: $f_c = 4.25$ ksi $E_s = 29000$ ksi

Code: ACI 318-89 $\beta_1 = 0.80$

Version: 2.10   Stress Profile : Block

                                     Reduction: $\phi_c = 0.70$ $\phi_b = 0.90$

Slenderness considered x-axis $k_b = 0.79$

**Fig. 13-15 PCACOL On-Screen Results for 16 in. x 16 in. Column
with 8#9 Bars (Exterior Columns)**

Example 13.2—Slenderness Effects by Moment Magnifier Method—
Columns not Braced Against Sidesway

Solve Example 13.1, assume columns are not braced against sidesway.

		Code
	Calculations and Discussion	**Reference**

1. Load data from frame analysis.

	Interior Columns	Exterior Columns
Service Loads (kips)	D = 363 L = 128 W = 0	D = 282 L = 64 W = 4
Service Moments (ft-kips)	Top: D = 16.5 L = 13.5 W = 50	Top: D = 35 L = 20 W = 25
	Btm: D = 43 L = 34 W = 50	Btm: D = 58 L = 30 W = 25

2. Factored load combinations to be considered.   9.2

For application in Eq. (10-6), it is helpful to distinguish between the factored "nonsway" moments due to gravity loads (M_{2b}) and the factored "sway" moments due to lateral loads (M_{2s}).

Example 13.2—Continued

Calculations and Discussion	Code Reference

Interior Columns

(a) gravity loads

$\quad$ U $\quad = 1.4D + 1.7L$ $\qquad\qquad\qquad\qquad\qquad\qquad\qquad$ Eq. (9-1)

$\quad$ P_u $\quad = 1.4(363) + 1.7(128) = 726$ kips

$\quad$ M_{1b} $\quad = 1.4(16.5) + 1.7(13.5) = 46$ ft-kips

$\quad$ M_{2b} $\quad = 1.4(43) + 1.7(34) = 118$ ft-kips

(b) gravity + wind loads

$\quad$ U $\quad = 0.75(1.4D + 1.7L + 1.7W)$ $\qquad\qquad\qquad\qquad$ Eq. (9-2)

$\quad$ P_u $\quad = 0.75(1.4 \times 363 + 1.7 \times 128 + 0) = 544$ kips

$\quad$ M_{2b} $\quad = 0.75(1.4 \times 43 + 1.7 \times 34) = 89$ ft-kips

$\quad$ M_{2s} $\quad = 0.75(1.7 \times 50) = 64$ ft-kips

or $\;$ U $\quad = 0.9D + 1.3W$ $\qquad\qquad\qquad\qquad\qquad\qquad$ Eq (9-3)

$\quad$ P_u $\quad = 0.9(363) = 327$ kips

$\quad$ M_{2b} $\quad = 0.9(43) = 39$ ft-kips

$\quad$ M_{2s} $\quad = 1.3(50) = 65$ ft-kips

Exterior Columns

(a) gravity loads

$\quad$ U $\quad = 1.4D + 1.7L$ $\qquad\qquad\qquad\qquad\qquad\qquad\qquad$ Eq. (9-1)

$\quad$ P_u $\quad = 1.4(282) + 1.7(64) = 504$ kips

$\quad$ M_{1b} $\quad = 1.4(35) + 1.7(20) = 83$ ft-kips

$\quad$ M_{2b} $\quad = 1.4(58) + 1.7(30) = 132$ ft-kips

(b) gravity + wind loads

$\quad$ U $\quad = 0.75(1.4D + 1.7L + 1.7W)$ $\qquad\qquad\qquad\qquad$ Eq. (9-2)

$\quad$ P_u $\quad = 0.75(1.4 \times 282 + 1.7 \times 64 + 1.7 \times 4) = 383$ kips

$\quad$ M_{2b} $\quad = 0.75(1.4 \times 58 + 1.7 \times 30) = 99$ ft-kips

$\quad$ M_{2s} $\quad = 0.75(1.7 \times 25) = 32$ ft-kips

or $\;$ U $\quad = 0.9D + 1.3W$ $\qquad\qquad\qquad\qquad\qquad\qquad$ Eq. (9-3)

$\quad$ P_u $\quad = 0.9(282) + 1.3(4) = 259$ kips

$\quad$ M_{2b} $\quad = 0.9(58) = 52$ ft-kips

$\quad$ M_{2s} $\quad = 1.3(25) = 33$ ft-kips

Example 13.2—Continued

	Code
Calculations and Discussion	**Reference**

3. Preliminary selection of column size and reinforcement.

   Use Design Aid EB9 (see Part 11). Note that the EB9 Design Tables do not include the strength reduction factor φ in the tabulated values; P_u/φ and M_u/φ must be used when designing with this aid.

   Base preliminary selection on gravity load combination, excluding slenderness effects.

   Interior Columns

   $$\text{required } P_n = \frac{P_u}{\varphi} = \frac{726}{0.7} = 1037 \text{ kips}$$

   $$M_n = \frac{M_u}{\varphi} = \frac{118}{0.7} = 169 \text{ ft--kips}$$

   Exterior Columns

   $$\text{required } P_n = \frac{P_u}{\varphi} = \frac{504}{0.7} = 720 \text{ kips}$$

   $$M_n = \frac{M_u}{\varphi} = \frac{132}{0.7} = 189 \text{ ft--kips}$$

   Try 18 in. x 18 in. Column Size (see Table 13.2 reproduced from EB9).

   To account for slenderness effects, try 4#9 bars for both interior and exterior columns. $A_s = 4.00 \text{ in.}^2, \rho = 1.23\%$.

   For $P_n = 1037$ kips, $M_n \approx 266$ ft--kips > 169 ft--kips   O.K.
        $P_n = 720$ kips, $M_n \approx 342$ ft--kips > 189 ft--kips   O.K.

   Estimate $\dfrac{k\ell_u}{r} \approx \dfrac{1.2\,(18 \times 12)}{0.3 \times 18} = 48$ 10.11.3

   $22 < \left(\dfrac{k\ell_u}{r} = 48\right) < 100$ 10.11.4

   Column slenderness effects must be included, and the Moment Magnifier Method is an   10.11.4.3
   acceptable procedure. See Fig. 13-4.

4. Compute properties of 18 in. x 18 in. section with 4#9 bars for slenderness evaluation.

   (a) Effective length factors (k)   10.11.2

Example 13.2—Continued

Table 13-2   Reproduced from page 12 of EB9

PCA—LOAD AND MOMENT STRENGTH TABLES FOR CONCRETE COLUMNS

COLUMN SIZE =18X18 INCHES
FPC=5,000   FY=60,000

NO A1	NO A2	SIZE NO	PCNT	0. PO	BALANCE PB	MB	0	40	80	120	160	260	360	460	560	660	760	860	960	1060	1160	1260	1360	1560
8	0	6	1.09	1573	540	355	133	156	179	202	223	272	312	341	353	342	327	307	282	249	208			
6	0	7	1.11	1578	538	356	135	159	181	203	225	273	313	342	354	343	328	309	283	250	210			
4	0	9	1.23	1600	533	366	148	171	193	215	236	283	323	353	363	351	336	316	291	259	219			
10	0	6	1.36	1622	538	383	163	187	209	231	252	300	341	370	380	366	350	329	303	271	232			
6	0	8	1.46	1641	533	391	174	196	218	240	261	309	348	378	387	373	356	335	310	278	240	194		
8	0	7	1.48	1645	535	395	176	199	221	243	264	312	352	382	391	376	359	338	312	281	242	197		
4	0	10	1.57	1660	528	398	183	206	227	249	269	316	355	386	393	378	361	341	316	285	247	202		
6	0	9	1.85	1711	528	429	215	237	259	280	300	347	387	417	424	406	387	365	339	309	272	229		
4	0	11	1.93	1725	523	432	220	242	263	284	304	350	389	420	425	408	389	367	342	312	276	234		
8	0	9	1.95	1729	530	441	226	249	270	292	312	359	399	429	436	417	397	375	348	318	282	239		
6	0	10	2.35	1802	523	477	266	288	309	330	350	396	435	467	469	449	427	403	377	347	312	272	226	
8	0	9	2.47	1823	524	492	281	302	324	344	364	411	451	481	484	462	440	415	388	358	323	284	238	
4	0	14	2.78	1879	512	510	305	326	346	366	385	430	468	499	500	477	454	430	404	375	342	304	261	
6	0	11	2.89	1899	516	528	321	342	362	383	402	447	487	518	518	494	470	445	418	388	354	316	273	
4	0	18	4.94	2269	468	689	506	525	543	562	579	620	657	687	671	648	619	590	561	530	499	465	430	349
0	0	0	6.00	2461	439	769	599	617	635	652	669	709	745	765	744	723	699	668	637	606	574	541	506	432
0	0	0	7.00	2641	410	840	683	701	718	735	752	790	825	829	808	787	766	740	708	676	644	611	577	505
0	0	0	8.00	2822	377	908	765	782	799	816	832	869	903	890	869	847	826	804	779	746	713	680	646	575
6	2	6	1.09	1573	544	326	134	157	179	200	221	263	292	314	325	318	307	291	268	237	199			
8	4	5	1.15	1584	547	326	141	164	186	207	225	264	297	315	326	319	308	293	271	241	203			
6	2	7	1.48	1645	541	356	177	199	220	241	260	297	324	345	354	344	332	315	293	264	229			
10	6	5	1.53	1654	547	353	183	205	224	242	260	297	327	344	352	343	331	315	294	267	232	190		
8	4	6	1.63	1671	543	361	194	215	236	253	269	306	335	352	360	350	337	321	300	273	239	198		
6	2	8	1.95	1729	537	391	226	248	268	288	306	336	361	381	388	375	361	343	321	295	262	223		
10	6	6	2.17	1769	542	398	249	267	285	302	319	351	378	391	396	383	369	352	332	307	277	241		
8	4	7	2.22	1778	539	403	256	275	291	306	321	357	381	395	401	387	373	356	335	310	280	243		
6	2	9	2.47	1823	532	429	280	300	320	338	353	378	401	420	425	409	393	374	353	327	297	262	220	
8	4	8	2.93	1906	533	453	322	337	353	367	382	416	434	446	448	432	416	397	377	353	325	293	255	
6	2	10	3.14	1943	526	477	346	365	384	396	406	429	452	470	471	453	435	415	393	368	340	308	271	
8	4	9	3.70	2046	527	506	389	404	419	433	447	474	490	501	500	482	463	444	422	399	373	343	310	
6	2	11	3.85	2073	520	528	415	432	442	451	460	483	504	522	520	500	480	458	436	412	385	354	321	240
4	4	6	1.09	1573	547	304	134	157	178	197	212	249	278	294	304	300	292	279	258	230	193			
4	2	7	1.11	1578	543	317	136	159	181	202	221	259	285	305	316	311	301	286	264	235	197			
4	6	6	1.36	1622	549	312	164	182	199	217	233	265	291	304	312	308	300	288	269	244	211			
4	4	8	1.46	1641	540	340	174	196	217	237	255	287	311	330	339	331	320	305	284	257	222			
4	4	7	1.48	1645	544	326	177	198	215	230	245	279	303	317	325	319	311	297	278	253	220			
4	2	9	1.85	1711	540	366	215	236	256	274	292	316	338	356	364	353	341	326	306	280	249	210		
4	4	8	1.95	1729	540	352	224	239	254	268	282	314	333	345	350	342	332	319	301	278	249	213		
4	2	10	2.35	1802	532	398	265	285	304	320	329	351	372	390	394	382	368	352	333	309	280	245	204	
4	4	9	2.47	1823	535	380	267	281	294	308	321	350	363	374	378	368	356	343	326	305	279	247	208	
4	2	11	2.89	1899	526	432	317	336	350	359	367	388	408	425	427	413	397	380	361	338	312	280	243	
10	2	5	1.15	1584	545	342	141	164	187	208	229	275	306	329	341	332	319	301	278	246	207			
8	2	6	1.36	1622	543	355	164	186	208	229	250	291	320	343	353	343	330	312	289	259	222			
10	2	6	1.63	1671	541	383	194	216	238	258	278	320	349	372	381	368	353	334	310	281	246	203		
8	2	7	1.85	1711	538	395	217	239	260	280	299	336	363	384	392	378	363	344	321	293	260	219		
8	2	8	2.44	1817	533	441	278	299	319	339	357	386	412	432	437	420	402	382	360	333	302	265	222	
8	2	9	3.09	1934	528	492	344	364	384	402	414	440	465	484	486	466	446	425	402	375	346	312	274	
10	4	5	1.34	1619	546	347	162	185	207	228	245	285	317	336	346	337	324	308	286	256	220			
6	4	6	1.36	1622	545	332	164	186	207	225	241	278	306	323	332	325	314	299	279	251	216			
6	4	7	1.85	1711	540	365	216	237	253	268	283	318	342	356	363	353	341	326	306	281	250	212		
10	4	6	1.90	1720	541	390	223	245	264	281	297	335	364	381	387	375	360	343	321	294	261	222		
6	4	8	2.44	1817	537	402	273	288	303	318	332	365	383	396	399	387	373	358	338	315	287	253	213	
6	4	9	3.09	1934	531	443	328	342	356	370	384	412	427	438	439	424	409	392	373	351	326	296	260	
6	6	5	1.15	1584	550	312	141	164	183	201	219	257	285	302	312	307	299	285	264	236	200			
8	6	5	1.34	1619	547	332	162	184	203	221	239	277	306	323	332	325	314	300	279	251	216			
6	6	6	1.63	1671	547	341	192	210	228	245	262	293	320	333	340	333	323	309	290	265	233	193		
8	6	6	1.90	1720	545	369	220	238	256	273	290	322	349	362	368	358	346	330	311	286	255	217		

Example 13.2—Continued

Calculations and Discussion	**Code Reference**

k for a braced condition must be taken as 1.0, unless an analysis shows that a lower value may be used. k for an unbraced condition must take into account the effect of cracking and reinforcement on relative stiffness, and must be greater than 1.0. With $k\,\ell_u/r < 60$, use $0.5EI_g$ for beams (to account for the effects of cracking and reinforcement on relative stiffness) and EI_g for columns to evaluate k factors. Determine k factors from Figs. 13-9 and 13-10.

$$I_g = \frac{18^4}{12} = 8{,}748 \text{ in.}^4$$

$$E_c = 57{,}000\sqrt{5000} = 4.03 \times 10^3 \text{ ksi}$$

For 18 ft long columns $\dfrac{EI_g}{\ell_c} = \dfrac{4.03 \times 10^3 \times 8748}{18 \times 12} = 163 \times 10^3 \text{ in.-kips}$

For 11 ft long columns $\dfrac{EI_g}{\ell_c} = \dfrac{4.03 \times 10^3 \times 8748}{11 \times 12} = 267 \times 10^3 \text{ in.-kips}$

Interior Columns

(top & btm) $\psi_a = \psi_b = \dfrac{\Sigma(EI/\ell_c)}{\Sigma(EI/\ell)} = \dfrac{163 + 267}{2(0.5 \times 460)} = 0.93$

From Fig. 13-9 (braced columns)$k_b = 0.78$
From Fig. 13-10 (unbraced columns)$k_s = 1.30$

Exterior Columns

(top & btm) $\psi_A = \psi_B = \dfrac{163 + 267}{0.5 \times 460} = 1.87$

From Fig. 13-9 (braced columns) $k_b = 0.85$
From Fig. 13-10 (unbraced columns) $k_s = 1.57$

(b) Critical load $P_c = \pi^2 EI/(k\,\ell_u)^2$
    Compute EI from Eq. (10-10)   Eq. (10-9)

$$I_{se} = 4.00(6.5)^2 = 169 \text{ in.}^4$$

$$E_s = 29 \times 10^3 \text{ ksi}$$ 8.5.2

For calculation of δ_b:

$$\beta_d = \frac{\text{factored dead load}}{\text{total factored load}} = \frac{1.4D}{1.4D + 1.7L}$$ 10.0

Example 13.2—Continued

Note: For ACI 318-89, β_d is a ratio of factored axial loads, not moments.

$$\beta_d \text{ (interior)} = \frac{1.4(363)}{726} = 0.70$$

$$\beta_d \text{ (exterior)} = \frac{1.4(282)}{504} = 0.78$$

$$EI = \frac{\frac{E_c I_g}{5} + E_s I_{se}}{1 + \beta_d}$$

Eq. (10-10)

$$EI_{(int)} = \frac{\frac{4.03 \times 10^3 \times 8748}{5} + 29 \times 10^3 \times 169}{1 + 0.70}$$

$$= \frac{11.95 \times 10^6}{1.70} = 7.03 \times 10^6$$

$$EI_{(ext)} = \frac{11.95 \times 10^6}{1 + 0.78} = 6.71 \times 10^6$$

For calculation of δ_s; $\beta_d = 0$:

$$EI_{(int \,\&\, ext)} = \frac{11.95 \times 10^6}{1.0} = 11.95 \times 10^6$$

Interior Columns

$$P_{cb} \text{ (braced)} = \frac{\pi^2 (7.03 \times 10^6)}{(0.78 \times 18 \times 12)^2} = 2444 \text{ kips}$$

$$P_{cs} \text{ (unbraced)} = \frac{\pi^2 (11.95 \times 10^6)}{(1.3 \times 18 \times 12)^2} = 1496 \text{ kips}$$

Exterior Columns

$$P_{cb} \text{ (braced)} = \frac{\pi^2 (6.71 \times 10^6)}{(0.85 \times 18 \times 12)^2} = 1965 \text{ kips}$$

Example 13.2—Continued

Calculations and Discussion	Code Reference

$$P_{cs} \text{ (unbraced)} = \frac{\pi^2 (11.95 \times 10^6)}{(1.57 \times 18 \times 12)^2} = 1026 \text{ kips}$$

5. Interior Columns—Final design including slenderness effects

(a) For gravity load combination   Eq. (9-1)

$P_u = 726$ kips
$M_{1b} = 46$ ft$-$kips
$M_{2b} = 118$ ft$-$kips

C_m may be taken as 1.0 unless computed by

$$C_m = 0.6 + 0.4 \left(\frac{M_{1b}}{M_{2b}} \right) \geq 0.4$$ Eq. (10-12)

Assuming member bent in single curvature:

$$C_m = 0.6 + 0.4 \left(\frac{46}{118} \right) = 0.76$$

$$\delta_b = \frac{C_m}{1 - \dfrac{P_u}{\varphi P_{cb}}} = \frac{0.76}{1 - \dfrac{726}{0.7 \times 2444}} = 1.32$$ Eq. (10-7)

Minimum moment for slenderness effects   10.11.5.4

$M_{2b} \geq P_u(0.6 + 0.03h)$

$$= \frac{726(0.6 + 0.03 \times 18)}{12} = 69 \text{ ft}-\text{kips} < 118 \text{ ft}-\text{kips}$$

$M_c = \delta_b M_{2b} = 1.32(118) = 156$ ft$-$kips

$$\text{required } P_n = \frac{P_u}{\varphi} = \frac{726}{0.7} = 1037 \text{ kips}$$

$$M_n = \frac{M_u}{\varphi} = \frac{156}{0.7} = 223 \text{ ft}-\text{kips}$$

Review of Table 13-2 indicates that the 4#9 bars are adequate for gravity loading:
($M_n = 266$ ft-kips > 223 ft-kips)

Example 13.2—Continued

Calculations and Discussion

18 in. x 18 in. —4#9 bars			
P_n (kips)	960	1037	1060
M_n (ft-kips)	291	266	259

Assuming member bent in double curvature, the more common condition for conventional cast-in-place construction:

$$C_m = 0.6 - 0.4\left(\frac{46}{118}\right) = 0.44$$

$$\delta_b = \frac{0.44}{1 - \frac{726}{0.7 \times 2444}} = 0.76 < 1.0$$

For double curvature bending, gravity load combination is not affected by slenderness effects.

$$\text{required } P_n = \frac{726}{0.7} = 1037 \text{ kips}$$

$$M_n = \frac{118}{0.7} = 169 \text{ ft}-\text{kips}$$

(b) For gravity + wind load combination

Eq. (9-2)

P_u = 544 kips
M_{2b} = 89 ft−kips
M_{2s} = 64 ft−kips

Assuming member bent in single curvature, $C_m = 0.76$

$$\delta_b = \frac{0.76}{1 - \frac{544}{0.7 \times 2444}} = 1.11$$

Eq. (10-7)

$$\delta_s = \frac{1.0}{1 - \frac{\Sigma P_u}{\varphi \Sigma P_{cs}}}$$

Eq. (10-8)

Example 13.2—Continued

	Code
Calculations and Discussion	**Reference**

For the 7x3 bay building, factored axial loads P_u and critical loads P_{cs} must be summed for the 20 exterior columns and 12 interior columns. Assume corner columns carry ½ the load of edge columns; use 18 exterior columns for the summation of P_u.

$\Sigma P_u \approx 18(383) + 12(544) = 6894 + 6528 = 13,422$ kips
$\Sigma P_{cs} = 20(1026) + 12(1496) = 20,520 + 17,952 = 38,472$ kips

$$\delta_s = \frac{1.0}{1 - \frac{13,422}{0.7 \times 38,472}} = 1.99$$

Minimum moment for slenderness effects   10.11.5.5

$M_{2s} \geq P_u(0.6 + 0.03h)$
$\quad = \frac{544(0.6 + 0.03 \times 18)}{12} = 52 \text{ ft}-\text{kips} < 64 \text{ ft}-\text{kips}$

$M_c = \delta_b M_{2b} + \delta_s M_{2s} = 1.11(89) + 1.99(64) = 226 \text{ ft}-\text{kips}$ Eq. (10-6)

required $P_n = \frac{544}{0.7} = 777$ kips

$\quad\quad M_n = \frac{226}{0.7} = 323 \text{ ft}-\text{kips}$

Review of Table 13-2 indicates that the 4#9 bars are adequate for gravity plus wind loading ($M_n = 333$ ft-kips > 323 ft-kips).

18 in. x 18 in. —4#9 bars			
P_n (kips)	760	777	860
M_n (ft-kips)	336	333	316

By inspection, Eq. (9-3) load combination does not govern.

Assuming member bent in double curvature, $C_m = 0.44$

$$\delta_b = \frac{0.44}{1 - \frac{544}{0.7 \times 2444}} = 0.65 < 1.0$$

Example 13.2—Continued

$\delta_s = 1.99$

$M_c = 1.0(89) + 1.99(64) = 216 \text{ ft}-\text{kips}$

required $P_n = \dfrac{544}{0.7} = 777 \text{ kips}$

$\quad\quad M_n = \dfrac{216}{0.7} = 309 \text{ ft}-\text{kips}$

6. Exterior Columns—Final design including slenderness effects

   (a) For gravity load combination .... Eq. (9-1)

   $P_u \quad = 504 \text{ kips}$
   $M_{1b} = 83 \text{ ft}-\text{kips}$
   $M_{2b} = 132 \text{ ft}-\text{kips}$

   For single curvature bending:

   $C_m = 0.6 + 0.4\left(\dfrac{83}{132}\right) = 0.85$ .... Eq. (10-12)

   $\delta_b = \dfrac{0.85}{1 - \dfrac{504}{0.7 \times 1965}} = 1.34$ .... Eq. (10-7)

   Minimum moment for slenderness effects 10.11.5.4

   $M_{2b} \geq P_u(0.6 + 0.03h)$
   $\quad = \dfrac{504(0.6 + 0.03 \times 18)}{12} = 48 \text{ ft}-\text{kips} < 132 \text{ ft}-\text{kips}$

   $M_c = 1.34(132) = 177 \text{ ft}-\text{kips}$

   required $\quad P_n = \dfrac{504}{0.7} = 720 \text{ kips}$

   $\quad\quad\quad\quad M_n = \dfrac{177}{0.7} = 253 \text{ ft}-\text{kips}$

   Review of Table 13-2 indicates that the 4#9 bars are more than adequate for gravity loading.

Example 13.2—Continued

For double curvature bending:

$$C_m = 0.6 - 0.4\left(\frac{83}{132}\right) = 0.35 < 0.40$$

$$\delta_b = \frac{0.40}{1 - \dfrac{504}{0.7 \times 1965}} = 0.63 < 1.0$$

For double curvature bending, gravity load combination is not affected by slenderness effects.

$$\text{required} \quad P_n = \frac{504}{0.7} = 720 \text{ kips}$$

$$M_n = \frac{132}{0.7} = 189 \text{ ft-kips}$$

(b) For gravity plus wind load combination $\qquad$ Eq. (9-2)

$$P_u = 383 \text{ kips}$$
$$M_{2b} = 99 \text{ ft-kips}$$
$$M_{2s} = 32 \text{ ft-kips}$$

Assuming member bent in single curvature, $C_m = 0.85$

$$\delta_b = \frac{0.85}{1 - \dfrac{383}{0.7 \times 1965}} = 1.18 \qquad \text{Eq. (10-7)}$$

$$\delta_s = \frac{1.0}{1 - \dfrac{\Sigma P_u}{\varphi \Sigma P_{cs}}} = 1.99 \quad \text{(Same as for interior columns)} \qquad \text{Eq. (10-8)}$$

Minimum moment for slenderness effects $\qquad$ 10.11.5.5

$$M_{2s} \geq P_u(0.6 + 0.03h)$$
$$= \frac{383(0.6 + 0.03 \times 18)}{12} = 36 \text{ ft-kips} > 32 \text{ ft-kips}$$

$$M_c = 1.18 (99) + 1.99 (36) = 188 \text{ ft-kips} \qquad \text{Eq. (10-6)}$$

Example 13.2—Continued

required $P_n = \dfrac{383}{0.7} = 547$ kips

$M_n = \dfrac{188}{0.7} = 269$ ft$-$kips

Assuming member bent in double curvature, $C_m = 0.40$

$$\delta_b = \dfrac{0.40}{1 - \dfrac{383}{0.7 \times 1965}} = 0.55 < 1.0$$

$\delta_s = 1.99$

$M_c = 1.0(99) + 1.99(36) = 171$ ft$-$kips

required $P_n = \dfrac{383}{0.7} = 547$ kips

$M_n = \dfrac{171}{0.7} = 244$ ft$-$kips

Review of Table 13-2 indicates that the 4#9 bars are more than adequate for gravity plus wind loading. For the exterior columns somewhat less reinforcement could be used. Such a redesign is left to the reader.

Use 18 in. x 18 in. section with 4#9 bars for all columns of the first story.

7. Using the investigative option of PCACOL Program, compare design results for the interior and exterior 18 in. x 18 in. columns with 4#9 bars.

The PCACOL program utilizes the moment magnifier method for column slenderness evaluation, similar to the method used in the long-hand calculations of this example.

The on-screen graphic results of PCACOL for the interior columns are reproduced in Fig. 13-16(a). Both the interaction diagram and a scaled column cross section showing vertical reinforcement are displayed. The required load combinations are plotted relative to the strength interaction curve for the 18 in. x 18 in. column size with 4#9 bars. The factored gravity load combinations [(Eq. (9-1)] is shown as point "1." The gravity + wind load combination [Eq. (9-2)] is shown as point "2." Dead + wind load combination [Eq. (9-3)] is also shown as point "3", considerably below the strength provided curve. For completeness, the PCACOL on-screen design options, design parameters, and design results are reproduced in Figs. 13-16(b) through (d).

Example 13.2—Continued

The on-screen graphic results for the exterior columns are reproduced in Fig. 13-17. The required load combinations are plotted relative to the strength interaction curve for the 18 in. x 18 in. column size with 4#9 bars. The factored gravity load combination [Eq. 9-1)] is shown as point "1." The gravity + wind load combinations, Eq. (9-2) and Eq. (9-3), are shown as points "2" and "3", respectively. For the exterior columns, somewhat less reinforcement could be used.

Example 13.2—Continued

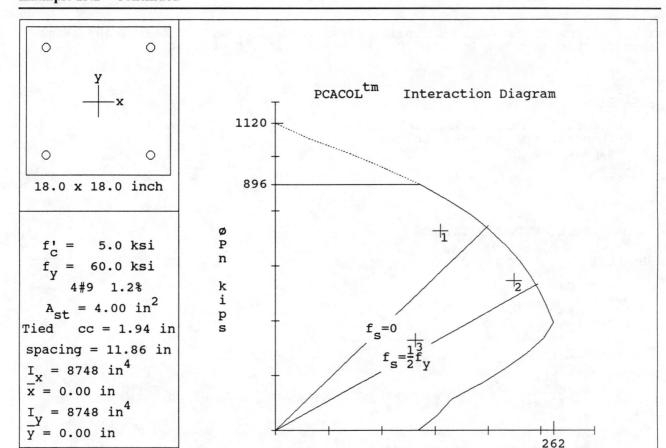

18.0 x 18.0 inch

f'_c = 5.0 ksi
f_y = 60.0 ksi
     4#9   1.2%
A_{st} = 4.00 in^2
Tied   cc = 1.94 in
spacing = 11.86 in
I_x = 8748 in^4
$\bar{x}$ = 0.00 in
I_y = 8748 in^4
$\bar{y}$ = 0.00 in

© 1989 PCA

PCACOL^{tm} Interaction Diagram

1120
896

ϕP_n kips

ϕM_{nx} (ft-k)

262

$f_s = 0$
$f_s = \frac{1}{2}f_y$

Licensed To: Portland Cement Association, Skokie, IL

Project: Example 13.2 Interior   File name: C:\PCACOL\DATA\EX13_2I.COL

Column Id: Slenderness Unbraced   Material Properties:

Engineer: E_c = 4030 ksi ϵ_u = 0.003 in/in

Date: Time: f_c = 4.25 ksi E_s = 29000 ksi

Code: ACI 318-89 β_1 = 0.80

Version: 2.10   Stress Profile : Block

                                        Reduction: ϕ_c = 0.70 ϕ_b = 0.90

Slenderness considered x-axis k_b = 0.78 k_s = 1.30

**Fig. 13-16(a)   PCACOL On-Screen Results for 18 in. x 18 in. Column
with 4#9 Bars (Interior Columns)**

Example 13.2—Continued

11/14/89 PCACOL(tm) V2.10 Proprietary Software of PORTLAND CEMENT ASN. Page 2
16:46:50 Licensed to: Portland Cement Association, Skokie, IL

Titles:

    Project Id.   Example 13.2 Interior
    Column Id.   Slenderness Unbraced
    Engineer
    Date
    Time

File:

    C:\PCACOL\DATA\EX13_2I.COL

Options:

    Investigation
    Slender
    US in-lbs
    ACI 318-89

Properties:

    fċ = 5.0 ksi
    Ec = 4030 ksi
    fc = 4.25 ksi
beta1 = 0.800
    eu = 0.0030 in/in
    fy = 60.0 ksi
    Es = 29000 ksi
 erup = 0.0000 in/in
    Stress Profile: Block

Geometry:

    Concrete:

        Rectangular: Width = 18.00 in
                     Depth = 18.00 in

    Reinforcement:

        Confinement: Tied   phic = 0.70
                               phib = 0.90
                                  a = 0.80

        Layout: Rectangular

        Pattern: All Sides Equal

                    4 #9   Cover = 1.94 in

                    Ast = 4.00 in^2 at 1.23%

Fig. 13-16(b)   PCACOL Design Input Data (Interior Columns)

Example 13.2—Continued

```
11/14/89 PCACOL(tm) V2.10   Proprietary Software of PORTLAND CEMENT ASN. Page 3
16:46:50 Licensed to: Portland Cement Association, Skokie, IL
```

Slenderness:

Sway Criteria:

X-Axis: Not Braced Against Sidesway
        Not Hinged at Either End

Columns:

Column	h (ft)	Width (in)	Depth (in)	I (in^4)	fċ (ksi)	Ec (ksi)
Design	18.0	18.00	18.00	8748	5.0	4030
Above	11.0	18.00	18.00	8748	5.0	4030
Below	11.0	18.00	18.00	8748	5.0	4030

Beams:

Beam Location	l (ft)	Width (in)	Depth (in)	I (in^4)	fċ (ksi)	Ec (ksi)
			X-Axis			
Above Left	30.0	12.00	34.50	41064	5.0	4030
Above Right	30.0	12.00	34.50	41064	5.0	4030
Below Left	30.0	12.00	34.50	41064	5.0	4030
Below Right	30.0	12.00	34.50	41064	5.0	4030

Effective Length Factors:

Axis	Ψ values Top	Bottom	k Values Braced	Sway	klu/r
X	0.000	0.000	0.780	1.300	54.0

Fig. 13-16(c)   PCACOL Design Parameters   (Interior Columns)

Example 13.2—Continued

11/14/89 PCACOL(tm) V2.10 Proprietary Software of PORTLAND CEMENT ASN. Page 4
16:46:50 Licensed to: Portland Cement Association, Skokie, IL

Moment Magnification Factors:

```
------------------------------- X-Axis -----------------------------------
                    Braced                                 Sway
        --------------------------------------    ---------------------------------
Load    Pc     Betad    EI         Cm     Del     Pc      EI          Del
Case   (kips)          (k-in^2)                   (kips)  (k-in^2)
****   ******  *****   **********  *****  *****   ******  **********  *****
1 U1    2443   0.700   7.026e+006  0.756  1.314    1495   1.195e+007   N/A
  U2                               0.756  1.109                        1.991
  U3                               0.753  1.000                        1.426
```

Sum of Pc = 25.70*Pc
Sum of Pu = 24.60*Pu

Service Loads:

```
------------------------------- X-Axis -----------------------------------
       Axial Loads              Moments at Top          Moments at Bottom
Dead    Live    Latl    Dead    Live    Latl    Dead    Live    Latl
(kips)  (kips)  (kips)  (ft-k)  (ft-k)  (ft-k)  (ft-k)  (ft-k)  (ft-k)
******  ******  ******  ******  ******  ******  ******  ******  ******
 363     128      0      17      14      50      -43     -34     -50
```

Factored Load Combinations:

```
U1 = 1.40*Dead + 1.70*Live
U2 = 0.75*(1.40*Dead + 1.70*Live + 1.70*Lateral)
U3 = 0.90*Dead + 1.30*Lateral
```

```
         Applied Loads    Computed Strength
Load      AP      AMX       UP      UMX      UP/AP
Case    (kips)  (ft-k)    (kips)  (ft-k)
****    ******  ******    ******  ******    ******
1 U1     726     155       798     179       1.101
  U2     544     225       571     239       1.051
  U3     327     131       579     238       1.777
```

Program completed as requested.

Fig. 13-16(d)   PCACOL Load Input Data (Interior Columns)

Example 13.2—Continued

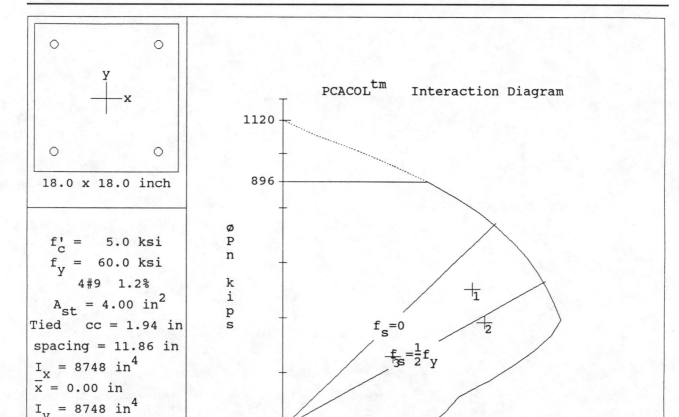

18.0 x 18.0 inch

f'_c = 5.0 ksi
f_y = 60.0 ksi
4#9 1.2%
A_{st} = 4.00 in^2
Tied cc = 1.94 in
spacing = 11.86 in
I_x = 8748 in^4
$\bar{x}$ = 0.00 in
I_y = 8748 in^4
$\bar{y}$ = 0.00 in

© 1989 PCA

PCACOL^{tm} Interaction Diagram

ϕP_n kips

$f_s=0$

$f_s=\frac{1}{2}f_y$

ϕM_{nx} (ft-k)

Licensed To: Portland Cement Association, Skokie, IL

Project: Example 13.2 Exterior   File name: C:\PCACOL\DATA\EX13_2E.COL

Column Id: Slenderness Unbraced   Material Properties:

Engineer: E_c = 4030 ksi ϵ_u = 0.003 in/in

Date: Time: f_c = 4.25 ksi E_s = 29000 ksi

Code: ACI 318-89 β_1 = 0.80

Version: 2.10   Stress Profile : Block

                                      Reduction: ϕ_c = 0.70 ϕ_b = 0.90

Slenderness considered x-axis k_b = 0.85 k_s = 1.57

**Fig. 13-17   PCACOL On-Screen Results for 18 in. x 18 in. Column
with 4#9 Bars (Exterior Columns)**

Shear

UPDATE FOR THE '89 CODE

For ACI 318-89, an upper limit has been placed on all shear strength equations based on concrete strength f_c' ... $\sqrt{f_c'}$ must not exceed 100 psi except in beams and joists where a higher value is permitted provided a minimum amount of shear reinforcement is used as specified in Section 11.1.2.1. Recent tests of concrete beams with concrete strengths exceeding 10,000 psi indicated that shear strength based on $\sqrt{f_c'}$ becomes less conservative as f_c' increases above about 10,000 psi.

GENERAL CONSIDERATIONS

The relatively abrupt nature of a "shear" failure, as compared to a ductile flexural failure, makes it desirable to design members so that strength in shear is equal to, or greater than, strength in flexure. To ensure a ductile flexural failure, the Code (1) limits the minimum and maximum amount of longitudinal reinforcement and (2) except for certain types of construction (Section 11.5.5.1), requires a minimum amount of shear reinforcement in all flexural members if the required shear strength exceeds 50 percent of the design shear strength contributed by the conrete ($\frac{1}{2}\varphi V_c$).

The determination of the amount of shear reinforcement is based on a modified form of the truss analogy. The truss analogy assumes that shear reinforcement resists the total transverse shear. Considerable research has indicated that shear strength provided by concrete V_c can be assumed equal to the shear causing inclined cracking; therefore, shear reinforcement need be designed to carry only the excess shear.

Only shear design for nonprestressed members with clear-span-to-effective-depth ratios greater than 5 is considered in Part 14, including horizontal shear design for composite concrete flexural members. Shear design for deep flexural members is presented in Part 19; for prestressed members, see Part 27.

11.1. SHEAR STRENGTH

Design provisions for shear are presented in terms of shear forces (rather than stresses) to be compatible with the other design conditions for the Strength Design Method, which are expressed in terms of loads, moments, and forces.

Accordingly, shear is expressed in terms of the factored shear force V_u, using the basic shear strength requirement:

$$\text{required shear strength} \leq \text{design shear strength}$$

$$V_u \leq \varphi V_n \qquad \text{Eq. (11-1)}$$

where $$V_n = V_c + V_s \qquad \text{Eq. (11-2)}$$

Substituting for V_n: $$V_u \leq \varphi V_c + \varphi V_s$$

where the design shear strength φV_n is simply the sum of the shear strength provided by concrete, φV_c, plus the shear strength provided by shear reinforcement, φV_s.

Shear strength at any section is computed using Eqs. (11-1) and (11-2), where the factored shear force V_u is obtained by applying the load factors specified in Section 9.2. For gravity loads, $V_u = 1.4\ V_d + 1.7\ V_l$. The strength reduction factor, $\varphi = 0.85$, is specified in Section 9.3.2.3.

The shear strength of a section is increased if a reaction produces compression in the end region of a member. For this condition, the provisions of Section 11.1.3.1 allow sections between the support and a distance "d" from the face of the support to be designed for the same shear force V_u as that computed at a distance "d." Typical support conditions where the factored shear force V_u at a distance "d" from the support can be used include: Members supported by bearing at the bottom of the member, as shown in Fig. 14-1(a); and members framing monolithically into another member, as illustrated in Figs. 14-1(b) and (c).

Support conditions where Section 11.1.3.1 can not be applied include members framing into a supporting member in tension, as shown in Fig. 14-1(d). In this case, the critical section for shear must be taken at the face of the supporting member and the shear within the connection should also be investigated. Also, Section 11.1.3.1 does not apply for shear in columns as shown in Fig. 14-1(e). Although the shear would generally be the same throughout the length of a column, the moment M_u at the face of support must be used if the strength provided by concrete V_c is computed from Eqs. (11-6) and (11-7).

With the 1983 Code, application of Section 11.1.3.1 was further restricted to cases where an abrupt change in shear does not occur between the face of support and distance "d". An example of such a condition is illustrated in Fig. 14-1(f), where a concentrated load is located close to the support. The shear between the support and "d" distance differs radically from that at distance "d." For this case, the maximum shear V_u must be taken at the face of support.

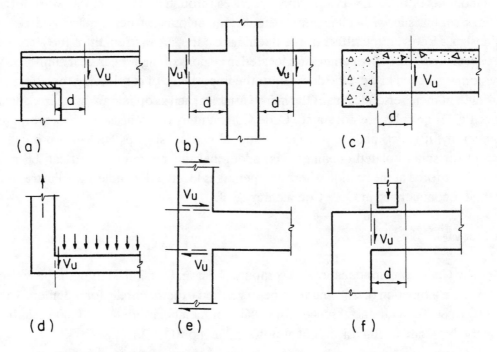

Fig. 14-1 Typical Support Conditions for Locating Factored Shear Force V_u

One other support condition is noteworthy. For brackets and corbels, the shear at the face of the support V_u must be considered, as shown in Fig. 14-2. However, the loading condition is such that the shear-friction provisions of Section 11.7 are applied more appropriately to investigate the shear strength at the face of the bracket support. See Part 17 for design of brackets and corbels.

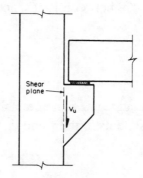

Fig. 14-2 Critical Shear Plane for Brackets

11.1.1.1 Web Openings - Often it is necessary to integrate mechanical and electrical service systems into the structural components of buildings. Passing these services through openings in the webs of the floor beams within the floor-ceiling sandwich eliminates a significant amount of dead space and results in a more economical design. However, the effect of the openings on the shear strength of the floor beams must be considered, especially when such openings are located in regions of high shear near supports. With the 1983 Code a provision was added to alert the designer to the importance of considering the effect of openings on the shear strength of members. Because of the many variables such as opening shape, size, and location along the span, specific design rules are not stated. Code Commentary references are, however, given for design guidance. Generally, it is desirable to provide additional vertical stirrups adjacent to both sides of a web opening, except for small isolated openings. The additional shear reinforcement can be proportioned to carry the total shear force at the section where an opening is located. Example 14.5 illustrates application of a design method recommended in Code Commentary Ref. 11.4.

11.1.2 Limit on $\sqrt{f_c'}$

Chapter 11 of the Code contains equations to compute shear and torsional strengths provided by concrete. These equations are a function of $\sqrt{f_c'}$ and have been verified experimentally for members with compressive stregths of 3000 to 8000 psi. In the absence of test data for members with $f_c' > 10,000$ psi, the values of $\sqrt{f_c'}$ are limited in the '89 Code to 100 psi, except as noted in Section 11.1.2.1.

Section 11.1.2 does not prohibit the use of concrete with $f_c' > 10,000$ psi. It merely directs the engineer not to count on any strength in excess of 10,000 psi when computing the shear and torsional strengths provided by conrete (V_c and T_c).

11.1.2.1 Minimum Web Reinforcement for Beams of High Strength Concrete

For reinforced or prestressed conrete beams and joists, based on some recent unpublished tests, the '89 Code permits $\sqrt{f_c'} > 100$ psi if a minimum amount of web reinforcement is provided. The required minimum amount of web reinforcement is provided. The required minimum amount of web reinforcement is that specified by Sections 11.5.5.3, 11.5.5.4, or 11.5.5.5 multiplied by $f_c'/5000$. The multiplier $f_c'/5000$ need not exceed a value of three.

For $f_c' = 10,000$ psi, minimum A_v computed by Eqs. (11-14), (11-15), or (11-16) is doubled. For $f_c' = 15,000$ psi or larger, minimum A_v is tripled. For f_c' between 10,000 and 15,000 psi, minimum A_v increases linearly as illustrated in Fig. 14-3.

11.2 LIGHTWEIGHT CONCRETE

Since the shear strength of lightweight aggregate concrete may be less than that of normal weight concrete with equal compressive strength, adjustments in the value of V_c, as computed for normal weight concrete, are necessary.

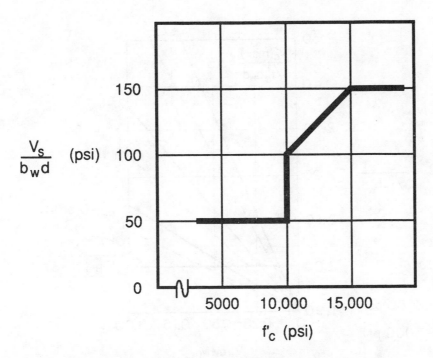

Fig. 14-3 Minimum Required Shear Reinforcement in Beams and Joists per Section 11.1.2.1, Based on Eq. 11-14.

When average splitting tensile strength f_{ct} is specified, $f_{ct}/6.7$ is substituted for $\sqrt{f_c'}$ in all equations for V_c; however, the value of $f_{ct}/6.7$ cannot be taken greater than $\sqrt{f_c'}$. When f_{ct} is not specified, $\sqrt{f_c'}$ is reduced using a multiplier of 0.75 for all-lightweight concrete or 0.85 for sand-lightweight concrete, with linear interpolation allowed when partial sand replacement is used.

11.3 SHEAR STRENGTH PROVIDED BY CONCRETE FOR NONPRESTRESSED MEMBERS

When computing the shear strength provided by concrete for members subject to shear and flexure only, designers have the option of using either the simplified form $V_c = 2\sqrt{f_c'}b_w d$ or the more elaborate expression given by Eq. (11-6). In computing V_c from Eq. (11-6), it should be noted that V_u and M_u are the values which occur simultaneously at the section considered. A maximum value of 1.0 is allowed for the ratio $V_u d/M_u$ for members not subject to axial compression to limit V_c near points of contraflexure (at these points M_u is zero or very small).

For members subject to shear and flexure with axial compression, axial tension, or torsion, simplified V_c expressions are given in Section 11.3.1, with optional more elaborate expressions for V_c available in Section 11.3.2.

Fig. 14-4 shows the variation of shear strength provided by concrete V_c with the ratio given by $V_u d/M_u$ for two values of concrete strength f_c' and reinforcement ratio $\rho_w = 0.5$, 1 and 2 percent.

14-5

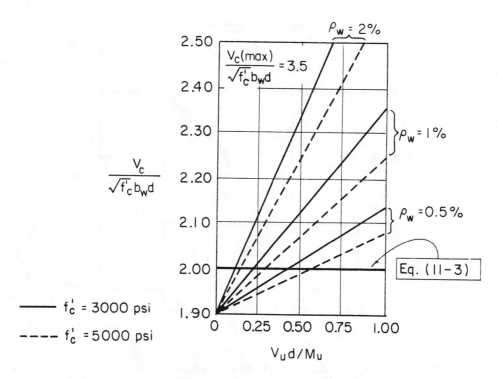

Fig. 14-4 Variation of $V_c/\sqrt{f_c'}b_w d$ with f_c', ρ_w, and Ratio $V_u d/M_u$, as per Eq. (11-6)

Fig. 14-5 shows the approximate range of values of V_c for sections under axial compression, as obtained from Eqs. (11-6) and (11-7). Values correspond to a 6 in. x 12 in. beam section with an effective depth of 10.8 in. The curves corresponding to the alternate expressions for V_c given by Eqs. (11-4) and (11-8), as well as that corresponding to Eq. (11-9) for members subject to axial tension, are also indicated.

Figure 14-6 shows the variation of V_c with N_u/A_g and f_c' for sections subject to axial compression, based on Eq. (11-4). For the range of N_u/A_g values shown, V_c varies from about 49% to 57% of the value of V_c as defined by Eq. (11-8).

Figure 14-7 is a plot of Eq. (11-5) giving the variations of V_c with the ratio $C_t T_u/V_u$ for sections subject to a factored torsional moment T_u greater than $\varphi(0.5\sqrt{f_c'}\Sigma x^2 y)$.

11.5. SHEAR STRENGTH PROVIDED BY SHEAR REINFORCEMENT

11.5.1 Types of Shear Reinforcement

Several types and arrangements of shear reinforcement permitted by Sections 11.5.1.1 and 11.5.1.2 are illustrated in Fig. 14-8. (Note that only vertical stirrups and wire fabric with wires located perpendicular to the axis of the member are permitted for prestressed members.) Vertical stirrups are the most common type of shear reinforcement. Inclined stirrups and longitudinal bent bars are rarely used as they require special care in field placing to locate them in proper positions.

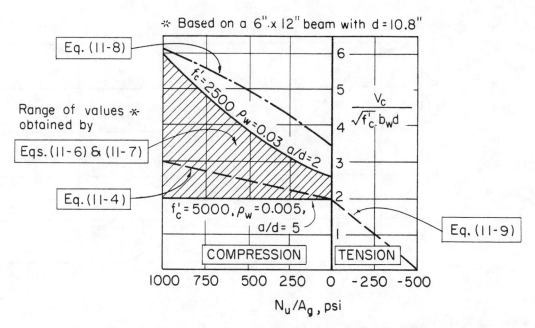

Fig. 14-5 Comparison of Design Equations for Shear and Axial Load

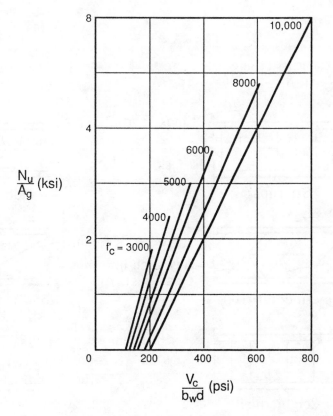

Fig. 14-6 Variation of $V_c/b_w d$ with f_c' and Ratio N_u/A_g as per Eq. (11-4)

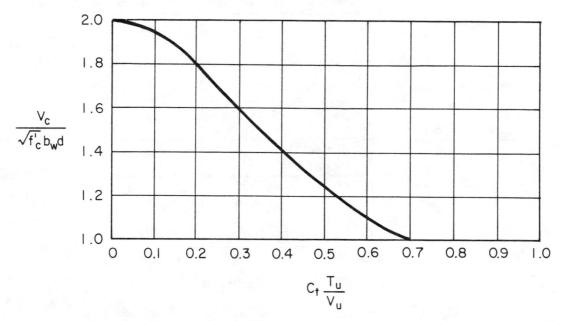

Fig. 14-7 Variation of $V_c/\sqrt{f_c'}$ with Ratio $C_t T_u/V_u$, as per Eq. (11-5)

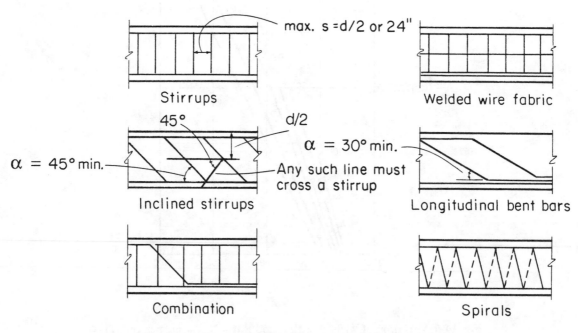

max. s = d/2 or 24"

Stirrups

Welded wire fabric

45°

d/2

$\alpha = 30°$ min.

$\alpha = 45°$ min.

Any such line must cross a stirrup

Inclined stirrups

Longitudinal bent bars

Combination

Spirals

Fig. 14-8 Types and Arrangements of Shear Reinforcement

11.5.3 Anchorage Details for Shear Reinforcement

To be fully effective, shear reinforcement must extend as close to full member depth as cover requirements and proximity of other reinforcement permit (Section 12.13.1), and be anchored at both ends to develop the design yield strength of the shear reinforcement. The anchorage details prescribed in Section 12.13 are presumed to satisfy this development requirement. For the '89 code, U-stirrup anchorage details have been significantly suimplified; see discussion for Section 12.13 in Part 4.

11.5.4 Spacing Limits for Shear Reinforcement

Spacing of vertical stirrups and wire spacing of wire fabric used as shear reinforcement must not exceed one-half the effective depth of the member (d/2), nor 24 in. When the quantity $(V_u - \varphi V_c)$ exceeds $\varphi 4\sqrt{f_c'}b_w d$, maximum spacing must be reduced by one-half (d/4), nor 12 in. Note also that the value of $(V_u - \varphi V_c)$ must not exceed $\varphi 8\sqrt{f_c'}b_w d$. For required shear strength above this value, a larger beam section must be provided, or possibly, the strength of concrete increased to increase the shear strength provided by concrete.

11.5.5 Minimum Shear Reinforcement

When the factored shear force V_u exceeds one-half the shear strength provided by concrete $(V_u > \varphi V_c/2)$, concrete flexural members must be provided with a minimum amount of shear reinforcement, except for slabs and footings, floor joists and wide, shallow beams (Section 11.5.5.1). For this condition, the required minimum shear reinforcement for nonprestressed members is

$$A_v = 50\ b_w s/f_y \qquad \text{Eq. (11-4)}$$

In essence, when minimum shear reinforcement is provided, the total design shear strength of a section is $\varphi V_n = \varphi(V_c + 50 b_w d)$.

Note that spacing of minimum shear reinforcement must not exceed d/2 or 24 in.

11.5.6 Design of Shear Reinforcement

When the factored shear force V_u exceeds the shear strength provided by concrete, φV_c, shear reinforcement must be provided to carry the excess shear. The Code equations expressing area of shear reinforcement A_v are presented in terms of shear strength V_s provided by shear reinforcement for direct application in Eqs. (11-1) and (11-2), rather than A_v directly. To assure correct application of the strength reduction factor,φ, equations for computing required shear reinforcement A_v directly are developed as follows.

When shear reinforcement perpendicular to axis of member is used (vertical stirrups), required area of shear reinforcement A_v spaced at a distance "s" is computed by

$$V_u \le \varphi V_n \qquad \text{Eq. (11-1)}$$

but $$V_n = V_c + V_s$$ Eq. (11-2)

and $$V_s = A_v f_y d/s$$ Eq. (11-17)

Substituting V_s in Eq. (11-2) and V_n in Eq. (11-1):

$$V_u \le \varphi V_c + \varphi A_v f_y d/s$$

Solving for A_v

$$A_v = \frac{(V_u - \varphi V_c)s}{\varphi f_y d}$$

Similarly, when inclined stirrups are used as shear reinforcement,

$$A_v = \frac{(V_u - \varphi V_c)s}{\varphi f_y (\sin\alpha + \cos\alpha)d}$$

where α is the angle between the inclined stirrup and longitudinal axis of member. See Fig. 14-8.

When shear reinforcement consists of a single bar or group of parallel bars, all bent-up at the same distance from the support,

$$A_v = \frac{(V_u - \varphi V_c)}{f_y \sin\alpha}$$

where α is the angle between the bent-up portion and longitudinal axis of member...not less than 30-deg. See Fig. 14-8. Note that the quantity $(V_u - \varphi V_c)$ must not exceed $\varphi 3\sqrt{f_c'} b_w d$.

DESIGN PROCEDURE FOR SHEAR REINFORCEMENT

Briefly, the design of a beam for shear involves the following steps:

1. Determine factored shear forces V_u at critical locations along the length of the member, starting at supports and or at "d" distance from supports per Fig. 14-1. For gravity loading, $V_u = 1.4V_d + 1.7 V_\ell$.

2. Determine shear strength provided by concrete φV_c. Using the simplified form, $V_c = 2\sqrt{f_c'} b_w d$ and $\varphi = 0.85$ for shear.

3. If $(V_u - \varphi V_c)$ exceed $\varphi 8\sqrt{f_c'} b_w d$, increase size of section; also determine distance from support where stirrups are not required $(V_u \le \varphi V_c/2)$.

4. Proportion stirrups to carry excess shear $(V_u - \varphi V_c)$. For vertical stirrups. ..$A_v = (V_u - \varphi V_c)s/\varphi f_y d$ but not less than $50\, b_w s/f_y$ where "s" cannot exceed d/2 or 24 in.

With practical bar sizes for shear reinforcement limited to #3, #4, and #5, it will be more expedient to select a bar size, usually #3 or #4 U-stirrup, and determine required stirrup spacing directly. For vertical stirrups... $s = \varphi A_v f_y d/(V_u - \varphi V_c)$ but not less than $(A_v f_y)/(50 b_w)$ nor d/2 or 24 in.

Generally, only a relatively few controlling sections along the length of a member need be considered; the required spacing of shear reinforcement at intermediate points is usually evident from the computed values at controlling points.

Minimum cost solutions for simple placing are usually limited to three spacings; first stirrup located 2 in. from face of support as a minimum clearance, an intermediate spacing, and finally a maximum spacing, usually d/2.[14.1]

The shear strength requirements are illustrated in Fig. 14-9.

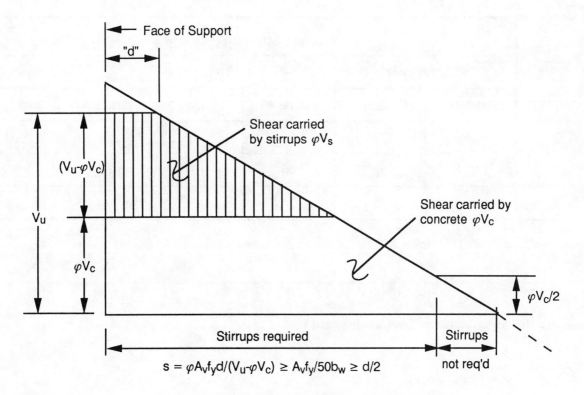

Fig. 14-9 Shear Strength Requirements Illustrated

The expression for shear strength provided by shear reinforcement φV_s can be assigned specific force values for a given stirrup size and strength of reinforcement. The selection and spacing of stirrups can be simplified if the spacing is expressed as a function of the effective depth "d" instead of numerical values. Practical limits of stirrup spacing generally vary from $s = d/2$ to $s = d/4$, since spacing closer than $d/4$ is not economical. With one intermediate spacing at $d/3$, a specific value of φV_s can be derived for each stirrup size and spacing as follows:

For vertical stirrups:
$$\varphi V_s = \varphi A_v f_y d / s \qquad\qquad \text{Eq. (11-7)}$$

substituting d/n for s, where $n = 2, 3,$ and 4

$$\varphi V_s = \varphi A_v f_y n$$

thus, for #3 U-stirrups @ $s = d/2$, $f_y = 60$ ksi and $\varphi = 0.85$

$$\varphi V_s = 0.85(2 \times 0.11)60 \times 2 = 22.44 \text{ kips, say 22 kips}$$

Values of φV_s given in Table 14-1 may be used to select shear reinforcement. Note that the φV_s values are independent of member size and concrete strength. Selection and spacing of stirrups using the design values for $\varphi V_s = (V_u - \varphi V_c)$ can be easily solved by numerical calculation or graphically. See Example 14.1.

Table 14-1 Shear Strength φV_s

Spacing	Shear Strength φV_s kips					
	#3 U-Stirrups*		#4 U-Stirrups*		#5 U-Stirrups*	
	Grade 40	Grade 60	Grade 40	Grade 60	Grade 40	Grade 60
d/2	15	22	27	40	42	63
d/3	22	33	40	61	63	95
d/4	30	45	54	81	84	126

* Stirrups with 2 legs (double values for 4 legs, etc.)

HORIZONTAL SHEAR IN COMPOSITE CONCRETE FLEXURAL MEMBERS

GENERAL CONSIDERATIONS

Horizontal shear in composite concrete flexural members is addressed in Chapter 17.

Horizontal shear forces act over the area of contact between the interconnected concrete surface of composite flexural members. These horizontal shear forces are due to moment gradient resulting from vertical shear forces. Full transfer of these horizontal shear forces at contact surfaces of interconnected elements is required by the Code. Horizontal shear strength must be investigated and provisions must be made to transfer horizontal shear to supporting elements. The Code considers that the strength of a composite member is the same whether or not the first element cast is shored during the casting and curing of the second element.

With publication of ACI 318-83, horizontal shear strength must be investigated in all composite flexural members. When the computed horizontal shear exceeds 350 psi, design for horizontal shear must be in accordance with the shear-friction procedures of Section 11.7.

17.5 HORIZONTAL SHEAR STRENGTH

The upper limits of horizontal shear strength depend upon the contact surface conditions. Horizontal shear strength V_{nh} may be evaluated on the basis of the provisions for intentional roughening or minimum ties, or both. The common requirement for all provisions is that interfacing surfaces must be clean and free of laitance. The horizontal shear strength may not be taken greater than $80b_v d$ for either intentionally roughened surfaces or minimum ties alone, or $350b_v d$ for both roughened surfaces and minimum ties together. Degree of roughness is specified only for the higher permissible shear strength (350 psi). Scoring the surface with a stiff bristled broom is common practice to satisfy the "intentionally roughened" requirement of Section 17.5.2.1. To permit the higher horizontal shear strength, a heavy raking or grooving of the surface is common practice to satisfy the "full 1/4 in. amplitude" requirement of Section 17.5.2.3.

If the horizontal shear exceeds $\varphi 350b_v d$, the section must be designed using the shear-friction method of Section 11.7, with shear-friction reinforcement provided as tie reinforcement for horizontal shear-transfer. For ties perpendicular to interfacing surfaces, required tie area is computed by Eq. (11-26). The friction coefficient μ is taken either as 1.0 for interface intentionally roughened or, 0.6 for interface not intentionally roughened. Note that the 0.6 factor increases the required tie area by 67 percent for unroughened surfaces. Note also that "intentionally roughened" by Section 11.7 means roughened to a full amplitude of 1/4 in., not just "roughened" as permitted by Section 17.5.2.1

17.6 TIES FOR HORIZONTAL SHEAR

According to Section 17.6.3, ties are required to be "fully anchored" into interconnected elements "in accordance with Section 12.13." Figure 14-9 shows some tie details that have been used successfully in testing and design practice. Figure 14-10 shows an extended stirrup detail used in tests of Ref. 14.2. Use of an embedded "hairpin" tie, as illustrated in Fig. 14-10, is common practice in the precast-prestressed industry.

Many precast products are manufactured in such a way that it is difficult to position tie reinforcement for horizontal shear before concrete is placed. Accordingly, the ties are embedded in the plastic concrete as permitted by Code Section 16.4.2.

Example 14.6 illustrates design for horizontal shear.

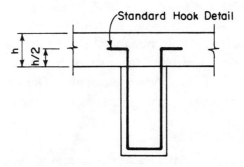

a) Extended Simple U- Stirrups

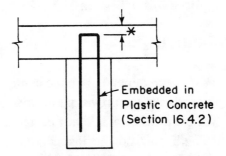

b) Embedded "Hairpin" Ties

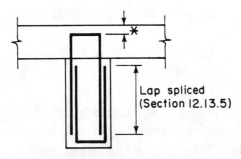

c) Extended Two-Piece U-Stirrups

*   Extended as close as cover and proximity of other reinforcement permit. A 3 in. minimum projection into cast-in-place segment is common practice.

Fig. 14-10 Ties for Horizontal Shear

REFERENCES

14.1 *Selection of Stirrups in Flexural Members for Economy*, PSI Bulletin 7901A, Concrete Reinforcing Steel Institute, Schaumburg, IL.

14.2. Hanson, N.W., *Precast-Prestressed Concrete Bridges 2. Horizontal Shear Connections,* Development Department Bulletin D35, Portland Cement Association, Skokie, 1960.

Example 14.1—Design for Shear - Members Subject to Shear and Flexure Only

Determine required size and spacing of vertical U-stirrups for a 30-foot span, simply supported beam.

b_w = 13 in.
 d = 20 in.
 f_c' = 3000 psi
 f_y = 40,000 psi
w_u = 4.5 kips/ft

Calculations and Discussion	Code Reference

For the purpose of this example, the live load will be assumed to be present on the full span, so that design shear at centerline of span is zero. (A design shear greater than zero at midspan is obtained by considering partial live loading of the span.) Using design procedure for shear reinforcement outlined in this part:

1.   Determine factored shear forces

     @ support: V_u = 4.5 (15) = 67.5 kips

     @ distance d from support:

     V_u = 6.75 – 4.5 (20/12) = 60 kips   11.1.2.1

2.   Determine shear strength provided by concrete

     $\varphi V_c = \varphi 2\sqrt{f_c'} b_w d$ Eq. (11-3)
         $= 0.85(2)\sqrt{3000}$
         $= 0.85(2)\sqrt{3000}\,(13)20 = 24.2$ kips

3.   Determine distance x_c from support beyond which concrete can carry total shear

     x_c = (67.5 – 24.2)/4.5 = 9.6 ft

4.   Determine distance x_m from support over which minimum shear reinforcement must be provided (i.e., up to where $V_u = \varphi V_c/2$).

     x_m = (67.5 – 12.1)/4.5 = 12.3 ft   11.5.5.1

Example 14.1—Continued

Calculations and Discussion	Code Reference

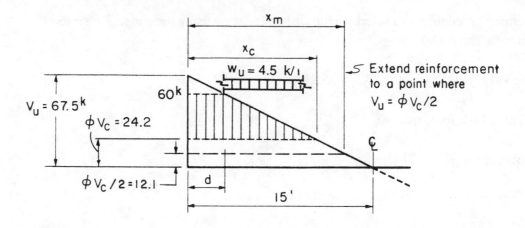

5. Determine required spacing of U-stirrups.   11.5.6.2

    s (req'd) = $\varphi A_v f_y d/(V_u - \varphi V_c)$

    Assuming #4 U-stirrups ($A_v = 0.40$ in.^{2}),

    @ distance d from support:

    s (req'd) = 0.85 (0.40) 40 (20)/(60 – 24.2) = 7.6 in.

    Since $(V_u - \varphi V_c)$ varies linearly between x = d and x = x_c and required spacing varies inversely with $(V_u - \varphi V_c)$, required spacing at any section between these points can be obtained directly from the value of s (req'd) corresponding to x = d.

    For instance, at a section x = d + [(x_c - d)/2] = 5.63 ft from support

    s (req'd) = (7.6) / (1/2) = 15.2 in.

6. Check maximum permissible spacing of stirrups.

    s (max) of vertical stirrups $\leq$ d/2 = 10 in.   11.5.4.1
             or $\leq$ 24 in.

    s (max) of #4 U-stirrups corresponding to minimum reinforcement area requirements

    s (max) = $A_v f_y/50\, b_w$ = 0.40(40,000)/50(13)   Eq. (11-14)
                  = 24.6 in.
    s (max) = 10 in.

Example 14.1—Continued

	Code
Calculations and Discussion	Reference

7. An alternate procdeure is to select stirrup size and spacing using the simplified method presented in Table 14-1.

$(V_u - \varphi V_c)$ at support $= 60 - 24.2 = 35.8$ kips

From Table 14-1, for Grade 40:

#4 U-Stirrups @ d/2 = 27 kips

#4 U-Stirrups @ d/3 = 40 kips

by interpolation, #4 U-Stirrups @ d/2.67 = 35.8 kips

length over which stirrups required $(67.5 - 12.1)/4.5$ = 12.3 ft

length over which #4 @ d/2.67 required $(67.5 - 24.2 - 27.0)/4.5$ = 3.6 ft

length over which #4 @ d/2 required   = 8.7 ft

Use #4 U-Stirrups:
6 @ 7 1/2 in.,
11 @ 10 in.,
from each end of beam.

Summary:

Stirrup spacing using #4 U-stirrups:

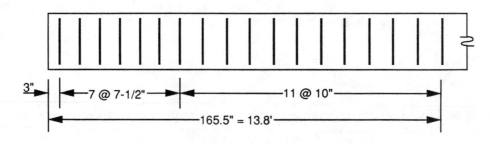

7 stirrups @ 7.5 in.
11 stirrups @ 10 in.

Example 14.2 —Design for Shear - with Axial Tension

Determine required spacing of vertical U-stirrups for a beam subject to axial tension.

f_c' = 3600 psi (sand−lightweight concrete, f_{ct} not specified)
f_y = 40,000 psi
M_d = 43.5 ft−kips
M_ℓ = 32.0 ft−kips
V_d = 12.8 kips
V_ℓ = 9.0 kips
N_d = −2.0 kips (tension)
N_ℓ = −15.2 kips (tension)

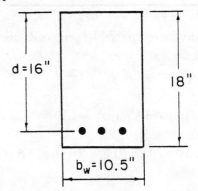

Calculations and Discussion	Code Reference

1. Determine factored loads

    Eq. (9-1)

    $M_u = 1.4(43.4) + 1.7(32.0) = 115.3$ ft−kips

    $V_u = 1.4(12.8) + 1.7(9.0) = 33.2$ kips

    $N_u = 1.4(−2.0) + 1.7(−15.2) = −28.6$ kips (tension)

2. Determine shear strength provided by concrete

    Since average splitting tensile strength f_{ct} is not specified, $\sqrt{f_c'}$ is reduced by a factor of 0.85 (sand-lightweight concrete)

    11.2.1.2

    $$\varphi V_c = 0.85\varphi 2\left[1 + \frac{N_u}{500A_g}\right]\sqrt{f_c'}\,b_w d$$

    $$= 0.85(0.85)2\left[1 + \frac{(-28,600)}{500(18)10.5}\right]\sqrt{3000}\,(10.5)16$$

    $$= 10.2 \text{ kips}$$

3. Check adequacy of cross-section.

    $$(V_u − \varphi V_c) \le 0.85\,(\varphi 8\sqrt{f_c'}\,b_w d)$$

    11.5.6.8

    23.0 kips ≤ 58.3 kips O.K.

Example 14.2 —Continued

	Code
Calculations and Discussion	**Reference**

4. Determine required spacing of U-stirrups

   $s \text{ (req'd)} = \varphi A_v f_y d / (V_u - \varphi V_c)$

   Assuming #3 U-stirrups ($A_v = 0.22$ in.^{2}),

   $s \text{ (req'd)} = 0.85 \, (0.22) \, 40 \, (16)/23.0 = 5.2$ in.

5. Check maximum permissible spacing of stirrups

   $(V_u - \varphi V_c) \leq 0.85 \, (\varphi 4 \sqrt{f_c'} b_w d)$ 11.5.4.3

   23.0 kips < 29.1 kips O.K.

   Provisions of Section 11.5.4.1 apply

   $s \text{ (max)}$ of vertical stirrups $\leq d/2 = 8$ in.   11.5.4.1
   or ≤ 24 in.

   $s \text{ (max)}$ of #3 U-stirrups corresponding to minimum reinforcement area requirements

   $s \text{ (max)} = A_v f_y / 50 \, b_w = 0.22(40,000)/50(10.5) = 16.8$ in.

   $s \text{ (max)} = 8$ in.

   Summary:

   Use #3 vertical stirrups @ 5.0 in. spacing.

Example 14.3 —Design for Shear - with Axial Compression

A tied compression member has been designed for the given load conditions. However, the original design did not take into account the fact that, under a reversal in the direction of lateral load (wind), the axial load, due to the combined effects of gravity and lateral loads, becomes $P_u = 10$ kips, with essentially no change in the values of M_u and V_u. Check shear reinforcement requirements for the column under (1) original design loads and (2) reduced axial load.

$M_u = 86$ ft−kips
$P_u = 160$ kips
$V_u = 20$ kips
$f_c' = 3000$ psi
$f_y = 40,000$ psi

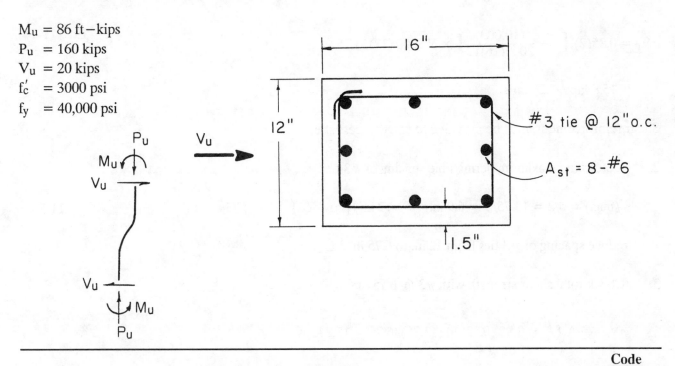

| | Code |
| Calculations and Discussion | Reference |

Condition 1: $P_u = N_u = 160$ kips

1. Determine shear strength provided by concrete

$$d = 16 - [1.5 + 0.375 + (0.750/2)] = 13.75 \text{ in.}$$

$$\varphi V_c = \varphi 2 \left[1 + \frac{N_u}{2000A_g}\right] \sqrt{f_c'} b_w d \qquad \text{Eq. (11-4)}$$

$$\varphi V_c = 0.85(2) \left[1 + \frac{(160,000)}{2000(16)12}\right] \sqrt{3000}\,(12)13.75$$

$$= 21.9 \text{ kips} > 20 \text{ kips}$$

Example 14.3 —Continued

Calculations and Discussion	**Code Reference**

Condition 2: $P_u = N_u = 10$ kips

1.   Determine shear strength provided by concrete.

$$\varphi V_c = 0.85(2) \left[1 + \frac{(10,000)}{2000(16)12}\right] \times \sqrt{3000}\,(12)13.75$$ Eq. (11-14)

$$= 15.8 \text{ kips} < 20 \text{ kips}$$

Shear reinforcement must be provided to carry excess shear.

2.   Determine maximum permissible spacing of #3 ties

s (max) = d/2 = 13.75/2 = 6.9 in.   12 in. (provided)   NG   11.5.4.1

reduce spacing of #3 ties from 12 in. to 6.75 in. o.c.

3.   Check total shear strength with #3 @ 6.75 in.

$$\varphi V_s = \varphi A_v f_y \frac{d}{s} = 0.85\,(0.22)40\,(13.75)/6.75 = 15.2 \text{ kips}$$

$$\varphi V_c + \varphi V_s = 15.8 + 15.2 \ = 31.0 \ \text{kips} \ > 20 \text{ kips}$$

Example 14.4 —Design for Shear - Concrete Floor Joist

Check shear requirements in the uniformly loaded floor joist shown below.

f'_c = 3600 psi

f_y = 40,000 psi

w_d = 58 lb/ft^2

w_l = 120 lb/ft^2

Assumed longitudinal reinforcement:

Two #5 bottom bars

#5 @ 9 in. top bars

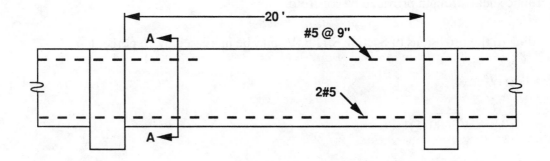

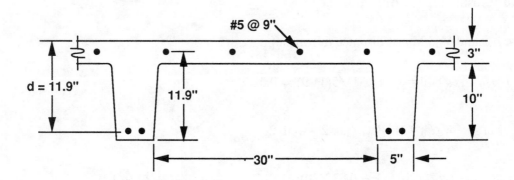

Example 14.4 —Continued

	Code
Calculations and Discussion	**Reference**

1. Determine factored load.

    $w_u = [1.4 (58) + 1.7 (120)] 35/12 = 832$ lb/ft Eq. (9-1)

2. Determine factored shear force.

    @ distance d from support: 11.1.2.1

    $V_u = 0.832 (10) - 0.832 (11.9/12) = 7.5$ kips 8.3.3

3. Determine shear strength provided by concrete.

    According to the provisions of Section 8.11.8, V_c may be increased by 10 percent.

    $\varphi V_c = 1.1\varphi 2\sqrt{f_c'}b_w d$ Eq. (11-3)

    $= 1.1(0.85)2\sqrt{3600}(5)11.9$

    $= 6.7$ kips < 7.5 kips N.G.

Calculate V_c using Eq. (11-6)

Compute ρ_w and $V_u d/M_u$ at distance d from support:

$\rho_w = A_s/b_w d = (4 \times 0.31)/5(11.9) = 0.0208$

$M_u = -w_u \ell_n^2/11 + w_u \ell_n d/2 - w_u d/2$ 8.3.3

$= -0.832(20)/11 + 0.832(20)(11.9/12)/2 - (0.832/2)(11.9/12)^2$

$= -30.3 + 8.3 - 0.4 = -22.4$ ft$-$kips

$V_u d/M_u = 7.5 (11.9/(22.4)(12) = 0.33 < 1$

Example 14.4 —Continued

	Code
Calculations and Discussion	Reference

$$\varphi V_c = 1.1(0.85) \left[1.9\sqrt{3600} + 2500\,(0.0208)0.33\right] \times \left[5(11.9)\right]$$

$$= 7.3 \text{ kips} < 7.5 \text{ kips} \quad \text{N.G.}$$

In accordance with the provisions of Section 8.11.8, the shear strength of concrete joist floor construction may be increased by use of shear reinforcement or by widening the end of ribs. Therefore, an increase in the joist section near the supports will be considered as follows.

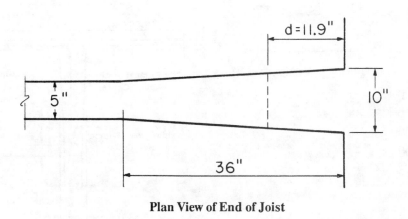

Plan View of End of Joist

Compute b_w at distance d from face of support

$$b_w = 5 + 5\left[(36 - 11.9)/36\right] = 8.3 \text{ in.}$$

Shear strength provided by concrete at distance d from support

$$\varphi V_c = 1.1(0.85)2\sqrt{3600}\,(8.3)11.9 \qquad \text{Eq. (11-3)}$$

$$= 11.1 \text{ kips} > 7.5 \text{ kips} \quad \text{O.K.}$$

Example 14.5 —Design for Shear - Shear Strength at Web Openings

The simply supported prestressed double tee beam shown below has been designed without web openings to carry a live load of 50 lb/ft^2 (w_u = 1520 lb/ft). Two 10-in.-deep by 36-in.-long web openings are required for passage of mechanical and electrical services. Investigate the shear strength of the beam at web opening A.

This design example is based on an experimental and analytical investigation reported in "Behavior and Design of Prestressed Concrete Beams with Large Web Openings," Research and Development Bulletin RD054D, Portland Cement Association, Skokie, IL. (Commentary Ref. 11.4.)

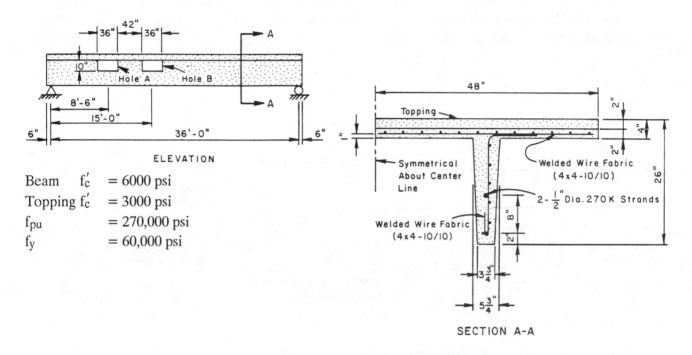

Beam f_c' = 6000 psi

Topping f_c' = 3000 psi

f_{pu} = 270,000 psi

f_y = 60,000 psi

	Code
Calculations and Discussion	**Reference**

This example treats only the shear strength considerations for the web opening. Other strength considerations need to be investigated, such as: to avoid slip of the prestressing strand, openings must be located outside the required strand development length, and strength of the struts to resist flexure and axial loads must be checked. The reader is referred to the complete design example in RD054D for such calculations. The design example in RD054D also illustrates procedures for checking service load stresses and deflections around the openings.

Example 14.5 —Continued

1.   Determine factored moment and shear at center of opening A. Since double tee
      is symmetric about centerline, consider one-half of double tee section.

w_u = 1520/2 = 760 lb/ft per tee

M_u = 0.760 (36/2)8.5– 0.760 (8.5^2)/2

    = 1066 in–kips

V_u = 0.760 (36/2) – 0.760 (8.5)

    = 7.2 kips

2.   Determine required shear reinforcement adjacent to opening. Vertical stirrups
      must be provided adjacent to both sides of web opening. The stirrups should be
      proportioned to carry the total shear force at the opening.

$$A_v = \frac{V_u}{\varphi f_y} = \frac{7200}{0.85 \times 60,000} = 0.14 \text{ in.}^2$$

Use #3 U-stirrup, one on each side of opening (A_v = 0.22 in.^{2})

3.   Using a simplified analytical procedure developed in reference RD054D, the axial
      and shear forces acting on the "struts" above and below opening A are calculated.
      Results are shown on the diagram below. The reader is referred to the complete
      design example in RD054D for the actual force calculations. Axial forces should be
      accounted for in the shear design of the struts.

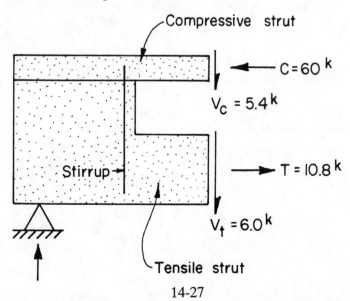

14-27

Example 14.5 —Continued

	Code
Calculations and Discussion	**Reference**

4. Investigate shear strength for tensile strut.

$V_u = 6.0$ kips
$N_u = -10.8$ kips   11.0
   $d = 0.8h = 0.8\,(12) = 9.6$ in.   11.0

b_w = average width of tensile strut = $[\,3.75 + (3.75 + 2 \times 12/22\,)\,]/2$
      = 4.3 in.

$$V_c = 2\Big(1 + \frac{N_u}{500\,A_g}\Big)\sqrt{f_c'}\,b_w d$$ Eq. (11-9)

$$= 2\Big(1 - \frac{10,800}{500 \times 4.3 \times 12}\Big)\sqrt{6000}\,(4.3)(9.6)$$

$$= 3.72 \text{ kips}$$

$\varphi V_c = 0.85\,(3.72) = 3.16$ kips   9.3.2.3

$V_u > \varphi V_c$

6.0 > 3.16 (shear reinforcement required in tensile strut)

$$A_v = \frac{(V_u - \varphi V_c)s}{\varphi f_y d}$$

$$= \frac{(6.0 - 3.16)9}{0.85 \times 60 \times 9.6} = 0.05 \text{ in.}^2$$

where, s = 0.75 h = 0.75 x 12 = 9 in.

Use #3 single leg stirrups at 9-in. centers in tensile strut, ($A_v = 0.11$ in.2). Anchor stirrups   11.5.4.1
around prestressing strands with 180° bend at each end.

5. Investigate shear strength for compressive strut.

$V_u = 5.4$ kips
$N_u = 60$ kips
   $d = 0.8h = 0.8\,(4) = 3.2$ in.
   $b_w = 48$ in.

Example 14.5 —Continued

Calculations and Discussion	**Code Reference**

$$V_c = 2\left(1 + \frac{N_u}{2000\,A_g}\right)\sqrt{f_c'}\,b_w d$$ Eq. (11-4)

$$= 2\left(1 + \frac{60{,}000}{2000 \times 192}\right)\sqrt{3000}\,(48)(3.2)$$

$$= 19.5 \text{ kips}$$ 17.2.3

$$\varphi V_c = 0.85\,(19.5) = 16.5 \text{ kips}$$

$$V_u < \varphi V_c$$

5.4 < 16.5   (shear reiforcement not required in compressive strut)

6.   Design Summary - See reinforcement details below.

(a)   Use U-shaped #3 stirrup adjacent to both edges of opening to contain cracking within the struts.

(b)   Use single-leg #3 stirrups at 9-in. centers as additional reinforcement in the tensile strut.

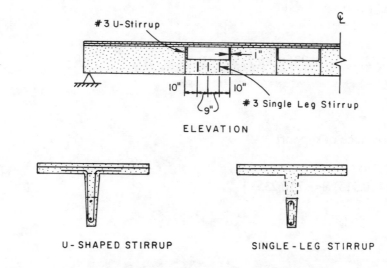

ELEVATION

U-SHAPED STIRRUP   SINGLE-LEG STIRRUP

DETAILS OF ADDITIONAL REINFORCEMENT

A similar design procedure is required for opening B.

Example 14.6 —Design for Horizontal Shear

For the composite slab and precast beam construction shown, design for transfer of horizontal shear at contact surface of beam and slab. Assume beam simply supported with a span of 30 feet.

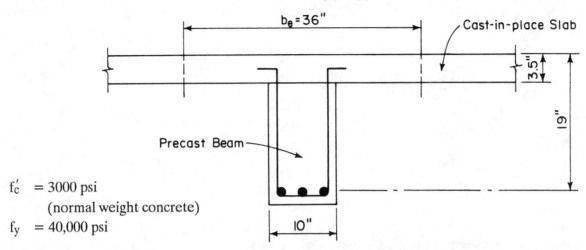

f_c' = 3000 psi
(normal weight concrete)
f_y = 40,000 psi

Calculations and Discussion	Code Reference

Case I: Service dead load = 315 lb/ft
Service live load = 235 lb/ft

1. Determine factored shear force V_u at span end. Eq. (9-1)

$V_u = 1.4D + 1.7L$

$= 1.4\,(0.315)(30/2) + 1.7(0.235)(30/2)$

$= 12.6$ kips

At a distance "d" from face of support 11.1.3.1

$V_u = 12.6 - \dfrac{19}{12}\left[1.4\,(0.315) + 1.7\,(0.235)\right]$

$= 11.3$ kips

2. Determine horizontal shear strength. 17.5.2

$V_u \le \varphi V_{nh}$ Eq. (17-1)

$\le \varphi\,(80 b_v d)$ 17.5.2.1

Example 14.6 —Continued

Calculations and Discussion	Code Reference

$\leq 0.85\ (80 \times 10 \times 19)/1000$ 17.5.2.2

11.3 kips $\leq$ 12.9 kips

Design in accordance with either Section 17.5.2.1 or 17.5.2.2:

If top surface of precast beam is intentionally roughened, no ties are required.   17.5.2.1

If top surface of precast beam is not intentionally roughened, minimum ties are required in accordance with Section 17.6.   17.5.2.2

Note: For either condition, top surface of precast beam must be cleaned and free of laitance prior to placing slab concrete.

Case II: Service dead load = 315 lb/ft
             Service live load   = 1000 lb/ft

1.   Determine factored shear force V_u at span end.

$V_u = 1.4\ (0.315)(15) + 1.7(1.0)(15)$ Eq. (9-1)

    $= 32.1$ kips

At a distance "d" from face of support   11.1.3.1

$V_u = 32.1 - \dfrac{19}{12}\ [1.4(0.315) + 1.7(1.0)]$

2.   Determine horizontal shear strength.   17.5.2

28.7 kips $> \big[\ 12.9$ kips $= (\varphi 80 b_v d)\ \big]$ 17.5.2.1 & 17.5.2.2

    $V_u < \varphi(350 b_v d)$

      $< 0.85\ (350 \times 10 \times 19)/1000$ 17.5.2.3

28.7 kips $\leq$ 56.5 kips

Example 14.6 —Continued

	Code
Calculations and Discussion	**Reference**

Design in accordance with Section 17.5.2.3:

Contact surface must be intentionally roughened to "a full amplitude of approximately 1/4-in.," and minimum ties provided in accordance with Section 17.6.

3. Determine required tie area.   17.6

$$A_v = \frac{50 b_w s}{f_y}$$ Eq. (11-14)

where s = 4(3.5) = 14 in. < 24 in.   17.6.1

$$A_v = \frac{50 \times 10 \times 14}{40,000} = 0.175 \text{ in.}^2 \text{ at 14 in. o.c.}$$
$$\text{or } 0.15 \text{ in.}^2/\text{ft}$$

4. Compare tie requirements with required vertical shear reinforcement at span end.

$V_u = 28.7 \text{ kips}$

$$V_c = 2\sqrt{f_c'} b_w d = 2\sqrt{3000} \times 10 \times 19/1000 = 20.8 \text{ kips}$$ Eq. (11-3)

$$V_u \le \varphi(V_c + V_s)$$ Eq. (11-1)

$$V_u \le \varphi V_c + \varphi A_v f_y \frac{d}{s}$$ Eq. (11-17)

$$\frac{A_v}{s} = \frac{(V_u \; \varphi V_c)}{\varphi f_y d} = \frac{28.7 - 0.85 \times 20.8}{0.85 \times 40 \times 19}$$

$$= 0.0171 \text{ in.}^2/\text{in.}$$

$s_{max} = 19/2 = 9.5 \text{ in.} < 24 \text{ in.}$ 11.5.4.1

$$A_v = 0.0171 \times 9.5 = 0.162 \text{ in.}^2$$

#3 U-stirrups @ 9.5 in. o.c. ($A_v = 0.28$ in.^{2}/ft) which exceeds that required for horizontal shear. Provide #3 U-stirrups @ 9.5 in. o.c. The ties must be adequately anchored into the slab by embedment or hooks. See Fig. 14-10.

Example 14.6 —Continued

Calculations and Discussion	Code Reference

Case III: Service dead load = 315 lb/ft
              Service live load   = 2270 lb/ft

Determine factored shear force V_u at span end.   Eq. (9-1)

$V_u = 1.4 \, (0.315)(15) + 1.7(2.27)(15)$

    $= 64.5$ kips

At distance "d" from support

$V_u = 64.5 - \dfrac{19}{12} \left[1.4 \, (0.315) + 1.7 \, (2.27) \right]$

    $= 57.6$ kips

57.6 kips > $\left[56.5 \text{ kips} = (\varphi 350 b_v d) \right]$ 17.5.2.4

Since V_u exceeds $\varphi \, (350 b_v d)$, design for horizontal shear must be in accordance with
Section 11.7 - Shear-Friction.   Shear along the contact surface between beam and slab
is resisted by shear-friction reinforcement across and perpendicular to the contact surface.

For the required area of tie reinforcement across the interface, the following three alternate methods are
suggested:

Alternative #1:

Using the shear-friction concept of Section 11.7, design for the total required shear-friction
reinforcement from centerline to span end. Converting to a unit stress, the factored horizontal
shear stress at span end is:

$V_{uh} = \dfrac{V_u}{b_v d} = \dfrac{57.6}{10 \times 19} = 303$ psi

Example 14.6— Continued

	Code
Calculations and Discussion	Reference

The shear "stress block" diagram may be shown as follows:

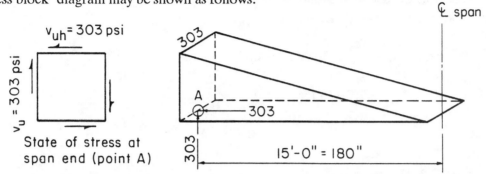

State of stress at span end (point A)

The total horizontal shear transfer between centerline and span end is:

$$V_{uh} = \frac{1}{2} \times 303 \times 180 \times 10 = 272.7 \text{ kips}$$

Required area of shear-friction reinforcement is computed by Eqs. (11-1) and (11-26):

$$V_{uh} \leq \varphi V_n \hspace{5cm} \text{Eq. (11-1)}$$

$$V_{uh} \leq \varphi A_{vf} f_y \mu \hspace{4.2cm} \text{Eq. (11-26)}$$

$$A_{vf} = \frac{V_{uh}}{\varphi f_y \mu}$$

If top surface of precast beam is intentionally roughened to approximately 1/4 in., $\mu = 1.0$.   11.7.4.3

$$A_{vf} = \frac{272.7}{0.85 \times 40 \times 1.0} = 8.02 \text{ in.}^2$$

$$s_{max} = 4 \times 3.5 = 14 \text{ in.} < 24 \text{ in.} \hspace{3cm} 17.6.1$$

Space ties uniformly from centerline of span to end. Use #5 double leg ties (U-stirrups).

$$A_{vf} = 0.31 \times 2 = 0.62 \text{ in.}^2$$

No. of ties required = 8.02/0.62 = 13

Use 13#5 U-stirrups @ 14 in. o.c.

Example 14.6 —Continued

If top surface of precast beam is intentionally roughened, $\mu = 0.6$.

$$A_{vf} = \frac{272.7}{0.85 \times 40 \times 0.6} = 13.37 \text{ in.}^2$$

No. of ties required = 13.37/0.62 = 22

Use 22#5 U-stirrups @ 8 in. o.c.

Note: Final section of ties will depend on beam (vertical) shear requirements.

Alternative #2:

If preferred, a varied tie spacing can be used, based on the actual shape of the horizontal shear distribution. The following method seems reasonable and has been used in the past:

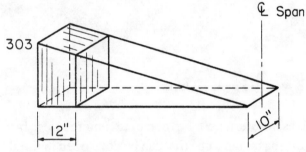

Referring to the horizontal shear stress block above, assume that the horizontal shear is uniform per foot of length, then the shear transfer force for the first foot is:

$$V_{uh} = 303 \times 10 \times 12 = 36.4 \text{ kips}$$

Using $\mu = 1.0$.

$$A_{vf} = \frac{36.4}{0.85 \times 40 \times 1.0} = 1.07 \text{ in.}^2/\text{ft}$$

With #5 double leg stirrups, $A_{vf} = 0.62$ in.^{2}.

$$s = \frac{0.62 \times 12}{1.07} = 6.95 \approx 7 \text{ in.}$$

Use #5 U-stirrups @ 7 in. o.c. for the first 14 in. from span end.

Example 14.6—Continued

	Code
Calculations and Discussion	**Reference**

This method can be used to determine the tie spacing for each successive one-foot length. The shear force will vary at each one-foot increment and the tie spacing can vary accordingly to a maximum of 14 in. toward the center of the span. This method will require more total ties than the 13 computed as based on a uniform spacing from span end to centerline.

Alternative #3:

Using the compressive force developed in the supported element: 17.5.3

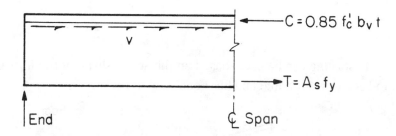

$$V_{uh} = C = 0.85 \times 3 \times 3.5\ 36 = 321.3 \text{ kips}$$

Use the shear-friction methods as in Alternate #1 to compute required ties. In actual design, use the smaller of either C or T. The 321.3 is the maximum force that can be developed in the slab element; the actual depth of the stress block may be less than the 3.5-in. slab depth.

Torsion

11.6 COMBINED SHEAR AND TORSION STRENGTH FOR NONPRESTRESSED MEMBERS WITH RECTANGULAR OR FLANGED SECTIONS

Design for torsion is analagous to that for shear. The factored torsional moment T_u is first computed; then the torsional moment strength T_c provided by the concrete is determined. If this value, modified by the strength reduction factor φ, is less than the factored torsional moment, torsion reinforcement must be provided to carry the stress. Torsion reinforcement, when required, must consist of both closed stirrups and longitudinal bars. Shear reinforcement requirements are added to the torsion requirements to determine size and spacing of combined closed stirrups. Similarly, longitudinal bars for torsion are added to those for flexure and axial force.

The interaction and design requirements for shear and torsion can be represented by elliptical interaction curves such as those shown in Fig. 15-1. Definitions of the Zones of the interaction diagram are given in Decision Table 15-1, which also gives the governing code sections and applicable code equations. In Fig. 15-1, the inner curve represents the strength provided by the concrete to resist combined shear and torsion. The outer curve is the maximum combination of shear and torsion strength permitted by the Code.

11.6.1 When Torsion Must Be Considered

Torsion effects may be neglected when the factored torsional moment T_u is equal to or less than $\varphi(0.5\sqrt{f_c'}\ \Sigma x^2 y)$, corresponding to a torsion stress of $1.5\sqrt{f_c'}$. This will include Zones (1), (2), and (3) in Fig. 15-1. See Fig. 15-2 for evaluation of $\Sigma x^2 y$ for typical member cross sections.

11.6.2 Torsional Moment to Maintain Equilibrium

As discussed in Commentary Sections 11.6.2 and 11.6.3, there are structural framing conditions in which the torsional moment must be resisted by the structural members to maintain equilibrium. In such cases the member must be designed for the full torsional moment, as shown in either Zone (6) or (7) of Fig. 15-1.

Table 15-1—Decision Table for Combined Shear and Torsion

Interaction Zone Fig. 14-1	Design Conditions	Code Reference	Required Reinforcement
(1)	1. $T_u < \varphi(0.5\sqrt{f'_c}\,\Sigma x^2 y)$ Torsion may be neglected 2. $V_u < \varphi V_c/2$	11.6.1 11.5.5.1	None
(2)	1. $T_u < \varphi(0.5\sqrt{f'_c}\,\Sigma x^2 y)$ Torsion may be neglected 2. $\varphi V_c > V_u > \varphi V_c/2$	11.6.1 11.5.5.3	Minimum Shear Only $A_v = \dfrac{50\,b_w s}{f_y}$ (11-14)
(3)	1. $T_u < \varphi(0.5\sqrt{f'_c}\,\Sigma x^2 y)$ Torsion may be neglected 2. $V_u > \varphi V_c$	11.6.1 11.5.6.1	Calculate shear only $A_v = \dfrac{(V_u - \varphi V_c)s}{\varphi f_y d}$ (11-17)
(4)	1. $T_u > \varphi(0.5\sqrt{f'_c}\,\Sigma x^2 y)$ 2. $V_u < \varphi V_c/2$	11.6.1 11.5.5.1	Minimum torsion only $2A_t = \dfrac{50\,b_w s}{f_y}$ (11-16) $A_\ell =$ Eq. (11-24) or (11-25)
(5)	1. $T_u > \varphi(0.5\sqrt{f'_c}\,\Sigma x^2 y)$ 2. $\varphi V_c > V_u > \varphi V_c/2$	11.6.1 11.5.5.5	Minimum combined shear and torsion $A_v + 2A_t = \dfrac{50\,b_w s}{f_y}$ (11-16) $A_\ell =$ Eq. (11-24) or (11-25)
(6) or (7)	1. $T_u > \varphi T_c$ 2. Torsional moment required for equilibrium 3. Design for T_u	11.6.9.1 11.6.2	Calculate combined shear and torsion $A_t = \dfrac{(T_u - \varphi T_c)s}{\varphi f_y\, \alpha_t\, x_1 y_1}$ (11-23) $A_\ell =$ Eq. (11-24) or (11-25)
(6)	1. $T_u > \varphi T_c$ 2. Uncracked section analysis for torsional moment T_u 3. Design for T_u or over design for cracking torque $T_u = \varphi(4\sqrt{f'_c}\,\Sigma x^2 y/3)$	11.6.9.1	Calculate combined shear and torsion $A_t = \dfrac{(T_u - \varphi T_c)s}{\varphi f_y\, \alpha_t\, x_1 y_1}$ (11-23) $A_\ell =$ Eq. (11-24) or (11-25)
(7)	1. $T_u > \varphi T_c$ 2. Redistribution of torsional moment after cracking 3. Design for cracking torque $T_u = \varphi(4\sqrt{f'_c}\,\Sigma x^2 y/3)$	11.6.9.1 11.6.3	Calculate combined shear and torsion $A_t = \dfrac{(T_u - \varphi T_c)s}{\varphi f_y\, \alpha_t\, x_1 y_1}$ (11-23) $A_\ell =$ Eq. (11-24) or (11-25)
(8)	1. $T_u > \varphi\, 5\, T_c$	11.6.9.4	Increase member section

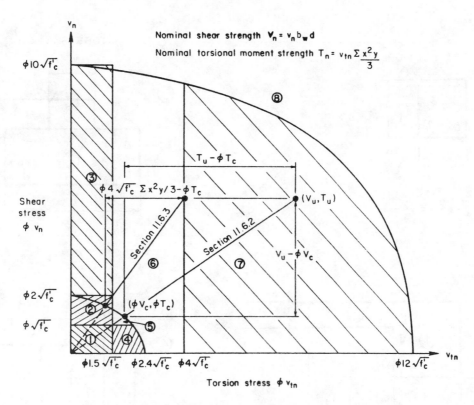

Fig. 15-1 Interaction of Shear and Torsion

11.6.3 Reduction of Torsional Moment

Where torsional moments occur in statically indeterminate framing conditions, the magnitude of the torsional moment will depend on the redistribution of loads between the member under consideration and the interacting structure. If the torsional moment (before redistribution) is greater than $\varphi(4\sqrt{f'_c}\ \Sigma x^2 y/3)$, torsional cracking is assumed. This will permit a large twist in the member for load redistribution while limiting the torsional moment to the cracking value. Thus, for Zones (6) and (7) in Fig. 15-1, the torsional moment for design may be taken conservatively as $\varphi(4\sqrt{f'_c}\ \Sigma x^2 y/3)$, corresponding to a torsional stress of $4\sqrt{f'_c}$. The reduced torsional moment, $\varphi(4\sqrt{f'_c}\ \Sigma x^2 y/3)$, is then used to determine adjusted shears and moment in the adjoining structural members. Note the Commentary caution concerning unusual framing conditions where cracking redistribution may not be realized.

If it is expected that the factored torsional moment will be less than the cracking torque, the structure may be analyzed using equilibrium and compatibility conditions (uncracked section analysis) to determine the torsional moment. If the torsional moment, so determined, is less than the cracking value, $\varphi(4\sqrt{f'_c}\ \Sigma x^2 y/3)$, the calculated value can be used to determine torsion reinforcement requirements. Such cases would fall in Zone (6) in Fig. 15-1.

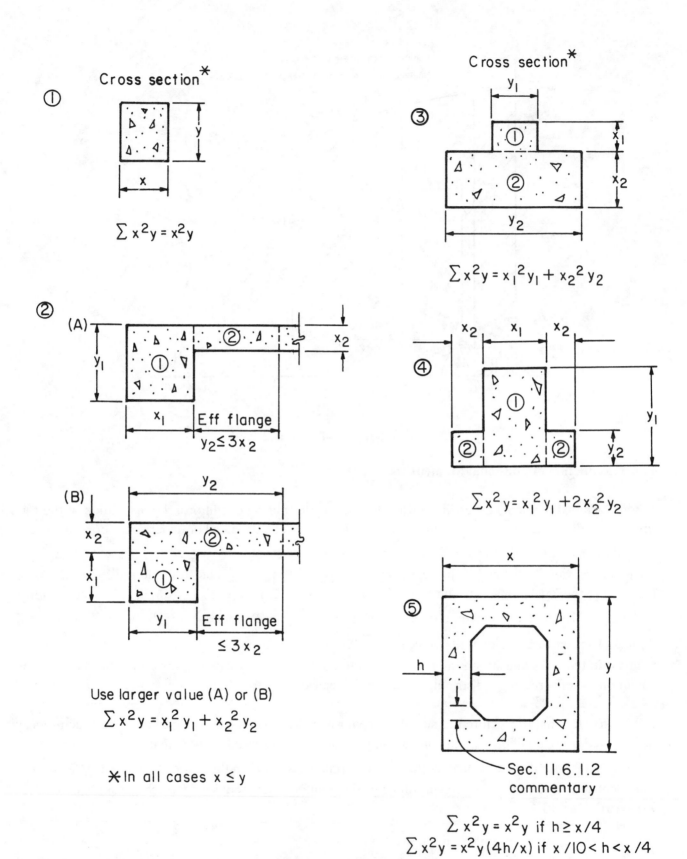

Fig. 15-2 Evaluation of $\Sigma(x^2y)$

15-4

11.6.4 Torsional Moment Near Support

Where torsional moment increases in the vicinity of the support, sections located within distance d from face of support may be designed for the torsional moment T_u at a distance d from face of support.

11.6.5 Torsional Moment Strength

The torsional strength of a section, provided by the combination of concrete and torsion reinforcement, reduced by the strength reduction factor for shear ($\varphi = 0.85$), must be equal to or greater than the factored torsional moment at that section.

The Code provisions for torsion are presented in the same format as that used for shear, with the code requirements expressed in terms of the factored torsional moment T_u directly, using the basic torsional moment strength relation:

required torsional moment strength ≤ design torsional moment strength

$$T_u \leq \varphi T_n \qquad \qquad \text{Eq. (11-20)}$$

$$\leq \varphi T_c + \varphi T_s \qquad \qquad \text{Eq. (11-21)}$$

where the design torsional moment strength φT_n is simply the sum of torsional moment strength provided by concrete φT_c plus torsional moment strength provided by torsion reinforcement φT_s.

11.6.6 Torsional Moment Strength Provided by Concrete

Eq. (11-22) is derived from the elliptical interaction curve for shear and torsion ($2\sqrt{f_c'}$ for shear only and $2.4\sqrt{f_c'}$ for torsion only) as shown in Fig. 15-1, and corresponds to Eq. (11-5) for shear. For a member subject to torsion only, the torsional moment strength provided by concrete is equivalent to a torsional stress of $2.4\sqrt{f_c'}$.

When axial tension is present in the member, the torsional moment strength provided by concrete should be neglected, or the values of Eq. (11-22) and Eq. (11-5) reduced by $(1 + N_u/500A_g)$, where N_u is negative for axial tension. There is no provision for increased torsional moment strength of the concrete when compression is present in the member.

11.6.7 Torsion Reinforcement Requirements

Torsion reinforcement, where required, must consist of closed stirrups, combined with longitudinal bars.

The stirrups required for torsion are added to those required for shear to give the total amount of stirrup reinforcement required for the combined loading. Closed stirrups **must** be used for torsion. U-shaped stirrups as commonly used for shear reinforcement are **not** suitable for use as torsion reinforcement. It is generally most economical to use a single type of stirrup in a particular beam and to combine the shear and torsion reinforcement. In this case all the stirrups must be closed stirrups. When the reinforcement is combined in this way, the most restrictive requirements for spacing and placement of both torsion and shear reinforcement must be met. In large beams, subject to heavy shear loads it may be convenient to provide multiple-legged stirrups to carry the shear and closed perimeter stirrups to carry the torsion. Detailing for closed stirrups used as torsion reinforcement is discussed on page 3-11, with recommended two-piece closed stirrup details illustrated in Figs. 3-9 and 3-10.

The longitudinal bars required for torsion must be distributed around the perimeter of the closed stirrups. At least one longitudinal bar must be placed in each corner of the closed stirrups. The longitudinal torsion bars required near the flexural tension and compression faces of a beam may be combined with the areas of flexural reinforcement required when the reinforcement in these locations is detailed. When detailing the longitudinal torsion bars, due care should be taken to ensure that it is effectively anchored so that its yield strength can be developed as assumed in design. This requires that the longitudinal bars have full development length provided beyond the $(d + b_t)$ distance called for in Section 11.6.7.6. Practically, this will usually mean continuous or lap spliced longitudinal reinforcement for the full length of the member. Note that the torsion reinforcement must be extended a distance equal to the effective depth "d" plus the width of that part of the cross section containing the closed stirrups resisting torsion "b_t."

11.6.8 Spacing Limits for Torsion Reinforcement

Maximum spacings of closed stirrups and longitudinal bars are specified to ensure proper performance of the torsion reinforcement cage. Where combined shear and torsion stirrups are used, the spacing limits for shear requirement given in Section 11.5.4 should be observed.

11.6.9 Design of Torsion Reinforcement

When the concrete section alone is not adequate to resist torsion, reinforcement must be provided in accordance with Eqs. (11-23), (11-24) and (11-25). Similar to shear, the code equation expressing area of torsion reinforcement A_t is presented in terms of torsional moment strength T_s for direct application in Eqs. (11-20) and (11-21), rather than A_t directly. For the usual case of combined shear and torsion, Eq. (11-23) will be more convenient for design in the form:

$$T_u \leq \varphi T_n \hspace{4cm} \text{Eq. (11-20)}$$

$$\leq \varphi (T_c + T_s)$$

$$\text{where } T_s = A_t\, \alpha_t\, x_1\, y_1 f_y/s \hspace{3cm} \text{Eq. (11-23)}$$

$$\text{Therefore } T_u \leq \varphi T_c + \varphi A_t\, \alpha_t\, x_1 y_1 f_y/s$$

which gives the required area, A_t, of **one leg** of a closed stirrup to resist torsion within a distance s. This area can then be combined with the similar shear reinforcement requirement $(A_t/s + A_v/2s)$ to give the required area of one leg of a combined stirrup per unit length of member. For a given stirrup size, the required stirrup spacing can be determined directly.

For flanged members in which closed stirrups are placed in more than one component of the section, Eq. (11-23) in combination with Eqs. (11-20) and (11-21) can be written:

$$\frac{A_t}{s} = \frac{T_u - \varphi T_c}{\varphi f_y \Sigma \alpha_t \, x_1 y_1}$$

assuming that all stirrups will be of the same bar size and spacing. The procedure for combining with shear reinforcement requirements would then be the same as above.

11.6.9.2 Minimum Torsion Reinforcement—Minimum closed stirrup requirements for combined shear and torsion are set in Section 11.5.5.5, and will include Zone (5) in the interaction diagram of Fig. 15-1. When shear force is negligible ($V_u \le \varphi V_c/2$), minimum area of closed stirrups for torsion can be determined from Eq. (11-16):

$$2A_t = \frac{50 b_w s}{f_y}$$

which applies to Zone (4) in Fig. 15-1. In more convenient form this becomes:

$$\frac{A_t}{s} = \frac{25 b_w}{f_y}$$

These minimum values may also apply in the lower strength levels of Zones (6) and (7). Stirrups (and longitudinal reinforcement to resist torsion) are not required in Zone (1), where both shear ($V_u \le \varphi V_c/2$) and torsion ($T_u \le \varphi 0.5\sqrt{f_c'} \, \Sigma x^2 y$) are minimal.

The contribution of the concrete to combined shear and torsion strength of a beam with web reinforcement varies between 40 and 100 percent of the torque at diagonal tension cracking. Therefore, in order that the strength of a cracked beam will be slightly greater than the cracking strength, a minimum amount of shear and torsion reinforcement must be provided to prevent failure at cracking.

In the case of shear without torsion, it has been found that a minimum amount of web reinforcement area equal to $50 \, b_w s/f_y$ will prevent a failure at shear cracking. For pure torsion, the contribution of the concrete to strength after cracking is much less than the cracking torque. Because of this, four times as much minimum web reinforcement is necessary to avoid failure at cracking in the case of pure torsion than in the case of shear, if equal volumes of web and longitudinal reinforcement are provided.

11.6.9.3 Longitudinal Bars Required for Torsion—For the small quantities of reinforcement corresponding to minimum reinforcement, the contribution of the torsion reinforcement to ultimate strength is proportional to the total volume of longitudinal and web reinforcement, and is essentially independent of the ratio of web reinforcement to longitudinal reinforcement. Because of this, the Code is able to specify the same minimum web reinforcement for any combination of torsion and shear; that is, $A_v + 2A_t = 50\ b_ws/f_y$. But this requires the use of more than an equal volume of longitudinal reinforcement for torsion, so that the total volume of stirrups plus longitudinal bars will be sufficient to ensure a ductile failure. This is achieved by use of Eq. (11-25), which also reduces the minimum amount of longitudinal torsion reinforcement as the ratio of torsion to shear decreases. Eq. (11-25) will govern for A_l rather than Eq. (11-24) if A_t is less than $\dfrac{100xs}{f_y}\left(\dfrac{T_u}{T_u + \dfrac{V_u}{3C_t}}\right)$. In Eq. (11-25), the x term is the length of the shorter side of the component rectangle of the cross-section which contains the torsion reinforcement. In a conventionally proportioned beam, x is the width of the web. In a wide, shallow beam, x is the overall depth of the beam.

If different yield strengths for the longitudinal bars and the closed stirrups are used, Eqs (11-24) and (11-25) are modified as follows:

$$A_l = 2A_t\left(\frac{x_1 + y_1}{s}\right) f_{vy}/f_y$$

and

$$A_l = \left[\frac{400sx}{f_{vy}}\left(\frac{T_u}{T_u + \dfrac{V_u}{3C_t}}\right) - 2A_t\ f_{vy}/f_y\right]\left(\frac{x_1 + y_1}{s}\right)$$

where
f_{vy} = yield strength of closed stirrups
f_y = yield strength of longitudinal bars

The validity of the provisions for design of torsion reinforcement depends on the development of the yield strength of the reinforcement at ultimate load. To ensure that yielding will occur, the maximum amount of reinforcement is set by an upper limit on the torsional moment strength of $5T_c$, or

$$\frac{12\sqrt{f_c'}\ \Sigma x^2 y}{3\left[\sqrt{1 + \left(\dfrac{0.4\ V_u}{C_t\ T_u}\right)^2}\ \right]}$$

The corresponding upper limit on shear strength is

$$\frac{10\sqrt{f_c'}\ b_wd}{\sqrt{1 + \left(2.5\ C_t\ \dfrac{T_u}{V_u}\right)^2}}$$

This interaction relationship, shown as the outer curve in Fig. 15-1, limits the shear stress to $10\sqrt{f_c'}$ when torsion is zero, and the torsion stress to $12\sqrt{f_c'}$ when shear is zero.

A DESIGN AID

In accordance with Section 11.6.1, torsion effects **may be neglected** when the factored torsional moment T_u does not exceed $\varphi(0.5\sqrt{f_c'}\ \Sigma x^2 y)$. Also, according to Section 11.6.3, in statically indeterminate framing conditions where reduction of torsional moment in a member can occur due to redistribution of internal forces, maximum factored torsional moment T_u need not exceed $\varphi(4\sqrt{f_c'}\ \Sigma x^2 y/3)$. The nomogram in Fig. 15-3 may be used to quickly check these two limiting torsional moment values.

Input data consists only of the overall dimensions (x and y) of the rectangular parts of the cross section resisting the torsional moment T_u. Input data for x and y, in inches, are located on the two outer vertical scales. A straight line drawn between x and y indicate values of $\varphi(0.5\sqrt{f_c'}\ \Sigma x^2 y)$ and $\varphi(4\sqrt{f_c'}\ \Sigma x^2 y/3)$ in ft-kips directly on the middle scale. When a cross section resisting torsion consists of component rectangles, the value of T_u for the section is simply the sum of the values for each component rectangle. The nomogram is based on $f_c' = 4000$ psi (normal weight concrete).

For f_c' other than 4000 psi and lightweight concrete, values of $\varphi(0.5\sqrt{f_c'}\ \Sigma x^2 y)$ and $\varphi(4\sqrt{f_c'}\ \Sigma x^2 y/3)$ should be multiplied by a correction factor listed in Table 15-2.

Table 15-2   Correction Factors for Concrete Strength and Type of Concrete

f_c'	Normal Weight Concrete	Lightweight Concrete	
		All Lightweight	Sand Lightweight
3000	0.866	0.650	0.736
4000	1.000	0.750	0.850
5000	1.118	0.838	0.950
6000	1.225	0.919	1.041

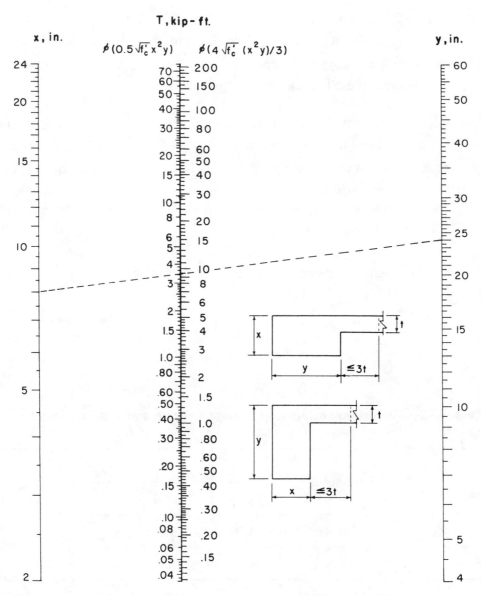

Fig. 15-3 Nomogram for Determining $\varphi(0.5\sqrt{f_c'}\ \Sigma x^2 y)$ and $\varphi(4\sqrt{f_c'}\ \Sigma x^2 y/3)$
(Reproduced from Design Aid EB106 — See Reference 15.1)

REFERENCES

15.1 "Design of Concrete Beams for Torsion," Portland Cement Association, Skokie, EB106D, 1983, 32 pages. Provides design aids in the form of graphs and tables that simplify torsion design requirements of the ACI 318 Code. Fully worked design examples are included to illustrate use of the design aids to effectively consider a number of possible solutions in a short time. Contains flow chart for programming on a hand-held calculator or microcomputer.

Example 15.1—Spandrel Beam Design for Combined Shear and Torsion

Design spandrel beam CI in parking garage second level for combined shear and torsion, assuming sections noted to be adequately sized for flexure and shear. The omission of columns in parking structures in many cases introduces appreciable torsion into spandrel members.

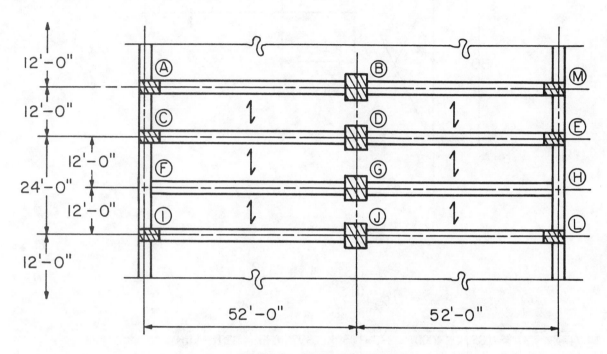

Partial Plan of Parking Garage

Note: Columns omitted at (F) and (H) for entry and exit

<u>Design Criteria:</u>

Typical bay = 12' x 52'
Slab thickness = 4½"
Live load = 50 lb/ft² uniform, or 2 kips concentrated
f'_c = 4000 psi (normal-weight concrete)
f_y = 60,000 psi

Height = 10' (floor to floor)
All beams = 15" x 30"
Exterior columns = 15" x 24"
Interior columns = 24" x 24"

Calculations and Discussion	Code Reference
1. The provisions of Section 11.6.3 greatly simplify the determination of the torsional moment in beam CI, since it is part of an indeterminate framing system in which redistribution of internal forces can occur following torsional cracking. The torsional moment for design can be assumed as $\varphi\left(4\sqrt{f'_c}\,\Sigma x^2 y/3\right)$. Find $\varphi\left(4\sqrt{f'_c}\,\Sigma x^2 y/3\right)$ for beam CI.	11.6.3

Example 15.1—Continued

	Code
Calculations and Discussion	**Reference**

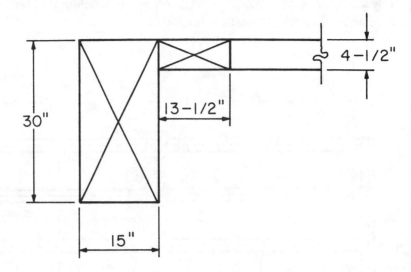

For slab portion
$$y = 3 \times 4.5 = 13.5 \text{ in.}$$

11.6.1.1

$$\Sigma x^2 y = (15^2 \times 30) + (4.5^2 \times 13.5) = 7023 \text{ in.}^3$$

9.3.2.3

$$\varphi = 0.85$$

$$\varphi \left(4\sqrt{f_c'} \Sigma x^2 y/3 \right) = 0.85 \left(4\sqrt{4000} \times \frac{7023}{3} \right) = 503{,}397 \text{ in. lb} = 42 \text{ ft} - \text{kips}$$

Also, using Design Aid—Fig. 15-3:

For x = 15, y = 30 $\Rightarrow$ 40.4
  x = 4.5, y = 13.5 $\Rightarrow$ __1.6__
  $\varphi \left(4\sqrt{f_c'} \Sigma x^2 y/3 \right) = 42.0 \text{ ft} - \text{kips}$

This value must be used in determining the redistribution of moment in beam FG. The resulting reaction at F will determine the shear in beam CI to be used in combination with the torsion.

2. Determine fixed end moments in beam FG.

Service DL $= \left(\dfrac{4.5}{12} \times 12 + \dfrac{25.5 \times 15}{144} \right) 150 = 1073 \text{ lb/ft}$

Service LL $= 50 \times 12 = 600 \text{ lb/ft}$

9.2.1

Factored load, U $= 1.4 \times 1073 + 1.7 \times 600 = 2522 \text{ lb/ft}$

Example 15.1—Continued

Code
References

Calculations and Discussion

$$\text{F.E.M.} \;=\; \frac{w\ell^2}{12} = \frac{2.522 \times 52^2}{12} = 568 \text{ ft}-\text{kips}$$

3. Apply reduced restraining torsional moment from step 1 = 42 ft-kips. Since this torsional moment is applied from both sides of F , the end moment on FG will be 2 × (42) = 84 ft-kips.

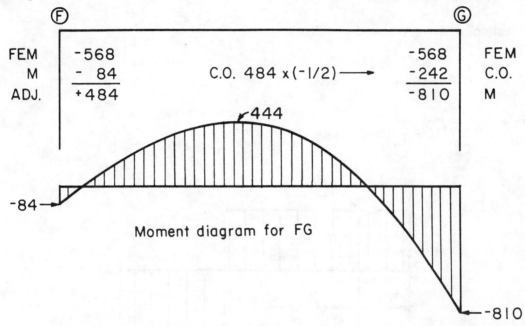

Moment diagram for FG

4. Find beam reaction at F and resulting shear in beam CI.

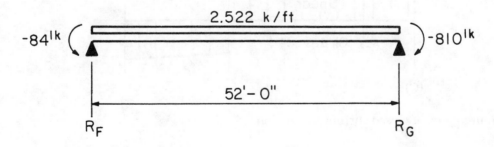

$$\Sigma M_G = 0$$

$$52\, R_F + 810 - 84 - 2.522 \times \left(\frac{52^2}{2}\right) = 0$$

Example 15.1—Continued

Calculations and Discussion	**Code Reference**

$$R_F = \frac{-810 + 84 + 3410}{52} = 51.6 \text{ kips}$$

$$DL \text{ in CI} = \left(\frac{30 \times 15}{144}\right) 150 = 469 \text{ lb/ft} = 0.469 \text{ kips/ft}$$

$$U = 1.4 \times 0.469 = 0.657 \text{ kips/ft} \qquad\qquad 9.2.1$$

Critical section in CI: (Assume d = 27.5 in.)

$$27.5 + \frac{15}{2} = 35 \text{ in. col. } \mathcal{L} \qquad\qquad 11.6.4$$

$$V_u = \frac{51.6}{2} + 0.657\left(12 - \frac{35}{12}\right) = 31.8 \text{ kips}$$

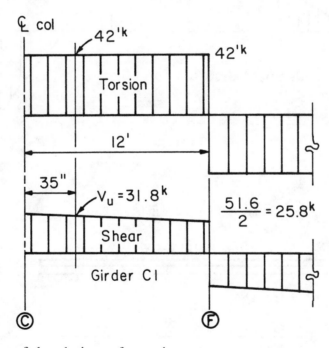

5. Calculate required area of closed stirrups for torsion

$$\frac{A_t}{s} = \frac{T_u - \varphi T_c}{\varphi f_y \, \alpha_t \, x_1 y_1} \qquad\qquad \begin{matrix} 11.6.5 \\ 11.6.9.1 \end{matrix}$$

T_u has been reduced to 42 ft-kips in step 1 above.

Example 15.1—Continued

Calculations and Discussion	**Code Reference**

$$T_c = \frac{0.8\sqrt{f_c'}\,\Sigma x^2 y}{\sqrt{1 + \left(\dfrac{0.4\,V_u}{C_t\,T_u}\right)^2}}$$

Eq. (11-22)

$$C_t = \frac{b_w d}{\Sigma x^2 y} = \frac{15 \times 27.5}{7023} = 0.0587$$

11.0

$$T_c = \frac{0.8\sqrt{4000} \times 7023}{\sqrt{1 + \left(\dfrac{0.4 \times 31.8}{0.0587 \times 42 \times 12}\right)^2}}$$

$$= 27.2 \text{ ft-kips}$$

Assuming $1\frac{1}{2}$ in. cover and #4 stirrup size

7.7.1

$$x_1 = 15 - 2(1.5 + 0.25) = 11.5 \text{ in.}$$

$$y_1 = 30 - 2(1.5 + 0.25) = 26.5 \text{ in.}$$

$$\alpha_t = 0.66 + 0.33\left(\frac{26.5}{11.5}\right) = 1.42$$

11.6.9.1

$$\frac{A_t}{s} = \frac{(42 - 0.85 \times 27.2)12}{0.85 \times 60 \times 1.42 \times 11.5 \times 26.5} = 0.0103 \text{ in.}^2/\text{in./leg}$$

6. Calculate required area of stirrups for shear.

$$V_c = \frac{2\sqrt{f_c'}\,b_w d}{\sqrt{\left(1 + 2.5C_t \dfrac{T_u}{V_u}\right)^2}} = \frac{2\sqrt{4000} \times 15 \times 27.5}{\sqrt{1 + \left(2.5 \times 0.0587 \times \dfrac{42 \times 12}{31.8}\right)^2}}$$

Eq. (11-5)

$$V_c = 20{,}610 \text{ lb} = 20.6 \text{ kips}$$

$$V_u = \varphi(V_c + V_s)$$

Eq. (11-1)
Eq. (11-2)

$$V_s = \frac{V_u}{\varphi} - V_c = \frac{31.8}{0.85} - 20.6 = 16.8 \text{ kips}$$

$$\frac{A_v}{s} = \frac{V_s}{f_y d} = \frac{16.8}{60 \times 27.5} = 0.0102 \text{ in.}^2/\text{in.}$$

Eq. (11-17)

Example 15.1—Continued

	Code
Calculations and Discussion	**Reference**

7. Determine combined shear and torsion stirrup requirements.

$$\frac{A_t}{s} + \frac{A_v}{2s} = 0.0103 + \frac{0.0102}{2} = 0.0154 \text{ in.}^2/\text{in./leg}$$

Try #3 bar, $A_b = 0.11 \text{ in.}^2$

$$s = \frac{0.11}{0.0154} = 7.14 \text{ in.}; \text{ Space #3 closed stirrups at 7 in.}$$

8. Check maximum stirrup spacing.

$$\frac{x_1 + y_1}{4} = \frac{11.5 + 26.5}{4} = 9.5 \ > \ 7 \qquad \text{O.K.}$$ 11.6.8.1

$$\frac{d}{2} = \frac{27.5}{2} = 13.75 \ > \ 7 \qquad \text{O.K.}$$ 11.5.4.1

9. Check requirements at center of span.

$$V_s \ = \ \frac{25.5}{0.85} - 20.6 = 9.75 \text{ kips}$$

$$\frac{A_v}{s} \ = \ \frac{9.75}{60 \times 27.5} = 0.0059$$

$$\frac{A_t}{s} \ = \ \frac{A_v}{2s} = 0.0103 + \frac{0.0059}{2} = 0.0133 \text{ in.}^2/\text{in./leg}$$

$$s \ = \ \frac{0.11}{0.0133} = 8.27 \text{ in.}; \text{ Use } 7-\text{in. spacing full length}$$

10. Check minimum stirrup area. 11.6.9.2

$$A_v + 2A_t = \frac{50 b_w s}{f_y} = \frac{50 \times 15 \times 7}{60,000} = 0.0875 \text{ in.}^2$$ Eq. (11-16)

Area provided $= 2 \times 0.11 = 0.22 \text{ in.}^2 \qquad \text{O.K.}$

11. Calculate longitudinal torsion reinforcement. 11.6.9.3

$$A_\ell = \frac{2A_t}{s}(x_1 + y_1) = 2 \times 0.0103(11.5 + 26.5) = 0.783 \text{ in.}^2$$ Eq. (11-24)

Example 15.1—Continued

Calculations and Discussion

Code Reference

$$A_l = \left[\frac{400 \times s}{f_y} \left(\frac{T_u}{T_u + \dfrac{V_u}{3C_t}} \right) - 2A_t \right] \left(\frac{x_1 + y_1}{s} \right)$$

(or substituting $\dfrac{50b_w s}{f_y}$ for $2A_t$)

$$\frac{50b_w s}{f_y} = 0.0875 \; < \; 2A_t = 2 \times 0.0103 \times 7 = 0.1442$$

Use $2A_t$

$$A_l = \left[\frac{400 \times 15 \times 7}{60,000} \left(\frac{42 \times 12}{42 \times 12 + \dfrac{31.8}{3 \times 0.0587}} \right) - 0.1442 \right] \left(\frac{11.5 + 26.5}{7} \right)$$

$$= 2.01 \text{ in.}^2$$

Provide $A_l = 2.01\text{-in.}^2$ Place longitudinal bars around perimeter of the closed stirrups, spaced at not more than 12 in., and locate one longitudinal bar in each corner of the closed stirrups. Longitudinal bars may be combined with the flexural reinforcement.

12. Flexural analysis of beam CI: (Ignoring flange action)
Consider column-beam joint at (C) :

$$\text{Col. stiffness} = \frac{4EI}{L} = \frac{4E \times 24 \times 15^3}{120 \times 12} = 225E$$
(above)

$$\text{Col. stiffness} = 225E$$
(below)

$$\text{Bm. CI stiffness} = \frac{4EI}{L} = \frac{4E \times 15 \times 30^3}{24 \times 12 \times 12} = 469E$$

$$\text{Bm. AC stiffness} = \frac{4EI}{L} = \frac{4E \times 15 \times 30^3}{12 \times 12 \times 12} = \underline{938E}$$

$$\Sigma = 1857E$$

Dist. Factors at (C) (and at (I)):

$$\text{Col. (above \& below)} = \frac{225}{1857} = 0.121$$

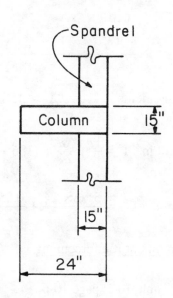

Plan joint (C)

Example 15.1—Continued

CA $\qquad = \dfrac{938}{1857} = 0.505$

CI $\qquad = \dfrac{469}{1857} = 0.253$

F.E.M.:

$$CA = \frac{w\ell^2}{12} = \frac{0.657 \times 12^2}{12} = 8 \text{ ft}-\text{kips}$$

$$CI = \frac{P\ell}{8} + \frac{w\ell^2}{12} = \frac{51.6 \times 24}{8} + \frac{0.657 \times 24^2}{12} = 186 \text{ ft}-\text{kips}$$

Two-step Moment Distribution: (Moments distributed to columns above
and below not shown)

		Ⓒ		Ⓘ	
D.F.	-.505	.253		.253	.505
F.E.M.	− 8	−186		−186	− 8
D.	− 90	+ 45		+ 45	− 90
C.O.	0	− 23		− 23	0
D.	− 12	+ 6		+ 6	− 12
	−110	−158		−158	−110

Final end moments in beam CI = 158 ft-kips

$$\text{Moment in CI } = \frac{P\ell}{8} + \frac{w\ell^2}{24} + (186 - 158)$$

$$= \frac{51.6 \times 24}{8} + \frac{0.657 \times 24^2}{24} + 28 = 199 \text{ ft}-\text{kips}$$

Negative moment reinforcement

Using Table 10-1, page 10-6

$$\text{Compute } \frac{M_u}{\varphi f'_c bd^2} = \frac{158 \times 12}{0.9 \times 4 \times 15 \times 27.5^2} = 0.0464$$

	Code
Calculations and Discussion	**Reference**

From Table 10-2, read $\omega \approx 0.048$

$$A_s = \rho bd = \frac{\omega f_c' \, bd}{f_y} = \frac{0.048 \times 4 \times 15 \times 27.5}{60} = 1.31 \text{ in.}^2$$

Positive moment reinforcement

Compute $\dfrac{M_u}{\varphi f_c' \, bd^2} = \dfrac{199 \times 12}{0.9 \times 4 \times 15 \times 27.5^2} = 0.0585$

From Table 10-2 read $\omega \approx 0.061$

$$A_s = \frac{\omega f_c' \, bd}{f_y} = \frac{0.061 \times 4 \times 15 \times 27.5}{60} = 1.67 \text{ in.}^2$$

13. Size combined longitudinal reinforcement. Eight longitudinal bars are required for torsion reinforcement to meet maximum spacing requirements.   11.6.8.2

Two corner bars (top and bottom) will be combined with the flexural reinforcement.

Positive moment section:

$$\frac{A\ell}{4} + A_s = \frac{2.01}{4} + 1.67 = 2.17 \text{ in.}^2$$

Use 6 – #6 bars, $A_s = 2.64 \text{ in.}^2$

Negative moment section:

$$\frac{A\ell}{4} + A_s = \frac{2.01}{4} + 1.31 = 1.81 \text{ in.}^2$$

Use 6 – #5 bars, $A_s = 1.86 \text{ in.}^2$

Extended positive moment bars: 12.11.1

$$\frac{A\ell}{4} + \frac{A_s}{4} = \frac{2.01}{4} + \frac{1.67}{4} = 0.92 \text{ in.}^2$$

Use 3 – #6 bars, $A_s = 1.32 \text{ in.}^2$

Example 15.1—Continued

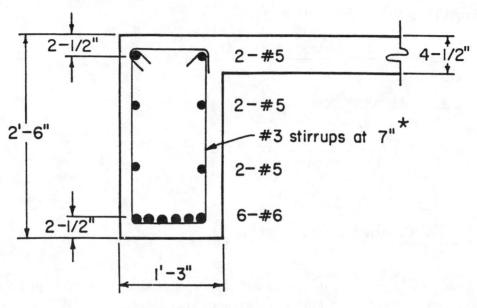

Positive Moment Section

Positive Moment Section

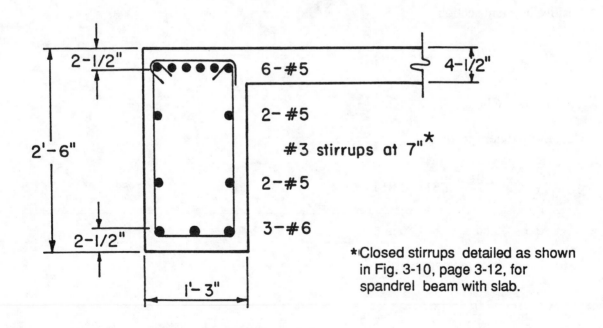

*Closed stirrups detailed as shown in Fig. 3-10, page 3-12, for spandrel beam with slab.

Negative Moment Section

Negative Moment Section

Example 15.1—Continued

Torsion bars in side of beam:

$$\frac{A\ell}{8} = \frac{2.01}{8} = 0.25 \text{ in.}^2$$

Use #5 bar, $A_b = 0.31 \text{ in.}^2$

Torsion bars in top corners of beam:

Extend two negative moment corner bars full length of beam.

Use #5 bar, $A_b = 0.31 \text{ in.}^2$

Example 15.2—Precast Spandrel Beam Design for Combined Shear and Torsion

Design a precast reinforced concrete spandrel beam for combined shear and torsion. Roof members are simply supported on spandrel ledge. Spandrel beams are connected to columns to transfer torsion. Continuity between spandrel beams is not provided.

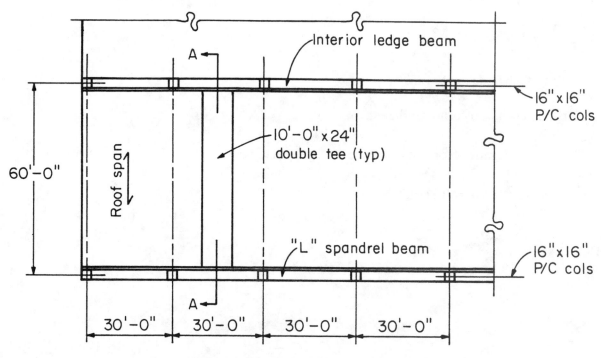

Partial plan of precast roof system

Design Criteria:
Live load   = 30 lb/ft^2
Dead Load = 64 lb/ft^2 (double tee + insulation + roofing)
$\qquad f_c' = $ 5000 psi
$\qquad f_y = $ 60,000 psi

Roof members are 10-ft-wide double-tee units, 24-in. deep. (Design of these units not included in this design example). For lateral support, alternate ends of roof members are fixed to supporting beams.

Calculations and Discussion	Code Reference

1. Assume double tee loading on spandrel beam as uniform. Calculate factored loading M_u, V_u, and T_u for spandrel beam.

Example 15.2—Continued

Calculations and Discussion

Code Reference

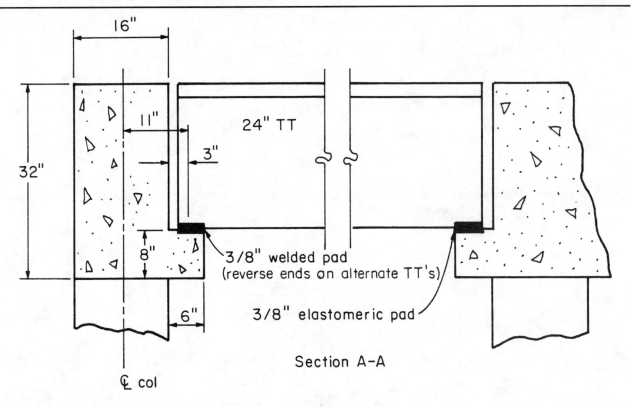

Section A–A

Dead Load:

Superimposed = $0.064 \times 60/2$ = 1.92

Spandrel = $(1.33 \times 2.67 + 0.5 \times 0.67)\, 0.150$ = 0.58

Total = 2.50 kips/ft

Live load = $0.030 \times 60/2 = 0.9$ kips/ft

Factored load = $1.4 \times 2.50 + 1.7 \times 0.9 = 5.03$ kips/ft 9.2.1

At center of span, $M_u = \dfrac{5.03 \times 30^2}{8} = 566$ ft–kips

End shear $V_u = 5.03 \times \dfrac{30}{2} = 75.45$ kips

Torsional factored load = $1.4 \times 1.92 + 1.7 \times 0.9 = 4.22$ kips/ft

End torsional moment $T_u = 4.22 \times \dfrac{30}{2} \times \dfrac{11}{12} = 58.0$ ft–kips

Critical section is at "d" distance from face of support. Assume d = 29.5 in.; critical section is at 29.5 + 8 = 37.5 in. from ₵ of column. 11.6.4, 11.1.3.1

Example 15.2—Continued

Calculations and Discussion	Code Reference

At critical section: $\left(15.0 - \dfrac{37.5}{12} = 11.88 \text{ ft from } \mathbb{C} \text{ of span}\right)$

$V_u = 75.45 \times \dfrac{11.88}{15} = 59.8 \text{ kips}$

$T_u = 58.0 \times \dfrac{11.88}{15} = 45.9 \text{ ft}-\text{kips}$

The spandrel beam must be designed for the full factored torsional moment since it is required to maintain equilibrium.   11.6.2

2. Determine $\Sigma x^2 y$ of spandrel beam section   11.6.1.1

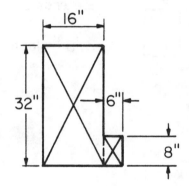

$\Sigma x^2 y = (16^2 \times 32) + 62 \times 8 = 8480 \text{ in.}^3$

3. Check if torsion may be neglected   11.6.1

$\varphi\left(0.5\sqrt{f_c'}\,\Sigma x^2 y\right) = 0.85\left(0.5\sqrt{5000} \times 8480\right)$
$= 254,841 \text{ in.}-\text{lb}$
$= 21.2 \text{ ft}-\text{kips} < T_u = 45.9 \text{ ft}-\text{kips}$

Also, using Design Aid — Fig. 15-3:

For $x = 16$, $y = 32$ $\Rightarrow$ 18.30
For $x = 6$, $y = 8$ $\Rightarrow$ .65
$\qquad \varphi\left(0.5\sqrt{f_c'}\,\Sigma x^2 y\right) \Rightarrow 18.95 \times 1.118 = 21.2 \text{ ft-kips}$

Torsion must be considered.

4. Calculate torsional moment strength provided by concrete.

Example 15-.2—Continued

Calculations and Discussion	Code Reference

$$T_c = \frac{0.8\sqrt{f_c'}\,\Sigma x^2 y}{\sqrt{1 + \left(\dfrac{0.4V_u}{C_t T_u}\right)^2}}$$

Eq. (11-22)

$$C_t = \frac{b_w d}{\Sigma x^2 y} = \frac{16 \times 29.5}{8480} = 0.0557$$

11.0

$$T_c = \frac{0.8\sqrt{5000} \times 8480}{\sqrt{1 + \left(\dfrac{0.4 \times 59.8}{0.0557 \times 45.9 \times 12}\right)^2}} = 378{,}203 \text{ in. lb} = 31.5 \text{ ft}-\text{kips}$$

5. Determine required area of closed stirrups for torsion.

$$\frac{A_t}{s} = \frac{T_u - \varphi T_c}{\varphi f_y\, \alpha_t\, x_1 y_1}$$

11.6.5
11.6.9.1

Assuming 1¼ in. cover and #4 stirrups for exterior exposure

7.7.2

$$x_1 = 16 - 2(1.25 + 0.25) = 13 \text{ in.}$$

$$y_1 = 32 - 2(1.25 + 0.25) = 29 \text{ in.}$$

$$\alpha_t = 0.66 + 0.33\left(\frac{29}{13}\right) = 1.40$$

$$\frac{A_t}{s} = \frac{(45.9 - 0.85 \times 31.5)12}{0.85 \times 60 \times 1.40 \times 13 \times 29} = 0.00851 \text{ in.}^2/\text{in./leg}$$

6. Calculate required area of stirrups for shear.

$$V_c = \frac{2\sqrt{f_c'}\,b_w d}{\sqrt{1 + \left(2.5 C_t \dfrac{T_u}{V_u}\right)^2}} = \frac{2\sqrt{5000} \times 16 \times 29.5}{\sqrt{1 + \left(2.5 \times 0.0557\, \dfrac{45.9 \times 12}{59.8}\right)^2}}$$

Eq. (11-5)

$$V_c = 41{,}062 \text{ lb} = 41.1 \text{ kips}$$

$$V_u \le \varphi(V_c - V_s)$$

Eq. (11-1)

$$V_s = \frac{V_u}{\varphi} - V_c = \frac{59.8}{0.85} - 41.1 = 29.3 \text{ kips}$$

Eq. (11-2)

Example 15.2—Continued

Calculations and Discussion	**Code Reference**

$$\frac{A_v}{s} = \frac{V_s}{f_y d} = \frac{29.3}{60 \times 29.5} = 0.0166 \text{ in.}^2/\text{in.}$$

Eq. (11-17)

7. Determine combined shear and torsion stirrup requirements.

$$\frac{A_t}{s} + \frac{A_v}{2s} = 0.00851 + \frac{0.0166}{2} = 0.0168 \text{ in.}^2/\text{in./leg}$$

Try #3, $A_b = 0.11$ in.^{2}

$$s = \frac{0.11}{0.01679} = 6.55 \text{ in.}$$

8. Check maximum stirrup spacing.

$$\frac{x_1 + y_1}{4} = \frac{13 + 29}{4} = 10.5 \text{ in. or } 12 \text{ in.}$$

11.6.8.1

$$\frac{d}{2} = \frac{29.5}{2} = 14.75 \text{ in. or } 24 \text{ in.}$$

11.5.4.1

$$4\sqrt{f'_c} b_w d = 4 \times \sqrt{5000} \times \frac{16 \times 29.5}{1000} = 133.5 \text{ kips} > V_s = 29.3 \text{ kips} \qquad \text{O.K.}$$

11.5.4.3

Use 6½-in. minimum and 10½-in. maximum spacing.

9. Check minimum stirrup area.

11.6.9.2

$$A_v + 2A_t \geq \frac{50 b_w s}{f_y} = \frac{50 \times 16 \times 6.5}{60,000} = 0.087 \text{ in.}^2$$

Eq. (11-16)

Area provided = $2 \times 0.11 = 0.22$ in.^{2} O.K.

10. Determine stirrup layout.
   Since both shear and torsion are zero at the center of span, and are assumed to vary linearly to the maximum value at the critical section, the start of maximum stirrup spacing can be determined by simple proportion.

$$\frac{s(\text{critical})}{s(\text{maximum})} \times 11.88 = \frac{6.5}{10.5} \times 11.88 = 7.35 \text{ ft}$$

Example 15.2—Continued

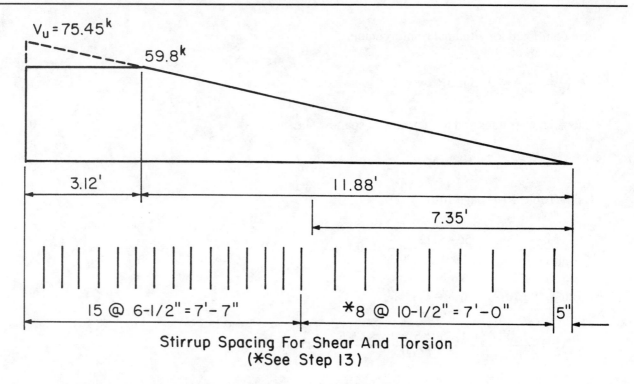

Stirrup Spacing For Shear And Torsion
(✳See Step 13)

11. Calculate longitudinal torsion reinforcement.   11.6.9.3

$$A_l = 2A_t\left(\frac{x_1 + y_1}{s}\right) = 2 \times 0.00851(13 + 29) = 0.715 \text{ in.}^2$$ Eq. (11-24)

$$A_l = \left[\frac{400xs}{f_y}\left(\frac{T_u}{T_u + \dfrac{V_u}{3C_t}}\right) - 2A_t\right]\left(\frac{x_1 + y_1}{s}\right)$$ Eq. (11-25)

(or substituting $\dfrac{50b_w s}{f_y}$ for $2A_t$)

$$\frac{50b_w s}{f_y} = 0.087 \;<\; 2A_t = 2 \times 0.00851 \times 6.5 = 0.1106$$

$$A_l = \left[\frac{400 \times 16 \times 6.5}{60,000}\left(\frac{45.9 \times 12}{45.9 \times 12 + \dfrac{59.8}{3 \times 0.0557}}\right) - 0.1106\right]\left(\frac{13 + 29}{6.5}\right) = 2.00 \text{ in.}^2$$

Provide $A_l = 2.00$ in.^{2} Place longitudinal bars around perimeter of the closed stirrups, spaced at not more than 12 in., and locate one longitudinal bar in each corner of the closed stirrups. Longitudinal bars may be combined with the flexural reinforcement.

Example 15.2—Continued

	Code
Calculations and Discussion	**Reference**

12. Size combined longitudinal reinforcement.

$$\frac{A_\ell}{8} = \frac{2.00}{8} = 0.25 \text{ in.}^2$$

Use #5 bar in sides and top corners of spandrel beam.

$$A_b = 0.31 \text{ in.}^2$$

Using Table 10-1, page 10-6

$$\frac{M_u}{\varphi f_c' bd^2} = \frac{566 \times 12}{0.9 \times 5 \times 16 \times 29.5^2} = 0.1084$$

From Table 10-2, read $\omega = 0.1165$

$$A_s = \frac{\omega f_c' bd}{f_y} = \frac{0.1165 \times 5 \times 16 \times 29.5}{60} = 4.58 \text{ in.}^2$$

At center of span:

$$\frac{A_\ell}{4} + A_s = \frac{2.00}{4} + 4.58 = 5.08 \text{ in.}^2$$

At end of span (extended reinforcement):   12.11.1

$$\frac{A_\ell}{4} + \frac{A_s}{3} = \frac{2.00}{4} + \frac{4.58}{3} = 2.03 \text{ in.}^2$$

Use 4 – #10 bars, $A_s = 5.08 \text{ in.}^2$

Extend 2 – #10 bars to end of girder

$$A_s = 2.54 \text{ in.}^2$$

13. Check required area of beam stirrups used as "hanger" reinforcement for beam ledge. Sufficient stirrups in beam section must be available to act also as hanger reinforcement for the beam ledge. See Part 17 for design of beam ledges.

Example 15.2—Continued

Code
Reference

Calculations and Discussion

Reaction from one double tee stem (5 ft between stems)

$$R_u = (1.4 \times 1.92 + 1.7 \times 0.9)5 = 21.1 \text{ kips/stem}$$

$$A_v(\text{oneleg}) = \frac{R_u}{\varphi f_y} = \frac{21.1}{0.85 \times 60} = 0.414 \text{ in.}^2/\text{stem}$$

Effective width of ledge over which hanger forces can be distributed may be evaluated from Reference 17.1. For the ledge loading dimensions of this example, $b_e = 26$ in.

$$\frac{A_v}{s} = \frac{0.414}{26} = 0.0160 \text{ in.}^2/\text{in.}$$

For #3 stirrups, $s_{max} = \frac{0.11}{0.0160} = 6.9$ in.

The #3 stirrups @ 6½ in. must be used for the full span length to act also as hanger reinforcement for the beam ledge.

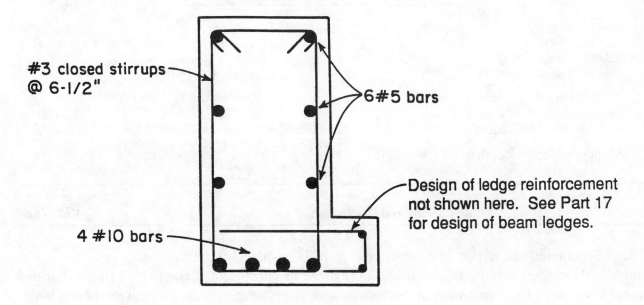

#3 closed stirrups
@ 6-1/2"

6#5 bars

Design of ledge reinforcement
not shown here. See Part 17
for design of beam ledges.

4 #10 bars

Reinforcement Details

Example 15.3—Spandrel Beam Analysis for Combined Shear and Torsion

For the floor slab framing system shown, develop the shear force and torsional moment diagrams for design of the spandrel beams.

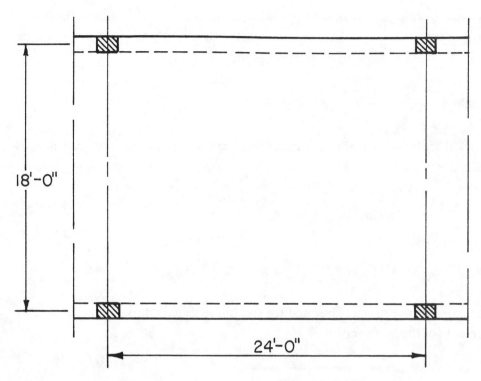

Design Data:

Slab thickness	= 7″ (87.5 lb/ft²)	Live load = 150 lb/ft² uniform	
Spandrel beams	= 24″ x 8″ (200 lb/ft²)	Dead load = 11 lb/ft² superimposed	
Columns	= 8″ x 12″		
f_c'	= 3000 psi		
Spandrel effective depth d = 21.5″			

Calculations and Discussion	Code Reference

In lieu of computing the torsional stiffness of the spandrel beam and the flexural stiffness of the slab to determine the magnitude of the torque to be applied to the spandrel beam, Section 11.6.3 greatly simplifies the determination of torsional moments in members of statically indeterminate framing systems where reduction of torsional moment in a member can occur due to redistribution of internal forces. A maximum factored torsional moment equal to $T_u = \varphi(4\sqrt{f_c'}\,\Sigma x^2 y/3)$ may be assumed at critical sections of such members.

Example 15.3—Continued

Calculations and Discussion	Code Reference

1. For the floor slab system shown, torsional loading from the slab may be assumed uniformly distributed along the spandrel beam (Section 11.6.3.2) with maximum torque $T_u = \varphi(4\sqrt{f_c'}\Sigma x^2 y/3)$ in the beam taken at distance "d" from face of column support (Section 11.6.4) and decreasing linearly to zero at midspan.

For the spandrel beam:

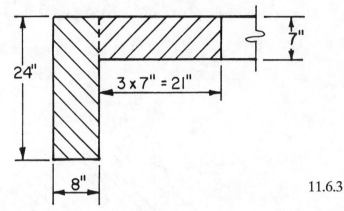

$$T_u = \varphi\left(4\sqrt{f_c'}\frac{\Sigma x^2 y}{3}\right)$$

$$= 0.85\left(4\sqrt{3000}\times\frac{2565}{3}\right)/1000$$

$$= 159.2 \text{ in.}-\text{kips}$$

$$= 13.3 \text{ ft}-\text{kips}$$

where $\Sigma x^2 y = 8^2 \times 24 + 7^2 \times 21 = 2565$ in.^{3}

11.6.3

11.6.1.1

Also, using Design Aid — Fig. 15-3:

For x = 8, y = 24   ⇒   9.4
     x = 7, y = 21   ⇒   6.0
       $\varphi(4\sqrt{f_c'}\Sigma x^2 y/3 = 15.4(0.866) = 13.3$ ft$-$kips

The torsional moment diagram is sketched as follows:

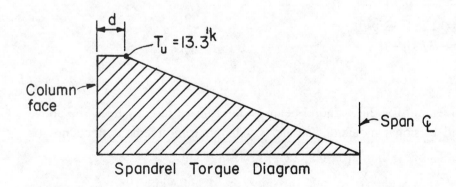

Spandrel Torque Diagram

Example 15.3—Continued

Code Reference

Calculations and Discussion

2 The shear force diagram is sketched as follows:

Slab loading w_u = 1.4D + 1.7L

$\qquad\qquad$ = 1.4(87.5 + 11) + 1.7(150)

$\qquad\qquad$ = 393 lb/ft^2

Beam loading $w_u = 1.4 \left(\frac{8\times24}{144} \times 150 \right) + 393 \times \frac{18}{2} = 3.82$ k/ft

$V_u = 3.82 \left(\frac{23}{2} - \frac{21.5}{12} \right) = 37.1$ kips @ a distance d from face

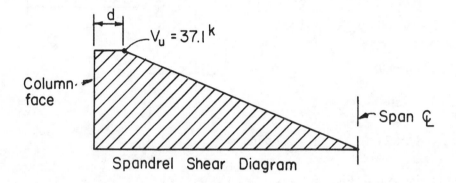

Spandrel Shear Diagram

Slab + $M_u = = \frac{w_u \ell_n^2}{8} - 1.37$

$\qquad\qquad = \frac{393 \, (16.67)^2}{8} - 1.37$

$\qquad\qquad = 12.28$ ft−kips/ft

$-M_u = 1.37$ kip−ft

Torsion reinforcement for the spandrel beam is determined directly from the above shear force and torsional moment diagrams. Torsion reinforcement need not be provided where T_u is less than $\varphi(0.5\sqrt{f_c'} \, \Sigma x^2 y)$.

$T_u < 0.85 \left(0.5\sqrt{3000} \times 2565 \right) \; < \; 59.7$ in.−kips $\; < \; 4.98$ ft−kips

Example 15.3—Continued

	Code
	Reference
Calculations and Discussion	

3. For slab design, the slab moments must be adjusted by the assumed torque from the   11.6.3.1
    spandrel beam.

    Torsional loading on spandrel beams, assumed as uniformly distributed, is approximated as

    $$\frac{2T_u}{\ell} = \frac{2 \times 13.3}{19.4} = 1.37 \text{ ft–kips/ft}$$

    where $\ell = 24 - (12 + 2 \times 21.5)/12 = 19.4 \text{ ft}$

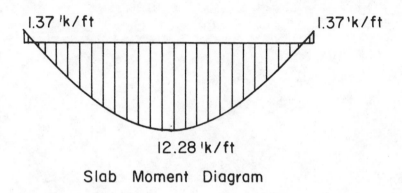

Slab Moment Diagram

Application of Section 11.6.3 for a spandrel beam with torque applied as a single
concentration is illustrated in Example 15.1.

Shear Friction

GENERAL CONSIDERATIONS

With the publication of ACI 318-83, Section 11.7 was completely rewritten to expand the shear-friction concept to include (1) applications where the shear-friction reinforcement is placed at an angle other than 90 degrees to the shear plane, (2) applications where concrete is cast against concrete not intentionally roughened, and (3) applications with lightweight concrete. In addition, a performance statement was added to allow "any other shear-transfer design methods" substantiated by tests. It is noteworthy that Section 11.9 refers to Section 11.7 for the direct shear-transfer in brackets and corbels; see Part 17.

11.7 SHEAR-FRICTION

The shear-friction concept provides a convenient tool for the design of members for direct shear where it is inappropriate to design for diagonal tension, as in precast connections, brackets and corbels. The concept is simple to apply and allows the designer to visualize the structural action within the member or joint. The approach is to assume that a crack has appeared at an unwanted, or unexpected location, as illustrated in Fig. 16-1. As slip begins to occur along the crack surface, the roughness of the crack forces the opposing faces of the crack to separate. This separation is resisted by reinforcement (A_{vf}) across the assumed crack (a separation of only 0.01 in. is sufficient to develop the yield strength of Grade 40 bars). The tensile force ($A_{vf}f_y$) developed in the reinforcement by this strain furnishes an equal and opposite normal clamping force, which in turn generates a frictional force parallel to the crack ($A_{vf}f_y\mu$) to resist further slip.

11.7.1 Applications

Shear-friction design is to be used where direct shear is being transferred across a given plane. Examples of application of the shear-friction concept are shown in Fig. 16-2, including potential crack locations. Successful application of the concept depends on proper selection of location of the assumed crack. Note that in end or edge bearing applications, the crack tends to occur at an angle of about 20 degrees to the direction of force application.

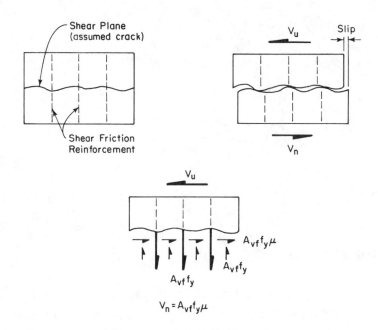

Fig. 16-1 Idealization of the Shear-Friction Concept

11.7.3 Shear-Transfer Design Methods

The shear-transfer design method presented in Section 11.7.4 is based on the simplest model of shear-transfer behavior, and results in a conservative prediction of shear-transfer strength. Other less simple shear-transfer relationships result in a closer prediction of shear-transfer strength. The performance statement of Section 11.7.3 "...any other shear-transfer design methods"... substantiated by tests includes the other methods within the scope and intent of Section 11.7. However, it should be noted that the provisions of Sections 11.7.5 through 11.7.10 apply to whatever shear-transfer method is used. One of the more exact methods is outlined in Commentary Section 11.7.3. Application of the "Modified Shear-Friction Method" is illustrated in Part 17, Example 17.2. Other acceptable methods are presented in publications by the Prestressed Concrete Institute.

11.7.4 Shear-Friction Design Method

As with the other shear design applications, the code provisions for shear-friction are presented in terms of shear-transfer strength V_n for direct application in the basic shear strength relation:

Required shear-transfer strength ≤ Design shear-transfer strength

$$V_u \leq \varphi V_n \qquad \text{Eq. (11-1)}$$

For shear-friction reinforcement perpendicular to shear plane:

$$V_u \leq \varphi A_{vf} f_y \mu \qquad \text{Eq. (11-26)}$$

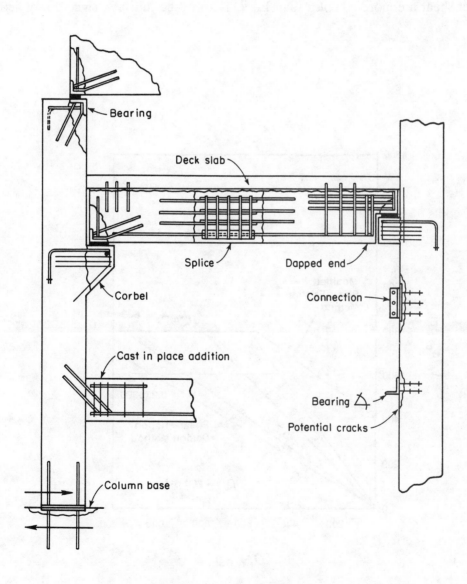

Bearing

Deck slab

Splice

Dapped end

Corbel

Connection

Cast in place addition

Bearing △

Potential cracks

Column base

**Fig. 16-2   Applications of the Shear-Friction Concept and
Potential Crack Locations**

The required area of shear-friction reinforcement can be computed directly from:

$$A_{vf} = \frac{V_u}{\varphi f_y \mu}$$

Or, for shear-friction reinforcement inclined to the shear plane, using Eq. (11-27):

$$A_{vf} = \frac{V_u}{\varphi f_y (\mu \sin\alpha_f + \cos\alpha_f)}$$

The basic Eq. (11-26) for shear-friction strength $V_n = A_{vf} f_y \mu$ reflects an "effective" coefficient of friction μ, representing a conservative lower bound to test data, as shown in Fig. 16-3. The actual mechanics of resistance to direct shear are more complex than Eq. (11-26) would indicate, since dowel action and the

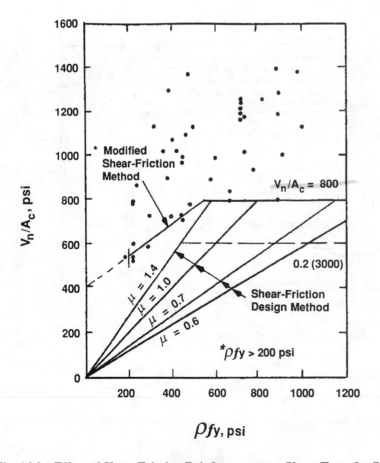

Fig. 16-3   Effect of Shear-Friction Reinforcement on Shear Transfer Strength
(dots indicate test results)

apparent cohesive strength of the concrete both contribute to direct shear strength. The high values for the basic coefficient of friction μ in Section 11.7.4.3 partially account for these discrepancies. The modified shear-friction strength given in Commentary Section 11.7.3 more closely approximates the effects of these factors.

Equation (11-27) is given for design applications where the shear-friction reinforcement crosses the shear-plane at an angle α_f other than 90 degrees, as illustrated in Fig. 16-4. Equation (11-27) resolves the total tensile force $A_{vf} f_y$ into two components: (1) a clamping component $A_{vf} f_y \sin\alpha_f$ with an associated frictional force $A_{vf} f_y \sin\alpha_f \mu$, plus (2) a slip-resisting component contributed directly by the inclination of the shear-friction reinforcement $A_{vf} f_y \cos\alpha_f$.

Note that Eq. (11-27) applies only when the shear force V_u produces tension in the shear-friction reinforcement.

11.7.4.3   Coefficient of Friction—The "effective" coefficients of friction, μ, for the various interface conditions include a parameter λ which accounts for the somewhat lower shear strength of all-lightweight and sand-lightweight concretes; for example, the μ value for all-lightweight concrete placed against hardened concrete not intentionally roughened is 0.6(0.75) = 0.45. Also, for lightweight concretes, Section 11.9.3.2.2 limits the shear-transfer strength V_n along the shear plane for design applications with low shear span-to-depth ratios a/d, such as brackets and corbels. This further restriction on lightweight concrete is illustrated in Example 16.1.

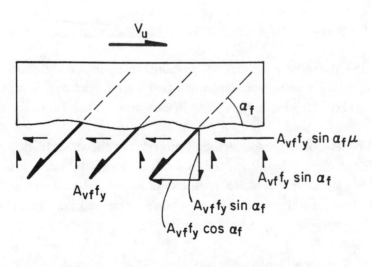

Fig. 16-4   Idealization of Inclined Shear-Friction Reinforcement

11.7.5 Maximum Shear-Transfer Strength

The shear-transfer strength V_n cannot be taken greater than $0.2f_c'$, nor 800 psi times the area of concrete section resisting shear transfer. This upper limit on V_n effectively limits the maximum reinforcement, as shown by Fig. 16-3.

11.7.7 Normal Forces

Equations (11-26) and (11-27) assume that there are no forces other than shear acting on the shear plane. A certain amount of moment is almost always present in brackets and corbels and other connections due to eccentricity of loads or applied moments at connections. Most joints also carry a significant amount of tension due to restrained shrinkage or thermal shortening of the connected members. A direct tensile force of at least $0.2V_u$ is required for design of connections such as brackets or corbels (Section 11.9.3.4), unless the actual force is accurately known. Friction of bearing pads, for example, can cause appreciable tensile forces on a corbel supporting a member subject to shortening. Reinforcement must be provided for direct tension according to Section 11.7.7, using $A_s = N_{uc}/\varphi f_y$, where N_{uc} is the factored tensile force, with φ taken as 0.85 (see Section 11.9.3.1).

Tensile reinforcement required for bending due to eccentricity of the loads, or other causes, is sized in the normal manner. Note that Section 11.9.3.1 specifies a φ of 0.85 for all design calculations in accordance with Section 11.9 (brackets and corbels). Special consideration should also be given to development of the full strength of the flexural reinforcement.

Since direct tension perpendicular to the shear plane (crack face) detracts from the shear-transfer strength, it follows that compression will add to the strength. This is now provided for in Section 11.7.7 which allows the use of "permanent net compression" as an additive force to the shear-friction clamping force. It is advisable, though not required, to use a load factor of 0.9 with such compressive loads.

11.7.8 - 11.7.10 Reinforcement Details

Section 11.7.8 requires that the shear-friction reinforcement be "appropriately placed" along the shear plane. Where no moment acts on the shear plane, uniform distribution of the bars is proper. Where a moment exists, the reinforcement should be distributed in the flexural tension portion of the shear plane.

Reinforcement should be adequately embedded on both sides of the shear plane to develop the full yield strength of the bars. Since space is limited to thin walls, corbels, and brackets, it is often necessary to use special anchorage details such as welded plates, angles, or cross bars. Reinforcement should be anchored in confined concrete. Confinement may be provided by beam or column ties, "external" concrete, or special added reinforcement.

In Section 11.7.9, the term "intentionally roughened" concrete at the interface is defined as "roughened to a full amplitude of approximately $\frac{1}{4}$ in." This can be accomplished by raking the plastic concrete or by bushhammering or chiseling hardened concrete surfaces.

A final requirement of Section 11.7.10, often overlooked, is that structural steel interfaces must be clean and free of paint. This requirement is based on tests to evaluate the friction coefficient for concrete anchored to unpainted structural steel by studs or rebars ($\mu = 0.7$). Data are not available for painted surfaces. If painted surfaces are to be used, a lower value of μ would be appropriate.

DESIGN EXAMPLES

In addition to Examples 16.1 and 16.2, shear-friction design is also illustrated for direct shear-transfer in brackets and corbels (see Part 17), and horizontal shear transfer between composite members (see Part 14) and at column/footing connections (see Part 24).

Example 16.1—Shear-Friction Design

A tilt-up wall panel is subject to seismic shear forces as shown. The wall also shortens due to temperature and shrinkage changes causing a strain of 0.0005 in./in. Design the shear connectors shown below. The connectors are detailed to minimize build-up of tensile forces in the connection due to wall shortening. Use lightweight concrete, w_c = 95 pcf. f_c' = 4000 psi and f_y = 60,000 psi. Panel width = 16 ft.

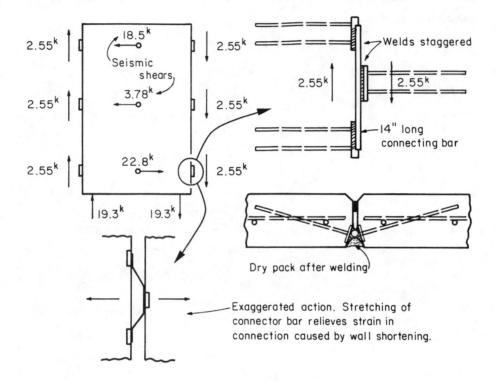

Calculations and Discussion	**Code Reference**

1. Calculate tension in "connecting" bar due to shear forces and wall shortening of 0.0005 in./in. Try #4 connecting bar (A_b = 0.20 in.^{2}).

   For 16-ft wide panel:

   Total strain at joint = 0.0005 (16)(12) = 0.096 in.

Example 16.1—Continued

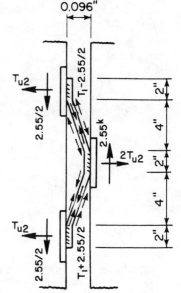

Length of 4-in. portion of bar between welds after straining =

$$\sqrt{4^2 + 0.096^2} = 4.00115183 \text{ in.}$$

$$\varepsilon = \frac{4.00115183 - 4.0}{4.0} = 0.00028796 \text{ in./in.}$$

$$T_1 = E_s \varepsilon_s A_b = 29 \times 10^6 \, (0.00028796)(0.20) = 1670 \text{ lb}$$

According to the general building code, based on the 1988 edition of the *Uniform Building Code,*

$$U = 1.4(D + L + E)$$

Shrinkage and temperature forces should be included as live load due to their unpredictable nature. The total shear force is assumed as divided between both ends of the connector bar.

$$T_{u1} = 1.4 \left(1670 + \frac{2550}{2} \right) = 4123 \text{ lb max.}$$

$$\varphi T_n = 0.85 \, (60,000)(0.20) = 10,200 \text{ lb}$$

$$T_{u1} = 4123 \text{ lb} \le T_n = 10,200 \text{ lb}$$

Therefore, #4 Connector bar is O.K. (#3 would also have been adequate.)

Force in the connection plates normal to the shear plane can be determined from geometry of detail.

$$T_{u2} = 1.4 \, (1670) \left(\frac{0.096}{4.0} \right) = 56 \text{ lb (negligible)}$$

2. Size end welds for $T_{u1} = 4123$ lb

$$T_{uw} = \varphi \, (25,000) \, \ell_w t_w = 0.70(25,000) \, \ell_w (0.25) = 4375 \, \ell_w$$

$$\ell_w = \frac{4123}{4375} = 0.94 \text{ in.} \qquad \text{Use } 1\frac{1}{4} \text{ in. welds at ends}$$

At center weld, 2.0 (2550) = 5100 lb must be developed.

$$\ell_w = \frac{5100}{4375} = 1.17 \text{ in.} \qquad \text{Use } 1\frac{1}{4} \text{ in. weld at center.}$$

Example 16.1—Continued

	Code
Calculations and Discussion	**Reference**

Check distance between welds.

$$\frac{14 - 1.25 - 1.25 - 1.25}{2} = 5.1 \text{ in.} > 4.0 \text{ in.} \quad \text{OK}$$

3. Design anchor plates using shear-friction method

    Center plate is most heavily loaded. Try 2 in. x 4 in. x ¼ in. plate.

    $V_u = 1.4E = 1.4(2550) = 3570 \text{ lb}$ Eq. (11-1)

    $V_u \leq \varphi V_n$ Eq. (11-6)

    $V_u \leq \varphi \, (A_{vf} \, f_y \, \mu)$

    Solving, $A_{vf} = \dfrac{V_u}{\varphi f_y \mu} = \dfrac{3570}{0.85(60,000)(0.525)} = 0.13 \text{ in.}^2$

    where, for lightweight concrete (95 pcf), $\mu = 0.75(0.7) = 0.525$ 11.7.4.3

    Add reinforcement for direct tensile force.

    $2T_{u2} = 2(56) = 112 \text{ lb}$ (negligible)

    Use 2 #3 bars per plate $(A_{vf} = 0.22 \text{ in.}^2)$

    Weld bars to plates to develop full f_y.

    Check maximum shear transfer strength permitted for connection. For lightweight   11.9.3.2.2
    aggregate concrete:

    $$V_{n(max)} \, k = \left[0.2 - 0.07\left(\frac{a}{d}\right)\right] f_c' A_c \text{ or } \left[800 - 280\left(\frac{a}{d}\right)\right] A_c$$

    Use $\dfrac{a}{d} \approx \dfrac{0.25}{2.5} = 0.1$ $A_c = 2 \times 4 = 8 \text{ in.}^2$

    $V_{n(max)} = [0.2 - 0.07(0.1)] \times 4000(8) = 6176 \text{ lb}$ Eq. (11-1)

    or $V_n = (800 - 280 \times 0.1)(8) = 6176 \text{ lb}$

    $\varphi V_{n(max)} = 0.85 (6176) = 5250 \text{ lb}$

    $V_u = 3570 \text{ lb} \leq \varphi V_{n(max)} = 5250 \text{ lb}$ OK

    Use 2 in. x 4 in. x ¼ in. plates

Example 17.3—Beam Ledge Design

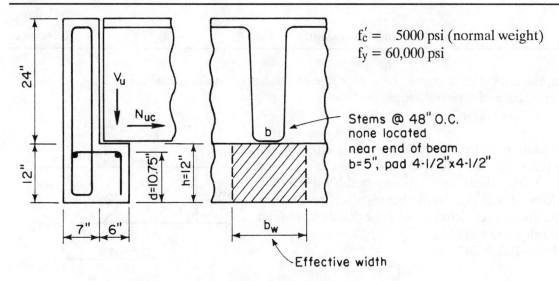

$f'_c = $ 5000 psi (normal weight)
$f_y = 60,000$ psi

Stems @ 48" O.C.
none located
near end of beam
b=5", pad 4-1/2"x4-1/2"

Effective width

The L-beam shown is to support a double-tee parking deck spanning 64 ft. Maximum service loads per stem are: DL = 11.1 kips; LL = 6.4 kips; total load = 17.5 kips. The loads may occur at any location on the L-beam ledge except near beam ends. The stems of the double-tees rest on 4.5 in. x 4.5 in. x 1/4 in. neoprene bearing pads (1000 psi maximum service load).

Design in accordance with the Code provisions for brackets and corbels may require a wider ledge than the 6 in. shown. To maintain the 6-in. width, one of the following may be necessary: (1) Use of a higher strength bearing pad (up to 2000 psi); or (2) Anchoring primary ledge reinforcement A_s to an armor angle.

This example will be based on the 6-in. ledge with 4.5-in.-square bearing pad. At the end of the example an alternative design will be shown.

Calculations and Discussion	Code Reference

1. Check 4.5 × 4.5 bearing pad size (1000 psi maximum service load).
   4.5 × 4.5(1) = 20.25 kips > 17.5 kips   O.K.

2. Check concrete bearing strength.

   V_u = 1.4(11.1) + 1.7(6.4) = 26.42 kips   Eq. (9-1)

   $\varphi P_{nb} = \varphi(0.85 f'_c A_1)$ 10.15.1
   $= 0.70(0.85 \times 5 \times 4.5 \times 4.5) = 60.24$ kips > 26.42 kips   O.K.   9.3.2.4

3. Determine shear-span a for both shear and flexure. The reaction is considered to be at the outer third point of the bearing pad.
   For shear, a = $4.5\left(\dfrac{2}{3}\right) + 1.0 = 4$ in.

Example 17.3—Continued

	Code
Calculations and Discussion	**Reference**

For flexure, the critical section is at the center line of the hanger reinforcement (A_v). Assume 1-in. cover and #4 bar stirrups.
a = 4 + 1 + 0.25 = 5.25 in.

4. Determine width of ledge to be considered as effective for each stem reaction. An effective width equal to width of bearing plus 4 times shear-span a is suggested.[17.1] (If a stem reaction is located near the end of a beam ledge, the effective width may be further restricted to twice the distance between center of bearing and end of ledge).
b + 4a = 4.5 + 4(4) = 20.5 in.

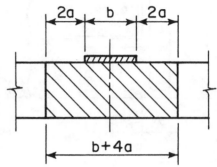

5. Check effective ledge section for shear-transfer strength V_n.   11.9.3.2.1

For $f_c' = 5000$ psi; $V_{n(max)} = 800 A_c$, where $A_c = (b + 4a)d$

$$V_n = \frac{800(20.5)(10.75)}{1000} = 176.3 \text{ kips}$$

$\varphi V_n = 0.85(176.3) = 149.8$ kips > 26.42 kips   O.K.

6. Determine shear-friction reinforcement A_{vf}.   11.9.3.2

$$A_{vf} = \frac{V_u}{\varphi f_y \mu} = \frac{26.42}{0.85(60)1.4} = 0.37 \text{ in.}^2/\text{stem}$$ 11.7.4.1
11.7.4.3

7. Determine reinforcement to resist direct tension A_n. Unless special provisions are made   11.9.3.4
to reduce direct tension, N_u should be taken not less than $0.2V_u$ to account for unexpected forces due to restrained long-time deformation of the supported member, or other causes. When the beam ledge is designed to resist specific horizontal forces, the bearing plate should be welded to the tension reinforcement A_s.

$N_u = 0.2V_u = 0.2(26.42) = 5.28$ kips

$$A_n = \frac{N_u}{\varphi f_y} = \frac{5.28}{0.85(60)} = 0.104 \text{ in.}^2/\text{stem}$$

8. Determine flexural reinforcement A_f.

$M_u = V_u a + N_u (h - d) = 26.42(5.25) + 5.28(12 - 10.75) = 145.3$ in.-kips

Find A_f using ordinary flexural design methods. For beam ledges, use $j_u d = 0.8d$.   11.9.3.1

Example 17-3—Continued

Calculations and Discussion	Code Reference

Also, φ should be taken as 0.85

$$A_f = \frac{145.3}{0.85(60)(0.8 \times 10.75)} = 0.331 \text{ in.}^2/\text{stem}$$

9. Determine primary tension reinforcement A_s.   11.9.3.5

$$\left(\frac{2}{3}\right)A_{vf} = \left(\frac{2}{3}\right)0.37 = 0.247 \ < \ A_f = 0.331; \ A_f \text{ controls design}$$

$$A_s = A_f + A_n = 0.331 + 0.104 = 0.435 \text{ in.}^2/\text{stem}$$

Check $A_{s(min)} = 0.04\left(\frac{f_c'}{f_y}\right)bd$ 11.9.5

$$= 0.04\left(\frac{5}{60}\right)20.5 \times 10.75 = 0.735 \text{ in.}^2/\text{stem} \ > \ 0.435 \text{ in.}^2/\text{stem}$$

For typical shallow ledge members, minimum A_s by Section 11.9.5 will almost always govern.

10. Determine shear reinforcement A_h.   11.9.4

$$A_h = 0.5(A_s - A_n) = 0.5A_f = 0.5(0.331) = 0.166 \text{ in.}^2/\text{stem}$$

11. Determine final size and spacing of ledge reinforcement.

Distribution of A_h depends on the magnitude of the punching shear strength around the bearing area:

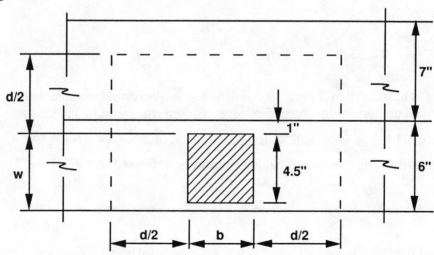

Example 17.3—Continued

Calculations and Discussion

$V_n = 4\sqrt{f_c'}\ b_o d = 4\sqrt{5000} \times 36 \times \dfrac{10.75}{1000} = 109.5$ kips

where $b_o = (b + 2w + 2d) = (4.5 + 2 \times 5 + 2 \times 10.75) = 36$ in.

$\varphi V_n = 0.85(109.5) = 93$ kips $>$ 26.42 kips

Combine $A_s + A_h$ in one layer

Since all required areas of reinforcement are computed for the reaction of one stem of the double-tee, divide areas by effective ledge width (20.5 in.) to obtain area per inch of ledge so that an adequate amount of reinforcement is provided along the effective width no matter where the tee stems are located on the ledge.

$A_s + A_h = \dfrac{0.735 + 0.166}{20.5} = 0.044$ in.^{2}/in. of ledge

Try #4 bar, $A_b = 0.20$ in.^{2} Try #5 bar, $A_b = 0.31$ in.^{2}

$s_{max} = \dfrac{0.20}{0.044} = 4.55$ in. $s_{max} = \dfrac{0.31}{0.044} = 7.05$ in.

Use #5 @ 7 in.

12. Check required area of beam stirrups used as "hanger" reinforcement.

A_v (one leg) $= \dfrac{V_u}{\varphi f_y} = \dfrac{26.42}{0.85(60)} = 0.518$ in.^{2}/leg

Use 20.5 in. effective ledge width (somewhat conservative) for distribution of hanger reinforcement.*

A_v (one leg) $= \dfrac{0.518}{20.5} = 0.0253$ in.^{2}/in. of ledge.

Sufficient stirrups in beam section for combined shear and torsion must be available to act also as hanger reinforcement for the beam ledge. Closed stirrups in beam section must be at least equal to $\dfrac{A_t}{s} + \dfrac{A_v}{2s} \geq 0.0253$ in.^{2}/in.; i.e., #3 @ 4 in. or #4 at 8 in.

If effective ledge width for distribution of hanger reinforcement is 37.5 in.*

---

* A more exact effective width (over which to distribute hanger forces) may be evaluated from Reference 17.1. For the ledge loading and dimensions of this example, $b_e = 37.5$ in.

Example 17.3—Continued

Code
Calculations and Discussion
Reference

A_v (one leg) $= \dfrac{0.518}{37.5} = 0.0138$ in.^{2}/in.

Thus, $\dfrac{A_t}{s} + \dfrac{A_v}{2s} \geq 0.0138$ in.^{2}/in.; i.e., #3 @ 8 in. or #4 @ 14 in.

13. Reinforcement Details

In accordance with Section 11.9.7, bearing area (4.5-in. pad) must not extend beyond straight portion of beam ledge reinforcement, nor beyond inside edge of transverse anchor bar. With a 4.5-in. bearing pad, this requires that the width of ledge be increased to 9 in. as shown below. Alternately a 6-in. ledge with a 3-in. medium strength pad (1500 psi) and the ledge reinforcement welded to an armor angle would satisfy the intent of Section 11.9.7.

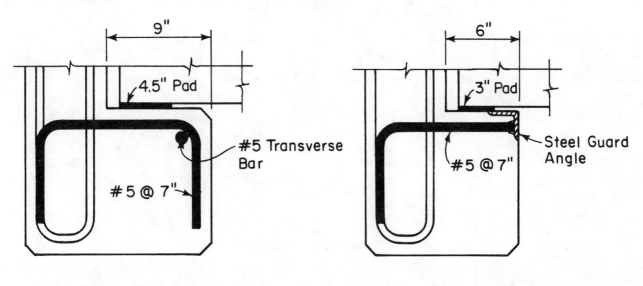

9 in. ledge detail   6 in. ledge detail

Shear in Slabs

UPDATE FOR THE '89 CODE

For ACI 318-89, several changes, both technical and editorial, have been made to the design provisions for shear in slabs. Code Sections 11.11 and 11.12, treating shear strength in slabs, have been completely reorganized in a more usable format:

11.11 - Transfer of Moments in Beam-Column Connections

11.12 - Shear Strength of Slabs and Footings

11.12.2 - Shear Strength Provided by Concrete

11.12.3 - Shear Strength Provided by Bars or Wires

11.12.4 - Shear Strength Provided by Steel I- or Channel-Shaped Sections (Shearheads)

11.12.5 - Openings in Slabs

11.12.6 - Transfer of Moments in Slab-Column Connections

Technically, the following has been accomplished within the text of new Code Sections 11.11 and 11.12:

(1)   Although it is customary to use a rectangular critical perimeter b_0 to define the critical section for two-way shear, a literal or legalistic interpretation of the definition of critical perimeter for a square or rectangular loaded area (column) would require round corners located at a distance d/2 from the corners of a square or rectangular column. The definition of b_0 has been revised to permit a rectangular critical section (Section 11.12.1.3).

(2)   While just implied in the '83 Code provisions, investigation of two-way shear at critical perimeters b_0 more than a distance d/2 away from the loaded area is now specifically required at edges of column capitals and drop panels and at outer limits of slab shear reinforcement (Section 11.12.1.2).

(3) Recent tests have indicated a decrease in shear strength provided by concrete as the ratio of the critical perimeter b_0 to the effective depth d increases, i.e, at larger and larger critical sections more distant from a loaded area or column. A new transition equation has been introduced to reflect this shear strength decrease [Eq. (11-37)].

(4) In the '83 Code, only "shearhead" reinforcement was permitted to increase the shear strength when both shear and moment transfer were required at slab-column connections. For the '89 Code, bars or wires used as shear reinforcement are also permitted for shear and moment transfer at slab-column connections.

One additional item relating to slabs: the two Code expressions, Eqs. (11-42) and (13-1), defining the transfer of unbalanced moment at slab-column connections, part by eccentricity of shear and part by flexure, have been modified to more correctly define the dimensions of the critical shear perimeter for edge and corner columns. See discussion on Section 11.12.6.

11.12   SHEAR STRENGTH OF SLABS AND FOOTINGS

Design of slabs for shear in the region of columns, concentrated loads or reactions is governed by the more severe of two conditions: wide-beam action and two-way action. For flexural members, such as one-way slabs in which the bending action is primarily in one direction, wide-beam action is the primary mode of behavior and the design requirements of Sections 11.1 through 11.5 must be satisfied. For two-way slab systems, such as flat plates and flat slabs, two-way action is the primary mode of behavior and the failure mechanism changes to that of punching, thus necessitating a different design approach. Even though wide-beam action shear rarely controls the shear strength of two-way slab systems, the designer must ensure that shear strength for beam action is not exceeded. Tributary areas and corresponding critical sections for wide-beam action shear strength and two-way shear strength at a slab and column connection are illustrated in Fig. 18-1. See Example 20.1.

Investigation of two-way shear at critical sections b_0 more than a distance d/2 away from the column is also required at edges of column capitals and drop panels, and at outer limits of slab shear reinforcement (Section 11.12.1.2). See Fig. 18-2.

Note that it is permissible to use a rectangular perimeter b_0 to define the critical section for square and rectangular columns (Section 11.12.1.3). A literal interpretation of the definition of perimeter b_0 for square or rectangular column or drop panel would require round corners located d/2 away from the corners of rectangular shapes. This clarification is added in the '89 Code. For shapes other than rectangular, the Commentary recommends that "β_c be taken as the ratio of the longest overall dimension of the effective loaded area to the largest overall dimension of the effective loaded area measured perpendicular thereto". β_c is the ratio of long side to short side of concentrated load or reaction area. See Example 18-2.

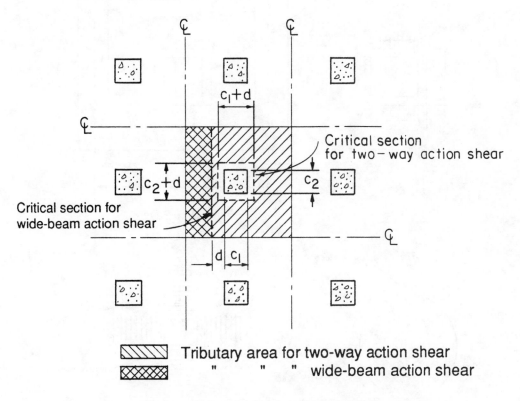

Fig. 18-1 Tributary Areas and Critical Sections for
Slab Shear at a Slab-Column Connection

11.12.2 Shear Strength Provided by Concrete

Two-way action shear strength of slabs (without shear reinforcement) is affected by the following five principal variables:

(1) concrete strength

(2) relationship between size of loaded area and slab thickness

(3) loaded area aspect ratio (shape of loaded area)

(4) perimeter area aspect ratio - new for '89 Code

(5) shear-to-moment ratio at slab-column connections

These variables are taken into account in the formulation of Eqs. (11-36) and (11-37):

$$V_c = (2 + 4/\beta_c)\sqrt{f_c'}\,b_o d \qquad\qquad\text{Eq. (11-36)}$$

$$V_c = (2 + \alpha_s/\beta_o)\sqrt{f_c'}\,b_o d \qquad\qquad\text{Eq. (11-37)}$$

18-3

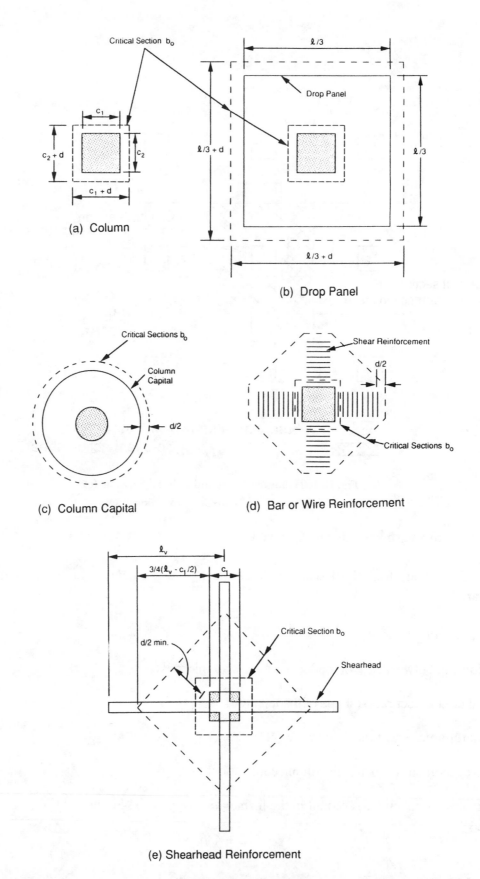

(a) Column

(b) Drop Panel

(c) Column Capital

(d) Bar or Wire Reinforcement

(e) Shearhead Reinforcement

Fig. 18-2 Critical Sections b_0 for Investigation of Two-Way Action Shear Strength

18-4

Shear strength affected by the loaded area aspect ratio β_c was first introduced in the 1977 Code Edition. The β_c variable provides a transition between two-way action shear ($4\sqrt{f_c'}$) and beam-action shear ($2\sqrt{f_c'}$) as the loaded area (support size) becomes more elongated. Shear strength variation as a function of β_c is shown in Fig. 18-3. For a support size less than 2 to 1 ($\beta_c \le 2$), Eq. (11-36) reduces to $V_c = 4\sqrt{f_c'}\,b_o d$.

For ACI 318-89, a new Eq. (11-37) has been introduced to account for a decrease in shear strength caused by an increase in the perimeter area aspect ratio β_o, where β_o is the ratio of the critical perimeter b_o to the effective depth d. Recent tests have indicated a decrease in shear strength as the ratio of the perimeter b_o to the effective depth d increases. The new Eq. (11-37) was therefore introduced. Shear strength variation as a function of β_o is shown in Fig. 18-4.

Equation (11-37) reduces to $V_c = 4\sqrt{f_c'}\,b_o d$ for:

Interior Columns (4 sides effective $\alpha_s = 40$) when $\beta_o \le 20$

Edge Columns (3 sides effective $\alpha_s = 30$) when $\beta_o \le 15$

Corner Columns (2 sides effective $\alpha_s = 20$) when $\beta_o \le 10$

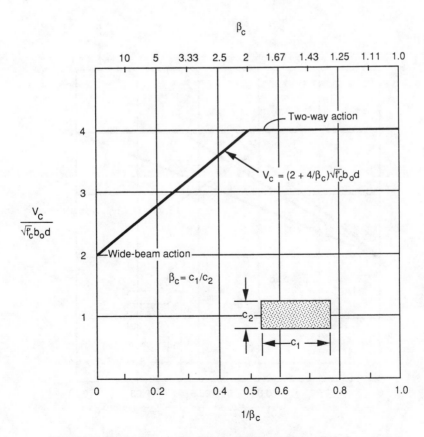

Fig. 18-3 Shear Strength of Slabs Without Shear Reinforcement
(loaded area aspect ratio affect β_c)

Note that two-way action shear strength V_c is the lesser of values given by Eqs. (11-36) and (11-37), but not greater than $4\sqrt{f_c'}b_o d$, where V_c is to be investigated at each of the critical sections b_o indicated in Fig. 18-2(a), (b) and (c) which depict slab-column connections without shear reinforcement. With the adoption of the new Eq. (11-37), shear strength for very thin slabs or at larger and larger critical sections more distant from a loaded area or column, such as at the edges of drop panels or column capitals, may be significantly reduced due to the larger perimeter area aspect ratio β_o. See Example 18.2.

The same shear strength V_c, the lesser of values given by Eqs. (11-36) and (11-37) but not greater than $4\sqrt{f_c'}b_o d$, applies for transfer of moment at slab-column connections without shear reinforcement. See discussion on Section 11.12.6.

Under certain design conditions, it may be necessary to increase the shear strength of slabs by use of shear reinforcement consisting of bars, wires, or steel I- or channel-shapes (shearheads). Note that both types of shear reinforcement are permitted in slabs for direct shear, and shear and moment transfer at slab-column connections. Use of bars or wires as shear reinforcement when both shear and moment transfer are required at slab-column connections is new with the '89 Code; previously only shearhead reinforcement was permitted for moment and shear transfer.

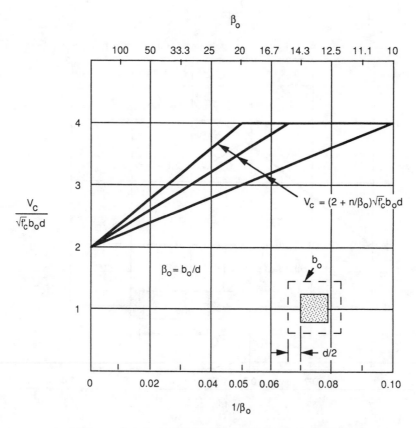

Fig. 18-4 Shear Strength of Slabs Without Shear Reinforcement
(Effect of Perimeter Area Aspect Ratio β_o)

11.12.3 Shear Strength Provided by Bar or Wire Reinforcement

Shear reinforcement is required when the factored shear force V_u exceeds the design shear strength φV_c available without reinforcement (Section 11.12.2.).

When bars or wires are used as shear reinforcement in slabs, the design procedure is similar to that for beams using the design provisions of Section 11.5. Required area of shear reinforcement A_v is computed from Eqs. (11-1), (11-2), and (11-17) as follows:

$$V_u \leq \varphi V_n \qquad\qquad\qquad \text{Eq. (11-1)}$$

$$\leq \varphi V_c + \varphi V_s \qquad\qquad\qquad \text{Eq. (11-2)}$$

$$\leq \varphi V_c + \varphi A_v f_y d / s$$

$$\text{Eq. (11-17)}$$

Solving, $\qquad A_v \geq (V_u - \varphi V_c) s / \varphi f_y d$

where $V_c = 2\sqrt{f_c'} b_o d$. A_v is the total area of required shear reinforcement to be extended from the sides of the column. For interior column supports, required area per side $= A_v/4$; for edge columns $= A_v/3$; and for corner columns $= A_v/2$. The shear reinforcement should be placed symmetrically about the column in location, number and spacing. Bar shear reinforcement usually takes the form of single or double vertical U-stirrups, uniformly spaced from each column side, see Fig. 18-5.

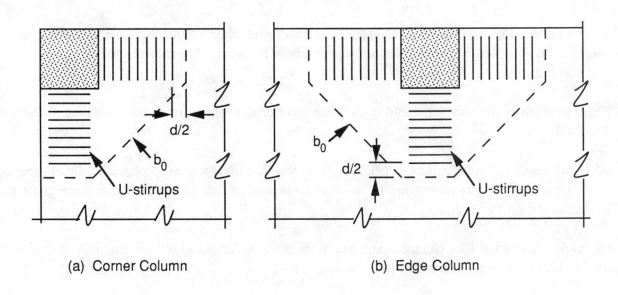

(a) Corner Column   (b) Edge Column

Fig. 18-5 Bar Shear Reinforcement Layout

The critical section for determining required area A_v is at the perimeter b_o located d/2 away from the column perimeter. Note that, as for beams, shear reinforcement must be designed to carry all shear in excess of $2\sqrt{f'_c}b_o d$, not $4\sqrt{f'_c}b_o d$ as used to evaluate shear strength of slabs without shear reinforcement. Also, as for beams, maximum spacing is d/2. Shear reinforcement must extend a distance from each column side to a perimeter b_o where concrete shear strength $V_c = 2\sqrt{f'_c}b_o d$ is adequate to carry the factored shear force V_u.

Shear strength of slabs with bar reinforcement is illustrated in Fig. 18-6. Note that the maximum shear strength V_n permitted with bars is $6\sqrt{f'_c}b_o d$. With $V_c = 2\sqrt{f'_c}b_o d$, the maximum shear strength provided by bar reinforcement is limited to $4\sqrt{f'_c}b_o d$. Application of the design criteria for bar shear reinforcement is illustrated in Example 18.3.

Note the warning in Commentary Section 11.12.3: "The importance of anchorage details for slab shear reinforcement cannot be over emphasized. Due to the small cover over slab reinforcement, stirrups with 90-deg hooks at the top corners have failed prematurely in tests by loss of anchorage when the hooks straightened out accompanied by loss of the cover concrete." Suggested anchorage details with ends of stirrups anchored by 135-deg bends plus $6d_b$ extensions at the free ends of the bar is illustrated in Fig. 18-6.

When moment, in addition to direct shear, is transferred between a slab and a supporting column, the provisions of Section 11.12.6 must be satisfied. See discussion on Section 11.12.6.

11.12.4   Shear Strength Provided by Steel I- or Channel-shaped Sections (Shearheads)

Shear strength may be provided by shearheads when the factored shear force V_u exceeds the shear strength φV_c available without shear reinforcement. V_c is the lesser of values given by Eqs. (11-36), (11-37), subject to a maximum of $4\sqrt{f'_c}b_o d$.

In the design of shearhead reinforcement, three important basic criteria must be considered as follows:

1. To ensure that the required shear strength of the slab is reached before the flexural strength of the shearhead is exceeded, a minimum flexural strength must be provided to the shearhead (Section 11.12.4.6).

2. Shear strength in the slab at the end of the shearhead reinforcement must not be exceeded (Section 11.12.4.8).

3. After the above requirements are satisfied, certain reduction in column strip negative moment reinforcement may be made in proportion to the contribution of the shearhead reinforcement to flexural strength (Section 11.12.4.9).

Structural details of steel I- or channel-shaped sections used as shearhead reinforcement is presented in sections 11.12.4.1 through 11.12.4.6. See Fig. 18-7.

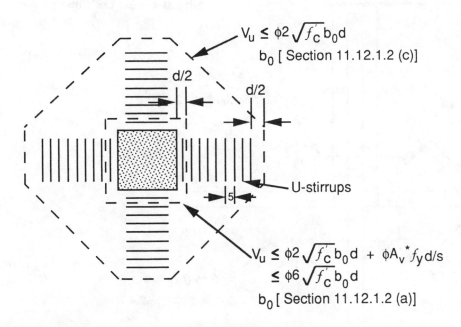

$V_u \leq \phi 2\sqrt{f'_c}\, b_0 d$

b_0 [Section 11.12.1.2 (c)]

d/2

d/2

U-stirrups

$V_u \leq \phi 2\sqrt{f'_c}\, b_0 d + \phi A_v{}^* f_y d/s$

$\leq \phi 6\sqrt{f'_c}\, b_0 d$

b_0 [Section 11.12.1.2 (a)]

* Total area of shear reinforcement
  on the four sides of the interior
  column support.

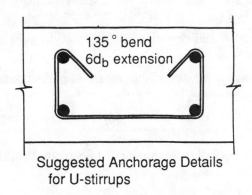

135° bend
6d$_b$ extension

Suggested Anchorage Details
for U-stirrups

Fig. 18-6 Shear Strength of Slabs with Bars or Wires Used as Shear Reinforcement (Section 11.12.3)

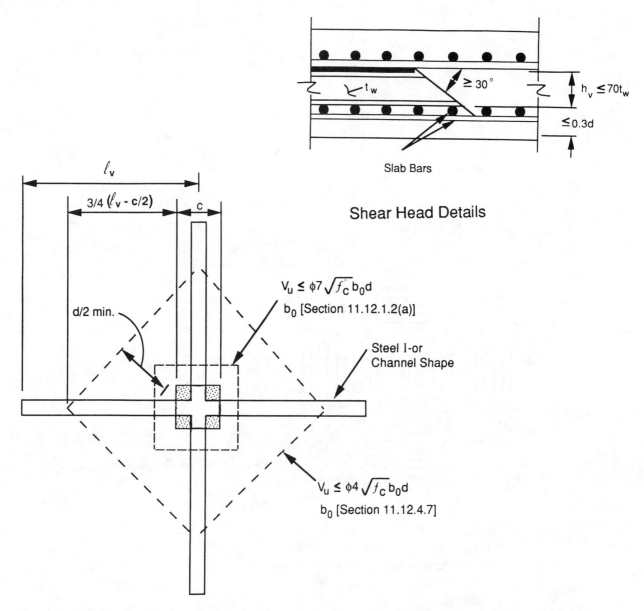

Fig. 18-7 Shear Strength of Slabs with "Shearheads" Used as Shear Reinforcement (Section 11.12.4)

1. Steel shapes must be fabricated by welding with a full penetration weld into identical arms at right angles.

2. Overall depth of steel shape must not exceed $70t_w$, where t_w is thickness of web of steel shape.

3. Ends of steel shape must be square cut or cut at an angle not exceeding 30-deg with the horizontal.

4. Compression flange (bottom flange) of steel shape must be located within 0.3d distance of the bottom of the slab.

5. Stiffness of steel shape must not be less than 15 percent of the stiffness of the composite cracked slab section of width $(c_2 + d)$. See Fig. 18-8.

Example 16.2—Shear-Friction Design (Inclined Shear Plane)

For the pilaster beam support shown, design for shear transfer across the potential crack plane. In edge bearing applications a crack tends to occur at an angle of about 20 degrees to the direction of force applications. Beam reactions are D = 25 kips, L = 30 kips, and T = 20 kips due to estimate of shrinkage and temperature change effects.

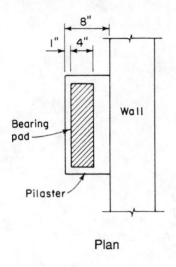

Plan

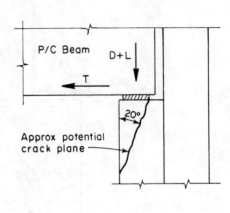

Sectional Elevation

	Code
Calculations and Discussion	**Reference**

1. Factored loads to be considered.

    Beam reaction R_u = 1.4D + 1.7L = 1.4(25) + 1.7(30) = 35 + 57 = 86 kips   Eq. (9-1)

    Shrinkage and temperature effects T_u = 1.7(20) = 34 kips
    but not less than $0.2(R_u)$ = 0.2(86) = 17.2 kips

    Note that the live load factor of 1.7 is used with T, due to the low confidence level in determining shrinkage and temperature effects occurring in service. Also, a minimum value of 20 percent of the beam reaction is considered (see Section 11.9.3.4 for corbel design).

2. Evaluate force conditions along potential crack plane

    Direct shear transfer force along shear plane:

    $V_u = R_u \sin \alpha_f + T_u \cos \alpha_f = 86(\sin 70°) + 34(\cos 70°)$
    $\quad = 80.8 + 11.6 = 92.4$ kips

Example 16.2—Continued

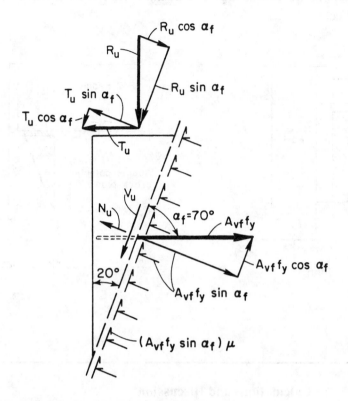

Net tension (or compression) across shear plane:

$$N_u = T_u \sin \alpha_f - R_u \cos \alpha_f = 34(\sin 70°) - 86(\cos 70°)$$
$$= 31.9 - 29.4 = 2.5 \text{ kips (net tension)}$$

If the load conditions were such as to result in net compression across the shear plane, it still should not have been used to reduce the required A_{vf}, because of the uncertainty in evaluating the shrinkage and temperature effects. Also, Section 11.7.7 permits a reduction in A_{vf} only for "permanent" net compression.

3. Shear-friction reinforcement to resist direct shear transfer

$$A_{vf} = \frac{V_u}{\varphi f_y (\mu \sin \alpha_f + \cos \alpha_f)}$$
Eq. (11-27)

$$= \frac{92.4}{0.85 \times 60(1.4 \sin 70° + \cos 70°)} = 1.09 \text{ in.}^2$$
11.7.4.3

Example 16.2—Continued

Code
Reference

Calculations and Discussion

4. Reinforcement to resist net tension

$$A_n = \frac{N_u}{\varphi f_y (\sin\alpha_f)} = \frac{2.5}{0.85 \times 60(\sin 70°)} = 0.05 \text{ in.}^2$$

Note that with failure primarily controlled by shear, $\varphi = 0.85$ is used (see Section 11.9.3.1 for corbel design).

5. Add A_{vf} and A_n for uniform distribution along potential crack plane.

$$A_s = 1.09 + 0.05 = 1.14 \text{ in.}^2$$

Use #3 closed ties (2 legs per tie)

Number required = $1.14/2(0.11) = 5.2$

Ties should be distributed along length of potential crack plane; approximate length = $5(\tan 70°) \approx 14$ in.

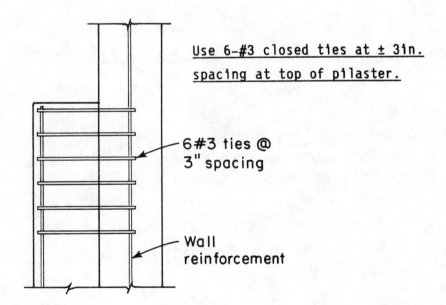

Use 6-#3 closed ties at ± 3in.
spacing at top of pilaster.

6#3 ties @
3" spacing

Wall
reinforcement

Example 16.2—Continued

	Code
Calculations and Discussion	**Reference**

6. Check reinforcement requirements for dead load only plus shrinkage and temperature effects. Use 0.9 load factor for dead load to maximize net tension across shear plane.

$R_u = 0.9D = 0.9(25) = 22.5$ kips, $T_u = 34$ kips

$V_u = 22.5(\sin 70°) + 34(\cos 70°) = 21.1 + 11.6 = 32.7$ kips

$N_u = 34(\sin 70°) - 22.5(\cos 70°) = 31.9 - 7.7 = 24.2$ kips

$$A_{vf} = \frac{32.7}{0.85 \times 60\,(1.4\sin 70° + \cos 70°)} = 0.39 \text{ in.}^2$$

$$A_n = \frac{24.2}{0.85 \times 60 \times \sin 70°} = 0.50 \text{ in.}^2$$

$A_s = 0.39 + 0.50 = 0.89 \text{ in.}^2 < 1.14 \text{ in.}^2$

Therefore, original design for full dead load + live load governs.

Brackets, Corbels and Beam Ledges

GENERAL CONSIDERATIONS

Design provisions for brackets and corbels were completely revised for the 1983 Code, eliminating the empirical equations of the 1977 Code and simplifying design by using the shear-friction method exclusively for shear-transfer strength V_n. No changes have been made from the 1983 to the 1989 edition of the Code.

11.9 BRACKETS, CORBELS AND BEAM LEDGES

The design procedure for brackets and corbels recognizes the deep beam or simple truss action of these short-shear-span members, as illustrated in Fig. 17-1. Four possible failure modes must be controlled: (1) Direct shear failure at the interface between bracket or corbel and supporting member; (2) Yielding of the tension tie due to moment and direct tension; (3) Crushing of the internal compression "strut;" and (4) Localized bearing or shear failure under the loaded area.

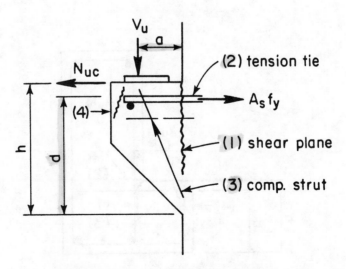

Fig. 17-1   Structural Action of Corbel

The design provisions of Section 11.9 apply only to members having a shear span-to-depth ratio of unity or less (a/d ≤ 1) since, for longer spans, diagonal tension cracks may form and the use of horizontal shear reinforcement may not suffice. Furthermore, the method has not been validated by tests for a/d > 1.

A second restriction limits the design method to cases where the factored shear force V_u exceeds the factored tensile force N_{uc}, again because no test data are available for load conditions with N_{uc} exceeding V_u.

Design of beam ledges is similar to that of a bracket or corbel with respect to loading conditions, design considerations, and reinforcing details. Accordingly, even though not specifically addressed by the Code, special design of beam ledges is included in Part 17. The four failure modes discussed above for brackets and corbels are also noted for beam ledges in Fig. 17-2. One additional failure mode needs to be considered for the beam ledge; (5) Separation between ledge and beam web near the top of the ledge in the vicinity of the ledge load. The vertical component of the inclined compression force must be picked up by the beam stirrups (stirrup legs A_v adjacent to the side face of the beam) acting as "hanger" reinforcement to carry the ledge load to the top of beam. Note that the critical section for moment is taken at center of beam stirrups, not at face of beam. Also, for beam ledges, the internal moment arm should not be taken greater than 0.8 for flexural strength. This reflects observed behavior in tests.[17.1, 17.2]

11.9.1 - 11.9.7   Design Procedure

The critical section for design of a bracket or corbel is taken at the face of the support; however, in the case of beam ledges (Fig. 17-2) the critical section for flexural design must be taken at the centerline of the beam stirrup, which acts as a "hanger" to carry vertical loads.

The critical section is "designed to resist simultaneously a shear V_u, a moment $[V_u a + N_{uc}(h - d)]$, and a horizontal tensile force N_{uc}." The value of N_{uc} must be no less than $0.2V_u$, unless some provision is made to avoid tensile forces. Since slip joints and flexible bearings do not always function as designed, good practice

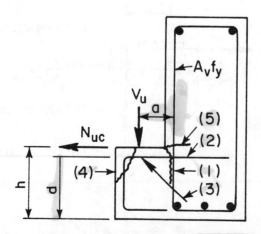

Fig. 17-2   Structural Action of Beam Ledge

dictates a value of at least $0.2V_u$ for N_{uc}, in any case. Since the tensile force N_{uc} is due to indeterminate causes such as restrained shrinkage or temperature stresses, a load factor of 1.7, as for live load, is required.

For normal weight concrete, shear strength V_n must not be taken greater than $0.2f_c' b_w d$ nor $800 b_w d$ (as in shear-friction design). For lightweight concrete, V_n is somewhat more restricted by Section 11.9.3.2.2. Tests show that for lightweight concrete, the maximum shear strength is a function of a/d as well as f_c'.

For design purposes, the total reinforcement required is divided into three parts, with each determined separately: A_{vf}, area of shear-friction reinforcement to resist direct shear V_u; A_f, area of flexural reinforcement to resist moment $V_u a + N_{uc}(h - d)$; and A_n, area of tensile reinforcement to resist direct tensile force N_{uc}. With behavior predominantly controlled by shear, a single value of $\varphi = 0.85$ should be used for all design conditions.

Once the separate areas of reinforcement A_{vf}, A_f, and A_n have been determined, the actual reinforcement to be provided, A_s and A_h, may be sized, where A_s will act as the primary tension reinforcement and A_h will act as shear reinforcement.

Since the flexural reinforcement A_f is balanced by a compression area, it also contributes to the direct shear-transfer strength of the section; thus, the total required reinforcement $(A_s + A_h)$ will be somewhat less than the total of $(A_{vf} + A_f + A_n)$.

The required reinforcement is distributed to conform to the results of tests. The total of $(A_s + A_h)$ must be the greater of two amounts: (1) the sum of A_{vf} and A_n or (2) the sum of $3A_f/2$ and A_n, thus a comparison of A_{vf} and $3A_f/2$ (or $2A_{vf}/3$ and A_f) will determine control.

With control determined, primary tension reinforcement A_s may be found from either $A_s = (2A_{vf}/3 + A_n)$ or $A_s = (A_f + A_n)$, whichever is larger. The required area of shear reinforcement parallel to A_s is then computed as $A_h = 0.5(A_s - A_n)$. This results in the total $(A_s + A_h)$ as required above. For brackets and corbels, A_h is to be distributed uniformly within two-thirds of the depth "d" next to A_s. For typical shallow ledge members, distribution of A_h depends on the magnitude of the punching shear strength around the bearing area. If the required shear strength V_u is greater than the design punching shear strength φV_n, A_h should be distributed in the upper third of the ledge depth; otherwise A_h may be added to A_s. See Example 17.3.

A minimum ratio of primary tension reinforcement $\rho = 0.04(f_c'/f_y)$ is required to assure a ductile failure after cracking under moment and direct tensile force. All reinforcement must be fully developed on both sides of the critical section. Anchorage within the support is usually accomplished by embedment or hooks. Within the bracket or corbel, the distance between load and support face is usually too short, so that special anchorage must be provided at the outer ends of both A_s and A_h. Anchorage of A_s is normally provided by welding a cross bar of equal size across the ends of A_s (Fig. 17-3) or welding to an armor angle. In the former case, the cross bar must be located beyond the edge of the loaded area. Where anchorage is provided by a hook or a loop in A_s, the load must not project beyond the straight portion of the hook or loop (Fig. 17-4). In beam ledges, anchorage may be provided by a hook or loop, with the same limitation on the load location

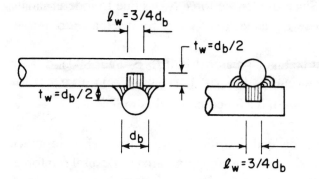

Fig. 17-3 Cross-Bar Weld Details

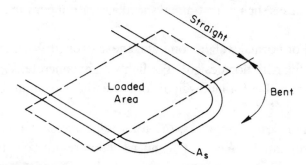

Fig. 17-4 Loaded Area Limitation with Loop Bar Detail

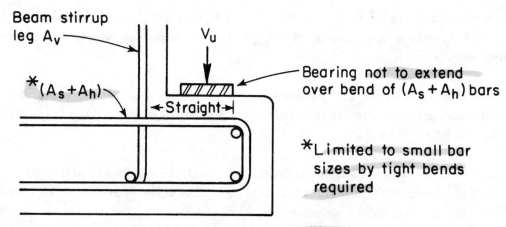

Fig. 17-5 Bar Details for Beam Ledge

(Fig. 17-5). Where a corbel or beam ledge is designed to resist specific horizontal forces, the bearing plate should be welded to A_s.

The closed stirrups or ties used for A_h must be similarly anchored, usually by bending around a "framing bar" of the same diameter as the closed stirrups or ties.

REFERENCES

17.1 Mirza, Sher Ali, and Furlong, Richard W., "Strength Criteria for Concrete Inverted T-Girder," *Journal of Structural Engineering* V. 109, No. 8, Aug. 1983, pp. 1836-1853.

17.2 Mirza, Sher Ali, and Furlong, Richard W., "Serviceability Behavior and Failure Mechanisms of Concrete Inverted T-Beam Bridge Bent Caps," *ACI Journal, Proceedings* V. 80, No. 4, July-August 1983, pp. 294-304.

Example 17.1—Corbel Design

Design a corbel with minimum dimensions to support a beam as shown below. The corbel is to project from a 14-in. square column. Restrained creep and shrinkage create a horizontal force of 20 kips at the welded bearing.

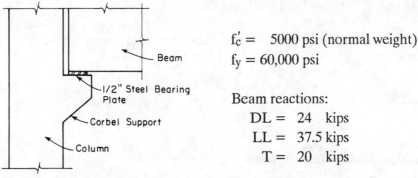

$f'_c = $ 5000 psi (normal weight)

$f_y = $ 60,000 psi

Beam reactions:

DL = 24 kips

LL = 37.5 kips

T = 20 kips

	Code
Calculations and Discussion	**Reference**

1. Size bearing plate based on bearing strength on concrete according to Section 10.15. Width of bearing plate = 14 in.

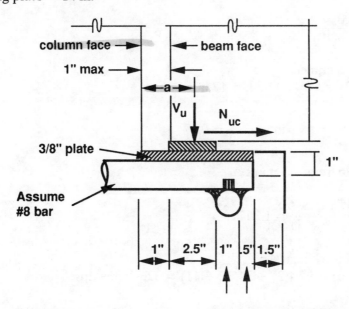

V_u = 1.4(24) + 1.7(37.5) = 97.4 kips

V_u $\leq \varphi P_{nb} = \varphi (0.85 f'_c A_1)$

97.4 = 0.70(0.85 × 5 × A_1) = 2.975A_1 9.3.2.4

$A_1 = \dfrac{97.4}{2.975} = 32.74$ in.^{2}

Bearing length $= \dfrac{32.74}{14} = 2.34$ in.

Use 2.5 in. x 14 in. bearing plate

Example 17.1—Continued

	Code
Calculations and Discussion	**Reference**

$N_{uc} = 1.7(20) = 34$ kips　　(treated as Live Load)　　　　　　　　　11.9.3.4

2. Determine a with 1 in. max. clearance at beam end. Beam reaction is assumed at third point at bearing plate.

$a = \dfrac{2}{3}(2.5) + 1.0 = 2.67$ in.

Use a = 3 in. maximum
Detail cross bar just outside outer bearing edge.

3. Determine total depth of corbel based on limiting shear-transfer strength V_n.

For $f'_c = 5000$ psi, $V_{n(max)} = 800b_wd$　　　　　　　　　　　　　　　11.9.3.2.1

$V_u \le \varphi V_n = \varphi(800b_wd)$

required $d = \dfrac{97,400}{0.85(800 \times 14)} = 10.23$ in.

Assuming #8 bar plus tolerance,
h = 10.23 + 1.0 = 11.23 in.　　　　Use h = 12 in.

For design, d = 12.0 – 1.0 = 11.0 in.　　　$\dfrac{a}{d} = 0.27$

4. Determine shear-friction reinforcement A_{vf}.　　　　　　　　　　　　11.9.3.2

$A_{vf} = \dfrac{V_u}{\varphi f_y \mu} = \dfrac{97.4}{0.85(60)(1.4 \times 1)} = 1.36$ in.^{2}　　　　　　　11.7.4.1
　　　　　　　　　　　　　　　　　　　　　　　　　　　　　　　　　11.7.4.3

5. Determine flexural reinforcement A_f.　　　　　　　　　　　　　　　11.9.3.3
    $M_u = V_ua + N_{uc}(h - d) = 97.4(3) + 34(12 - 11) = 326.2$ in.–kips

Find A_f using ordinary flexural design methods or conservatively use $j_ud = 0.9d$

$A_f = \dfrac{326.2}{0.85(60)(0.9 \times 11)} = 0.646$ in.^{2}

Note that for all design calculations, $\varphi = 0.85$　　　　　　　　　　　11.9.3.1

6. Determine direct tension reinforcement A_n.　　　　　　　　　　　　11.9.3.4

Example 17-1—Continued

Calculations and Discussion	**Code References**

$$A_n = \frac{N_{uc}}{\varphi f_y} = \frac{34}{0.85(60)} = 0.667 \text{ in.}^2$$

7. Determine primary tension reinforcement A_s.   11.9.3.5

$$\frac{2}{3}A_{vf} = \frac{2}{3}1.36 = 0.907 \quad A_f = 0.646; \quad \frac{2}{3}A_{vf} \text{ controls design}$$

$$A_s = \frac{2}{3}A_{vf} + A_n = 0.907 + 0.667 = 1.57 \text{ in.}^2$$

Use 2 #8 bars, $A_s = 1.58 \text{ in.}^2$ 11.9.5

Check minimum reinforcement:

$$\rho_{min} = 0.04\left(\frac{f_c'}{f_y}\right) = 0.04\left(\frac{5}{60}\right) = 0.003$$

$$A_{s(min)} = 0.003(14)(11) = 0.462 \text{ in.}^2 \quad < \quad A_s = 1.58 \text{ in.}^2 \qquad \text{O.K.}$$

8. Determine shear reinforcement A_h.   11.9.4

$$A_h = 0.5(A_s - A_n) = 0.5(1.57 - 0.667) = 0.454 \text{ in.}^2$$

Use 3 #3 stirrups, $A_h = 0.66 \text{ in.}^2$
Distribute stirrups in two-thirds of effective corbel depth adjacent to A_s.

Example 17.1—Continued

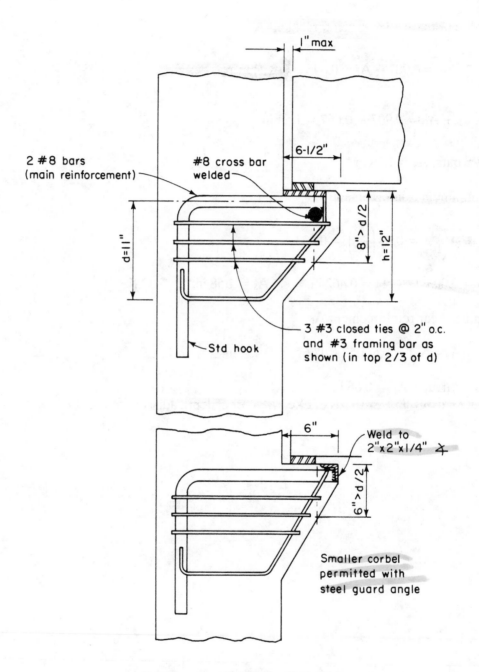

1" max

2 #8 bars
(main reinforcement)

#8 cross bar
welded

6-1/2"

8" > d/2

h=12"

d=11"

3 #3 closed ties @ 2" o.c.
and #3 framing bar as
shown (in top 2/3 of d)

Std hook

6"

Weld to
2"x2"x1/4" ∠

6" > d/2

Smaller corbel
permitted with
steel guard angle

Reinforcement Details for Corbel

Design a corbel to project from a 14-in.-square column to support the following beam reactions:

     dead load = 32 kips
     live load = 30 kips
     horizontal force = 24 kips
     $f_c' =$ 4000 psi (all lightweight)
     f_y = 60,000 psi

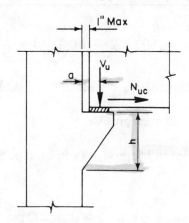

	Calculations and Discussion	**Code Reference**

1. Size bearing plate

   $V_u = 1.4(32) + 1.7(30) = 95.8$ kips   Eq. (9-1)

   $V_u \le \varphi P_{nb} = \varphi(0.85 f_c' A_1)$ 10.15.1

   $95.8 = 0.70(0.85 \times 4 \times A_1)$ 9.3.2.4

   Solving, $A_1 = 40.3$ in.^{2}

   Length of bearing required $= \dfrac{40.3}{14} = 2.9$ in.

   Use 14 in. x 3 in. bearing plate

2. Determine a.

   Assume beam reaction to act at outer third point of bearing plate.
   Assume 1-in. gap between back edge of bearing plate and column face. Hence,

   $a = 1 + \dfrac{2}{3}(3) = 3$ in.

3. Determine total depth of corbel based on limiting shear-transfer strength V_n. For easier placement of reinforcement and concrete, try h = 15 in. Assuming #8 bar with 1-in. cover, d = 15 − 1 − 0.5 = 13.5 in., $\dfrac{a}{d} = 0.22$

   For lightweight concrete and $f_c' \ge 4000$ psi: 11.9.3.2.2

Example 17.2—Continued

$$V_n = \left(800 - 280\frac{a}{d}\right) b_w d \text{ lbs.} = (800 - 280 \times 0.22)14 \times \frac{13.5}{1000} = 139.6 \text{ kips}$$

$$\varphi V_n = 0.85(139.6) = 118.7 \text{ kips} > V_u = 95.8 \text{ kips} \qquad \text{O.K.}$$

4. Determine shear-friction reinforcement A_{vf}.   11.9.3.2

   Using a Modified Shear-Friction Method as permitted by Section 11.7.3 (see commentary Section 11.7.3):

   $$V_n = 0.8A_{vf} f_y + K_1 b_w d, \text{ with } \frac{A_{vf} f_y}{b_w d} \text{ not less than 200 psi}$$
   For all lightweight concrete, $K_1 = 200$ psi

   $$V_u \le \varphi V_n \le \varphi(0.8A_{vf} f_y + 0.2b_w d)$$

   Solving for A_{vf}:

   $$A_{vf} = \frac{V_u - \varphi 0.2b_w d}{\varphi 0.8 f_y}, \text{ but not less than } 0.2b_w \frac{d}{f_y}$$

   $$= \frac{95.8 - 0.85 \times 0.2 \times 14 \times 13.5}{0.85(0.8)60} = 1.56 \text{ in.}^2$$

   but not less than $0.2 \times 14 \times \frac{13.5}{60} = 0.63 \text{ in.}^2$

   For comparison, compute A_{vf} by Code Eq. (11-26):

   For all lightweight concrete,   11.7.4.3
   $\mu = 1.4\lambda = 1.4(0.75) = 1.05$

   $$A_{vf} = \frac{V_u}{\varphi f_y \mu} = \frac{95.8}{0.85 \times 60 \times 1.05} = 1.79 \text{ in.}^2$$

5. Determine flexural reinforcement A_f.   11.9.3.3

   $$M_u = V_u a + N_{uc}(h - d) = 95.8(3) + 40.8(15 - 13.5) = 348.6 \text{ in.-kips}$$

   where $N_{uc} = 1.7(24) = 40.8$ kips   11.9.3.4

   Find A_f using ordinary flexural design methods, or conservatively use $j_u d = 0.9d$

Example 17.2—Continued

$$A_f = \frac{M_u}{\varphi f_y \, j_u d} = \frac{348.6}{0.85 \times 60 \times 0.9 \times 13.5} = 0.56 \text{ in.}^2$$

Note that for all design calculations, $\varphi = 0.85$ 11.9.3.1

6. Determine direct tension reinforcement A_n.   11.9.3.4

$$A_n = \frac{N_{uc}}{\varphi f_y} = \frac{40.8}{0.85 \times 60} = 0.80 \text{ in.}^2$$

7. Determine primary tension reinforcement A_s.   11.9.3.5

$$\left(\frac{2}{3}\right) A_{vf} = \left(\frac{2}{3}\right) 1.56 = 1.04 \text{ in.}^2 \; > \; A_f = 0.56 \text{ in.}^2; \quad \left(\frac{2}{3}\right) A_{vf} \text{ controls design}$$

$$A_s = \left(\frac{2}{3}\right) A_{vf} + A_n = 1.04 + 0.80 = 1.84 \text{ in.}^2$$

Use 3 #7 bars, $A_s = 1.80 \text{ in.}^2$

Check $A_{s(min)} = 0.04 \left(\frac{4}{60}\right) 14 \times 13.5 = 0.504 \text{ in.}^2 \; < \; A_s = 1.80 \text{ in.}^2$ O.K.

8. Determine shear reinforcement A_h.   11.9.4

$A_h = 0.5(A_s - A_n) = 0.5 \, (1.84 - 0.80) = 0.52 \text{ in.}^2$

Use 3 #3 stirrups, $A_h = 0.66 \text{ in.}^2$

The shear reinforcement is to be placed within two thirds of the effective corbel depth adjacent to A_s.

$$s_{max} = \left(\frac{2}{3}\right) \frac{13.5}{3} = 3 \text{ in.} \qquad \text{Use 3 in. stirrup spacing}$$

9. Corbel Details

Corbel will project $(1 + 3 + 2) = 6$ in. from column face.

Use 6-in. depth at outer face of corbel, then depth at outer edge of bearing plate will be   11.9.2
greater than $\frac{d}{2}$.

Example 17.2—Continued

$6 + \dfrac{7.5}{3} = 8.5 \text{ in.} > \dfrac{13.5}{2}$ O.K.

A_s to be anchored at front face of corbel by welding a #7 bar transversely across ends of A_s bars.

11.9.6

A_s must be anchored within column by standard hook.

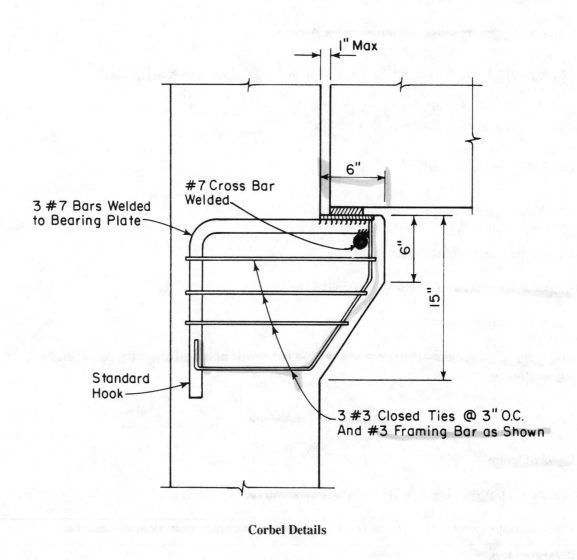

Corbel Details

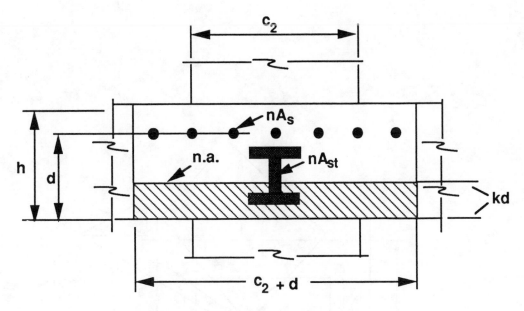

**Fig. 18-8 Stiffness of Cracked Composite Slab Section
(Steel Transformed to Equivalent Concrete)**

6. Plastic moment strength M_p of steel shape must not be less than the value computed by Eq. (11-40).

Shear strength of slabs with shearheads is illustrated in Fig. 18-7. The critical section for shear is located at 3/4 the length of the shearhead from the edge of the column $[3/4(\ell_v - c/2)]$, but need not be located closer than d/2 from the column perimeter. The length of the shearhead must be such that the factored shear force V_u to be transferred from slab to column does not exceed $\varphi 4\sqrt{f_c'}\,b_o d$ on this critical section. Note that the maximum shear strength V_n permitted with shearheads is $7\sqrt{f_c'}\,b_o d$; with $V_c = 4\sqrt{f_c'}\,b_o d$, without shear reinforcement, use of shearheads can increase the shear strength by $3\sqrt{f_c'}\,b_o d$.

Application of the design criteria for shearhead reinforcement is illustrated in Example 18.3. When moment, in addition to direct shear, is transferred between a slab and a column, the provisions of Section 11.12.6 must be satisfied.

11.12.5 Openings in Slabs

The effect of openings (vertical holes through slabs) on the shear strength of slabs must be investigated when the openings are located within the column strip areas of slabs or within middle strip areas when the openings are located closer than 10 times the slab thickness (10h) from a column. A reduction in shear strength is made by considering as ineffective that portion of the critical section b_o which is enclosed by straight lines projecting from the column centroid to the edges of the opening. Ineffective portions of critical sections b_o are illustrated in Fig. 18-9. For slabs with shear reinforcement, the ineffective portion of the perimeter b_o is one-half of that without shear reinforcement. The one-half factor is interpreted to apply equally to shearhead reinforcement and bar or wire reinforcement.

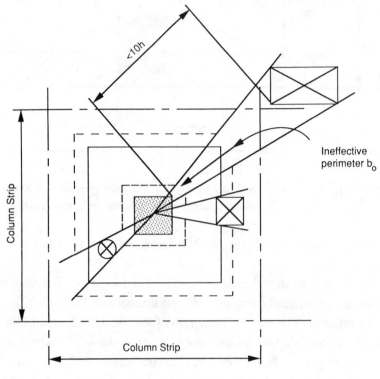

(a) Slab with Drop Panel

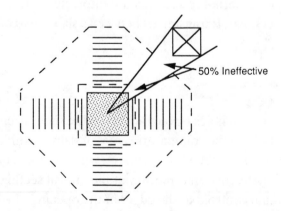

(b) Slab with Bar Reinforcement

Fig. 18-9 Effect of Slab Openings on Shear Strength

13.5 OPENINGS IN SLAB SYSTEMS

Openings of any size are permitted in slab systems without beams if special analysis indicates that both strength and serviceability of the slab system, considering the effects of the opening, are satisfied. Without special analysis, openings up to a certain size are permitted as illustrated in Fig. 18-10. The size of openings located within intersecting middle strip areas is unlimited. Within the area of the slab common to intersecting column strips, size of openings is the most restrictive due to their effect on slab shear strength or load transfer near slab-column connections. See discussion on effect of slab openings on shear strength (Section 11.12.5) and Fig. 18-9. Without special analysis, size of openings within intersecting column strips is limited to one-sixteenth of the slab span length in either direction $[1/8(\ell/2) = \ell/16]$. Within the slab area common to one column and one middle strip, opening size is limited to one-eighth the span length in either direction $[1/4(\ell/2) = \ell/8]$.

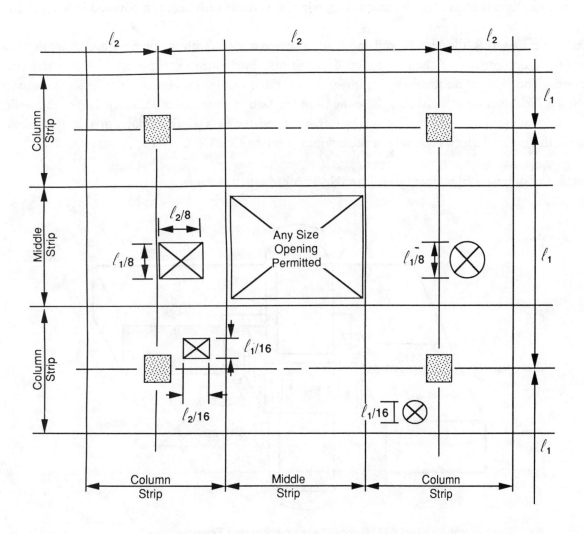

Fig. 18-10 Openings in Slab Systems without Beams

18-13

The total amount of reinforcement required for the panel without openings, in both directions, must be maintained; reinforcement interrupted by any opening must be replaced, one-half on each side of the opening.

11.12.6 Transfer of Moments at Slab-Column Connections

Load transfer between a slab and a column directly, without intermediate load transfer through a beam, is one of the more critical design conditions for two-way slab systems without beams between column supports. Shear strength at an exterior slab-column connection (without spandrel beams) is especially critical, because the total exterior negative slab moment must be transferred directly to the column. This aspect of two-way slab design should not be taken lightly by the designer. Two-way slab systems usually are fairly "forgiving" in the event of an error in the distribution or even in the amount of flexural reinforcement; however, little or no forgiveness is to be expected if a critical error in the provision of shear strength is made.

Note that the provisions of Section 11.12.6 (or Section 13.3.3) do not apply to slab systems with beams framing into the column support. With beams, load transfer from the slab, through the beams, to the columns is considerably less critical. Shear strength in slab systems with beams is covered in Section 13.6.8.

The Code specifies that the unbalanced moment between a slab (without beams) and column must be transferred by eccentricity of shear (Section 11.12.6) and by flexure (Section 13.3.3) at a slab-column connection. The general mechanism of transfer is illustrated in Fig. 18-11. Shear transfer is assumed to occur on a critical section at a distance $d/2$ away from the face of the column (Section 11.12.1.2), while the fraction of unbalanced moment transferred by flexure is resisted by a width of slab equal to the transverse column width c_2, plus $1.5h$ on each side of the column (Section 13.3.3.2).

The fraction of unbalanced moment transferred by eccentricity of shear is:

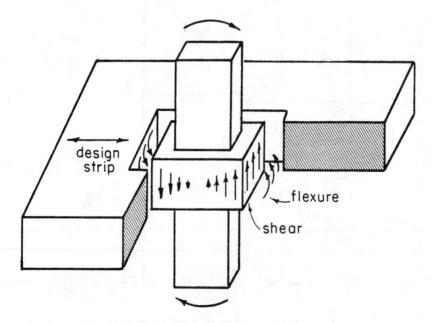

Fig. 18-11 Direct Shear and Moment Transfer

$$\gamma_v = 1 - \frac{1}{1 + (2/3)\sqrt{b_1/b_2}}$$

Eq. (11-42)

and that transferred by flexure is:

$$\gamma_f = \frac{1}{1 + (2/3)\sqrt{b_1/b_2}}$$

Eq. (13-1)

where b_1 and b_2 are the lengths of the sides of the critical shear perimeter for moment transfer, see Fig. 18-12. For the '89 Code, the parameters b_1 and b_2 were introduced to more correctly define the dimensions of the critical shear perimeter for edge and corner columns. The previous parameters, $c_1 + d$ and $c_2 + d$, used in Eqs. (11-42) and (13-1), correctly defined the side lengths for interior columns but were generally not correct for edge and corner columns. Figure 18-13 gives a graphical solution to Eqs. (11-42) and (13-1).

The unbalanced moment transferred by eccentricity of shear is $\gamma_v M_u$, where M_u is the unbalanced moment at the centroid of the critical section. The unbalanced moment M_u at an exterior support of an end span will generally not be computed at the centroid of the critical transfer section in the frame analysis. By the Direct

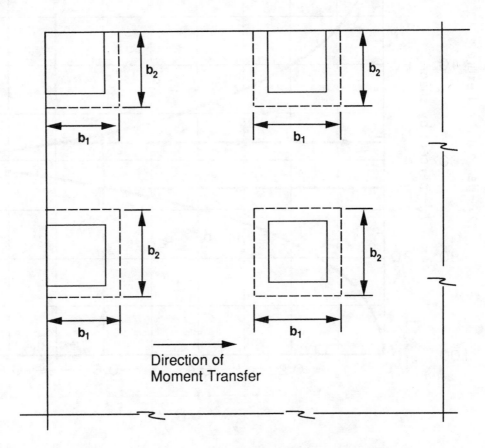

Fig. 18-12 Parameters b₁ and b₂ for Eqs. (11-42) and (13-1)

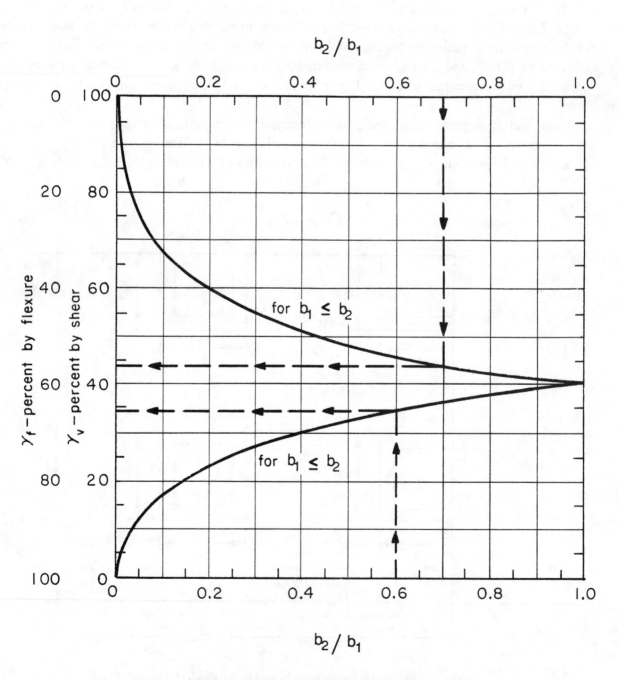

Fig. 18-13 Solution to Eqs. (11-42) and (13-1)

Design Method of Chapter 13, moments are computed at the face of support. Considering the approximate nature of the procedure used to evaluate the stress distribution due to moment-shear transfer, it seems unwarranted to consider a change in moment to the transfer centroid; use of the moment values from frame analysis (centerline of support) or from Section 13.6.3.3 (face of support) directly are accurate enough.

Unbalanced moment transfer between slab and an edge column (without edge beams) requires special consideration when slabs are analyzed for gravity loads using the moment coefficients of the Direct Design Method. To assure adequate shear strength when using the approximate end-span moment coefficient (Section 13.6.3.6), the full nominal moment strength M_n provided by the column strip must be taken as the transfer moment in computing the fraction of unbalanced moment transferred by eccentricity of shear ($\gamma_v M_n$) in accordance with Section 11.12.6. See Part 20 for further discussion of this special shear strength requirement and its application in Example 20.1.

For two-way slabs analyzed for gravity loads by the Equivalent Frame Method, the computed frame moment at the exterior support of an end span is used directly as the unbalanced transfer moment.

The factored shear stress on the critical transfer section is the sum of stresses caused by direct shear and unbalanced moment transfer, as follows:

$$v_{u1} = V_u/A_c + \gamma_v M_u c/J$$

and
$$v_{u2} = V_u/A_c - \gamma_v M_u c'/J$$

The Code-defined distribution of shear stress at an edge column (bending perpendicular to edge) and an interior column is shown in Fig. 18-14. Expressions for A_c, c, c', J/c and J/c' are developed in Tables 18-1 and 18-2.

For slabs without shear reinforcement, the maximum shear stress (V_{u1}) must not exceed the shear strength provided by concrete V_c given in Section 11.12.2. For the moment transfer design condition, Eqs. (11-36) and (11-37) must be expressed in terms of shear stress rather than shear force:

$$v_c = \varphi(2 + 4/\beta_c)\sqrt{f_c'} \qquad\qquad \text{Eq. (11-36)}$$

or
$$v_c = \varphi(2 + \alpha_s/\beta_0)\sqrt{f_c'} \qquad\qquad \text{Eq. (11-37)}$$

Note that for usual design conditions (slab thicknesses and column sizes), the governing shear stress will be $\varphi 4\sqrt{f_c'}$.

For slabs with shear reinforcement consisting of bars or wires, the maximum shear stress (v_{u1}) must not exceed that corresponding to the shear strength ($V_c + V_s$) given in Section 11.12.3:

$$v_c + v_s = \varphi(2\sqrt{f_c'} + A_v f_y/b_o s)$$

18-17

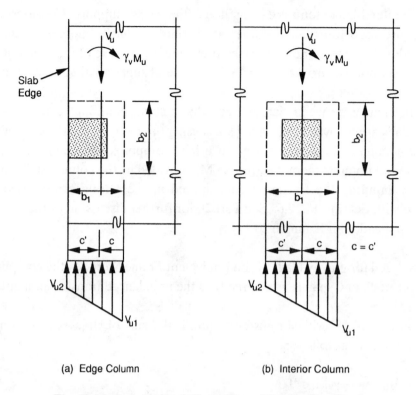

(a) Edge Column (b) Interior Column

**Fig. 18-14 Code-Defined Shear Stress Distribution due to
Moment-Shear Transfer at Slab-Column Connection**

but not greater than $\varphi 6\sqrt{f_c'}$, where A_v is the total area of shear reinforcement provided on the column sides, and b_o is the perimeter of the critical section located at d/2 away from the column perimeter.

To account for shear stress variation around the column, shear reinforcement to be extended from each column side may be determined separately by using an effective "beam" width (c + d):

$$A_v(\text{per side}) = (v_{u1} - \varphi v_c)\frac{(c + d)s}{\varphi f_y}$$

where v_{u1} is the maximum slab shear stress on a column side and $v_c = 2\sqrt{f_c'}$. Note, however, that Commentary Section 11.12.3 recommends that shear reinforcement be placed around the column "...as symmetrical as possible in location, number and spacing." Also, with symmetrical shear reinforcement on all column sides, the reinforcement extending from the column sides with less computed shear stress provides torsional resistance in the strip of slab (slab-beam) perpendicular to the direction of moment-shear transfer. Alternatively, with symmetrical shear reinforcement on all sides of the column, required area A_v may be computed by:

$$A_v = (v_{u1} - \varphi v_c)\frac{b_o s}{\varphi f_y}$$

where A_v is the total area of required shear reinforcement to be extended from the sides of the column, and b_o is the perimeter of the critical section located at d/2 away from the column perimeter.

Shear reinforcement may be terminated at a perimeter b_o where $v_c = 2\sqrt{f_c'}$ equals or exceeds v_{u1} where v_{u1} is based on perimeter b_o defined in Section 11.12.1.2(a). Application of the design criteria for transfer of shear and moment between a slab and column support is illustrated in Example 18.4.

For slabs with shear reinforcement consisting of steel I- or channel-shaped sections (shearheads), the sum of the shear stresses due to (1) direct shear V_u acting on the critical section defined in Section 11.12.4.7 and (2) moment transferred by eccentricity of shear $\gamma_v M_u$ about the centroid of the critical section defined in Section 11.12.1.2(a) must not exceed $\varphi 4\sqrt{f_c'}$.

$$v_u = V_u/b_o d + \gamma_v M_u c/J \le \varphi 4\sqrt{f_c'}$$

where perimeter b_o is defined in Section 11.12.4.7, and c and J are the section properties of the critical section located d/2 from the column perimeter (see Tables 18-1 and 18-2). Note that this seemingly inconsistent stress summation on different critical sections is a conservative estimation of maximum shear stress. See Commentary Section 11.12.6.3. If $v_u > \varphi 4\sqrt{f_c'}$, the length of the shearhead must be increased to reduce the direct shear stress component. Note that the maximum shear stress (v_{u1}) due to combined direct shear and moment transfer must not exceed $\varphi 7\sqrt{f_c'}$ on the critical section at d/2 from the column perimeter; see Fig. 18-7.

INTERIOR COLUMN

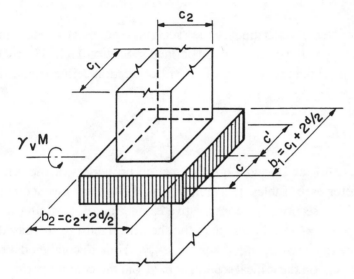

Concrete area of critical section:

$$A_c = 2(a + b)d$$

Modulus of critical section:

$$\frac{J}{c} = \frac{J}{c'} = [ad(a + 3b) + d^3]/3$$

where

$$c = c' = a/2$$

CORNER COLUMN

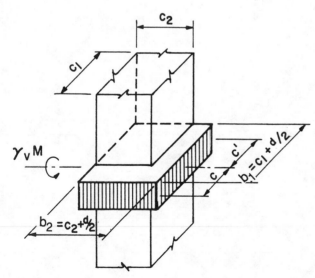

Concrete area of critical section:

$$A_c = (a + b)d$$

Modulus of critical section:

$$\frac{J}{c} = [ad(a + 4b) + d^3(a + b)/a]/6$$

$$\frac{J}{c'} = [a^2d(a + 4b) + d^3(a + b)]/6(a + 2b)$$

where

$$c = a^2/2(a + b)$$

$$c' = a(a + 2b)/2(a + b)$$

EDGE COLUMN (Bending Parallel to edge)

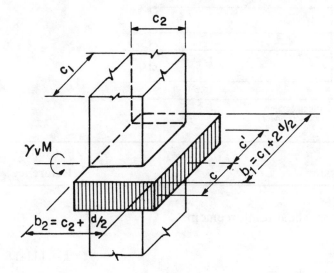

Concrete area of critical section:

$$A_c = (a + 2b)d$$

Modulus of critical section:

$$\frac{J}{c} = \frac{J}{c'} = [ad(a + 6b) + d^3]/6$$

where

$$c = c' = a/2$$

EDGE COLUMN (Bending perpendicular to edge)

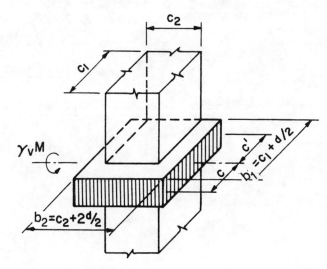

Concrete area of critical section:

$$A_c = (2a + b)d$$

Modulus of critical section:

$$\frac{J}{c} = [2ad(a + 2b) + d^3(2a + b)/a]/6$$

$$\frac{J}{c'} = [2a^2d(a + 2b) + d^3(2a + b)]/6(a + b)$$

where

$$c = a^2/2(a + b)$$
$$c' = a(a + b)/(2a + b)$$

Example 18.1—Shear Strength of Slab at Column Support

Determine two-way action shear strength at an interior column support of a flat plate slab system for the following design conditions.

Column dimensions = 48 in. x 8 in.
Slab effective depth d = 6.5 in.
Specified concrete strength f'_c = 3000 psi

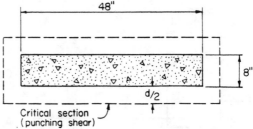

Calculations and Discussion	Code Reference

1. Two-way action shear (punching shear) without shear reinforcement:

    $V_u \leq \varphi V_n$ Eq. (11-1)

    $\leq \varphi V_c$ 11.12.2

2. Effect of loaded area aspect ratio β_c:

    $\varphi V_c = \varphi(2 + 4/\beta_c)\sqrt{f'_c}\, b_o d$ Eq. (11-36)

    $= 0.85(2 + 4/6)\sqrt{3000} \times 138 \times 6.5 = 111.4$ kips

    where $\beta_c = 48/8 = 6$ 11.12.2.2

    $b_o = 2(48 + 6.5 + 8 + 6.5) = 138$ in.   11.12.1.2

    $\varphi = 0.85$ 9.3.2.3

3. Effect of perimeter area aspect ratio β_o:

    $\varphi V_c = \varphi(2 + \alpha_s/\beta_o)\sqrt{f'_c}\, b_o d$ Eq. (11-37)

    $= 0.85(2 + 40/21.2)\sqrt{3000} \times 138 \times 6.5 = 162.3$ kips

    where $\alpha_s = 40$ for interior column support   11.12.2.1

    $\beta_o = b_o/d = 138/6.5 = 21.2$

4. Excluding effect of β_c and β_o:

Example 18.1—Continued

$$\varphi V_c = \varphi 4 \sqrt{f_c'} b_o d$$

$$= 0.85 \times 4\sqrt{3000} \times 138 \times 6.5 = 167 \text{ kips}$$

5. $\varphi V_n = 111.4$ kips.   The aspect ratio β_c results in a 34% reduction in two-way action shear strength.

Example 18.2—Shear Strength for Non-Rectangular Support

For the L-shaped column support shown, check punching shear strength for a factored shear force transfer of $V_u = 105$ kips. Use $f'_c = 3000$ psi. Effective slab depth = 5.5 in.

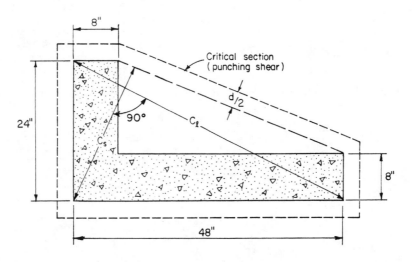

	Code
Calculations and Discussion	**Reference**

1. For shapes other than rectangular, the Commentary recommends that β_c be taken as the ratio of the longest overall dimension of the effective loaded area c_ℓ to the largest overall dimension of the effective loaded area c_s, measured perpendicular to c_ℓ. As shown above, $\qquad$ 11.12.2.1

    $\beta_c = c_\ell/c_s = 54/25* = 2.16$

    For the critical section shown, $\qquad$ 11.12.1.2
    b_0 (perimeter of critical section) = 141 in.*

    *Scaled dimensions are accurate enough

2. Two-way action shear (punching shear) without shear reinforcement:

    $V_u \le \varphi V_n$ $\qquad$ Eq. (11-1)

    $\quad \le \varphi V_c$ $\qquad$ 11.12.2

3. Punching shear strength V_c without shear reinforcement is the lesser of values given by Eqs. (11-36) and (11-37), but not greater than $4\sqrt{f'_c}\,b_0 d$.

Example 18.2—Continued

	Code
Calculations and Discussion	**Reference**

$V_c = (2 + 4/\beta_c)\sqrt{f'_c}\, b_o d$ Eq. (11-36)

$\quad = (2 + 4/2.16)\sqrt{3000} \times 141 \times 5.5 = 163.6$ kips

$V_c = (2 + \alpha_s/\beta_o)\sqrt{f'_c}\, b_o d$ Eq. (11-37)

$\quad = (2 + 40/25.6)\sqrt{3000} \times 141 \times 5.5 = 151.3$ kips

where $\alpha_s = 40$ for interior column support   11.12.2.1

$\beta_o = b_o/d = 141/5.5 = 25.6$

$\quad V_c = 4\sqrt{f'_c}\, b_o d$

$\quad = 4\sqrt{3000} \times 141 \times 5.5 = 169.9$ kips

$\varphi V_c = 0.85(151.3) = 128.6$ kips   9.3.2.3

4. $V_u \leq \varphi V_c$

   105 kips < 128.6 kips O.K.

Example 18.3—Shear Strength of Slab with Shear Reinforcement

Consider an interior panel of a flat plate slab system supported by a 12 in. square column. Panel size $\ell_1 = \ell_2$ = 21 ft. Determine shear strength of slab at column support, and if not adequate, increase the shear strength by shear reinforcement. Overall slab thickness h = 7.5 in. (d = 6 in.).

$f_c' = 3000$ psi
$f_y = 60,000$ psi (bar reinforcement)
$f_y = 36,000$ psi (structural steel)

Superimposed factored load   = 150 psf
Column strip negative moment, $M_u = 175$ ft$-$kips

Calculations and Discussion	Code Reference
A. Two-way action shear (punching shear) without shear reinforcement.	11.12.2
$V_u \leq \varphi V_n$	Eq. (11-1)
$\leq \varphi V_c$	Eq. (11-2)

1. Since there are no shear forces at the center line of adjacent panels (see Fig. 18-1), tributary areas and critical sections for slab shear are shown below.

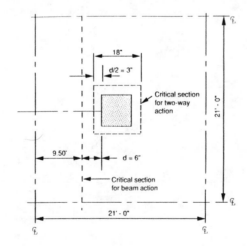

For 7.5 in. slab, factored dead load $w_d = 1.4(94) = 132$ psf   9.2.1
$w_u = 132 + 150 = 282$ psf

18-26

Example 18.3—Continued

	Code
Calculations and Discussion	**Reference**

2. Wide-Beam Action Shear:

Investigation of wide-beam action shear strength is made at the critical section at a distance d from face of column support.　　　　　11.1.3.1

$V_u = 0.282(9.5 \times 21) = 56.3$ kips

$V_c = 2\sqrt{f'_c}b_w d = 2\sqrt{3000}(21 \times 12)\,6 = 165.6$ kips　　　　Eq. (11-3)

$\varphi V_c = 0.85(165.6) = 140.8$ kips　　　　9.3.2.3

56.3 kips < 140.8 kips　　Wide-beam action will rarely control
　　　　　　　　　　　　the shear strength of two-way slab systems.

3. Two-Way Action Shear:

Investigation of two-way action shear strength is made at　　　11.12.1.2(a)
the critical section b_o located at d/2 from the column perimeter.
Total factored shear force to be transferred from slab to column.

$V_u = 0.282(21^2 - 1.5^2) = 123.7$ kips

Shear strength V_c without shear reinforcement　　　　11.12.2.2

$b_o = 4(18) = 72$ in.　　　　11.12.1.2(a)

$\beta_c = 12/12 = 1.0 < 2$ (see Fig. 18−3)　　　　Eq. (11-36)

$\beta_o = b_o/d = 72/6 = 12 < 20$ (see Fig. 18−4)　　　　Eq. (11-37)

$V_c = 4\sqrt{f'_c}b_o d = 4\sqrt{3000} \times 72 \times 6 = 94.6$ kips

$\varphi V_c = 0.85(94.6) = 80.4$ kips

123.7 kips > 80.4 kips

Example 18.3—Continued

	Code
Calculations and Discussion	Reference

Shear strength of slab is not adequate to transfer the
factored shear force V_u = 123.7 kips from slab to column
support. Shear strength may be increased by:

(1) increasing concrete strength f_c'

(2) increasing slab thickness at column support, using drop panel

(3) providing shear reinforcement (bars, wires, or steel I- or channel-shapes)

B.   Increase shear strength by increasing strength of slab concrete.

$$V_u \le \varphi V_n$$ Eq. (11-1)

$$123,700 \le 0.85(4\sqrt{f_c'} \times 72 \times 6)$$

Solving, f_c' = 7000 psi

C.   Increase shear strength by increasing slab thickness at column support
with drop panel. Provide drop panel in accordance with Sections 9.5.3.4
and 13.4.7 (see Fig. 18-15). Minimum overall slab thickness at drop panel
= 1.25(7.5) = 9.375 in. Check 9.5 in. slab thickness (2 in. projection below
slab); d ≈ 8 in. Size of drop panel = 21/6 = 3.50 ft. Try 7 ft × 7 ft drop panel.

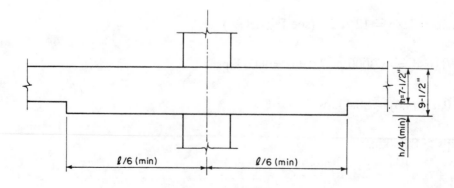

Fig. 18-15 Drop Panel Details

Example 18.3—Continued

Code
Reference

Calculations and Discussion

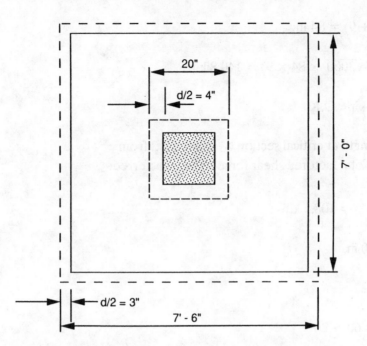

1. Investigate shear strength at critical section b_o located at d/2 from column perimeter. Total factored shear force to be transferred:

   For 2 in. drop panel projection, $w_d = 1.4(25) = 35$ psf

   $$V_u = 0.282(21^2 - 1.67^2) + 0.035(7^2 - 1.67^2)$$

   $$= 123.6 + 1.6 = 125.2 \text{ kips}$$

   $b_o = 4(12 + 8) = 80$ in.   11.12.1.2(a)

   $\beta_c = 1.0 < 2$ Eq. (11-36)

   $\beta_o = b_o/d = 80/8 = 10.0 < 20$ Eq. (11-37)

   $\varphi V_c = \varphi 4\sqrt{f_c'}b_o d = 0.85(4\sqrt{3000} \times 80 \times 8) = 119.2$ kips

   125.2 kips > 119.2 kips   Increase slab thickness at drop panel
   to 10.5 in. (3 in. projection below slab);
   $d \approx 9$ in.

Example 18.3—Continued

$$b_o = 4(12 + 9) = 84 \text{ in.}$$

$$\varphi V_c = 0.85(4\sqrt{3000} \times 84 \times 9) = 140.8 \text{ kips}$$

125.2 kips < 140.8 kips O.K.

2. Investigate shear strength at critical section b_o located d/2 from edge of drop panel. Total factored shear force to be transferred:

$$V_u = 0.282(21^2 - 7.5^2) = 108.5 \text{ kips}$$

$$b_o = 4(84 + 6) = 360 \text{ in.}$$

11.12.1.2(b)

$$\beta_c = 84/84 = 1.0 < 2$$

Eq. (11-36)

$$\beta_o = b_o/d = 360/6 = 60 > 20$$

Eq. (11-37)

$$\varphi V_c = \varphi(2 + \alpha_s/\beta_o)\sqrt{f_c'}\,b_o d$$

Eq. (11-37)

$$= 0.85(2 + 40/60)\sqrt{3000} \times 360 \times 6 = 268.2 \text{ kips}$$

108.5 kips < 268.2 kips O.K.

Note the significant decrease in shear strength at
edge of drop panel due to large perimeter area aspect ratio β_o:

$$V_c = 2.67\sqrt{f_c'}\,b_o d$$

A 7 ft × 7 ft drop panel with a 3 in. projection below slab will provide
adequate shear strength for superimposed factored loads of 150 psf.

D. Increase shear strength by bar reinforcement. See Fig. 18-6.

11.12.3

1. Check maximum shear strength permitted with bars.

11.12.3.2

Example 18.3—Continued

	Code
Calculations and Discussion	**Reference**

$V_u \le \varphi V_n$

Eq. (11-1)

$\quad \le \varphi(6\sqrt{f_c'}\,b_o d)$

$\quad \le 0.85(6\sqrt{3000} \times 72 \times 6) = 120.7$ kips

$V_u = 0.282(21^2 - 1.5^2) = 123.7$ kips

123.7 kips > 120.7 kips   Say O.K.

A more exact calculation of "d" would result in approximately equal shear values

2. Determine shear strength provided by concrete with bar shear reinforcement.

11.12.3.1

$V_c = 2\sqrt{f_c'}\,b_o d = 2\sqrt{3000} \times 72 \times 6 = 47.3$ kips

$\varphi V_c = 0.85(47.3) = 40.2$ kips

3. Design shear reinforcement in accordance with Section 11.5.
   Required area of shear reinforcement A_v is computed by:

$A_v = (V_u - \varphi V_c)s/\varphi f_y d$

Assume s = 3 in. (maximum spacing permitted = d/2)

11.5.4.1

$A_v = (123.7 - 40.2)3/0.85 \times 60 \times 6 = 0.82$ in.^{2}

where A_v is total area of shear reinforcement required
on the four sides of the column (see Fig. 18-6).

A_v(per side) = 0.82/4 = 0.21 in.^{2}

A check on the calculations can easily be made; referring to Fig. 18-6,

$V_u \le \varphi 2\sqrt{f_c'}\,b_o d + \varphi A_v f_y d/s$

$\quad \le 0.85(2\sqrt{3000} \times 72 \times 6)/1000 + 0.85(4 \times 0.22 \times 60 \times 6/3)$

Example 18.3—Continued

	Code
Calculations and Discussion	**Reference**

$\le 40.2 + 89.8 = 130$ kips

123.7 kips < 130 kips O.K.

4. Determine distance from sides of column where stirrups
    may be terminated. (See Fig. 18-6)

11.12.1.2(c)

$V_u \le \varphi V_c$

Eq. (11-1)

$\le \varphi 2\sqrt{f'_c}\, b_o d$

For square column (see sketch below),

$b_o = 4(12 + a\sqrt{2})$

$123,700 \le 0.85 \times 2\sqrt{3000} \times 4(12 + a\sqrt{2})6$

Solving, a = 30.7 in.

Stirrups may be terminated d/2 = 3" inside the critical perimeter b_o.

Use 10 - #3 U-stirrups @ 3 in. spacing along each column line as shown below.

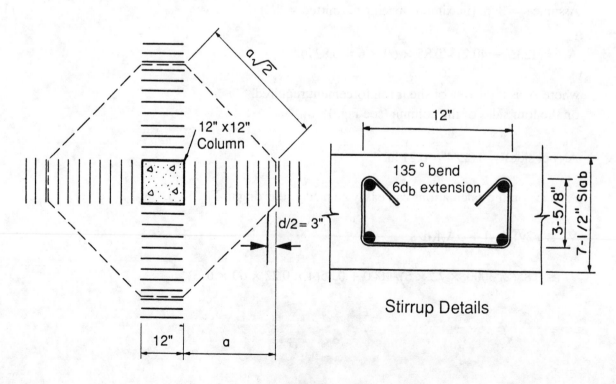

Stirrup Details

Example 18.3—Continued

	Code
Calculations and Discussion	**Reference**

E. Increase shear strength by steel I shapes (shearheads) 11.12.5

1. Check maximum shear strength permitted with steel shapes. See Fig. 18-7. 11.12.4.8

$V_u = 0.282(21^2 - 1.5^2) = 123.7$ kips

$V_u \le \varphi V_n$ Eq. (11-1)

$\le \varphi(7\sqrt{f_c'}\,b_o d)$ 11.12.1.2(a)

$\le 0.85(7\sqrt{3000} \times 72 \times 6) = 140.8$ kips

123.7 kips < 140.8 kips   O.K.

2. Determine minimum required perimeter b_o at critical section 11.12.4.8
to limit shear strength to $V_n = 4\sqrt{f_c'}\,b_o d$. See Fig. 18-7.

$V_u \le \varphi V_n$ Eq. (11-1)

$123,700 \le 0.85\,(4\sqrt{3000}\,b_o \times 6)$

Solving, $b_o = 111$ in. 11.12.4.7

3. Determine required length of shearhead arm ℓ_v 11.12.4.7
to satisfy $b_o = 111$ in. at 3/4 the distance $(\ell_v - c_1/2)$

$$b_o \approx 4\sqrt{2}\left[\frac{c_1}{2} + \frac{3}{4}(\ell_v - c_1/2)\right]$$

with $b_o = 111$ in. and $c_1 = 12$ in., solving, $\ell_v = 24.2$ in.

4. To ensure that premature flexural failure of shearhead
does not occur before shear strength of slab is reached,
required plastic moment strength M_p of each shearhead arm:

$$\varphi M_p = \frac{V_u}{2\eta}\left[h_v + \alpha_v(\ell_v - c_1/2)\right]$$ Eq. (11-40)

Example 18.3 —Continued

	Code
Calculations and Discussion	**Reference**

For a four (identical) arm shearhead, $\eta = 4$;
and assuming $h_v = 4$ in. and $\alpha_v = 0.25$:

11.12.4.5

$$\varphi M_p = \frac{123.7}{2(4)}\left[4 + 0.25(24.2 - 12/2)\right] = 132.2 \text{ in.} - \text{kips}$$

Required $M_p = 132.2/0.9 = 146.9$ in.-kips

Try S4 x 9.5 (plastic modulus $Z_x = 4.04$ in.^{3})

$$M_p = Z_x f_y = 4.04(36) = 145.4 \text{ in.} - \text{kips} \approx 146.9 \text{ in.} - \text{kips say O.K.}$$

5. Check depth limitation of S4 x 9.5 shearhead

11.12.4.2

$70t_w = 70(0.326) = 22.82$ in. $> h_v = 4$ in. O.K.

6. Determine location of compression flange of steel
    shape with respect to compression surface of slab.

11.12.4.4

$0.3d = 0.3(6) = 1.8$ in. $< 0.75 + 2(0.625) = 2$ in. N.G.

(3/4 in. cover + 2#5 bars).

Therefore, the #5 bars in the lower layer must be cut; see Fig. 18-16

7. Determine relative stiffness ratio α_v.

(S4 × 9.5)
$A_{st} = 2.79$ in.^{2}
$I_s = 6.79$ in.^{4}

11.12.4.5

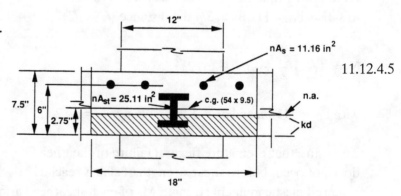

A_s provided for $M_u = 175$ ft-kips #5 @ 5 in.

c.g. of S4 x 9.5 from compression face $= 0.75 + 2 = 2.75$ in.
Effective slab width $= c_2 + d = 12 + 6 = 18$ in.

Example 18.3—Continued

Transformed Section Properties:

For $f_c' = 3000$ psi, use $n = E_s/E_c = 9$
Steel transformed to equivalent concrete,

$$nA_s = 9(4 \times 0.31) = 11.16 \text{ in.}^2$$

$$nA_{st} = 9(2.79) = 25.11 \text{ in.}^2$$

Neutral axis of composite cracked slab section may be obtained
by equating the static moments of the transformed areas.

$$18(kd)^2/2 = 25.11(2.75 - kd) + 11.16(6 - kd)$$

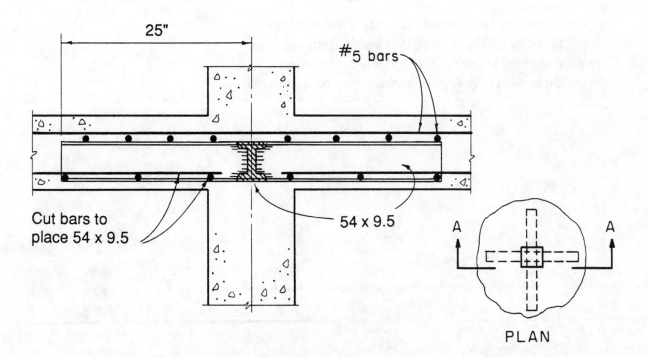

Fig. 18-16 Details of Shearhead Reinforcement

Example 18.3—Continued

	Code
Calculations and Discussion	Reference

Solving , kd = 2.36 in.

Composite $I = 18(2.36)^3/3 + n(I_s \text{ steel shape}) + 11.16(3.64)^2 + 25.11 (0.39)^2$

$$= 78.7 + 9(6.79) + 147.9 + 3.8 = 291.5 \text{ in.}^4$$

$\alpha_v = nI_s/I \text{ composite} = 9(6.79)/291.5 = 0.21 > 0.15 \text{ O.K.}$

Therefore, S4 × 9.5 section satisfies all code requirements for shearhead reinforcement. See Fig. 18-18 for details.

8. Determine contribution of shearhead to   11.12.4.9
    negative moment strength of column strip.

$$M_v = \frac{\varphi \alpha_v V_u}{2\eta} (\ell_v - c_1/2)$$ Eq. (11-41)

$$= \frac{0.9 \times 0.21 \times 123.7}{2 \times 4}(25 - 6) = 55.5 \text{ in.} - \text{kips}$$

However, M_v must not exceed either $M_p = 146.9$ in.-kips
or 0.3(175)12 = 630 in.- kips, or the change in column strip
moment over the length ℓ_v. For this design, approximately
3% of the column strip negative moment may be considered
resisted by the shearhead reinforcement.

Example 18.4—Shear Strength of Slab with Transfer of Moment

Consider an exterior (edge) panel of a flat plate slab system supported by a 16 in. square column. Determine shear strength for transfer of direct shear and moment between slab and column support. Overall slab thickness h = 7 in. (d ≈5.75 in.). Consider two loading conditions:

(1) Total factored shear force V_u = 30 kips
Column strip negative moment capacity M_n = 37.5 ft-kips

(2) Alternatively, V_u = 65 kips
M_n = 37.5 ft-kips

f'_c = 3000 psi
f_y = 60,000 psi

	Code
Calculations and Discussion	Reference

A. Section properties for shear stress computations

Referring to Fig. 18-16: edge column-bending perpendicular to edge,

$b_1 = c_1 + d/2 = 16 + 5.75/2 = 18.88$ in.
$b_2 = c_2 + d = 16 + 5.75 = 21.75$ in.
$c = b_1^2/(2b_1 + b_2)$
$\quad = 18.88^2/(2 \times 18.88 + 21.75)$
$\quad = 5.99$ in.
$A_c = (2b_1 + b_2)d = 342.2$ in.^{2}

$J/c = \left[2b_1d(b_1 + 2b_2) + d^3(2b_1 + b_2)/b_1\right]/6 = 2358$ in.^{3}
$c' = b_1 - c = 18.88 - 5.99 = 12.89$ in.
$J/c' = (J/c)(c/c') = 2358(5.99/12.89)$
$\quad\quad = 1096$ in.^{3}

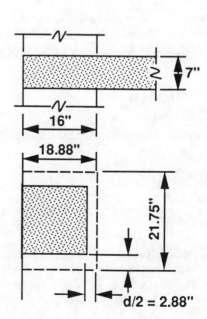

Example 18.4—Continued

Calculations and Discussion

B. Loading condition (1), $V_u = 30$ kips
    $M_u = 37.5$ ft-kips

1. Portion of total column strip negative moment M_u
    to be transferred by eccentricity of shear.
    For $b_1/b_2 = 18.88/21.75 = 0.87$

11.12.6.1

$$\gamma_v = 1 - \frac{1}{1 + 2/3\sqrt{b_1/b_2}}$$

Eq. (11-42)

$$= 1 - \frac{1}{1 + 2/3\sqrt{0.87}} = 0.38$$

$\gamma_v M_u = 0.38(37.5) = 14.3$ ft−kips

Note: Considering the approximate nature of the
moment transfer procedure, assume the transfer
moment $\gamma_v M_u$ is at centroid of critical transfer section.

2. Check shear strength of slab without shear reinforcement.

Combined shear stress at inside edge of critical transfer section.

$v_{u1} = V_u/A_c + \gamma_v M_u/(J/c)$

$= 30,000/342.2 + 14.3 \times 12,000/2358 = 87.7 + 72.8$

$= 160.5$ psi

Combined shear stress at outside edge of critical transfer section:

$v_{u2} = V_u/A_c - \gamma_v M_u/(J/c')$

$= 30,000/342.2 - 14.3 \times 12,000/1096 = 87.7 - 156.6$

$= 68.9$ psi

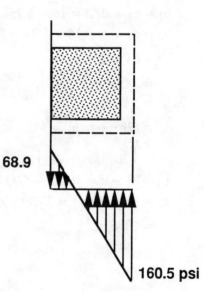

68.9

160.5 psi

Example 18.4—Continued

Calculations and Discussion

Code
Reference

Permissible shear stress

$\varphi V_n = \varphi 4\sqrt{f_c'} = 0.85(4\sqrt{3000}) = 186.2$ psi

160.5 psi < 186.2 psi

Slab shear strength is adequate
for the required shear and moment
transfer between slab and column.

Design for the portion of unbalanced moment transferred by flexure,
$\gamma_f M_u = (1 - \gamma_v)M_u = (1 - 0.38)37.5 = 23.3$ ft−kips, must also be considered.
See Examples 20.1 and 21.1.

C. Loading condition (2), $V_u = 65$ kips
    $M_u = 37.5$ ft-kips

1. Check shear strength of slab without shear reinforcement.

    Combined shear stress at inside edge of critical transfer section:

    $v_{u1} = 65,000/342.2 + 14.3 \times 12,000/2358$

    $\quad = 189.9 + 72.8 = 262.7$ psi

    At outside edge:

    $v_{u2} = 189.9 - 156.6 = 33.3$ psi

    Permissible shear stress = 186.2 psi

    $\varphi v_n = 262.7$ psi $> v_{u1} = 186.2$ psi

    Shear reinforcement must be provided to carry excess shear stress,
    either bar reinforcement or steel I- or channel-shapes (shearheads).

    Increase slab shear strength by bar reinforcement.

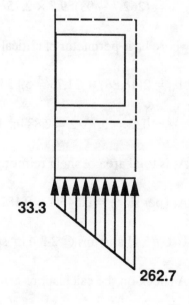

Example 18.4—Continued

	Code
Calculations and Discussion	**Reference**

2. Check maximum shear stress permitted with bar reinforcement. 11.12.3.2

$$v_{u1} \leq \varphi 6\sqrt{f_c'}$$ 11.12.1.2(a)
$$\leq 0.85(6\sqrt{3000}) = 279.3 \text{ psi}$$

   262.7 psi < 279.3 psi   O.K.

3. Determine shear stress carried by concrete with bar reinforcement. 11.12.3.1

$$\varphi v_c = \varphi 2\sqrt{f_c'} = 0.85(2\sqrt{3000}) = 93 \text{ psi}$$

4. With symmetrical shear reinforcement on all sides of column, required A_v is computed by

$$A_v = (v_{u1} - \varphi v_c)b_o s/\varphi f_y$$

$$= (262.7 - 93)59.5 \times 2.75/0.85 \times 60{,}000 = 0.545 \text{ in.}^2$$

   where b_o is perimeter of critical section located at d/2 from column perimeter

   $$b_o = 2(18.88) + 21.75 = 59.5 \text{ in.}$$

   $$s = d/2 = 5.75/2 = 2.88 \text{ in.} \text{Use } s = 2{-}3/4 \text{ in.}$$

   A_v is total area of shear reinforcement required on the three sides of the column.

   $$A_v \text{ (per side)} = 0.545/3 = 0.182 \text{ in.}^2$$

   Use #3 U-stirrups @ 2-3/4 in. spacing ($A_v = 0.22 \text{ in.}^2$)

   A check on the calculations can easily be made; for #3 U-stirrups @ 2-3/4 in:

   $$v_c + v_s = \varphi(2\sqrt{f_c'} + A_v f_y/b_o s)$$

   $$= 0.85\left[2\sqrt{3000} + (3 \times 0.22)60{,}000/59.5 \times 2.75\right]$$

Example 18.4—Continued

	Code
Calculations and Discussion	Reference

$$= 0.85(109.5 + 242.0) = 298.8 \text{ psi}$$

262.7 psi < 298.8 psi O.K.

5. Determine distance from sides of column where stirrups may be terminated.

$$V_u \leq \varphi V_c$$

$$\leq \varphi 2\sqrt{f_c'} b_o d$$

where $b_o = (2a\sqrt{2} + 16 \times 3)$

$$65,000 \leq 0.85 \times 2\sqrt{3000}\,(2a\sqrt{2} + 48)5.75$$

Solving, $a = 25.95$ in.

Stirrups required = $(25.95 - d/2)/2.75 = 8.4$
(Stirrups may be terminated at $d/2 = 2.88$ in. inside perimeter b_o)

Use 9 - #3 U-stirrups @ 2-3/4 in. spacing along
the three sides of the column. Use similar stirrup
detail as for Example 18.3.

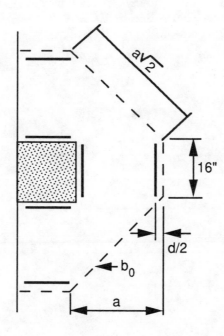

Deep Flexural Members

UPDATE FOR THE '89 CODE

For ACI 318-89, design of **continuous** deep flexural members for shear strength must be based on the regular beam design procedures of Sections 11.1 through 11.5 and must also satisfy the provisions of Sections 11.8.4, 11.8.9 and 11.8.10. The special shear strength provisions of Section 11.8 are intended to apply only to **simply supported** deep beams. Recent tests of continuous deep beams have indicated that the special shear provisions of Section 11.8 are inadequate for continuous members. A new Section 11.8.3 has been introduced directing the engineer to base shear strength of continuous members on the design provisions of Sections 11.1– 11.5. Thus, Section 11.8 is basically limited to simply supported deep beams.

GENERAL CONSIDERATIONS

The Code gives two definitions for "deep" members. For **flexure**, members with overall depth-to-clear-span ratios greater than 2/5 for continuous spans or 4/5 for simple spans are defined as "deep" (Section 10.7.1). For **shear**, a "deep" member is one with an effective depth-to-span ratio of 1/5 or greater (Section 11.8.1).

No specific provisions for designing deep members for flexure are found in the Code, but such members must be designed "taking into account nonlinear distribution of strain and lateral buckling" (Section 10.7.1). Appropriate references for the design of deep beams for flexure are given in the Commentary and at the end of Part 19.

Information on lateral buckling is more difficult to find. Fortunately, most walls and beams receive lateral support from supported floor or roof members, so lateral buckling of the compression flange is rarely a problem; see Fig. 19-1(a). Some form of lateral support is required at intervals not exceeding 50 times the least width of the compression flange (Section 10.4.1), even if the member is free-standing; see Fig. 19-1(b). For free-standing walls, a lateral stability check should be made and an adequate margin of safety against lateral buckling provided.

Lateral buckling in a vertical direction (Fig. 19-2), particularly near concentrated loads and at supports, can be checked by the column moment magnifier, or by numerical or energy methods. A simplified procedure

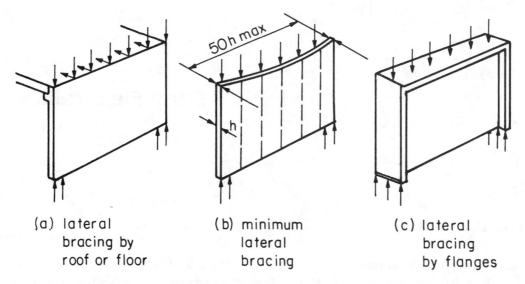

Fig. 19-1   Lateral Support for Deep Flexural Members

(a) lateral bracing by roof or floor

(b) minimum lateral bracing

(c) lateral bracing by flanges

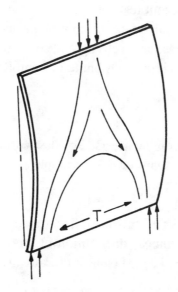

Fig. 19-2   Lateral Buckling of Deep Flexural Members

for wall-like beams (tilt-up panels) is provided in Reference 19.1. If the height-to-thickness ratio of a member is limited to 25, buckling should not be a problem.

10.7   DEEP FLEXURAL MEMBERS

The Code requires that "nonlinear distribution of strain" be taken into account in flexural design of deep members. Elastic analysis by Dischinger and others (19.2, 19.3) has shown that the shape of the elastic stress curve can be quite different from the linear distribution usually assumed. At midspan, the neutral axis moves away from the loaded face of the member as the depth to clear span increases. Over the supports, the resultant elastic tensile forces can be within a third of the member depth from the top fiber. Although tensile stresses found by Dischinger are usually less than would be expected from a linear analysis, such stresses can be as much as 31 percent higher.

Nonlinear distribution of strains and stresses assumes an uncracked, homogeneous cross section and, therefore, does not apply to design at the ultimate moment strength (nominal moment strength M_n for design), since cracking usually occurs before the moment strength can be developed. This would imply that the tensile reinforcement required to develop the moment strength M_n could be placed near the extreme tensile fiber as is customary for ordinary flexural members. Reference 19.3, however, recommends that tensile reinforcement be distributed throughout the tensile area and centered at or near the resultant of the tensile forces, so that, when cracking occurs, there will not be a sudden shift in the location of the resultant tensile force. Both methods of sizing and placing reinforcement are illustrated in Design Example 19.1 and it is left to the judgment of the designer to choose the more appropriate method.

Development of horizontal tensile reinforcement in single-span simply-supported deep members requires special consideration. Since moments increase rapidly from zero at the face of the support, the reinforcement may not have sufficient anchorage length to develop the required moment strength near the support. Tensile bars may be anchored by development length (if available), standard hooks, or by special anchorage devices.

The most radical departure from a linear strain and stress distribution is in compression areas at or near supports of continuous members. Compressive forces may be confined to the bottom 5 or 10 percent of the member depth and compressive stresses may be as high as 14 times those indicated by linear strain and stress distribution.[19.2] In these cases, reinforcing details require special consideration. If service load compressive stresses approach about $0.45f'_c$, it may be necessary to treat the compression area as an axially loaded member, using laterally tied reinforcement to carry the compressive forces as the moment strength is approached.

11.8   SHEAR STRENGTH OF DEEP FLEXURAL MEMBERS

The special shear strength provisions for deep flexural members apply only to members having a clear-span-to-effective-depth ratio (ℓ_n/d) less than 5. The deep members must be loaded at the top face as shown in Fig. 19-3. Since the principal tensile forces in deep members are primarily horizontal (vertical cracking), horizontal shear reinforcement is effective in resisting the tensile forces. Truss bars are, therefore, not recommended as shear reinforcement in deep members.

For the '89 Code, different shear design procedures are prescribed for simply supported and continuous deep flexural members. Design of simply supported members for shear must be based on the special provisions of Section 11.8. Design of continuous members for shear must be based on the regular beam design procedures of Sections 11.1 through 11.5 as well as 11.8.4, 11.8.9 and 11.8.10. Also, when loads are applied through the sides or bottom of the member, simply supported or continuous, the shear design provisions of Section 11.1 through 11.5 must be used.

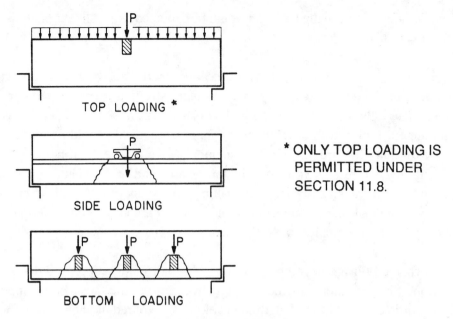

Fig. 19-3   Loading of Deep Flexural Members

11.8.2 Simply Supported Deep Flexural Members

For shear design of simply supported deep members (Section 11.8.2), the maximum factored shear force V_u is calculated at a distance from the face of the support defined as 0.15 times the clear span for uniformly loaded beams or 0.50 times the shear span "a" for beams with concentrated loads, but in no case greater than "d" distance from face of support. (Section 11.8.5).

The factored shear force V_u must not exceed the shear strength provided by $\varphi V_n = \varphi(V_c + V_s)$, where V_c is the shear strength provided by concrete and V_s is the shear strength provided by shear reinforcement, both horizontal and vertical. V_c may be computed from either the more complex Eq. (11-30), which takes into account the effects of the tensile reinforcement and $M_u/V_u d$ at the critical section, or may be determined from the simpler Eq. (11-29), $V_c = 2\sqrt{f_c'} b_w d$. Eq. (11-30) is illustrated in Fig. 19-4.

The first step in design is to check if V_u is less than φV_c, with V_c equal to $2\sqrt{f_c'} b_w d$. If shear strength provided by concrete is not adequate to carry the factored shear force V_u, calculate φV_s for minimum shear reinforcement and add to φV_c. Using the minimum shear reinforcement requirements of Section 11.8.9 ($A_v = 0.0015 b_w s$) and Section 11.8.10 ($A_{vh} = 0.0025 b_w s_2$), shear strength Eq. (11-31) reduces to

$$V_s = (0.029d - 0.001\ell_n) \, b_w f_y/12$$

Shear strength with minimum shear reinforcement becomes

$$V_u \leq \varphi(V_c + V_s)$$
substituting for V_c from Eq. (11-29)
$$V_u \leq \varphi[2\sqrt{f_c'} b_w d + (0.029d - 0.001\ell_n) b_w f_y/12]$$

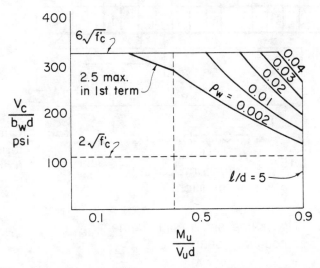

$$V_c = [3.5 - (2.5)(M_u/V_u d)]$$
$$\times \left[1.9\sqrt{f'_c} + 2500\,\rho_w \,\frac{V_u d}{M_u}\right] b_w d$$

PLOTTED FOR 3,000 psi
CONCRETE, SIMPLE SPAN
AND UNIFORM LOAD.

Fig. 19-4   Shear Strength of "Simply Supported" Deep Flexural members

If shear strength with minimum shear reinforcement is still not adequate, the more complex Eq. (11-30) can be used to calculate a higher concrete shear strength, or additional shear reinforcement, A_v and A_{vh}, may be added to increase the shear strength of the section. Shear reinforcement required at the critical section must be provided throughout the span in all cases (Section 11.8.11).

For design convenience, required area of shear reinforcement A_v and A_{vh} in terms of factored shear force V_u can be computed directly by

$$\frac{A_v}{s}\left(\frac{1 + \ell_n/d}{12}\right) + \frac{A_{vh}}{s2}\left(\frac{11 - \ell_n/d}{12}\right) = \frac{V_u - \varphi V_c}{\varphi f_y d}$$

Shear strength $V_n = V_c + V_s$ must not be taken greater than:

$$\text{for}\,\ell_n/d \le 2.....V_n = 8\sqrt{f'_c}\,b_w d \quad \text{and}$$
$$\text{for}\,\ell_n/d > 2.....V_n = \frac{2}{3}(10 + \ell_n/d)\sqrt{f'_c}\,b_w d$$

At the upper limit of $\ell_n/d = 5$, $V_n = 10\sqrt{f'_c}\,b_w d$...the same as for ordinary beams.

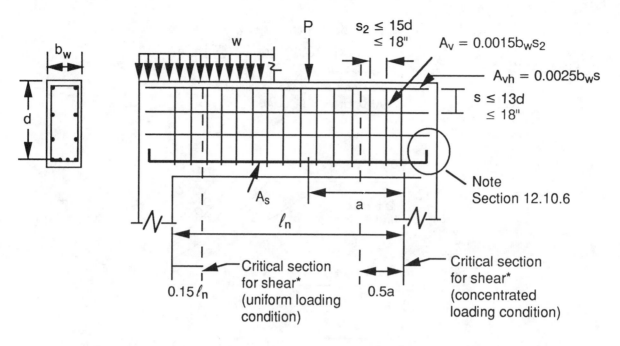

Fig. 19-5   Design Details for Simply-Supported Deep Beams ($l_n/d < 5$)

A strict reading of Code Section 11.8.8 would appear to suggest that no shear reinforcement is needed in a simply supported deep beam unless V_u exceeds V_c, which may be as high as $\varphi 6\sqrt{f_c'}$. However, a deep beam without shear reinforcement is not recommended. It would be more appropriate to conform with Sections 11.8.9 and 11.8.10 for the design of simply supported deep beams.

Design details for simply-supported deep members is illustrated in Fig. 19-5.

11.8.3 Continuous Deep Flexural Members

For shear design of continuous deep members (Section 11.8.3), the design procedure is the same as for ordinary beams. The maximum factored shear force V_u is calculated at "d" distance from face of support. The factored shear force V_u must not exceed the shear strength provided by the section $\varphi(V_c + V_s)$, where V_c may be computed from either the more complex Eq. (11-6), or may be determined from the simpler Eq. (11-3), $V_c = 2\sqrt{f_c'}b_w d$. Section 11.8.3 also specifies that design of continuous deep flexural members must also satisfy Sections 11.8.4, 11.8.9 and 11.8.10. Section 11.8.4 sets an upper limit to V_n. Sections 11.8.9 and 11.8.10 specify minimum vertical and horizontal shear reinforcement, respectively.

The first step in design is to check if V_u is less than φV_c, with V_c equal to $2\sqrt{f_c'}b_w d$. If shear strength provided by concrete is not adequate to carry the factored shear force V_u, calculate φV_s for minimum shear reinforcement and add to φV_c. Using the minimum shear reinforcement of Section 11.8.9 ($A_v = 0.0015b_ws$), shear strength Eq. (11-17) reduces to

$$V_s = 0.0015f_yb_wd$$

For Grade 40 bars...$V_s = 60b_wd$
For Grade 60 bars...$V_s = 90b_wd$

Note that minimum shear reinforcement by Section 11.8.9 is greater than that required by Eq. (11-14). Shear strength with minimum shear reinforcement becomes

$$V_u \leq \varphi(2\sqrt{f_c'}\,b_wd + 0.0015f_yb_wd)$$

If shear strength with minimum shear reinforcement is still not adequate, the more complex Eq. (11-6) can be used to calculate a higher concrete shear strength, or additional shear reinforcement A_v may be added to increase the shear strength of the section, where

$$\frac{A_v}{s} = \frac{V_u - \varphi V_c}{\varphi f_y d}$$

Shear reinforcement may be varied along the length of span as for ordinary beams; however, a minimum area of both horizontal and vertical reinforcement, A_v and A_{vh}, in accordance with Sections 11.8.9 and 11.8.10 must be provided full span length irrespective of shear force conditions. Note that the spacing "s" of vertical shear reinforcement A_v must not exceed $d/5$, nor 18 in....somewhat closer maximum spacing than that permitted for ordinary beams. Note also that the horizontal shear reinforcement A_{vh} does not contribute to shear strength V_s for continuous deep members.

As for simply supported deep members, in continuous deep members, shear strength V_n must not be taken greater than:

$$\ell_n/d \leq 2.....V_n = 8\sqrt{f_c'}\,b_wd$$
$$\ell_n/d > 2.....V_n = \frac{2}{3}(10 + \ell_n/d)\sqrt{f_c'}\,b_wd$$

REFERENCES

19.1   "Tilt-Up Load-Bearing Walls," A Design Aid, Portland Cement Association, Skokie, IL, EB074D, 1980, 28 pp. A "column model" (a panel considered hinged along loaded edges and free along vertical edges) is used to compute load capacities of reinforced concrete tilt-up wall panels (with a central curtain of reinforcement) that rest on continuous footings. An approximate but rational means of evaluating effects of isolated footings and sustained loads on capacity of these slender walls is included, as well as load-moment interaction charts and tables and design applications.

19.2   "Design of Deep Girders," Portland Cement Association, Skokie, IL, IS079D, 10 pp. Presents analysis of deep girders according to elastic theory of Franz Dischinger, with special studies and numerical examples added. Data and procedures illustrated apply to design of deep wall-like members such as in bins, hoppers, and foundation walls.

19.3   Chow, Li., Conway, H., and Winter, G., "Stresses in Deep Beams," *Transactions,* ASCE, Vol 118, 1953, pp. 686-708.

Example 19.1—Design of Deep Flexural Members

This design example has been adapted from the PCA publication, "Design of Deep Girder," and modified in accordance with the ACI Code and the strength design method. The publication may be used directly to design deep members by the Alternate Design Method, or it may be used to locate the tensile resultants and check cracking under the Strength Design Method of the Code.

An interior span of a continuous deep girder is shown below.
Width of beam and support, b_w = 15 in.
Uniform loads: Live load = 10 kips/ft   Dead load = 10 kips/ft

f_c' =   3000 psi

f_y = 40,000 psi

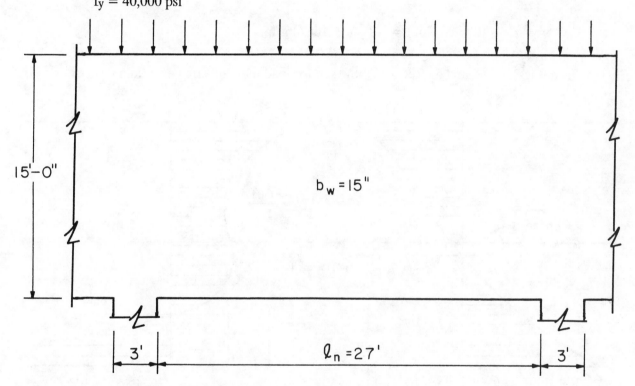

A.   Determine if deep beam provisions apply

$$\frac{\ell_n}{h} = \frac{27}{15} = 1.8$$

For flexure.....$\frac{\ell_n}{h} < 2.5$ 10.7.1

For shear......$\frac{\ell_n}{d} < 5$ 11.8.1

Design for flexure and shear must satisfy deep beam provisions of Sections 10.7 and 11.8.

Example 19.1—Continued

	Code
Calculations and Discussion	**Reference**

B. Design for Flexure   10.7

   1. Determine moment stresses (at service loads)

      Refer to Reference 19.2, "Design of Deep Girders," for design constants as follows:

$$\varepsilon = \frac{C}{L} = \frac{3}{30} = \frac{1}{10} \qquad \beta = \frac{H}{L} = \frac{15}{30} = \frac{1}{2}$$

$$w = 10 + 10 = 20 \text{ kip/ft} = \frac{20{,}000}{12} = 1667 \text{ lbs/in.}$$

      From Figs. 2, 3, 4, and 5 (Ref. 19.2), the service load moment stresses at mid-span and support are:

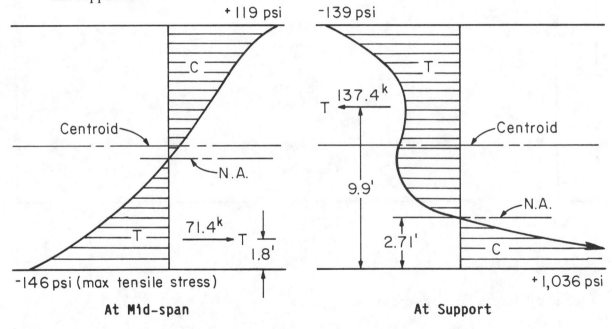

      To avoid cracking at service loads, tensile stresses should not exceed the modulus of rupture.

$$f_r = 7.5\sqrt{f'_c} = 7.5\sqrt{3000} = 411\text{psi} > 146 \quad \text{O.K.}$$ Eq. (9-9)

      Designers using the Alternate Design Method of Appendix B may proceed directly with the flexural design as outlined in Reference 19.2, calculating the required reinforcement from the tensile resultants (T) and distributing the reinforcement appropriately. The following procedure is in accordance with the Strength Design Method of the Code.

Example 19.1—Continued

2. Determine required moment strengths

$$U = 1.4(10) + 1.7(10) = 31.0 \text{ k/ft}$$ Eq. (9-1)

@ mid-span (Ref. 19.2):

$$M_u = \frac{w\ell_u^2(1 - \varepsilon^2)}{24}$$

$$= \frac{31 \times 30^2(1 - 0.1^2)}{24} = 1151 \text{ ft–kip}$$

@ support (Ref. 19.2):

$$M_u = \frac{w\ell_u^2(1 - \varepsilon)(2 - \varepsilon)}{24}$$

$$= \frac{31 \times 30^2(1 - 0.1)(2 - 0.1)}{24} = 1988 \text{ ft–kips}$$

Factored moment and shear diagrams are determined as follows:

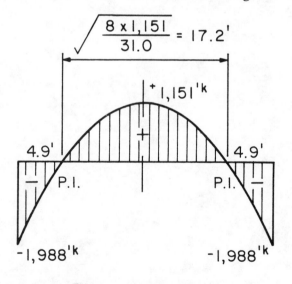

Factored moments

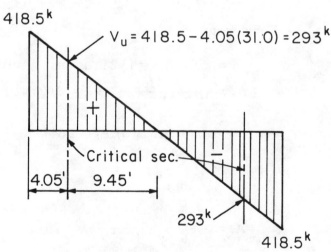

Factored shear force

3. Determine flexural reinforcement

Method 1—(using full effective depth d)

Example 19.1—Continued

$d = 15.0 - \dfrac{3}{12} = 14.75$ ft; assume $j_u = 0.9$

$A_s = \dfrac{M_u}{\varphi f_y j_u d}$

@ mid-span:

$A_s = \dfrac{1151}{0.9 \times 40 \times 0.9 \times 14.75} = 2.41$ in.^{2}

Use 4 #7 bars, $A_s = 2.40$ in.^{2}

@ support:

$A_s = \dfrac{1988}{0.9 \times 40 \times 0.9 \times 14.75} = 4.16$ in.^{2}

Use 2 #10 and 2 #9 bars, $A_s = 4.54$ in.^{2}

Locate primary reinforcement A_s (top and bottom) as close to tension face as cover and other reinforcement allow.

Method 2—(using depth to tensile resultant)

@ mid-span:

$d = 15.0 - 1.8 = 13.2$ ft; assume $j_u = 0.9$

$A_s = \dfrac{1151}{0.9 \times 40 \times 0.9 \times 13.2} = 2.69$ in.^{2}

Use 6 #6 bars (3 each face), $A_s = 2.64$ in.^{2}

@ support:

$d = 9.9$ ft.

$A_s = \dfrac{1988}{0.9 \times 40 \times 0.9 \times 9.9} = 6.20$ in.^{2}

Use 14 #6 bars (7 each face), $A_s = 6.16$ in.^{2}

Reinforcement determined by this method should be distributed in the total tensile area, approximately centered on the resultant tensile force.

Example 19.1—Continued

Calculations and Discussion	Code Reference

4. Determine minimum horizontal and vertical reinforcement in side faces of girder. The minimum "wall" type reinforcement will be used in addition to the primary tensile reinforcement.

Horizontal reinforcement:

$$A_{vh} = 0.0025 b_w s_2$$
$$= 0.0025 \times 15 \times 12 = 0.45 \text{ in.}^2/\text{ft}$$

$$s_2 \le \frac{d}{3}, 3b_w, \text{ or } 18 \text{ in.}$$

Use #5 bars @ 16 in. (each face), $A_{vh} = 0.46$ in.^{2}/ft.

10.7.4

14.3.3
11.8.10

Vertical reinforcement:

$$A_v = 0.0015 b_w s$$
$$= 0.0015 \times 15 \times 12 = 0.27 \text{ in.}^2/\text{ft}$$

$$s \le \frac{d}{5}, 3b_w, \text{ or } 18 \text{ in.}$$

Use #4 bars @ 17.5 in. (each face), $A_v = 0.27$ in.^{2}/ft

14.3.2
11.8.9

Example 19.1—Continued

	Code
Calculations and Discussion	**Reference**

Space all horizontal bars at 8 in. or 16 in. for simplicity. Extension of the negative moment reinforcement beyond the point of inflection is normally the largest of d, $12d_b$, or $\ell_n/16$. For deep members, this extension requirement (d controls) would not allow any bar cut-off. For deep members with minimum reinforcement A_{vh} provided throughout the span in addition to the flexural reinforcement, the $\ell_n/16$ extension seems adequate. This excessive d extension is unnecessary. $\ell_n/16 = 1.7$ ft is used in this example. Cut-off location = 4.9 + 1.7 = 6.6 ft.

An alternate distribution of reinforcement might be to use the #6 bars @ 9½ in. throughout, since the area of reinforcement is the same as alternating #5 and #6 bars @ 8 in. as suggested above.

C. Design for Shear

Design of continuous deep flexural members for shear is based on the regular beam design procedures of Sections 11.1 through 11.5. For this example, taking maximum factored shear force V_u at "d" distance from support is not reasonable; design for shear will be based on V_u at $0.15\ell_n$ from face of support.   11.8.3

$0.15\ell_n = 0.15(27) = 4.05$ ft, $V_u = 293$ kips

1. Determine shear strength without shear reinforcement

$$\varphi V_c = \varphi(2\sqrt{f_c'}\, b_w d) \qquad\qquad \text{Eq. (11-3)}$$

$$= \frac{0.85(2\sqrt{3000} \times 15 \times 14.83 \times 12)}{1,000}$$

$\varphi V_c = 248$ kips

$V_u = 293$ kips $>$ 248 kips   N.G.

Shear strength provided by concrete φV_c is not adequate to carry the factored shear force V_u.

2. Determine shear strength with minimum shear reinforcement

$$\varphi V_s = \varphi(0.0015 f_y b_w d) \qquad\qquad \text{11.8.9}$$
$$= 0.85(0.0015 \times 60 \times 15 \times 14.83 \times 12) = 204 \text{ kips} \qquad \text{9.3.2.3}$$

$\varphi(V_c + V_s) = 248 + 204 = 452$ kips $>$ 293 kips   O.K.

Check maximum shear strength permitted:

Example 19.1—Continued

	Code
Calculations and Discussion	**Reference**

$$\frac{\ell_n}{d} = \frac{27}{14.83} = 1.82 \ < \ 2$$

$$V_n = 8\sqrt{f_c'}\,b_w d = 8\sqrt{3000} \times 15 \times 14.83 \times 12 = 1170 \text{ kips}$$ 11.8.4

$$\varphi V_n = 0.85(1170) = 995 \text{ kips} \ > \ 293 \text{ kips} \qquad \text{O.K.}$$

Shear strength with minimum shear reinforcement will usually be adequate for deep beams.

$$A_v = 0.0015 b_w s = 0.0015 \times 15 \times 12 = 0.27 \text{ in.}^2/\text{ft.}$$ 11.8.9

Use #4 bars @ 17.5 in. (each face)

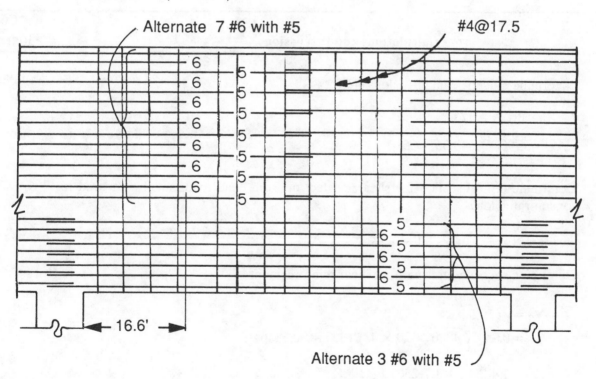

Example 19.2—Design of Deep Flexural Members

Design required shear reinforcement for the essentially simply supported transfer girder supporting the single column shown below. Column loads: dead load = 200 kips, live load = 250 kips.

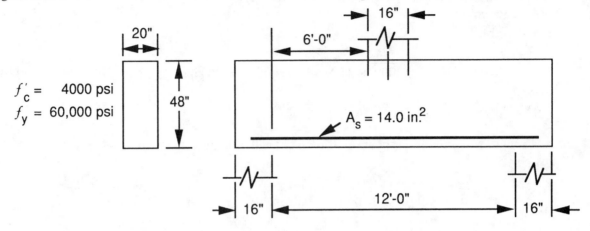

	Code
Calculations and Discussion	**Reference**

1. Determine if deep beam provisions apply

    $d \approx 48 - 5 = 43$ in.

    $\dfrac{\ell_n}{d} = \dfrac{12 \times 12}{43} = 3.35 \ < \ 5$ 11.8.1

2. Determine critical section for shear; neglect uniform dead load...small compared to   11.8.5
    concentrated loads.

    $0.5a = 0.5(6) = 3$ ft. $\ < \ d = 3.58$ ft.

3. Determine shear strength without shear reinforcement   11.8.6

    $$\varphi V_c = \varphi(2\sqrt{f'_c}\,b_w d)$$ Eq. (11-29)
    $$= 0.85(2\sqrt{4000} \times 20 \times 43)/1000 = 92.5 \text{ kips}$$ 9.3.2.3

    $$V_u = \frac{1.4(200) + 1.7(250)}{2} = 352.5 \text{ kips} \ > \ 92.5 \qquad \text{N.G.}$$

    Shear strength provided by concrete φV_c is not adequate to carry the factored shear force V_u.

4. Check maximum shear strength permitted.   11.8.4

    For $\dfrac{\ell_n}{d} = 3.35$, $V_n = \dfrac{2}{3}\left(10 + \dfrac{\ell_n}{d}\right)\sqrt{f'_c}\,b_w d$ Eq. (11-28)

Example 19.2—Continued

	Code
Calculations and Discussion	**Reference**

$$V_n = \frac{2}{3}(10 + 3.35)\sqrt{4000} \times 20 \times 43/1000 = 484 \text{ kips}$$

$$\varphi V_n = 0.85(484) = 411 \text{ kips} \quad > \quad 352.5 \qquad \text{O.K.}$$

5. Determine shear strength with minimum shear reinforcement

11.8.9
11.8.10

$$\varphi V_s = \varphi(0.029d - 0.001\ell_n)b_w \frac{f_y}{12}$$

$$= 0.85 \, (0.029 \times 43 - 0.001 \times 12 \times 12)20 \times 60/12$$

$$\varphi V_s = 94 \text{ kips}$$

$$\varphi(V_c + V_s) = 92.5 + 94 = 186.5 \text{ kips} \quad < \quad 352.5 \qquad \text{N.G.}$$

6. Determine shear strength φV_c using more complex Eq. (11-30) at critical section.

11.8.7

$$V_c = \left(3.5 - 2.5 \frac{M_u}{V_u d}\right)\left(1.9 \sqrt{f_c'} + 2500 \rho_w \frac{V_u d}{M_u}\right)b_w d$$

At critical section:

Eq. (11-30)

$$\frac{M_u}{V_u d} = \frac{352.5 \times 3}{352.5 \times 3.58} = 0.84$$

$$(3.5 - 2.5 \times 0.84) = 1.4 \quad < \quad 2.5$$

$$\rho_w = \frac{A_s}{b_w d} = \frac{14.0}{20 \times 43} = 0.0163$$

$$V_c = 1.4\left[1.9\sqrt{4000} + \frac{2500(0.0163)}{0.84}\right]20 \times 43 = 203 \text{ kips}$$

$$\varphi V_c = 0.85(203) = 173 \text{ kips}$$

Check $\varphi V_c \leq \varphi(6\sqrt{f_c'} b_w d = 0.85(6\sqrt{4000} \times 20 \times 43) = 277 \text{ kips} \quad > \quad 173 \qquad \text{O.K.}$

11.8.7

$$\varphi(V_c + V_s) = 173 + 94 = 267 \text{ kips} \quad < \quad 352.5 \qquad \text{N.G.}$$

Greater than minimum shear reinforcement must be provided.

7. Determine required shear reinforcement using Eq. (11-31) modified to reflect A_v and A_{vh} directly.

11.8.8

$$\frac{A_v}{s}\left(\frac{1 + \ell_n/d}{12}\right) + \frac{A_{vh}}{s2}\left(\frac{11 - \ell_n/d}{12}\right) = \frac{V_u - \varphi V_c}{\varphi f_y d}$$

Eq. (11-31)

Example 19.2—Continued

$$\frac{V_u - \varphi V_c}{\varphi f_y d} = \frac{352.5 - 173}{0.85 \times 60 \times 43} = 0.0819 \text{ in.}^2/\text{in.}$$

Use minimum horizontal reinforcement:

11.8.10

$$A_{vh} = 0.0025 b_w s_2 = 0.0025 \times 20 \times 12 = 0.60 \text{ } in.^2/\text{ft}$$

$$s_2 \le \frac{d}{3} = \frac{43}{3} = 14.3 \text{ in.} \quad < \quad 18 \text{ in.}$$

Use #5 @ 12 in. (each face), $A_{vh} = 0.62$ in.^{2}/ft

$$\frac{A_{vh}}{s_2} = \frac{2 \times 0.31}{12} = 0.0517 \text{ in.}^2/\text{in.}$$

$$\frac{A_v}{s}\left(\frac{1 + 3.35}{12}\right) + 0.0517\left(\frac{11 - 3.35}{12}\right) = 0.0819$$

Eq. (11-31)

Solving for $\frac{A_v}{s} = 0.1349$ in.^{2}/ft

$$A_v = 0.1349(12) = 1.619 \text{ in.}^2/\text{ft}$$

$$s < \frac{d}{5} = \frac{43}{5} = 8.6 \text{ in.} \quad < \quad 18 \text{ in.}$$

Use #5 @ 4½ in. (each face), $A_v = 1.65$ in.^{2}/ft

Alternatively, "tighten-up" the spacing of the horizontal bars to #5 @ 9 in. (each face), $\frac{A_{vh}}{s_2} = \frac{2 \times 0.31}{9} = 0.0689$ in.^{2}/in.

$$\frac{A_v}{s}\left(\frac{1 + 3.35}{12}\right) + 0.0689\left(\frac{11 - 3.35}{12}\right) = 0.0819$$

Solving for $\frac{A_v}{s} = 0.1048$ in.^{2}/in.

$$A_v = 0.1048(12) = 1.257 \text{ in.}^2/\text{ft.}$$

Use #5 @ 6 in. (each face), $A_v = 1.24$ in.^{2}/ft.   O.K.

Example 19.2—Continued

8. Check shear strength provided using #5 @ 9 in. (each face) for horizontal and #5 @ 6 in. (each face) for vertical shear reinforcement.

$$V_s = \left[\frac{A_v}{s} \left(\frac{1 + \ell_n /d}{12} \right) + \frac{A_{vh}}{s_2} \left(\frac{11 - \ell_n /d}{12} \right) \right] f_y d$$

Eq. (11-31)

$$= \left[0.1033 \left(\frac{1 + 3.35}{12} \right) + 0.0689 \left(\frac{11 - 3.35}{12} \right) \right] 60 \times 43 = 210.1 \text{ kips}$$

$\varphi V_s = 0.85(210.1) = 178.6$ kips

$\varphi(V_c + V_s) = 173 + 178.6 = 351.6$ kips ≈ 352.5 O.K.

9. Both horizontal and vertical shear reinforcement required at the critical section must be provided throughout the span. See reinforcement details below.

11.8.11

Use #5 @ 9 in. (each face) for horizontal
and #5 @ 6 in. (each face) for vertical shear reinforcement.

Note: The main flexural reinforcement must be extended into the supports as far as cover requirements and proximity of other steel permit...and if need be, bent upwards to obtain adequate embedment.

12.10.6

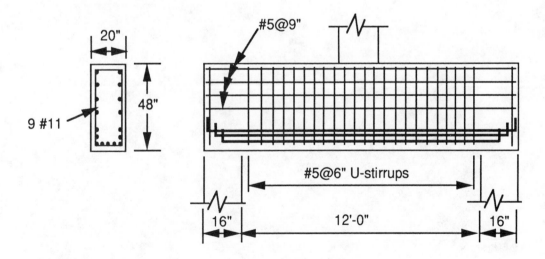

Reinforcement Details

Two-Way Slab Systems

13.1 SCOPE

Figure 20-1 shows the various types of two-way reinforced concrete slab systems in use at the present time, that may be designed according to Chapter 13.

A solid slab supported on beams on all four sides [Fig. 20-1(a)] was the original slab system in reinforced concrete. With this system, if the ratio of the long to the short side of a slab panel is two or more, load transfer is predominantly by bending in the short direction and the panel essentially acts as a one-way slab. As the ratio of the sides of a slab panel approaches unity (square panel), significant load is transferred by bending in both orthogonal directions, and the panel should be treated as a two-way rather than a one-way slab.

As time progressed and technology evolved, the column-line beams gradually began to disappear. The resulting slab system consisting of solid slabs supported directly on columns is called the flat plat [Fig. 20-1(b)]. The two-way flat plate is very efficient and economical and is currently the most widely used slab system for multistory construction, such as motels, hotels, dormitories, apartment buildings, and hospitals. In comparison to other concrete floor/roof systems, flat plates can be constructed in less time and with minimum labor costs because the system utilizes the simplest possible formwork and reinforcing steel layout. The use of flat plate construction also has other significant economic aspects. For instance, because of the shallow thickness of the floor system, story heights are automatically reduced, resulting in smaller overall height of exterior walls and utility shafts; shorter floor-to-ceiling partitions; reductions in plumbing, sprinkler, and duct risers; and a multitude of other items of construction. In cities like Washington, D.C., where the maximum height of buildings is restricted, the thin flat plate permits the construction of the maximum number of stories on a given plan area. Flat plates also provide for the most flexibility in the layout of columns, partitions, small openings, etc. An additional advantage of flat plate slabs that should not be overlooked is their inherent fire resistance. Slab thickness required for structural purposes will, in most cases, provide the fire resistance required by the general building code, without having to apply spray-on fire proofing, or install a suspended ceiling. This is of particular importance where job conditions allow direct application of the ceiling finish to the flat plate soffit, eliminating the need for suspended ceilings. Additional cost and construction time savings are then possible as compared to other structural systems.

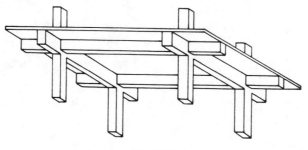

(a) Two-way Slab

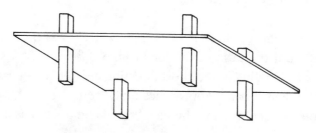

(b) Flat Plate

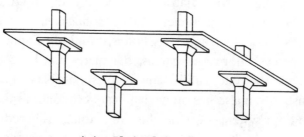

(c) Flat Slab

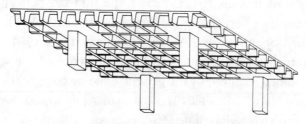

(d) Waffle Slab (Two-Way Joist Slab)

Fig. 20-1   Types of Two-Way Slab Systems

The principal limitation on the use of flat plate construction is imposed by shear around the columns (Section 13.3.4). For heavy loads or long spans, the flat plate is often thickened locally around the columns creating what are known as drop panels. When a flat plate incorporates drop panels, it is called a flat slab [Fig. 20-1(c)]. Also for reasons of shear around the columns, the column tops are sometimes flared, creating column capitals. For purposes of design, a column capital is part of the column, whereas a drop panel is part of the slab (Sections 13.7.3 and 13.7.4).

Waffle slab construction [Fig. 20-1(d)] consists of rows of concrete joists at right angles to each other with solid heads at the column (needed for shear strength). The joists are commonly formed by using standard square "dome" forms. The domes are omitted around the columns to form the solid heads. For design purposes, waffle slabs are considered as flat slabs with the solid heads acting as drop panels (Section 13.1.3). Waffle slab construction allows a considerable reduction in dead load as compared to conventional flat slab construction. Thus, it is particularly advantageous where the use of long span and/or heavy loads is desired without the use of deepened drop panels or support beams. The geometric shape formed by the joist ribs is often architecturally desirable.

13.1.4 Deflection Control—Minimum Slab Thickness

Minimum thickness/span ratios enable the designer to avoid extremely complex deflection calculations in routine designs. Deflections of two-way slab systems need not be computed if the overall slab thickness meets the minimum requirements specified in Section 9.5.3. Minimum slab thicknesses for flat plates, flat slabs and (waffle slabs) based on Table 9.5(c), and two-way beam-supported slabs based on Eqs. (9-11), (9-12), and (9-13) are summarized in Table 20-1, where ℓ_n is the clear span length in the long direction of a two-way slab panel. The tabulated values are the controlling minimum thicknesses governed by interior, side, or corner panels assuming a constant slab thickness for all panels making up a slab system. Practical

Table 20-1—Minimum Thickness for Two-Way Slab Systems
(Grade 60 Reinforcement)

Two-Way Slab Systems		Minimum h
Flat Plat Flat Plate with Edge Beams[1]	$[h_{min} = 5 \text{ in.}]$	$\ell_n/30$ $\ell_n/33$
Flat Slab[2] Flat Slab[2] with Edge Beams	$[h_{min} = 4 \text{ in.}]$	$\ell_n/33$ $\ell_n/36$
Two-Way Slab[3] (square panels) Two-Way Slab[3] (rectangular 2:1 panels)	$[h_{min} = 3\frac{1}{2} \text{ in.}]$	$\ell_n/40.7$ $\ell_n/49.3$

[1]Edge beam-to-slab stiffness ratio $\alpha > 0.8$ (Section 9.5.3.5)
[2]Drop panel length $> \ell/3$, depth ≥ 1.25 h (Section 9.5.3.4)
[3]Average beam-to-slab stiffness ratio $\alpha_m = 2.0$ [Section 9.5.3.3(b)]

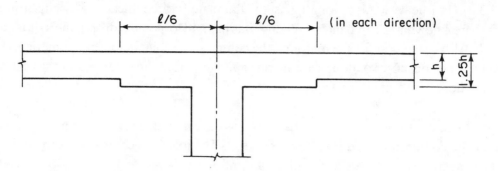

Fig. 20-2   Drop Panel Details (Section 9.5.3.4)

edge beam sizes will usually provide beam-to-slab stiffness ratios α greater than the minimum specified value of 0.8, so that a 10% increase in slab thickness for flat plate and flat slab panels with discontinuous edges would not be required. A "Standard" size drop panel that would allow a 10% reduction in the minimum required thickness of a flat slab floor system is illustrated in Fig. 20-2. Note that a drop of larger size and depth may be used if required for shear strength; however, a corresponding lesser slab thickness is not permitted unless deflections are computed. The values for two-way slabs are based on an average beam-to-slab stiffness ratio α_m = 2.0; a lesser slab thickness may be used with higher beam-to-slab stiffness ratios per Eq. (9-11).

For design convenience, minimum thicknesses for the six types of two-way slab systems listed in Table 20-1 are plotted in Fig. 20-3.

Refer to Part 8 for a general discussion on control of deflections for two-way slab systems, including design examples of deflection calculations for two-way slabs.

13.2   DEFINITIONS

13.2.1   Design Strip

For analysis of a two-way slab system by either the Direct Design Method (Section 13.6) or the Equivalent Frame Method (Section 13.7), the slab system is divided into design strips consisting of a column strip and half middle strip(s) as defined in Sections 13.2.1 and 13.2.2, and as illustrated in Fig. 20-4. The column strip is defined as having a width equal to one-half the transverse or longitudinal span, whichever is smaller. The middle strip is bounded by two column strips. Some judgment is required in applying the definitions given in Section 13.2.1 for column strips with varying span lengths along the design strip.

The reason for specifying that the column strip width be based on the shorter of ℓ_1, or ℓ_2 is to account for the tendency for moment to concentrate about the column line when the span length of the design strip is less than its width.

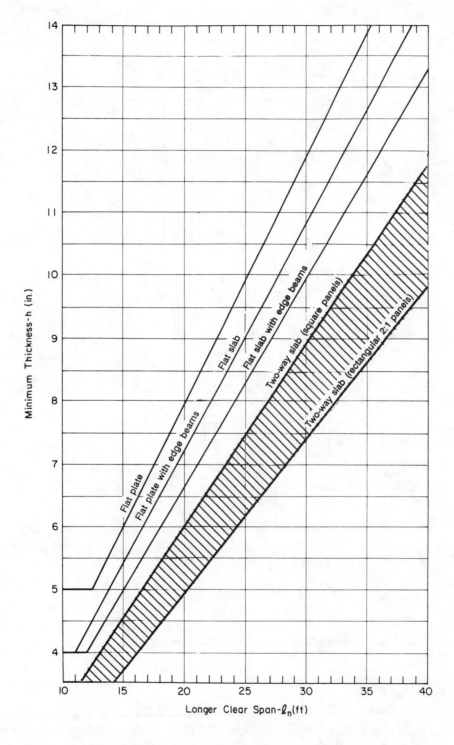

Fig. 20-3 Minimum Slab Thickness for Two-Way Slab Systems
(See Table 20-1)

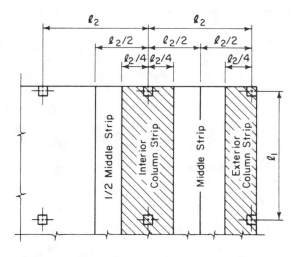

(a) Column Strip for $\ell_2 \leq \ell_1$

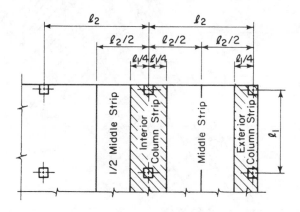

(b) Column Strip for $\ell_2 > \ell_1$

Fig. 20-4   Definition of Design Strips

13.2.4   Effective Beam Section

For slab systems with beams between supports, the beams include portions of the slab as flanges, as shown in Fig. 20-5. Design constants and stiffness parameters used with the Direct Design and Equivalent Frame analysis methods are based on the effective beam section as shown.

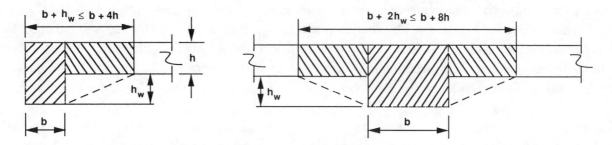

Fig. 20-5   Effective Beam Section

13.3   DESIGN PROCEDURES

Section 13.3.1 permits design (analysis) of two-way slab systems by "any" method that satisfies Code-defined strength requirements (Sections 9.2 and 9.3), and all applicable Code serviceability requirements, including specified limits on deflections (Section 9.5.3).

13.3.1.1   Gravity Load Analysis—For designer "convenience," two methods of analysis of two-way slab systems under gravity loads are addressed in Chapter 13: the simpler Direct Design Method of Section 13.6, and the more complex Equivalent Frame Method of Section 13.7. The Direct Design Method is an approximate method using moment coefficients directly, while the Equivalent Frame (elastic analysis) Method is more exact. The approximate analysis procedure of the Direct Design Method will give reasonably conservative moment values for the stated design conditions for slab systems within the limitations of Section 13.6.1.

Both methods are for analysis under gravity loads only, and are limited in application to buildings with columns and/or walls laid out on a basically orthogonal grid; i.e., where column lines taken longitudinally and transversely through the building are mutually perpendicular. Both methods are applicable to slabs with or without beams between supports. Note that neither method applies to slab systems with beams spanning between other beams; the beams must be located along column lines and be supported by columns or other essentially nondeflecting supports at the corners of the slab panels.

13.3.1.2   Lateral Load Analysis—For lateral load analysis of frames, effects of slab cracking and reinforcement on stiffness of frame members must be taken into account. During the life of a structure, ordinary occupancy loads and volume changes due to shrinkage and temperature effects will cause cracking of slabs. To ensure that lateral drift caused by wind or earthquakes is not underestimated, cracking of slabs must be considered in stiffness assumptions for lateral drift calculations. Note the "performance statement" on lateral load analysis of frames in Section 13.3.1.2. See Commentary Section 13.3.1.2 for guidance on stiffness assumption for lateral load analysis.

While the equivalent frame defined in Section 13.7 is limited to gravity load analysis, it can be used for lateral load analysis, if modified to account for reduced stiffness of the slab-beams. The stiffness of slab members is affected not only by cracking, but also by other parameters such as ℓ_2/ℓ_1, c_1/ℓ_1, c_2/c_1, and on concentration of reinforcement in the slab width defined in Section 13.3.3.2 for unbalanced moment transfer by flexure. This added concentration of reinforcement increases stiffness by preventing premature yielding and softening in the slab near the column supports. Consideration of actual stiffness as affected by cracking and other factors is important for lateral load analysis because lateral displacement can significantly affect the moments in the columns, especially in tall unbraced frame buildings. Also, actual lateral displacement for a single story, or for the total height of a building is an important consideration for building stability and performance.

Cracking reduces stiffness of the slab-beams as compared with that of an uncracked floor. Such stiffness reductions are particularly important when lateral loads are considered on unbraced frames. The magnitude of the loss of stiffness due to cracking will depend on the type of slab system and reinforcement details. For example, prestressed slab systems with reduced slab cracking due to prestressing, and slab systems with large beams between columns will lose less stiffness than a conventional reinforced flat plate framing system.

Since it is difficult to evaluate the effect of cracking on stiffness, it is usually sufficient to use a lower bound value. On the assumption of a fully cracked slab with minimum reinforcement at all locations, a stiffness for the slab-beam equal to one-fourth that based on the gross area of concrete ($K_{sb}/4$) should be reasonable. A detailed evaluation of the effect of cracking may also be made. Since slabs normally have more than minimum reinforcement and are not fully cracked, except under very unusual conditions, the one-fourth value should be expected to provide a safe lower bound for stiffness under lateral loads.

For both vertical and lateral load analyses, moments at critical sections of the slab-beams are transversely distributed in accordance with Sections 13.6.4 (column strips) and 13.6.6 (middle strips).

Moments from an Equivalent Frame (or Direct Design) analysis for gravity loading may be combined with moments from a lateral load analysis (Section 13.3.1.3).

13.3.4   Shear in Two-Way Slab Systems

If two-way slab systems are supported by beams or walls, the slab shear is seldom a critical factor in design, as the shear force at factored loads is generally well below the shear capacity of the concrete.

In contrast, when two-way slabs are supported directly by columns as in flat plates or flat slabs, shear around the columns is of critical importance. Shear strength at an exterior slab-column connection (without edge beams) is especially critical because the total exterior negative slab moment must be transferred directly to the column. This aspect of two-way slab design should not be taken lightly by the designer. Two-way slab systems will normally be found to be quite "forgiving" if an error in the distribution or even in the amount of flexural reinforcement is made, but there will be no forgiveness if a critical lapse occurs in providing the required shear strength.

For slab systems supported directly by columns, it is advisable at an early stage in design to check the shear strength of the slab in the vicinity of columns as illustrated in Fig. 20-6.

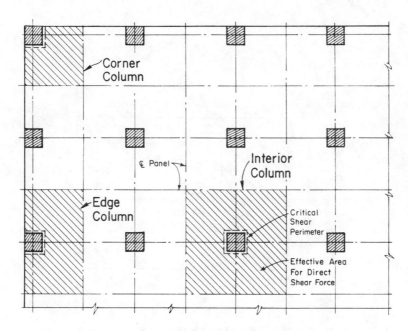

Fig. 20-6   Critical Locations for Slab Shear Strength

Two types of shear need to be considered in the design of flat plates or flat slabs supported directly on columns. The first is the familiar one-way or beam-type shear, which may be critical in long narrow slabs. Analysis for beam shear considers the slab to act as a wide beam spanning between the columns. The critical section is taken at a distance d from the face of the column. Design against beam shear consists of checking for satisfaction of the requirement indicated in Fig. 20-7(a). Beam shear in slabs is seldom a critical factor in design, as the shear force is usually well below the shear capacity of the concrete.

Two-way or "punching" shear is generally the more critical of the two types of shear in slab systems supported directly on columns. Punching shear considers failure along the surface of a truncated cone or pyramid around a column. The critical section is taken perpendicular to the slab at a distance d/2 from the perimeter of a column. The shear force V_u to be resisted can be easily calculated as the total factored load on the area bounded by panel centerlines around the column, less the load applied within the area defined by the critical shear perimeter, see Fig. 20-6.

In the absence of a significant moment transfer from the slab to the column, design against punching shear consists of making sure that the requirement of Fig. 20-7(b) is satisfied. For practical design, only direct shear (uniformly distributed around the perimeter b_o) occurs around interior slab-column supports where no (or insignificant) moment is to be transferred from the slab to the column. Significant moments may have to be carried when unbalanced gravity loads on either side of an interior column or horizontal loading due to wind must be transferred from the slab to the column. At exterior slab-column supports, the total exterior slab moment from gravity loads (plus any lateral load moments due to wind or earthquake) must be transferred directly to the column.

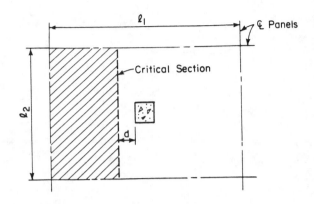

(a) Beam Shear

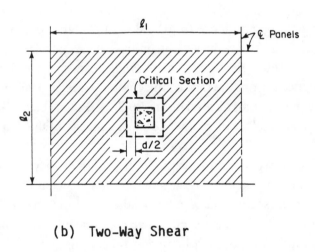

(b) Two-Way Shear

Fig. 20-7   Direct Shear at an Interior Slab-Column Support (see Fig. 20-6).

13.3.3   Transfer of Moment in Slab-Column Connections

Transfer of moment between a slab and a column takes place by a combination of flexure (Section 13.3.3) and eccentricity of shear (Section 11.12.6.1). Shear due to moment transfer is assumed to act on a critical section at a distance $d/2$ from the face of the column—the same critical section around the column as that used for direct shear transfer [Fig. 20-7(b)]. The portion of the moment transferred by flexure is assumed to be transferred over a width of slab equal to the transverse column width c_2, plus 1.5 times the slab thickness (1.5h) on either side of the column (Section 13.3.3.2). Concentration of negative reinforcement is to be used to resist moment on this effective slab width. The combined shear stress due to direct shear and moment transfer often governs the design, especially at the exterior slab-column supports.

The portions of the total unbalanced moment M_u to be transferred by eccentricity of shear and by flexure are given by Eqs. (11-42) and (13-1), where $\gamma_v M_u$ is considered transferred by eccentricity of shear, and $\gamma_f M_u$ is considered transferred by flexure. Referring to Fig. 18-13, at an interior square column with $b_1 = b_2$, 40% of the moment is transferred by eccentricity of shear ($\gamma_v M_u = 0.40\ M_u$), and 60% by flexure ($\gamma_f M_u = 0.60 M_u$), where M_u is the transfer moment at the centroid of the critical section. The moment M_u at the exterior slab-column support will generally not be computed at the centroid of the critical transfer section. In the Equivalent Frame analysis, moments are computed at the column centerline. In the Direct Design Method, moments are computed at the face of support. Considering the approximate nature of the procedure used to evaluate the stress distribution due to moment transfer, it seems unwarranted to consider a change in moment to the critical section centroid; use of the moment values at column centerline (EFM) or at face of support (DDM) directly would usually be accurate enough.

The factored shear stress on the critical transfer section is the sum of the direct shear and the shear caused by moment transfer,

$$v_u = \frac{V_u}{A_c} + \gamma_v\, M_u\, \frac{c}{J}$$

or

$$v_u = \frac{V_u}{A_c} - \gamma_v\, M_u\, \frac{c'}{J}$$

For slabs supported on square columns, shear stress v_u must not exceed $\varphi 4\sqrt{f_c'}$.

Computation of the combined shear stress involves the following properties of the critical transfer section:

A_c = area of critical section

c or c' = distance from centroid of critical section to face of section where stress is being computed

J = property of critical section analogous to polar moment of inertia

The above properties are given in terms of formulas in Figs. 18-15 and 18-16 of Part 18. Note that in the case of flat slabs, two different critical sections need to be considered in punching shear calculations, as shown in Fig. 20-8.

Unbalanced moment transfer between slab and an edge column (without spandrel beams) requires special consideration when slabs are analyzed by the Direct Design Method for gravity loads. See discussion on Section 13.6.3.6 in Part 21.

For an in-depth discussion of shear in slabs, including consideration of shear reinforcement in slabs, refer to Part 18.

13.4 SLAB REINFORCEMENT

- Minimum area of reinforcement "in each direction" for two-way slab systems = 0.0018bh (b = slab width, h = total thickness) for Grade 60 bars for either top or bottom steel ... Section 13.4.1

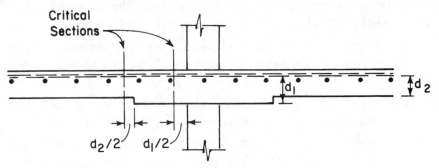

Fig. 20-8   Critical Shear-Transfer Sections for Flat Slabs

- Maximum bar spacing is 2h, but not more than 18 in. ... Section 13.4.2

- Minimum extensions for reinforcement in slabs without beams (flat plates and flat slabs) are prescribed in Fig. 13.4.8 ... Section 13.4.8.1

Note that the reinforcement details of Code Fig. 13.4.8 do not apply to two-way slabs with beams between supports or to slabs in frames resisting lateral loads. For those slabs, a general analysis must be made according to Chapter 12 of the Code to determine bar lengths based on the moment variation. For the '89 Code, reinforcement details for bent bars have been deleted from Fig. 13.4.8 in view of their rare usage in today's construction. Designers who wish to use bent bars in two-way slabs (without beams) should refer to Fig. 13.4.8 of the '83 Code.

SEQUEL

The Direct Design Method and the Equivalent Frame Method for gravity load analysis of two-way slab systems are treated in detail in the following Parts 21 and 22, respectively.

Two-Way Slabs —
Direct Design Method

UPDATE FOR THE '89 CODE

For ACI 318-89, three revisions relating to two-way slab design warrant mention...

(1) The Code expression giving the fraction of unbalanced moment transferred by flexure between a slab and a column, Eq. (13-1), has been modified to more clearly define the dimensions of the critical shear perimeter for edge and corner columns. This is a companion change to Eq. (11-42) for the moment fraction transferred by eccentricity of shear. See discussion on Section 11.12.6 in Part 18.

(2) Reinforcement details for bent bars have been eliminated from Fig. 13.4.8 in consideration of their rare usage in today's construction. Designers who wish to use bent bars in two-way slabs (without beams) should refer to Fig. 13.4.8 of the '83 Code.

(3) As a companion revision to new Code Section 7.13 providing special reinforcement for structural integrity, a new Section 13.4.8.5 has been introduced requiring a portion of the column strip bottom bars in two-way slabs without beams to be made continuous or to be spliced at column supports. A portion of the middle strip bottom bars must also be extended at column lines. These two reinforcement details for improved structural integrity of flat plate and flat slab systems are illustrated in Code Fig. 13.4.8. See discussion on Section 7.13 in Part 3.

GENERAL CONSIDERATIONS

The Direct Design Method is an approximate procedure for analyzing two-way slab systems subject to loads due to gravity only. Since it is an approximate procedure, the method is limited to slab systems meeting the limitations specified in Section 13.6.1. Two-way slab systems not meeting these limitations must be analyzed by more accurate procedures such as the Equivalent Frame Method, as specified in Section 13.7. See Part 22 for discussion and examples using the Equivalent Frame Method.

With the publication of ACI 318-83, the Direct Design Method for moment analysis of two-way slab systems was greatly simplified by eliminating all stiffness calculations for determining design moments in an end span. A table of moment coefficients for distribution of the total span moment in an end span (Section 13.6.3.3) replaced the expressions for distribution as a function of the stiffness ratio α. As a companion change, the approximate Eq. (13-4) for unbalanced moment transfer between the slab and an interior column was also simplified with elimination of the α (concrete not intentionally roughened) term. With these changes, the Direct Design Method became a truly direct design procedure, with all design moments determined directly from moment coefficients. Also, a new Section 13.6.3.6 was added addressing a special provision for moment transfer between a slab and an edge column when the approximate moment coefficients of Section 13.6.3.6 are used. See discussion on Section 13.6.3.6. Commentary Section 13.6.3.3 includes a "Modified Stiffness Method" reflecting the original distribution method, and confirming that existing design aids and computer programs based on the original distribution as a function of the stiffness ratio α are still equally applicable for usage.

PRELIMINARY DESIGN

Before proceeding with the Direct Design Method analysis, a preliminary slab thickness h needs to be determined for control of deflections according to the minimum thickness requirements of Section 9.5.3. Table 20-1 and Fig. 20-3 of Part 20 can be used to simplify minimum thickness computations.

For slab systems without beams, it is advisable at this stage in the design to check the shear strength of the slab in the vicinity of columns or other support locations in accordance with the special shear provision for slabs of Section 11.11. See discussion on Section 13.3.4 in Part 20.

Once a slab thickness has been selected, the Direct Design Method, which is essentially a three-step analysis procedure, involves: (1) determining the total factored static moment for each span, (2) dividing the total factored static moment between negative and positive moments within each span, and (3) distributing the negative and positive moments to the column and middle strips within each span.

For analysis, the slab system is divided into design strips consisting of a column strip and half middle strip(s) as defined in Sections 13.2.1 and 13.2.2, and as illustrated in Fig. 21-1. Some judgement is required in applying the definitions given in Section 13.2.1 for column strips with varying span lengths along the design strip.

13.6.1 Limitations

The Direct Design Method applies within the following conditions illustrated in Fig. 21.2:

1. There must be three or more continuous spans in each direction;

2. Slab panels must be rectangular with a ratio of longer to shorter span (centerline-to-centerline of supports) not greater than 2;

3. Successive span lengths (centerline-to-centerline of supports) in each direction must not differ by more than 1/3 of the longer span;

4. Columns must not be offset more than 10% of the span (in direction of offset) from either axis between centerlines of successive columns;

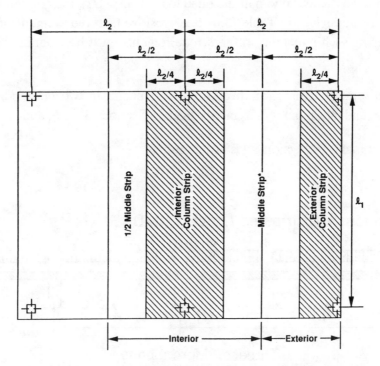

(a) Column strip for $l_2 \leq l_1$

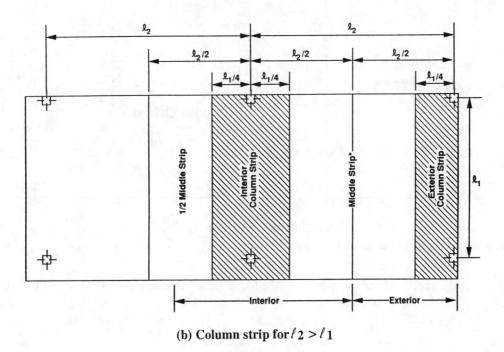

(b) Column strip for $l_2 > l_1$

*When edge of exterior design strip is supported by a wall, the factored moment resisted by this middle strip is defined in Section 13.6.6.3

Fig. 21-1 Definition of Design Strips

5. Loads must be uniformly distributed, with the live load not more than 3 times the dead load (L/D ≤ 3). Note that if the live load exceeds one-half the dead load (L/D >1/2), the column-to-slab stiffness ratios must exceed certain values given in Table 13.6.10, or positive factored moments in panels supported by columns not meeting such minimum stiffness requirements must be magnified by a coefficient computed by Eq. (13-5);

6. For two-way beam-supported slabs, relative stiffness of beams in two perpendicular directions must satisfy the minimum and maximum requirements given in Section 13.6.1.6; and

7. Redistribution of moments by Section 8.4 is not permitted.

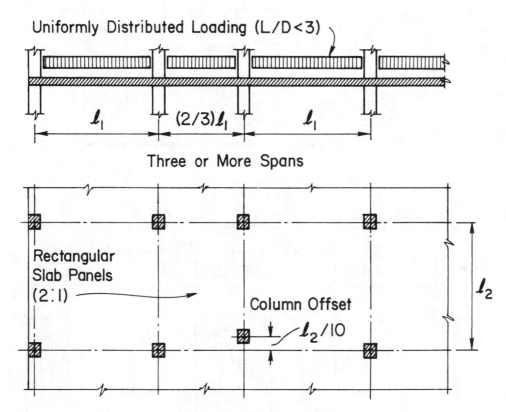

Fig. 21-2 Conditions for Analysis by Coefficients

13.6.2 Total Factored Static Moment for a Span

For uniform loading, the total design moment M_o for a panel is calculated by the simple static moment expression:

$$M_o = w_u \ell_2 \ell_n^2/8 \qquad \text{Eq. (13-3)}$$

where w_u is the factored combination of dead and live loads (psf), $w_u = 1.4 w_d + 1.7 w_\ell$ The clear span ℓ_n (in the direction of analysis) is defined in a straightforward manner for columns or other supporting elements of rectangular cross section. The clear span starts at the face of support. Face of support is defined as shown

21-4

in Fig. 21-3. One limitation requires that the clear span not be taken as less than 65% of the span center-to-center of supports (Section 13.6.2.5). ℓ_2 is simply the span (centerline-to-centerline of supports) transverse to ℓ_n; however, when the span adjacent and parallel to an edge is being considered, the distance from edge of slab to panel centerline is used for ℓ_2 in calculation of M_0.

13.6.3 Negative and Positive Factored Moments

The total static moment for a span is divided into negative and positive design moments as shown in Fig. 21.4. End span moments in Fig. 21-4 are shown for a "flat plate or flat slab" without spandrels (slab system without beams between interior supports and without an edge beam). For other end span conditions, the total static moment M_0 is distributed as shown in Table 21-1.

13.6.3.6 Special Provision for Load Transfer Between Slab and an Edge Column—For columns supporting a slab **without** beams, load transfer directly between the slab and supporting columns (without intermediate load transfer through beams) is one of the more critical design conditions for the flat plate or flat slab system. Shear strength of the slab-column connection is critical. This aspect of two-way slab design should not be taken lightly by the designer. Two-way slab systems are fairly "forgiving" of an error in the distribution or even in the amount of flexural reinforcement; however, there is little or no forgiveness if a critical error in the provision of strength is made. See part 18 for special shear provisions for direct shear and moment transfer at slab-column connections.

Table 21-1 Distribution of Total Static Moment for an End Span

Factored Moment	(1) Slab Simply Supported on Concrete or Masonry Wall	(2) Two-Way Slabs	(3) Flat Plates and Flat Slabs Without Edge Beam	(4) With Edge Beam	(5) Slab Monolithic with Concrete Wall
Interior Negative	0.75	0.70	0.70	0.70	0.65
Positive	0.63	0.57	0.52	0.50	0.35
Exterior Negative	0	0.16	0.26	0.30	0.65

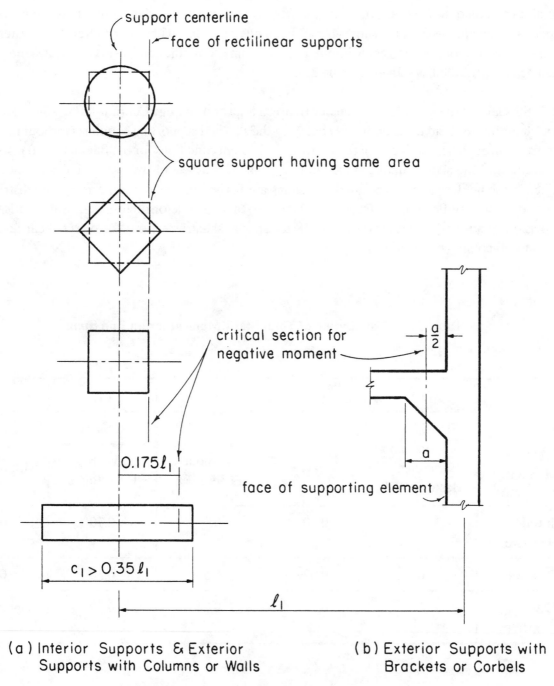

support centerline

face of rectilinear supports

square support having same area

critical section for negative moment

$0.175\ell_1$

$c_1 > 0.35\ell_1$

ℓ_1

$\frac{a}{2}$

a

face of supporting element

(a) Interior Supports & Exterior Supports with Columns or Walls

(b) Exterior Supports with Brackets or Corbels

Fig. 21-3 Critical Sections for Negative Design Moment

For exterior columns supporting a slab without spandrel beams, the load transfer condition is more critical, because the total exterior negative moment from the slab must be transferred to the columns. Section 13.6.3.6 addresses this potentially critical moment transfer between a slab and an edge column. To ensure adequate shear strength when using the approximate end-span moment coefficients of Section 13.6.3.3, the full nominal moment strength M_n provided by the column strip must be used in determining the fraction of unbalanced moment transferred by eccentricity of shear ($\gamma_v M_n$) in accordance with Section 11.12.6. (For end spans without edge beams, the column strip is proportioned to resist the total exterior negative factored moment.) The M_n requirement for evaluating slab shear strength required at the exterior column for moment transfer by eccentricity of shear is illustratred in Fig. 21-5. The total reinforcement provided in the column strip includes the additional reinforcement concentrated over the column to resist the fraction of unbalanced moment transferred by flexure, $\gamma_f M_u = \gamma_f (0.26 M_o)$, where the moment coefficient (0.26) is from Section 13.6.3.3, and γ_f is given by Eq. (13-1).

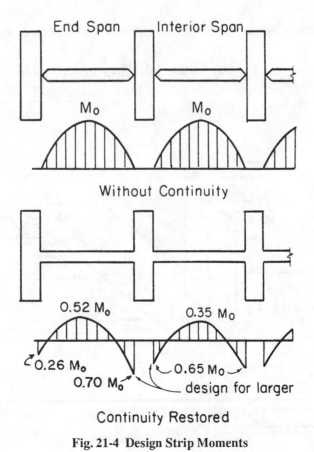

Fig. 21-4 Design Strip Moments

13.6.4 Factored Moments in Column Strips

The amounts of negative and positive factored moments to be resisted by a column strip, as defined in Fig. 21-1, depends on the relative beam-to-slab stiffness ratio and the panel width-to-length ratio in the direction of analysis. An exception to this is when a support has a large transverse width.

The column strip at the exterior of an end span is required to resist the total factored negative moment in the design strip unless edge beams are provided.

When the transverse width of a support is equal to or greater than three fourths (3/4) of the design strip width, Section 13.6.4.3 requires that the negative factored moment be uniformly distributed across the design strip.

The percentage of total negative and positive moments to be resisted by a column strip may be determined from the tables in Section 13.6.4.1 (interior negative), Section 13.6.4.2 (exterior negative) and Section 13.6.4.4 (positive), or from the following expressions:

Percentage of negative factored moment at interior support to be resisted by column strip

$$= 75 + 30(\alpha_1 \ell_2/\ell_1)(1 - \ell_2/\ell_1) \qquad (1)$$

Percentage of negative factored moment at exterior support to be resisted by column strip

$$= 100 - 10\beta_t + 12\beta_t(\alpha_1 \ell_2/\ell_1)(1 - \ell_2/\ell_1) \qquad (2)$$

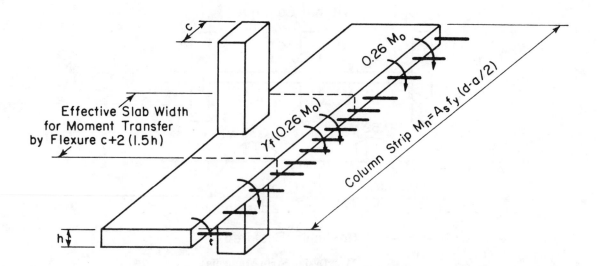

Fig. 21-5 Nominal Moment Strength of Column Strip for Evaluation of $\gamma_v M_n = (1 - \gamma_f) M_n$

Percentage of positive factored moment to be resisted by column strip

$$= 60 + 30(\alpha_1 \ell_2 / \ell_1)(1.5 - \ell_2 / \ell_1) \qquad (3)$$

Note: When $\alpha_1 \ell_2 / \ell_1 > 1.0$, use 1.0 in above equations. When $\beta_t > 2.5$, use 2.5 in Eq. (2) above.

For slabs without beams between supports ($\alpha = 0$) and without edge beams ($\beta_t = 0$), the distribution of total negative moments to column strips is simply 75 and 100 percent for interior and exterior supports, respectively, and the distribution of total positive moment is 60 percent. For slabs with beams between supports, distribution depends on the beam-to-slab stiffness ratio; when edge beams are present, the ratio of torsional stiffness of edge beam to flexural stiffness of slab also influences distribution. Figures 21-6, 21-7, and 21-8 simplify evaluation of the beam-to-slab stiffness ratio α. To evaluate β_t, stiffness ratio for edge beams, Table 21-2 simplifies calculation of the torsional constant C.

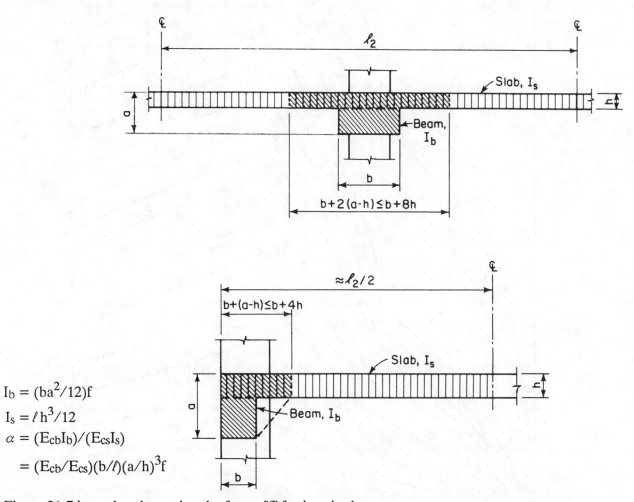

$$I_b = (ba^2/12)f$$
$$I_s = \ell h^3/12$$
$$\alpha = (E_{cb}I_b)/(E_{cs}I_s)$$
$$= (E_{cb}/E_{cs})(b/\ell)(a/h)^3 f$$

Figure 21-7 is used to determine the factor "f" for interior beams.
Figure 21-8 is used to determine the factor "f" for edge beams.

Fig. 21-6 Effective Beam and Slab Sections for Stiffness Ratio α

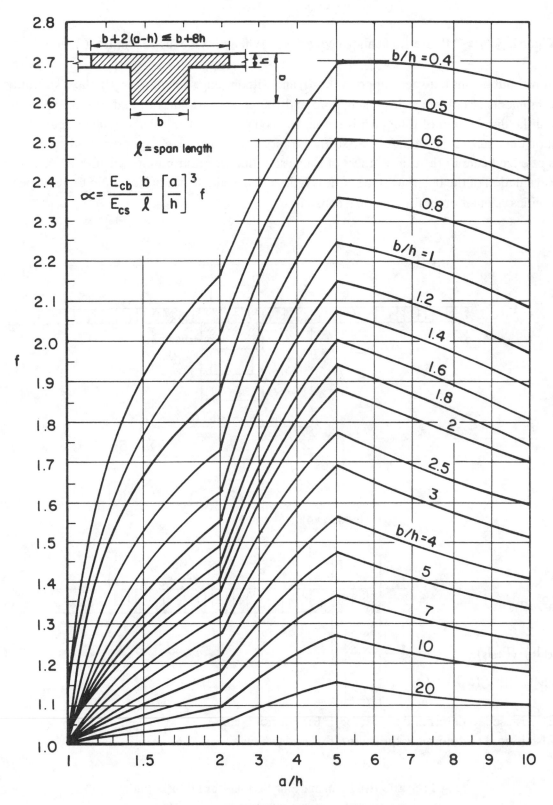

Fig. 21-7 Beam-to-Slab Stiffness Ratio α (Interior Beams)

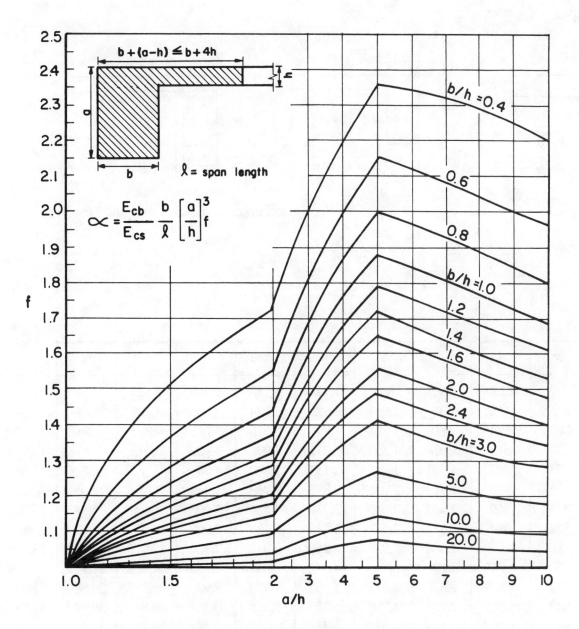

Fig. 21-8 Beam to Slab Stiffness Ratio α **(Edge Beams)**

Table 21-2—Design Aid for Computing Torsional Section Constant C

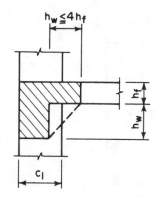

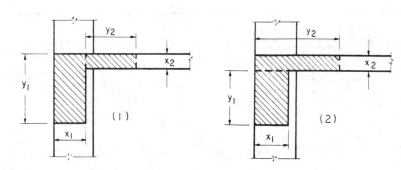

Edge Beam

Use Larger Value of C computed from (1) or (2)

(Section 13.2.4)

y \ x*	4	5	6	7	8	9	10	12	14	16
12	202	369	592	868	1,118	1,538	1,900	2,557		
14	245	452	736	1,096	1,529	2,024	2,566	3,709	4,738	
16	288	534	880	1,325	1,871	2,510	3,233	4,861	6,567	8,083
18	330	619	1,024	1,554	2,212	2,996	3,900	6,013	8,397	10,813
20	373	702	1,167	1,782	2,553	3,482	4,567	7,165	10,226	13,544
22	416	785	1,312	2,011	2,895	3,968	5,233	8,317	12,055	16,275
24	458	869	1,456	2,240	3,236	4,454	5,900	9,469	13,885	19,005
27	522	994	1,672	2,583	3,748	5,183	6,900	11,197	16,628	23,101
30	586	1,119	1,888	2,926	4,260	5,912	7,900	12,925	19,373	27,197
33	650	1,243	2,104	3,269	4,772	6,641	8,900	14,653	22,117	31,293
36	714	1,369	2,320	3,612	5,284	7,370	9,900	16,381	24,860	35,389
42	842	1,619	2,752	4,298	6,308	8,828	11,900	19,837	30,349	43,581
48	970	1,869	3,183	4,984	7,332	10,286	13,900	23,293	35,836	51,773
54	1,098	2,119	3,616	5,670	8,356	11,744	15,900	26,749	41,325	59,965
60	1,226	2,369	4,048	6,356	9,380	13,202	17,900	30,205	46,813	68,157

* Small side of a rectangular cross section with dimensions x and y.

13.6.5 Factored Moments in Beams

When a design strip contains beams between columns, the factored moment assigned to the column strip must be distributed between the slab and the beam portions of the column strip. The amount of the column strip factored moment to be resisted by the beam varies linearly between zero and 85 percent as $\alpha_1 l_2 / l_1$ varies between zero and 1.0. When $\alpha_1 l_2 / l_1$ is equal to or greater than 1.0, 85 percent of the total column strip moment must be resisted by the beam. In addition, the beam section must resist the effects of loads applied directly to the beam, including weight of beam stem projecting above or below the slab.

13.6.6 Factored Moments in Middle Strips

Factored moments not assigned to column strips must be resisted by middle strips. An exception to this is a middle strip adjacent to and parallel with an edge supported by a wall, where the moment to be resisted is twice the factored moment assigned to the half middle strip corresponding to this first row of interior supports. (See Fig. 21-1.)

13.6.9 Factored Moments in Columns and Walls

Supporting columns and walls must resist any negative moments transferred from the slab system.

For interior columns (or walls), the approximate Eq. (13-4) may be used to determine the unbalanced moment transferred by gravity loading, unless an analysis is made considering the effects of pattern loading and unequal adjacent spans. The transfer moment is computed directly as a function of span length and gravity loading. For the more usual case with equal transverse and adjacent spans, Eq. (13-4) reduces to

$$M_u = 0.07(0.5 w_l l_2 l_n^2) \tag{4}$$

where, w_l = factored live load, psf
              l_2 = span length transverse to l_n
              l_n = clear span length in the direction of analysis

At exterior column or wall supports, the total exterior negative factored moment from the slab system (Section 13.6.3.3) is transferred directly to the supporting members. Due to the approximate nature of the moment coefficients, it seems unwarranted to consider the change in moment from face of support to centerline of support; use the moment values from Section 13.6.3.3 directly.

Columns above and below the slab must resist a portion of the support moment based on the relative column stiffnesses--generally, in proportion to column lengths above and below the slab. Again, due to the approximate nature of the moment coefficients of the Direct Design Method, the refinement of considering the change in moment from centerline of slab-beam to top or bottom of column seems unwarranted.

13.6.10 Provisions for Effects of Pattern Loadings

When the unfactored dead-to-live load ratio is less than 2 (high live-to-dead load ratio), the effect of increased moment due to pattern loading can be neglected in the analysis, if sufficiently stiff columns are provided. If the columns above and below the slab do not meet the minimum stiffness α_{min} of Code Table 13.6.10, and the dead-to-live load ratio is less than 2, the positive factored moments must be increased by the coefficient computed from Eq. (13-5).

DESIGN AID - DIRECT DESIGN MOMENT COEFFICIENTS

Distribution of the total free-span moment M_o into negative and positive moments, and then, column and middle strip moments, involves direct application of moment coefficients to the total moment M_o. The moment coefficients are a function of span (interior or exterior) and slab support conditions (type of two-way slab system.) For design convenience, moment coefficients for typical two-way slab systems are given in Tables 21-3 through 21-7. Tables 21-3 through 21-6 apply to flat plates or flat slabs with differing end support conditions. Table 21-7 applies to two-way slabs supported on beams on all four sides. Final moments for the column strip and the middle strip are computed by directly using the tabulated values.

The moment coefficients of Table 21-4 (flat plate with edge beams) are valid for $\beta_t \geq 2.5$. The coefficients of Table 21-7 (two-way beam-supported slabs) apply for $\alpha_1 l_2 / l_1 \geq 1.0$ and $\beta_t \geq 2.5$. Many practical beam sizes will provide beam-to-slab stiffness ratios such that $\alpha_1 l_2 / l_1$ and β_t will be greater than these limits, allowing moment coefficients to be taken directly form the tables, without further conderation of stiffnesses

Table 21-3—Design Moment Coefficients for Flat Plate or Flat Slab Supported Directly on Columns

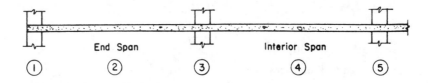

Slab Moments	End Span			Interior Span	
	(1)	(2)	(3)	(4)	(5)
	Exterior Negative	Positive	First Interior Negative	Positive	Interior Negative
Total Moment	0.26 M_o	0.52 Mo	0.70 Mo	0.35 Mo	0.65 Mo
Column Strip	0.26 M_o	0.31M_o	0.53 M_o	0.21 Mo	0.49 Mo
Middle Strip	0 M_o	0.21M_o	0.17 M_o	0.14 Mo	0.16 Mo

Notes: (1) All negative moments are at face of support

and interpolation for moment coeffcients. However, if beams are present, the two stiffness parameters α_1 and β_t will need to be evaluated. For two-way slabs, and for $E_{cb} = E_{cs}$, the stiffness parameter α_1 is simply the ratio fo the moments of inertia of the effective beam and slab sections in the direction of analysis, $\alpha_1 = I_b/I_s$ as illustrated in Fig. 21-6. Fig. 21-7 and 21-8 simplify evaluation of the α term.

For $E_{cb} = E_{cs}$, relative stiffness provided by an edge beam is reflected by the parmater $\beta_t = C/2I_s$, where I_s is the moment of inertia of the effective slab section spanning in the direction of ℓ_1 and having a width equal to ℓ_2, $I_s = \ell_2 h^3/12$. The constant C pertains to the torsional stiffness of the effective edge beam cross section. It is found by dividing the beam section into its component rectangles, each having a smaller dimension x and a larger dimension y, and by summing the contributions of all the parts by means of the equation:

$$C = \Sigma(1 - 0.63x/y)(x^3y/3)$$

Eq. (13-7)

The subdivision can be done in such a way as to maximize C. Table 21-2 simplifies calculation of the torsional constant C.

Table 21-4—Design Moment Coefficients for Flat Plate or Flat Slab with Edge Beams

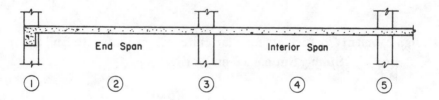

Slab Moments	End Span			Interior Span	
	(1)	(2)	(3)	(4)	(5)
	Exterior Negative	Positive	First Interior Negative	Positive	Interior Negative
Total Moment	0.30 M_o	0.50 Mo	0.70 Mo	0.35 Mo	0.65 M_o
Column Strip	0.23 M_o	0.30 M_o	0.53 M_o	0.21 M_o	0.49 M_o
Middle Strip	0.07 M_o	0.20 M_o	0.17 M_o	0.14 M_o	0.16 M_o

Notes: (1) All negative moments are at face of support
      (2) Torsional stiffness of spandrel beams $\beta_t \geq 2.5$. For values of β_t less than 2.5, exterior negative column strip moment increases to $(0.30 - 0.03\beta_t) M_o$.

Table 21-5—Design Moment Coefficients for Flat Plate or Flat Slab with End Span Integral with Wall

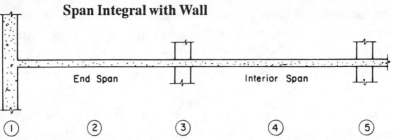

Slab Moments	End Span			Interior Span	
	(1)	(2)	(3)	(4)	(5)
	Exterior Negative	Positive	First Interior Negative	Positive	Interior Negative
Total Moment	$0.65 M_O$	$0.35 M_O$	$0.65 M_O$	$0.35 M_O$	$0.65 M_O$
Column Strip	$0.49 M_O$	$0.21 M_O$	$0.49 M_O$	$0.21 M_O$	$0.49 M_O$
Middle Strip	$0.16 M_O$	$0.14 M_O$	$0.16 M_O$	$0.14 M_O$	$0.16 M_O$

Note: (1) All negative moments are at face of support.

Table 21-6—Design Moment Coefficients for Flat Plate or Flat Slab with End Span Simply Supported on Wall

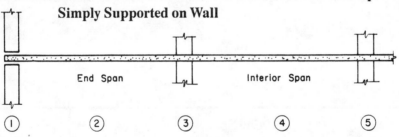

Slab Moments	End Span			Interior Span	
	(1)	(2)	(3)	(4)	(5)
	Exterior Negative	Positive	First Interior Negative	Positive	Interior Negative
Total Moment	0	$0.63 M_O$	$0.75 M_O$	$0.35 M_O$	$0.65 M_O$
Column Strip	0	$0.38 M_O$	$0.56 M_O$	$0.21 M_O$	$0.49 M_O$
Middle Strip	0	$0.25 M_O$	$0.19 M_O$	$0.14 M_O$	$0.16 M_O$

Note: (1) All negative moments are at face of support.

Table 21-6 Design Moment Coefficients for Two-Way Beam-Supported Slab

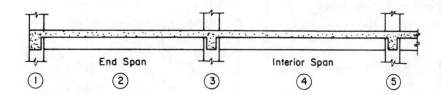

Span ratio l_2/l_1	Slab and Beam Moments		End Span			Interior Span	
			(1)	(2)	(3)	(4)	(5)
			Exterior Negative	Positive	First Interior Negative	Positive	Interior Negative
	Total Moment		$0.16 M_0$	$0.57 M_0$	$0.70 M_0$	$0.35 M_0$	$0.65 M_0$
0.5	Column Strip	Beam	$0.12 M_0$	$0.43 M_0$	$0.54 M_0$	$0.27 M_0$	$0.50 M_0$
		Slab	$0.02 M_0$	$0.08 M_0$	$0.09 M_0$	$0.05 M_0$	$0.09 M_0$
	Middle Strip		$0.02 M_0$	$0.06 M_0$	$0.07 M_0$	$0.03 M_0$	$0.06 M_0$
1.0	Column Strip	Beam	$0.10 M_0$	$0.37 M_0$	$0.45 M_0$	$0.22 M_0$	$0.42 M_0$
		Slab	$0.02 M_0$	$0.06 M_0$	$0.08 M_0$	$0.04 M_0$	$0.07 M_0$
	Middle Strip		$0.04 M_0$	$0.14 M_0$	$0.17 M_0$	$0.09 M_0$	$0.16 M_0$
2.0	Column Strip	Beam	$0.06 M_0$	$0.22 M_0$	$0.27 M_0$	$0.14 M_0$	$0.25 M_0$
		Slab	$0.01 M_0$	$0.04 M_0$	$0.05 M_0$	$0.02 M_0$	$0.04 M_0$
	Middle Strip		$0.09 M_0$	$0.31 M_0$	$0.38 M_0$	$0.19 M_0$	$0.36 M_0$

Example 21.1—Two-Way Slab without Beams Analyzed by Direct Design Method

Using the Direct Design Method, determine design moments for the slab system shown, in the transverse direction, for an intermediate floor.

Story height   = 9 ft
Column dimensions = 16 in. x 16 in.
Lateral loads to be resisted by shear walls
No edge beams
Partition weight   = 20 psf
Service live load   = 40 psf
$f_c' = 3000$ psi (for slab)
$f_c' = 5000$ psi (for column)
$f_y = 60,000$ psi

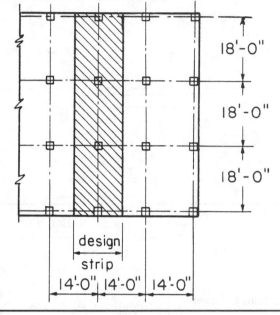

design strip

14'-0" | 14'-0" | 14'-0"

Calculations and Discussion	Code Reference

1. Preliminary design for slab thickness h

   a. Control of deflection:
     For slab systems without beams (flat plate),   9.5.3.2
     the minimum overall thickness h with Grade   Table 9.5(b)
     60 reinforcement is (see Table 20.1):

     $h = \ell_n/30 = 200/30 = 6.67$ in.

     where ℓ_n is the length of clear span in the long direction = 216 - 16 = 200 in.

     This is larger than the 5 in. minimum specified   9.5.3.2(a)
     for slabs without drop panels.

   b. Shear strength of slab:

     Use an average effective depth, d ≈ 5.75 in. (3/4-in. cover and #4 bar)

     Factored dead load, $w_d = (87.5 + 20)1.4 = 150.5$ psf
     Factored live load, $w_\ell = 40 \times 1.7$ = 68.0 psf
     Total factored load, w_u = 218.5 psf

Example 21.1—Continued

Calculations and Discussion	Code Reference

Investigation for wide-beam action is made on a 12-in. wide strip at d distance from face of support in the long direction (see Fig. 21-9).

11.11.1.1

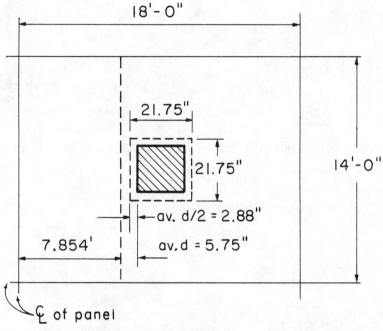

Fig. 21-9 Critical Sections for One-Way and Two-Way Shear

$V_u = 0.2185 \times 7.854 = 1.72$ kips

$V_c = 2\sqrt{f_c'} b_w d$
Eq. (11-3)

$\quad = 2\sqrt{3000} \times 12 \times 5.75/1000 = 7.56$ kips

$\varphi V_c = 0.85 \times 7.56 = 6.43$ kips

$V_u < \varphi V_c$

Since there are no shear forces at the centerline of adjacent panels (see Fig. 21-9), the shear strength in two-way action at "d/2" distance around a support is computed as follows:

Example 21.1—Continued

$V_u = 0.2185(18 \times 14 - 1.81^2) = 54.3$ kips

$V_c = 4\sqrt{f_c'}b_o d$ (for square columns)

Eq. (11-36)

$\quad = 4\sqrt{3000} \times (4 \times 21.75) \times 5.75/1000 = 109.6$ kips

$\varphi V_c = 0.85 \times 109.6 = 93.2$ kips

$V_u < \varphi V_c$

Therefore, preliminary design indicates that a 7 in. slab is adequate for control of deflection and shear strength.

2. Check applicability of Direct Design Methods — 13.6.1

There is a minimum of three continuous spans in each direction, — 13.6.1.1

Long-to-short span ratio is 1.29 < 2.0, — 13.6.1.2

Successive span lengths are equal, — 13.6.1.3

Columns are not offset, — 13.6.1.4

Loads are uniformly distributed with live-to-dead load ratio of 0.37 < 3.0, — 13.6.1.5

Slab system is without beams — 13.6.1.6

3. Factored moments in slab

a. Total factored moment per span: — 13.6.2

$\quad M_o = w_u \ell_2 \ell_n^2/8$ — Eq. (13-3)

$\quad\quad = 0.2185 \times 14 \times 16.67^2/8 = 106.3$ ft-kips

Example 21.1—Continued

Calculations and Discussion

Code
Reference

b. Distribution of the total factored moment M_o per span into negative and positive moments, and then into column and middle strip moments, involves direct application of the moment coefficients to the total moment M_o. Referring to Table 21-3 (flat plate without edge beams):

13.6.3
13.6.4
13.6.6

	Total Moment (ft-kips)	Column Strip Moment (ft-kips)	Moment (ft-kips) in Two Half-Middle Strips*
End Span:			
Exterior Negative	$0.26\,M_o = 27.6$	$0.26\,M_o = 27.6$	0
Positive	$0.52\,M_o = 55.3$	$0.31\,M_o = 33.0$	$0.21\,M_o = 22.3$
Interior Negative	$0.70\,M_o = 74.4$	$0.53\,M_o = 56.3$	$0.17\,M_o = 18.1$
Interior Span:			
Positive	$0.35\,M_o = 37.2$	$0.21\,M_o = 22.3$	$0.14\,M_o = 14.9$
Negative	$0.65\,M_o = 69.1$	$0.49\,M_o = 52.1$	$0.16\,M_o = 17.0$

*That portion of the total moment M_o not resisted by the column strip is assigned to the two half-middle strips.

Note: The factored moments may be modified by 10 percent, provided the total factored static moment in any panel is not less than that computed from Eq. (13-3). This modification is omitted here.

13.6.7

4. Check for effects of pattern loading

13.6.10

Dead-to-live load ratio:

$\beta_a = 107.5/40 = 2.69$

When $\beta_a \geq 2.0$ pattern loading effect may be neglected.

5. Factored moments in columns

13.6.9

a. Interior columns (with equal transverse and adjacent spans):

Example 21.1—Continued

Calculations and Discussion	Code Reference

$M = 0.07(0.5 w_\ell \ell_2 \ell_n^2)$ Eq. (13-4)

$\quad = 0.07(0.5 \times 1.7 \times 0.04 \times 14 \times 16.67^2)$

$\quad = 9.3$ ft$-$kips

With the same column size and length above and below the slab:

$M_c = 9.3/2 = 4.65$ ft$-$kips

This moment is combined with the factored axial load (for each story) for design of the interior columns.

b.   Exterior columns:

Total exterior negative moment from slab must be transferred directly to the columns: $M_u = 27.6$ ft-kips. With same column size and length above and below the slab:

$M_c = 27.6/2 = 13.8$ ft$-$kips

This moment is combined with the factored axial load (for each story) for design of the exterior column.

6.   Transfer of gravity load shear and moment at exterior column.

Check slab shear and flexural strength at edge column   11.12.6
due to direct shear and unbalanced moment transfer.   13.3.3

a.   Factored shear force transfer at exterior column:

$V_u \approx w_u \ell_1 \ell_2 /2$

$\quad \approx 0.2185 \times 14 \times 18/2 = 27.5$ kips

Example 21.1—Continued

	Code
Calculations and Discussion	**Reference**

b. Unbalanced moment transfer at exterior column: 13.6.3.6
                                                                      13.3.3.2

When the end span moments are determined using the
approximate moment coefficients of Section 13.6.3.3,
the special provision of Section 13.6.3.6 (moment transfer
between slab and an edge column) requires that the fraction
of unbalanced moment transferred by eccentricity of shear
must be based on the full column strip nominal moment
strength, M_n provided. The total reinforcement provided in
the column strip includes the additional reinforcement
concentrated over the column to resist the fraction of
unbalanced moment transferred by flexure $\gamma_f M_u$, where M_u
is the exterior negative factored moment from the slab.
For a slab without an edge beam, the total $M_u = 27.6$ ft-kips
is resisted by the column strip. (Minimum reinforcement per
Section 13.4.1 is provided in the middle strip.)

For both middle strip and column strip: 13.4.1

$A_{s(min)} = 0.0018bh \times 84 \times 7 = 1.06$ in.^{2} 7.12.2.1(b)
where b is width of design strip = 14/2 = 7 ft = 84 in.

For $s_{max} = 2h = 2 \times 7 = 14$ in., total bars required   13.4.2
$\qquad = 84/14 = 6$ bars

Check total reinforcement required for column strip
negative moment $M_u = 27.6$ ft-kips.

$$\frac{M_u}{\varphi f'_c bd^2} = \frac{27.6 \times 12}{0.9 \times 3 \times 84 \times 5.75^2} = 0.0442$$

where d = 7.0 - 1.25 = 5.75 in. (3/4 in. cover and #4 bar)

From Table 10-2,

$\omega = 0.0454$
$\rho = \omega f'_c / f_y = 0.0454 \times 3/60 = 0.00227$
$A_s = \rho bd = 0.00227 \times 84 \times 5.75 = 1.10$ in.^{2}

Example 21.1—Continued

Calculations and Discussion	Code Reference

For #4 bars, total bars required = 1.10/0.20 = 5.5 bars

6 bars for s_{max} = 14 in. governs.

Use #4 bars @ ±14-in. spacing in middle strip and in column strip

Additional reinforcement required over column within effective slab width of c + 2 (1.5h) = 16 + 2 (1.5 × 7) = 37 in. to resist fraction of unbalanced moment transferred by flexure is computed from Eq. (13-1).

13.3.3.2

Referring to Figs. 18-12 and 18-13, for (square) edge columns:

$b_1 = c_1 + d/2 = 16 + 5.75/2 = 18.88$ in.
$b_2 = c_2 + d = 16 + 5.75 = 21.75$ in.
$b_1/b_2 = 0.87$

$$\gamma_f = \frac{1}{1 + (2/3)\sqrt{b_1/b_2}} = \frac{1}{1 + (2/3)\sqrt{0.87}} = 0.62$$

Eq. (13-1)

$\gamma_f M_u = 0.62(27.6) = 17.1$ ft−kips must be transferred within the effective slab width of 37 in. Use 2 additional #4 bars over column. Check moment strength for 4#4 bars within 37-in. slab width. See sketch below.

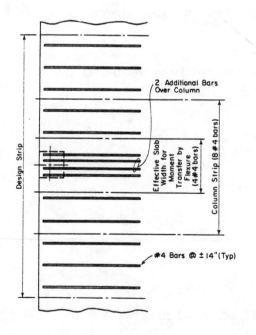

Example 21.1—Continued

	Code
Calculations and Discussion	Reference

For 4 #4 bars:

$A_s = 4(0.20) = 0.80$ in.^{2}

$\omega = A_s f_y / f'_c \, bd = 0.80 \times 60/(3 \times 37 \times 5.75) = 0.0752$

From Table 10-2, $M_n/f'_c \, bd^2 = 0.0719$

$M_n = 0.0719 \times 3 \times 37 \times 5.75^2/12 = 22.0$ ft–kips

$\varphi M_n = 0.9(22.0) = 19.8$ ft–kips > 17.1 ft–kips O.K.

Fraction of unbalanced moment transferred by
eccentricity of shear must be based on full nominal
moment strength M_n provided in column strip.

13.6.3.6

For $6 + 2 = 8$ # 4 bars:

$A_s = 8(0.20) = 1.60$ in.^{2}

$\omega = 1.60 \times 60/(3 \times 84 \times 5.75) = 0.0663$

From Table 10-2, $M_n/f'_c \, bd^2 = 0.0637$

$M_n = 0.0637 \times 3 \times 84 \times 5.75^2/12 = 44.2$ ft–kips

Assume transfer moment M_n at centroid of critical transfer section.

c.   Combined shear stress at inside face of critical transfer section.
For shear strength equations, see Part 18.

11.12.6.2

$v_u = V_u/A_c + \gamma_v M_n/(J/c)$

$= 27,500/342.2 + 0.38 \times 44.2 \times 12,000/2358$

$= 80.4 + 85.5 = 165.9$ psi

Example 21.1—Continued

	Code
Calculations and Discussion	Reference

where (referring to Fig. 18-16, edge column bending perpendicular to edge), with b_1 = 18.88 in. and b_2 = 21.75 in.

$$c = b_1^2/(2b_1 + b_2) = 18.88^2/(2 \times 18.88 + 21.75)$$

$$= 5.99 \text{ in.}$$

$$A_c = (2b_1 + b_2)d = 342.2 \text{ in.}^2$$

$$J/c = \left[2b_1d(b_1 + 2b_2) + d^3(2b_1 + b_2)/b_1\right]/6$$

$$= 2358 \text{ in.}^3$$

$$\gamma_v = 1 - \frac{1}{1+(2/3)\sqrt{b_1/b_2}}$$

$$= 1 - \frac{1}{1+(2/3)\sqrt{18.88/21.75}} = 0.38$$ Eq. (11-42)

d.   Permissible shear stress   11.12.6.2

$$\varphi V_n = \varphi 4\sqrt{f_c'} = 0.85 \times 4\sqrt{3000}$$

$$= 186.2 \text{ psi} > 165.9 \text{ psi} \quad \text{O.K.}$$

Using the Direct Design Method, determine design moments for the slab system shown in the transverse direction, for an intermediate floor.

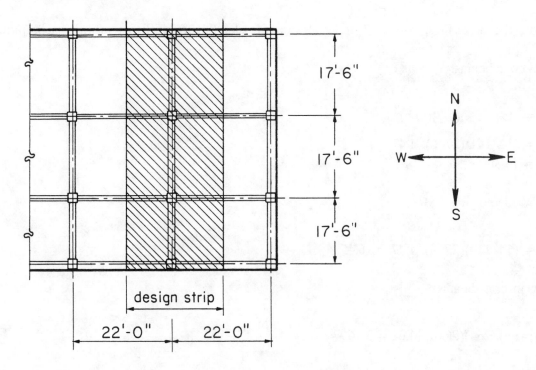

Story height	=	12 ft
Edge beam dimensions	=	14 in. × 27 in.
Interior beam dimensions	=	14 in. × 20 in.
Column dimensions	=	18 in. × 18 in.
Slab thickness	=	6 in.
Service live load	=	100 psf

f'_c = 4000 psi (for all members), normal weight concrete
f_y = 60,000 psi

Example 21.2 —Continued

	Code
Calculations and Discussion	**Reference**

1. Preliminary design for slab thickness, h ,   9.5.3
   Control of deflection

   With the aid of Figs. 21-6, 21-7, and 21-8, beam-to-slab flexural stiffness ratio α is computed as follows:

   NS edge beams:

   $$\alpha = (E_{cb}/E_{cs})(b/\ell)(a/h)^3 f$$
   $$= (14/141)(27/6)^3\, 1.47$$
   $$= 13.30$$

   EW edge beams:

   $$\alpha = (14/114)(27/6)^3\, 1.47 = 16.45$$

   NS interior beams:

   $$\alpha = (14/264)(20/6)^3\, 1.61 = 3.16$$

   EW interior beams:

   $$\alpha = (14/210)(20/6)^3\, 1.61 = 3.98$$

   Since all $\alpha > 2.0$, (see Fig. 8-2), Eq. (9-12) will control for minimum thickness.

   Therefore,

   $$h = \frac{\ell_n\,(0.8 + f_y/200{,}000)}{36 + 9\beta}$$ Eq. (9-12)

   $$= \frac{246(0.8 + 60{,}000/200{,}000)}{36 + 9(1.28)} = 5.69 \text{ in.}$$

   where $\beta = 20.5/16 = 1.28$

Example 21.2 —Continued

	Code
Calculations and Discussion	Reference

l_n = clear span in long direction measured face to face
of columns = 20 ft 6 in. = 246 in.
Use 6 in. slab thickness

2. Check applicability of Direct Design Method   13.6.1

There is a minimum of three continuous spans in each direction,   13.6.1.1
Long-to-short span ratio is 1.26 < 2.0,   13.6.1.2
Successive span lengths are equal,   13.6.1.3
Columns are not offset   13.6.1.4
Loads are uniformly distributed with live-to-dead ratio of 1.33 < 3.0   13.6.1.5

13.6.1.6

Interior panel: $\alpha_1 l_2^2 / \alpha_2 l_1^2 = 1.25$

Exterior panel: $\alpha_1 l_2^2 / \alpha_2 l_1^2 = 0.30$

3. Factored moments in slab

    a. Total factored moment per span: 13.6.2

Eq. (13-3)

$M_0 = w_u l_2 l_n^2 / 8$

$\quad = 0.288 \times 22 \times 16^2 / 8 = 202.8$ ft$-$kips
where $w_u = w_d + w_l = 1.4(75 + 9.3) + 1.7(100) = 288$ psf
(9.3 psf is weight of beam stem per foot divided by l_2)

    b. Negative and positive factored moments   13.6.3

Interior span:

Negative moment = 0.65 M_0 = 131.8 ft$-$kips
Positive moment = 0.35 M_0 = 71.0 ft$-$kips

End span (Two-way slabs):   13.6.3.3

Exterior negative moment = 0.16 M_0 = 32.4 ft$-$kips
Positive moment   = 0.57 M_0 = 115.6 ft$-$kips
Interior negative moment = 0.70 M_0 = 142.0 ft$-$kips

Example 21.2 —Continued

	Code
Calculations and Discussion	**Reference**

Note: The factored moments may be modified by 10 percent,   13.6.7
provided the total factored static moment in any panel is not
less than that computed from Eq. (13.3). This modification
is omitted here.

4.   Distribution of factored moments in column and middle strips   13.6.4

     13.6.6

    a.   Percentage of total negative and positive moments to column strip.

        At interior support: using Eq. (1)

$$75 + 30(\alpha_1 \ell_2/\ell_1)(1 - \ell_2/\ell_1) = 75 + 30(1 - 1.26) = 67\%$$

        where α_1 = (in the direction of ℓ_1) is computed with the aid of Fig. 21-6.

$$\alpha_1 = (b/\ell)(a/h)^3 f$$
$$= (14/264)(20/6)^3 \, 1.61 = 3.16$$
$$\alpha_1 \ell_2/\ell_1 = 3.16 \times 22/17.5 = 3.98 > 1.0 \ \text{Use } 1.0$$

        At exterior support: using Eq. (2)

$$100 - 10\beta_t + 12\beta_t(\alpha_1 \ell_2/\ell_1)(1 - \ell_2/\ell_1) =$$
$$100 - 10(1.88) + 12(1.88)(1 - 1.26) = 75\%$$
$$\text{where } \beta_t = c/(2I_s) = 17{,}868 / (2 \times 4752) = 1.88$$
$$I_s = \ell_2 h^3/12 = 264 \times 6^3/12 = 4752 \text{ in.}^4$$

Example 21.2 —Continued

Calculations and Discussion

Code
Reference

C is taken as the larger value computed (with the aid of Fig. 21-9)
for the torsional member shown below.

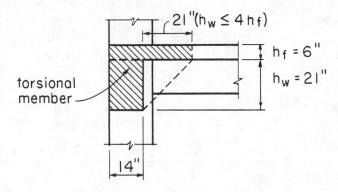

$x_1 = 14$ in. $x_2 = 6$ in.	$x_1 = 14$ in. $x_2 = 6$ in.
$y_1 = 21$ in. $y_2 = 35$ in.	$y_1 = 27$ in. $y_2 = 21$ in.
$C_1 = 11{,}141$ in.^{4} $C_2 = 2248$ in.^{4}	$C_1 = 16{,}628$ in.^{4} $C_2 = 1240$ in.^{4}
$\Sigma C = 11{,}141 + 2248 = 13{,}389$ in.^{4}	$\Sigma C = 16{,}628 + 1240 = 17{,}868$ in.^{4}

Positive moment: using Eq. (3)

$$60 + 30(\alpha_1 \ell_2/\ell_1)(1.5 - \ell_2/\ell_1) = 60 + 30(1.5 - 1.26) = 67\%$$

Example 21.2—Continued

Factored moments in column strips and middle strips are summarized as follows:

	Factored Moment (ft-kips)	Column Strip		Moment (ft-kips) in Two Half-Middle Strips[2]
		Percent	Moment[1] (ft-kips)	
End Span:				
Exterior Negative	32.4	75	24.3	8.1
Positive	115.6	67	77.5	38.1
Interior Negative	142.0	67	95.1	46.9
Interior Span:				
Negative	131.8	67	88.3	43.5
Positive	71.0	67	47.6	23.4

[1] Since $\alpha_1 l_2 / l_1 > 1.0$, beams must be proportioned to resist 85 percent of column strip moment as per Section 13.6.5.1.

[2] That portion of the factored moment not resisted by the column strip is assigned to the half-middle strips.

5. Check for effects of pattern loading   13.6.10

Dead-to-live load ratio:

$\beta_a = 75/100 = 0.75 < 2.0$

Since $\alpha_1 > 2.0$, required α_{min} by Table 13.6.10 is zero.

Therefore, pattern loading effect may be neglected.

Example 21.2—Continued

Calculations and Discussion	Code References

6. Factored moments in columns 13.6.9

   a. Interior columns (with equal transverse and adjacent spans): 13.6.9

$$M = 0.07(0.5 w_l \ell_2 \ell_n^2)$$ Eq. (13-4)
$$= 0.07(0.5 \times 1.7 \times 0.1 \times 22 \times 16^2)$$
$$= 33.5 \text{ ft-kips}$$

With the same column size and length above and below the slab:

$$M_c = 33.5/2 = 16.75 \text{ ft-kips}$$

This moment is combined with the factored axial load (for each story) for design of the interior columns.

   b. Exterior columns:

The total exterior negative moment from the slab beam is transferred to the exterior columns; with the same column size and length above and below the slab system:

$$M_c = 32.4/2 = 16.2 \text{ ft-kips}$$

7. Shear strength

   a. Beams.   Since $\alpha_1 \ell_2/\ell_1 = 3.98$, beams must resist total shear ($b_w = 14$ in., $d = 17$ in.)   13.6.8.1

NS Beams:

$$V_u = w_u \ell_1^2/4 = 0.288(17.5)^2/4 = 22.1 \text{ kips}$$

$$\varphi V_c = \varphi 2\sqrt{f_c'}\, b_w d$$ Eq. (11-3)

Example 21.2 —Continued

$$= 0.85 \times 2\sqrt{4000} \times 14 \times 17 = 25.6 \text{ kips}$$

$$V_u < \varphi V_c$$

Only minimum shear reinforcement is required
where $V_u > \varphi V_c$

EW Beams:

$$V_u = w_u \ell_1 (2\ell_2 - \ell_1)/4$$

$$= 0.288 \times 17.5(2 \times 22 - 17.5)/4 = 33.4 \text{ kips}$$

$$V_u > \varphi V_c$$

Required shear strength to be provided by shear
reinforcement $V_s = (V_u - \varphi V_c)/\varphi = (33.4 - 25.6)/0.85$

$$= 9.2 \text{ kips}$$

b.   Slabs. ($b_w = 12$ in., $d = 5.5$ in.)   13.6.8.4

$$V_u = w_u \ell_1/2 = 0.275 \times 17.5/2 = 2.4 \text{ kips}$$

$$\varphi V_c = \varphi 2\sqrt{f_c} b_w d$$

$$= 0.85 \times 2\sqrt{4000} \times 12 \times 5.5 = 7.1 \text{ kips}$$

$$V_u < \varphi V_c$$

Shear strength of slab is adequate without shear reinforcement.

9.   Edge beams must be designed to resist moment not transferred to exterior
columns by parallel beams, in accordance with Section 11.6.

Two-Way Slabs — Equivalent Frame Method

GENERAL CONSIDERATIONS

The Equivalent Frame Method of analysis for gravity loading converts a three-dimensional frame system with two-way slabs into a series of two-dimensional frames (slab-beams and columns), with each frame extending the full height of the building, as illustrated in Fig. 22-1. The width of each equivalent frame extends to mid-span between column centerlines. The complete analysis of the two-way slab system for a building consists of analyzing a series of equivalent interior and exterior frames spanning longitudinally and transversely through the building. For gravity loading, the slab-beams of each floor or roof (level) may be analyzed separately, with the far ends of attached columns considered fixed (Section 13.7.2.5).

The equivalent frame method of elastic analysis applies to buildings with columns laid out on a basically orthogonal grid, with column lines extending longitudinally and transversely through the building. The analysis method is applicable to slabs with or without beams between supports.

The Equivalent Frame Method may also be used for lateral load analysis if stiffness of frame members are modified to account for cracking and other relevant factors. See discussion on Section 13.3.1.2 in Part 20.

PRELIMINARY DESIGN

Before proceeding with Equivalent Frame analysis, a preliminary slab thickness h needs to be determined for control of deflections, according to the minimum thickness requirements of Section 9.5.3. Table 20-1 and Fig. 20-3 of Part 20 may be used to simplify minimum thickness computations. For slab systems without beams, it is advisable at this stage of design to check the shear strength of the slab in the vicinity of columns or other support locations, according to the special provisions for slabs of Section 11.11. See discussion on Section 13.3.4 in Part 20.

13.7.2 Equivalent Frame

Application of the frame definitions given in Sections 13.7.2, 13.2.1, and 13.2.2 is illustrated in Fig. 22-2. Some judgement is required in applying the definitions given in Section 13.2.1 for column strips with varying span lengths along the design strip.

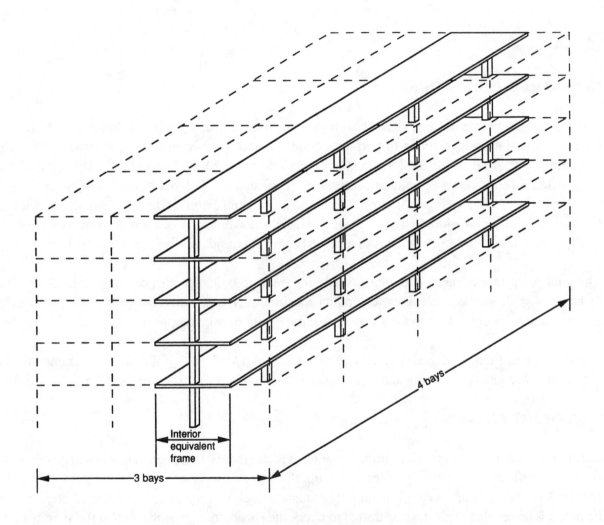

Interior
equivalent
frame

3 bays

4 bays

Fig. 22-1 Equivalent Frames for 5-Story Building

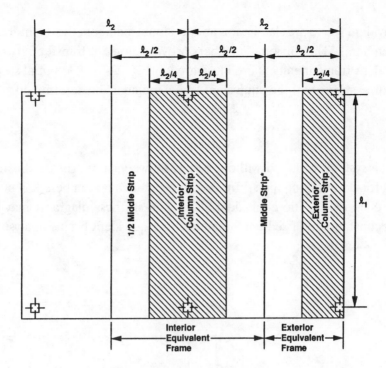

(a) Column Strip for $l_2 \leq l_1$

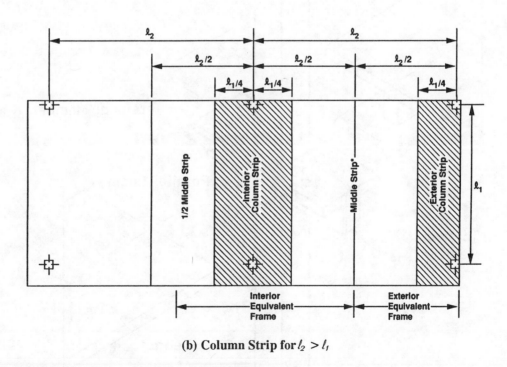

(b) Column Strip for $l_2 > l_1$

*When edge of exterior design strip is supported by a wall, the factored moment resisted by this middle strip is defined in Section 13.6.6.3.

Fig. 22-2 Design Strips of Equivalent Frame

Members of the equivalent frame are slab-beams and torsional members (horizontal members) supported by columns (vertical members). The torsional members provide moment transfer between the slab-beams and columns. The equivalent frame members are illustrated in Fig. 22-3. The initial step in the frame analysis requires that the flexural stiffness of the equivalent frame members be determined.

13.7.3 - Slab-Beams

Common types of slab systems with and without beams between supports are illustrated in Figs. 22-4 and 22-5. Cross sections for determining the stiffness of the slab-beam members K_{sb} between support center-lines are shown for each type. The equivalent slab-beam stiffness diagrams may be used to determine moment distribution constants and fixed-end moments for Equivalent Frame analysis.

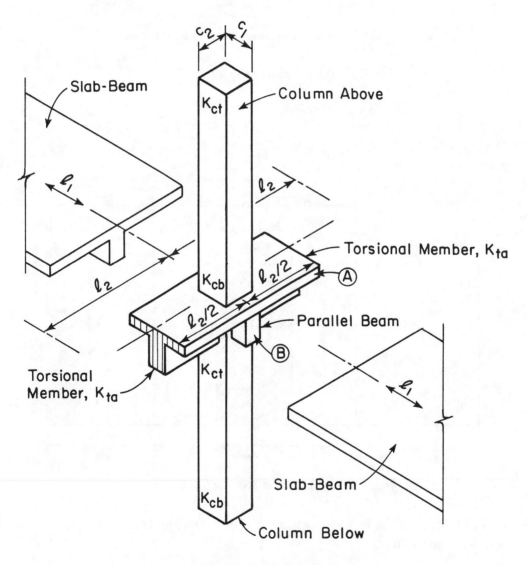

Fig. 22-3 Equivalent Frame Members

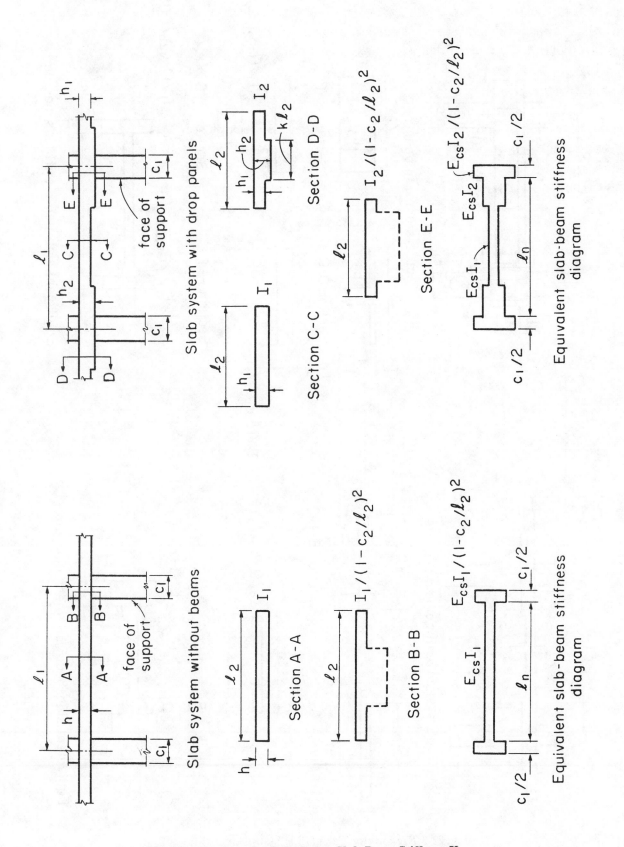

Fig. 22-4 Sections for Calculating Slab-Beam Stiffness K_{sb}

22-5

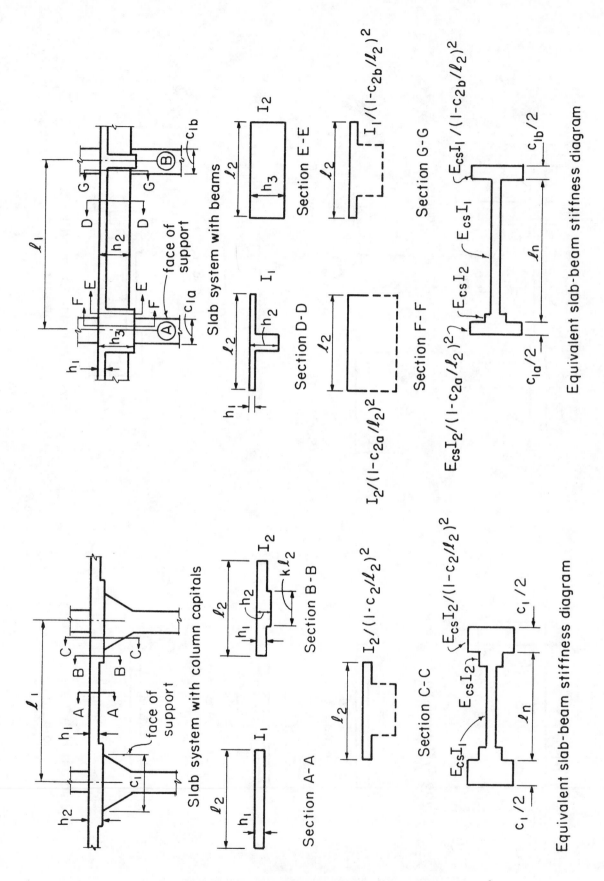

Fig. 22-5 Sections for Calculating Slab-Beam Stiffness K_{sb}

Stiffness calculations are based on the following considerations:

(a) The moment of inertia of the slab-beam between faces of supports is based on the gross cross-sectional area of the concrete. Variation in the moment of inertia along the axis of the slab-beam between supports is taken into account. (Section 13.7.3.2.)

(b) A support is defined as a column, capital, bracket or wall. Note that a beam is not considered a supporting member for the equivalent frame. (Section 13.7.3.3.)

(c) The moment of inertia of the slab-beam from the face of support to the centerline of support is assumed equal to the moment of inertia of the slab-beam at the face of support, divided by the quantity $(1 - c_2/\ell_2)^2$. (Section 13.7.3.3.)

The magnification factor $1/(1 - c_2/\ell_2)^2$ applied to the moment of inertia between support face and support centerline, in effect, makes each slab-beam at least a haunched member within its length. Consequently, stiffness and carryover factors and fixed-end moments based on the usual assumptions of uniform prismatic members cannot be applied to the slab-beam members.

Tables A1 through A6 in Appendix 22A give stiffness coefficients, carry-over factors, and fixed-end moment coefficients for different geometric and loading configurations. A wide range of column-to-span ratios in both longitudinal and transverse directions is covered in the tables. Table A1 can be used for flat plates and two-way slabs with beams. Tables A2 through A5 are intended to be used for flat slabs and waffle slabs with various drop (solid head) depths. Table A6 covers the unusual case of a flat plate combined with a flat slab. Fixed-end moment coefficients are provided for both uniform and partially uniform loads. Partial load coefficients were developed for loads distributed over a length of span equal to $0.2\ell_1$. However, loads acting over longer portions of span may be considered by summing the effects of loads acting over each $0.2\ell_1$ interval. For example, if the partial loading extends over $0.6\ell_1$, then the coefficients corresponding to three consecutive $0.2\ell_1$ intervals are to be added. This provides flexibility in the arrangement of loading. For concentrated loads, a high intensity of partial loading may be considered at the appropriate location, and assumed to be distributed over $0.2\ell_1$. For parameter values in between those listed, interpolation may be made. Stiffness diagrams are shown on each table. With appropriate engineering judgment, different span conditions may be considered with the help of information given in these tables.

13.7.4 Columns

Common types of column end support conditions for slab systems are illustrated in Fig. 22-6. The column stiffness is based on a height of column ℓ_c measured from the mid-depth of the slab above to the mid-depth of the slab below. The column stiffness diagrams may be used to determine column flexural stiffness, K_c. The stiffness diagrams are based on the following considerations:

(a)The moment of inertia of the column outside the slab-beam joint is base on the gross cross-sectional area of the concrete. Variation in the moment of inertia along the axis of the column between slab-beam joints is taken into account. For columns with capitals, the moment of inertia is assumed to vary linearly from the base of the capital to the bottom of the slab-beam (Sections 13.7.4.1 and 13.7.4.2).

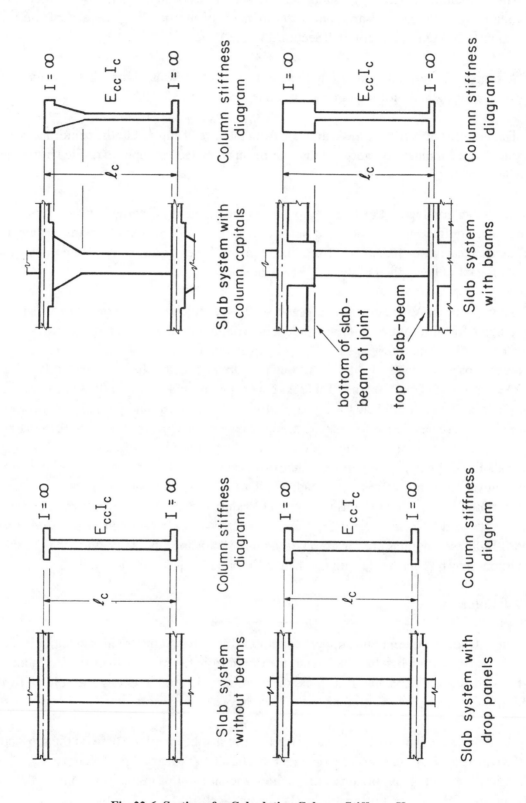

Fig. 22-6 Sections for Calculating Column Stiffness K_c

(b)The moment of inertia is assumed infinite ($I = \infty$) from the top to the bottom of the slab-beam at the joint. As with the slab-beam members, the stiffness factor K_c for the columns cannot be based on the assumption of uniform prismatic members (Sections 13.7.4.3).

Table A7 in Appendix 22A can be used to determine the actual column stiffnesses and carry-over factors.

13.7.5 Torsional Members

Torsional members for common slab-beam joints are illustrated in Fig. 22-7. The cross section of a torsional member is the largest of those defined by the three conditions given in Section 13.7.5.1. The governing condition (a), (b) or (c) is indicated below each illustration in Fig. 22-7.

The stiffness K_t of the torsional member is calculated by the following expression:

$$K_t = \Sigma \left[\frac{9E_{cs}C}{\ell_2 \left[1 - (c_2/\ell_2)\right]^3} \right] \qquad \text{Eq. (13-6)}$$

where the summation extends over torsional members framing into a joint: two (2) for interior frames, and one (1) for exterior frames.

The term C is a cross-sectional constant that defines the torsional properties of each torsional member framing into a joint:

$$C = \Sigma \left[1 - 0.63 \, (x/y)\right] x^3 y/3 \qquad \text{Eq. (13-7)}$$

where x = the shorter dimension of a rectangular part and y = the longer dimension of a rectangular part

The value of C is computed by dividing the cross section of a torsional member into separate rectangular parts and summing the C values for the component rectangles. It is appropriate to subdivide the cross section in a manner that results in the highest possible value of C. Application of the C expression is illustrated in Fig. 22-8.

If beams frame into the support in the direction moments are being determined, the torsional stiffness K_t given by Eq. (13-6) needs to be increased as follows:

$$K_{ta} = K_t \, I_{sb}/I_s$$

where K_{ta} = increased torsional stiffness due to the parallel beam (note parallel beam shown in Fig. 22-3)

I_s = moment of inertia of a width of slab equal to the full width between panel centerlines, ℓ_2, excluding that portion of the beam stem extending above and below the slab (note part A in Fig. 22-3).

$$= \ell_2 \, h^3/12$$

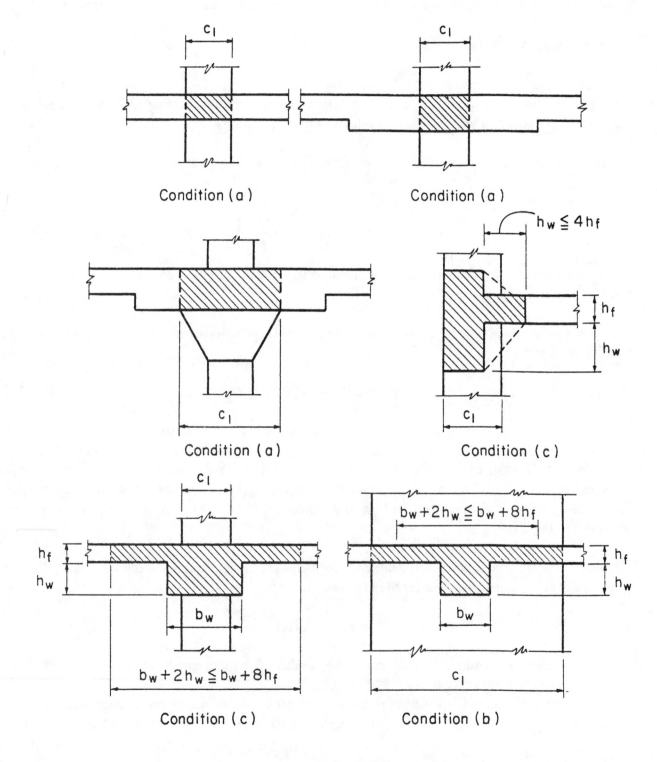

Fig. 22-7 Torsional Members

22-10

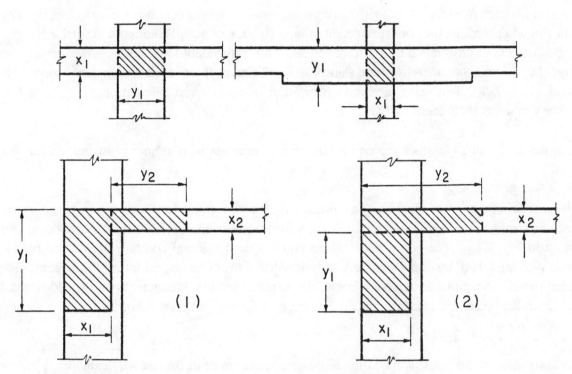

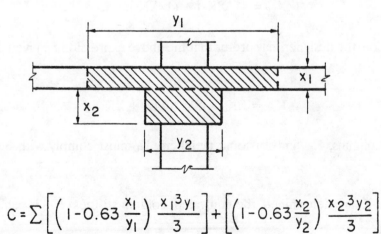

Use larger value of C computed from (1) or (2)

$$C = \sum \left[\left(1 - 0.63 \frac{x_1}{y_1}\right) \frac{x_1^3 y_1}{3} \right] + \left[\left(1 - 0.63 \frac{x_2}{y_2}\right) \frac{x_2^3 y_2}{3} \right]$$

**Fig. 22-8 Cross Sectional Constant C , Defining Torsional
Properties of a Torsional Member**

I_{sb} = moment of inertia of the slab section specified for I_s including that portion of the beam stem extending above and below the slab (for the parallel beam illustrated in Fig. 22-3, I_{sb} is for the full tee section shown).

Equivalent Columns (Commentary Section 13.7.4)

With the publication of ACI 318-83, the "equivalent column" concept of defining a single-stiffness element consisting of the actual columns above and below the slab-beams plus an attached transverse torsional member was eliminated from the code. With the increasing use of computers for two-way slab analysis by the Equivalent Frame procedure, the concept of combining stiffnesses of actual columns and torsional members into a single-stiffness has lost much of it attractiveness. The "Equivalent Column" is, however, retained in the Commentary as an aid to analysis where slab-beams at different floor levels are analyzed separately for gravity loads, especially when using moment distribution or other hand calculation procedures for the analysis. See Commentary Section 13.7.4.

Both Examples 22.1 and 22.2 utilize the equivalent column concept with moment distribution for gravity load analysis.

The equivalent column concept modifies the column stiffness to account for the torsional flexibility of the slab-to-column connection which reduces its efficiency for transmission of moments. An equivalent column is illustrated in Fig. 22-3. The equivalent column consists of the actual columns above and below the slab-beams, plus "attached" torsional members on both sides of the columns, extending to the centerlines of the adjacent panels. Note that for an edge frame, the attached torsional member is on one side only. The presence of parallel beams will also influence the stiffness of the equivalent column.

The flexural stiffness of the equivalent column K_{ec} is given in terms of its inverse, or flexibility, as follows:

$$1/K_{ec} = (1/\Sigma K_c) + (1/\Sigma K_t)$$

For purposes of computation, the designer may prefer that the above expression be given directly in terms of stiffness as follows:

$$K_{ec} = \Sigma K_c \times \Sigma K_t / (\Sigma K_c + \Sigma K_t)$$

Stiffnesses of the actual columns, K_c, and torsional members, K_t, must comply with Sections 13.7.4 and 13.7.5.

After the values of K_c and K_t are determined, the equivalent column stiffness K_{ec} is computed. Using Fig. 22-3 for illustration,

$$K_{ec} = [(K_{ct} + K_{cb})(K_{ta} + K_{ta})] / [(K_{ct} + K_{cb}) + (K_{ta} + K_{ta})]$$

where $\quad$ K_{ct} = flexural stiffness at top of lower column framing into joint,

$\quad\quad\quad\quad\quad K_{cb}$ = flexural stiffness at bottom of upper column framing into joint,

$\quad\quad\quad\quad\quad K_{ta}$ = torsional stiffness of each torsional member, one on each side of the column, increased due to the parallel beam (if any).

13.7.6 Arrangement of Live Load

In the usual case where the exact loading pattern is not known, the maximum factored moments are developed with loading conditions illustrated by the three-span partial frame in Fig. 22-9, and described as follows:

(a) When the service live load does not exceed three-quarters of the service dead load, only loading pattern (1) with full factored live load on all spans need be analyzed for negative and positive factored moments.

(b) When the service live-to-dead load ratio exceeds three-quarters, the five loading patterns shown need to be analyzed to determine all factored moments in the slab-beam members. Loading patterns (2) through (5) consider partial factored live loads for determining factored moments. However, with partial live loading, the factored moments cannot be taken less than those occurring with full factored live load on all slab-beam spans; hence load pattern (1) needs to be included in the analysis.

For slab systems with beams, loads supported directly by the beams (such as the weight of the beam stem or a wall supported directly by the beams) may be inconvenient to include in the frame analysis for the slab loads, $w_d + w_\ell$. An additional frame analysis may be required with the beam section designed to carry these loads in addition to the portion of the slab moments assigned to the beams.

13.7.7 Factored Moments

Moment distribution is probably the most convenient hand calculation method for analyzing partial frames involving several continuous spans with the far ends of upper and lower columns fixed. The mechanics of the method will not be described here, except for a brief discussion of the following two points: (1) the use of the equivalent column concept to determine joint distribution factors; and (2) the proper procedure to distribute the equivalent column moment obtained in the frame analysis to the actual columns above and below the slab-beam joint. See Examples 22.1 and 22.2.

A frame joint with stiffness factors K shown for each member framing into the joint is illustrated in Fig. 22-10. Expressions are given below for the moment distribution factors DF at the joint, using the equivalent column stiffness, K_{ec}. These DF factors are used directly in the moment distribution procedure.

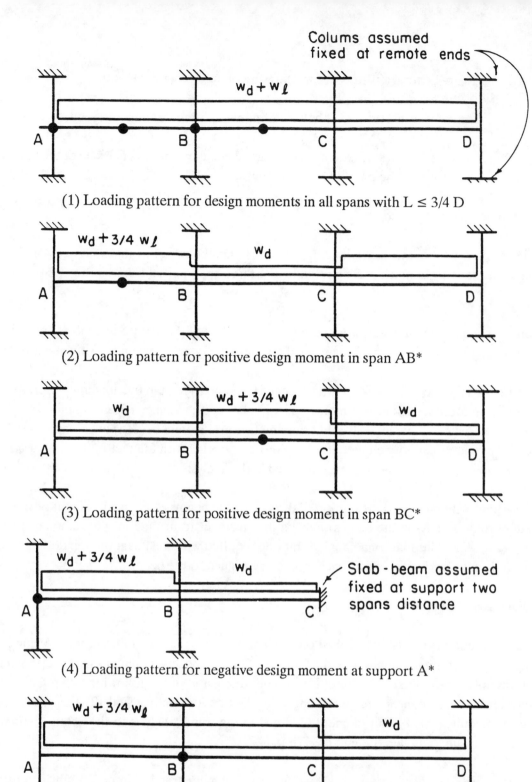

(1) Loading pattern for design moments in all spans with $L \leq 3/4 D$

(2) Loading pattern for positive design moment in span AB*

(3) Loading pattern for positive design moment in span BC*

(4) Loading pattern for negative design moment at support A*

(5) Loading pattern for negative design moment at support B*

Fig. 22-9 Partial Frame Analysis for Vertical Loading.

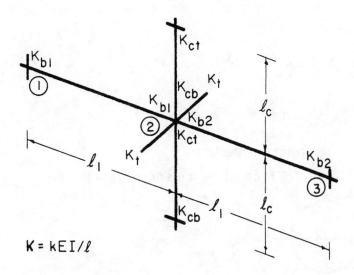

$$K = kEI/\ell$$

Fig. 22-10 Moment Distribution Factors DF

Equivalent column stiffness,

$$K_{ec} = \Sigma K_c \times \Sigma K_t / (\Sigma K_c + \Sigma K_t)$$

$$= [(K_{ct} + K_{cb})\,(K_t + K_t)] \,/\, [(K_{ct} + K_{cb}) + (K_t + K_t)]$$

Slab-beam distribution factor,

$$DF \text{ (span 2-1)} = K_{b1} / (K_{b1} + K_{b2} + K_{ec})$$

$$DF \text{ (span 2-3)} = K_{b2} / (K_{b1} + K_{b2} + K_{ec})$$

Equivalent column distribution factor (unbalanced moment from slab-beam),

$$DF = K_{ec} / (K_{b1} + K_{b2} + K_{ec})$$

The unbalanced moment determined for the equivalent column in the moment distribution cycles is distributed to the actual columns above and below the slab-beam in proportion to the actual column stiffnesses at the joint. Referring to Fig. 22-10:

Portion of unbalanced moment to upper column,

$$= K_{cb} / (K_{cb} + K_{ct})$$

Portion of unbalanced moment to lower column,

$$= K_{ct} / (K_{cb} + K_{ct})$$

The "actual" columns are then designed for these moments.

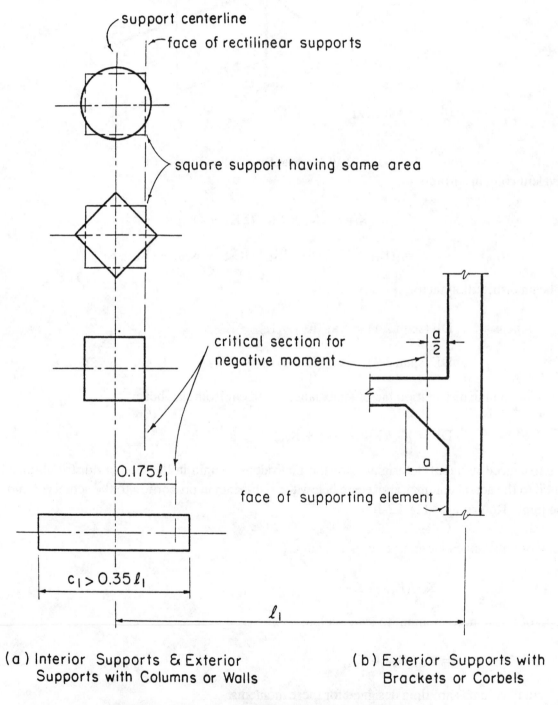

(a) Interior Supports & Exterior
    Supports with Columns or Walls

(b) Exterior Supports with
    Brackets or Corbels

Fig. 22-11 Critical Sections for Negative Factored Moment

22-16

13.7.7.1 - 13.7.7.3 Negative Factored Moments - Negative factored moments for design must be taken at faces of rectilinear support, but not at a distance greater than $0.175\ell_1$ from the center of support. This absolute value is a limit on long narrow supports in order to prevent undue reduction in design moment. The support member is defined as a column, capital, bracket or wall. Non-rectangular supports should be treated as square supports having the same cross-sectional area. Note that for slab systems with beams, the faces of beams are not considered a face-of-support location. Support conditions for locating the negative factored moment are illustrated in Fig. 22-11. Note the special requirements illustrated for exterior supports.

13.7.7.4 Moment Redistribution - Should a designer choose to use the Equivalent Frame Method to analyze a slab system that meets the limitations of the Direct Design Method, the factored moments may be reduced so that the total static factored moment (sum of the average negative and positive moments) need not exceed M_0 computed by Eq. (13-3). This permissible reduction is illustrated in Fig. 22-12.

Since the Equivalent Frame Method of analysis is not an approximate method, the moment redistribution allowed in Section 8.4 may be used. Excessive cracking may result if these provisions are imprudently applied. The burden of judgement is left to the designer as to what, if any, redistribution is warranted.

13.7.7.5 - Factored Moments in Column Strips and Middle Strips - Negative and positive factored moments may be distributed to the column strip and the two half middle strips of the slab-beam in accordance with Sections 13.6.4, 13.6.5 and 13.6.6, provided the requirement of Section 13.6.1.6 is satisfied.

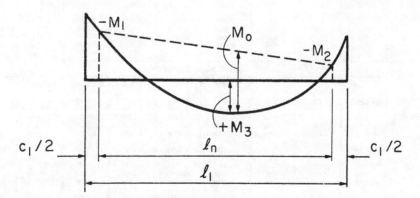

$$M_0 = \left[(M_1 + M_2)/2\right] + M_3 \text{ need not be greater than } w\ell_2\ell_n^2/8$$

Permissible reduction for moments M_1, M_2, and $M_3 =$

$$\left[w\ell_2\ell_n^2/8\right]/\left[(M_1 + M_2)/2 + M_3\right]$$

Fig. 22-12 Total Static Design Moment for a Span

APPENDIX 22A

DESIGN AIDS FOR MOMENT DISTRIBUTION CONSTANTS

Table A1 - Moment Distribution Constants for Slab-Beam Members

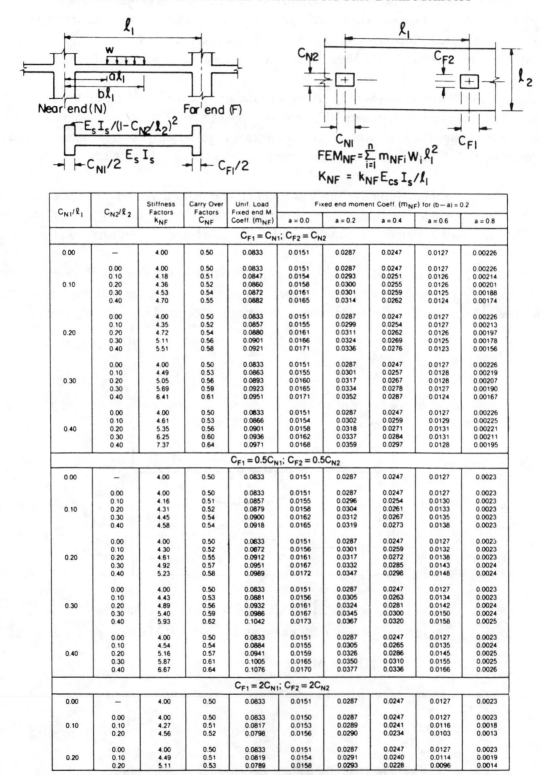

$$FEM_{NF} = \sum_{i=1}^{n} m_{NFi} W_i \ell_1^2$$

$$K_{NF} = k_{NF} E_{cs} I_s / \ell_1$$

C_{N1}/ℓ_1	C_{N2}/ℓ_2	Stiffness Factors k_{NF}	Carry Over Factors C_{NF}	Unif. Load Fixed end M. Coeff. (m_{NF})	Fixed end moment Coeff. (m_{NF}) for (b−a) = 0.2				
					a = 0.0	a = 0.2	a = 0.4	a = 0.6	a = 0.8
				$C_{F1} = C_{N1}$; $C_{F2} = C_{N2}$					
0.00	—	4.00	0.50	0.0833	0.0151	0.0287	0.0247	0.0127	0.00226
	0.00	4.00	0.50	0.0833	0.0151	0.0287	0.0247	0.0127	0.00226
	0.10	4.18	0.51	0.0847	0.0154	0.0293	0.0251	0.0126	0.00214
0.10	0.20	4.36	0.52	0.0860	0.0158	0.0300	0.0255	0.0126	0.00201
	0.30	4.53	0.54	0.0872	0.0161	0.0301	0.0259	0.0125	0.00188
	0.40	4.70	0.55	0.0882	0.0165	0.0314	0.0262	0.0124	0.00174
	0.00	4.00	0.50	0.0833	0.0151	0.0287	0.0247	0.0127	0.00226
	0.10	4.35	0.52	0.0857	0.0155	0.0299	0.0254	0.0127	0.00213
0.20	0.20	4.72	0.54	0.0880	0.0161	0.0311	0.0262	0.0126	0.00197
	0.30	5.11	0.56	0.0901	0.0166	0.0324	0.0269	0.0125	0.00178
	0.40	5.51	0.58	0.0921	0.0171	0.0336	0.0276	0.0123	0.00156
	0.00	4.00	0.50	0.0833	0.0151	0.0287	0.0247	0.0127	0.00226
	0.10	4.49	0.53	0.0863	0.0155	0.0301	0.0257	0.0128	0.00219
0.30	0.20	5.05	0.56	0.0893	0.0160	0.0317	0.0267	0.0128	0.00207
	0.30	5.69	0.59	0.0923	0.0165	0.0334	0.0278	0.0127	0.00190
	0.40	6.41	0.61	0.0951	0.0171	0.0352	0.0287	0.0124	0.00167
	0.00	4.00	0.50	0.0833	0.0151	0.0287	0.0247	0.0127	0.00226
	0.10	4.61	0.53	0.0866	0.0154	0.0302	0.0259	0.0129	0.00225
0.40	0.20	5.35	0.56	0.0901	0.0158	0.0318	0.0271	0.0131	0.00221
	0.30	6.25	0.60	0.0936	0.0162	0.0337	0.0284	0.0131	0.00211
	0.40	7.37	0.64	0.0971	0.0168	0.0359	0.0297	0.0128	0.00195
				$C_{F1} = 0.5C_{N1}$; $C_{F2} = 0.5C_{N2}$					
0.00	—	4.00	0.50	0.0833	0.0151	0.0287	0.0247	0.0127	0.0023
	0.00	4.00	0.50	0.0833	0.0151	0.0287	0.0247	0.0127	0.0023
	0.10	4.16	0.51	0.0857	0.0155	0.0296	0.0254	0.0130	0.0023
0.10	0.20	4.31	0.52	0.0879	0.0158	0.0304	0.0261	0.0133	0.0023
	0.30	4.45	0.54	0.0900	0.0162	0.0312	0.0267	0.0135	0.0023
	0.40	4.58	0.54	0.0918	0.0165	0.0319	0.0273	0.0138	0.0023
	0.00	4.00	0.50	0.0833	0.0151	0.0287	0.0247	0.0127	0.0023
	0.10	4.30	0.52	0.0872	0.0156	0.0301	0.0259	0.0132	0.0023
0.20	0.20	4.61	0.55	0.0912	0.0161	0.0317	0.0272	0.0138	0.0023
	0.30	4.92	0.57	0.0951	0.0167	0.0332	0.0285	0.0143	0.0024
	0.40	5.23	0.58	0.0989	0.0172	0.0347	0.0298	0.0148	0.0024
	0.00	4.00	0.50	0.0833	0.0151	0.0287	0.0247	0.0127	0.0023
	0.10	4.43	0.53	0.0881	0.0156	0.0305	0.0263	0.0134	0.0023
0.30	0.20	4.89	0.56	0.0932	0.0161	0.0324	0.0281	0.0142	0.0024
	0.30	5.40	0.59	0.0986	0.0167	0.0345	0.0300	0.0150	0.0024
	0.40	5.93	0.62	0.1042	0.0173	0.0367	0.0320	0.0158	0.0025
	0.00	4.00	0.50	0.0833	0.0151	0.0287	0.0247	0.0127	0.0023
	0.10	4.54	0.54	0.0884	0.0155	0.0305	0.0265	0.0135	0.0024
0.40	0.20	5.16	0.57	0.0941	0.0159	0.0326	0.0286	0.0145	0.0025
	0.30	5.87	0.61	0.1005	0.0165	0.0350	0.0310	0.0155	0.0025
	0.40	6.67	0.64	0.1076	0.0170	0.0377	0.0336	0.0166	0.0026
				$C_{F1} = 2C_{N1}$; $C_{F2} = 2C_{N2}$					
0.00	—	4.00	0.50	0.0833	0.0151	0.0287	0.0247	0.0127	0.0023
	0.00	4.00	0.50	0.0833	0.0150	0.0287	0.0247	0.0127	0.0023
0.10	0.10	4.27	0.51	0.0817	0.0153	0.0289	0.0241	0.0116	0.0018
	0.20	4.56	0.52	0.0798	0.0156	0.0290	0.0234	0.0103	0.0013
	0.00	4.00	0.50	0.0833	0.0151	0.0287	0.0247	0.0127	0.0023
0.20	0.10	4.49	0.51	0.0819	0.0154	0.0291	0.0240	0.0114	0.0019
	0.20	5.11	0.53	0.0789	0.0158	0.0293	0.0228	0.0096	0.0014

Table A2—Moment Distribution Constants for Slab-Beam
Members (Drop thickness = 0.25h)

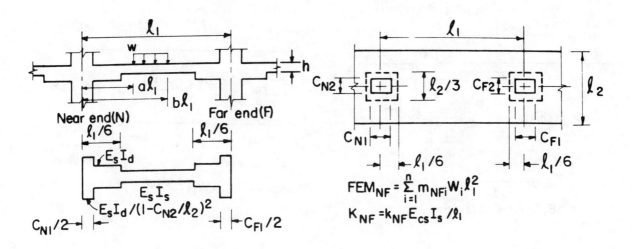

C_{N1}/ℓ_1	C_{N2}/ℓ_2	Stiffness Factors k_{NF}	Carry Over Factors C_{NF}	Unif. Load Fixed end M. Coeff. (m_{NF})	Fixed end moment Coeff. (m_{NF}) for (b—a) = 0.2				
					a = 0.0	a = 0.2	a = 0.4	a = 0.6	a = 0.8
$C_{F1} = C_{N1}$; $C_{F2} = C_{N2}$									
0.00	—	4.79	0.54	0.0879	0.0157	0.0309	0.0263	0.0129	0.0022
0.10	0.00	4.79	0.54	0.0879	0.0157	0.0309	0.0263	0.0129	0.0022
	0.10	4.99	0.55	0.0890	0.0160	0.0316	0.0266	0.0128	0.0020
	0.20	5.18	0.56	0.0901	0.0163	0.0322	0.0270	0.0127	0.0019
	0.30	5.37	0.57	0.0911	0.0167	0.0328	0.0273	0.0126	0.0018
0.20	0.00	4.79	0.54	0.0879	0.0157	0.0309	0.0263	0.0129	0.0022
	0.10	5.17	0.56	0.0900	0.0161	0.0320	0.0269	0.0128	0.0020
	0.20	5.56	0.58	0.0918	0.0166	0.0332	0.0276	0.0126	0.0018
	0.30	5.96	0.60	0.0936	0.0171	0.0344	0.0282	0.0124	0.0016
0.30	0.00	4.79	0.54	0.0879	0.0157	0.0309	0.0263	0.0129	0.0022
	0.10	5.32	0.57	0.0905	0.0161	0.0323	0.0272	0.0128	0.0021
	0.20	5.90	0.59	0.0930	0.0166	0.0338	0.0281	0.0127	0.0019
	0.30	6.55	0.62	0.0955	0.0171	0.0354	0.0290	0.0124	0.0017
$C_{F1} = 0.5C_{N1}$; $C_{F2} = 0.5C_{N2}$									
0.00	—	4.79	0.54	0.0879	0.0157	0.0309	0.0263	0.0129	0.0022
0.10	0.00	4.79	0.54	0.0879	0.0157	0.0309	0.0263	0.0129	0.0022
	0.10	4.96	0.55	0.0900	0.0160	0.0317	0.0269	0.0131	0.0022
	0.20	5.12	0.56	0.0920	0.0164	0.0325	0.0276	0.0134	0.0022
0.20	0.00	4.79	0.54	0.0879	0.0157	0.0309	0.0263	0.0129	0.0022
	0.10	5.11	0.56	0.0914	0.0162	0.0323	0.0275	0.0133	0.0022
	0.20	5.43	0.58	0.0950	0.0167	0.0337	0.0286	0.0138	0.0022
$C_{F1} = 2C_{N1}$; $C_{F2} = 2C_{N2}$									
0.00	—	4.79	0.54	0.0879	0.0157	0.0309	0.0263	0.0129	0.0022
0.10	0.00	4.79	0.54	0.0879	0.0157	0.0309	0.0263	0.0129	0.0022
	0.10	5.10	0.55	0.0860	0.0159	0.0311	0.0256	0.0117	0.0017

Table A3—Moment Distribution Constants for Slab-Beam Members (Drop thickness = 0.50h)

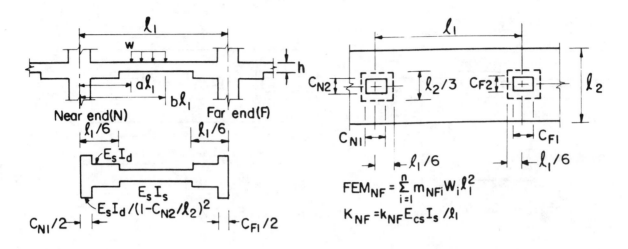

$$FEM_{NF} = \sum_{i=1}^{n} m_{NFi} W_i \ell_1^2$$

$$K_{NF} = k_{NF} E_{cs} I_s / \ell_1$$

C_{N1}/ℓ_1	C_{N2}/ℓ_2	Stiffness Factors k_{NF}	Carry Over Factors C_{NF}	Unif. Load Fixed end M. Coeff. (m_{NF})	Fixed end moment Coeff. (m_{NF}) for (b—a) = 0.2				
					a = 0.0	a = 0.2	a = 0.4	a = 0.6	a = 0.8
$C_{F1} = C_{N1}$; $C_{F2} = C_{N2}$									
0.00	—	5.84	0.59	0.0926	0.0164	0.0335	0.0279	0.0128	0.0020
0.10	0.00	5.84	0.59	0.0926	0.0164	0.0335	0.0279	0.0128	0.0020
	0.10	6.04	0.60	0.0936	0.0167	0.0341	0.0282	0.0126	0.0018
	0.20	6.24	0.61	0.0940	0.0170	0.0347	0.0285	0.0125	0.0017
	0.30	6.43	0.61	0.0952	0.0173	0.0353	0.0287	0.0123	0.0016
0.20	0.00	5.84	0.59	0.0926	0.0164	0.0335	0.0279	0.0128	0.0020
	0.10	6.22	0.61	0.0942	0.0168	0.0346	0.0285	0.0126	0.0018
	0.20	6.62	0.62	0.0957	0.0172	0.0356	0.0290	0.0123	0.0016
	0.30	7.01	0.64	0.0971	0.0177	0.0366	0.0294	0.0120	0.0014
0.30	0.00	5.84	0.59	0.0926	0.0164	0.0335	0.0279	0.0128	0.0020
	0.10	6.37	0.61	0.0947	0.0168	0.0348	0.0287	0.0126	0.0018
	0.20	6.95	0.63	0.0967	0.0172	0.0362	0.0294	0.0123	0.0016
	0.30	7.57	0.65	0.0986	0.0177	0.0375	0.0300	0.0119	0.0014
$C_{F1} = 0.5C_{N1}$; $C_{F2} = 0.5C_{N2}$									
0.00	—	5.84	0.59	0.0926	0.0164	0.0335	0.0279	0.0128	0.0020
0.10	0.00	5.84	0.59	0.0926	0.0164	0.0335	0.0279	0.0128	0.0020
	0.10	6.00	0.60	0.0945	0.0167	0.0343	0.0285	0.0130	0.0020
	0.20	6.16	0.60	0.0962	0.0170	0.0350	0.0291	0.0132	0.0020
0.20	0.00	5.84	0.59	0.0926	0.0164	0.0335	0.0279	0.0128	0.0020
	0.10	6.15	0.60	0.0957	0.0169	0.0348	0.0290	0.0131	0.0020
	0.20	6.47	0.62	0.0987	0.0173	0.0360	0.0300	0.0134	0.0020
$C_{F1} = 2C_{N1}$; $C_{F2} = 2C_{N2}$									
0.00	—	5.84	0.59	0.0926	0.0164	0.0335	0.0279	0.0128	0.0020
0.10	0.00	5.84	0.59	0.0926	0.0164	0.0335	0.0279	0.0128	0.0020
	0.10	6.17	0.60	0.0907	0.0166	0.0337	0.0273	0.0116	0.0015

Table A4—Moment Distribution Constants for Slab-Beam Members (Drop thickness = 0.75h)

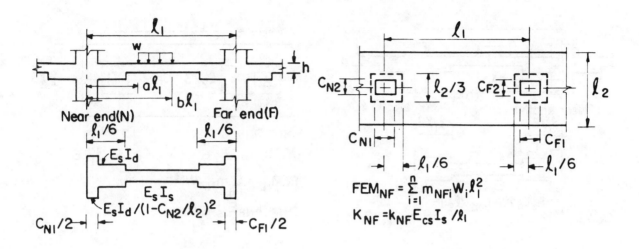

$$FEM_{NF} = \sum_{i=1}^{n} m_{NFi} W_i \ell_1^2$$

$$K_{NF} = k_{NF} E_{cs} I_s / \ell_1$$

C_{N1}/ℓ_1	C_{N2}/ℓ_2	Stiffness Factors k_{NF}	Carry Over Factors C_{NF}	Unif. Load Fixed end M. Coeff. (m_{NF})	Fixed end moment Coeff. (m_{NF}) for (b—a) = 0.2				
					a = 0.0	a = 0.2	a = 0.4	a = 0.6	a = 0.8
$C_{F1} = C_{N1}$; $C_{F2} = C_{N2}$									
0.00	—	6.92	0.63	0.0965	0.0171	0.0360	0.0293	0.0124	0.0017
0.10	0.00	6.92	0.63	0.0965	0.0171	0.0360	0.0293	0.0124	0.0017
	0.10	7.12	0.64	0.0972	0.0174	0.0365	0.0295	0.0122	0.0016
	0.20	7.31	0.64	0.0978	0.0176	0.0370	0.0297	0.0120	0.0014
	0.30	7.48	0.65	0.0984	0.0179	0.0375	0.0299	0.0118	0.0013
0.20	0.00	6.92	0.63	0.0965	0.0171	0.0360	0.0293	0.0124	0.0017
	0.10	7.12	0.64	0.0977	0.0175	0.0369	0.0297	0.0121	0.0015
	0.20	7.31	0.65	0.0988	0.0178	0.0378	0.0301	0.0118	0.0013
	0.30	7.48	0.67	0.0999	0.0182	0.0386	0.0304	0.0115	0.0011
0.30	0.00	6.92	0.63	0.0965	0.0171	0.0360	0.0293	0.0124	0.0017
	0.10	7.29	0.65	0.0981	0.0175	0.0371	0.0299	0.0121	0.0015
	0.20	7.66	0.66	0.0996	0.0179	0.0383	0.0304	0.0117	0.0013
	0.30	8.02	0.68	0.1009	0.0182	0.0394	0.0309	0.0113	0.0011
$C_{F1} = 0.5C_{N1}$; $C_{F2} = 0.5C_{N2}$									
0.00	—	6.92	0.63	0.0965	0.0171	0.0360	0.0293	0.0124	0.0017
0.10	0.00	6.92	0.63	0.0965	0.0171	0.0360	0.0293	0.0124	0.0017
	0.10	7.08	0.64	0.0980	0.0174	0.0366	0.0298	0.0125	0.0017
	0.20	7.23	0.64	0.0993	0.0177	0.0372	0.0302	0.0126	0.0016
0.20	0.00	6.92	0.63	0.0965	0.0171	0.0360	0.0293	0.0124	0.0017
	0.10	7.21	0.64	0.0991	0.0175	0.0371	0.0302	0.0126	0.0017
	0.20	7.51	0.65	0.1014	0.0179	0.0381	0.0310	0.0128	0.0016
$C_{F1} = 2C_{N1}$; $C_{F2} = 2C_{N2}$									
0.00	—	6.92	0.63	0.0965	0.0171	0.0360	0.0293	0.0124	0.0017
0.10	0.00	6.92	0.63	0.0965	0.0171	0.0360	0.0293	0.0124	0.0017
	0.10	7.26	0.64	0.0946	0.0173	0.0361	0.0287	0.0112	0.0013

Table A5—Moment Distribution Constants for Slab-Beam
Members (Drop thickness = h)

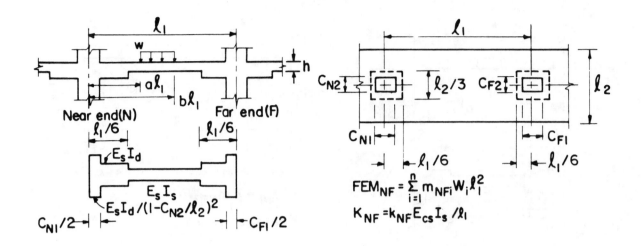

$$FEM_{NF} = \sum_{i=1}^{n} m_{NFi} w_i \ell_1^2$$

$$K_{NF} = k_{NF} E_{cs} I_s / \ell_1$$

C_{N1}/ℓ_1	C_{N2}/ℓ_2	Stiffness Factors k_{NF}	Carry Over Factors C_{NF}	Unif. Load Fixed end M. Coeff. (m_{NF})	Fixed end moment Coeff. (m_{NF}) for (b—a) = 0.2				
					a = 0.0	a = 0.2	a = 0.4	a = 0.6	a = 0.8
$C_{F1} = C_{N1};\ C_{F2} = C_{N2}$									
0.00	—	7.89	0.66	0.0993	0.0177	0.0380	0.0303	0.0118	0.0014
0.10	0.00	7.89	0.66	0.0993	0.0177	0.0380	0.0303	0.0118	0.0014
	0.10	8.07	0.66	0.0998	0.0180	0.0385	0.0305	0.0116	0.0013
	0.20	8.24	0.67	0.1003	0.0182	0.0389	0.0306	0.0115	0.0012
	0.30	8.40	0.67	0.1007	0.0183	0.0393	0.0307	0.0113	0.0011
0.20	0.00	7.89	0.66	0.0993	0.0177	0.0380	0.0303	0.0118	0.0014
	0.10	8.22	0.67	0.1002	0.0180	0.0388	0.0306	0.0115	0.0012
	0.20	8.55	0.68	0.1010	0.0183	0.0395	0.0309	0.0112	0.0011
	0.30	9.87	0.69	0.1018	0.0186	0.0402	0.0311	0.0109	0.0009
0.30	0.00	7.89	0.66	0.0993	0.0177	0.0380	0.0303	0.0118	0.0014
	0.10	8.35	0.67	0.1005	0.0181	0.0390	0.0307	0.0115	0.0012
	0.20	8.82	0.68	0.1016	0.0184	0.0399	0.0311	0.0111	0.0011
	0.30	9.28	0.70	0.1026	0.0187	0.0409	0.0314	0.0107	0.0009
$C_{F1} = 0.5 C_{N1};\ C_{F2} = 0.5 C_{N2}$									
0,00	—	7.89	0.66	0.0993	0.0177	0.0380	0.0303	0.0118	0.0014
0.10	0.00	7.89	0.66	0.0993	0.0177	0.0380	0.0303	0.0118	0.0014
	0.10	8.03	0.66	0.1006	0.0180	0.0386	0.0307	0.0119	0.0014
	0.20	8.16	0.67	0.1016	0.0182	0.0390	0.0310	0.0120	0.0014
0.20	0.00	7.89	0.66	0.0993	0.0177	0.0380	0.0303	0.0118	0.0014
	0.10	8.15	0.67	0.1014	0.0181	0.0389	0.0310	0.0120	0.0014
	0.20	8.41	0.68	0.1032	0.0184	0.0398	0.0316	0.0121	0.0013
$C_{F1} = 2 C_{N1};\ C_{F2} = 0.5 C_{N2}$									
0.00	—	7.89	0.66	0.0993	0.0177	0.0380	0.0303	0.0118	0.0014
0.10	0.00	7.79	0.66	0.0993	0.0177	0.0380	0.0303	0.0118	0.0014
	0.10	8.20	0.67	0.0981	0.0179	0.0382	0.0297	0.0113	0.0010

Table A6—Moment Distribution Constants for Slab-Beam Members
(Column dimensions assumed equal
at near end and far end – $c_{F1} = c_{N1}$, $c_{F2} = c_{N2}$)

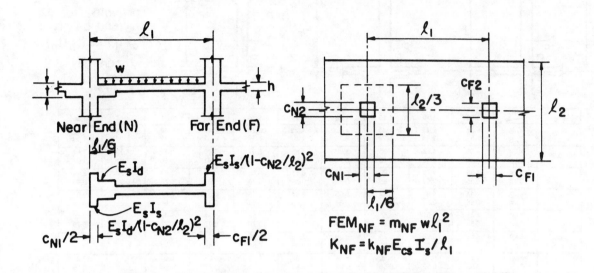

$$FEM_{NF} = m_{NF} w \ell_1^2$$
$$K_{NF} = k_{NF} E_{cs} I_s / \ell_1$$

C_1/ℓ_1	C_2/ℓ_2	t = 1.5h						t = 2h					
		k_{NF}	C_{NF}	m_{NF}	k_{FN}	C_{FN}	m_{FN}	k_{NF}	C_{NF}	m_{NF}	k_{FN}	C_{FN}	m_{FN}
0.00	—	5.39	0.49	0.1023	4.26	0.60	0.0749	6.63	0.49	0.1190	4.49	0.65	0.0676
0.10	0.00	5.39	0.49	0.1023	4.26	0.60	0.0749	6.63	0.49	0.1190	4.49	0.65	0.0676
	0.10	5.65	0.52	0.1012	4.65	0.60	0.0794	7.03	0.54	0.1145	5.19	0.66	0.0757
	0.20	5.86	0.54	0.1012	4.91	0.61	0.0818	7.22	0.56	0.1140	5.43	0.67	0.0778
	0.30	6.05	0.55	0.1025	5.10	0.62	0.0838	7.36	0.56	0.1142	5.57	0.67	0.0786
0.20	0.00	5.39	0.49	0.1023	4.26	0.60	0.0749	6.63	0.49	0.1190	4.49	0.65	0.0676
	0.10	5.88	0.54	0.1006	5.04	0.61	0.0826	7.41	0.58	0.1111	5.96	0.66	0.0823
	0.20	6.33	0.58	0.1003	5.63	0.62	0.0874	7.85	0.61	0.1094	6.57	0.67	0.0872
	0.30	6.75	0.60	0.1008	6.10	0.64	0.0903	8.18	0.63	0.1093	6.94	0.68	0.0892
0.30	0.00	5.39	0.49	0.1023	4.26	0.60	0.075	6.63	0.49	0.1190	4.49	0.65	0.0676
	0.10	6.08	0.56	0.1003	5.40	0.61	0.085	7.76	0.62	0.1087	6.77	0.67	0.0873
	0.20	6.78	0.61	0.0996	6.38	0.63	0.092	8.49	0.66	0.1055	7.91	0.68	0.0952
	0.30	7.48	0.64	0.0997	7.25	0.65	0.096	9.06	0.68	0.1047	8.66	0.69	0.0991

Table A7—Stiffness and Carry-Over Factors for Columns

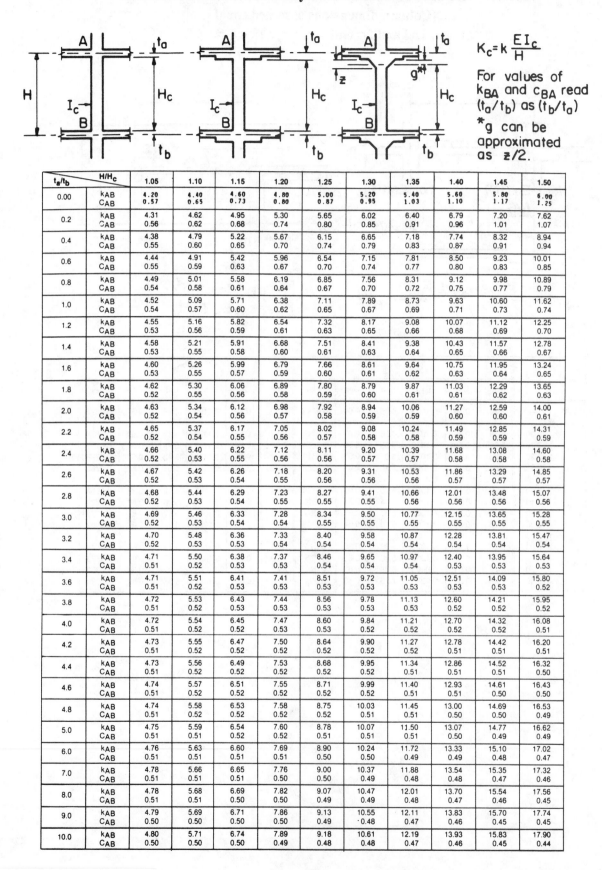

$$K_c = k \frac{EI_c}{H}$$

For values of k_{BA} and c_{BA} read (t_a/t_b) as (t_b/t_a)

*g can be approximated as $z/2$.

t_a/t_b	H/H$_c$	1.05	1.10	1.15	1.20	1.25	1.30	1.35	1.40	1.45	1.50
0.00	kAB	4.20	4.40	4.60	4.80	5.00	5.20	5.40	5.60	5.80	6.00
	CAB	0.57	0.65	0.73	0.80	0.87	0.95	1.03	1.10	1.17	1.25
0.2	kAB	4.31	4.62	4.95	5.30	5.65	6.02	6.40	6.79	7.20	7.62
	CAB	0.56	0.62	0.68	0.74	0.80	0.85	0.91	0.96	1.01	1.07
0.4	kAB	4.38	4.79	5.22	5.67	6.15	6.65	7.18	7.74	8.32	8.94
	CAB	0.55	0.60	0.65	0.70	0.74	0.79	0.83	0.87	0.91	0.94
0.6	kAB	4.44	4.91	5.42	5.96	6.54	7.15	7.81	8.50	9.23	10.01
	CAB	0.55	0.59	0.63	0.67	0.70	0.74	0.77	0.80	0.83	0.85
0.8	kAB	4.49	5.01	5.58	6.19	6.85	7.56	8.31	9.12	9.98	10.89
	CAB	0.54	0.58	0.61	0.64	0.67	0.70	0.72	0.75	0.77	0.79
1.0	kAB	4.52	5.09	5.71	6.38	7.11	7.89	8.73	9.63	10.60	11.62
	CAB	0.54	0.57	0.60	0.62	0.65	0.67	0.69	0.71	0.73	0.74
1.2	kAB	4.55	5.16	5.82	6.54	7.32	8.17	9.08	10.07	11.12	12.25
	CAB	0.53	0.56	0.59	0.61	0.63	0.65	0.66	0.68	0.69	0.70
1.4	kAB	4.58	5.21	5.91	6.68	7.51	8.41	9.38	10.43	11.57	12.78
	CAB	0.53	0.55	0.58	0.60	0.61	0.63	0.64	0.65	0.66	0.67
1.6	kAB	4.60	5.26	5.99	6.79	7.66	8.61	9.64	10.75	11.95	13.24
	CAB	0.53	0.55	0.57	0.59	0.60	0.61	0.62	0.63	0.64	0.65
1.8	kAB	4.62	5.30	6.06	6.89	7.80	8.79	9.87	11.03	12.29	13.65
	CAB	0.52	0.55	0.56	0.58	0.59	0.60	0.61	0.61	0.62	0.63
2.0	kAB	4.63	5.34	6.12	6.98	7.92	8.94	10.06	11.27	12.59	14.00
	CAB	0.52	0.54	0.56	0.57	0.58	0.59	0.59	0.60	0.60	0.61
2.2	kAB	4.65	5.37	6.17	7.05	8.02	9.08	10.24	11.49	12.85	14.31
	CAB	0.52	0.54	0.55	0.56	0.57	0.58	0.58	0.59	0.59	0.59
2.4	kAB	4.66	5.40	6.22	7.12	8.11	9.20	10.39	11.68	13.08	14.60
	CAB	0.52	0.53	0.55	0.56	0.56	0.57	0.57	0.58	0.58	0.58
2.6	kAB	4.67	5.42	6.26	7.18	8.20	9.31	10.53	11.86	13.29	14.85
	CAB	0.52	0.53	0.54	0.55	0.56	0.56	0.56	0.57	0.57	0.57
2.8	kAB	4.68	5.44	6.29	7.23	8.27	9.41	10.66	12.01	13.48	15.07
	CAB	0.52	0.53	0.54	0.55	0.55	0.55	0.56	0.56	0.56	0.56
3.0	kAB	4.69	5.46	6.33	7.28	8.34	9.50	10.77	12.15	13.65	15.28
	CAB	0.52	0.53	0.54	0.54	0.55	0.55	0.55	0.55	0.55	0.55
3.2	kAB	4.70	5.48	6.36	7.33	8.40	9.58	10.87	12.28	13.81	15.47
	CAB	0.52	0.53	0.53	0.54	0.54	0.54	0.54	0.54	0.54	0.54
3.4	kAB	4.71	5.50	6.38	7.37	8.46	9.65	10.97	12.40	13.95	15.64
	CAB	0.51	0.52	0.53	0.53	0.54	0.54	0.54	0.53	0.53	0.53
3.6	kAB	4.71	5.51	6.41	7.41	8.51	9.72	11.05	12.51	14.09	15.80
	CAB	0.51	0.52	0.53	0.53	0.53	0.53	0.53	0.53	0.53	0.52
3.8	kAB	4.72	5.53	6.43	7.44	8.56	9.78	11.13	12.60	14.21	15.95
	CAB	0.51	0.52	0.53	0.53	0.53	0.53	0.53	0.52	0.52	0.52
4.0	kAB	4.72	5.54	6.45	7.47	8.60	9.84	11.21	12.70	14.32	16.08
	CAB	0.51	0.52	0.52	0.53	0.53	0.52	0.52	0.52	0.52	0.51
4.2	kAB	4.73	5.55	6.47	7.50	8.64	9.90	11.27	12.78	14.42	16.20
	CAB	0.51	0.52	0.52	0.52	0.52	0.52	0.52	0.51	0.51	0.51
4.4	kAB	4.73	5.56	6.49	7.53	8.68	9.95	11.34	12.86	14.52	16.32
	CAB	0.51	0.52	0.52	0.52	0.52	0.52	0.51	0.51	0.51	0.50
4.6	kAB	4.74	5.57	6.51	7.55	8.71	9.99	11.40	12.93	14.61	16.43
	CAB	0.51	0.52	0.52	0.52	0.52	0.52	0.51	0.51	0.50	0.50
4.8	kAB	4.74	5.58	6.53	7.58	8.75	10.03	11.45	13.00	14.69	16.53
	CAB	0.51	0.52	0.52	0.52	0.52	0.51	0.51	0.50	0.50	0.49
5.0	kAB	4.75	5.59	6.54	7.60	8.78	10.07	11.50	13.07	14.77	16.62
	CAB	0.51	0.51	0.52	0.52	0.51	0.51	0.51	0.50	0.49	0.49
6.0	kAB	4.76	5.63	6.60	7.69	8.90	10.24	11.72	13.33	15.10	17.02
	CAB	0.51	0.51	0.51	0.51	0.50	0.50	0.49	0.49	0.48	0.47
7.0	kAB	4.78	5.66	6.65	7.76	9.00	10.37	11.88	13.54	15.35	17.32
	CAB	0.51	0.51	0.51	0.50	0.50	0.49	0.48	0.48	0.47	0.46
8.0	kAB	4.78	5.68	6.69	7.82	9.07	10.47	12.01	13.70	15.54	17.56
	CAB	0.51	0.51	0.50	0.50	0.49	0.49	0.48	0.47	0.46	0.45
9.0	kAB	4.79	5.69	6.71	7.86	9.13	10.55	12.11	13.83	15.70	17.74
	CAB	0.50	0.50	0.50	0.50	0.49	0.48	0.47	0.46	0.45	0.45
10.0	kAB	4.80	5.71	6.74	7.89	9.18	10.61	12.19	13.93	15.83	17.90
	CAB	0.50	0.50	0.50	0.49	0.48	0.48	0.47	0.46	0.45	0.44

Using the equivalent frame method, determine design moments for the slab system shown in the transverse direction, for an intermediate floor.

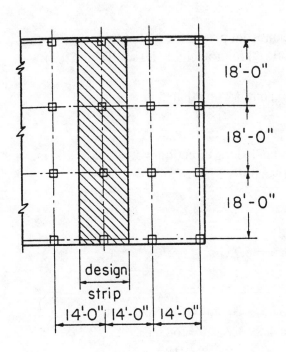

Story height = 9 ft

Column dimensions = 16 in. × 16 in.

Lateral loads to be resisted by shear walls

No edge beams

Partition weight = 20 psf

Service live load = 40 psf

$f_c' = 3000$ psi (for slab)

$f_c' = 5000$ psi (for column)

$f_y = 60,000$ psi

Example 22.1--—Continued

Calculations and Discussion	Code Reference

1. Preliminary design for slab thickness h.

   (a) Control of deflections:

   For slab systems without beams (flat plate), the minimum overall thickness h with Grade 60 reinforcement is (see Table 20-1):

   9.5.3.2

   $h = \ell_n/30 = 200/30 = 6.67$ in., but not less than 5 in.

   Table 9.5(b)
   9.5.3.2(a)

   where ℓ_n = length of clear span in long direction = 216 - 16 = 200 in.
   Try 7 in. slab for all panels (weight = 87.5 psf)

   (b) Shear strength of slab:

   Use average effective depth d = 5.75 in.
   (3/4 in. Cover + #4 bar)

   Factored dead load, $w_d = (87.5 + 20)\ 1.4 = 150.5$ psf

   9.2.1

   Factored live load, $w_\ell = 40 \times 1.7$ $=\ \underline{68.0\ \text{psf}}$
   Total factored load $=218.5$ psf

   Wide-beam action: Wide beam action is investigated for a 12 in. wide strip taken at d distance from the face of support in the long direction (See Fig. 22-13).

   11.12.1.1

   $V_u = 0.2185 \times 7.854 = 1.72$ kips
   $V_c = 2\sqrt{f_c'}\,b_w d$
   $\varphi V_c = 0.85 \times 2\sqrt{3000} \times 12 \times 5.75/1000 = 6.43$ kips
   $V_u < \varphi V_c$ O.K.

Example 22.1-—Continued

Calculations and Discussion

Two-way action: Since there are no shear forces at the centerlines
of adjacent panels (see Fig. 22-13), the shear strength at d/2 distance
around the support is computed as follows:

$V_u = 0.2185 \, (18 \times 14 - 1.81^2) = 54.3$ kips

$V_c = 4\sqrt{f_c'} \, b_o d$ (for square column)

Eq. (11-36)

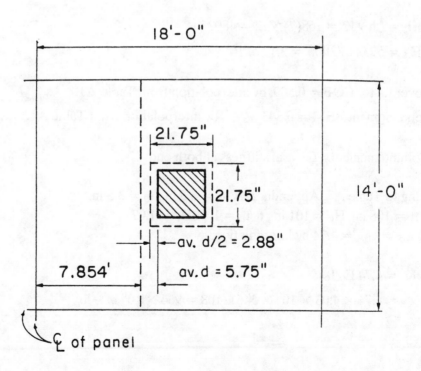

Fig. 22-13 Critical Sections for Shear for Example Problem

Example 22.1—Continued

Preliminary design indicates that a 7 in. overall slab thickness
is adequate for control of deflections and shear strength.

2. Frame members of equivalent frame.

Determine moment distribution constants and fixed-end moments for the equivalent frame members. The
moment distribution procedure will be used to analyze the partial frame. Stiffness factors k, carry over
factors COF, and fixed-end moment factors FEM, for the slab-beams and column members are determined
by conjugate beam procedures. These calculations are shown here.

(a) Slab-beam, flexural stiffness at both ends K_{sb}:

$c_{N1}/\ell_1 = 16/(18 \times 12) = 0.07$, $c_{N2}/\ell_2 = 16/(14 \times 12) = 0.1$

By interpolation, form Table A1, Appendix 22A, $k_{NF} = k_{FN} = 4.13$
Thus, $K_{sb} = 4.13\, E_{cs}I_s/\ell_1$

$$= 4.13 \times 3.12 \times 10^6 \times 4802/216 = 286 \times 10^6 \text{ in.}-\text{lb}$$

where $I_s = \ell_2 h^3/12 = 168(7)^3/12 = 4802 \text{ in.}^4$

$$E_{cs} = 57{,}000\sqrt{3000} = 3.12 \times 10^6 \text{ psi}$$

8.5.1

Carry-over factor COF = 0.509, by interpolation from Table A1.
Fixed-end moment FEM = $0.843\, w_{\ell_2} \ell_1^2$, by interpolation from Table A1.

(b) Column members, flexural stiffness at both ends K_c:

Referring to Table A7, Appendix 22A, $t_a = 3.5$ in., $t_b = 3.5$ in.,
H = 9 ft = 108 in., H_c = 101 in., t_a/t_b = 1, H/H_c = 1.07
Thus, $k_{AB} = k_{BA} = 4.74$ by interpolation.

$K_c = 4.74\, E_{cc}I_c/\ell_c$

$$= 4.74 \times 4.03 \times 10^6 \times 5461/108 = 966 \times 10^6 \text{ in.}-\text{lb}$$

Example 22.1—Continued

	Code
Calculations and Discussion	Reference

where, $I_c = c^4/12 = (16)^4/12 = 5461$ in.^{4}

$\quad E_{cc} = 57,000\sqrt{5000} = 4.03 \times 10^6$ psi

$\quad \ell_c = 9$ft $= 108$ in.

8.5.1

(c) Torsional members, torsional stiffness K_t:

$K_t = 9\,E_{cs}\,C/\left[\ell_2(1 - c_2/\ell_2)^3\right]$

$\quad = 9 \times 3.12 \times 10^6 \times 1325/\left[168(0.905)^3\right]$

$\quad = 299 \times 10^6$ in.$-$lb

Eq. (13-6)

where $C = \Sigma(1 - 0.63\,x/y)(x^3y/3)$

$\quad = (1 - 0.63 \times 7/16)(7^3 \times 16/3) = 1325$ in.^{4}

$c_2 = 16$ in. and $\ell_2 = 14$ ft $= 168$ in.

Eq. (13-7)

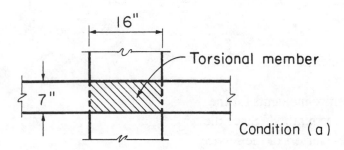

(d) Equivalent column stiffness K_{ec}:

$K_{ec} = \Sigma K_c \times \Sigma K_t/(\Sigma K_c + \Sigma K_t)$

$\quad = (2 \times 966)(2 \times 299)/\left[(2 \times 966) + (2 \times 299)\right]$

$\quad = 457 \times 10^6$ in.$-$lb

where ΣK_t is for two torsional members, one on each side
of column, and ΣK_c is for the upper and lower columns
at the slab-beam joint of an intermediate floor.

Example 22.1—Continued

	Code
Calculations and Discussion	**Reference**

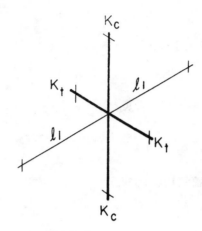

(e) Slab-beam joint distribution factors DF:
At exterior joint:
DF = 286/(286 + 457) = 0.385

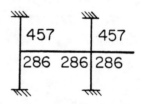

At interior joint:
DF = 286/(286 + 286 + 457) = 0.278
COF for slab-beam = 0.509

3. Partial frame analysis of equivalent frame:

Determine maximum negative and positive moments for the
slab-beams using the moment distribution method. Since the
service live load does not exceed three-quarters of the service
dead load, design moments are assumed to occur at all critical
sections with full factored live load on all spans.   13.7.6.2

L/D = 40/(87.5 + 20) = 0.37 < 3/4

(a) Factored load and fixed-end moments:

Factored dead load w_d = 1.4(87.5 + 20)
            = 150.5 psf

Factored live load w_ℓ = 1.7(40)
            = 68 psf

Example 22.1—Continued

Factored load $w_d + w_l$ = 218.5 psf

FEM's for slab-beams $= 0.0843 w_d l_2 l_1^2$ (Table A1, Appendix 22A)

FEM's due to $w_d + w_l = 0.0843(0.2185 \times 14)18^2$
$$= 83.6 \text{ ft}-\text{kips}$$

(b) Moment distribution is shown in Table 22-1. Counterclockwise rotational moments acting on the member ends are taken as positive. Positive span moments are determined from the following equation:

$$M(\text{midspan}) = M_o - 1/2 (M_L + M_R)$$
where M_o is the moment at midspan for a simple beam.

When the end moments are not equal, the maximum moment in the span does not occur at midspan, but its value is close to that at midspan.

Positive moment in span 1-2:

$$+M = (0.2185 \times 14)18^2/18 - 1/2(53.0 + 95.1)$$
$$= 49.8 \text{ ft}-\text{kips}$$

Positive moment in span 2-3:

$$+M = (0.2185 \times 14)18^2/18 - 1/2(86.4 + 86.4)$$
$$= 37.5 \text{ ft}-\text{kips}$$

4. Design moments.

Positive and negative factored moments for the slab system in the transverse direction are shown in Fig. 22-14. The negative design moments are taken at the faces of rectilinear supports but not at distances greater than $0.175 \, l_1$ from the center of supports.

13.7.7.1

0.67 ft < 0.175 × 18 ft (Use face of support location).

Example 22.1—Continued

Table 22-1—Moment Distribution for Partial Frame
(Transverse Direction)

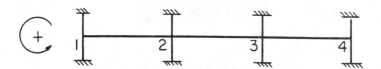

Joint	1	2		3		4
Member	1-2	2-1	2-3	3-2	3-4	4-3
DF COF	0.385 0.509	0.278 0.509	0.278 0.509	0.278 0.509	0.278 0.509	0.385 0.509
FEM COM*	+83.6 0 +2.3 +0.3	-83.9 -16.4 0 -0.5	+83.6 0 -2.3 -0.3	-83.6 0 +2.3 +0.3	+83.6 +16.4 0 +0.5	-83.6 0 -2.3 -0.3
Σ DM**	+86.2 -33.2	-100.5 +5.4	+81.0 +5.4	-81.0 -5.4	+100.5 -5.4	-86.2 +33.2
Neg. M ***	+53.0 +52.6	-95.1 -93.6	+86.4 +85.0	-86.4 -85.0	+95.1 +93.6	-53.0 -52.6
M @ £ of span	49.8 49.5***		37.5 37.5***		49.8 49.5***	

* Carry-over moment, COM, is the negative product of the distribution factor, carry-over factor and unbalanced joint moment carried to the opposite end of span.

** Distributed moment, DM, is the negative product of the distribution factor and the unbalanced joint moment.

*** Moments from ADOSS, the computer program for the analysis and design of slab systems by the Portland Cement Association. See accompanying Example 22.1 solved by ADOSS and reproduced at the end of this example.

Example 22.1—Continued

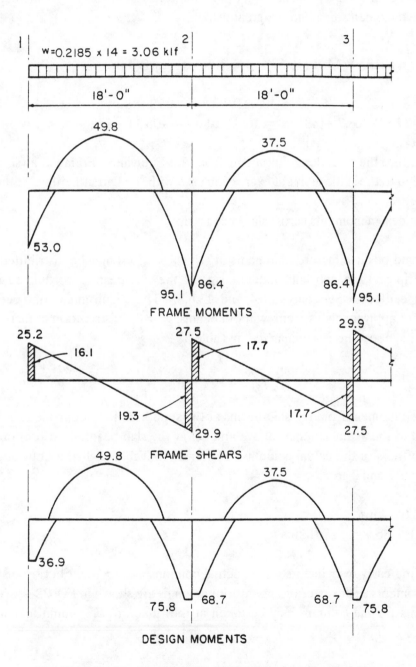

**Fig. 22-14 Positive and Negative Design Moments for Slab-Beam
(All Spans Loaded with Full Factored Live Load)**

Example 22.1—Continued

	Code
Calculations and Discussion	**Reference**

5.   Total factored moment per span

Slab systems within the limitations of Section 13.6.1 may have the resulting analytical moments reduced in such proportion that the numerical sum of the positive and average negative moments need not be greater than,

13.7.7.4

$$M_O = w \ell_2 \ell_n^2 / 8 = 0.2185 \times 14 \times (16.67)^2 / 8 = 106.3 \text{ ft}-\text{kips}$$

End spans: $49.8 + (36.9 + 75.8)/2 = 106.2 \text{ ft-kips} \approx 106.3 \text{ ft-kips}$
Interior span: $37.5 + (68.7 + 68.7)/2 = 106.2 \text{ ft-kips} \approx 106.3 \text{ ft-kips}$

It may be seen that the total design moments from the Equivalent Frame analysis are very close indeed to the static moment expression used with the Direct Design Method.

6.   Distribution of design moments across slab-beam strip.

13.7.7.5

The negative and positive factored moments at critical sections may be distributed to the column strip and the two half middle strips of the slab-beam according to the proportions specified in Sections 13.6.4 and 13.6.6.   The requirement of Section 13.6.1.6 does not apply for slab systems without beams, $\propto = 0$. Distribution of factored moments at critical sections is summarized in Table 22-2.

7.   Column moments.

The unbalanced moment from the slab-beams at the supports of the equivalent frame are distributed to the actual columns above and below the slab-beam in proportion to the relative stiffness of the actual columns.   Referring to Table 22-1, the unbalanced moment at joints 1 and 2 are:

Joint 1 = +53.0 ft-kips
Joint 2 = -95.1 + 86.4 = - 8.7 ft-kips

The stiffness and carry-over factors of the actual columns and the distribution of the unbalanced moments to the exterior and interior columns are shown in Fig. 22-15.   The design moments for the columns may be taken at the juncture of column and slab. Summarizing:

Example 22.1—Continued

Table 22-2 — Distribution of Factored Moments

	Factored Moment (ft-kips)	Column Strip		Moment (ft-kips) in Two Half-Middle Strips[**]
		Percent[*]	Moment (ft-kips)	
End Span:				
Exterior Negative	36.9	100	36.9	0.0
Positive	49.8	60	29.9	19.9
Interior Negative	75.8	75	56.9	18.9
Interior Span:				
Negative	68.7	75	51.5	17.2
Positive	37.5	60	22.5	15.0

* For slab systems without beams

** That portion of the Factored momnet not resisted by the column strip is assigned to the two half-middle strips.

Example 22.1—Continued

Design moment in exterior column = 25.2 ft-kips
Design moment in interior column = 4.13 ft-kips

8. Transfer of gravity load shear and moment at exterior column.

    Check slab shear and flexural strength at edge column   11.12.6
    due to direct shear and unbalanced moment transfer   13.3.3

    (a) Factored shear force transfer at exterior column:

    $V_u \approx w_u \ell_1 \ell_2/2$
    $\approx 0.2185 \times 14 \times 18/2 = 27.5$ kips

    (b) Unbalanced moment transfer at exterior column:

    When factored moments are determined by a more accurate method of framing
    analysis, considering actual member stiffnesses, such as the Equivalent Frame Analysis
    procedure, transfer moment is taken directly from the results of the frame analysis.
    Unbalanced moment at exterior column (Table 22-2) is M_u - 36.9 ft-kips.

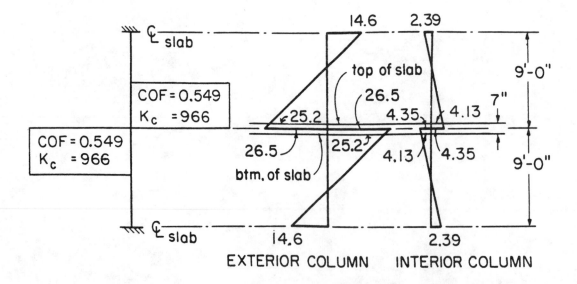

Fig. 22-15 Column Moments (Unbalanced Moments from slab-beam)

Example 22.1—Contined

	Code
Calculations and Discussion	**Reference**

(Note the special provision of Section 13.6.3.6 for unbalanced moment transfer between slab and an edge column when the approximate moment coefficients of the Direct Design Method are used. See Example 21.1). Considering the approximate nature of the moment transfer analysis procedure, assume the transfer moment M_u is at centroid of critical transfer section.

(c) Combined shear stress at inside face of critical transfer section.

For shear strength equations, see Part 18.   11.12.6.2

$$v_u = V_u/A_c + \gamma_v M_u/(J/c)$$
$$= 27,500/342.2 + 0.38 \times 36.9 \times 12,000/2358$$
$$= 80.4 + 71.4 = 151.8 \text{ psi}$$

where fraction of unbalanced moment transferred by eccentricity of shear at edge slab-column connection, $\gamma_v = 0.38$ [Eq. (11-42)]. For section properties A_c and (J/c), see Example 21.1.

(d) Combined shear stress at outside face of critical transfer section.   11.12.6.2

$$V_u = 27,500/342.2 - 0.38 \times 36.9 \times 12,000/1096$$
$$= 80.4 - 153.5 = 73.1 \text{ psi}$$

(e) Permissible shear stress   11.12.6.1.1

$$\varphi V_n = \varphi 4\sqrt{f_c'} = 0.85 \times 4\sqrt{3000}$$
$$= 186.2 \text{ psi} > 151.8 \text{ psi} \text{ O.K.}$$

(f) Design for unbalanced moment transfer by flexure for both   13.3.3
   middle strip and column strip:

$$A_{s(min)} = 0.0018bh = 0.0018 \times 84 \times 7 = 1.06 \text{ in.}^2$$ 13.4.17.12.2.1(b)

where b is width of design strip = 14/2 = 7 ft. = 84 in.

For #4 bars, total bars required = 1.06/0.20 = 5.3 bars   13.4.2
For $s_{max} = 2h = 2 \times 7 = 14$ in., total bars required = 84/14 = 6 bars

Example 22.1—Continued

Check total reinforcement required for column strip negative moment
M_u = 36.9 ft kips. Using Table 10-1, Page 10-6,

$$\frac{M_u}{\varphi f'_c bd^2} = \frac{36.9 \times 12}{0.9 \times 3 \times 84 \times 5.75^2} = 0.591$$

where d $\approx$ 7.0 - 1.25 = 5.75 in.
From Table 10-1; ω = 0.0613

$A_s = \omega f'_c/fy = 0.613 \times 3 \times 84 \times 5.75/60 = 1.48$ in.^{2}

For #4 bars, total bars required = 1.48/0.20 = 7.4 bars.

Use 6 #4 bars @ 14 in. spacing in middle strip and portion of column strip outside unbalanced moment transfer band having width equal to c+2 (1.5h)=16+2 (1.5 × 7) = 37 in.

13.3.3.2

Additional reinforcement required over column within effective slab width of 37 in. to resist fraction of unbalanced moment transferred by flexure is computed from Eq. (13-1). For (square) edge column, γ_f =62% [Eq. (13-1)] (See Example 21.1)

$\gamma_f M_u$ = 0.62 (36.9) = 22.9 ft-kips must be transferred within the effective slab width of 37 in. Try 2 additional bars over column. Check moment strength for 4 #4 bars within 37 in. width.

For 4 #4 bars: A_s = 4(0.20) = 0.80 in.^{2}
$\omega = A_s fy/f'_c bd = 0.80 \times 60/3 \times 37 \times 5.75$
  = 0.0752

From Table 10-1, $M_n/f'_c bd^2$ = 0.0719
$M_n = 0.0719 \times 3 \times 37 \times 5.75^2/12 = 22.0$ ft$-$kips
φM_n = 0.9(22.0) = 19.8 ft$-$kips < 22.9 ft$-$kips N.G.

Try 3 additional bars. Moment strength for 5 #4 bars within 37 in. width:

A_s = 5(0.20) = 1.00 in.^{2}
$\omega = 1.00 \times 60/3 \times 37 \times 5.75 = 0.0928$

Example 22.1—Continued

Calculations and Discussion

Code
Reference

From Table 10-1, $M_n/f_c'bd^2 = 0.0877$

$\varphi M_n = 0.90(0.0877 \times 3 \times 37 \times 5.75^2/12)$

$= 24.1 \text{ ft}-\text{kips} > 22.9 \text{ ft}-\text{kips}$ O.K.

Detail bars as shown below. Total bars in column strip
$= 6 + 3 = 9 \text{ bars} > 7.4$ required for total column strip
negative moment. O.K.

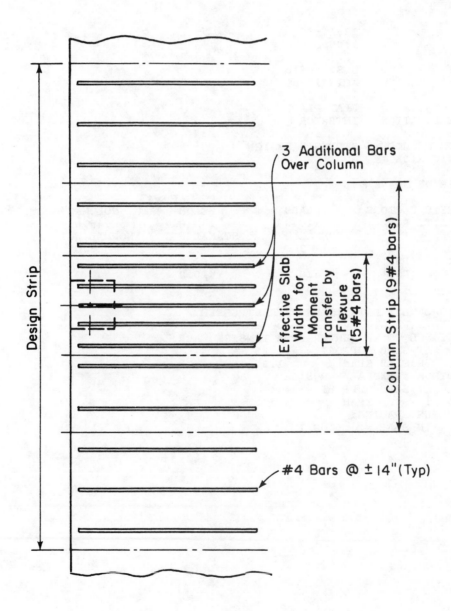

Example 22.1—Continued

Printout of Example 22.1 Solved by Computer Program ADOSS

```
11-1-89 ADOSS(tm) 5.20  Proprietary Software of PORTLAND CEMENT ASSN.
 1:39:53 PM Licensed to: Portland Cement Association, Skokie, IL

FILE NAME         DATA\EX22-1.ADS

PROJECT ID.       PCA Notes on ACI 318 Example 22.1
                  --------------------------------------
SPAN ID.          Typ. Interior
                  -------------------

ENGINEER

DATE              12-11-89
TIME              1:05pm

UNITS             U.S. in-lb
CODE              ACI 318-89

SLAB SYSTEM       FLAT PLATE
FRAME LOCATION    INTERIOR

DESIGN METHOD     STRENGTH DESIGN
SEISMIC RISK      NONE OR LOW

NUMBER OF SPANS   5

CONCRETE FACTORS       SLABS        BEAMS        COLUMNS
  DENSITY(PCF  )       145.0        145.0        145.0
  TYPE             NORMAL WGT   NORMAL WGT   NORMAL WGT
  f'c    (KSI)          3.0          3.0          5.0
  fct    (PSI)        367.0        367.0        473.8
  fr     (PSI)        410.8        410.8        530.3


REINFORCEMENT DETAILS: NON-PRESTRESSED
  YIELD STRENGTH Fy  = 60.00 KSI
  DISTANCE TO RF CENTER FROM TENSION FACE:
        AT SLAB SUPPORT =    1.25 IN
        IN SLAB SPAN    =    1.25 IN
  MINIMUM FLEXURAL BAR SIZE:
        AT SLAB SUPPORT =  # 4
        IN SLAB SPAN    =  # 4
  MINIMUM SPACING:
        IN SLAB =    6.00 IN
```

Example 22.1—Continued

SPAN/LOADING DATA
****

SPAN NUMBER	LENGTH L1 (FT)	Tslab (IN)	WIDTH LEFT (FT)	L2*** RIGHT (FT)	SLAB SYSTEM	DESIGN STRIP (FT)	COLUMN STRIP** (FT)	UNIFORM_LOADS S. DL (PSF)	LIVE (PSF)
1*	.7	7.0	7.0	7.0	1	14.0	.0	20.0	40.0
2	18.0	7.0	7.0	7.0	1	14.0	7.0	20.0	40.0
3	18.0	7.0	7.0	7.0	1	14.0	7.0	20.0	40.0
4	18.0	7.0	7.0	7.0	1	14.0	7.0	20.0	40.0
5*	.7	7.0	7.0	7.0	1	14.0	.0	20.0	40.0

```
*   -Indicates cantilever span information.
**  -Strip width used for positive flexure.
***-L2 widths are 1/2 dist. to transverse column.
"E"-Indicates exterior strip.
```

PARTIAL LOADING DATA
****
PARTIAL LOADINGS ARE NOT SPECIFIED

COLUMN/TORSIONAL DATA
****

COLUMN NUMBER	COLUMN_ABOVE_SLAB C1 (IN)	C2 (IN)	HGT (FT)	COLUMN_BELOW_SLAB C1 (IN)	C2 (IN)	HGT (FT)	CAPITAL** EXTEN. (IN)	DEPTH (IN)	COLUMN STRIP* (FT)	MIDDLE STRIP* (FT)
1	16.0	16.0	9.0	16.0	16.0	9.0	.0	.0	7.0	7.0
2	16.0	16.0	9.0	16.0	16.0	9.0	.0	.0	7.0	7.0
3	16.0	16.0	9.0	16.0	16.0	9.0	.0	.0	7.0	7.0
4	16.0	16.0	9.0	16.0	16.0	9.0	.0	.0	7.0	7.0

```
Columns with zero "C2" are round columns.
*  -Strip width used for negative flexure.
**-Capital extension distance measured from face of column.
```

COLUMN NUMBER	TRANSVERSE BEAM WIDTH (IN)	DEPTH (IN)	ECCEN (IN)	DROP PANEL/SOLID HEAD X1_LEFT (FT)	X1_RIGHT (FT)	X2_WIDTH (FT)	THICK (IN)	SUPPORT FIXITY* %
1	.0	.0	.0	.0	.0	.0	.0	100%
2	.0	.0	.0	.0	.0	.0	.0	100%
3	.0	.0	.0	.0	.0	.0	.0	100%
4	.0	.0	.0	.0	.0	.0	.0	100%

```
*  -Support fixity of 0% denotes pinned condition.
   Support fixity of 999% denotes fixed end condition.
```

Example 22.1—Continued

```
                     LATERAL LOAD/OUTPUT DATA
                     *************************

LATERAL LOADS ARE NOT SPECIFIED



OUTPUT DATA
PATTERN LOADINGS:                  1 THRU 4
PATTERN LIVE LOAD FACTOR (1-3)  =     75%

LOAD FACTORS:
  U = 1.40*D + 1.70*L
  U =  .75( 1.40*D + 1.70*L + 1.70*W)
  U =  .90*D + 1.30*W


OUTPUT OPTION(S) FOR ANALYSIS:
  1. Data Check
  2. CL Moments & Shears
  3. Design Moments & Shears
  4. Reinforcing required

OUTPUT OPTION(S) FOR DESIGN:
  2. Bar Sizing, Deflection, & Material Quantities

**TOTAL UNFACTORED DEAD LOAD =        80.417 KIPS
                   LIVE LOAD =        30.987 KIPS



      ---- STATICS  PRINT-OUT  FOR  GRAVITY  LOAD ANALYSIS ----
          ***************************************************

                J O I N T   M O M E N T S  ( FT - KIPS )
                ------------------------------------------
JOINT                   PATTERN-1                        PATTERN-2
NUMBER    LEFT     RIGHT     TOP     BOTTOM     LEFT     RIGHT     TOP     BOTTOM
----------------------------------------------------------------------------------
    1      -.6      50.1    -24.8    -24.8      -.6      33.9    -16.7    -16.7
    2     -87.1     80.8      3.2      3.2     -69.8     74.4     -2.3     -2.3
    3     -80.8     87.1     -3.2     -3.2     -74.4     69.8      2.3      2.3
    4     -50.1       .6     24.8     24.8     -33.9       .6     16.7     16.7
JOINT                   PATTERN-3                        PATTERN-4
NUMBER    LEFT     RIGHT     TOP     BOTTOM     LEFT     RIGHT     TOP     BOTTOM
----------------------------------------------------------------------------------
    1      -.4      50.4    -25.0    -25.0      -.6      52.6    -26.0    -26.0
    2     -80.3     61.9      9.2      9.2     -93.6     85.0      4.3      4.3
    3     -61.9     80.3     -9.2     -9.2     -85.0     93.6     -4.3     -4.3
    4     -50.4       .4     25.0     25.0     -52.6       .6     26.0     26.0

                J O I N T   S H E A R S  ( KIPS )
                ---------------------------------
JOINT       PATTERN-1          PATTERN-2          PATTERN-3          PATTERN-4
NUMBER    LEFT     RIGHT     LEFT     RIGHT     LEFT     RIGHT     LEFT     RIGHT
----------------------------------------------------------------------------------
    1      -1.7     23.0      -1.7     16.3      -1.3     23.1      -1.9     24.6
    2     -26.9     25.3     -20.3     24.8     -26.4     18.3     -29.2     26.9
    3     -25.3     26.9     -24.8     20.3     -18.3     26.4     -26.9     29.2
    4     -23.0      1.7     -16.3      1.7     -23.1      1.3     -24.6      1.9
```

Example 22.1—Continued

NEGATIVE FLEXURE
****

COLUMN NUMBER	PATTERN NUMBER	LOCATION @COL. FACE		TOTAL DESIGN (FT-K)	COLUMN STRIP (FT-K)	MIDDLE STRIP (FT-K)	BEAM IN COL.STRIP (FT-K)
1	4		R	36.8	36.3	.5	Not applicable
2	4	L		-74.8	-56.1	-18.7	Not applicable
3	4		R	74.8	56.1	18.7	Not applicable
4	4	L		-36.8	-36.3	-.5	Not applicable

POSITIVE FLEXURE
****

SPAN NUMBER	PATTERN NUMBER	LOCATION FROM LEFT (FT)	TOTAL DESIGN (FT-K)	COLUMN STRIP (FT-K)	MIDDLE STRIP (FT-K)	BEAM IN COL. STRIP (FT-K)
2	4	8.6	49.5	29.7	19.8	Not applicable
3	2	8.6	37.5	22.5	15.0	Not applicable
4	4	9.4	49.5	29.7	19.8	Not applicable

SHEAR ANALYSIS
****

NOTE--Allowable shear stress in slabs = 219.09 PSI when ratio
of col. dim. (long/short) is less than 2.0.
Allowable shear stress in beams = 109.54 PSI (see "CODE").

--Wide beam shear (see "CODE") is not computed, check manually.

--After the column numbers, C = Corner, E = Exterior, I = Interior.

DIRECT SHEAR WITH TRANSFER OF MOMENT
--------------- AROUND COLUMN ---------------

COL. NO.	ALLOW. STRESS (PSI)	PATT NO.	REACTION (KIPS)	SHEAR STRESS (PSI)	PATT NO.	REACTION (KIPS)	UNBAL. MOMENT (FT-K)	SHEAR TRANSFR (FT-K)	SHEAR STRESS (PSI)
1E	219.09	4	26.1	89.84	4	26.1	41.1	15.7	184.09
2I	219.09	4	55.6	130.78	4	55.6	-8.6	-3.4	143.89
3I	219.09	4	55.6	130.78	4	55.6	8.6	3.4	143.89
4E	219.09	4	26.1	89.84	4	26.1	-41.1	-15.7	184.09

Example 22.1—Continued

N E G A T I V E   R E I N F O R C E M E N T
****

COLUMN NUMBER	PATT NO.	LOCATION @COL FACE	TOTAL DESIGN (FT-K)	COLUMN STRIP AREA (SQ.IN)	COLUMN STRIP WIDTH (FT)	MIDDLE STRIP AREA (SQ.IN)	MIDDLE STRIP WIDTH (FT)	COL.STRIP.BM AREA (SQ.IN)	COL.STRIP.BM WIDTH (IN)
1	4	‖ R	36.8	1.45	7.0	1.06	7.0	Not applicable	
2	4	L ‖	-74.8	2.30	7.0	1.06	7.0	Not applicable	
3	4	‖ R	74.8	2.30	7.0	1.06	7.0	Not applicable	
4	4	L ‖	-36.8	1.45	7.0	1.06	7.0	Not applicable	

P O S I T I V E   R E I N F O R C E M E N T
****

SPAN NUMBER	PATT NO.	LOCATION FROM LEFT (FT)	TOTAL DESIGN (FT-K)	COLUMN STRIP AREA (SQ.IN)	COLUMN STRIP WIDTH (FT)	MIDDLE STRIP AREA (SQ.IN)	MIDDLE STRIP WIDTH (FT)	COL.STRIP.BM AREA (SQ.IN)	COL.STRIP.BM WIDTH (IN)
2	4	8.6	49.5	1.18	7.0	1.06	7.0	Not applicable	
3	2	8.6	37.5	1.06	7.0	1.06	7.0	Not applicable	
4	4	9.4	49.5	1.18	7.0	1.06	7.0	Not applicable	

I N F L E C T I O N   P O I N T S
****

COLUMN NUMBER	GOVERNING PATTERN NUMBER	DISTANCE LEFT (FT)	GOVERNING PATTERN NUMBER	DISTANCE RIGHT (FT)
1	4	.67	4	2.70
2	2	4.50	3	4.50
3	3	4.50	2	4.50
4	4	2.70	4	.67

D E S I G N   R E S U L T S
****

NOTE--The schedule given below is a guide for proper reinforcement
      placement and is based on reasonable engineering judgement.
      Unusual boundary and/or loading conditions may require
      modification of this schedule.

   --Bars are distributed uniformly across column and middle strips.

N E G A T I V E   R E I N F O R C E M E N T
****

COLUMN NUMBER	COLUMN STRIP LONG BARS NO	COLUMN STRIP LONG BARS SIZE	COLUMN STRIP LONG BARS LENGTH LEFT (FT)	COLUMN STRIP LONG BARS LENGTH RIGHT (FT)	COLUMN STRIP SHORT BARS NO	COLUMN STRIP SHORT BARS SIZE	COLUMN STRIP SHORT BARS LENGTH LEFT (FT)	COLUMN STRIP SHORT BARS LENGTH RIGHT (FT)	MIDDLE STRIP LONG BARS NO	MIDDLE STRIP LONG BARS SIZE	MIDDLE STRIP LONG BARS LENGTH LEFT (FT)	MIDDLE STRIP LONG BARS LENGTH RIGHT (FT)
1	4	# 4	.67	5.67	3	# 4	.67	4.00	6	# 4	.67	4.33
2	6	# 4	5.67	5.67	6	# 4	4.00	4.00	6	# 4	5.54	5.54
3	6	# 4	5.67	5.67	6	# 4	4.00	4.00	6	# 4	5.54	5.54
4	4	# 4	5.67	.67	3	# 4	4.00	.67	6	# 4	4.33	.67

Example 22.1—Continued

```
                    P O S I T I V E    R E I N F O R C E M E N T
                 **********************************************
          *      C O L U M N        S T R I P    *   M I D D L E    S T R I P
          *    LONG   BARS     *  SHORT   BARS    *  LONG   BARS   *  SHORT   BARS
  SPAN    * ---- B A R ----  * ---- B A R ----  * ---- B A R ---- * ---- B A R ----
  NUMBER  * NO  SIZE LENGTH  * NO  SIZE LENGTH  * NO  SIZE LENGTH * NO  SIZE LENGTH
          *          (FT)    *          (FT)    *          (FT)   *          (FT)
          ------------------------------------------------------------------------
     2      3   # 4  17.58     3   # 4  15.58     3   # 4  18.08    3   # 4  15.13
     3      3   # 4  17.50     3   # 4  13.50     3   # 4  18.50    3   # 4  12.60
     4      3   # 4  17.58     3   # 4  15.58     3   # 4  18.08    3   # 4  15.13
```

```
     ADDITIONAL INFORMATION AT SUPPORTS
     ------------------------------------
          * REINF. SUMMARY* ADD'L R/F REQ'D DUE TO UNBALANCED (U.) MOMENT TRANSFER
  COLUMN  * --------------* -----------------------------------------------------
  NUMBER  * W/O U. MOMENT * MAX.U. *GAMMA* FLEXURAL  *PATT* CRITICAL     SECTION
          * REQ'D - PROV'D* MOMENT * -f  * TRANSFER  *NO. * SLABW - AREA -   R/F
          *(SQ.IN) (SQ.IN)* (FT-K) *     * (FT-K)    *    * (FT)   (SQ.IN)
          --------------------------------------------------------------------------
     1      2.51    2.60     51.9    .62     32.0       4     3.1    1.34   4  # 4
     2      3.35    3.60    -18.4    .60    -11.0       3     3.1     .44   0  # 4
     3      3.35    3.60     18.4    .60     11.0       3     3.1     .44   0  # 4
     4      2.51    2.60    -51.9    .62    -32.0       4     3.1    1.34   4  # 4
```

NOTE: Zero transfer "CRITICAL SLABW" indicates no support dimensions
      given for transfer.
      If beam(s) are present, transfer mode may be due to beam shear
      and/or torsion, check manually.

```
          ADDITIONAL INFORMATION FOR IN-SPAN CONDITIONS
          ---------------------------------------------
          *  REINF.  SUMMARY  *
  SPAN    * ----------------- *  TOTAL FACTORED SPAN
  NUMBER* *    AT MIDSPAN      *  STATIC DESIGN MOMENT
          * REQ'D. - PROV'D.  *  (W/O PARTIAL LOADS)
          * (SQ.IN)  (SQ.IN)  *      (FT-K)
          ----------------------------------------------------
     2        2.24     2.40            104.2
     3        2.12     2.40            104.2
     4        2.24     2.40            104.2
```

Example 22.1—Continued

```
             D E F L E C T I O N     A N A L Y S I S
             ************************************

NOTES--The deflections below must be combined with those of
       the analysis in the perpendicular direction. Consult
       users manual for method of combination and limitations.

     --Spans 1 and  5 are cantilevers.

     --Time-dependent deflections are in addition to those
       shown and must be computed as a multiplier of the dead
       load(DL) deflection. See "CODE" for range of multipliers.

     --Deflections due to concentrated or partial loads may be larger
       at the point of application than those shown at the centerline.
       Deflections are computed as from an average uniform loading
       derived from the sum of all loads applied to the span.

     --Modulus of elasticity of concrete, Ec =  3156. KSI
```

SPAN NUMBER	DEAD LOAD Ieff. (IN^4)	COLUMN STRIP DEFLECTION DUE TO: DEAD (IN)	LIVE (IN)	TOTAL (IN)	MIDDLE STRIP DEFLECTION DUE TO: DEAD (IN)	LIVE (IN)	TOTAL (IN)
1	4802.	-.005	-.002	-.007	-.005	-.002	-.007
2	4802.	.087	.045	.133	.045	.021	.066
3	4802.	.055	.039	.094	.024	.018	.041
4	4802.	.087	.045	.133	.045	.021	.066
5	4802.	-.005	-.002	-.007	-.005	-.002	-.007

```
             Q U A N T I T Y     E S T I M A T E S
             ************************************

TOTAL QUANTITIES
----------------
CONCRETE           ....      16.7 CU.YD
FORMWORK           ....      775. SQ.FT
REINFORCEMENT (IN THE DIRECTION OF ANALYSIS)
   (NEGATIVE)      ....      336. LBS
   (POSITIVE)      ....      391. LBS

SUMMARY OF QUANTITIES
---------------------
CONCRETE           ....       .58 CU.FT/SQ.FT
FORMWORK           ....      1.00 SQ.FT/SQ.FT
REINFORCEMENT**    ....       .94 LBS./ SQ.FT

    **(IN THE DIRECTION OF ANALYSIS)

          * Program completed as requested *
```

Example 22.2 —Two-Way Slab with Beams Analyzed by Equivalent Frame Method

Using the equivalent frame method, determine design moments for the slab system shown, in the transverse direction, for an intermediate floor.

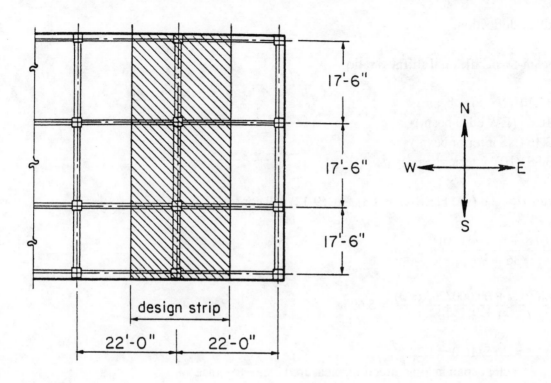

Story height	= 12 ft
Edge beam dimensions	= 14 in. × 27 in.
Interior beam dimensions	= 14 in. × 20 in.
Column dimensions	= 18 in. × 18 in.
Slab thickness	= 6 in.
Service live load	= 100 psf

f'_c = 4000 psi (for all members), normal weight concrete

f_y = 60,000 psi

Example 22.2 —Continued

	Code
Calculations and Discussion	Reference

1. Preliminary design for slab thickness h.

    Control of deflection: 0.5.3

    The beam-to-slab flexural stiffness ratio,

    α = 13.30 (NS edge beam)*
       = 16.45 (EW edge beam)*
       = 3.16 (NS interior beam)*
       = 3.98 (EW interior beam)*

    Since all $\alpha >$ 2.0 (see Fig. 8-2, part 8), Eq. (9-12) will control. Therefore,

    $$h = \frac{\ell_n(0.8 + f_y/200,000)}{36 + 9\beta}$$ Eq. (9-12)

    $$= \frac{246(0.8 + 60,000/200,000)}{36 + 9(1.28)} = 5.69 \text{ in.}$$

    where β = 20.5/16.0 = 1.28
           ℓ_n = clear span in long direction, measured face$-$to$-$face
                of columns = 20 ft 6 in. = 246 in.

    Use 6" slab thickness.

2. Frame members of equivalent frame.

    Determine moment distribution constants and fixed-end moment coefficients
    for the equivalent frame members. The moment distribution procedure will be
    used to analyze the partial frame for vertical loading. Stiffness factors k, carry
    over factors COF, and fixed-end moment factors FEM, for the slab-beams
    and column members are determined by conjugate beam procedures. These
    calculations are not shown here.

---

*Calculations for α terms are given in Example 22.1, Part 21.

Example 22.2—Continued

(a) Slab-beams, flexural stiffness at both ends K_{sb}:
Referring to Table A1, Appendix 22A,

$$K_{sb} = 4.11 \ E_c I_{sb}/\ell_1 = 4.11 \times 25{,}387 \ E_c/210 = 497 \ E_c$$

where I_{sb} = moment of inertia of slab-beam section shown in Fig. 22-16 and
computed with the aid of Fig. 22-21
$$= 2.72 \ (14 \times 20^3) = 25{,}387 \ in.^4$$
ℓ_1 = 17 ft 6 in. = 210 in.

Carry-over factor COF = 0.507

Fixed-end moment, FEM = $0.0842 \ w\ell_2\ell_1^2$

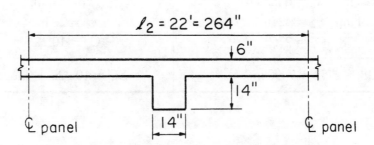

Fig. 22-16 Cross-Section of Slab-Beam

(b) Column members, flexural stiffness K_c:
Referring to Table A7, Appenidx 22A,

For interior columns:

$$K_{ct} = 6.82 \ E_c I_c/\ell_c = 6.82 \times 8748 E_c/144 = 414 E_c$$
$$K_{cb} = 4.99 \ E_c I_c/\ell_c = 4.99 \times 8748 E_c/144 = 303 E_c$$

Example 22.2 —Continued

Calculations and Discussion	Code Reference

For exterior columns:

$K_{ct} = 8.57 \, E_c I_c/\ell_c = 8.57 \times 8748E_c/144 = 521E_c$
$K_{cb} = 5.31 \, E_c I_c/\ell_c = 5.31 \times 8748E_c/144 = 323E_c$
where $I_c = (c)^4/12 = (18)^4/12 = 8748 \text{ in.}^4$
  $\ell_c = 12 \text{ ft} = 144 \text{ in.}$

(c)   Torsional members, torsional stiffness K_t:

$K_t = 9E_cC/\ell_2(1 - c_2/\ell_2)^3$

Eq. (13-6)

where $C = \Sigma(1 - 0.63 \, x/y)(x^3y/3)$

Eq. (13-7)

For interior columns:

$K_t = 9E_c \times 11{,}698/[264 \, (0.932)^3] = 493E_c$

$x_1 = 14$ in. $x_2 = 6$ in.		$x_1 = 14$ in. $x_2 = 6$ in.	
$y_1 = 14$ in. $y_2 = 42$ in.		$y_1 = 20$ in. $y_2 = 14$ in.	
$C_1 = 4738$ $C_2 = 2752$		$C_1 = 10{,}226$ $C_2 = 736$	
$\Sigma C = 4738 + 2752 = 7490 \text{ in.}^4$		$\Sigma C = 10{,}226 + 736 \times 2 = 11{,}698 \text{ in.}^4$	

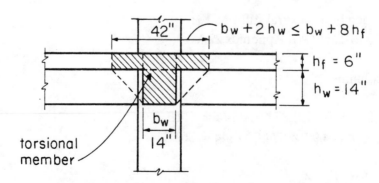

Fig. 22-17 Torsional Member at Interior Column

where C is taken as the larger value computed with the aid of Table 21.2 for the torsional member shown in Fig. 22-17.

Example 22.2—Continued	
Calculations and Discussion	Reference

For exterior columns:

$$K_t = 9E_c \times 17,868/[264 \, (0.932)^3] = 752E_c$$

where C is taken as the larger value computed (with the aid of Table 21-2) for the torsional member shown in Fig. 22-18.

$x_1 = 14$ in. $x_2 = 6$ in. $y_1 = 21$ in. $y_2 = 35$ in. $C_1 = 11,141$ $C_2 = 2248$	$x_1 = 14$ in. $x_2 = 6$ in. $y_1 = 27$ in. $y_2 = 21$ in. $C_1 = 16,628$ $C_2 = 1240$
$\Sigma C = 11,141 + 2248 = 13,389$ in.^{4}	$\Sigma C = 16,628 + 1240 = 17,868$ in.^{4}

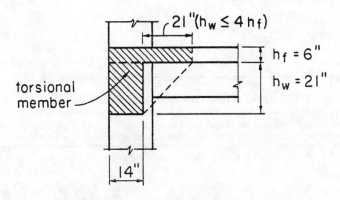

Fig. 22-18 Torsional Member at Exterior Column

22-51

Example 22.2—Continued

Code
Reference

Calculations and Discussion

(d) Parallel beams, increased torsional stiffness Kta:

For interior columns:
$K_{ta} = K_t I_{sb}/I_s = 493E_c \times 25,387/4752 = 2634E_c$

For exterior columns:
$K_{ta} = 752E_c \times 25,387/4752 = 4017E_c$
where I_s = moment of inertia of slab-section shown in Fig. 22-19.
   $= 264 (6)^3/12 = 4752$ in.^{4}
   I_{sb} = moment of inertia of full "T" section shown in Fig. 22-19
       and computed with the aid of Fig. 22-21
       $= 2.72 (14 \times 20^3/12) = 25,387$ in.^{4}

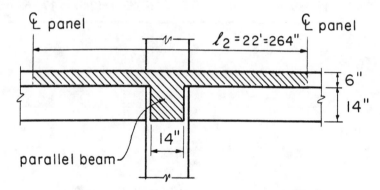

Fig. 22-19 Slab-Beam in the Direction of Analysis

(e) Equivalent column stiffness, K_{ec}:

$K_{ec} = \Sigma K_c \times \Sigma K_{ta}/(\Sigma K_c + \Sigma K_{ta})$

where ΣK_{ta} is for two torsional members, one on each side of column,
and ΣK_c is for the upper and lower columns at the slab-beam joint
of an intermediate floor.

Example 22.2—Continued

Calculations and Discussion

Code
Reference

For interior columns:

$$K_c = \frac{(303E_c + 414E_c)(2 \times 2634E_c)}{(303E_c + 414E_c) + (2 \times 2634E_c)}$$

$$= 631E_c$$

For exterior columns:

$$K_c = \frac{(323E + 521E)(2 \times 4017E)}{(323E + 521E) + (2 \times 4017E)}$$

$$= 764E$$

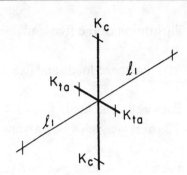

(f) Slab-beam joint distribution factors DF:

At exterior joint:

$$DF = 497E_c/(497E_c + 764E_c)$$
$$= 0.394$$

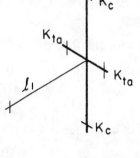

At interior joint:

$$DF = 497E_c/(408E_c + 497E_c + 631E_c)$$
$$= 0.306$$
COF for slab-beam = 0.507

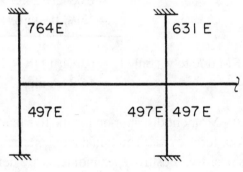

3. Partial frame analysis of equivalent frame.

Determine maximum negative and positive moments for the
slab-beams using the moment distribution method.

With a service live-to-dead load ratio:

$$L/D = 100/75 = 1.33 \ > 3/4$$

the frame will be analyzed for five loading conditions with pattern loading
and partial live load as allowed by Section 13.7.6.3. (See Fig. 22-9 for an

13.7.6.3

Example 22.2—Continued

illustration of the five load patterns considered.)

(a)   Factored loads and fixed-end moments:

Factored dead load, w_d = 1.4 (75 + 9.3) = 118 psf
9.3 psf is weight of beam stem per foot divided by ℓ_2

Factored live load , w_ℓ = 1.7(100) = 170 psf

Factored load , w_d + w_ℓ = 288 psf

FEM's for slab-beams   = 0.842 $w_{\ell_2}\ell_1^2$
FEM due to w_d + w_ℓ = 0.0842(0.288 × 22)17.5^2
                = 163.4 ft−kips

FEM due to w_d + 3/4 w_ℓ

                = 0.0842(0.2455 × 22)17.5^2
                = 139.3 ft−kips

FEM due to w_d only   = 0.084(0.118 × 22)17.5^2
                = 66.9 ft−kips

(b)   Moment distribution for the five loading conditions is shown in Table 22.3.
Counterclockwise rotational moments acting on the member ends are taken
as positive. Positive span moments are determined from the equation:

$M_{(midspan)}$ = M_o - 1/2 (M_L + M_R)
where M_o is the moment at midspan for a simple beam.

When the end moments are not equal, the maximum moment in the span
does not occur at midspan, but its value is close to that at midspan.

Example 22.2—Continued

Calculations and Discussion

Positive moment in span 1-2 for loading (1):

$$+M = (0.288 \times 22)17.5^2/8 - 1/2(102.8 + 185.0)$$
$$= 98.7 \text{ ft-kips}$$

The following moment values for the slab-beams may be obtained by summarizing the results in Table 22-3. Note that according to Section 13.7.6.3, the design moments shall be taken not less than those occurring with full factored live load on all spans.

Maximum positive moment in end span
= the larger of 98.7 or 92.2 = 98.7 ft-kips

Maximum positive moment in interior span*
= the larger of 73.6 or 78.4 = 78.4 ft-kips

Maximum negative moment at end support
= the larger of 102.8 or 94.0 = 102.8 ft-kips

Maximum negative moment at interior support of exterior span
= the larger of 185.0 or 161.5 = 185.0 ft-kips

Maximum negative moment at interior support of interior span
= the larger of 169.4 or 154.1 = 169.4 ft-kips

---

*This is the only moment governed by the pattern loading with partial live load. All other maximum moments occur with full factored live load on all spans

Example 22.2 - Continued

	Code
Calculations and Discussion	**Reference**

Table 22-3 —Moment Distribution for Partial Frame
(Transverse Direction)

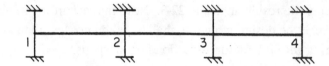

Joint	1	2		3		4
Member	1-2	2-1	2-3	3-2	3-4	4-3
DF COF	0.394 0.507	0.306 0.507	0.306 0.507	0.306 0.507	0.306 0.507	0.394 0.507

(1) All spans loaded with full factored live load

	1-2	2-1	2-3	3-2	3-4	4-3
FEM COM*	+163.4	-163.4 -32.6	+163.4	-163.4	+163.4 +32.6	-163.4
"	+5.1		-5.1	+5.1		-5.1
"	+0.8	-1.0	-0.8	+0.8	+1.0	-0.8
"	+0.3	-0.2	-0.3	+0.3	+0.2	-0.3
Σ DM**	+169.6 +66.8	-197.2 +12.2	+157.2 +12.2	-157.2 -12.2	+197.2 -12.2	-169.6 -66.8
Total	+102.8	-185.0	+169.4	-169.4	+185.0	-102.8

Example 22.2—Continued

Table 22-3—Continued

(2) First and third spans loaded with 3/4 factored live load

FEM	+139.3	+139.3	+66.9	-66.9	+139.3	-139.3
COM*	+11.2	-27.8	-11.2	+11.2	+27.8	-11.2
"	+6.1	-2.2	-6.1	+6.1	+2.2	-6.1
"	+1.3	-1.2	-1.3	+1.3	+1.2	-1.3
"	+0.4	-0.3	-0.4	+0.4	+0.3	-0.4
Σ	+158.3	-170.8	+47.9	-47.9	+170.8	-158.3
DM**	-62.4	+37.6	+37.6	-37.6	-37.6	+62.4
Total	+95.9	-133.3	+85.5	-85.5	+133.3	-95.9
Midspan M	92.2				92.2	

(3) Center span loaded with 3/4 factored live load

FEM	+66.9	-66.9	+139.3	-139.3	+66.9	-66.9
COM*	-11.2	-13.4	-11.2	-11.2	+13.4	+11.2
"	+0.3	+2.2	+0.3	+0.3	-2.2	-0.3
"	-0.3	-0.1	-0.3	-0.3	+0.1	+0.3
Σ	+55.7	-78.2	+150.5	-150.5	+78.2	-55.7
DM**	-21.9	-22.1	-22.1	+22.1	+22.1	+21.9
Total	+33.8	-100.3	+128.4	-128.4	+100.3	-33.8
Midspan M			78.4			

Example 22.2—Continued

		Code
	Calculations and Discussion	Reference

Table 22-3—Continued

(4) First span loaded with 3/4 factored live load and beam-slab
assumed fixed at support two spans away

FEM	+139.3	-139.3	+66.9	-66.9		
COM*	+11.2	-27.8		+11.2		
"	+4.3	-2.2		+4.3		
"	+0.3	-0.9		+0.3		
Σ	+155.1	-170.2	+66.9			
DM**	-61.1	+31.6	+31.6			
Total	+94.0	-138.6	+98.5	-51.1		

(5) First and second span loaded with 3/4 factored live load

FEM	+139.3	-139.3	+139.3	-139.3	+66.9	-66.9
COM*		-27.8	+11.2		+13.4	+11.2
"	+2.6		-2.1	+2.6	-2.2	-2.1
"	+0.3	-0.2	-0.1	+0.3	+0.4	-0.1
Σ	+142.2	-167.3	+148.3	-136.4	+78.5	-57.9
DM**	-56.0	+5.8	+5.8	+17.7	+17.7	+22.8
Total	+86.2	-161.5	+154.1	-118.7	+96.2	-35.1

*Carry-over moment, COM, is the negative product of the distribution factor, carry-over and unbalanced joint moment carried to the opposite end of span.

**Distributed moment, DM, is the negative product of the distribution factor and the unbalanced joint moment.

Example 22.2—Continued

	Code
Calculations and Discussion	**Reference**

4. Design Moments.

   Positive and negative factored moments for the slab system in the transverse direction are shown in Fig. 22-20. The negative factored moments are taken at the face of rectilinear supports at distances not greater than $0.175\ell_1$ from the center of supports.

   13.7.7.1

   $0.75 < 0.175 \times 17.5$ (Use face of support location).

5. Total factored moment per span.

   13.7.7.4

   Slab systems within the limitations of Section 13.6.1 may have the resulting analytical moments reduced in such proportion that the numerical sum of the positive and average negative moments are not greater than the total static moment M_o given by Eq. (13-3). Check limitations of Section 13.6.1.6 for relative stiffness of beams in two perpendicular directions.

   For interior panel:

   $\alpha_1\ell_2^2/\alpha_2\ell_1^2 = 3.16\,(22)^2/3.98(17.5)^2 = 1.25$
   $0.2 < 1.25 < 5.0$ O.K.

   13.6.1.6

   For exterior panel:

   $3.16(22)^2/16.45(17.5)^2 = 0.30$
   $0.2 < 0.30 < 5.0$ O.K.

   All limitations of Section 13.6.1 are satisfied and the provisions of Section 13.7.7.4 may be applied.

   $M_o = w\ell_2\ell_n^2/8 = 0.288 \times 22 \times 16^2/8 = 202.8$ ft-kips

   Eq. (13-3)

Example 22.2 —Continued

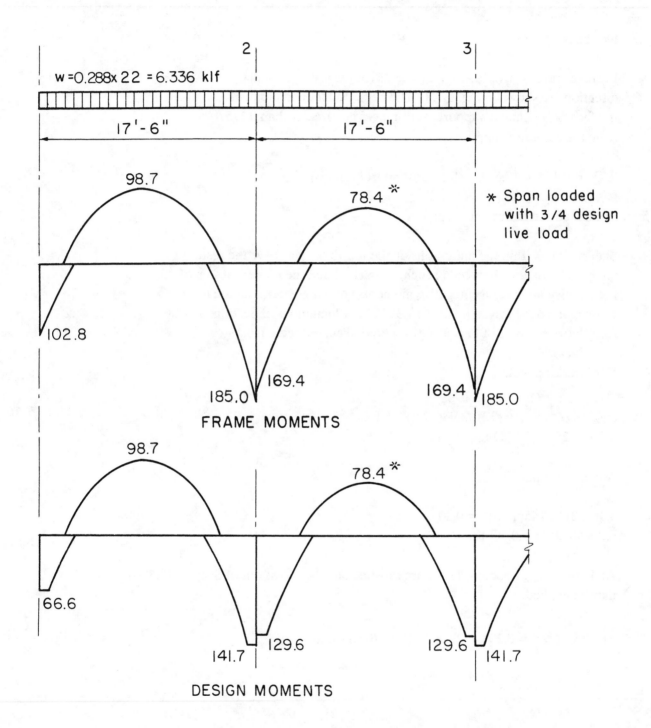

**Fig. 22-20 Positive and Negative Design Moments for Slab-Beam
(All Spans Loaded With Full Factored Live Load Except As Noted)**

Note: See Section 13.6.8 for distribution of shear forces.

22-60

Example 22.2 —Continued

Calculations and Discussion	Code Reference

End span: $98.7 + (66.6 + 141.7)/2 = 202.9$ ft-kips

Interior span: $78.4 + (129.6 + 129.6)/2 = 208.0$ ft-kips

To illustrate proper procedure, the interior span factored moments may be reduced as follows:

Permissible reduction $= 202.8/208.0 = 0.98$
Adjusted negative design moment $= 129.6 \times 0.98 = 126.4$ ft$-$kips
Adjusted positive design moment $= 78.4 \times 0.98 \underline{\ = \ \ 76.4}$ ft$-$kips
$$M_O = 202.8 \text{ ft}-\text{kips}$$

6. Distribution of design moments across slab-beam strip   13.7.7.5

    Negative and positive factored moments at critical sections may be distributed
    to the column strip, beam and two-half middle strips of the slab-beam according to
    the proportions specified in Sections 13.6.4, 13.6.5 and 13.6.6, if requirements
    of Section 13.6.1.6 are satisfied.

(a) Since the relative stiffnesses of beams are between 0.2 and 5.0 (see Step No. 5),
    the moments can be distributed across slab-beams as specified in Section 13.6.4,
    13.6.5 and 13.6.6.

(b) Distribution of factored moments at critical section:

$l_2/l_1 \quad = 22/17.5 = 1.257$
$\alpha_1 l_2/l_1 \quad = 3.16 \times 1.257 = 3.97$
$\beta_t = C/2I_s = 17,868\,/(2 \times 4752) = 1.88$
where $\quad I_s = 22 \times 12 \times 6^3/12 = 4752$ in.^{4}
$\qquad C = 17,868$ in.^{4} (see Fig. 22$-$18)

Example 22.2—Continued

Factored moments at critical sections are summarized in Table 22-4.

Table 22-4 Distribution of Design Moments

	Factored Moment (ft-kips)	Column Strip		Moment (ft-kips) in Two Half-Middle Strips**
		Percent*	Moment (ft-kips)	
End Span:				
Exterior Negative	66.6	75	50.0	16.6
Positive	98.7	67	66.1	32.6
Interior Negative	141.7	67	94.9	46.8
Interior Span:				
Negative	129.6	67	86.8	42.8
Positive	78.4	67	52.5	25.9

* Since $\alpha_1 \ell_2 / \ell_1 > 1.0$ beams must be proportioned to resist 85 percent of column strip moment as per Section 13.6.5.1.

** That portion of the factored moment not resisted by the column strip is assigned to the two half-middle strips.

7.　Calculations for shear in beams and slab are performed in Example 21.2, Part 21.

Example 22.2—Continued

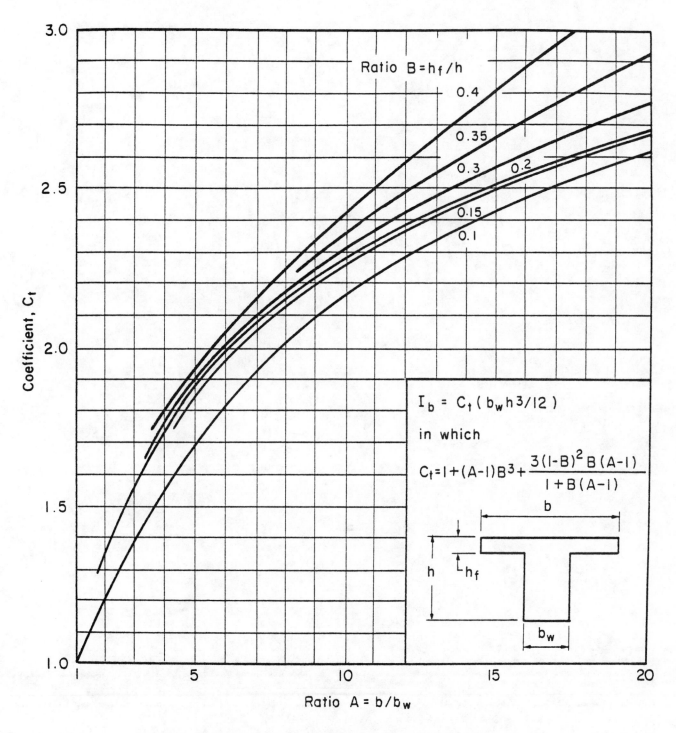

**Fig. 22-21 Coefficient C_t for Gross Moment of Inertia of Flanged
Sections (Flange on One or Two Sides)**

WALLS

GENERAL CONSIDERATIONS

Section 14.2.2 allows the designer two options in designing walls: walls may be designed as compression members (Section 14.4) using the strength design provisions for flexure and axial loads of Chapter 10; or they may be designed by the Empirical Design Method of Section 14.5. The provisions of Section 14.2 and 14.3 apply to walls designed by either method. No minimum wall thicknesses are prescribed for walls designed as compression members (Section 14.4).

Note that the Empirical Design Method applies to load bearing walls, and only to walls of solid rectangular cross section. Load bearing walls of nonrectangular cross section, such as ribbed wall panels, must be designed by Section 14.4. Cantilever retaining walls are designed by the flexural design provisions of Chapter 10.

Section 14.2.3 alerts the designer not to overlook the special shear provisions for walls of Section 11.10. Shear forces must be considered in the design of walls; in some cases the required shear reinforcement may exceed the minimum wall reinforcement of Section 14.3. Example 23.4 illustrates shear design of walls.

One additional item concerning walls: the Code provisions for force transfer at footings (Section 15.8) specifically address force transfer between a wall and footing, with Section 15.8.2.2 requiring a minimum amount of reinforcement, not less than the minimum wall reinforcement, to be provided across the interface between a wall and a supporting footing.

14.4 WALLS DESIGNED AS COMPRESSION MEMBERS

Where wall geometry and loading conditions do not satisfy the limitations of Section 14.5, usually where lateral loads are present, walls must be designed as compression members by the strength design provisions for flexure and axial loads of Chapter 10. Minimum reinforcement requirements of Section 14.3 apply to walls designed by strength design provisions of Chapter 10. Especially note that the vertical wall reinforcement need not be enclosed by lateral ties (as for columns) if the conditions of Section 14.3.6 are satisfied. All other Code provisions for compression members apply to walls designed by Chapter 10.

As with columns, design of walls is difficult without the use of design aids. Wall design is further complicated by the fact that slenderness is a consideration in the design of practically all walls. Two methods for slenderness consideration are specified in the Code. The so-called rigorous analysis, which takes into account variable wall stiffness, is specified in Section 10.10.1. In lieu of that procedure, the approximate evaluation of slenderness effects prescribed in Section 10.11 may be used.

Considering the approximate slenderness method, note that Eqs. (10-10) and (10-11) for EI were not originally derived for members with a single layer of reinforcement.

An EI equation for walls with a single layer of reinforcement has been suggested by MacGregor:[23.1]

$$EI = \frac{E_cI_g}{\beta}\left(0.5 - \frac{e}{h}\right) \geq 0.1\frac{E_cI_g}{\beta} \tag{1}$$

$$\leq 0.4\frac{E_cI_g}{\beta}$$

where $\beta = 0.9 + (0.5\,\beta_d{}^2) - 12\rho \geq 1.0$
β_d = ratio of dead load to total load, and
ρ = ratio of area of vertical reinforcement to area gross concrete area

For ACI 318-89, the definition of the creep effect factor β_d, included with Eqs. (10-10) and (10-11) for EI, has been modified. It is now defined in terms of a sustained load ratio, rather than a sustained moment ratio; see discussion on Section 10.11.5.2 in Part 13. For consistency, the same sustained load ratio seems appropriate for the EI expression for walls.

Comparison of EI by Eq. (1) and Code Eq. (10-11) is shown in Fig. 23-1. Values of EI in terms of E_cI_g are plotted as a function of eccentricity e/h for several values of β_d, with reinforcement ratio ρ constant at 0.0015. Values obtained using Eq. (10-11) are constant for each value of β_d, since e/h is not a variable. For walls with higher load eccentricity, Code Eq. (10-11) appears to overestimate wall stiffness. For walls designed by Chapter 10 with slenderness evaluation by Section 10.11, Eq. (1) is recommended in lieu of Code Eq. (10-11) for wall stiffness. Example 23.2 illustrates application of Code Section 10.11 using Eq. (1) for wall stiffness.

When wall slenderness exceeds the limit for application of the approximate slenderness evaluation method of Section 10.11 ($k\ell_u/h \geq 30$), a more detailed evaluation of wall slenderness effects is required, as defined in Section 10.10.1. The slender load-bearing concrete wall panels currently used in some building systems, especially in tilt-up wall construction, are in this high slenderness category. The more detailed slenderness analysis should account for the influence of variable wall stiffness, the effects of deflections on the moments and forces, and the effects of duration of loads. Such an analysis is presented in PCA design aid EB074D *Tilt-up Load-Bearing Walls.*[23.2] It presents load capacities of slender wall panels ($20 \leq k\ell_u/h \leq 50$) with thicknesses varying from 5-1/2 to 9-1/2 in. and having single or double layers of reinforcement. Design

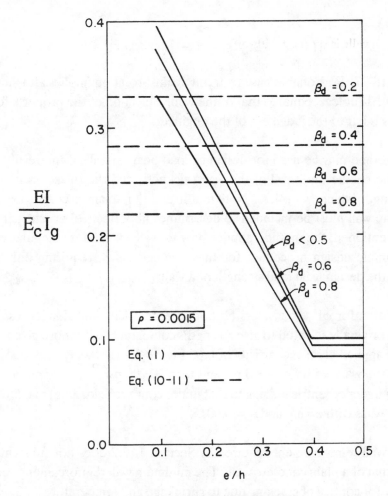

Fig. 23-1 Stiffness EI of Walls

assistance is given in the form of load capacity coefficient tables. A description of how the design tables were developed is included in the publication. The design aid is simple to use, requiring only a minimum amount of calculations. Example 23.3 illustrates application of EB074D.

14.5 EMPIRICAL DESIGN METHOD

With the publication of ACI 318-83, the empirical wall design Eq. (14-1) was modified to reflect the general range of end conditions encountered in wall design, and to allow for a wider range of design applications. The wall strength equation in previous codes was based on the assumption of a wall with top and bottom ends restrained against lateral movement, and with rotation restraint at one end, corresponding to an effective length factor between 0.8 and 0.9. Axial load strength values determined from the original equation could be unconservative unless at least one end of the wall was restrained against rotation. This pinned-pinned end condition can exist in certain walls, particularly of precast tilt-up construction, and in some large panel wall systems. In addition, in some other types of wall construction, the top end of the wall is free, and not braced against translation. In these cases, it is necessary to reflect the proper effective length in the design equation. Values of effective length factor k are now given in Section 14.5.2 for commonly occurring wall end conditions. Revised Eq. (14-1) will give the same results as the 1977 Code Eq. (14-1) for walls braced against translation of both ends and with reasonable base restraint against rotation. Using an effective length factor k = 0.8, the slenderness function becomes

$$1 - [(0.8\ell/32h)^2] = 1 - [(\ell/40h)^2]$$

as in Eq. (14-1) of ACI 318-77. Reasonable base restraint against rotation implies attachment to a member having a flexural stiffness EI/ℓ at least equal to that of the wall. Selection of the proper k for a particular set of support end conditions is left to the judgment of the engineer.

The Empirical Design Method may be used for design of load-bearing walls if the resultant of the vertical loads is located within the middle one-third of the wall thickness, and the thickness is at least 1/25th the unsupported height or length, whichever is less. Note that in addition to any eccentric axial loads, the effect of any lateral loads on the wall must be included to determine the "effective" eccentricity of the resultant vertical load. Also, as mentioned, the method applies only to walls of solid rectangular cross-section. The empirical method is a simple design procedure for these limited cases, requiring only a single strength calculation to determine the design axial load strength of a wall.

The single strength equation for φP_{nw} considers both load eccentricity and slenderness effects. The eccentricity factor 0.55 was originally selected to give strengths comparable with those given by Chapter 10 for members with axial load applied at an eccentricity of h/6. Figure 23-2 shows typical load-moment strength curves for 8-, 10-, and 12-in. walls with $f_c'= 4000$ psi and $f_y = 60,000$ psi. [23.3] The curves yield eccentricity factors (ratios of strength under eccentric loading to that under concentric loading) of 0.562, 0.568, and 0.563 for the 8-, 10-, and 12-in. walls with e=h/6 and $\rho = 0.0015$.

Note that the minimum wall reinforcement required by Section 14.3.2 does not substantially increase the strength of a wall above that of a plain concrete wall. The minimum wall reinforcement required by Section 14.3 is provided primarily for control of cracking due to shrinkage and temperature stresses.

For slender walls with different end restraint conditions, Eq. (14-1) results in strengths comparable with those given by the slenderness evaluation procedure of Section 10.11, as illustrated in Fig. 23-3. Principal

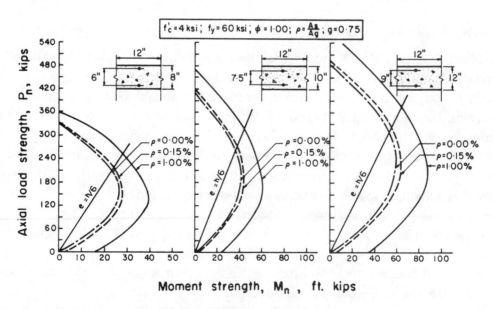

Fig. 23-2 Typical Load-Moment Strength Curves for 8-, 10-, and 12-in. Walls

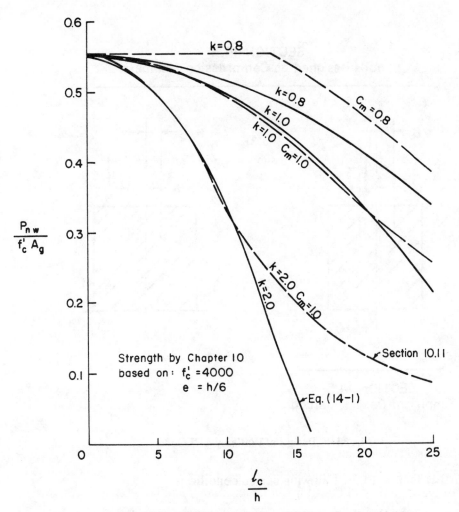

Fig. 23-3 Wall Strength: Eq. (14-1) vs. Section 10.11

application of the Empirical Design Method is for relatively short walls spanning vertically, and subject to vertical loads only, such as those resulting from the reaction of floor or roof systems supported on a wall. Application becomes extremely limited when lateral loads must be considered, because of the "effective" load eccentricity limitation of h/6. Walls other than short walls carrying "reasonably concentric" loads should be designed as compression members for axial load and flexure by provisions of Chapter 10 (Section 14.4). Figure 23-4 compares design by Eq. (14-1) versus that by Chapter 10.

14.5.2 Empirical Design Procedure

When the eccentricity e of the "effective" load does not exceed h/6 the design is performed considering P_u as a concentric load. The factored axial load P_u must be less than the design axial load strength φP_{nw} computed by Eq. (14-1):

$$P_u \leq \varphi\, P_{nw}$$

$$\leq 0.55\, \varphi f_c' A_g \left[1 - \left(\frac{k\ell_c}{32h} \right)^2 \right] \qquad \text{Eq. (14-1)}$$

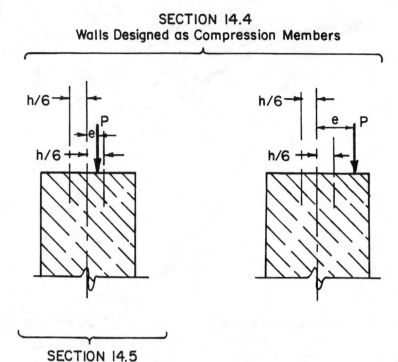

SECTION 14.4
Walls Designed as Compression Members

SECTION 14.5
Empirical Design Method

Fig. 23-4 Design of Walls by ACI Code

Use of Eq. (14-1) is further limited to the following design conditions:

(1) Wall thickness h must not be less than $\ell_c/25$ nor 4 in. Basement walls and foundation walls must be at least 7-1/2 in. thick (Section 14.5.3.2).

(2) Walls must contain both horizontal and vertical reinforcement. The area of horizontal reinforcement must be not less than 0.0025 times the area of gross concrete section, and that of the vertical reinforcement not less than 0.0015 times the area of gross concrete section. These ratios may be reduced to 0.0020 and 0.0012 respectively, when bars #5 or smaller having $f_y \geq 60,000$ psi, or welded wire fabric with W31 or D31 or smaller wires are used. In walls greater than 10 in. thick (except basement walls) the reinforcement in each direction must be placed in two layers (Section 14.3).

(3) Length of wall to be considered as effective for each beam reaction must not exceed center-to-center distance between reactions, nor width of bearing plus 4h (14.2.4.)

(4) The wall must be anchored to the floors or to columns and other structural elements of the building (Section 14.2.6).

Example 23.1 illustrates application of the Empirical Design Method to a bearing wall supporting precast floor beams. It should be noted that the reinforcement and minimum thickness requirements of Sections 14.3 and 14.5.3 may be waived where structural analysis shows adequate strength and wall stability (Section 14.2.7). This required condition may be satisfied by a design using *Building Code Requirements for Structural Plain Concrete*, ACI 318.1-89. [23.4.]

DESIGN SUMMARY

A trial procedure for wall design is suggested: first assume a wall thickness h and a reinforcement ratio ρ, then check the trial wall for the applied loading conditions.

It is not within the scope of Part 23 to include design aids for a broad range of wall and loading conditions. The intent is to present examples of design options and aids. Using programmable calculators, the designer can, with reasonable effort, produce design aids to fit the range of conditions usually encountered in practice and in a form of his choice. For example, strength interaction diagrams such as those plotted in Fig. 23-5 (ρ = 0.0015) and Fig. 23-6 (ρ = 0.0025) can be helpful design aids for evaluation of wall strength. "Blow-ups" of the lower portions of the strength interaction diagrams are shown for specific walls (h = 6.5 in.). Load

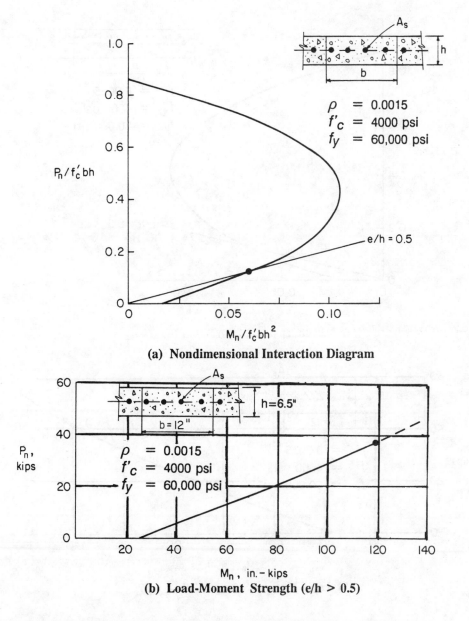

(a) Nondimensional Interaction Diagram

(b) Load-Moment Strength (e/h > 0.5)

Fig. 23-5 Load-Moment Interaction Diagram for Wall (Single Layer Reinforcement, ρ = 0.0015)

charts, such as Fig. 23-7 can also be developed for specific walls. Design aids such as the one shown in Figure 23-8 may facilitate selection of wall reinforcement. The designer can tailor his design aids to the "normal construction practice" in his area.

Prestressed walls are not covered specifically in Part 23. Prestressing of walls is advantageous for handling (precast panels) and for increased buckling resistance. For design of prestressed walls the designer should consult Ref. 23.5.

11.10 SPECIAL "SHEAR" PROVISIONS FOR WALLS

For most low-rise buildings, horizontal shear forces acting in the plane of walls are small, and can usually be neglected in design. Such in-plane forces, however, become a design consideration in major structures

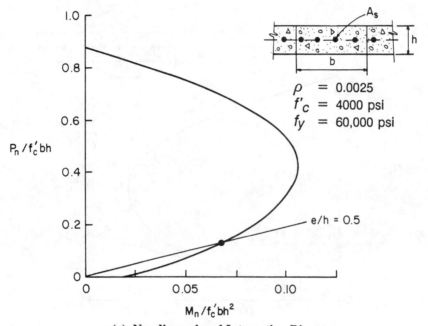

(a) Nondimensional Interaction Diagram

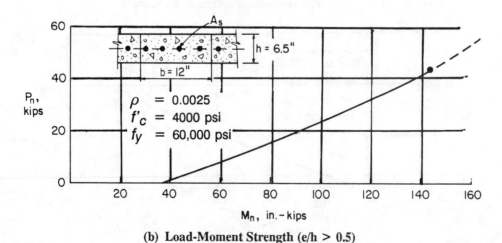

(b) Load-Moment Strength (e/h > 0.5)

Fig. 23-6 Load-Moment Interaction Diagram for Wall (Single Layer Reinforcement, $\rho = 0.0025$)

23-8

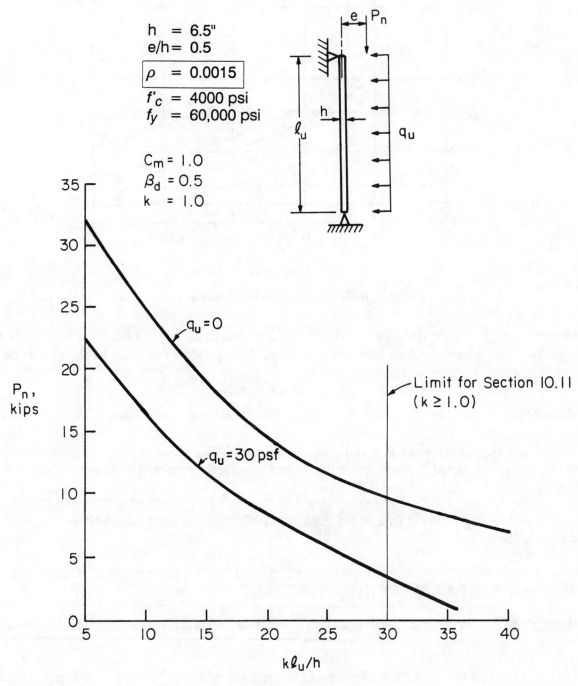

Fig. 23-7 Design Chart for 6.5-in. Wall

23-9

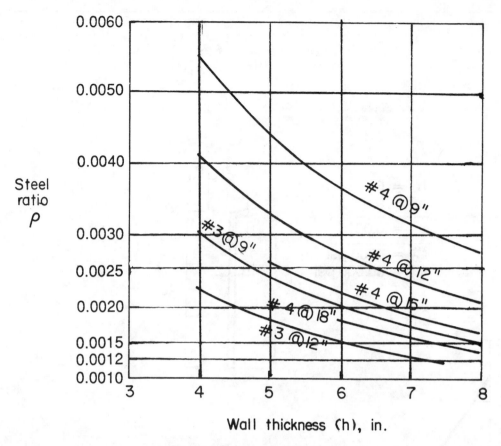

Fig. 23-8 Design Aid for Wall Reinforcement

where a limited number of walls resist the total lateral load, such as in high-rise buildings. Flexural strength must also be considered when in-plane loads are significant. Example 23.4 illustrates in-plane shear design of walls, including design for flexural strength.

REFERENCES

23.1 MacGregor, J.G., "Design and Safety of Reinforced Concrete Compression Members," paper presented at International Association for Bridge and Structural Engineering Symposium, Quebec, 1974.

23.2 *Tilt-Up Load Bearing Walls - A Design Aid,* Publication EB074D, Portland Cement Association, Skokie, IL, 1980.

23.3 Kripanaryanan, K.M., "Interesting Aspects of the Empirical Wall Design Equation," *ACI Journal, Proceedings* Vol. 74, No. 5, May 1977, pp. 204-207.

23.4 *Building Code Requirements for Structural Plain Concrete*, ACI 318.1-89, American Concrete Institute, Detroit, MI, 1989.

23.5 *PCI Design Handbook - Precast and Prestressed Concrete,* 3rd Edition, Prestressed Concrete Institute, Chicago, IL, 1985.

Example 23.1—Design of Bearing Wall by Empirical Design Method (Section 14.5)

A concrete bearing wall supports a floor system of precast single tees spaced at 8 ft on centers. The stem of each tee section is 8 in. wide. The tees have full bearing on the wall. The height of wall is 15 ft, and the wall is considered laterally restrained at top. Design of the wall is required.

Design Data:

    Floor beam reactions due to dead load = 28 kips
    live load = 14 kips
    $f_c' =$ 4000 psi
    $f_y =$ 60,000 psi
    Neglect self weight of wall

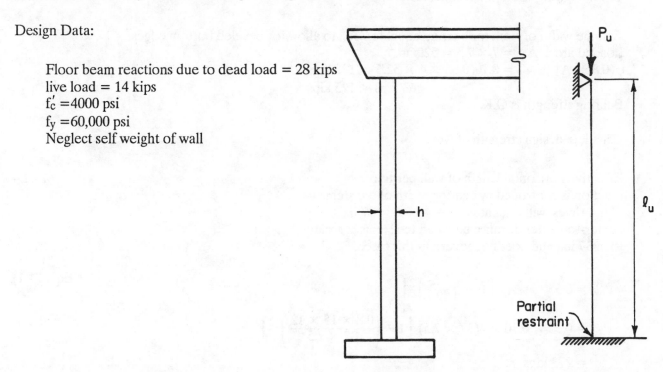

	Code
Calculations and Discussions	**Reference**

The general design procedure is to select a trial wall thickness h, then check the trial wall for the applied loading conditions.

1.   Select trial wall thickness h.

    $h > \ell_u/25$ but not less than 4 in.   14.5.3.1
      $> 15 \times 12/25 = 7.2$ in.   14.2.6
    Try h = 7.5 in.

2.   Calculate factored loading.

    $P_u = 1.4D + 1.7L$ Eq. (9-1)
        $= 1.4(28) + 1.7(14)$
        $= 39.2 + 23.8 = 6$ kips

Example 23.1—Continued

	Code
Calculations and Discussion	**Reference**

3. Check bearing strength on concrete.   10.15

Assume width of stem for bearing equal to 7 in., to allow for beveled bottom edges.
Loaded area, $A_1 = 7 \times 7.5 = 52.5$ in.^{2}
$\varphi(0.85 f_c' A_1) = 0.70(0.85 \times 4 \times 52.5) = 125$ kips
                        63 kips < 125 kips
Bearing strength is O.K.

4. Calculate design strength of wall.

Effective horizontal length of wall per tee   14.2.4
reaction is controlled by bearing width of tee stem
plus 4 times wall thickness. $7 + 4 \times 7.5 = 37$ in.
center-to-center distance between tee stems is greater
than 37 in., and does not govern in this case.

$$\varphi P_{nw} = 0.55 \; \varphi f_c' A_g \left[1 - \left(\frac{k\ell_c}{32h}\right)^2\right]$$ Eq. (14-1)

$$= 0.55 \times 0.70 \times 4(37 \times 7.5) \left[1 - \left(\frac{0.8 \times 15 \times 12}{32 \times 7.5}\right)^2\right]$$

$$= 273 \text{ kips}$$

$P_u \le \varphi P_{nw}$

63 kips < 273.5 kips O.K.   7.5-in.-thick wall is adequate,
                                 with sufficient margin for
                                 possible effect of load eccentricity.

5. Select reinforcement. Provide single layer of reinforcement.

Based on 1-ft width of wall, and Grade 60 reinforcement:

horizontal $A_s = 0.0020 \times 12 \times 7.5 = 0.180$ in.^{2}/ft   14.3.3
vertical $A_s = 0.0012 \times 12 \times 7.5 = 0.108$ in.^{2}/ft   14.3.2
Spacing $= 3h$, but not greater than 18 in.   7.6.5
         $= 3 \times 7.5 = 22.5$ in. (18 in. governs)
Horizontal A_s - Use #4 bars @ 12-in. on center ($A_s = 0.20$ in.^{2}/ft)
Vertical A_s - Use #4 bars @ 18-in. on center ($A_s = 0.13$ in.^{2}/ft)

Design aids such as Fig. 23-8 may be used to select reinforcement directly.

Example 23.2 —Design of Tilt-up Wall Panel by Chapter 10 (Section 14.4)

Design of the wall shown is required.

Roof dead load (8 ft double tees) = 50 psf
Roof live load = 20 psf
Wind load = 35 psf

ℓ_u = 16 ft.
k = 1.0
f_c' = 4000 psi
f_y = 60,000 psi
w_c = 150 pcf

Wall is considered laterally restrained at top.

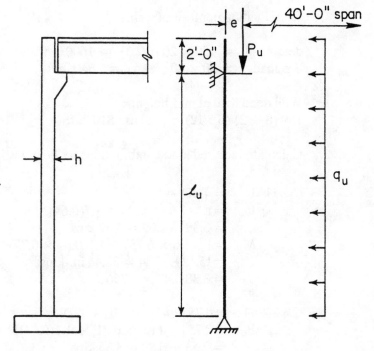

	Code
Calculations and Discussions	**Reference**

1.   Select trial wall section and reinforcement.

Try h = 6.5 in. (e = 6.75 in.)
A_s = #4 @ 12 in. = 0.20 in.^{2}/ft
$\rho = A_s/bh = 0.20/12 \times 6.5 = 0.00256$

[Design for 1-ft (b = 12 in.) wall section]

Effective length of wall for roof reaction:

(4-in. tee stems @ 4 ft o.c.)   14.2.4
4 in. + 4 (6.5) = 30 in. (governs)
distance between stems = 48 in.

Example 23.2 —Continued

Roof loading per foot of wall:

dead load = $50 \times 20\,(4/2.5) = 1600$ lb/ft
live load = $20 \times 20\,(4/2.5) = 640$ lb/ft

Wall dead load at mid-height:
$150\,(8 + 2)\,(6.5/12) = 813$, use 810 lb/ft

2. Calculate factored load combinations to be investigated.

Case l: U $= 1.4D + 1.7L$ Eq. (9-1)
    $P_u = 1.4\,(1.6 + 0.81) + 1.7(0.64)$
    $= 3.38 + 1.09 = 4.47$ kips
    $M_u = 1.4(1.6 \times 6.75) + 1.7(0.64 \times 6.75)$
    $= 15.12 + 7.34 = 22.46$ in.-kips
    $\beta_d = 3.38/4.47 = 0.756$

Case 2: U $= 0.75\,(1.4D + 1.7L + 1.7W)$ Eq. (9-2)
    $P_u = 0.75\,[1.4(1.6 + 0.81) + 1.7(0.64)]$
    $= 2.53 + 0.82 = 3.35$ kips
    $M_u = 0.75\,[5.12 + 7.34 + 1.7\,(0.035 \times 16^2 \times 12/8)]$
    $= 11.34 + 5.51 + 17.13 = 33.98$ in.-kips
    $\beta_d = 2.53/3.35 = 0.756$

Case 3: U $= 0.9D + 1.3W$ Eq. (9-3)
    $P_u = 0.9\,(1.6 + 0.81) = 2.17$ kips
    $M_u = 0.9\,(1.6 \times 6.75) + 1.3\,(0.035 \times 16^2 \times 12/8)$
    $= 9.72 + 17.47 = 27.19$ in.-kips
    $\beta_d = 2.17/2.17 = 1.00$

3. Check wall slenderness.   10.11.4

$k\ell_u/r = 1.0\,(16 \times 12)/0.3 \times 6.5 = 98.5$ 10.11.3
$98.5 < 100$ Approximate evaluation of   10.11.4.3
    slenderness effects by Section 10.11 may be used.

4. Calculate moment magnification by Section 10.11.5 using Eq. (1) for EI.

$$EI = \frac{E_c I_g}{\beta}\left(0.5 - \frac{e}{h}\right) \geq 0.1\frac{E_c I_g}{\beta}$$ Eq. (1)

$$\leq 0.4\frac{E_c I_g}{\beta}$$

Example 23.2—Continued

	Code
Calculations and Discussion	Reference

$E_c = 57{,}000 \sqrt{4000} = 3.605 \times 10^6$ psi

$I_g = \dfrac{12(6.5)^3}{12} = 274.6$ in.^{4}

$e/h = \dfrac{6.75}{6.5} = 1.04$

8.5.1

$EI = 0.1 \left(\dfrac{3.605 \times 10^6 \times 274.6}{\beta} \right) = \dfrac{99 \times 10^6}{\beta}$ lb-in.^{2}

$P_c = \dfrac{\pi^2 EI}{(k\ell_u)^2} = \dfrac{\pi^2}{(16 \times 12)^2} \left(\dfrac{99 \times 10^3}{\beta} \right) = \dfrac{26.5}{\beta}$ kips

Eq. (10-9)

$\beta = 0.9 + 0.5\beta_d^2 - 12\rho \geq 1.0$

$\beta = 0.9 + 0.5\beta_d^2 - 12(0.00256)$

$ = 0.869 + 0.5\beta_d^2 \geq 1.0$

$\delta_b = \dfrac{1.0}{1 - (P_u/\varphi P_c)} \geq 1.0$

Eq. (10-7)

Calculate increased φ factor for largest $P_u = 4.47$ kips

9.3.2.2(b)

$\varphi = 0.9 - 0.2P_u/(0.1f_c'A_g) \geq 0.70$

$ = 0.9 - 0.2 \times 4.47/(0.1 \times 4 \times 12 \times 6.5) = 0.87$

Although slight variations from this value will occur for different load combinations, this single value is considered adequate.

$M_c = \delta_b M_u$

Eq. (10-6)

Moment Magnification

Load Case	P_u kips	M_u in.-kips	β_d	$\beta \geq 1.0$	EI lb-in.^{2}	P_c kips	δ_b	M_c in.-kips
1	4.47	22.46	0.756	1.15	86×10^6	23.0	1.29	28.97
2	3.35	33.98	0.756	1.15	86×10^6	23.0	1.20	40.78
3	2.17	27.19	1.00	1.37	72×10^6	19.3	1.15	31.27

Example 23.2 —Continued

Calculations and Discussion	Code Reference

5. Compare strength required vs. strength provided.

    Referring to load-moment interaction curve, Fig. 23-6(b):

    Case 1: Required nominal strength P_u/φ = 4.47/0.87 = 5.14 kips
    M_c/φ = 28.97/0.87 =   33.3 in.-kips
    From Fig. 23-6(b) at P_n = 5.14 kips, read M_n = 51 in.-kips > 33.3 in.-kips   O.K.

    Case 2: Required nominal strength P_u/φ = 3.35/0.87 = 3.85 kips
    M_c/φ = 40.78/0.87 = 46.9 in.-kips
    From Fig. 23-6(b) at P_n = 3.85 kips, read M_n = 46 in.-kips $\approx$ 46.9 in.-kips say O.K.

    Case 3: Required nominal strength P_u/φ = 2.17/0.87 = 2.49 kips
    M_c/φ = 31.27/0.87 = 35.9 in.-kips
    From Fig. 23-6(b) at P_n = 2.49 kips, read M_n = 42 in.-kips > 35.9 in.-kips O.K.

Example 23.3—Design of Tilt-up Wall Panel Using Design Aid EB074D

Design of the wall shown is required.

Roof dead load (8 ft. double tees) = 50 psf

Roof live load = 20 psf
Wind load = 20 psf
Unsupported length ℓ_u = 18 ft.

k = 1.0
f'_c = 4000 psi
f_y = 60,000 psi

Wall is considered laterally restrained at top.

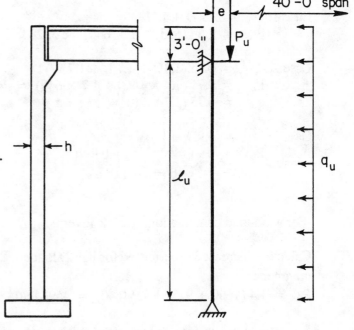

	Code
Calculations and Discussion	**Reference**

1. Select trial wall section and reinforcement.

    Try h = 6.5 in. (e = 6.75 in.)
    A_s = #4 @ 12 in. = 0.20 in.^{2}/ft
    ρ = A_s/bh = 0.20/12 × 6.5 = 0.00256
    (Design for 1-ft (b = 12 in.) wall section)
    Effective length of wall for roof reaction:
    (4-in. tee stems @ 4 ft o.c.)
    4 in. + 4 (6.5) = 30 in. (governs)   14.2.4
    distance between stems = 48 in.
    Roof loading per foot of wall:
    dead load = 50 × 20 (4/2.5) = 1600 lb/ft
    live load = 20 × 20 (4/2.5) = 640 lb/ft

Example 23.3 —Continued

	Code
Calculations and Discussion	Reference

2. Calculate factored load combinations to be investigated.*

Case 1: U $\quad$ = 1.4D + 1.7L $\qquad\qquad\qquad\qquad\qquad\qquad$ Eq. (9-1)

$\quad\quad\quad$ P_u $\quad$ = 1.4 (1600) + 1.7 (640) = 3328 lb/ft

Case 2: U $\quad$ = 0.75 (1.4D + 1.7 L + 1.7 W) $\qquad\qquad\qquad\qquad$ Eq. (9-2)

$\quad\quad\quad$ P_u $\quad$ = 0.75 (1.4 × 1600 + 1.7 × 640) = 2496 lb/ft

$\quad\quad\quad$ q_u $\quad$ = 0.75 (1.7 × 20) = 25.5 psf

Case 3: U $\quad$ = 0.9D + 1.3W $\qquad\qquad\qquad\qquad\qquad\qquad$ Eq. (9-3)

$\quad\quad\quad$ P_u $\quad$ = 0.9 (1600) = 1440 lb/ft

$\quad\quad\quad$ q_u $\quad$ = 1.3 (20) = 26 psf

For wind load combinations, case 2 governs.

Calculate increased φ permitted for lightly loaded $\qquad\qquad$ 9.3.2.2(b)
members:

P_u = 1.4 (1600 + 975) + 1.7 (640) = 4693 lb/ft

where wall dead load at mid-height = 150 (9+3) (6.5/12) = 975 lb/ft

φ = 0.9 – 0.2 P_u /(0.1f_c' A_g) ≥ 0.70

$\quad$ = 0.9 – 0.2 x 4.69/(0.1 × 4 × 12 × 6.5) = 0.87

Although slight variations from this value will occur for
different load combinations, this single value is considered adequate.

3. Check Strength, Using Design Aid EB074D.

The tabulated strength values in EB074D are nominal strengths P_n. When designing with this aid, the factored loads P_u must be divided by the φ factor to enter the strength tables directly.

Case 1: Required P_u /φ= 3328/0.87 = 3825 lb/ft

$\quad\quad\quad$ q_u /φ =0

From Strength Table A5 (see Table 23-1), for h = 6-1/2 in.,
ρ ≈ 0.0025, e = 6.75 in., q_u/φ = 0, and
kℓ_u /h = 1.0 × 18 × 12/6.5 = 33

* Strength tables of Design Aid EB074D include weight of wall.

Example 23.3 —Continued

Code
Reference

Calculations and Discussion

By interpolation, coeff. = 0.0184

P_n = 0.0184 x 4000 x 12 x 6.5 = 5740 lb/ft

$P_n \geq P_u/\varphi$

5740 lb/ft > 3825 lb/ft   O.K.

Case 2: Required P_u /φ = 2496/0.87 = 2870 lb/ft

q_u /φ = 25.5/0.87 = 29.3 ≈ 30 psf

From Strength Table A6 (see Table 23-2), for h = 6-1/2 in.,

$\rho \approx 0.0025$, e = 6.75 in., q_u/φ = 30 psf, and $k\ell_u/h$ = 33

By interpolation, coeff. = 0.011

P_n = 0.011 x 4000 x 12 x 6.5 = 3432 lb/ft

$P_n \geq P_u /\varphi$

3432 lb/ft > 2870 lb/ft   O.K.

Therefore, h = 6-1/2 in. and A_s = #4 @ 12 in. is adequate for

load combinations investigated.

Table 23-1—Reproduced from Page 15 of EB074D

$\rho = \dfrac{A_s \times 100}{b_1 \times h}$	End eccentricity, e, in.	q_u/φ = 0 psf Slenderness ratio, $k\ell_u/h$ =				q_u/φ = 15 psf Slenderness ratio, $k\ell_u/h$ =			
		20	30	40	50	20	30	40	50
0.15	1.00	0.498	0.347	0.227	0.155	0.468	0.331	0.191	0.085
	3.25	0.094	0.042	0.018	0.013	0.087	0.021	0.005	**
	6.75	0.018	0.014	0.005	0.003	0.017	0.009	0.003	**
0.25	1.00	0.498	0.347	0.227	0.155	0.468	0.331	0.191	0.090
	3.25	0.110	0.050	0.026	0.018	0.105	0.037	0.011	0.003
	6.75	0.029	0.022	0.010	0.006	0.025	0.015	0.006	0.002
0.50	1.00	0.498	0.347	0.227	0.155	0.483	0.331	0.191	0.100
	3.25	0.128	0.066	0.034	0.022	0.124	0.055	0.023	0.011
	6.75	0.049	0.034	0.020	0.012	0.045	0.029	0.016	0.009
0.75	1.00	0.498	0.347	0.227	0.155	0.498	0.331	0.191	0.110
	3.25	0.146	0.082	0.042	0.026	0.142	0.073	0.035	0.019
	6.75	0.069	0.046	0.030	0.018	0.065	0.044	0.026	0.016

f_y = 60 ksi, f_c' ≤ 4,000 psi, w = 150 pcf

$h = 6½''$

P_u/φ = (coeff.) $b_1 h f_c'$

Table 23-2—Reproduced from Page 16 of EB074D

$h = 6\frac{1}{2}''$

$P_u/\varphi = (\text{coeff.})\, b_1 h f_c'$

$\rho = \dfrac{A_s \times 100}{b_1 \times h}$	End eccentricity, e, in.	$q_u/\varphi = 30$ psf Slenderness ratio, $k\ell_u/h =$				$q_u/\varphi = 45$ psf Slenderness ratio, $k\ell_u/h =$			
		20	30	40	50	20	30	40	50
0.15	1.00	0.468	0.316	0.035	–	0.438	0.110	–	–
	3.25	0.079	0.011	**	–	0.067	**	–	–
	6.75	0.016	0.005	**	–	0.014	**	–	–
0.25	1.00	0.468	0.316	0.151	0.030	0.438	0.301	0.065	–
	3.25	0.101	0.026	0.006	**	0.092	0.016	**	–
	6.75	0.024	0.014	0.004	**	0.023	0.009	**	–
0.50	1.00	0.483	0.316	0.151	0.040	0.453	0.301	0.070	0.010
	3.25	0.121	0.046	0.016	0.004	0.114	0.036	0.010	0.003
	6.75	0.042	0.028	0.013	0.003	0.040	0.024	0.009	0.002
0.75	1.00	0.498	0.316	0.151	0.050	0.468	0.301	0.070	0.020
	3.25	0.141	0.066	0.026	0.009	0.137	0.056	0.021	0.006
	6.75	0.061	0.042	0.023	0.007	0.059	0.039	0.020	0.005

$f_y = 60$ ksi $f_c' \leqslant 4,000$ psi $w = 150$ pcf

Example 23.4—Shear Design of Wall

Investigate the shear and flexural strengths for the wall shown.

h = 8 in.
$f_c' = 3000$ psi
$f_y = 60,000$ psi

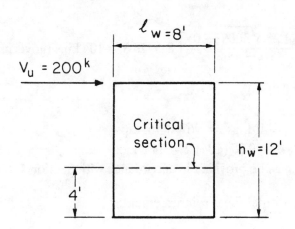

	Code
Calculations and Dsicussion	**Reference**

1. Calculate maximum shear strength permitted.

    $V_u \le \varphi V_n$ Eq. (11-1)
         $\le \varphi 10 \sqrt{f_c'}\ hd$ 11.10.3
         $\le 0.85 \times 10\ \sqrt{3000} \times 8\ (0.8 \times 96) = 286$ kips
                    200 kips $\le$ 286 kips

    8-in. wall section is adequate

2. Calculate shear strength provided by concrete V_c.   11.10.6

    Critical section for shear: 11.10.7
    $\ell_w/2 = 8/2\ = 4$ ft (governs)
    $h_w/2 = 12/2 = 6$ ft

    $$V_c = 3.3 \sqrt{f_c'}\ hd + \frac{N_u d}{4\ell_w} \qquad \text{Eq. (11-32)}$$

        $= 3.3 \sqrt{3000}\ (8)\ (76.8) + 0 = 111$ kips

    $$\text{or } V_c = d \left[0.6\sqrt{f_c'} + \frac{\ell_w \left(1.25 \sqrt{f_c'} + 0.2 N_u/\ell_w\ h\right)}{\dfrac{M_u}{V_u} - \dfrac{\ell_w}{2}} \right] hd \qquad \text{Eq. (11-33)}$$

Example 23.4—Continued

$$= \left[0.6\sqrt{3000} + \frac{96\,(1.25\sqrt{3000} + 0)}{96 - 48} \right] 8 \times 76.8 = 104 \text{ kips (governs)}$$

where, $d = 0.8\ell_w = 0.8(96) = 76.8$ in.

11.10.4

$N_u = 0$

$M_u = (12 - 4)\,V_u = 8V_u \text{ ft-kips} = 96\,V_u \text{ in.-kips}$

$V_u = 200 \text{ kips} > \varphi V_c /2 = 0.85\,(104)/2 = 44.2 \text{ kips}$

11.10.8

Shear reinforcement must be provided in accordance with Section 11.10.9.

3. Calculate required horizontal shear reinforcement.

11.10.9.1

$$V_u \;\leq\; \varphi V_n \qquad\qquad\qquad\qquad \text{Eq. (11-1)}$$
$$+\;\; \leq\; \varphi\,(V_c + V_s) \qquad\qquad \text{Eq. (11-2)}$$
$$\leq\; \varphi\,V_c + \varphi\frac{A_v\,f_y\,d}{s_2} \qquad \text{Eq. (11-34)}$$

Solving for $\dfrac{A_v}{s_2} = \dfrac{(V_u - \varphi V_c)}{\varphi f_y d}$

$$= \frac{(200 - 0.85 \times 104)}{0.85 \times 60 \times 76.8} = 0.0285$$

For
$$2\ \#3:\ s_2 = \frac{2 \times 0.11}{0.0285} = 7.72 \text{ in.}$$

$$2\ \#4:\ s_2 = \frac{2 \times 0.20}{0.0285} = 14.04 \text{ in.}$$

$$2\ \#5:\ s_2 = \frac{2 \times 0.31}{0.0285} = 21.76 \text{ in.}$$

Maximum spacing not greater than the smaller of:

11.10.9.3

$\ell_w /5 = 8 \times 12/5 = 19.2$ in.

$3h = 3 \times 8 = 24.0$ in.

or 18 in.

Use 2 #4 @ 14 in.

Check $\rho_h \geq 0.0025$

11.10.9.2

$$\rho_h = \frac{A_v}{A_g} = \frac{2 \times 0.20}{8 \times 14} = 0.0036 > 0.0025 \text{ O.K.}$$

Example 23.4 —Continued

Calculations and Discussion	Code Reference

4. Calculate vertical shear reinforcement

$$\rho_n = 0.0025 + 0.5(2.5 - \frac{h_w}{\ell_w})\,(\rho_h - 0.0025)$$ Eq. (11-35)

$$= 0.0025 + 0.5(2.5 - 1.5)\,(0.0036 - 0.0025)$$

$$= 0.0031$$

Use 2 #4 @ 14 in.

5. Design for flexure.

$$M_u = V_u h_w = 200 \times 12 = 2400 \text{ ft-kips}$$

Using Table 10-2:

Compute, $\dfrac{M_u}{\varphi f_c' b d^2} = \dfrac{2400 \times 12}{0.9 \times 3 \times 8 \times 76.8^2} = 0.2261$

where $d = 0.8\ell_w = 0.8 \times 96 = 76.8$ in.   11.10.4

A larger value of d could be used if determined by
a strain compatibility analysis.

From Table 10-2, read $\omega \approx 0.269$

$$A_s = \rho h d = \frac{\omega f_c' h d}{f_y} = \frac{0.269 \times 3 \times 8 \times 76.8}{60} = 8.26 \text{ in.}^2$$

Use 11 #8 bars each side

$$A_s = 8.69 \text{ in.}^2$$

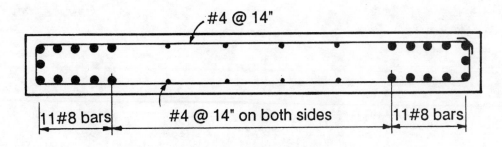

#4 @ 14"

11#8 bars   #4 @ 14" on both sides   11#8 bars

Reinforcement Details

24

FOOTINGS

GENERAL CONSIDERATIONS

Provisions of Chapter 15 apply primarily for design of footings supporting a single column (isolated footings). Chapter 15 does not provide specific design provisions for footings supporting more than one column (combined footings) other than to state that combined footings must be proportioned to resist the factored loads and induced reactions in accordance with the appropriate design requirements of the code...and where applicable the provisions of Chapter 15 for isolated footings. As a general design approach, combined footings are designed as for beams in the longitudinal direction and as an isolated footing in the transverse direction over a defined width on each side of the supported columns. Detailed discussion of combined footing design is beyond the scope of Part 24. Commentary references 15.1 and 15.2 are suggested for detailed design recommendations for combined footings.

15.2 LOADS AND REACTIONS

The first step in design is to determine the required footing base area on the basis of permissible soil bearing pressures or pile bearing loads and actual unfactored service loads in whatever combination governs the design. For the usual isolated footing, the footing is assumed to be rigid with the soil pressure considered as uniform for concentric loading and triangular or trapizoidal for eccentric loading (combined effect of axial load and bending).

When the plan dimensions of the footing have been established, the depth and required reinforcement follow. For this purpose, the contact pressures and all loads are increased by the appropriate load factors specified in Section 9.2 - Required Strength. The factored loads or related internal moments and shears are then used to proportion the footing for required shear and moment strength.

15.5 SHEAR IN FOOTINGS

Shear strength of a footing in the vicinity of the supported member (column or wall) must be determined for the more severe of the two conditions stated in Section 11.12. See discussion for Section 11.12 in Part 18.

Both wide-beam action (Section 11.12.1.1) and two-way action (Section 11.12.1.2) for the footing must be checked to determine required footing depth. Beam action assumes the footing acts as a wide beam with a critical section across its entire width. If this condition is the more severe, design for shear proceeds in accordance with Sections 11.1 through 11.5. Even though wide-beam action rarely controls the shear strength of footings, the designer must ensure that shear strength for beam action is not exceeded. Two-way action for the footing checks "punching" shear strength. The critical section for punching shear is a perimeter b_0 around the supported member with the shear strength computed in accordance with Eq. (11-36). Tributary areas and corresponding critical sections for wide-beam action and two-way action for an isolated footing are illustrtated in Fig. 24-1. Note that it is permissible to use a rectangular perimeter b_0 to define the critical section for square or rectangular columns (Section 11.12.1.3).

For ACI 318-89, a new Eq. (11-37) has been introduced to account for a decrease in shear strength affected by the ratio of the critical perimeter b_0 to the effective depth d. Rarely, however, will Eq. (11-37) govern for isolated footings supporting a single column. For typical isolated footings, the perimeter area aspect ratio $\beta_o = b_0/d$ will be considerably less than the limiting value to reduce the shear strength below the $4\sqrt{f_c'}\, b_0 d$ upper limit. (For an isolated footing supporting a single column, the $4\sqrt{f_c'}\, b_0 d$ upper limit [Eq. (11-38)] governs for $\beta_o \leq 20$) See Fig. 18-4 (Part 18) and Example 24.2.

The shear strength for two-way action [Eq. (11-36)] is a function of support size β_c, with a reduction in shear strength from $4\sqrt{f_c'}\, b_0 d$ depending upon the β_c ratio. β_c is the ratio of long-to-short side of the column or support area. Fig. 24-2 to $2\sqrt{f_c'}\, b_0 d$ illustrates the shear strength reduction as a function of β_c.

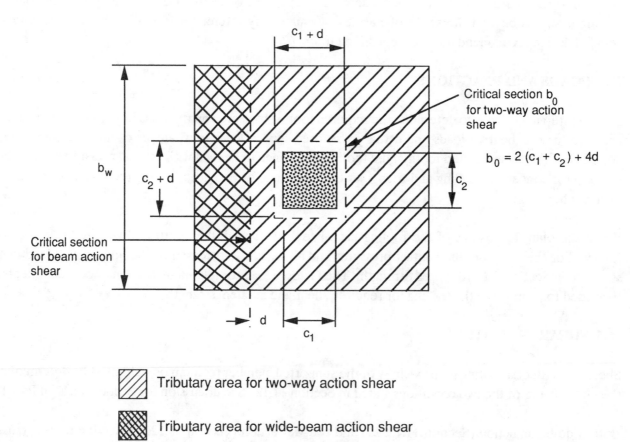

Fig. 24-1 Tributary Areas and Critical Sections for Shear in Footings

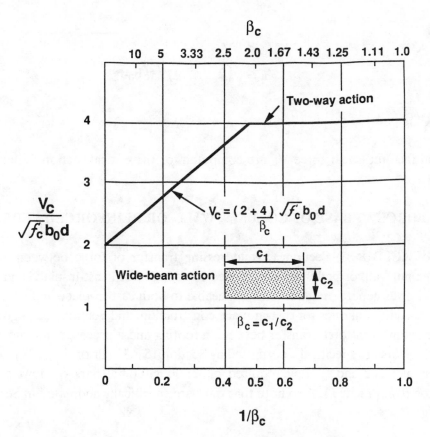

Fig. 24-2 Shear Strength of Footings

If the factored shear force V_u at the critical section exceeds the shear strength φV_c given by Eq. (11-36), shear reinforcement must be provided. If shear reinforcement consisting of bars or wires is used, the shear strength may be increased to a maximum value of $6\sqrt{f_c'}\, b_o d$. However, shear reinforcement must be designed to carry shear in excess of $2\sqrt{f_c'}\, b_o d$. This limit is one-half that permitted by Eq. (11-36) with a β_c ratio of 2 or less.

For footing design (without shear reinforcement), the shear strength equations may be summarized as follows:

Wide-Beam Action for Footing

$$V_u \leq \varphi V_n \qquad\qquad\qquad \text{Eq. (11-1)}$$
$$V_u \leq \varphi\left(2\sqrt{f_c'}\, b_w d\right) \qquad \text{Eq. (11-3)}$$

where footing width b_w and factored shear force V_u are computed for the critical section defined in Section 11.12.1.1. See Fig. 24-1.

Two-Way Action for Footing

$$V_u \leq \varphi V_n \qquad \text{Eq. (11-1)}$$
$$V_u \leq \varphi\left(2 + 4/\beta_c\right)\sqrt{f_c'}\,b_o d \qquad \text{Eq. (11-36)}$$

but not greater than $\varphi 4\sqrt{f_c'}\,b_o d$

where perimeter b_o and factored shear force V_u are computed for the critical section defined in Section 11.12.1.2. See Fig. 24-1.

15.8. TRANSFER OF FORCE AT BASE OF COLUMN, WALL, OR REINFORCED PEDESTAL

With the publication of ACI 318-83, Section 15.8 addressing transfer of force between a footing and supported member (column, wall, or pedestal) was revised to address both cast-in-place and precast construction. Section 15.8.1 gives general requirements applicable to both cast-in-place and precast construction. Section 15.8.2 gives additional rules for cast-in-place construction; and Section 15.8.3 gives additional rules for precast construction. For force transfer between a footing and a precast column or wall, anchor bolts or mechanical connectors are specifically permitted by Section 15.8.3. (Prior to the '83 Code, connection between a precast member and footing required either longitudinal bars or dowels crossing the interface, contrary to common practice.) Also note that walls are specifically addressed in Section 15.8 for force transfer to footings.

All forces applied at the base of a column or wall (supported member) must be transferred to the footing (supporting member) by bearing on concrete and/or by reinforcement. Tensile forces must be resisted entirely by reinforcement. Bearing on concrete for both supported and supporting member must not exceed the concrete bearing strength permitted by Section 10.15. See discussion on Section 10.15 in Part 6.

For a supported column

$$\varphi P_{nb} = \varphi(0.85 f_c' A_1)$$

where f_c' is strength of column concrete.

For the usual case of a supporting footing with a total area considerably greater than the supported column
($\sqrt{A_2/A_1} \geq 2$)

$$\varphi P_{nb} = 2\left[\varphi(0.85 f_c' A_1)\right]$$

where f_c' is strength of footing concrete.

When bearing strength is exceeded, reinforcement must be provided to transfer the excess load. A minimum area of reinforcement must be provided across the interface of column or wall and footing, even where concrete bearing strength is not exceeded. With the force transfer provisions addressing both cast-in-place and precast construction, including force transfer between a wall and footing, the minimum reinforcement requirements are based on the type of supported member.

CIP Columns:

$$A_s = 0.005A_s$$

CIP Walls:

$$A_s = \text{minimum vertical wall reinforcement (Section 14.3.2)}$$

P/C Columns:

$$A_s = 200A_g/f_y$$

P/C Walls:

$$A_s = 50A_g/f_y$$

For cast-in-place (CIP) construction, reinforcement may consist of extended reinforcing bars or dowels. For precast construction, reinforcement may consist of anchor bolts or mechanical connectors. Unfortunately, the Code does not give any specific data for design of anchor bolts or mechanical connectors. See Example 24.9. Note that for ACI 318-89, Section 15.8.2.3 of ACI 318-83 is deleted. This section, which limited the diameter of dowels from footing to supported member, was deleted because it was considered unduly restrictive and unnecessary. The Code does not limit dowel size at other construction joints where dowels may be used for force transfer.

The Shear-Friction Design Method of Section 11.7.4 should be used for horizontal force transfer between column and footing. See Examples 24.4 and 24.5. Consideration of some of the lateral force being transferred by shear through a formed shear key is questionable. Considerable slip is required to develop a shear key. Shear keys, if provided, should be considered as an added mechanical factor of safety only, with no design shear force assigned to the shear key. See Example 24.6.

PLAIN CONCRETE PEDESTALS AND FOOTINGS

Plain concrete pedestals and footings are designed in accordance with *ACI Building Code Requirements for Structural Plain Concrete* (ACI 318.1-89). See Example 24.8.

REFERENCE

24.1 *PCI Design Handbook—Precast and Prestressed Concrete*, 3rd Edition, Prestressed Concrete Institute, Chicago, IL, 1985.

Example 24.1—Design for Base Area of Footing

Determine the base area A_f required for a square spread footing with the following design conditions:

Service dead load = 350 kips

Service live load = 275 kips

Service surcharge = 100 psf

Assume average weight of soil and concrete above footing base = 130 pcf

Permissible soil pressure = 4.5 ksf

Column dimension = 30 in. x 12 in.

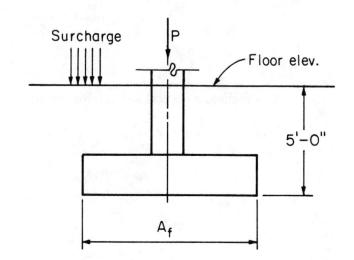

Calculations and Discussion	Code Reference

1. Total weight of surcharge:

   $0.130 \times 5 + 0.100 = 0.750$ ksf

2. Net permissible soil pressure:

   $4.5 - 0.75 = 3.75$ ksf

3. Footing base area required: 15.2.2

   $$A_f = \frac{350 + 275}{3.75} = 167 \text{ ft}^2$$

   Use 13 ft x 13 ft square footing ($A_f = 169$ ft^2)

   Note that base area of footing is determined using service loads (unfactored loads) with the permissible soil pressure.

4. Factored loads and soil reaction due to factored loading:

   $U = 1.4 (350) + 1.7 (275) = 957.5$ kips   (Eq. (9-1)

Example 24.1—Continued

Calculations and Dsicussion	Code Reference

$$q_s = \frac{U}{A_f} = \frac{957.5}{169} = 5.70 \text{ ksf}$$

To proportion the footing for strength (depth and required reinforcement) factored loads must be used.

15.2.1

Example 24.2—Design for Depth of Footing

For the design conditions of Example 24.1, determine the overall thickness of footing required.

$f_c' = 3000$ psi

$P_u = 957.5$ kips

$q_s = 5.70$ ksf

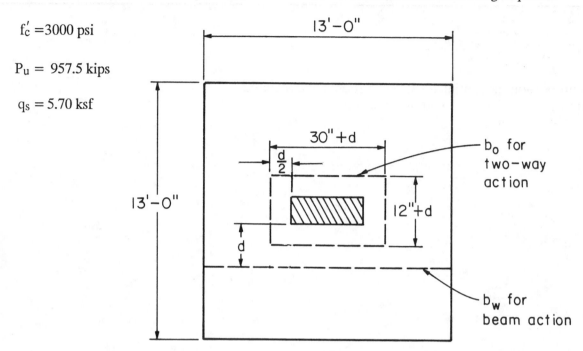

Calculations and Discussion	Code Reference

Determine depth based on shear strength without shear reinforcement. Depth required for shear usually controls footing thickness. Both wide-beam action and two-way action for the footing must be investigated. Assume overall thickness of 33 in. Average d = 28 in. | 11.12

1. Wide-beam action for footing: | 11.12.1.1

$V_u \leq \varphi V_n$ Eq. (11-1)

$\leq \varphi(2\sqrt{f_c'} b_w d)$ | Eq. (11-3)

$\leq 0.85 \quad (2\sqrt{3000} \times 156 \times 28)/1000 = 407$ kips

where $b_w = 13$ ft = 156 in.

$V_u = 5.70(6.0 - 2.33)13 = 272$ kips

272 kips < 407 kips wide-beam action shear strength is O.K.

2. Two-way action for footing: | 11.12.1.2

$V_u \leq \varphi V_n$ | Eq. (11-1)

Example 24.2—Continued

$\leq \varphi \left[(2 + 4/\beta_c)\sqrt{f_c'}\, b_o d \right]$

Eq. (11-36)

but not greater than $\varphi(4\sqrt{f_c'}\ b_o d)$

Eq. (11-38)

$\leq \varphi \left[(2 + 4/2.5)\sqrt{f_c'}\, b_o d \right] = \varphi(3.6\sqrt{f_c'}\, b_o d)$

$= 0.85\,(3.6\sqrt{3000} \times 196 \times 28)/1000 = 920$ kips

where $\beta_c = 30/12 = 2.5$

$b_o = 2(30 + 28) + 2(12 + 28) = 196$ in.

$V_u = 5.70(169 - 4.83 \times 3.33) = 872$ kips

872 kips $<$ 920 kips two-way action "shear strength" is O.K.

check Eq. (11-37):

$\varphi V_n = \varphi \left[(2 + \alpha_s/\beta_o)\sqrt{f_c'}\, b_o d \right]$

Eq. (11-37)

$\qquad = \varphi \left[(2 + 40/7)\sqrt{f_c'}\, b_o d \right]$

$\qquad = \varphi(7.7\sqrt{f_c'}\, b_o d) > \varphi(4\sqrt{f_c'}\, b_o d)$

where $\alpha_s = 40$ for interior columns, and

$\beta_o = b_o/d = 196/28 = 7$

Rarely will Eq. (11-37) govern for isolated footings supporting a single column.

Example 24.3 —Design for Footing Reinforcement

For the design conditions of Example 24.1, determine required footing reinforcement.

$f_c' = 3000$ psi

$f_y = 60,000$ psi

$P_u = 957.5$ kips

$q_s = 5.70$ ksf

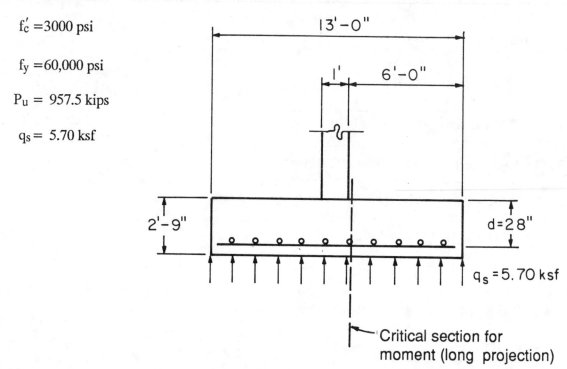

13'-0"

1' 6'-0"

2'-9" d=28"

$q_s = 5.70$ ksf

Critical section for
moment (long projection)

	Calculations and Discussion	Code Reference

1. Critical section for moment is at face of column: 15.4.2

$M_u = 5.70 \times 13 \times 6^2 /2 = 1334$ ft-kips

2. Compute A_s required using formula (4) of Part 10:

Required $R_u = \dfrac{M_u}{\varphi b d^2} = \dfrac{1334 \times 12 \times 1000}{0.9 \times 156 \times 28^2} = 145$ psi

$$\rho = \frac{0.85 f_c'}{f_y}\left(1 - \sqrt{1 - \frac{2R_u}{0.85 f_c'}}\right)$$

$$= \frac{0.85 \times 3}{60}\left(1 - \sqrt{1 - \frac{2 \times 145}{0.85 \times 3000}}\right) = 0.0025$$

Example 24.3 —Continued

	Code
Calculations and Discussion	**Reference**

Check minimum A_s required for structural slabs of uniform
thickness; for Grade 60 reinforcement:

10.5.3

$\rho_{min} = 0.0018 < 0.0025$ O.K.

7.12.2

Earlier editions of the Code were subject to interpretation regarding whether the minimum reinforcement
requirements of Section 10.5 should apply to footings. A minimum amount of reinforcement is required to
guard against a mode of failure that can occur in very lightly reinforced members; if the moment strength of
a cracked section is less than the moment strength of the uncracked section, the member will fail immediately
upon formation of a crack. It seems reasonable that this mode of failure applies to footings as well as beams.
The lesser value for slabs (Section 10.5.3) is considered adequate because of the lateral distribution of any
overload possible with slab-like members. Commentary Section 10.5.3 is more explicit, stating that slabs
(footings) which help support the structure vertically should meet the requirements of Section 10.5.3.

Required $A_s = \rho bd$

$$A_s = 0.0025 \times 156 \times 28 = 10.92 \text{ in.}^2$$

Use 14 #8 bars ($A_s = 11.06 \text{ in.}^2$) each way

Note that a lesser amount of reinforcement is required in the perpendicular direction. For ease of placement use same reinforcement each way.

3.   Check development of reinforcement.

15.6

Critical section for development is the same
as that for moment (at face of column).

15.6.3

For #8 bars: $\ell_{db} = 0.04 A_b f_y / \sqrt{f_c'} = 0.04 \, (0.79) \, (60,000) / \sqrt{3000} = 34.6 \text{in.}$

12.2.2
(Table 4-2)

Modification factor for bar spacing and cover:

Clear spacing $\approx [156 - 2(6) - 14(1.0)]/13 = 10 \text{ in.} = 10d_b$

Clear spacing $> 5d_b$

Cover $> 2d_b$

Side Cover $> 2.5 d_b \, \ell_d = 0.8 \ell_{db}$

12.2.3.4

Example 24.3—Continued

$\ell_d = 0.8(34.6) = 27.7$ in.

Check minimum development length:

$0.03 d_b f_y / \sqrt{f_c'} = 0.03 (1.00)(60,000) / \sqrt{3000} = 32.9$ in. (Table 4-4)

32.9 in. > 27.7 in.

$\ell_d = 32.9$ in. < 60 in. (short projection) O.K.

Example 24.4 —Design for Transfer of Force at Base of Column

For the design conditions of Example 24.1, check force transfer at interface of column and footing.

f_c' (column) = 5000 psi

f_c' (footing) = 3000 psi

f_y = 60,000 psi

P_u = 957.5 kips

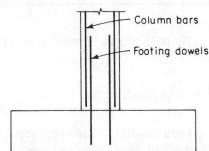

Column bars

Footing dowels

Calculations and Discussion	Code Reference

1. Bearing strength on column concrete, $f_c' = 5000$ psi: | 15.8.1.1

$\varphi P_{nb} = \varphi(0.85 f_c' A_1) = 0.70(0.85 \times 5 \times 12 \times 30) = 1071$ kips | 10.15.1

$\qquad\qquad\qquad\qquad$ 1071 kips > 957.5 kips O.K. | 9.3.2.4

2. Bearing strength on footing concrete, $f_c' = 3000$ psi | 15.8.1.1

For bearing on footing concrete, the bearing strength is increased due to the large footing area permitting a greater distribution of the column load. The increase permitted varies between 1 and 2 in accordance with the expression | 10.15.1

$$\sqrt{A_2/A_1} \le 2,$$

where A_1 is the column area (loaded area) and A_2 is the area of the lower base of the largest frustum of a pyramid, cone, or tapered wedge contained wholly within the support and having for its upper base the loaded area, and having side slopes of 1 vertical to 2 horizontal For the 30 in. x 12 in. column supported on the 13 ft x 13 ft square footing,

$$\sqrt{A_2/A_1} = \sqrt{138 \times 156/30 \times 12} = 7.7 > 2$$

Note: When the loaded area A_1 is one-fourth or less of the supporting area A_2, as for footings, the bearing strength will be increased by a factor of 2.

Also, bearing on the column concrete will always govern until the strength of the column concrete exceeds twice that of the footing concrete.

Example 24.4—Continued

	Code
Calculations and Discussion	**Reference**

$\varphi P_{nb} = 2 \left[\varphi(0.85 f_c' A_1) \right] = 2 \left[0.70 \left(0.85 \times 3 \times 12 \times 30 \right) \right] = 1285 \text{ kips}$

$> 957.5 \text{ kips} \quad \text{O.K.}$

3. Required dowel bars between column and footing: 5.8.2

Even though bearing strength on column and footing concrete   15.8.2.1
is adequate to transfer the factored loading, a minimum area
of reinforcement is required across the interface.

$A_s \text{ (min)} = 0.005 (30 \times 12) = 1.80 \text{ in.}^2$

Provide 4 #6 bars as dowels ($A_s = 1.76 \text{ in.}^2$)

4. Development of dowel reinforcement:

$\ell_d = 0.02 \, d_b f_y / \sqrt{f_c'}$ 12.3.2

but not less than $0.0003 d_b f_y$

For #6 bars   (Table 4-7):

Development length within the column,

$\ell_d = 0.02 \times 0.75 \times 60,000 / \sqrt{5000} = 12.7 \text{ in.}$

$\ell_{d \text{ (min)}} = 0.0003 \times 0.75 \times 60,000 = 13.5 \text{ in. (governs)}$

Development within the footing,

$\ell_d = 0.02 \times 0.75 \times 60,000 / \sqrt{3000} = 16.4 \text{ in. (governs)}$

$\ell_{d(\text{min})} = 0.0003 \times 0.75 \times 60,000 = 13.5 \text{ in.}$

Available length for development above footing reinforcement,

$= 33 - 3 \text{ (cover)} - 2 \times 1.0 \text{ (footing bar dia.)} - 0.75 \text{ (dowel bar dia.)} = 27.3 \text{ in.}$

$27.3 \text{ in.} > 16.4 \text{ in} \quad \text{O. K.}$

Example 24.5—Design for Transfer of Force by Reinforcement

For the design conditions given below, provide for transfer of force between column and footing.

12 in. x 12 in. tied reinforced column
with 4 #14 longitudinal bars

$f_c' = 4000$ psi (column and footing)

$f_y = 60,000$ psi

$P_d = 200$ kips

$P_l = 100$ kips

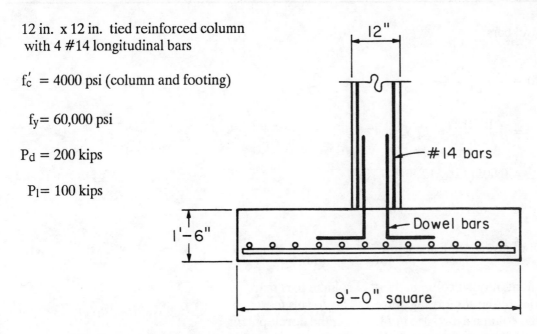

	Calculations and Discussion	Code Reference

1. Bearing strength on column concrete: 15.8.1.1

$\varphi P_{nb} = \varphi(0.85 f_c' A_1) = 0.70\ (0.85 \times 4 \times 12 \times 12)$ 10.15.1

$\qquad = 342.7$ kips

$P_u \quad = 1.4 \times 200 + 1.7 \times 100 = 450$ kips   Eq. (9-1)

450 kips $\quad > 342.7$ kips N.G.

The column load cannot be transferred by bearing on   15.8.1.2
concrete alone. The excess load $= 450 - 342.7 = 107.3$ kips
must be transferred by reinforcement.

2. Bearing strength on footing concrete: 15.8.1.1

$\sqrt{A_2/A_1}\ > 2$

$\varphi P_{nb} = 2\ (342.7) = 685.4$ kips > 450 kips O.K.

Example 24.5—Continued

	Code
Calculations and Discussion	Reference

3. Required dowel bars: 15.8.1.2

$$A_s \text{ (required)} = \frac{(P_u - \varphi P_{nb})}{\varphi f_y}$$ 9.3.2.4

$$= \frac{107.3}{0.70 \times 60} = 2.55 \text{ in.}$$

$A_s \text{ (min)} \quad = 0.005 (12 \times 12) = 0.72 \text{ in.}$ 15.8.2.1

Try 4 # 8 bars as dowels ($A_s = 3.16$ in.^{2})

4. Development of dowel reinforcement:

For development into the column, the #14 column bars may   15.8.2.4
be lap spliced with the #8 footing dowels. The dowels must
extend into the column a distance not less than the development
length of the #14 column bars or the lap splice length of the #8
footing dowels, whichever is greater.

For #14 bars: $\ell_d = 0.02 \, d_b f_y / \sqrt{f_c'}$ 12.3.2

but not less than $0.0003 d_b f_y$

$= 0.02 \times 1.693 \times 60,000 / \sqrt{4000} = 32.1 \text{ in.}$ (governs)

$\ell_{d(min)} = 0.0003 \times 1.693 \times 60,000 = 30.5 \text{ in.}$

For #8 bars:

lap length $= 0.0005 \ d_b f_y$ 12.16.1

$= 0.0005 \times 1.0 \times 60,000 = 30.0 \text{ in.} > 12 \text{ in.}$

The #8 dowel bars must extend not less than 32 in. into the column.

For development into the footing, the #8 dowels must   15.8.2.4
extend a full development length.

$\ell_d = 0.02 \times 1.0 \times 60,000 / \sqrt{4000} = 19.0 \text{ in.}$ 12.3.2

Example 24.5—Continued

| Calculations and Discussion | Code Reference |

$\ell_{d(min)}$ $0.0003 \times 1.0 \times 60,000 = 18.0$ in.

ℓ_d may be reduced to account for excess area.   12.3.3.1

A_s required / A_s provided $= 2.55/3.16 = 0.81$

$\ell_d = 19 \times 0.81 = 15.4$ in.

If the footing dowels are bent for placement on top of the footing reinforcement (as   12.1
shown in the sketch), the bent portion cannot be considered effective for developing the
bars in compression. Available length for development above footing reinforcement
$\approx 18 - 6 = 12$ in. < 15.4 in. required. Either the footing depth must be increased or a
larger number of smaller-sized dowels used.

Increase footing depth to 1'-9" and provide 4 #8 dowels, extended 32 in. into the column
and bent 90-deg. for placement on top of the footing reinforcement. Total length of #8
dowels $= 32 + 15 = 47$ in., say 4 ft.

Example 24.6 - Design for Transfer of Horizontal Force at Base of Column

For the design conditions of Example 24.5, provide for transfer of a horizontal factored force of 95 kips acting at the base of the column.

$f_c' = 4000$ psi

$f_y = 60,000$ psi

	Code
Calculations and Discussion	**Reference**

1. The shear-friction design method of Section 11.7.4   15.8.1.4

   $V_u \leq \varphi V_n$ Eq. (11-1)

   $V_n = (A_{vf} f_y \mu)$ Eq. (11-26)

   Use $\mu = 0.6$ (concrete not intentionally roughened)   11.7.4.3

   and $\varphi = 0.85$ (shear)

   $A_{vf} = \dfrac{95}{0.85 \times 60 \times 0.6} = 3.10$ in.^{2}

2. The 4 #8 footing dowels ($A_s = 3.16$ in.^{2}) provided for
   vertical force transfer may also act as shear-friction
   reinforcement. Therefore, no additional reinforcement
   is required for the horizontal force transfer. If the 4 #8
   dowels were not adequate, a 40% reduction in required
   A_{vf} is possible if the footing concrete in contact with the
   column concrete is roughened "to an amplitude of
   approximately 1/4 in." With the roughened surface, $\mu = 1.0$,
   and required $A_{vf} = 95/0.85 \times 60 \times 1.0 = 1.86$ in.^{2}
   Tensile development length of the dowels needs to be checked.

   Development within the column,

   For #8 bars: $\ell_{db} = 30$ in.   (Table 4-2)

   Clear spacing $> 2d_b$ 12.2.3.3

   Cover $> d_b$

   $\ell_d = 1.4 \ell_{db}$

   $\ell_d = 1.4(30) = 42$ in. > 28.5 in.   (Table 4-4)

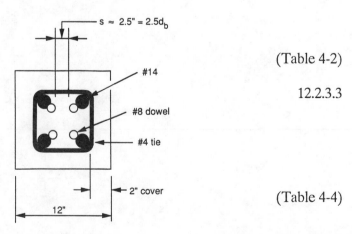

Example 24.6 —Continued

Calculations and Discussion	Code Reference

Check A_{tr} required to permit $\ell_d = 1.0\,\ell_{db}$ 12.2.3.1(b)

$A_{tr} / s = d_b\, N/40 = 1.0 \times 4/40 = 0.10$ in.^{2}

#5 ties @ 6 in. are required within development   (Table 4-3)
length to permit $\ell_d = 1.0\,\ell_{db}$

Use normal column ties (#4 @ 12 in.) with $\ell_d = 1.4\,\ell_{db} = 42$ in.
for development of #8 dowels within column.

Development within footing (use standard end hook),

For #8 bar, $\ell_{dh} = 19$ in.   (Table 4-8)

However, ℓ_{dh} can be reduced due to   12.5.3.2
favorable confinement conditions
provided by increased cover— $\ell_{dh} = 0.7\,(19) = 13.3$ in.

With 1'-9" footing depth, available depth for hook
development $= 21 - 6 = 15$ in. > 13.3 in. O.K.

Use 15 in. hook embedment into footing to secure hook
on top of footing reinforcement for placement.

Total length of #8 dowels $= 42 + 15 = 57$ in., use 4 ft - 9 in. long dowels

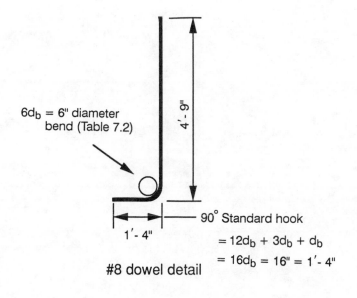

6d_b = 6" diameter
bend (Table 7.2)

4' - 9"

1' - 4"

90° Standard hook
= 12d_b + 3d_b + d_b
= 16d_b = 16" = 1'- 4"

#8 dowel detail

Example 24.6—Continued

Calculations and Discussion	Code Code Refer

3. Check maximum shear transfer strength permitted.   11.7.5

$V_u \le \varphi(0.2f_c'A_c)$ but not greater than $\varphi\,(800\,A_c)$

$\le 0.85\,(0.2 \times 4 \times 12 \times 12) = 97.9\,\text{kips},$

but not greater than $0.85\,(800 \times 12 \times 12)\,/\,1000 = 97.9\,\text{kips}$

97.9 kips < 97.9 kips O.K.

The top of the footing at the interface between column   11.7.9
and footing must be clean and free of laitance before
placement of the column concrete.

Example 24.7 —Design for Depth of Footing on Piles

For the footing supported on piles shown,
determine the required thickness of footing (pile cap).

pile diameter = 12 in.

column dimensions = 16 in. x 16 in.

f_c' = 4000 psi

Load per pile:

P_d = 20 kips

P_l = 10 kips

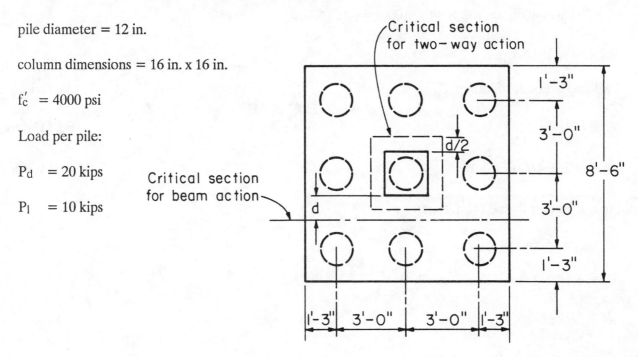

	Calculations and Discussion	Code Reference

A. Depth required for shear usually controls footing thickness. Both wide-beam action and two-way action for the footing must be investigated. Assume an overall thickness of 1 ft-9 in. Average d ≈ 14 in.

11.11

15.7

Factored pile loading:

$P_u = 1.4\,(20) + 1.7\,(10) = 45$ kips

Eq. (9-1)

1. Wide-beam action for footing:

11.12.1.1

$V_u \le \varphi V_n$

Eq. (11-1)

$V_u \le \varphi(2\sqrt{f_c'}\,b_w d)$

Eq. (11-3)

V_u (neglecting footing wt.) = 3 × 45 = 135 kips

b_w = 8 ft−6 in. = 102 in.

Example 24.7 —Continued

Calculations and Discussion	Code Reference

$V_u \le 0.85(2\sqrt{4000} \times 102 \times 14)/1000 = 153.5$ kips

135 kips < 153.5 kips   O.K.

2. Two-way action for footing: 11.12.1.2

$V_u \le \varphi V_n$ Eq. (11-1)

$V_u \le \varphi(2 + 4/\beta_c)\sqrt{f_c'}\,b_o d$ Eq. (11-36)

but not greater than $\varphi\,(4\sqrt{f_c'}\,b_o d)$

$V_u = 8 \times 45$ kips $= 360$ kips

$b_o = 4(16 + 14) = 120$ in.

       ↰ $d/2 \times 2 = \underline{\underline{d}}$

$\beta_c = 1.0$

$\varphi V_n = \varphi\,4(\sqrt{f_c'}\,b_o d)$

$V_u \le 0.85(4\sqrt{4000} \times 120 \times 14)/1000 = 361.3$ kips

360 kips < 361.3 kips O.K.

Check Eq. (11-37): $\varphi V_n = \varphi\left[(2 + \alpha_s/\beta_o)\sqrt{f_c'}\,b_o d\right]$

$\qquad\qquad\qquad = \varphi\left[(2 + 40/8.6)\sqrt{f_c'}\,b_o d\right]$ Eq. (11-37)

$\qquad\qquad\qquad = \varphi\left[(6.65\sqrt{f_c'}\,b_o d) > \varphi(4\sqrt{f_c'}\,b_o d)\right]$ Eq. (11-38)

Where $\alpha_s = 40$ for interior columns, and

$\beta_o = b_o\,/d = 120\,/\,14 = 8.6$

Rarely will Eq. (11-37) govern for isolated footings supporting a single column. Therefore, the 14 in. effective depth is adequate for shear.

Example 24.7—Continued

Calculations and Discussion

B. Check "punching" shear strength at piles.
   With piles spaced at 3 ft-0 in. on center, critical
   perimeters do not overlap.

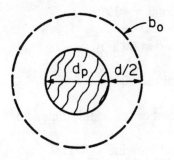

V_u = 45 kips per pile

$b_o = \pi(12 + 14) = 81.7$ in.

$V_u \le \varphi \, (4\sqrt{f_c'} \, b_o d)$ 11.11.2

$V_u \le 0.85 \times 4 \sqrt{4000} \times 81.7 \times 14 / 1000 = 45$ kips

45 kips < 246 kips O.K.

Example 24.8—Design of Plain Concrete Footing

Proportion a plain concrete square footing for the following design conditions:

Service dead load = 40 kips

Service live load = 60 kips

Service surcharge = 0

Supported member (pedestal) dimensions = 12 in. x 12 in.

Permissible soil pressure = 4.0 ksf

f_c' = 3000 psi (footing and pedestal)

Design is to be in accordance with ACI 318.1-89

Calculations and Discussion	Code Reference

1. Footing base area:

$$A_f = \frac{40 + 60}{4.0} = 25 \text{ ft}^2$$

7.2.2

Use 5 ft x 5 ft square footing ($A_f = 25 \text{ ft}^2$)

Note that the base area is determined using unfactored service loads with the permissible soil pressure. To proportion footing for strength, factored loads must be used.

U = 1.4 (40) + 1.7 (60) = 158 kips

7.2.1

$$q_s = \frac{U}{A_f} = \frac{158}{25} = 6.32 \text{ ksf}$$

2. Determine footing depth. For plain concrete, flexural strength will usually control thickness. Referring to sketch, the critical section for moment is at face of pedestal.

7.2.5

$$M_u = q_s \frac{b}{2} \left(\frac{b-c}{2}\right)^2$$

$$= 6.32 \times 2.5 \, (2)^2 = 6.32 \text{ ft-kips}$$

24-24

Example 24.8—Continued

Calculations and Discussion	**Code Reference**

$$f_t \geq \frac{M_c}{I} = \frac{6\,M_u}{bh^2}$$

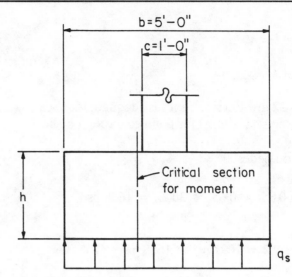

b=5'-0"

c=1'-0"

Critical section for moment

h

q_s

Permissible flexural stress:

6.2.1(a)

$$f_t = 5\,\varphi\,\sqrt{f_c'}$$

$$= 5 \times 0.65\sqrt{3000} = 178 \text{ psi}$$

6.2.2

$$178 \geq \frac{6\,(\,63.2 \times 12 \times 1000\,)}{(\,5 \times 12\,)h^2}$$

Solving for h = 20.6 in.

For concrete cast against soil, bottom 2 in. of concrete
cannot be considered for strength computations.
(The reduced overall thickness is to allow for unevenness
of excavation and for some contamination of the concrete
adjacent to the soil.) Use overall footing depth of 24 in.

6.3.5

3. Check shear strength for 24 in. footing depth.
    Use effective depth for shear $h_{eff} = 24 - 2 = 22$ in.

    The critical section for beam action is located (a distance equal to
    effective depth away from face of pedestal)
    only 2.0 − 1.83 = 0.17 ft from edge of footing; therefore,
    beam action shear is not critical.

7.2.6.2(a)

    Two-way action for footing:

7.2.6.2(b)

$$v_u = \frac{3V_u}{2b_oh} = \frac{3(107.4)1000}{2 \times 136 \times 22} = 54 \text{ psi}$$

Eq. (7-2)

where,

$$V_u = 6.32\,(\,5^2 - 2.83^2\,) = 107.4 \text{ kips}$$

$$b_o = 4\,(12 + 22) = 136 \text{ in.}$$

7.0

$$\varphi V_c = \varphi(2 + 4/\beta_c)\sqrt{f_c'} \quad \text{but not greater than } \varphi\,4\sqrt{f_c'}$$

6.2.1(c)

$$= 0.65 \times 4\sqrt{3000} = 142 \text{ psi}$$

6.2.2

Example 24.8—Continued

	Code
Calculations and Discussion	**Reference**

$v_u \leq \varphi v_c$

54 psi < 142 psi O.K.

Therefore, the 22 in. effective depth is adequate for shear.
Shear stress rarely will control for plain concrete members.

4. Bearing stress on pedestal

7.3.3

$f_b = 0.85\varphi f_c' = 0.85 \times 0.65 \times 3000 = 1658$ psi

6.2.1(d)
6.2.2

158 kips / 12 × 12 = 1097 psi < 1658 psi O.K.

Example 24.9—Design for Transfer of Force at Base of Precast Column

For the 18 in. x 18 in. precast column and base plate detail shown, check force transfer between column and pedestal for a factored load P_u = 1050 kips.

f_c' = 5000 psi (column)

f_c' = 3000 psi (pedestal)

f_y = 60,000 psi

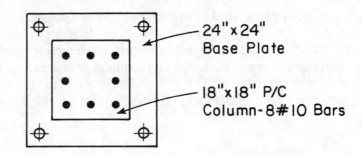

24"x24" Base Plate

18"x18" P/C Column-8#10 Bars

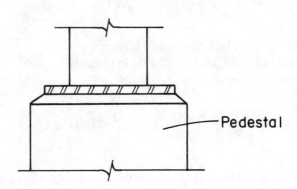

Pedestal

	Code
Calculations and Discussion	**Reference**

1. Bearing strength on column concrete (between precast column and base plate), f_c' = 5000 psi

   $$\varphi P_{nb} = \varphi(0.85 f_c' A_1)$$ 10.15.1

   $$= 0.07 \,(0.85 \times 5 \times 18 \times 18) = 964 \text{ kips} < 1050 \text{ kips}$$ 9.3.2.4

2. Bearing strength on pedestal concrete (between base plate and pedestal), f_c' = 3000 psi

   $$\varphi P_{nb} = 0.70 \,(0.85 \times 3 \times 24 \times 24) = 1028 \text{ kips} < 1050 \text{ kips}$$ 10.15.1

Example 24.9 —Continued

3. Column load cannot be transferred by bearing on concrete alone
    for either column or pedestal. The excess load between column
    and base plate (1050 – 964 = 86 kips), and between base plate and
    pedestal (1050 – 1028 = 22 kips) must be transferred by reinforcement.
    In the manufacture of precast columns it is common practice to cast
    the base plate with the column. The base plate is secured to the column
    either by deformed bar anchors (dowels) or column bars welded to the
    base plate.

    Required area of dowel bars

$$A_s \text{ (required)} = \frac{(P_u - \varphi P_{nb})}{\varphi f_y} = \frac{(1050 - 964)}{0.7 \times 60} = 2.04 \text{ in.}^2$$ 9.3.2.4

    Also, connection between precast column and base plate   15.8.3.1
    must have a tensile strength not less than 200 A_g in pounds,
    where A_g is area of precast column

$$A_s \text{ (min)} \quad = \frac{200 A_g}{f_y} = \frac{200 \times 18 \times 18}{60,000} = 1.08 \text{ in.}^2 < 2.04 \text{ in.}^2$$

    Required #5 deformed bar anchors $= \dfrac{2.04}{0.31} = 6.6$

    Development of bar anchors:

$$\ell_d = 0.02 d_b f_y / \sqrt{f_c'} = 0.02 \times 0.625 \times 60,000 / \sqrt{5000}$$ 12.3.2

$$= 10.6 \text{ in.}$$

    but not less than $0.003 d_b f_y = 0.003 \times 0.625 \times 60,000 = 11.25$ in.

    Considering excess steel area provided (say 8 #5 bar anchors)   12.3.3

$$\ell_d = 11.25 \times 2.04 / (8 \times 0.31) = 9.25 \text{ in.}$$

    Use 8 #5 x 10 in. deformed bar anchors. Anchors are
    automatically welded (similar to headed studs) to base plate.
    The base plate and bar anchor assembly is then cast with the column.

Example 24.9—Continued

4. Excess load between base plate and pedestal (1050 – 1028 = 22 kips) must also be transferred by reinforcement, with an area not less than $200A_g/f_y$.

Check 4 anchor bolts, ASTM A36 steel.

$$A_s \text{ (required)} = \frac{(1050 - 1028)}{0.7 \times 36} = 0.873 \text{ in.}^2$$

but not less than A_s (min) = 200 x 18 x 18/36,000 = 1.80 in.^{2}

Note: The code minimum of $200 A_g/f_y$ applies also to the connection between base plate and pedestal.

Required number of 3/4 in. anchor bolts = 1.80/0.44 = 4.09

Use four 3/4 in. anchor bolts

The anchor bolts must be embedded into the pedestal to develop their design strength in bond. Determine embedment length of smooth anchor bolt as 2 times the embedment length for a deformed bar.

$$\ell_d = 2 (0.02 \times 0.75 \times 36,000 / 3000) = 19.7 \text{ in.}$$

12.3.2

but not less than 2 (0.0003 × 0.75 × 36,000) = 16.2 in.

Use four 3/4 in. x 1 ft-9 in. anchor bolts. Enclose anchor bolts with 4#3 ties at 3 in. centers. See Connection Detail.

Note: The reader should refer to the PCI Design Handbook [24.1] for an in-depth treatise on design and construction details for precast column connections. Design for the base plate thickness is also addressed in the PCI Design Handbook.

Example 24.9—Continued

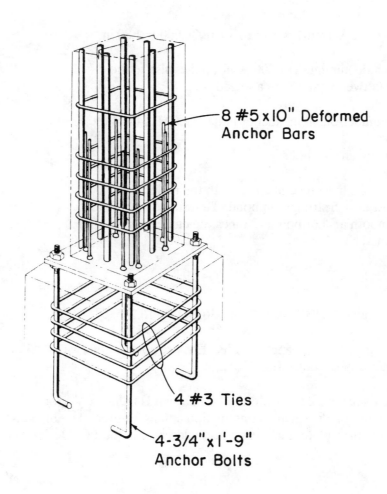

8 #5 x 10" Deformed
Anchor Bars

4 #3 Ties

4-3/4"x1'-9"
Anchor Bolts

Connection Detail

25

Precast Concrete

GENERAL CONSIDERATIONS

A concrete element is considered to be precast when the element is cast somewhere other than in its final position in the structure. With the 1983 Code Edition, the scope of Chaper 16 was revised to require that all precast concrete, whether produced in a plant or on site, meet the requirements of the Chapter. As the scope originally read, the provisions of Chapter 16 applied only to precast concrete members manufactured under plant-controlled conditions. The provisions of Chapter 16 are such that there are no extra benefits to be gained by meeting "plant-controlled conditions." Thus, the 1983 revision was actually a clarification of intent.

Design and detailing of joints and connections of precast concrete structures can be a very specialized task. Sections 16.2.2 and 16.2.4 provide typical design considerations related to joints and connections. In many cases, connections in precast concrete structures represent a discontinuity in the elastic properties of the structure. Typically, the precast members themselves are much stiffer than the joints connecting them, resulting in an abrupt change in behavior at the joints. This effect can be significant in determining the forces transferred through the joints from one member to another, as well as in computing deflections of the structure. Moment connections for rigid frame buildings often fall into this category. Figure 25-1 shows a diagram which models the above effect.

Design and use of brackets and corbels in precast concrete structures are common. The special shear provisions for brackets and corbels (Section 11.9) should be followed.

When connections are subject to repeated loads, stress reversals, or seismic conditions, the joint should be designed to maintain a ductile behavior if possible. Although connecting plates or reinforcing bars may be ductile in their behavior, the welds between them may not be. For example, the welds indicated in the typical connection detail in Fig. 25-1 may fail in a sudden mode before the top plate yields. Such welds should be increased in size to preclude brittle failure.

Since standard connection methods for typical joints may vary from one manufacturer to another, the designer may consider indicating the general scheme for connecting the elements, specifying the required forces to be transferred through the joint. If this is done, the designer should request details and calculations from the manufacturer to verify project requirements. Industry standards, such as Refs. 25.1 through 25.3, should be followed.

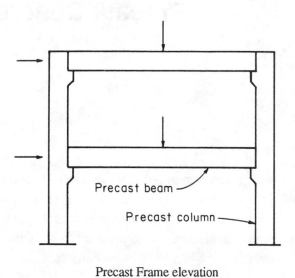

Precast Frame elevation

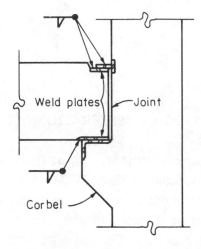

Typical Connection Detail

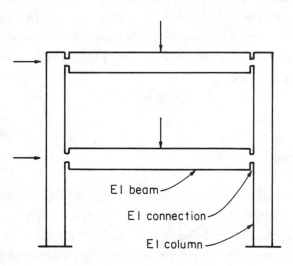

Equivalent Stiffness (EI) Diagram

Fig 25-1 Precast Frame with Moment Connections

The design of precast concrete wall panels should follow the requirements of Chapters 10 and 14 of the Code, and the recommendations given in Refs. 25.1 to 25.5. Minimum reinforcement as specified in Chapter 14 is necessary. Walls acting as deep flexural members should be analyzed as such, and should conform to all applicable provisions of the Code, especially Sections 10.7 and 11.8.

Shop drawings should include information related to the fabrication, storage, shipping, and erection of the precast elements. In addition to those items mentioned in Sections 16.4.1 and 16.6.1, the designer may require that the following items be included on the shop drawings:

(1)   Concrete strengths at removal from forms, at the time of prestressing, and at 28 days;

(2)   Handling methods and methods of support during storage, and erection; and

(3)   Initial and final camber for prestressed concrete members.

REFERENCES

25.1 *PCI Design Handbook—Precast and Prestressed Concrete,* 3rd Edition, Prestressed Concrete Institute, Chicago, IL., 1985.

25.2 *PCI Manual for Structural Design of Architectural Precast Concrete,* Prestressed Concrete Institute, Chicago, IL., 1977.

25.3 *Design and Typical Details of Connections for Precast and Prestressed Concrete,* 2nd Edition, Prestressed Concrete Institute, Chicago, IL., 1988.

25.4 *Connections for Precast-Prestressed Concrete Buildings-Including Earthquake Resistance,* Publication TR-2-82, Prestressed Concrete Institute, Chicago, IL., 1982.

25.5 *Design Connections for Precast Prestressed Concrete Buildings for the Effects of Earthquake,* Publication TR-5-86, Prestressed Concrete Institute, Chicago, IL., 1986.

Prestressed Concrete – Flexure

UPDATE FOR THE '89 CODE

For ACI 318-89, several changes, both technical and editorial, have been made to the design provisions for prestressed concrete, namely...

- 18.0 and Eq. (18-3)...For use in Eq. (18-3), an additional factor γ_p for Type II high strength deformed bars for prestressed concrete (ASTM A722) with a yield to tensile strength ratio of 0.80 has been added..."$\gamma_p = 0.55$ for f_{py}/f_{pu} not less than 0.80." Factors for all prestressing tendon materials are now included with the γ_p notation for determining the flexural strength of members with bonded prestressing tendons per Eq. (18-3). Note also the clarification in the commentary regarding use of Eq. (18-3).

- 18.1.3...Section 7.6.5 addressing maximum spacing of reinforcement for conventionally reinforced (non-prestressed) walls and slabs has been added to the list of code sections that do not apply to prestressed concrete. Maximum spacing of bonded reinforcement for prestressed two-way flat plates is addressed in Section 18.9.3.3..."spacing of bonded reinforcement shall not exceed 12 in." Maximum spacing for unbonded tendons in prestressed slab systems is addressed in Section 18.12.4..."spacing of prestressing tendons or groups of tendons in one direction shall not exceed 8 times the slab thickness, nor 5 ft."

- 18.5...The limiting tensile stress in prestressing tendons due to tendon jacking force has been lowered from $0.85f_{pu}$ to $0.80f_{pu}$, and the wording clarified to indicate the **lesser** of $0.80f_{pu}$ or the maximum value recommended by the manufacturer. The reduction from 0.85 to 0.80 f_{pu} will affect only low-relaxation strands. For low-relaxation strand, the original $0.85f_{pu}$ maximum stress due to tendon jacking force was not compatible with the tendon stress immediately after prestress transfer of $0.74f_{pu}$. With stressing equipment now in use for unbonded tendons, initial stressing to $0.85f_{pu}$ results in a tendon stress significantly greater than the $0.74f_{pu}$ value permitted immediately after prestress transfer. Using a maximum stress at jacking of $0.80f_{pu}$ will result in tendon stress at transfer approximating the limiting value of $0.74f_{pu}$.

- 18.7 and Eq. (18-3)...Use of Eq. (18-3) to compute the moment strength of a prestressed flexural member with bonded tendons is clarified in Commentary Section 18.7.2 to apply only when all of the prestressed reinforcement is located in the tension zone. When part of the prestressed reinforcement is located in the compression zone of a member cross-section, Eq. (18-3), involving d_p, is not valid. Moment strength for such a condition must be computed by a general analysis based on stress and strain compatibility, using the stress-strain properties of the prestressing tendons and the assumptions given in Section 10.2. See Example 26.4.

- 18.14...A new Section 18.14.3 has been added to specifically require corrosion protection for unbonded single strand tendons, particularly when used in corrosive environments, in accordance with the Post-Tensioning Institute (PTI) "Specification for Unbonded Single Strand Tendons."[26.3] Performance specifications are presented for tendons in normal (noncorrosive) and aggressive (corrosive) environments. The specifications are presented in a format for direct adoption as part of a project specifications with changes, additions, or deletions as necessary to adapt the specification to the specific job conditions and job specification format.

- 18.18...A seven (7) percent tolerance between force determination from elongation measurements [18.18.1(a)] and gage pressure [18.18.1(b)] has been added for post-tensioned construction. The 5 percent tolerance has been retained for pretensioned members.

The original 5 percent tolerance reflected experience primarily with pretensioned concrete elements. Subsequent experience with post-tensioned construction has indicated that, due to effects of friction loss variation along the length of post-tensioned tendons, a 7 percent value is a more reasonable tolerance between gage pressure and elongation measurements for post-tensioned construction.

- 18.19...The widely used wedge-type anchorages and couplers for strand post-tensioning tendons will not develop 100 percent f_{pu} as implied by '83 Code Section 18.9.1. New Section 18.19.1 addresses both unbonded and bonded tendons. It requires tendon anchorage and tendon coupler assemblies for both bonded and unbonded tendons to develop 95 percent of the specified ultimate strength of the tendon when tested in an unbonded condition. Note also that Section 18.19.4 is clarified to require permanent corrosion protection for not only anchorages and end fittings but also couplers.

GENERAL CONSIDERATIONS

In prestressed members, compressive stresses are introduced into the concrete to reduce tensile stresses resulting from applied loads including their own weight (dead load). Prestressing tendons such as wire, strands, or bars impart compressive stresses to the concrete. Pretensioning is a method of prestressing in which the tendons are tensioned before concrete is placed. Post-tensioning is a method of prestressing in which the tendons are tensioned after the concrete has hardened.

The act of prestressing a member introduces "prestressing loads" to the member. The induced prestressing loads, acting in conjunction with the externally applied loads, must provide serviceability and strength to the member beginning immediately after prestress force transfer and continuing throughout the life of the member.

Prestressed structures must be analyzed taking into account prestressed loads, service loads, temperature, creep, shrinkage and the structural properties of all materials involved. However, the Code indicates that empirical or simplified analytical methods should be avoided since such approximate methods may not adequately account for prestressing forces. Thus, various sections of the Code are excluded although the "entire Code applies to prestressed concrete structures" (Sections 18.1.2 and 18.1.3).

For the '89 Code, one additional code section has been added to the list of exempted sections listed in Section 18.1.3...Section 7.6.5 addressing maximum spacing of reinforcement for conventionally reinforced (nonprestressed) walls and slabs. Maximum spacing for bonded reinforcement in prestressed two-way flat plates and for unbonded tendons in prestressed slab systems is provided in Sections 18.9 and 18.12, respectively.

18.2 GENERAL

The Code specifies strength and serviceability requirements that are basic to any structure, prestressed or nonprestressed. In particular, both strength and allowable stresses must be checked. It also calls attention to certain structural aspects more common in prestressed concrete structures, such as stress concentrations at anchorages (Section 18.2.3), compatibility of deformation with the adjoining structure (Section 18.2.4), and the possibility of buckling of thin flanges as well as of any part of the member between points where concrete and prestressing tendons are in contact (Section 18.2.5). Regarding the effect of prestressing on adjoining parts of the structure, it is frequently necessary to calculate column moments due to axial shortening of prestressed floors. For the analysis of buckling, the engineer should refer to recognized texts on the subject. At the present time minimum "width to thickness ratios" are not given by the Code. In computing section properties, Section 18.2.6 requires that effect of loss of area due to open ducts for post-tensioning must be considered in design.

18.3 DESIGN ASSUMPTIONS

In applying fundamental structural principles (equilibrium, stress-strain relations, and geometric compatibility) to prestressed structures, certain simplifying assumptions can be made. For computation of strength (Section 18.3.1), the basic assumptions are the same as for nonprestressed members. An exception is that Section 10.2.4 refers only to nonprestressed reinforcement. For behavior at service conditions, the "straight-line theory" (referring to the straight line variation of stress with strain) may be used. For analysis at service conditions, the modulii of elasticity for concrete and reinforcement (nonprestressed and prestressed) are given in Section 8.5.

18.4 and 18.5 PERMISSIBLE STRESSES

Both concrete and prestressing tendon stresses are limited to ensure proper behavior at service loads and immediately after prestress transfer. The Code allows a concrete tension of $6\sqrt{f_c'}$ at the ends of simply supported members. Concrete stress limitations at service loads are based on the definition of a "precompressed" tensile concrete zone. A precompressed tensile zone is that portion of the member cross section in which flexural tension occurs under dead and live loads. Under Section 18.4.2, for a tensile stress between $6\sqrt{f_c'}$ and $12\sqrt{f_c'}$, a 50 percent increase in concrete cover is required by Section 7.7.3.2. For this stress, deflection requirements must comply with Section 9.5.4.

Deflections of prestressed members calculated according to Section 9.5.4 should not exceed the values listed in Table 9.5(c), a table new to the '89 Code. According to Section 9.5.1, prestressed concrete members, like any other concrete members, should be designed to have adequate stiffness to prevent deformations which may adversely affect the strength or serviceability of the structure.

With the 1983 Code edition, permissible tendon stresses were revised to recognize the higher yield strength of low-relaxation tendons. In terms of the specified minimum tensile strength f_{pu}, the permissible stress due to tendon jacking force for low-relaxation wire and strands was limited to $0.85f_{pu}$, with ordinary wire, strands, and bars limited to $0.80\ f_{pu}$. Subsequently, the higher permissible stress for low-relaxation tendons was determined to be incompatible with the tendon stress immediately after prestress transfer of $0.74f_{pu}$. With stressing equipment currently in use for unbonded tendons, initial stressing to $0.85f_{pu}$ results in a tendon stress at prestress transfer significantly greater than the limiting value of $0.74f_{pu}$. Consequently, for the '89 Code, the limiting stress due to tendon jacking force has been lowered from $0.85f_{pu}$ to $0.80f_{pu}$. Using a maximum stress at jacking of $0.80f_{pu}$ will result in a tendon stress at transfer approximating the limiting value of $0.74f_{pu}$. In essence, the permissible stress at jacking is now the same for both low-relaxation and ordinary tendons. Permissible stresses in prestressing tendons, in terms of the specified minimum tensile strength f_{pu}, are summarized as follows:

(a) Due to tendon jacking force:
low-relaxation wire and strands ($f_{py} = 0.90\ f_{pu}$) . . . . $0.80\ f_{pu}$*
stress-relieved wire and strands, and plain bars (ASTM A722) ($f_{py} = 0.85\ f_{pu}$) . . . . $0.80\ f_{pu}$
deformed bars (ASTM A722) ($f_{py} = 0.80\ f_{pu}$) . . . . $0.75\ f_{pu}$

(b) Immediately after prestress transfer:
low-relaxation wire and strands ($f_{py} = 0.90\ f_{pu}$) . . . . $0.74\ f_{pu}$
stress-relieved wire and strands, and plain bars ($f_{py} = 0.85\ f_{pu}$) . . . . $0.70\ f_{pu}$
deformed bars ($f_{py} = 0.80\ f_{pu}$) . . . . $0.66\ f_{pu}$

(c) Immediately after tendon anchorage . . . . $0.70\ f_{pu}$
(at anchorages and couplers)

Note that the permissible stresses given in Sections 18.5.1(a) and (b) apply to both pretensioned and post-tensioned tendons.

18.6   LOSS OF PRESTRESS

Another factor which must be considered in design of prestressed members is the loss of prestress due to various causes. These losses can drastically affect the behavior of a member at service loads. Although calculation procedures and certain values of creep strain, friction factors, etc., may be recommended, they are at best only an estimate. When designing members whose behavior (deflection in particular) is sensitive

---

*reduced from $0.85\ f_{pu}$ to $0.80\ f_{pu}$ for ACI 318-89.

to prestress losses, the engineer should establish through tests the time-dependent properties of materials to be used in the analysis/design of the structure. Refined analyses should be performed to assess the prestress losses. Specific provisions for computing friction loss in post-tensioning tendons are provided in Section 18.6.2. Allowance for other types of prestress losses are discussed in Ref. 26.1. Note that the designer is required to show on the design drawings the magnitude and location of prestressing forces as required by Section 1.2.1(g). Prior to the '83 Code edition, the acceptable ranges of tendon jacking forces and tendon elongations were also required to be placed on design drawings. Generally, the designer does not know which materials will be used, thus information on jacking forces and tendon elongations are usually given on shop drawings rather than on the design drawings.

ESTIMATING PRESTRESS LOSSES

Lump sum values of prestress losses that were widely used as a design estimate of prestress losses prior to the '83 Code edition (35,000 psi for pretensioning and 25,000 psi for post-tensioning) are now considered obsolete. Also, the lump sum values may not be adequate for some design conditions.

Reference 26.1 offers guidance to compute prestress losses and it is adaptable to computer programs. It allows step-by-step computation of losses which is necessary for rational analysis of deformations. The method is too tedious for hand calculations.

Reference 26.2 presents a reasonably accurate and easy procedure for estimating prestress losses due to various causes for pretensioned and post-tensioned members with bonded and unbonded tendons. The procedure is intended for practical design applications under normal design conditions. The various sources of loss of prestress and equations for computing each effect (taken from Ref. 26.2) are summarized below. The simple equations enable the designer to estimate the various types of prestress loss rather than using a lump sum value. The reader is referred to Ref. 26.2 for an in-depth discussion of the procedure, including sample computations for typical prestressed concrete beams.

COMPUTATION OF LOSSES

Elastic Shortening of Concrete (ES)

For members with bonded tendons:

$$ES = K_{es}E_s\frac{f_{cir}}{E_{ci}} \tag{1}$$

in which

K_{es} = 1.0 for pretensioned members

K_{es} = 0.5 for post-tensioned members where tendons are tensioned in sequential order to the same tension. With other post-tensioning procedures, the value for K_{es} may vary from 0 to 0.5.

$$f_{cir} = K_{cir}f_{cpi} - f_g \tag{2}$$

in which

$K_{cir} = 1.0$ for post-tensioned members

$K_{cir} = 0.9$ for pretensioned members.

For members with unbonded tendons:

$$ES = K_{es}E_s\frac{f_{cpa}}{E_{ci}} \tag{1a}$$

in which

f_{cpa} = average compressive stress in the concrete along the member length at the center of gravity of the tendons immediately after the prestress has been applied to the concrete.

Creep of Concrete (CR)

For members with bonded tendons:

$$CR = K_{cr}\frac{E_s}{E_c}(f_{cir} - f_{cds}) \tag{3}$$

in which

$K_{cr} = 2.0$ for pretensioned members

$K_{cr} = 1.6$ for post-tensioned members

For members made of sand lightweight concrete the foregoing values of K_{cr} should be reduced by 20 percent.

For members with unbonded tendons:

$$CR = K_{cr}\frac{E_s}{E_c}f_{cpa} \tag{3a}$$

Shrinkage of Concrete (SH)

$$SH = 8.2 \times 10^{-6} K_{sh} E_s \left(1 - 0.06\frac{V}{S}\right)(100 - RH) \tag{4}$$

26-6

in which

K_{sh} = 1.0 for pretensioned members

or

K_{sh} is taken from Table 26-1 for post-tensioned members.

Table 26-1—Values of K_{sh} for Post-Tensioned Members

Time, days*	1	3	5	7	10	20	30	60
K_{sh}	0.92	0.85	0.80	0.77	0.73	0.64	0.58	0.45

*Time after end of moist curing to application of prestress

Relaxation of Tendon Stress (RE)

$$RE = [K_{re} - J(SH + CR + ES)]C \qquad (5)$$

in which the values of K_{re}, J, and C are taken from Tables 26-2 and 26-3.

Table 26-2—Values of K_{re} and J

Type of tendon	K_{re}	J
270 Grade stress-relieved strand or wire	20,000	0.15
250 Grade stress-relieved strand or wire	18,500	0.14
240 or 235 Grade stress-relieved wire	17,600	0.13
270 Grade low-relaxation strand	5,000	0.040
250 Grade low-relaxation wire	4,630	0.037
240 or 235 Grade low-relaxation wire	4,400	0.035
145 or 160 Grade stress-relieved bar	6,000	0.05

Table 26-3—Values of C

f_{pi}/f_{pu}	Stress relieved strand or wire	Stress-relieved bar or low relaxation strand or wire
0.80		1.28
0.79		1.22
0.78		1.16
0.77		1.11
0.76		1.05
0.75	1.45	1.00
0.74	1.36	0.95
0.73	1.27	0.90
0.72	1.18	0.85
0.71	1.09	0.80
0.70	1.00	0.75
0.69	0.94	0.70
0.68	0.89	0.66
0.67	0.83	0.61
0.66	0.78	0.57
0.65	0.73	0.53
0.64	0.68	0.49
0.63	0.63	0.45
0.62	0.58	0.41
0.61	0.53	0.37
0.60	0.49	0.33

Friction

Computation of friction losses is covered in Section 18.6.2. When the tendon is tensioned, the friction losses computed can be checked with reasonable accuracy by comparing the measured elongation and the prestressing force applied by the tensioning jack.

SUMMARY OF NOTATION

A_c = area of gross concrete section at the cross section considered

A_{ps} = total area of prestressing tendons

CR = stress loss due to creep of concrete

e = eccentricity of center of gravity of tendons with respect to center of gravity of concrete at the cross section considered

E_{ci} = modulus of elasticity of concrete at time prestress is applied

E_c = modulus of elasticity of concrete at 28 days

E_s = modulus of elasticity of prestressing tendons. Usually 28,000,000 psi

ES = stress loss due to elastic shortening of concrete

f_{cds} = stress in concrete at center of gravity of tendons due to all superimposed permanent dead loads that are applied to the member after it has been prestressed

f_{cir} = net compressive stress in concrete at center of gravity of tendons immediately after the prestress has been applied to the concrete.

f_{cpa} = average compressive stress in the concrete along the member length at the center of gravity of the tendons immediately after the prestress has been applied to the concrete

f_{cpi} = stress in concrete at center of gravity of tendons due to P_{pi}

f_g = stress in concrete at center of gravity of tendons due to weight of structure at time prestress is applied

f_{pi} = stress in tendon due to P_{pi}, $f_{pi} = P_{pi}/A_{ps}$

f_{pu} = specified tensile strength of prestressing tendon, psi

I_c = moment of inertia of gross concrete section at the cross section considered

M_d = bending moment due to dead weight of member being prestressed and to any other permanent loads in place at time of prestressing

M_{ds} = bending moment due to all superimposed permanent dead loads that are applied to the member after it has been prestressed

P_{pi} = prestressing force in tendons at critical location on span after reduction for losses due to friction and seating loss at anchorages but before reduction for ES, CR, SH, and RE

RE = stress loss due to relaxation of tendons

RH = average relative humidity surrounding the concrete member (see Fig. 26-1)

SH = stress loss due to shrinkage of concrete

V/S = volume to surface ratio, usually taken as gross cross-sectional area of concrete member divided by its perimeter

18.7 FLEXURAL STRENGTH

The flexural strength of prestressed members can be calculated using the same assumptions as for non-prestressed members. Prestressing tendons, however, do not have a well defined yield point as does mild reinforcement. As a prestressed cross section reaches its flexural strength (defined by a maximum compressive concrete strain of 0.003), stress in the prestressed reinforcement at nominal strength f_{ps} will vary depending on the amount of prestressing. The value of f_{ps} can be obtained using the conditions of equilibrium, stress-strain relations, and strain compatibility. However, the analysis is quite cumbersome, especially in the case of unbonded prestressing. For bonded prestressing, the compatibility of strains can be considered at an individual section, while for unbonded prestressing, compatibility relations can be written only at the anchorage points and will depend on the entire cable profile and member loading. To avoid such lengthy calculations, the Code allows f_{ps} to be obtained by the approximate Eqs. (18-3), (18-4), and (18-5).

For members with bonded prestressing tendons, an approximate value of f_{ps} given by Eq. (18-3) may be used for flexural members reinforced with a combination of prestressed and nonprestressed reinforcement (partially prestressed members), taking into account effects of any nonprestressed tension reinforcement (ω), any compression reinforcement (ω'), concrete strength (β_1), and an appropriate factor for type of prestressing material used (γ_p). For a fully prestressed member, Eq. (18-3) reduces to:

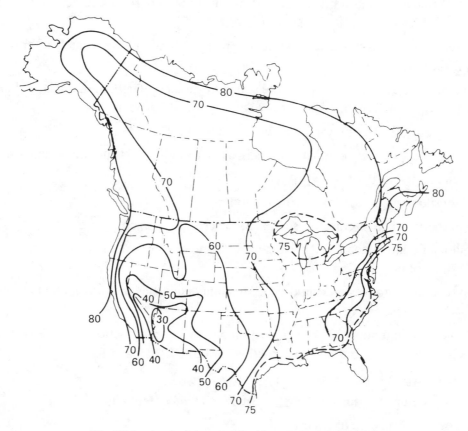

Fig. 26-1   Annual Average Ambient Relative Humidity

$$f_{ps} = f_{pu}\left(1 - \frac{\gamma_p}{\beta_1}\rho_p\frac{f_{pu}}{f_c'}\right)$$

where γ_p = 0.55 for deformed bars

$\qquad$ = 0.40 for stress−relieved wire and strands, and plain bars

$\qquad$ = 0.28 for low−relaxation wire and strands

and β_1, as defined in Section 10.2.7.3,

$\qquad \beta_1$ = 0.85 for $f_c' \leq$ 4000 psi

$\qquad$ = 0.80 for f_c' = 5000 psi

$\qquad$ = 0.75 for f_c' = 6000 psi

$\qquad$ = 0.70 for f_c' = 7000 psi

$\qquad$ = 0.65 for $f_c' \geq$ 8000 psi

Eq. (18-3) can be written in nondimensional form as follows:

$$\omega_p = \omega_{pu}\left(1 - \frac{\gamma_p}{\beta_1}\omega_{pu}\right) \qquad (6)$$

where

$$\omega_{pu} = \frac{A_{ps} f_{pu}}{bd_p f_c'} \tag{7}$$

For the '89 Code, use of Eq. (18-3) to compute the moment strength of a prestressed member with bonded tendons has been clarified in Commentary Section 18.7.2 to apply only when all of the prestressed reinforcement is located in the tension zone. When part of the prestressed reinforcement is located in the compression zone of a member cross section, Eq. (18-3), involving d_p, is not valid. Flexural strength for such a condition must be computed by a general analysis based on stress and strain compatibility, using the stress-strain properties of the prestressing tendons and the assumptions given in Section 10.2.

For members with unbonded prestressing tendons, an approximate value of f_{ps} given by Eqs. (18-4) and (18-5) may be used. Eq. (18-5) applies to members with high span-to-depth ratios, such as post-tensioned one-way slabs, flat plates and flat slabs.

Section 18.7.2 also permits a more accurate determination of f_{ps} based on a strain compatibility analysis. Design Example 26-4 illustrates the procedure.

With the value of f_{ps} known, the nominal moment strength can be calculated as follows:

$$M_n = A_{ps} f_{ps} \left(d_p - 0.59 \frac{A_{ps} f_{ps}}{bf_c'} \right) \tag{8}$$

or in nondimensional terms:

$$R_n = \omega_p (1 - 0.59 \, \omega_p) \tag{9}$$

where

$$R_n = \frac{M_n}{bd_p^2 f_c'} \text{ and } \omega_p = \frac{A_{ps} f_{ps}}{bd_p f_c'} \tag{10}$$

18.8 LIMITS FOR REINFORCEMENT OF FLEXURAL MEMBERS

The requirements for percentage of reinforcement are illustrated in Fig. 26-2. Note that reinforcement can be added to provide a reinforcement index higher than $0.36\beta_1$; however, this added reinforcement cannot be assumed to contribute to the moment strength.

Section 18.8.3 requires the total amount of prestressed and nonprestressed reinforcement of flexural members to be adequate to develop a design moment strength at least equal to 1.2 times the cracking moment strength ($\varphi M_n \geq 1.2 M_{cr}$), where M_{cr} is computed by elastic theory using a modulus of rupture equal to $7.5\sqrt{f_c'}$. The provisions of Section 18.8.3 are analogous to Section 10.5 for nonprestressed members; a

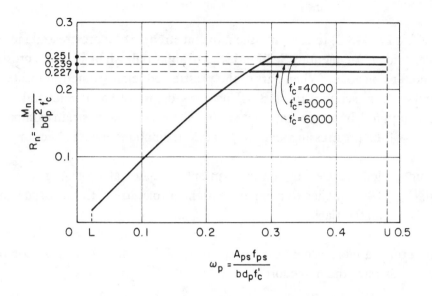

Note: The lower limit (L) is determined by the cracking moment (Section 18.8.3) and the upper limit (U) is determined by the permissible serviceability stresses (Section 18.4). Actual values for these limits differ for each problem.

Fig. 26-2   Permissible Limits of Prestressed Reinforcement and Influence on Moment Strength

precaution against abrupt flexural failure resulting from rupture of the prestressing tendons when failure occurs immediately after cracking. The provision ensures that cracking will occur before flexural strength is reached, and by a large enough margin so that significant deflection will occur to warn that the ultimate capacity is being approached. The typical prestressed member will have a fairly large margin between cracking strength and flexural strength, but the designer must be certain by checking it.

Cracking moment for a prestressed member is determined by summing all the moments that will cause a stress in the bottom fiber equal to the modulus of rupture f_r. Referring to Fig. 26-3, for a prestressed composite member

$$-\left(\frac{P_{se}}{A_c}\right) - \left(\frac{P_{se}\,e}{S_b}\right) + \left(\frac{M_d}{S_b}\right) + \left(\frac{M_a}{S_c}\right) = + f_r$$

Solving for

$$M_a = \left(f_r + \frac{P_{se}}{A_c} + \frac{P_{se}\,e}{S_b}\right) S_c - M_d\left(\frac{S_c}{S_b}\right)$$

Since

$$M_{cr} = M_d + M_a$$

$$M_{cr} = \left(f_r + \frac{P_{se}}{A_c} + \frac{P_{se}\,e}{S_b} \right) S_c - M_d \left(\frac{S_c}{S_b} - 1 \right)$$

For a prestressed member alone (without composite slab), $S_c = S_b$, therefore, M_{cr} reduces to

$$M_{cr} = \left(f_r + \frac{P_{se}}{A_c} \right) S_b + P_{se}\,e$$

Examples 26.6 and 26.7 illustrate computation of the cracking moment strength of prestressed members.

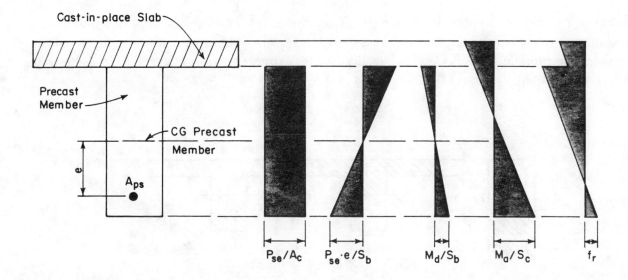

A_{ps} = area of prestressed reinforcement in tensile zone

A_c = area of precast member

S_b = section modulus for bottom of precast member

S_c = section modulus for bottom of composite member

P_{se} = effective prestress force

e = eccentricity of prestress force

M_d = dead load moment of composite member

M_a = additional moment to cause a stress in bottom fiber equal to modulus of rupture f_r

Fig. 26-3   Stress Conditions for Evaluating Cracking Moment Strength

Note that an exception waives the 1.2 times the cracking strength requirement for those cases where the strength provided is at least twice the flexural and shear strength required by Section 9.2.

For flexural strength: $\varphi M_n \geq 2M_u \geq 2(1.4M_d + 1.7M_l)$

For shear strength: $\varphi V_n \geq 2V_u \geq 2(1.4V_d + 1.7V_l)$

The $1.2M_{cr}$ provision often requires excessive reinforcement for certain prestressed flexural members specially for short span hollow-core members. The exception is intended to limit the amount of additional reinforcement required to amounts comparable to the similar requirements for nonprestressed members in Section 10.5.2.

18.9   MINIMUM BONDED REINFORCEMENT

A minimum amount of bonded reinforcement is desirable in members with unbonded tendons. Reference to Code Commentary discussion for Section 18.9 is suggested.

For all flexural members with unbonded prestressing tendons, except two-way solid slabs of uniform thickness, a minimum area of bonded reinforcement computed by Eq. (18-6) must be uniformly distributed over the precompressed tensile zone as close as practical to the extreme tension fiber. Fig. 26-4 illustrates application of Eq. (18-6).

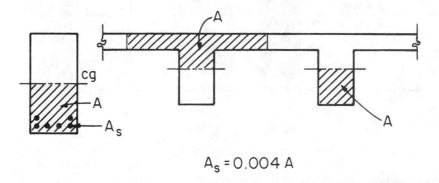

$$A_s = 0.004 \, A$$

Fig. 26-4   Bonded Reinforcement for Flexural Members

For solid slabs of uniform thickness, the special provisions of Section 18.9.3 apply. Depending on the tensile stress in the concrete at service loads, the requirements for positive moment areas of solid slabs are illustrated in Fig. 26-5(a).

Fig. 26-5(b) illustrates the minimum bonded reinforcement requirements for the negative moment areas at column supports. The bonded reinforcement must be located within the width $C_2 + 2(1.5h)$ as shown, with 4 bars minimum spaced not greater than 12 inches.

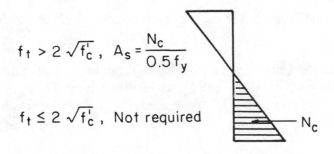

$$f_t > 2\sqrt{f_c'}, \quad A_s = \frac{N_c}{0.5 f_y}$$

$$f_t \leq 2\sqrt{f_c'}, \quad \text{Not required}$$

(a) Positive Moment Areas

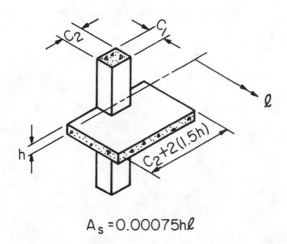

$$A_s = 0.00075h\ell$$

(b) Negative Moment Areas

Fig. 26-5 Bonded Reinforcement for Flat Plates

18.11 COMPRESSION MEMBERS—COMBINED FLEXURE AND AXIAL LOADS

Provisions of the Code for calculating the strength of prestressed compression members are the same as for members without prestressing. The only two additional considerations are (1) accounting for prestressing strains, and (2) using an appropriate stress-strain relation for the prestressing tendons. Example 26.7 illustrates the calculation procedure.

For compression members with an average concrete stress due to prestressing of less than 225 psi, minimum nonprestressed reinforcement must be provided.

REFERENCES

26.1 PCI Committee on Prestress Loses, "Recommendations for Estimating Prestress Losses," *PCI Journal,* Vol. 20, No. 4, July-August 1975, pp. 43-75.

26.2 Zia, Paul, et al., "Estimating Prestress Losses," *Concrete International: Design and Construction,* Vol. 1, No. 6, June 1979, pp. 32-38.

26.3 PTI Ad-Hoc Committee for Unbonded Single Strand Tendons, "Specification for Unbonded Single Strand Tendons," *PCI Journal,* Vol. 30, No. 2, March-April 1985, pp. 22-39.

Example 26.1—Estimating Prestress Losses

For the simply supported double-tee shown below, estimate loss of prestress using the procedures of Ref. 26.2, as outlined earlier under "Computation of Losses." Assume the unit is manufactured in Green Bay, WI.

live load = 20 psf
roof load = 20 psf
dead load = 47 psf
span = 32 ft
f'_{ci} = 3500 psi
f'_c = 5000 psi
f_{pu} = 270,000 psi (low-relaxation strands)
f_{py} = 0.90 f_{pu}
jacking stress = 0.74 f_{pu} = 200 ksi
stress at transfer = 0.70 f_{pu} = 189 ksi
force at transfer = P_i = 4(0.0845) × 189 = 63.9 kips

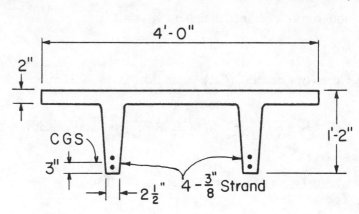

Assume the following for loss computations:
E_{ci} = 3587 ksi
E_c = 4287 ksi
E_s = 28,000 ksi

Section Properties
A_c = 180 in.^{2}
I_c = 2864 in.^{4}
y_b = 10.0 in.
y_t = 4.0 in.
A_{ps} = 0.32 in.^{2}
e = 7.0 in.

Calculations and Discussion	Code Reference

1. Elastic Shortening of Concrete (ES); using Eq. (1) of Part 26

$$ES = K_{es} \, E_s \, \frac{f_{cir}}{E_{ci}} = 1.0(28,000) \, \frac{0.6}{3587} = 4.68 \text{ ksi}$$

where $f_{cir} = K_{cir} \, f_{cpi} - f_g$

$$= K_{cir}\left(\frac{P_i}{A_c} + \frac{P_i e^2}{I_c}\right) - \frac{M_d \, e}{I_c}$$

$$= 0.9\left(\frac{63.9}{180} + \frac{63.9 \times 7^2}{2864}\right) - \frac{288.8 \times 7}{2864} = 0.60 \text{ ksi}$$

$$M_d = 0.047 \times 4 \times 32^2 \times \frac{12}{8} = 288.8 \text{ in.-kips (dead load of unit)}$$

Example 26.1—Continued

and for pretensioned members, $K_{es} = 1.0$

$$K_{cir} = 0.9$$

2. Creep of Concrete (CR); using Eq. (3)

$$CR = K_{cr}\frac{E_s}{E_c}(f_{cir} - f_{cds}) = 2.0 \times \frac{28,000}{4287}(0.60 - 0.30) = 3.92 \text{ ksi}$$

where $f_{cds} = M_{ds}\frac{e}{I} = 122.9 \times \frac{7}{2864} = 0.30 \text{ ksi}$

$$M_{ds} = 0.02 \times 4 \times 32^2 \times \frac{12}{8} = 122.9 \text{ in.–kips (roof load only)}$$

and $K_{cr} = 2.0$ for pretensioned members.

3. Shrinkage of Concrete (SH); using Eq. (4)

$$SH = 8.2 \times 10^{-6} K_{sh} E_s \left(1 - 0.06 \frac{V}{S}\right)(100 - RH)$$

$$= 8.2 \times 10^{-6} \times 28,000 (1 - 0.06 \times 1.22)(100 - 75) = 5.32 \text{ ksi}$$

where $\frac{V}{S} = \frac{180}{(4 \times 12 \times 2) + (2 \times 2) + (12 \times 4)} = 1.22$

RH = average relative humidity surrounding the concrete member from Fig. 26–1. For Green Bay, Wisconsin, RH = 75%

and $K_{sh} = 1.0$ for pretensioned members.

4. Relaxation of Tendon Stress (RE); using Eq. (5)

$$RE = [K_{re} - J(SH + CR + ES)] C$$

$$= [5 - 0.04 (5.32 + 3.92 + 4.68)] 0.95 = 4.22 \text{ ksi}$$

where, for 270 Grade low-relaxation strand:

Example 26.1—Continued

Calculations and Discussion

Code Reference

$K_{re} = 5$ ksi (Table 26−2)

$J = 0.04$ (Table 26−2)

$C = 0.95$ (Table26−3 for $\frac{f_{pi}}{f_{pu}} = 0.74$)

5. Total allowance for loss of prestress
   ES + CR + SH + RE = 4.68 + 3.92 + 5.32 + 4.22 = 18.14 ksi

   18.6.1

6. Effective prestress stress f_{se} and effective prestress force P_e

   $f_{se} = 0.74\, f_{pu} -$ allowance for all prestress losses

   $= 0.74\,(270) - 18.14 = 181.7$ ksi

   $P_e = f_{se}\, A_{ps} = 181.7 \times (4 \times 0.0845) = 61.4$ kips

For the simply supported double-tee shown below, check all permissible concrete stresses immediately after prestress transfer and at service load assuming the unit is used for roof framing.

Live load (20) + roofing (20) = 40 psf
Dead load = 47 psf
Span = 32 ft
f'_{ci} = 3500 psi
f'_c = 5000 psi
f_{pu} = 270,000 psi
(low-relaxation strands
f_{py} = 0.9 f_{pu})

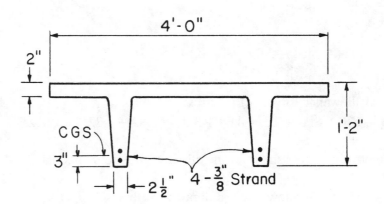

<u>Section Properties</u>
A = 180 in.^{2}
I = 2864 in.^{4}
y_b = 10.0 in.
y_t = 4.00 in.
A_{ps} = 0.32 in.^{2}

Calculations and Discussion	Code Reference
1. Calculate permissible stresses in concrete.	18.4
<u>At prestress transfer</u>	18.4.1
Tension: $6\sqrt{f'_{ci}}$ = 354 psi*	
Compression: 0.60 f'_{ci} = 2100 psi	
<u>At service load</u>	18.4.2
Tension: $6\sqrt{f'_c}$ = 424 psi	
Compression: 0.45 f'_c = 2250 psi	

2. Calculate service load moments at mid span:

$$M_d = \frac{w_d \ell^2}{8} = \frac{0.047 \times 4 \times 32^2}{8} = 24.0 \text{ ft}-\text{kips}$$

*At ends of simply supported members; otherwise $3\sqrt{f'_{ci}}$

Example 26.2—Continued

| | **Calculations and Discussion** | **Code Reference** |

$$M_{d+l} = \frac{w_{d+l}\,\ell^2}{8} = \frac{0.087 \times 4 \times 32^2}{8} = 44.6 \text{ ft}-\text{kips}$$

3. Calculate prestress force and eccentricity:
   At transfer: $P_i = 0.74 f_{pu} A_{ps} = 0.74 \times 270 \times 0.32 = 63.9$ kips   18.5.1

   At service load:
       Allowance for loss of prestress = 18.1 ksi   18.6
       (estimate of loss of prestress calculated by method of Ref. 26.2. See Example 26.1
       for calculations of prestress loss for this Example)

   Effective prestress stress: $f_{se} = 0.74\, f_{pu} -$ prestress loss
   $$= (0.74 \times 270) - 18.1 = 181.7 \text{ ksi}$$
   $P_e = f_{se} A_{ps} = 181.7 \times 0.32 = 58.1$ kips
   Eccentricity: $e = y_b - 3 = 10 - 3 = 7$ in.

4. Calculate extreme fiber stresses by "straight line theory" which leads to the following well known formulas:
   $$f_t = \frac{P}{A} + \frac{My_t}{I} - \frac{Pey_t}{I}$$
   $$f_b = \frac{P}{A} - \frac{My_b}{I} + \frac{Pey_b}{I}$$

Stresses at Prestress Transfer (psi)

	Support		Mid Span	
	Top	Bottom	Top	Bottom
P_i/A	+ 355	+ 355	+ 355	+ 355
$P_i ey/I$	− 623	+1562	− 623	+1562
My/I	—	—	+ 402	− 1006
Total	− 268 (O.K.)	+1917 (O.K.)	+ 134 (O.K.)	+ 911 (O.K.)
Permissible	− 354	+2100	+2100	+2100

Compression (+)
Tension (−)

Example 26.2—Continued

	Code
Calculations and Discussion	**Reference**

Stresses at Prestress Transfer (psi)

	Support		Mid Span	
	Top	Bottom	Top	Bottom
P_e/A	+ 323	+ 323	+ 323	+ 323
$E_e ey/I$	– 568	+1420	– 568	+1420
My/I	—	—	+ 747	– 1869
Total	– 245	+1743 (O.K.)	+ 502 (O.K.)	– 126 (O.K.)
Permissible	No limit	+2250	+2250	– 424

Compression (+)
Tension (–)

Notes:

1. The violation of permissible stresses is allowed under Section 18.4.3 in certain cases.

2. The tension stresses computed at the support will always be less than the calculated values since the assumption of plane sections remaining plane cannot hold at this point. As a rough guide, a distance d (section depth) from the end of the member is required for this assumption to apply with a good degree of accuracy.

3. In computing stresses at the support, reduction of prestress to account for the bond transfer length of the wire can be used.

Example 26.3—Flexural Strength of Prestressed Member Using Approximate Value for f_{ps}

Calculate the nominal moment strength of the prestressed member shown below.

$f'_c = 5000$ psi

$f_{pu} = 270,000$ psi (stress-relieved strands $f_{py} = 0.85\, f_{pu}$)

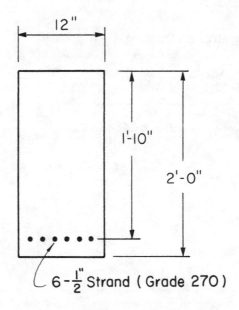

12"

1'-10"

2'-0"

$6 - \frac{1}{2}$" Strand (Grade 270)

	Code
Calculations and Discussion	**Reference**

1. Calculate stress in prestressed reinforcement at nominal strength using approximate value for f_{ps}. For a fully prestressed member, Eq. (18-3) reduces to:

$$f_{ps} = f_{pu}\left(1 - \frac{\gamma_p}{\beta_1}\rho_p\,\frac{f_{pu}}{f'_c}\right)$$

Eq. (18-3)

$$= 270\left(1 - \frac{0.40}{0.80} \times 0.00348 \times \frac{270}{5}\right) = 245 \text{ ksi}$$

where $\gamma_p = 0.40$ for $\dfrac{f_{py}}{f_{pu}} = 0.85$

18.0

$\beta_1 = 0.80$ for $f'_c = 5000$ psi

10.2.7.3

$$\rho_p = \frac{A_{ps}}{bd_p} = \frac{6 \times 0.153}{12 \times 22} = 0.00348$$

Example 26.3—Continued

2. Calculate reinforcement index ω_p from Eq. (10) of Part 26   18.8.1

$$\omega_p = \frac{A_{ps}\,f_{ps}}{b d_p\, f'_c} = \frac{0.918 \times 245}{12 \times 22 \times 5} = 0.170 \; < \; 0.36\beta_1 = 0.36(0.80) = 0.288 \quad \text{O.K.}$$

3. Calculate nominal moment strength from Eq. (3) of Part 26

$$M_n = A_{ps}\,f_{ps}\left(d_p - 0.59\frac{A_{ps}\,f_{ps}}{b\,f'_c}\right)$$

$$M_n = \frac{0.918 \times 245}{12}\left(22 - 0.59\frac{0.918 \times 245}{12 \times 5}\right) = 371 \text{ ft}-\text{kips}$$

Example 26.4—Flexural Strength of Prestressed Member Based on Strain Compatibility

The rectangular beam section shown below is reinforced with a combination of prestressed and non-prestressed strands. Calculate the nominal moment strength.

f'_c = 5000 psi
f_{pu} = 270,000 psi (low-relaxation strand; f_{py} = 0.9 f_{pu})
E_{ps} = 28,000 ksi
losses = 31.7 ksi (calculated by method of Ref. 26.2, excluding allowance for shortening of concrete. See Section 18.6—Loss of Prestress for procedure.

Calculations and Discussion	Code Reference

1. Calculate effective strain in prestressing steel.

    ε = (0.74 f_{pu} – losses)/E_{ps} = (0.74 × 270 – 31.7)/28,000 = 0.0060

    Note: This value of ε should not include losses due to shortening of concrete.

2. Draw strain diagram at nominal moment strength (defined by a maximum concrete compressive strain of 0.003). For f'_c = 5000, β_1 = 0.80.

    18.3.1

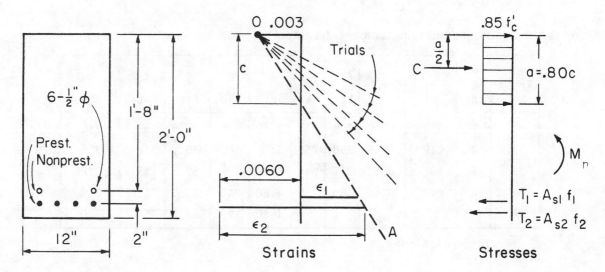

3. Obtain equilibrium of horizontal forces.

    The "strain line" drawn above from point 0, must be such that one has equilibrium of horizontal forces.

    $C = T_1 + T_2$

Example 26.4—Continued

In computing T_1 and T_2, the strains ε_1, ε_2 and stresses f_1, f_2 must satisfy the stress-strain relation for the strand (see Fig. 26-6). Equilibrium can be obtained by a trial-and-error procedure as follows:

1. assume c (location of neutral axis)

2. compute ε_1 and ε_2 (a graphical procedure is very convenient)

3. obtain f_1 and f_2 from stress-strain curve (see Fig. 26-6)

4. compute $a = \beta_1 c$

5. compute $C = 0.85 f'_c ab$

6. compute T_1 and T_2

7. check equation $C = T_1 + T_2$

8. if $C < (T_1 + T_2)$, increase C, or vice versa. Return to step 2 above and repeat until convergence.

Using this procedure the following table may be obtained:

Trial No.	c in.	ε_1	ε_2	f_1 ksi	f_2 ksi	a in.	C kips	T_1 kips	T_2 kips	$T_1 + T_2$ kips
1	3	.017	.0250	258	263	2.40	122	78.9	161	240
2	4	.012	.0195	250	259	3.20	163	76.5	158.5	235
3	5	.009	.0162	228	248	4.00	205	69.8	151.8	222
4	6	.007	.0140	180	253	4.80	245	55.1	155	210
*						4.40	220	65.0	155	220

*By interpolation

4. Calculate nominal moment strength.

Using C = 220 kips, T_1 = 65 kips and T_2 = 155 kips, the nominal moment strength can be calculated as follows:

Taking moments about T_2:

$M_n = (19.8 \times 220 - 2 \times 65)/12 = 352$ ft−kips

Example 26.4—Continued

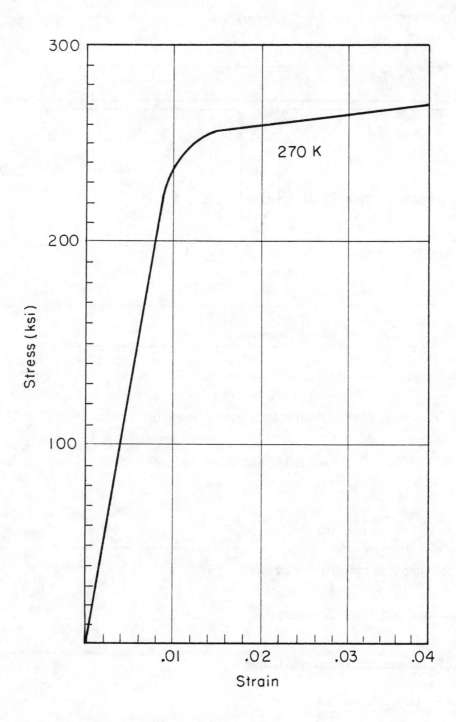

Fig. 26-6   Stress-Strain Relation for 270k Low-Relaxation Strand

Example 26.5—Limits for Reinforcement of Prestressed Flexural Member

For the single tee section shown below, check limits for the prestressed reinforcement provided.

$f_c' = 5000$ psi

$f_{pu} = 270,000$ psi (stress-relieved strands; $f_{py} = 0.85\ f_{pu}$)

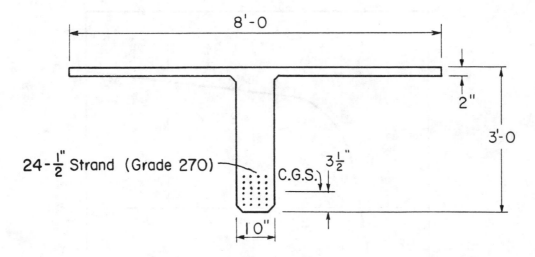

	Code
Calculations and Discussion	

<u>Example No. 26.5.1</u>

1. Calculate stress in prestressed reinforcement at nominal strength.

$$\omega_{pu} = \frac{A_{ps}f_{pu}}{bd_pf_c'} = \frac{24 \times 0.153 \times 270}{96 \times 32.5 \times 5} = 0.0636$$

$$f_{ps} = f_{pu}\left(1 - \frac{\gamma_p}{\beta_1}\omega_{pu}\right) = 270\left(1 - \frac{0.4}{0.8} \times 0.0636\right) = 261 \text{ ksi}$$

Eq. (18-3)

2. Calculate required depth of concrete stress block.

$$a = \frac{24 \times 0.153 \times 261}{96 \times 0.85 \times 5} = 2.35 \text{ in.} \quad > \quad h_f = 2 \text{ in.}$$

3. Calcuate area of reinforcement to develop flange.

$$A_{pf} = \frac{0.85h_f(b - b_w)f_c'}{f_{ps}} = \frac{0.85 \times 2(96 - 10)5}{261} = 2.8 \text{ in.}^2$$

Example 26.5—Continued

	Code
Calculations and Discussion	**Reference**

4. Calculate area of reinforcement to develop web.

$$A_{pw} = A_{ps} - A_{pf} = 24 \times 0.153 - 2.8 = 0.88 \text{ in.}^2$$

5. Check $\omega_{pw} \le 0.36\beta_1 = 0.36(0.8) = 0.288$ 18.8.1

$$\omega_{pw} = \frac{A_{pw}f_{ps}}{b_w d_p f'_c} = \frac{0.88 \times 261}{10 \times 32.5 \times 5} = 0.142 \; < \; 0.288 \qquad \text{O.K.}$$

Example No. 26.5.2

In the previous problem check the limits of reinforcement assuming a 3 in. thick flange.

1. $f_{ps} = 261$ ksi

2. $a = 2.35 \; < \; h_f = 3$ in.

   Since the stress block is entirely within the flange, the section acts effectively as a rectangular section.

3. Check limit $\omega_p \le 0.36\beta_1$ 18.8.1

$$\omega_p = \frac{A_{ps}f_{ps}}{bd f'_c} = \frac{24 \times 0.153 \times 261}{96 \times 32.5 \times 5} = 0.0615 \; < \; 0.288 \qquad \text{O.K.}$$

Note that the limitation on ω_{pw} used in Example No. 26.5.1 needs to be checked only when the depth of the equivalent stress block is larger than the depth of the flange (see Fig. 26-7).

Example 26.5—Continued

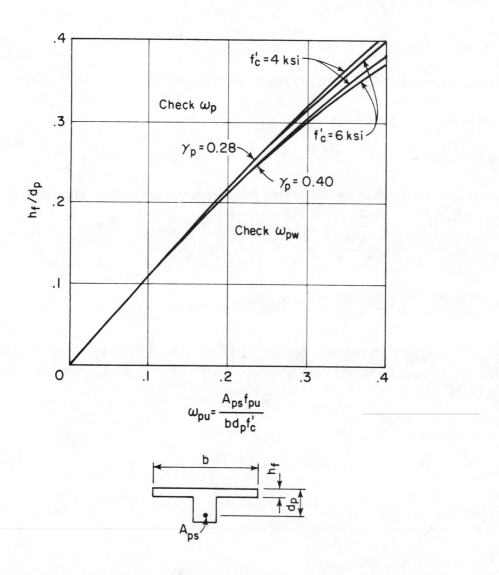

Fig. 26-7   Relation Between Flange Thickness and Steel Percentage for Controlling ω_{pw}. Based on Depth of Rectangular Stress Block

Example 26.6—Cracking Moment Strength of Prestressed Member

For the prestressed member of Example 26.3, calculate the cracking moment strength and compare it with the design moment strength.

$f_c' = 5000$ psi
$f_{pu} = 270,000$ psi
losses = assume 20%

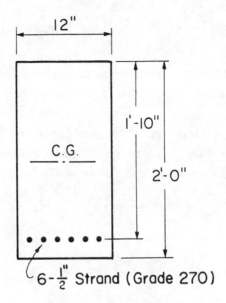

12"

1'-10"

C.G.

2'-0"

$6 - \frac{1}{2}"$ Strand (Grade 270)

Calculations and Discussion	Code Reference

1. Calculate cracking moment strength using equation developed on page 26-13.

$$M_{cr} = \left(f_r + \frac{P_{se}}{A_c} \right) S_b + \left(P_{se} \times e \right)$$

$$f_r = 7.5 \sqrt{f_c'} = 530 \text{ psi}$$ Eq. (9-9)

$$P_{se} = 0.8 \times 6 \times 0.153 \times 0.7 \times 270 = 138 \text{ kips (assuming 20\% losses)}$$

$$S_b = \frac{bh^2}{6} = \frac{12 \times 24^2}{6} = 1152 \text{ in.}^3$$

$$A_c = bh = 12 \times 24 = 288 \text{ in.}^2$$

$$e = 12 - 2 = 10 \text{ in.}$$

Example 26.6—Continued

$$M_{cr} = \left(0.530 + \frac{138}{288}\right) 1152 + (138 \times 10) = 2543 \text{ in.} - \text{kips} = 212 \text{ ft} - \text{kips}$$

Note that cracking moment strength needs to be determined for checking maximum reinforcement per Section 18.8.3.

2. Section 18.8.3 requires that the total reinforcement (prestressed and nonprestressed) must be adequate to develop a design moment strength at least equal to 1.2 times the cracking moment strength. From Example 26.3, M_n = 371 ft-kips.

$\varphi M_n \geq 1.2 M_{cr}$ 18.8.3

$0.9(371) > 1.2(212)$

$334 > 254$ O.K.

Example 26.7—Cracking Moment Strength of Prestressed Composite Member

For the 6 in. solid flat slab with 2 in. composite topping, calculate the cracking moment strength. The slab is supported on bearing walls with 15 ft span.

Section properties
per foot of width

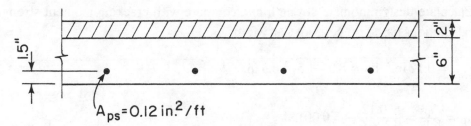

$A_c = 72$ in.^{2}

$S_b = 72$ in.^{3}

$S_c = 132.7$ in.^{3}

$w_c = 83$ psf

$A_{ps} = 0.12$ in.^{2}/ft

$f'_c = 5000$ (lightweight concrete)

$f_{pu} = 250,000$ psi

assume 25% losses

	Code
Calculations and Discussion	**Reference**

1. Calculate cracking moment strength using equation developed on page 26-13. All calculations based on per foot slab width.

$$M_{cr} = \left(f_r + \frac{P_{se}}{A_c} + \frac{P_{se}\,e}{S_b} \right) S_c - (M_{db} + M_{ds}) \left(\frac{S_c}{S_b} - 1 \right)$$

$$f_r = 0.75 \left(7.5\sqrt{5000} \right) = 398 \text{ psi}$$ 9.5.2.3

$$P_{se} = 0.75(0.12 \times 0.7 \times 250) = 15.75 \text{ kips}$$

$$e = 3 - 1.5 = 1.5 \text{ in.}$$

$$M_{db+ds} = \frac{w\,\ell^2}{8} = \frac{0.083 \times 15^2}{8} = 2.33 \text{ ft-kips} = 28.0 \text{ in.-kips}$$

$$M_{cr} = \left(0.398 + \frac{15.75}{72} + \frac{15.75 \times 1.5}{72} \right) 132.7 - 28.0 \left(\frac{132.7}{72} - 1 \right)$$

$$= 125.4 - 23.6 = 101.8 \text{ in.-kips}$$

Example 26.7—Continued

	Code
Calculations and Discussion	**Reference**

2. Calculate design moment strength and compare with cracking moment strength. All calculations based on per foot slab width.

$A_{ps} = 0.12$ in.^{2}, $d_p = 8.0 - 1.5 = 6.5$ in.

18.0

$$\rho_p = \frac{A_{ps}}{bd_p} = \frac{0.12}{12 \times 6.5} = 0.00154$$

For $f_{pu} = 250$ ksi and $f_c' = 5000$ psi, Eq. (18-3) reduces to:

$$f_{ps} = f_{pu}\left(1 - 0.5\rho_p\frac{f_{pu}}{f_c'}\right) = 250\left(1 - 0.5 \times 0.00154 \times \frac{250}{5}\right) = 240.4 \text{ ksi}$$

$$a = \frac{A_{ps}f_{ps}}{0.85f_c'b} = \frac{0.12 \times 240.4}{0.85 \times 5 \times 12} = 0.57 \text{ in.}$$

$$M_n = A_{ps}f_{ps}(d_p - a/2) = 0.12 \times 240.4(6.5 - 0.57/2) = 179.3 \text{ in.-kips}$$

$$\varphi M_n = 0.9(179.3) = 161.4 \text{ in.-kips}$$

$$\varphi M_n \geq 1.2(M_{cr})$$

18.8.3

$$161.4 > 1.2(101.8)$$

$$161.4 > 122.2 \qquad \text{O.K.}$$

Example 26.8—Prestressed Compression Member

For the short column shown below, calculate the nominal strength M_n for a nominal axial load $P_n = 30$ kips.

$f_c' = 5000$ psi
$f_{pu} = 270,000$ psi (stress-relieved strand)
losses = assume 10%

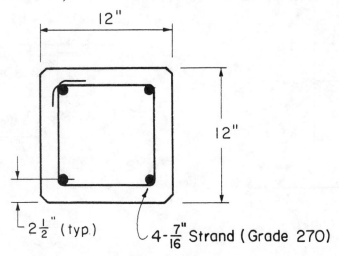

| | | Code |
| Calculations and Discussion | | Reference |

The same "strain compatibility" procedure used for flexure must be used here. The only difference is that for columns the load P_n must be included in the equilibrium of axial forces.

1. Calculate effective prestress.

$f_{se} = 0.9 \times 0.7 f_{pu} = 0.9 \times 0.7 \times 270 = 170$ ksi

$P_e = A_{ps} f_{se} = 4 \times 0.115 \times 170 = 78.2$ kips

2. Calculate average prestress on column section.   18.11.2.1

$f_{pc} = \dfrac{P_e}{A_g} = \dfrac{78.2}{12^2} = 0.542 \; > \; 0.225$ ksi

Minimum reinforcement as per Section 10.9.1 not required.

3. Calculate effective strain in prestressing steel.

$\varepsilon = \dfrac{f_{se}}{E_{ps}} = \dfrac{170}{28,000} = 0.0061$

Example 26.8—Continued

4. Draw strain diagram at nominal moment strength (defined by a maximum concrete compressive strain of 0.003). For $f_c' = 5000$ psi, $\beta_1 = 0.80$.

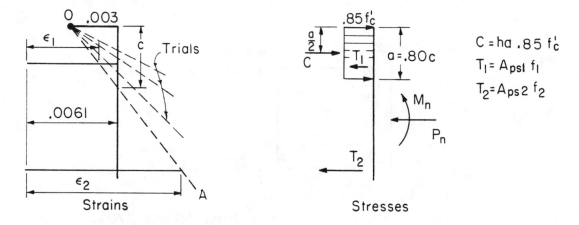

5. Obtain equilibrium of axial forces. The strain line OA drawn above, must be such that equilibrium of axial forces exists.

$C = T_1 + T_2 + P_n$

This can be done by trial-and-error as outlined in Example 26.4. Assuming different values of c, the following trial table is obtained:

6. Calculate nominal moment strength.

Using $C = 123$ kips, $P_n = 30$ kips, $T_1 = 35$ kips, and $T_2 = 58$ kips, the moment strngth can be calculated as follows:

Taking moments about P_n,

$M_n = (4.80 \times 123 + 58 \times 3.5 - 35 \times 3.5)/12 = 55.9$ ft-kips

Prestressed Concrete – Shear

UPDATE FOR THE '89 CODE

One clarification...for prestressed members, "area of shear reinforcement shall not be less than the **smaller** A_v from Eq. (11-14) and (11-15)." The '83 Code did not indicate which equation, (11-14) or (11-15), governed for minimum area of shear reinforcement. Design practice has been to use the smaller amount from these two equations, now clarified for ACI 318-89.

11.1 SHEAR STRENGTH FOR PRESTRESSED MEMBERS

Shear is expressed in terms of the factored shear force V_u directly, using the basic shear strength requirement:

$$\text{Required Shear Strength} \leq \text{Design Shear Strength}$$

$$V_u \leq \varphi V_n \qquad\qquad \text{Eq. (11-1)}$$

$$\text{where } V_n = V_c + V_s \qquad\qquad \text{Eq. (11-2)}$$

$$\text{Therefore } V_u \leq \varphi V_c + \varphi V_s$$

The nominal shear strength, V_n, is simply the sum of the shear strength provided by concrete (Section 11.4) plus the shear strength provided by shear reinforcement (Section 11.5). Beginning with the 1977 Code, shear design provisions are presented in terms of shear forces ($V_n = V_c + V_s$) to better clarify application of the material strength reduction factor φ for shear design. In force format, the φ factor is correctly applied to the material strength φV_c and φV_s directly, with only the load factor U applied to the required shear strength V_u.

11.4   SHEAR STRENGTH PROVIDED BY CONCRETE FOR PRESTRESSED MEMBERS

Similar to that of nonprestressed concrete, shear strength provided by concrete in prestressed members is presented in terms of nominal shear strength V_n, with the material understrength factor φ included with the basic shear strength Eq. (11-1). Substituting Eq. (11-2), the contribution of concrete to design shear strength is equal to φV_c where V_c is taken from the appropriate expressions of Section 11.4.

Section 11.4 is arranged with the simplified procedure (Section 11.4.1) followed by the more complex option (Section 11.4.2):

Section 11.4.1—Simplified Option

$$V_c = \left(0.6\sqrt{f_c'} + 700\,\frac{V_u d}{M_u}\right) b_w d \qquad \text{Eq. (11-10)}$$

but need not be less than
$$V_c = \left(2\sqrt{f_c'}\,b_w d\right) \qquad \text{Eq. (11-3)}$$

and
$$V_c \text{ should not exceed } \left(5\sqrt{f_c'}\,b_w d\right)$$

Section 11.4.2—More Complex Option

The lesser of
$$V_{ci} = \left[0.6\sqrt{f_c'}\,b_w d + V_d + \frac{V_i M_{cr}}{M_{max}}\right] \qquad \text{Eq. (11-11)}$$

or
$$V_{cw} = \left[\left(3.5\sqrt{f_c'} + 0.3\,f_{pc}\right)b_w d + V_p\right] \qquad \text{Eq. (11-13)}$$

11.4.1—The simplified V_c expressions of Section 11.4.1 are limited to members with an effective total prestress force at least equal to 40% of the tensile strength of the flexural reinforcement. Note that the limitation on "d" for Eq. (11-10) applies only to the $(V_u d/M_u)$ term of Eq. (11-10).

The value of "d" in the term $(b_w d)$ is as defined in Section 11.0. Note also the additional limits on Eq. (11-10) in the end regions of pretensioned members as provided in Sections 11.4.3 and 11.4.4. Actually, Section 11.4.4 is to ensure that the effect on shear strength of reduced prestress is properly taken into account when bonding of some of the tendons is prevented (debonding) near the ends of a pretensioned member, as permitted by Section 12.9.3.

11.4.2—The optional V_c expressions of Section 11.4.2 are difficult to apply without design aids, and should be used only when V_c by Section 11.4.1 is not adequate. Shear strength by Section 11.4.2 is governed by the lesser value resulting from either flexural-shear cracking (V_{ci}) or web-shear cracking (V_{cw}).

11.4.2.1—V_{ci} usually governs for members subject to uniform loading. The total shear strength V_{ci} is the sum of three parts: (1) the shear force required to transform a flexural crack into an inclined crack—

$0.6\sqrt{f_c'}\ b_w d$; (2) the **unfactored** dead load shear force—V_d; and (3) the portion of the remaining **factored** shear force that will cause a flexural crack to initially occur—$V_i\ M_{cr}/M_{max}$. Note that V_{ci} need not be taken less than $1.7\sqrt{f_c'}\ b_w d$.

For non-composite members, V_d is the shear force caused by the **unfactored** dead load. For composite members, V_d is computed using the **unfactored** selfweight plus **unfactored** superimposed dead load. The value V_i is the **factored** shear force resulting from the externally applied loads occuring simultaneously with M_{max}. V_i is determined by subtracting V_d from the shear force resulting from the total factored loads, V_u. Similarly, $M_{max} = M_u - M_d$. The load combination used to determine V_i and M_{max} is the one that causes maximum moment at the section under consideration. When calculating the cracking moment M_{cr}, the load used to determine f_d is the same unfactored load used to compute V_d.

11.4.2.2—The expression for web shear strength V_{cw} usually governs for heavily prestressed beams with thin webs, especially when the beam is subject to large concentrated loads near simple supports. Eq. (11-13) predicts the shear strength at first web-shear cracking. An alternate value of V_{cw} can be computed as the shear force corresponding to dead load plus live load that results in a principal tensile stress of $4\sqrt{f_c'}$ at the centroid of the axis of the member, or at the interface of web and flange when the centroidal axis is located in the flange. This alternate method may be advantageous when designing members where shear is critical. Note the limitation on V_{cw} in the end regions of pretensioned members as provided in Sections 11.4.3 and 11.4.4.

11.5   SHEAR STRENGTH PROVIDED BY SHEAR REINFORCEMENT FOR PRESTRESSED MEMBERS

Where the factored shear force V_u exceeds the shear strength $\varphi V_c/2$, shear reinforcement must be provided to satisfy the basic shear strength Eqs. (11-1) and (11-2). For the usual case with shear reinforcement perpendicular to the axis of the member, the design shear strength provided by shear reinforcement is equal to

$$\varphi V_s = \varphi(A_v\ f_y\ d/s) \qquad \text{Eq. (11-17)}$$

where A_v is the area of shear reinforcement within a distance s.

For design of shear reinforcement, required area A_v is computed directly from the basic shear strength Eqs. (11-1) and (11-2),

$$V_u \le \varphi V_n \qquad \text{Eq. (11-1)}$$

$$\le \varphi V_c + \varphi V_s \qquad \text{Eq. (11-2)}$$

$$V_u \le \varphi V_c + \varphi A_v\ f_y\ \frac{d}{s} \qquad \text{Eq. (11-17)}$$

Solving for

$$A_v = \frac{(V_u - \varphi V_c)s}{\varphi f_y d}$$

Note that the φ factor is correctly applied to the strength provided by concrete and shear reinforcement.

Putting it all together, the shear strength equality for the load combination of dead load and live load can be summarized as follows:

$$\text{Required Shear Strength} \le \text{Design Shear Strength}$$

$$V_u \le \varphi V_n$$

$$\le \varphi(V_c + V_s)$$

$$1.4\,V_d + 1.7\,V_\ell \le \varphi\left[\left(0.6\sqrt{f_c'} + 700\,\frac{V_u d}{M_u}\right)b_w d + A_v\,f_y\,\frac{d}{s}\right]$$

11.5.5.4—For prestressed members, minimum shear reinforcement may be computed as the smaller of either Eq. (11-14) or (11-15); however, Eq. (11-14) will generally give a higher minimum than Eq. (11-15). Note that Eq. (11-15) may not be used for members with an effective prestress force less than 40 percent of the tensile strength of the flexural reinforcement.

As permitted by Section 11.5.5.2, shear reinforcement may be omitted in any member if shown by physical tests that the required strength can be developed without shear reinforcement. Section 11.5.5.2 clarifies conditions for appropriate tests. Also, Commentary discussion gives further guidance on appropriate tests to meet the intent of Section 11.5.5.2. The Commentary also calls attention to the need for sufficient stirrups in all thin-web, post-tensioned members to support the tendons in the design profile, and to provide reinforcement for tensile stresses in the webs resulting from local deviations of the tendons from the design tendon profile.

Example 27.1—Design for Shear (Section 11.4.1)

For the prestressed single tee shown, determine shear requirements using V_c by Eq. (11-10):

Precast concrete: f'_c = 5000 psi (sand lightweight)
Topping concrete: f'_c = 4000 psi (normal weight)
Prestressing steel: Thirteen ½-in. dia. 270 ksi strands (single depression at midspan)
Span = 60 ft (simple)
Dead load = 723 lb/ft (includes topping)
Live load = 600 lb/ft
f_{se} (after all losses) = 125 ksi

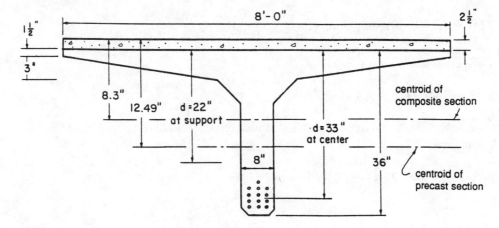

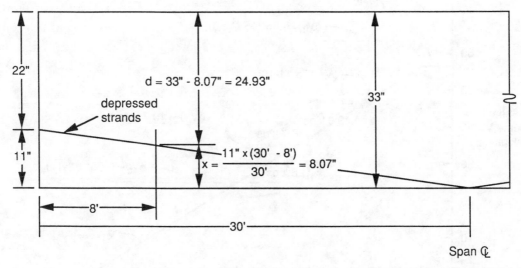

Strand Profile

Example 27.1—Continued

	Code
Calculations and Discussion	**Reference**

1. Determine factored shear force V_u at various locations along the span. The results are shown in Fig. 27-1.

2. Determine shear strength provided by concrete V_c using Eq. (11-10). The effective prestress f_{se} is greater than 40 percent of f_{pu} (125 ksi > 0.40 × 270 = 108 ksi). Note that the value of "d" need not be taken less than 0.8h for shear strength computations. Typical computations using (Eq. 11-10) for a section 8 feet from support are as follows, assuming the shear is entirely resisted by the web of the precast section:

   11.4.1

   11.0

   $w_u = 1.4 (0.723) + 1.7 (0.600) = 2.03$ kips/ft   Eq. (9-1)

   $V_u = [(60/2) - 8] 2.03 = 44.66$ kips

   $M_u = 30 \times 2.03 \times 8 - 2.03 \times 8 \times 4 = 422$ ft-kips

   For non-composite section, at 8 ft from support, distance d (centroid of tendons) = 24.93 in.; 0.8h = 28.8 in.

   For composite section, d = 27.43 in.; 0.8h = 30.8 in.

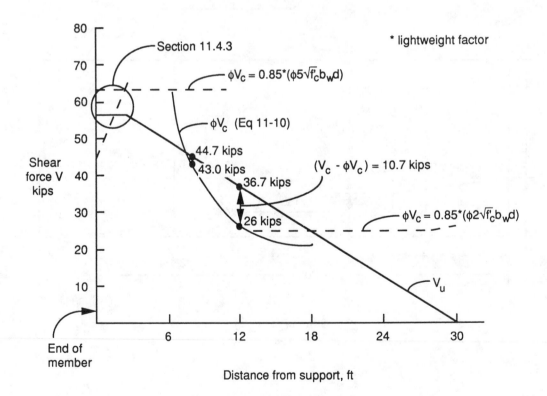

Fig. 27-1   Shear Force Variation Along Member

27-6

Example 27.1—Continued

Calculations and Discussion	Code Reference

$$V_c = \left(0.6\sqrt{f_c'} + 700\ V_u d/M_u\right) b_w d$$ Eq. (11-10)

but not less than $2\sqrt{f_c'}\ b_w d$

nor greater than $5\sqrt{f_c'}\ b_w d$

$V_c = (0.6 \times 0.85^* \sqrt{5000} + 700 \times 44.66 \times 27.43^{**}/422 \times 12)\ 8 \times 30.8$

$\quad = (36 + 169)\ 8 \times 30.8 = 50.6$ kips (governs)

$\quad \geq 2 \times 0.85\sqrt{5000} \times 8 \times 30.8 = 29.6$ kips

$\quad \leq 5 \times 0.85\sqrt{5000} \times 8 \times 30.8 = 74.0$ kips

$\varphi V_c = 0.85 \times 50.6 = 43.0$ kips (see Fig. 27-1) 11.2.1.2

Note: For members simply supported and subject to uniform loading, $V_u d/M_u$ in. Eq. (11-10) becomes a simple function of d/ℓ, where ℓ is the span length,

$$V_c = \left[0.6\sqrt{f_c'} + 700\ d\ (\ell - 2x)\ /x\ (\ell - x)\right] b_w d$$ Eq. (11-10)

where x is the distance from section being investigated to support. At 8 feet from support,

$$V_c = \left[0.6 \times 0.85\sqrt{5000} + 700 \times 27.43(60 - 16)/8(60 - 8)12\right] 8 \times 30.8 = 50.6 \text{ kips}$$

3. In the end regions of pretensioned members, the shear stength provided by concrete V_c may be limited by the provisions of Section 11.4.3. For this design, Section 11.4.3 does not apply because the section at h/2 is farther out into the span than the bond transfer length (see Fig. 27-2). The following will, however, illustrate typical calculations to satisfy Section 11.4.3. Compute V_c at 10 in. from end of member.

Bond transfer length for $\frac{1}{2}$-in. diameter strand = 50 (0.5) = 25 in.
Prestress force at 10-in. location = (10/25)125 × 0.153 × 13 = 99.4 kips

---

*factor for sand-lightweight concrete	11.2.1.2
**must use total effective d in the term $V_u d/M_u$	11.4.1

Example 27.1—Continued

Vertical component of prestress force at 10-in. location, V_p = 3.0 kips.

Distance d = 24.8 in., use 0.8h = 30.8 in.

11.4.2.3

M_d (unfactored weight of precast unit + topping) = 215.8 in.-kips

Distance of composite section centroid from the centroid of precast unit, c = 4.19 in.

Tendon eccentricity, e = 24.8 – 12.49 =

= 12.31 in. below the centroid of the precast section

f_{pc} (see notation definition) = $\dfrac{P_h}{A_g} - (P_he)\dfrac{c}{I_g} + M_d\dfrac{c}{I_g}$ = 112.1 psi

where A_g and I_g are for the precast section alone.

$V_{cw} = \left(3.5\sqrt{f_c'} + 0.3f_{pc}\right)b_wd + V_p$

Eq. (11-3)

$\quad = \left(3.5 \times 0.85\sqrt{5000} + 0.3 \times 112.1\right)8 \times 30.8 + 3000 = 63.1$ kips

$\varphi V_{cw} = 0.85 \times 63.1 = 53.6$ kips

The results of such an analysis are shown graphically in Fig. 27-2.

4. Compare factored shear V_u with shear strength provided by concrete φV_c. Where V_u > φV_c, shear reinforcement must be provided to carry the excess, otherwise provide minimum shear reinforcement.

Shear reinforcement required at 11.9 ft from support is calculated as follows:

$V_u = \left[\left(\dfrac{60}{2}\right) - 11.9\right]2.03 = 36.74$ kips

$\varphi V_c = 26.0$ kips $\quad$ (see Fig. 27–1)

Example 27.1—Continued

Calculations and Discussion

Code Reference

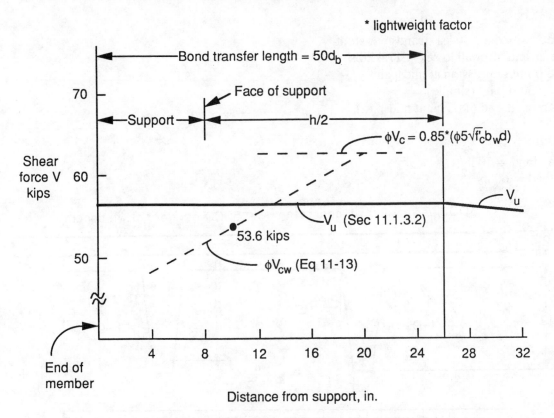

Fig. 27-2   Shear Force Variation at End of Member

$$A_v = \frac{(V_u - \varphi V_c)s}{\varphi f_y d} = \frac{(36.74 - 26.0)12}{0.85 \times 60 \times 30.8} = 0.082 \text{ in.}^2/\text{ft}$$

Check minimum required by Section 11.5.5.4. Use Eq. (11-15) since it generally requires a smaller amount for typical building members.

(Eq. (11-15)

$$A_v^{(min)} = \frac{A_{ps}}{80} \frac{f_{pu}}{f_y} \frac{s}{d} \sqrt{\frac{d}{b_w}}$$

$$= \frac{1.99}{80} \times \frac{270}{60} \times \frac{12}{30.8} \sqrt{\frac{30.8}{8}} = 0.086 \text{ in.}^2/\text{ft} \text{ (governs)}$$

Use #3 stirrups @ 18 in. for entire member length. ($A_v = 0.147$ in.^{2}/ft)

Example 27.2—Design for Shear (Section 11.4.2)

For the pretensioned double tee shown, determine shear requirements using V_c by Eqs. (11-11) and (11-13).

Prestress:
>   Three 3/8-in. dia. 250 ksi strands per stem
>   Force/strand after all losses = 11.4 kips
>   Single point depression at midspan
>   Span = 30 ft 0 in. (simple)
>   d = 8.5 in. at end (11.5 in. at midspan)

Loading:
>   Ceiling load = 10 lb/ft^2
>   Live load = 60 lb/ft^2

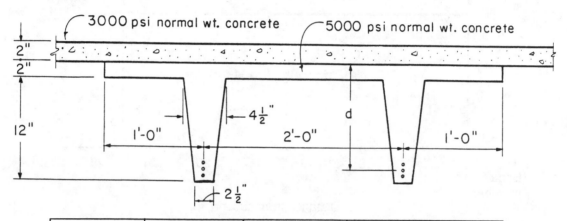

Properties	Area, in.^{2}	Weight psf	I, in.^{4}	Y_b, in.
Precast	180	47	2,864	10.0
Composite		25	4,203*	11.45*

* Corrected for difference in concrete strengths.

	Code
Calculations and Discussion	Reference

1. Determine factored shear V_u at various locations along the span. The results are shown in Fig. 27-3.

2. Determine shear strength provided by concrete V_{ci} using Eq. (11-11). Note that the value of "d" need not be taken less than 0.8h for shear strength computations. Values of V_d and M_d [f_d in Eq. (11-12)] are for **unfactored** dead load (member dead load plus   11.4.2.3

Example 27.2—Continued

Calculations and Discussion

Code Reference

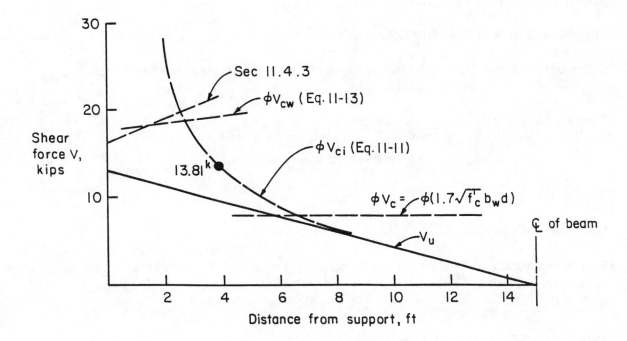

Fig. 27-3   Shear Force Variation Along Member

superimposed dead load for composite members). The value of V_i is equal to the shear caused by total factored load minus V_d. Likewise, $M_{max} = M_u - M_d$. For f_d in Eq. (11-12), the designer needs to know the construction sequence. For an unshored composite member, the beam and slab dead loads are resisted by the precast unit while the ceiling loads are resisted by the composite section. Typical computations for a section 4 feet from support are as follows. Assume the shear is resisted by the web of the precast section.

$w_d = 47 + 25 + 10 = 82 \ lb/ft^2$

$w_u = 1.4 \ (82 \times 4) + 1.7 \ (60 \times 4) = 867 \ lb/ft$ Eq. (9-1)

Unfactored dead load shear $V_d = 0.082 \times 4 \ (15 - 4) = 3.61$ kips

Total factored shear $V_u = 0.867 \ (15 - 4) = 9.54$ kips

$V_i = V_u - V_d = 9.54 - 3.61 = 5.93$ kips

Total factored moment $M_u = 0.867 \times 15 \times 4 - 0.867 \times 4^2/2 = 45.1$ ft-kips

Example 27.2—Continued

Unfactored dead load moment

$M_d = (0.082 \times 4)15 \times 4 - (0.082 \times 4) \times 4^2/2 = 17.1 \, \text{ft−kips}$

$M_{max} = M_u - M_d = 45.1 - 17.1 = 28.0 \, \text{ft-kips}$

Note that both V_i and M_{max} result from the total factored loads less the unfactored dead loads (member plus superimposed dead load).

Dead load moment carried by precast section:

$M_{d1} = (47 + 25)4 \times 15 \times 4 - (47 + 25)4 \times 4^2/2 = 14.9 \, \text{ft−kips}$

Dead load stress:

$f_{d1} = M_{d1} \, Y_b/I_g = 14.9 \times 12 \times 10/2864 = 0.627 \, \text{ksi}$

For composite sections, dead load moments carried by precast unit at any section consists of dead load of slab and beam only. Dead load moment carried by composite section consists of ceiling dead load:

$M_{d2} = (10 \times 4)15 \times 4 - (10 \times 4)4^2/2 = 2.08 \, \text{ft−kips}$

Dead load stress:

$f_{d2} = M_{d2} \, Y_{bc}/I_{gc} = 2.08 \times 12 \times 11.45/4203 = 0.068 \, \text{ksi}$

Therefore, dead load stress $f_d = f_{d1} + f_{d2} = 0.695 \, \text{ksi}$.

Eccentricity of prestressing tendons:

$e = 10 - [14 - (4 \times 3/15 + 8.5)] = 5.3 \, \text{in.}$

Horizontal component of prestressing force $P_h = 68 \, \text{kips}$

Concrete stress due to prestress:

$f_{pe} = \dfrac{P_h}{A_g} + P_h e \dfrac{Y_b}{I_g} = \dfrac{68}{180} + 68 \times 5.3 \times \dfrac{10}{2864} = 1.64 \, \text{ksi}$

Cracking moment:

$$M_{cr} = \left(\frac{I}{Y_t}\right)\left(6\sqrt{f_c'} + f_{pe} - f_d\right)$$
Eq. (11-12)

Example 27.2—Continued

$$= \left(\frac{4203}{11.45}\right) \frac{(6\sqrt{5000} + 1640 - 695)}{12} = 41.75 \text{ ft-kips}$$

Note I and Y_t are properties of composite section.

Average width of two stems $b_w = 7$ in.

Effective depth $d = 4 \times \dfrac{3}{15} + 8.5 + 2 = 11.3$ in.

Distance $0.8 h = 12.8$ in. (governs)   11.4.2.3

$$V_{ci} = 0.6\sqrt{f_c'}\, b_w d + V_d + \frac{V_i M_{cr}}{M_{max}}$$ Eq. (11-11)

but not less than $1.7\sqrt{f_c'}\, b_w d$

$$V_{ci} = 0.6\sqrt{5000} \times 7 \times 12.8 + 3.61 + \frac{5.93 \times 41.75}{28.0}$$

$$= 3.80 + 3.61 + 8.84 = 16.25 \text{ kips} \quad \text{(governs)}$$

$$\geq 1.7\sqrt{5000} \times 7 \times 12.8 = 10.77 \text{ kips}$$

$\varphi V_{ci} = 0.85 \times 16.25 = 13.81$ kips (see Fig. 27-3)

3. Determine shear strength provided by concrete V_{cw} using Eq. (11-13). Note the use of a reduced prestress force in the end regions (Section 11.4.3). Computations using Eq. (11-13) are similar to those presented for Design Example 27.1. The results of such an analysis are shown graphically in Fig. 27-3.

4. Compare factored shear force V_u with shear strength provided by concrete φV_c. Where $V_u > \varphi V_c$, shear reinforcement must be provided to carry the excess, otherwise provide minimum shear reinforcement. Referring to Fig. 27-3, only minimum shear reinforcement by Section 11.5.5.4 is required for this design. Using Eq. (11-15):

Eq. (11-15)

$$A_v = \frac{A_{ps}}{80} \frac{f_{pu}}{f_y} \frac{s}{d} \sqrt{\frac{d}{b_w}}$$

$$= \frac{6 \times 0.08}{80} \times \frac{250}{60} \times \frac{12}{12.8} \sqrt{\frac{12.8}{7}} = 0.032 \text{ in.}^2/\text{ft} = 0.016 \text{ in.}^2/\text{ft/stem}$$

Use 6 x 6 W1.4 x W1.4 W.W.F. $(A_v = 0.028 \text{ in.}^2/\text{ft})$

17.5.2.1

Example 27.2—Continued

	Code
Calculations and Discussion	**Reference**

5. Check horizontal shear strength between precast unit and topping slab.

$V_u = 0.867 (15 - 12.8/12) = 12.1$ kips

$\varphi V_{nh} = \varphi(80b_w d)$ 17.5.2.1

     $= 0.85(80 \times 48 \times 12.8)/1000 = 41.8$ kips

$V_u \leq \varphi V_{nh}$ Eq. (17-1)

12.1 kips < 41.8 kips   O.K.

Contact surface must be clean, free of laitance, and intentionally roughened (broom finish).

Prestressed Slab Systems

INTRODUCTION

Four Code sections are significant with respect to analysis and design of prestressed slab systems:

Section 11.12.2 – Shear strength of prestressed slabs.

Section 11.12.6 – Shear strength of prestressed slabs with moment transfer.

Section 18.7.2 – f_{ps} for calculation of flexural strength.

Section 18.12   – Prestressed Slab systems.

Discussion of each of these Code sections is presented below, followed by Example 28.1 on a post-tensioned flat plate. The design example illustrates application of the above Code sections as well as general applicability of the Code to analysis and design of post-tensioned flat plates.

11.12.2 Shear Strength

Section 11.12.2 includes specific provisions for calculation of shear strength in two-way prestressed concrete systems. At columns of two-way prestressed slabs (and footings) utilizing unbonded tendons and meeting the bonded reinforcement requirements of Section 18.9.3, the shear strength V_n must not be taken greater than the shear strength V_c computed in accordance with Sections 11.12.2.1 or 11.12.2.2, unless shear reinforcement is provided in accordance with Section 11.12.3 or 11.12.4. Section 11.12.2.2 gives the following value of the shear strength V_c at columns of two-way prestressed slabs:

$$V_c = (\beta_p\sqrt{f_c'} + 0.3f_{pc})b_od + V_p \qquad \text{Eq. (11-39)}$$

For ACI 318-89, Eq. (11-39) includes the term β_p as the smaller of 3.5 or $(1.5 + \alpha_s/\beta_o)$. The term α_s/β_o is to account for a decrease in shear strength affected by the perimeter area aspect ratio β_o, where β_o is the ratio of the critical perimeter b_o to effective depth d and α_s is to be taken as 40 for interior columns, 30 for edge columns, and 20 for corner columns. f_{pc} is the average value of f_{pc} for the two directions, and V_p is the vertical component of all effective prestress forces crossing the critical section. If shear strength is computed by Eq. (11-39), the following must be satisfied; otherwise, Section 11.12.2.1 for nonprestressed slabs applies:

(a) no portion of the column cross section can be closer to a discontinuous edge than 4 times the slab thickness,

(b) f_c' must not be taken greater than 5000 psi, and

(c) f_{pc} in each direction must not be less than 125 psi, nor be taken greater than 500 psi.

The revised Eq. (11-39) is a modified form of the new shear strength Eq. (11-37) for nonprestressed slabs. Discussion on Eq. (11-37) is presented in Part 18.

In accordance with the above limitations, shear strength Eqs. (11-36) and (11-37) for nonprestressed slabs are applicable to columns closer to the discontinuous edge than 4 times the slab thickness. Shear strength V_c is the lesser of values given by Eq. (11-36) and (11-37), but not greater than $4\sqrt{f_c'}\,b_o d$. For usual design conditions (slab thicknesses and column sizes), the controlling shear strength at edge columns will be $4\sqrt{f_c'}\,b_o d$.

11.12.6 Shear Strength with Moment Transfer

For moment transfer calculations, the controlling shear stress at columns of two-way prestressed slabs with bonded reinforcement in accordance with Section 18.9.3 is governed by Eq. (11-39) as follows:

$$v_c = (\beta_p\sqrt{f_c'} + 0.3f_{pc} + V_p/b_o d) \qquad \text{Eq. (11-39)}$$

If permissible shear stress is computed by Eq. (11-39), the following must be satisfied;

(a) no portion of column cross section can be closer to a discontinuous edge than 4 times the slab thickness,

(b) f_c' must not be taken greater than 5000 psi, and

(c) f_{pc} in each direction can not be less than 125 psi, nor be taken greater than 500 psi.

For edge columns under moment transfer conditions, the controlling shear stress will be as permitted for nonprestressed slabs. For usual design conditions, the governing shear stress at edge columns will be $\varphi\,4\sqrt{f_c'}$.

18.7.2 f_{ps} for Unbonded Tendons

In prestressed elements with unbonded tendons having a span/depth ratio greater than 35, the stress in prestressing steel at nominal strength is given by:

$$f_{ps} = f_{se} + 10{,}000 + f_c'/300\rho_p \qquad \text{Eq. (18-5)}$$

but not greater than f_{py}, nor (f_{se} + 30,000).

Nearly all prestressed one-way slabs and flat plates will have span/depth ratios greater than 35. Equation (18-5) provides values of f_{ps} which are generally 15,000 to 20,000 psi lower than the values of f_{ps} given by Eq. (18-4) which was devised primarily from results of beam tests. These lower values of f_{ps} are more compatible with values of f_{ps} obtained in more recent tests of prestressed one-way slabs and flat plates. Application of Eq. (18-5) is illustrated in Example 28.l.

18.12 SLAB SYSTEMS

Section 18.12 incorporates analysis and design procedures for two-way prestressed slab systems as follows:

(1)   Use of the Equivalent Frame Method of Section 13.7, or more detailed analysis procedures, is required for determination of factored moments and shears in prestressed slab systems.

(2)   Spacing of tendons in one direction must not exceed 8 times the slab thickness nor 5 ft. Spacing of tendons must also provide a minimum average prestress, after allowance for all prestress losses, of 125 psi on the slab section tributary to the tendon or tendon group. Special consideration must be given to tendon spacing in slabs with concentrated loads.

(3)   A minimum of two tendons must be provided in each direction through the critical shear section over columns. This provision, in conjunction with the limits on tendon spacing outlined in item 2 above, provides specific guidance for distributing tendons in prestressed flat plates in accordance with the

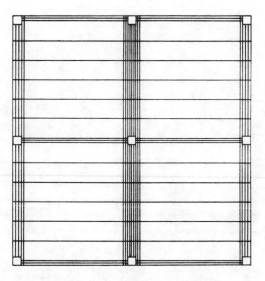

Fig. 28-1 Banded Tendon Distribution

"banded" pattern illustrated in Fig. 28-1. This method of tendon installation greatly simplifies detailing and installation procedures.

Calculation of equivalent frame properties is illustrated in Example 28.1. Tendon distribution is also discussed in Example 28.l.

Reference 28.1 illustrates application of ACI 318 requirements for design of one-way and two-way post-tensioned slabs. Detailed design examples are also presented in Ref. 28.1.

REFERENCE

28.1 *Design of Post-Tensioned Slabs*, Post-Tensioning Institute, Phoenix, AZ, 1977.

Example 28.1— Two-Way Prestressed Slab System

Design a typical transverse strip of the prestressed flat plate with partial plan and section shown below.

f'_c = 4000 psi (slabs and columns)

f_y = 60,000 psi

f_{pu} = 270,000 psi

live load = 40 psf
partition load = 15 psf

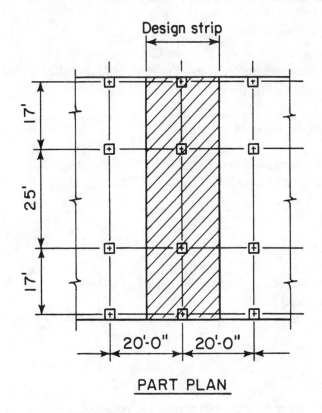

PART PLAN

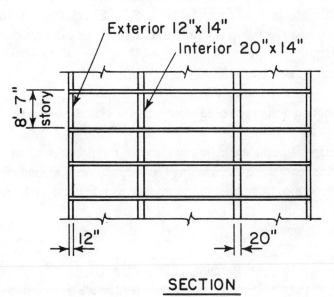

SECTION

Example 28.1—Continued

1.   Slab Thickness

For two-way prestressed slabs, a span/depth ratio of 45 has been found to result in the best overall economy and provide satisfactory structural performance. [28.1]

    Slab thickness at $\ell/45$:

    Longitudinal span - 20 x 12/45 = 5.3 in.
    Transverse span - 25 x 12/45 = 6.7 in.

    Use 6-1/2 in. slab

Slab weight	= 81 psf
Partition load	= 15 psf
Factored dead load	= 96 × 1.4 = 134 psf
Factored live load	= 40 × 1.7 = 68 psf
Total load	= 136 psf, unfactored
	= 202 psf, factored

2.   Design Procedure

Assume a set of loads to be balanced by parabolic tendons. Analyze an equivalent frame subjected to the net downward loads, according to Section 13.7. Check flexural stresses at critical sections, and revise load balancing tendon forces as required to obtain net flexural tension stresses according to Section 18.4.1.

When final forces are determined, obtain frame moments for factored dead and live loads. Calculate secondary moments induced in the frame by post-tensioning forces, and combine with factored load moments to obtain design factored moments. Provide minimum reinforcement in accordance with Section 18.9.

Check design flexural strength and increase nonprestressed reinforcement if required by strength criteria. Investigate shear strength, including shear due to vertical load and due to moment transfer, compare total to permissible values calculated in accordance with Section 11.12.2.

3.   Load Balancing

Arbitrarily, a force corresponding to an average compressive stress of 175 psi, with a parabolic tendon profile of maximum permissible sag, will be used for the initial estimate of balanced load.

Example 28.1— Continued

Then $F_e = 0.175 \times 6.5 \times 12$

$\quad\quad = 13.65$ kips/ft

Assuming 1/2 in. diameter, 270 ksi strand tendons and 30 ksi long term losses, effective force per tendon is $0.153 \times (0.7 \times 270 - 30) = 24.33$ kips.

For a 20 foot bay, $20 \times 13.65/24.33 = 11.2$ tendons.

Use eleven ½ in. diameter (270 ksi) tendons/bay

Then $F_e = 11 \times 24.33/20 = 13.38$ kips/ft

$\quad F_e/A = 13.38/78 = 0.172$ ksi

4.   Tendon Profile

For spans 1 and 3: $a = (3.25 + 5.5)/2 - 1.75 = 2.625$ in.

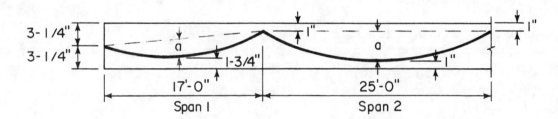

$w_{bal} = 8F_e a/12l^2 = 8 \times 13.38 \times 2.625/(12 \times 17^2) = 0.081$ ksf

Net load causing bending

$\quad w_{net} = 0.136 - 0.081 = 0.055$ ksf

Example 28.1— Continued

	Code
Calculations and Discussion	Reference

For span 2:   a = 6.5 − 1 − 1 = 4.5 in.

w_{bal} = 0.064 ksf

w_{net} = 0.072 ksf

5.   Equivalent Frame Properties   13.7

(a) Column Stiffness: 13.7.4

Column stiffness, including effects of "infinite" stiffness within the slab-column joint, may be calculated by classical methods or by simplified methods which are in close agreement. The following approximate stiffness K_c will give results within five percent of "exact" values. [28.1]

$$K_c = 4EI/(\ell - 2h)$$

where ℓ = center to center column height, and

h = slab thickness.

For exterior columns (14 in. × 12 in.):

$$I = 14 \times 12^3/12 = 2016 \text{ in.}^4$$

E_{col}/E_{slab} = 1.0

$$K_c = (4 \times 1.0 \times 2016)/(103 - 2 \times 6.5) = 90 \text{ in.}^3$$

$$\Sigma K_c = 90 \times 2 = 180 \text{ in.}^3 \text{ (joint total)}$$

Stiffness of torsional members is calculated as follows: 13.7.5

$$C = (1 - 0.63 \, x/y) \, x^3 y/3$$ Eq. (13-7)

$$= (1 - 0.63 \times 6.5/12)6.5^3 \times 12/3 = 724 \text{ in.}^4$$

Example 28.1— Continued

Calculations and Discussion	Code Reference

$$K_t = \frac{\Sigma 9CE}{\ell_2(1 - c_2/\ell_2)^3}$$ Eq. (13-6)

$$= \frac{9 \times 724 \times 1}{20 \times 12\,(1 - 1.17/20)^3} = 32.5 \text{ in.}^3$$

$\Sigma K_t = 32.5 \times 2 = 65 \text{ in.}^3(\text{joint total})$

Equivalent column stiffness,

$$1/K_{ec} = 1/\Sigma K_t + 1/\Sigma K_c$$ Com. 13.7.4

$$K_{ec} = (1/65 + 1/180)^{-1} = 48 \text{ in.}^3$$

For interior columns (14 x 20):

$$I = 14 \times 20^3/12 = 9333 \text{ in.}^4$$

$$K_c = (4 \times 1.0 \times 9333)/(103 - 2 \times 6.5) = 415 \text{ in.}^3$$

$$\Sigma K_c = 415 \times 2 = 830 \text{ in.}^3(\text{joint total})$$

$$C = (1 - 0.63 \times 6.5/20)6.5^3 \times 20/3 = 1456 \text{ in.}^4$$

$$K_t = \frac{9 \times 1456}{240(1 - 1.17/20)^3} = 65 \text{ in.}^3$$

$$\Sigma K_t = 65 \times 2 = 130 \text{ in.}^3 \text{ (joint total)}$$

$$K_{ec} = (1/830 + 1/130)^{-1} = 112 \text{ in.}^3$$

(b)   Slab-Beam Stiffness: 13.7.3

Slab stiffness, including effects of infinite stiffness within slab-column joint, can be calculated by the following approximate expression [28.1]

$$K_s = 4EI/(\ell_1 - c_1/2)$$

Example 28.1— Continued

	Code
Calculations and Discussion	**Reference**

where $\qquad \ell_t$ = span center$-$to$-$center of supports, and

$\qquad\qquad\qquad\qquad c_1$ = column dimension in direction of ℓ_t

At exterior column:

$$K_s = (4 \times 1 \times 20 \times 6.5^3)/(12 \times 17 - 12/2) = 111 \text{ in.}^3$$

At interior column (spans 1 & 3):

$$K_s = (4 \times 1 \times 20 \times 6.5^3)/(12 \times 17 - 20/2) = 110 \text{ in.}^3$$

For simplicity, use single value of 111 for both ends of spans 1 and 3.

At interior column (span 2):

$$K_s = (4 \times 1 \times 20 \times 6.5^3)/(12 \times 25 - 20/2) = 76 \text{ in.}^3$$

(c)   Distribution factors for analysis by moment distribution.

Slab distribution factor

At exterior joints $= 111/(111 + 48) = 0.70$

At interior joints for spans 1 and 3 $= 111/(111 + 76 + 111) = 0.37$

At interior joints for span 2 $= 76/298 = 0.25$

6.   Moment Distribution - Net Loads

Since the nonprismatic section causes only very small effects
on fixed-end moments and carryover factors, fixed-end moments
will be calculated from FEM $= w\ell^2/12$ and carryover factors taken as COF $= 1/2$.

For spans 1 and 3, net load FEM $= 0.055 \times 17^2/12 = 1.32ft-$kips

For span 2, net load FEM $\qquad = 0.072 \times 25^2/12 = 3.75ft-$kips

Example 28.1— Continued

	Code
Calculations and Discussion	**Reference**

Moment Distribution - Net Loads
(all moments are in ft-kips)

DF	0.70	0.37	0.25
FEM	-1.32	-1.32	-3.75
Distribution	+0.92	-0.90	+0.61
Carry-over	+0.45	+0.46	-0.30
Distribution	-0.32	+0.06	-0.05
Final	-0.27	-2.62	-3.49

7. Check Net Tensile Stresses

    (a) At face of column (interior face of interior column):

    Moment at column face is centerline moment $+ V_c/3$

    $$-M_{max} = -3.49 + 20/(3 \times 12)\,(12.5 \times 0.072)$$

    $$= -2.99 \text{ ft}-\text{kips}$$

    $$S = bh^2/6 = 12 \times 6.5^2/6 = 84.5 \text{ in.}^3$$

    $$f_t = M_{max}/S - F_e/A$$

    $$= 2.99 \times 12/84.5 - 0.172 = 0.425 - 0.172 = +0.253 \text{ ksi}$$

    $6\sqrt{f_c'} = 0.380 \text{ ksi} > 0.253 \text{ ksi} \quad \text{O.K.}$ \hfill 18.4.2

    (b) At midspan:

    $$+M_{max} = 0.072 \times 25^2/8 - 3.49 = +2.13 \text{ ft}-\text{kips}$$

Example 28.1— Continued

Calculations and Discussion	Code Reference

$$f_t = 2.13 \times 12/84.5 - 0.172 = +0.130 \text{ ksi}$$

$$2\sqrt{f_c'} = 2\sqrt{4000} = 0.126 \text{ ksi} < 0.130 \text{ ksi}$$

When tensile stress of $2\sqrt{f_c'}$ is exceeded, Section 18.9.3 requires that the total tensile force be carried by bonded reinforcement at a stress of $f_y/2$.   18.9.3.2

$$f_c = -2.13 \times 12/84.5 - 0.172 = 0.474 < 0.45f_c'$$ 18.4.2

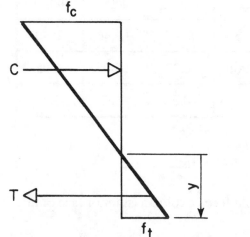

$$y = 6.5 \times 0.130/(0.130 + 0.474) = 1.4 \text{ in.}$$

$$T = 0.130 \times 1.40 \times 12/2 = 1.09 \text{ kips/ft}$$

$$A_s = 1.09/(60/2) = 0.036 \text{ in.}^2/\text{ft}$$ Eq. (18-7)

Use 4#4 bars at 60 in. on center at bottom
in the midspan region of span 2

$$(A_s = 0.04 \text{ in.}^2/\text{ft})$$

This completes the service load portion of the design. The design
strength in flexure and shear must still be verified to complete the design.

8. Flexural Strength

    (a) Calculation of design moments

    Design moments for statically indeterminate post-tensioned members
    are determined by combining frame moments due to factored dead and live
    loads with secondary moments induced into the frame by the tendons. The
    load balancing approach directly includes both primary and secondary effects,
    so that for service conditions only "net loads" need be considered.

Example 28.1— Continued

Calculations and Discussion

**Code
Reference**

At design flexural strength, the balanced load moments are used to determine secondary moments by subtracting the primary moment, which is simply F × e, at each support. For multistory buildings where typical vertical load design is combined with varying moments due to lateral loading, an efficient design approach would be to analyze the equivalent frame under each case of dead, live, balanced, and lateral loads, and combine the cases for each design condition with appropriate load factors. For this example the balanced load moments are determined by moment distribution as follows:

For spans 1 and 3, balanced load FEM $= 0.081 \times 17^2/12 = 1.95 \text{ft}-\text{kips}$

For span 2, balanced load FEM $\qquad = 0.064 \times 25^2/12 = 3.33 \text{ft}-\text{kips}$

Moment Distribution - Balanced Loads
(all moments are in ft-kips)

DF	0.70	0.37	0.25
FEM	+1.95	+1.95	+3.33
Distribution	-1.37	+0.51	-0.35
Carry-over	-0.25	+0.68	+0.17
Distribution	+0.18	-0.19	+0.13
Final	+0.51	+2.95	+3.28

Since load balancing accounts for both primary and secondary moments directly, secondary moments can be found from the following relationhip: $M_{bal} = M_1 + M_2$, or $M_2 = M_{bal} - M_1$
The primary moment M_1 equals F × e at each support.

Thus, the secondary moments are:

At exterior column:

$$M_2 = 0.51 - 13.38 \times 0/12 = 0.51 \text{ ft}-\text{kips}$$

Example 28.1—Continued

Calculations and Discussion

Code Reference

At interior column:

Spans 1 and 3,

$$M_2 = 2.95 - 13.38\,(3.25 - 1.0)/12 = 0.44 \text{ ft}-\text{kips}$$

Span 2,

$$M_2 = 3.28 - 13.38 \times 2.25/12 = 0.77 \text{ ft}-\text{kips}$$

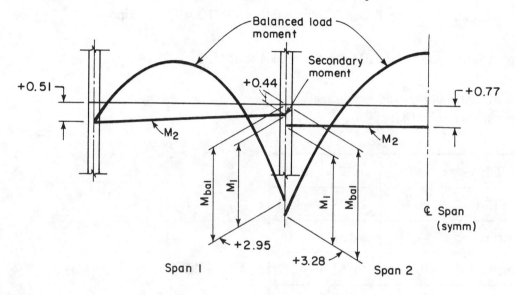

Factored load moments: ($w_u = 202$ psf)

For spans 1 and 3, factored load FEM

$$= 0.202 \times 17^2/12 = 4.86 \text{ ft}-\text{kips}$$

For span 2, factored load FEM

$$= 0.202 \times 25^2/12 = 10.52 \text{ ft}-\text{kips}$$

Example 28.1—Continued

Moment Distribution - Factored Loads
(all moments are in ft-kips)

DF	0.70	0.37	0.25
FEM	-4.86	-4.86	-10.52
Distribution	+3.40	-2.09	+1.42
Carry-over	+1.05	-1.70	-0.71
Distribution	-0.74	+0.37	-0.25
Final	-1.15	-8.28	-10.06

Combine factored load and secondary moments to obtain design moments:

	Span1		Span2
Factored load moments Secondary moments (ft-kips)	-1.15 +0.51	-8.28 +0.44	-10.06 +0.77
Moments at column centerline (ft-kips) Moment reduction to face of column, Vc/3 (ft-kips)	-0.64 +0.42	-7.84 +1.19	-9.29 +1.40
Design moments at face of column (ft-kips)	-0.22	-6.65	-7.89

Calculate design moments at midspan:

For span 1,

$$V_{ext} = 0.202 \times 17/2 - (8.28 - 1.15)/17$$

$$= 1.72 - 0.42 = 1.30 \text{ kips/ft}$$

$$V_{center} = 1.72 + 0.42 = 2.14 \text{ kips/ft}$$

Example 28.1—Continued

Calculations and Discussion	Code Reference

Distance of location of zero shear and maximum moment,

= 1.30/0.202 = 6.45 ft from exterior column.

End span positive moment:

$$1.30 \times 6.45 = +8.39 \text{ ft–kips/ft}$$

$$-0.202(6.45)^2/2 = -4.20 \text{ ft–kips/ft}$$
$$\text{end moment} = \underline{-1.14 \text{ ft–kips/ft}}$$
$$= +3.05 \text{ ft–kips/ft}$$
$$M_2 = \underline{+0.43 \text{ ft–kips/ft}}$$
$$M_{max} = +3.48 \text{ ft–kips/ft}$$

For span 2,

$$V = 0.202 \times 25/2 = 2.52 \text{ kips/ft}$$

$$+\text{Moment} = 0.202 \times 25^2/8 = +15.78 \text{ ft–kips/ft}$$
$$\text{end moment} = \underline{-10.06 \text{ ft–kips/ft}}$$
$$= +5.72 \text{ ft–kips/ft}$$
$$M_2 = \underline{+0.77 \text{ ft–kips/ft}}$$
$$M_{max} = +6.49 \text{ ft–kips/ft}$$

(b) Calculation of Flexural Strength

Check interior support section. Section 18.9.3 requires a minimum amount of bonded reinforcement in negative moment areas at column supports regardless of service load stress conditions or strength, unless more than the minimum is required for flexural strength. This minimum amount is to help ensure the integrity of the punching zone so that full shear strength can be developed.

$$A_s = 0.00075h\ell$$

Eq. (18-7)

$$= 0.00075 \times 6.5 \times (17 + 25)/2 \times 12 = 1.22 \text{ in.}^2$$

Example 28.1—Continued

Code Reference

Calculations and Discussion

Say 6#4 bars × 9 ft. Space at maximum 6 in. on centers, so that bars are placed within column width plus 1.5 times slab thickness on either side of column.

For average one foot strip:

$$A_s = 6 \times 0.20/20 = 0.06 \text{ in.}^2/\text{ft}$$

Initial check of flexural strength will be made considering this reinforcement.

Calculate stress in tendons at nominal strength:

$$f_{ps} = f_{se} + 10{,}000 + \frac{f_c'}{300\rho_p}$$

Eq. (18-5)

With 11 tendons in 20 ft bay:

$$\rho_p = A_{ps}/bd = 11 \times 0.153/20 \times 12 \times 5.5 = 0.00127$$

$$f_{se} = 0.7 \times 270 - 30 \text{ ksi (losses)} = 159 \text{ ksi}$$

$$f_{ps} = 159 + 10 + 4/300(0.00127) = 169 + 10.5 = 179.5 \text{ ksi}$$

Example 28.1—Continued

$A_{ps}f_{ps}$ = 179.5 × 0.153 × 11/20 = 15.10 kips/ft
A_sf_y = 0.06 × 60   = 3.60
                                      ———————
                                     18.70 kips/ft

$$a = \frac{A_{ps}f_{ps} + A_sf_y}{0.85f_c'b} = \frac{18.70}{0.85 \times 4 \times 12} = 0.46 \text{ in.}$$

Since bars and tendons are in same layer:

$$(d - a/2) = (5.5 - \frac{0.46}{2})/12 = 0.44 \text{ ft}$$

At column centerline:

$$\varphi M_n = 0.9 \times 0.44 \times 18.7 = 7.41 \text{ ft}-\text{kips/ft} < 7.89 \text{ ft}-\text{kips/ft}$$ 9.3.2.1

Calculate available strength at midspan and permissible moment
redistribution at column according to Section 18.10.4

$$\text{permissible redistribution} = 20\left(1 - \frac{w_p + w}{0.30}\right)$$

$$\Sigma_w = 18.7/5.5 \times 12 \times 4 = 0.071 < 0.20$$ 18.10.4.3

(18.7 kips/ft = $A_{ps}f_{ps}$ + A_sf_y, 5.5 in. = d, 12 in. = b, 4 ksi = f_c')

percent redistribution = 20(1 - 0.071/0.30) = 15.2%

redistribution moment = 0.152 × 7.41 = 1.13 ft-kips

Since the midspan of span 2 required 4#4 bars from service load
considerations, the flexural strength is:

                     $A_{ps}f_{ps}$ = 15.10 kips/ft
           A_sf_y = 0.04 × 60 =   2.40 kips/ft
                         = 17.50 kips/ft

$$a = 17.50/0.85 \times 4 \times 12 = 0.43 \text{ in.}$$

$$(d - a/2) = (5.5 - \frac{0.43}{2})/12 = 0.44 \text{ ft}$$

Example 28.1—Continued

At center of span,

$$\varphi M_n = 0.9 \times 0.44 \times 17.50 = 6.93 \text{ ft-kips/ft}$$

The required moment strength = 6.49 ft-kips/ft, which leaves
0.44 ft-kips/ft available to accommodate moment redistributed
from the support section. If 0.44 is redistributed,

$$-M = -7.89 + 0.44 = -7.45 \text{ ft-kips/ft} > 7.41 \text{ ft-kips/ft N.G.}$$

$$+M = +6.49 + 0.44 = +6.93 \text{ ft-kips/ft required} = 6.93 \text{ ft-kips/ft available O.K.}$$

Thus, minimum rebars plus tendons are not adequate for flexural strength at column support. Addition of 2#4 bars at midspan will make midspan strength identical to strength at column (which also has 6#4 bars) = 7.41 ft-kips/ft. For this case, 7.41 - 6.49 or 0.92 ft-kips/ft are available for redistribution. Redistributing 0.50 ft-kips,

$$-M = 7.89 + 0.50 = -7.39 \text{ ft-kips/ft} < 7.41 \text{ ft-kips/ft O.K.}$$

$$+M = +6.49 + 0.50 = +6.99 \text{ ft-kips/ft} < 7.41 \text{ ft-kips/ft O.K.}$$

With 2#4 bars added at midspan and redistribution of 0.50 ft-kips/ft from the negative moment section to midspan, both negative and positive moment sections are adequate. Midspan sections of spans 1 and 3 ave more than adequate strength by comparison with span 2.

The flexural strength at exterior columns is governed by moment transfer requirements. Since moment transfer also involves shear, the two aspects will be treated under shear strength considerations.

9. Shear and Moment Transfer Strength at Exterior Column.   11.12.6
                                                                                            13.3.3

(a) Shear and moment transferred at exterior column

$$V_u = 0.202 \times 17/2 - (8.28 - 1.15)/17 = 1.30 \text{ kips/ft}$$

Assume building enclosure is masonry and glass, weighing 0.40 kips/ft

Example 28.1—Continued

Total slab shear at exterior column:

$$V_u = (1.4 \times 0.40 + 1.30)20 = 37.2 \text{ kips}$$

Transfer moment $= 20(0.64) = 12.8$ ft-kips
(factored moment of exterior column centerline $= 0.64$ ft-kips/ft)

(b)   Combined shear stress at inside face of critical transfer section.

For shear strength equations, see Part 18, Fig. 18-16.

$$V_u = V_u/A_c + \gamma_v M_u/(J/c)$$

$$= 37,200/252 + 0.37 \times 12.8 \times 12,000/1419$$

$$= 148 + 40 = 188 \text{ psi}$$

where (referring to Fig. 18-16: edge column-bending perpendicular to edge)

assuming $d = 0.8 \times 6.5 = 5.2$ in.

$c_1 = 12$ in.

$c_2 = 14$ in.

$b_1 = c_1 + d/2 = 14.6$ in.

$b_2 = c_2 + d = 19.2$ in.

$c = b_1^2/(2b_1 + b_2) = 4.40$ in.

$A_c = (2b_1 + b_2)d = 252 \text{ in.}^2$

$J/c = [2b_1 d(b_1 + 2b_2) + d^3(2b_1 + b_2)/b_1]/6$

$$= 1419 \text{ in.}^3$$

Example 28.1—Continued

Calculations and Discussion	Code Reference

For $b_1/b_2 = 14.6/19.2 = 0.76$

Eq. (11-42)

$$\gamma_v = 1 - \frac{1}{1 + (2/3)\sqrt{b_1/b_2}} = 0.37$$

Eq. (11-42)

(c) Permissible Shear Stress

11.12.2.1

For edge columns:

$$v_c = \varphi 4\sqrt{f_c'} = 0.85 \times 4\sqrt{4000} = 215 \text{ psi} > 188 \text{ psi O.K.}$$

(d) Check Moment Transfer Strength

13.3.3

Although the transfer moment is small, for illustrative purposes, calculate the moment strength of the effective slab width for moment transfer (width of column plus 1.5 times the slab thickness on each side). Assume that of the 11 tendons required for the 20 ft bay width, 3 tendons are anchored within the column and are bundled together across the building. This amount should be noted on the design drawings. Besides providing flexural strength, this prestress will act directly on the critical section for shear and improve shear strength. As previously shown, a minimum amount of bonded reinforcement is required at all columns. For the exterior column, the required area is:

13.3.3.2

$$A_s = 0.00075h\ell = 0.00075 \times 6.5 \times 17 \times 12 = 1.0 \text{ in.}^2$$

Eq. (18-8)

Say 5#4 bars x 5 ft (including standard end hook).

Calculate stress in tendons:

Effective slab width = $14 + 2(1.5 \times 6.5) = 33.5$ in.

Within effective slab width: $f_{ps} = \dfrac{33.5 \times 3.25 \times 4}{300 \times 0.153 \times 3} + 169 = 172$ ksi

Eq. (18-5)

$$\rho_p = \frac{0.153 \times 3}{33.5 \times 3.25}$$

Example 28.1—Continued

	Code
Calculations and Discussion	**Reference**

Corresponding prestress force $= 172 \times 0.153 \times 3 = 79.0$ kips

$$A_s f_y = 5 \times 0.20 \times 60 = 60.0 \text{ kips}$$

$$A_{ps} f_{ps} + A_s f_y = 139.0 \text{ kips}$$

$$a = \frac{139.0}{0.85 \times 4 \times 33.5} = 1.22 \text{ in.}$$

tendon $j_u d = (3.25 - 1.22/2)/12 = 0.22$ ft

rebar $j_u d = (5.5 - 1.22/2)/12 = 0.41$ ft

$\varphi M_n = 0.9(79 \times 0.22 + 60 \times 0.41) = 37.4$ ft-kips

$$\gamma_f = -\frac{1}{1 + 2/3\sqrt{b_1/b_2}} = 0.63$$

Eq. (13-1)

$\gamma_f M_u = 0.63(12.8) = 8.1$ ft-kips $<< 37.4$ ft-kips O.K.

10. Shear and Moment Transfer Strength at Interior Column.

11.12.6

13.3.3

(a) Shear and moment transferred at interior column.

Direct shear and moment to the left and right of interior columns is calculated in Step 8 above.

$V_u = (2.14 + 2.52)20 = 93.2$ kips

Transfer moment $= 20(9.29 - 7.84) = 29.0$ ft-kips

(b) Combined shear stress at face of critical transfer section. For shear strength equations, see Part 18, Fig. 18-15.

$v_{u1} = V_u/A_c + \gamma_v M_u/(J/c)$

$\quad = 93,200/462 + 0.43 \times 29.0 \times 12,000/3664$

Example 28.1—Continued

$= 202 + 41 = 243$ psi

where (referring to Fig. 18-15: interior column)

$d \approx 0.8 \times 6.5 = 5.2$ in.

$c_1 = 20$ in.

$c_2 = 14$ in.

$b_1 = c_1 + d = 25.2$ in.

$b_2 = c_2 + d = 19.2$ in.

$A_c = 2(b_1 + b_2)d = 462$ in.^{2}

$J/c = [b_1 d(b_1 + 3b_2) + d^3]/3 = 3664$ in.^{3}

For $b_1/b_2 = 25.2/19.2 = 1.31$

$$\gamma_v = 1 - \frac{1}{1 + 2/3\sqrt{b_1/b_2}} = 0.43$$

Eq. (11-42)

(c) Permissible shear stress

For interior columns, Eq. (11-39) applies:

$$v_c = \varphi(\beta_p\sqrt{f_c'} + 0.3f_{pc} + V_p/b_o d)$$

Eq. (11-39)

V_p is the shear carried through critical transfer section by tendons.
For thin slabs, the V_p term must be carefully evaluated, as field placing
practices can have a great effect on the profile of the tendons through
the critical section. Conservatively, this term may be taken as zero.

Example 28.1 Continued

	Code
Calculations and Discussion	**Reference**

$$v_c = 0.85(3.5 \sqrt{4000} + 0.3 \times 172) = 232 \text{ psi} < 245 \text{ psi}$$

The V_p component of tendon force crossing the critical transfer section will make up the slight deficiency.

where $\quad \beta_p = (1.5 + \alpha_s/\beta_o) \qquad$ but not greater than 3.5

$\quad \beta_o = b_o/d.$

$\quad b_o = 2[(14 + 5.2) + (20 + 5.2)] = 88.8 \text{ in.}$

$\quad \alpha_s = 40$ for interior columns.

$\quad d = 5.2 \text{ in.}$

$\quad \beta_o = 88.8/5.2 = 17.1$

$(1.5 + 40/17.1) = 3.8 > 3.5$

(d) Check moment transfer strength   13.3.3

$$\gamma_f = 1 - \gamma_v = 1 - 0.43 = 0.57 \qquad\qquad\qquad \text{Eq. (13-1)}$$

Moment transferred by flexure within width of column plus 1.5 times   13.3.3.2
slab thickness on each side = 0.57(30.2) = 17.2 ft-kips

Effective slab width = $14 + 2(1.5 \times 6.5) = 33.5$ in.
Say $A_{ps}f_{ps} = 79$ kips (same as exterior column)

$$A_s = 0.00075h\ell = 0.00075 \times 6.5 \times 21 \times 12 = 1.22 \text{ in.}^2 \qquad \text{Eq. (18-8)}$$

Use 6#4 bars ($A_s = 1.20 \text{ in.}^2$)

$$A_s f_y = 1.20 \times 60 = 72 \text{ kips}$$

$A_{ps}f_{ps} + A_s f_y = 79 + 72 = 151$ kips

Example 28.1 Continued

$$a = \frac{151}{0.85 \times 4 \times 33.5} = 1.33 \text{ in.}$$

$$(d - a/2) = (5.5 - 1.33/2)/12 = 0.40 \text{ ft}$$

$$\varphi M_n = 0.9(151 \times 0.40) = 54.4 \text{ ft-kips}$$

Moment transfer strength in flexure of 54.4 ft-kips is much greater than the required transfer moment of 17.2 ft-kips.

11. Distribution of tendons

In accordance with Section 18.12, the 11 tendons per 20 ft bay will be distributed in a group of 3 tendons directly through the column with the remaining 8 tendons spaced at 2 ft 3 in. centers (about 4 times slab thickness). Tendons in the perpendicular direction will be placed in a narrow band through and immediately adjacent to the columns.

29

Shells and Folded Plate Members

INTRODUCTION

Chapter 19, addressing shell and folded plate members, was completely updated for ACI 318-83. In its present form, it reflects the current state-of-the-knowledge in analysis and design of folded plates and shells. Chapter 19 includes guidance on analysis methods appropriate for different types of shell structures, and provides specific direction as to design and proper placement of shell reinforcement. The Commentary on Chapter 19 should be helpful to designers; its contents reflect current information—including an extended reference listing.

GENERAL CONSIDERATIONS

Code requirements for shells and folded plates must, of necessity, be somewhat general in nature as compared to the provisions for other types of structures where the practice of design has been firmly established. Chapter 19 is specific in only a few critical areas peculiar to shell design; otherwise, it refers to standard provisions of the code. It should be noted that strength design is permitted for shell structures, even though most of the shells in this country have been built using working stress design procedures.

The Code, the Commentary, and the list of references are an excellent source of information and guidance on shell design. The list of references, however, does not exhaust the possible sources of design assistance.

(1) Chapter 19 covers the design of a large class of concrete structures that are quite different from the ordinary slab, beam and column construction. Structural action varies from shells with considerable bending in the shell portions (folded plates and barrel shells) to those with very little bending except at the junction of shell and support (hyperbolic paraboloids and domes of revolution). The problems of shell design, therefore, cannot be lumped together, as each type has its own peculiar attributes that must be thoroughly understood by the designer. Even shells classified under one type, such as the hyperbolic paraboloid, vary greatly in their structural action. Studies have shown that gabled hyperbolic paraboloids, for example, are much more complex than the simple membrane theory would indicate. This is one explanation for the lack of a rigid set of rules in the Code for the design of shells and folded plate structures.

(2) For the reasons given above, design of a shell requires considerable lead time to gain an understanding of the design problems for the particular type of shell. An attempt to design a shell without proper study may invite poor performance. Design of shell structues requires the ability to think in terms of three-dimensional space; this is only gained by study and experience. The conceptual stage is the most critical period in shell design, since this is when vital decisions on form and dimensions must be made.

(3) Strength of shell structures is inherent in their shape and is not created by boosting the performance of materials to their limit as in the case of other types of concrete structures such as conventional and prestressed concrete beams. Therefore, the design stresses in the concrete should not be raised to their highest acceptable values, except where required for very large structures. Deflections are normally not a problem if the stresses are low.

(4) Shell size is a very important determinant in the analytical precision required for its design. Short spans (up to 60 ft) can be designed using approximate methods such as the beam method for barrel shells, provided the exterior shell elements are properly supported by beams and columns. However, the limits and approximations of any method must be thoroughly understood. Large spans may require much more elaborate analyses. For example, a large hyperbolic paraboloid (150-ft span or more) may require a finite element analysis.

Application of the following Code provisions warrants further explanation

19.2 ANALYSIS AND DESIGN

19.2.6 Prestressed Shells

The components of force produced by prestressing tendons draped in a thin shell must be taken into account in the design. In the case of a barrel shell, it should be noted that the tendon does not lie in one plane, as shown in Fig. 29-1.

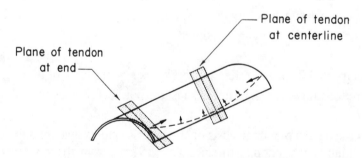

Fig. 29-1 Draped Prestressing Tendon in Barrel Shell

19.2.7 Design Method

The Strength Design Method is permitted for the design of shells, but it should be noted that for slab elements intersecting at an angle, and having high tensile stresses at inside corners, the ultimate strength is greatly reduced from that at the center of a concrete slab. Therefore, special attention should be given to the reinforcement used in these areas, and thickness should be greater than the minimum allowed by the strength method.

19.4 SHELL REINFORCEMENT

19.4.6 Membrane Reinforcement

For shells with essentially membrane forces, such as hyperbolic paraboloids and domes of revolution, it is usually convenient to place the reinforcement in the direction of the principal forces. Even though folded plates and barrel shells act essentially as longitudinal beams (traditionally having vertical stirrups as shear reinforcement), an orthogonal pattern of reinforcement (diagonal bars) is much easier to place and also assures end anchorage in the barrel or folded plate. With diagonal bars, five layers of reinforcement may be required at some points.

The direction of principal stresses near the supports is usually about 45 degrees, so that equal areas of reinforcement are needed in each direction to satisfy the requirements of Section 19.4.4. For illustration, Fig. 29-2 shows a plot of the principal membrane forces in a barrel shell with a span of 60 ft, a rise of 6.3 ft, a thickness of 3.5 in. and a snow load of 25 psf, and a roof load of 10 psf. Forces are shown in kips per linear foot.

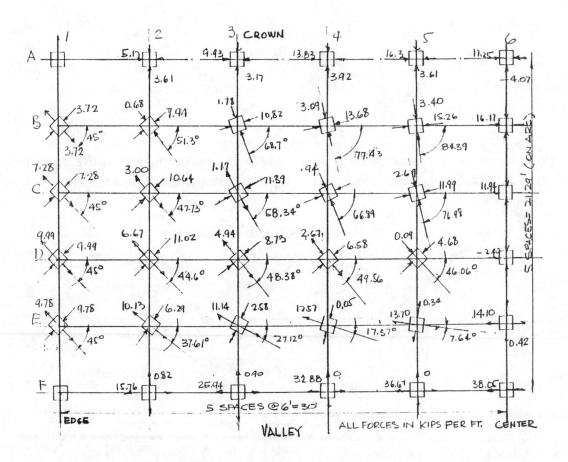

Fig. 29-2 Membrane Forces and Direction for 60-ft span Barrel Shell

19.4.8 Concentration of Reinforcement

In the case of long barrel shells (or domes) it is often desirable to concentrate tensile reinforcement near the edges rather than distribute the reinforcement over the entire tensile zone. When this is done, a minimum amount of reinforcement equal to 0.0035 bh must be distributed over the remaining portion of the tensile zone, as shown in Fig. 29-3. This amount in practical terms is twice the minimum steel requirement for shrinkage and temperature stresses.

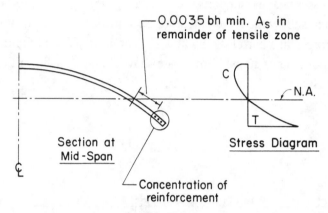

Fig. 29-3 Concentration of Shell Reinforcement

19.4.10 Spacing of Reinforcement

Spacing of reinforcement up to 5 times the shell thickness or 18 in. is the maximum permissible. Therefore, for shells less than 3.6 in. thick, 5 times the thickness controls. For thicker shells, the spacing of bars must not exceed 18 in.

30

Strength Evaluation of
Existing Structures

20.1 STRENGTH EVALUATION - GENERAL

Strength evaluation of existing structures requires experience and sound engineering judgment. Chapter 20 sets specific criteria for testing and evaluation of flexural members only. Other members may be evaluated by analysis or load tests, or a combination of both.

Load and Span Definitions

D= Service dead load supported by the member being tested, as defined by the general building code (without load factors).

L= Service live load supported by the member being tested, as specified by the general building code (without load factors). L may include live load reductions if permitted by the general code.

ℓ_t= Span of member under load test (shorter span of flat slabs supported on four sides). Span of member, except for cantilevers, is distance between centers of supports or clear distance between supports plus depth of member, whichever is less. Span for cantilevers is to be taken as two times the distance from support to cantilever end.

20.4 LOAD TESTS OF FLEXURAL MEMBERS

The load test procedures for flexural members and conditions of acceptance or rejection are summarized below. The code should be consulted for more detailed information.

Criteria

(1) Portion of structure to load test must be at least 56 days old . . . (20.3.2).

(2) Forty-eight hours prior to load test, apply full service dead load, D . . . (20.3.4).

(3) Immediately prior to application of test load, take data readings for measurements of deflections . . . (20.4.2).

(4)   In addition to full service dead load, D, apply test load equal to (0.2D + 1.45L) in four equal increments. Total load acting must equal 0.85 (1.4D + 1.7L) . . . (20.4.3 and 20.4.4).

(5)   Take deflection readings after 24 hours with load in place . . . (20.4.5).

(6)   Remove test load . . . (20.4.6).

(7)   Take final deflection readings 24 hours after removal of test load . . (20.4.6).

Acceptance

If visible evidence of failure has occured, the test portion has failed, and no re-testing is permitted. If the structure shows no visible evidence of failure, the following conditions must be satisfied:

(1)   When maximum deflection exceeds ℓ_t^2 /20,000h, the percentage recovery must be at least 75% after 24 hours (80% for prestressed members) . . . (20.4.8).

(2)   When maximum deflection is less than ℓ_t^2 /20,000h, recovery requirement is waived . . . (20.4.8).

Figures 30-1 and 30-2 illustrate application of the limiting deflection criteria.

(3)   Members failing to meet the 75% recovery criterion may be re-tested . . . (20.4.10).

(4)   Before re-testing, 72 hours must have elapsed after load removal, and on re-test the recovery must be 80% . . . . (20.4.10).

(5)   Prestressed members shall not be re-tested . . . (20.4.11).

20.5 MEMBERS OTHER THAN FLEXURAL MEMBERS

An analytical method to investigate these members is preferred. In testing compression members, generally a wall or column, the test would have to be carried out to destruction.

20.6 PROVISIONS FOR LOWER LOAD RATING

A lower load rating may be approved by the building official when a structure does not satisfy the analytical or percentage recovery tests. It should be recognized that the lower load rating will generally prohibit the use of the structure for the purpose originally intended.

SUMMARY

Chapter 20 does not cover load testing for the approval of novel design or construction methods, or for the demonstration of quality of prefabricated units, unless these units already form a part of the structure under test, nor is the chapter written to settle private disputes or litigation over construction quality.

References 30.1 to 30.4 published by the Concrete Reinforcing Steel Institute (CRSI) are suggested as additional guides in an investigation or strength evaluation where there is doubt concerning the load-carrying capacity of an element or portion of an existing building structure, or in evaluating the strength or load-carrying capacity of old structures.

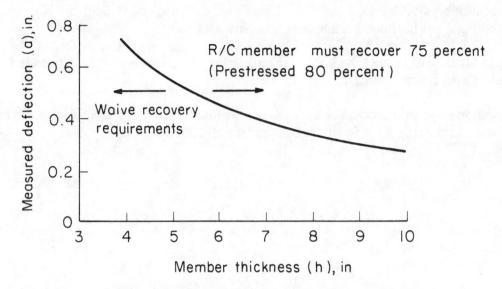

**Fig. 30-1 Load Testing Acceptance Criteria for Members
with Span Length, l_t = 20 ft**

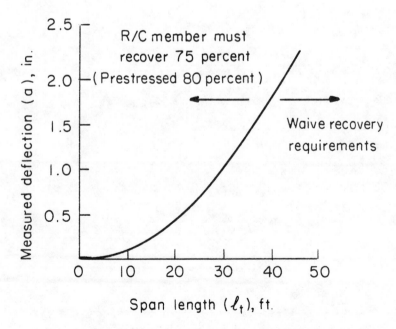

**Fig. 30-2 Load Testing Criteria for Members
with Overall Thickness, h = 6-½ in.**

REFERENCES

30.1   "Evaluation of Reinforcing Steel in Old Reinforced Concrete Structures", Engineering Data Report No. 11, Concrete Reinforcing Steel Institute, Schaumburg, IL.

30.2   "Applications of ACI 318 Load Test Requirements", Structural Bulletin No. 16, Concrete Reinforcing Steel Institute, Schaumburg, IL, Nov. 1987.

30.3   "Proper Load Tests Protect the Public", Engineering Data Report No. 27, Concrete Reinforcing Steel Institute, Schaumburg, IL.

30.4   "Evaluation of Reinforcing Steel Systems in Old Reinforced Concrete Structures", 1st Edition, Concrete Reinforcing Steel Institute, Schaumburg, IL, 1981.

Special Provisions for Seismic Design

BACKGROUND

Special provisions for earthquake resistance were first introduced into the 1971 edition of the ACI Code into Appendix A, and were included without revision in ACI 318-77. The original provisions of Appendix A were intended to apply only to reinforced concrete structures located in regions of high seismicity, and designed with a substantial reduction in total lateral seismic forces (as compared with the elastic response forces), in anticipation of inelastic structural behavior. Also, with publication of the 1971 Code edition, several changes were incorporated into the main body of the Code specifically to improve toughness, in order to increase the resistance of concrete structures to earthquakes or other catastrophic loads. While Appendix A was meant for application to lateral load resisting frames and walls in regions of high seismicity, the main body of the Code was supposed to be sufficient for regions where there is a probability of only moderate or light earthquake damage.

For the 1983 Code edition, the special provisions of Appendix A were extensively revised, to reflect then current knowledge and practice of the design and detailing of monolithic reinforced concrete structures for earthquake resistance. Appendix A to ACI 318-83 for the first time included special detailing for frames in zones of moderate seismic risk.

The special provisions for earthquake resistance form Chapter 21 of the 1989 Code edition. This move away from an appendix location into the main body of the Code has been made for reasons discussed in Part 1 of this publication.

For buildings located in regions of low seismic risk, no special design or detailing is required; the general requirements of Chapters 1 through 20 of the Code apply. Concrete structures proportioned by Chapters 1 through 20 of the Code are expected to provide a level of toughness adequate for low earthquake intensities.

For buildings located in regions of moderate seismic risk, reinforced concrete moment frames proportioned to resist earthquake effects require some special reinforcing details, as were specified for the first time in Section A.9 of ACI 318-83. The special details apply only to frames (beams, columns, and slabs) to which earthquake-induced forces have been assigned in design. The special reinforcing details serve to accommodate a suitable level of inelastic behavior if the frame is subjected to an earthquake of such intensity as to require it to perform inelastically. There are no special requirements for shearwalls provided to resist lateral effects of wind and earthquakes, or for nonstructural components of buildings located in regions of moderate seismic risk. Shearwalls proportioned by the non-seismic provisions of the Code are considered to have sufficient toughness at drift levels anticipated in regions of moderate seismicity.

For buildings located in regions of high seismic risk, where damage to construction has a high probability of occurrence, all building components, structural and nonstructural, must satisfy the requirements of Chapter 21 excluding those of Section 21.9. The special proportioning and detailing provisions of Chapter 21 are intended to provide a monolithic reinforced concrete structure with adequate toughness to respond inelastically under severe earthquake motions.

Seismic risk level is usually designated by zones or areas of equal risk or probability of damage, such as Zone 0—no damage; Zone 1—minor damage; Zone 2—moderate damage; and Zone 3—major damage. Areas within Zone 3 that are close to major fault systems are assigned to Seismic Zone 4. Seismic risk levels (Seismic Zone Maps) are under the jurisdiction of general building codes rather than ACI 318. In the absence of a general building code that addresses earthquake loads and seismic zoning, it is intended that local authorities (engineers, geologists, and building officials) should decide on the need and proper application of the special provisions for seismic design.

GENERAL CONSIDERATIONS

Economical earthquake-resistant design should aim at providing appropriate dynamic characteristics in structures so that acceptable response levels would result under the design earthquake(s). The structural properties over which the designer exercises some degree of control, which he can modify to achieve the desired results, are the magnitude and distribution of stiffness and mass, relative strengths of members and their deformabilities.

In some structures, such as slender free-standing towers or smoke stacks which depend for their stability on the stiffness of the single element making up the structure, or in nuclear containment buildings where a more-than-usual conservatism in design is required, yielding of the principal elements in the structure cannot be tolerated. In such cases, the design needs to be based on an essentially elastic response to moderate-to-strong earthquakes, with the critical stresses limited to the range below yield.

In most buildings, particularly those consisting of frames and other multiple-redundant systems, however, economy is achieved by allowing yielding to take place in some members under moderate-to-strong earthquake motion.

The performance criteria implicit in most earthquake code provisions require that a structure be able to:[31.1]

1. Resist earthquakes of minor intensity without damage; a structure would be expected to resist such frequent but minor shocks within its elastic range of stresses;

2. Resist moderate earthquakes with minor structural and some nonstructural damage; with proper design and construction, it is expected that structural damage due to the majority of earthquakes will be repairable; and

3. Resist major catastrophic earthquakes without collapse.

The above performance criteria allow only for the effects of a typical ground shaking. The effects of slides, subsidence or active faulting in the immediate vicinity of the structure, which may accompany an earthquake, are not considered.

While no clear quantitative definition of the above earthquake intensity ranges has been given, their use implies the consideration not only of the actual intensity level but also of their associated probability of occurrence with reference to the expected life of a structure.

The principal concern in earthquake-resistant design is the provision of adequate strength and ductility to assure life safety, i.e., prevention of collapse under the most intense earthquake that may reasonably be expected at a site during the life of a structure. Observations of building behavior in recent earthquakes, however, have made engineers increasingly aware of the need to ensure that buildings housing facilities essential to post-earthquake operations, such as hospitals, power plants, fire stations and communication centers, not only survive without collapse but remain operational after an earthquake. This means that such buildings should suffer a minimum amount of damage. Thus, damage control has been added to life safety as a second design criterion.

Often, damage control becomes desirable from a purely economic point of view. The extra cost of preventing severe damage to the nonstructural components of a building, such as partitions, glazing, ceiling, elevators and other mechanical systems, may be justified by the savings realized in replacement costs and from continued use of a building after a strong earthquake.

The principal steps involved in the earthquake resistant design of a typical concrete structure according to building code provisions are as follows:

1. Determination of design earthquake forces:

    (a) calculation of base shear corresponding to computed or estimated fundamental period of vibration of the structure (a preliminary design of the structure is assumed here);

    (b) distribution of the base shear over the height of the building.

2. Analysis of the structure under the (static) lateral forces calculated in Step 1, as well as under gravity and wind loads, to obtain member design forces.

3. Designing members and joints for the most unfavorable combination of gravity and lateral loads, and detailing them for ductile behavior.

The design base shear represents the total horizontal seismic force (service load level) that may be assumed acting parallel to the axis of the structure considered. The force in the other horizontal direction is assumed to act non-concurrently. Vertical seismic forces are not considered because the building is already designed for the vertical acceleration of gravity; the additional vertical acceleration due to an earthquake would normally be only a fraction of g. It can be accommodated by the factor of safety on gravity loads.

The code-specified design lateral forces have the same general distribution as the typical envelope of maximum horizontal shears indicated by an elastic dynamic analysis. However, the code forces, which are assumed to be resisted by a structure within its elastic (working) range of stresses, are substantially smaller than those which would be developed in a structure subjected to an earthquake of intensity equal to that of the 1940 El Centro, if the structure were to respond elastically to such ground excitation. Thus, buildings designed under the present codes would be expected to undergo fairly large deformations (four to six times the lateral displacements resulting from the code-specified, statically applied shears) when subjected to an earthquake with the intensity of the 1940 El Centro. These large deformations will be accompanied by yielding in many members of the structure, and in fact, such is the intent of the codes. The acceptance of the fact that it is economically unwarranted to design buildings to resist major earthquakes elastically and the recognition of the capacity of structures possessing adequate strength and ductility to withstand major earthquakes by responding inelastically to these, lies behind the relatively low forces specified by the codes. These reduced force levels must be and are coupled with additional requirements for the design and detailing of members and their connections in order to ensure sufficient deformation capacity in the inelastic range.

The capacity of a structure to deform in a ductile manner, that is to deform beyond the yield limit without significant loss of strength, allows such a structure to absorb a major portion of the energy from an earthquake without serious damage. Laboratory tests have demonstrated that reinforced concrete members and their connections, designed and detailed by the present codes, do possess the necessary ductility to allow a structure to respond inelastically to earthquakes of major intensity without significant loss of strength.

Because of the relatively large inelastic deformations which a building designed by the present codes may undergo during a strong earthquake, proper provisions must be made to ensure that the structure does not become unstable under the vertical loads. The codes thus prescribe the so-called strong column-weak beam design with the intent of confining yielding to the beams while the columns remain elastic throughout their seismic response. It is required that the sum of the moment strengths of the columns meeting at a joint, under the design axial loads, be greater than the sum of the moment strengths of the beams framing into the joint in the same plane.

The design provisions contained in the main body of the ACI Building Code as well as the regular provisions in most other codes do in fact provide some ductility which should be sufficient for structures subjected only to minor earthquakes that may occur frequently, such as those associated with UBC[31.2] or ANSI[31.3] Zone 1 areas. For structures that may be subjected to earthquakes of moderate intensity (Zone 2), some additional confinement, anchorage, and shear reinforcement details may be required. Some such details for lateral load resisting frames, for the first time, were given in Appendix A to the 1983 edition of the ACI Code (Section A.9, currently Section 21.9). For structures that may be subjected to strong intensity earthquakes (Zones 3 and 4), appreciable inelastic deformations can be expected, so that substantial ductility is required. The design provisions contained in Chapter 21 of ACI 318-89 (excluding Section 21.9) are primarily intended to provide this additional ductility.

UPDATE FOR THE '89 CODE

The following are significant changes from Appendix A of ACI 318-83 to Chapter 21 of ACI 318-89:

1. In Section 21.0 - Notation, the definition of A_j, effective cross-sectional area within a joint, has been revised; see discussion on Section 21.6.

    A new notation M_{pr}, probable flexural strength, has been added. See Figs. 31-3, 31-8, and the associated discussions.

2. In Section 21.1 - Definitions, the definitions of crosstie and hoop have been revised by requiring a six-diameter (but no less than 3 in.), rather than a ten-diameter, extension beyond 135-degree hooks.

3. Modified strength reduction factors for regions of high seismic risk have been moved from the former Section A.2.3 to Section 9.3.4 of the Code.

    A new Section 21.2.6 provides a more explicit restriction of reinforcement welding in earthquake areas than is given in Section 7.4.

4. An editorial change has been made to Section 21.3.1 to avoid misinterpretation that members not meeting the requirements of Sections 21.3.1.1 through 21.3.1.4 might be used, without having to comply with the requirements of Section 21.3.

    A change has been made to Section 21.3.3.5, to reflect the new six-diameter (but not less than 3 in.), rather than ten-diameter extension beyond the 135-degree hooks.

5. An editorial change has been made to Section 21.4.1 to avoid misinterpretation that members not meeting the requirements of Sections 21.4.1.1 and 21.4.1.2 might be used, without having to comply with the requirements of Section 21.4.

    A change has been made to Section 21.4.4.5 requiring a minimum extension of the special transverse reinforcement in columns supporting discontinued stiff elements beyond the discontinuity into the stiff element, and into the elements supporting the columns.

6. A revision has been made to Section 21.5.3.4 to point out that the forces in boundary members may be either tensile or compressive. Both must be considered in design, not only compression, as Section 4.5.3.4 of ACI 318-83 appears to imply.

    A new Section 21.5.3.6 has been introduced to require anchorage of transverse reinforcement terminating at the edges of shearwalls without boundary elements, except when V_u in the plane of the wall is less than $A_{cv} \sqrt{f_c'}$.

    A new Section 21.5.5 has been introduced to point out that columns supporting discontinuous walls must be reinforced in accordance with Section 21.4.4.5.

7. For joints confined on all four faces, a 50% reduction in the amount of confinement reinforcement is allowed by Section 21.6.2.2. The '89 Code further allows that at these locations, the spacing specified in Section 21.4.4.2(b) may be increased to 6 in.

   A change has been made in Section 21.6.3.1, to provide a conservative value of nominal shear strength for corner column joints, in concurrence with Committee 352 recommendations.

8. A change has been introduced in Section 21.7.1 to point out that the strength reduction factor of 1.0 and the allowance for over-strength reinforcement should apply to Section 21.7.1.2 as well as to Section 21.7.1.1. Also, it has been made clear that there is no need to calculate shear based on the full column moment strength when the upper limit on moment is controlled by flexural members framing into the column.

9. There had been considerable disagreement among engineers as to the application of Section A.8 of ACI 318-83. The new Section 21.8 specifically states that frame members assumed not to contribute to lateral resistance shall be detailed according to Section 21.8.1.1 or Section 21.8.1.2, depending on the magnitude of moments induced in those members when subjected to twice the lateral displacement under the factored lateral forces.

21.2 GENERAL REQUIREMENTS

21.2.2 Analysis and Proportioning of Structural Members

The interaction of all structural and nonstructural components affecting linear and nonlinear structural response are to be considered in analysis (21.2.2.1). Consequences of failure of structural and nonstructural components not forming part of the lateral force resisting system should also be considered (21.2.2.2). The intent of Sections 21.2.2.1 and 21.2.2.2 is to draw attention to the influence of nonstructural components on structural response and to hazards from falling objects.

Section 21.2.2.3 alerts the designer to the fact that the base of the structure as defined in analysis may not necessarily correspond to the foundation or ground level. It requires that structural members below base, which transmit forces resulting from earthquake effects to the foundation, shall also comply with the requirements of Chapter 21.

Even though some element(s) of a structure may not be considered part of the lateral force resisting system, the effect on all elements of displacements several times those caused by Code forces should be considered (21.2.2.4). The only exception should be when complete failure of the element would not result in loss of the vertical load carrying capacity of the structure.

21.2.3 Strength Reduction Factors

The strength reduction factors of Section 9.3.2 are not based on the observed behavior of reinforced concrete members under load or displacement cycles simulating earthquake effects. Some of those factors have been modified in Section 9.3.4 in view of the effects on strength of large reversing displacements into the inelastic range of response. Note that Sections 9.3.4.1 and 9.3.4.2 were Sections A.2.3.1 and A.2.3.2, respectively, in ACI 318-83. The provisions have been moved to Chapter 9 in ACI 318-89.

Section 9.3.4.1 refers to brittle members such as low-rise walls or portions of walls between openings which are proportioned such as to make it impractical to raise their nominal shear strength above the shear corresponding to nominal flexural strength for the pertinent loading conditions. The provision does not apply to beam-column joints.

Section 9.3.4.2 is intended to discourage the use of columns, not conforming to the strict transverse reinforcement requirements of Section 21.4.4, for resisting earthquake induced forces.

21.2.4, 21.2.5 Limitations on Materials

A minimum specified concrete strength f_c' of 3000 psi and a maximum specified reinforcement yield strength f_y of 60,000 psi are mandated. These limits are imposed as reasonable bounds on the variation of material properties, particularly with respect to their unfavorable effects on the sectional ductility of members in which they are used. A decrease in the concrete strength and an increase in the yield strength of the tensile reinforcement tend to decrease the ultimate curvature and hence the sectional ductility of a member subjected to flexure. Also, an increase in the yield strength of reinforcement is generally accompanied by a decrease in the ductility—as measured by the maximum deformation—of the material itself.

There is evidence suggesting that lightweight concrete ranging in strength up to 12,500 psi can attain adequate ultimate strain capacities. Testing to examine the behavior of high strength, lightweight concrete under high intensity, cyclic shear loads, including a critical study of bond characteristics, has not been extensive in the past. However, there are test data showing that properly designed lightweight concrete columns, with concrete strength ranging up to 6200 psi, maintained ductility and strength when subjected to large inelastic deformations from load reversals. It was felt by Committee 318 that a limit of 4000 psi on the strength of lightweight concrete was advisable, pending further testing of high strength lightweight concrete members under reversed cyclic loading. Note that lightweight concrete with a higher design compressive strength is allowed if it can be demonstrated by experimental evidence that structural members made with that lightweight concrete possess strength and toughness equal to or exceeding those of comparable members made with normal weight concrete of the same strength.

Chapter 21 requires that reinforcement for resisting flexure and axial forces in frame members and wall boundary elements be ASTM A 706. Grade 60 low alloy steel intended for application where welding or bending, or both, are important. However, ASTM A 615 billet steel bars of Grade 40 or 60 may be used in these members if the following two conditions are satisfied:

$$\text{actual } f_y \leq \text{specified } f_y + 18,000 \text{ psi}$$

$$\frac{\text{actual ultimate tensile stress}}{\text{actual } f_y} \geq 1.25$$

The first requirement helps to limit the magnitude of the actual shears that can develop in a flexural member above that computed on the basis of the specified yield value when plastic hinges form at the ends of a beam. The second requirement is intended to ensure steel with a sufficiently long yield plateau.

In the "strong column-weak beam" frame intended by the Code, the relationship between the moment capacities of columns and beams may be upset if the beams turn out to have much greater moment capacity than intended by the designer. Thus, the substitution of Grade 60 steel of the same area for specified Grade 40 steel in beams can be detrimental. The shear strength of beams and columns, which is generally based on the condition of plastic hinges forming at the ends of the members, may become inadequate if the moment capacity of member ends should be greater than intended as a result of the steel having a substantially greater yield strength than specified.

ACI 318-89 includes a new Section 21.2.6, combining the last sentence of the previous Section A.2.5.1 with an added sentence that states:"... welding of stirrups, ties, inserts, or other similar elements to longitudinal reinforcement required by design shall not be permitted." The new sentence is intended to provide a more explicit restriction of reinforcement welding in earthquake areas than is given in Section 7.5.4. The wording is similar to that included in Refs. 30.1 and 30.2. Welding or tack-welding of crossing reinforcing bars can lead to local embrittlement of the steel. If such welding will facilitate fabrication or field installation, it must be done only on bars added expressly for construction.

21.3 FLEXURAL MEMBERS OF FRAMES

These include members, having a clear span greater than 4 times the effective depth, that are subjected to a factored axial compressive force not exceeding $(A_g f_c'/10)$, where A_g is the gross cross-sectional area (21.3.1.1, 21.3.1.2). For the '89 Code, the first paragraph of Section 21.3.1 (former Section A.3.1) has been revised, without changing the intent of the original wording. It was felt that the original wording of Section A.3.1 could be misinterpreted to imply that members which did not meet the requirements of Sections 21.3.1.1 through 21.3.1.4 might be used, but did not have to comply with Section 21.3. The significant provisions relating to flexural members are

1. Limitations on section dimensions (21.3.1.3, 21.3.1.4):

   width-to-depth ratio $\geq$ 0.3
   width $\geq$ 10 in.

                             $\leq$ [width of supporting column +1.5 (depth of beam)]

   These limitations have been guided by experience with test specimens subjected to cyclic inelastic loading.

2. Limitations on flexural reinforcement ratio (21.3.2.1 and Fig. 31-1):

   $\rho_{min} = 200/f_y$ with at least two continuous bars at both top and bottom of member

   $\rho_{max} = 0.025$

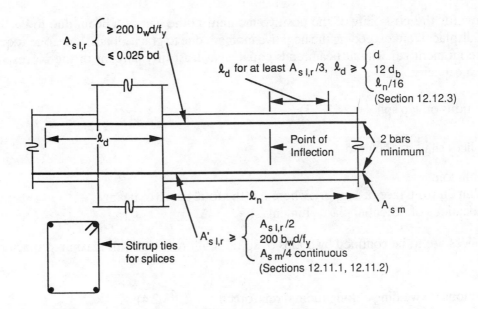

Subscripts l, r, and m indicate left support, right support, and midspan, respectively.

Fig. 31-1 Reinforcement Requirements for Flexural Members

Because the ductility of a flexural member decreases with increasing values of the reinforcement ratio, Chapter 21 limits the maximum reinforcement ratio to 0.025. The use of a limiting value based on the "balanced reinforcement ratio," as given in the main body of the Code, while applicable to members subjected to monotonically increasing loads, fails to address conditions in a flexural member subjected to reversals of inelastic deformation. The limiting ratio of 0.025 is based mainly on considerations of steel congestion and also on limiting shear stresses in beams of typical proportions. From a practical standpoint, lower steel ratios should be used whenever possible. The requirement of at least two bars, top and bottom, is dictated by construction rather than behavioral requirements.

The selection of the size, number, and arrangement of flexural reinforcement should be made with full consideration of construction requirements. This is particularly important in relation to beam-column connections, where construction difficulties can arise as a result of reinforcement congestion. The preparation of large-scale drawings of the connection, showing all beam, column, and joint reinforcement, will help eliminate unanticipated problems in the field. Such large-scale drawings will pay dividends in terms of lower bid prices and a smooth-running construction job.

3. Moment capacity requirements (Section 21.3.2.2):
    (Subscripts l, r, and m indicate left support, right support, and midspan, respectively.)

    At beam ends, $M_{n,l,r}^{+} \geq 0.50 M_{n,l,r}^{-}$

    At any point along beam span,

$$M_n^+ \text{ or } M_n^- \geq 0.25\, M_{n,\ell\, r}^+ \quad \text{(whichever is larger)}$$

To allow for the possibility of the positive moment at the end of a beam due to earthquake-induced lateral displacement exceeding the negative moment due to gravity loads, the Code requires a minimum positive moment capacity at beam ends equal to at least fifty percent of the corresponding negative moment capacity.

4.   Restrictions on lap splices (21.3.2.3):

Lap splices shall not be used

   -within joints
   -within 2h from face of support, where h is the total depth of beam
   -at locations of potential plastic hinging

Lap splices are to be confined by hoops or spiral reinforcement with maximum spacing or pitch of d/4 or 4 in.

5.   Restrictions on welding of longitudinal reinforcement (21.3.2.4):

Welded splices and mechanical connectors may be used provided

   -they are used only on alternate bars in each layer at any section
   -the center-to-center distance between splices of adjacent bars is at least 24 in.

Welded splices and mechanical connectors shall conform to the requirements given in the main body of the Code (Sections 12.14.3.1 through 12.14.3.4). A major requirement is that the splice should develop at least 125 percent of the specified yield strength of the bar.

6.   Transverse reinforcement requirements for confinement and shear (21.3.3):

Transverse reinforcement in beams must satisfy requirements associated with their dual function as confinement reinforcement and shear reinforcement (Fig. 31-2).

Confinement reinforcement in the form of hoops is required

   -over a distance 2h from faces of support (where h is the total depth of the member)
   -over distances 2h on both sides of sections where flexural yielding may occur due
     to earthquake loading.

Hoop spacing must satisfy the following requirements

   -first hoop at 2 in. from face of support
   -maximum spacing $\leq$ d/4
                       $\leq 8 \times$ (diameter of smallest longitudinal bar)
                       $\leq 24 \times$ (diameter of hoop bar)
                       ≤ 12 in.

Where hoops are not required, stirrup spacing $\leq$ d/2

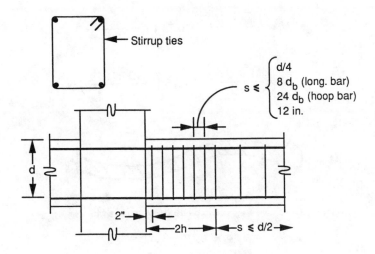

Fig. 31-2 Limitations on Transverse Reinforcement in Beams

Using continuous closed ties for hoop reinforcement may cause numerous construction problems. Section 21.3.3.5 provides that hoops in flexural members may be made up of two pieces of reinforcement: a U-stirrup having 135-degree hooks with six-diameter (but not less than 3 in.) extension anchored in the confined core and a crosstie to make a closed hoop. A ten-diameter extension beyond the 135-degree hook used to be required in the '83 Code.

Shear reinforcement is to be provided so as to preclude shear failure prior to the development of plastic hinges at beam ends. Design shears for determining shear reinforcement are to be based on a condition where plastic hinges occur at the beam ends due to the combined effects of lateral displacements and factored gravity loads (Fig. 31-3). The "probable flexural strength," M_{pr}, associated with plastic hinging, is to be computed using a strength reduction factor $\varphi = 1.0$, and assuming a stress in the tensile reinforcement of $f_s = 1.25\ f_y$.

In determining the required shear reinforcement, the contribution of the concrete, V_c, is to be neglected if the earthquake-induced shear is greater than one-half of the total design shear and the factored axial compressive force including earthquake effects is less than $A_g f'_c/20$. (Section 21.7.2.1).

Shear reinforcement shall be in the form of hoops in regions where confinement is also required, as indicated above for confinement reinforcement (see Fig. 31-2). Otherwise, stirrups or ties may be used (Section 21.7.2.2).

The transverse reinforcement provided shall satisfy the requirement for confinement or shear, whichever is larger.

$$V_l = \frac{M_{prl}^- + M_{prr}^+}{\ell} + 0.75\left(\frac{1.4D + 1.7L}{2}\right)$$

$$V_r = \frac{M_{prl}^+ + M_{prr}^-}{\ell} - 0.75\left(\frac{1.4D + 1.7L}{2}\right)$$

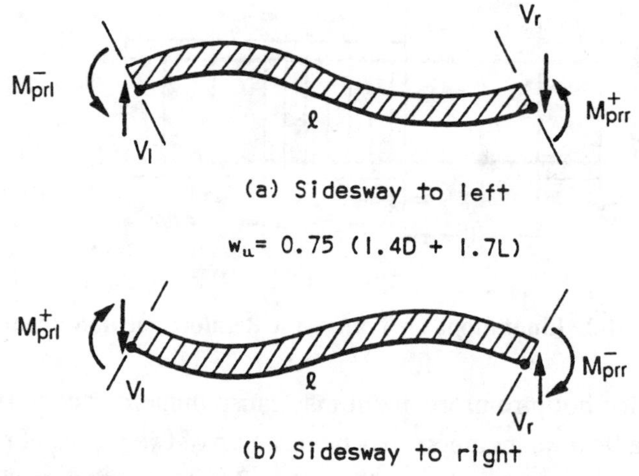

(a) Sidesway to left

$w_u = 0.75 (1.4D + 1.7L)$

(b) Sidesway to right

Fig. 31-3 Loading Cases for Design of Shear Reinforcement in Beams - Uniform Gravity Loads (Section 21.7.1.1)

Because the ductile behavior of earthquake-resistant frames designed to current codes is premised on the ability of the beams to develop plastic hinges with adequate rotational capacity, it is essential to ensure that shear failure does not occur before the flexural capacity of the beams has developed. Transverse reinforcement is required for two related functions: (a) to provide sufficient shear strength so that the full flexural capacity of a member can be developed, and (b) to help ensure adequate rotation capacity in plastic hinging regions by confining the concrete in the compression zone and providing lateral support to the compression steel. To be equally effective with respect to both functions under load reversals, the transverse reinforcement should be placed perpendicular to the longitudinal reinforcement.

Shear reinforcement in the form of stirrups or stirrup-ties is designed for the shear due to the factored gravity loads and the shear corresponding to plastic hinges forming at both ends. Plastic end moments associated with lateral displacements in either direction should be considered (Fig. 31-3). It is important to note that the required shear strength in beams (as in columns) is determined by the flexural strength of the frame member (as well as the factored loads acting on the member) rather than by the factored shear force calculated from a lateral load analysis.

Because of the direct dependence of the required web reinforcement on the yield strength of the flexural reinforcement, any unintended substantial overstrength in the latter could result in a nonductile shear failure preceding the development of the full flexural capacity of a member. The limitations on the actual strength of steel reinforcement mentioned earlier, as well as the use of $\varphi = 1.0$ and $f_s = 1.25\ f_y$ in calculating the probable strength, M_{pr}, of a beam end, are all designed to reduce the chances of a shear failure preceding flexural yielding. The use of $f_s = 1.25\ f_y$ reflects the strong likelihood of the deformation in the tensile reinforcement entering the strain-hardening range.

To allow for load combinations unaccounted for in design, a minimum amount of web reinforcement is required throughout the length of all flexural members. Within regions of potential hinging, stirrup-ties or hoops are required. Hoops may be made of two pieces of reinforcement, as mentioned earlier (Fig. 31-4). Consecutive crossties should have their 90-degree hooks on opposite faces of the flexural member.

21.4 FRAME MEMBERS SUBJECTED TO BENDING AND AXIAL LOAD

Chapter 21 makes a distinction between columns or beam-columns and flexural members on the basis of the magnitude of the factored axial load imposed on a member. Thus, when the factored axial load does not exceed $(A_g f_c'/10)$, the member falls under the category of flexural members, as discussed in the preceding section. When the factored axial load on a member resisting earthquake-induced forces exceeds $(A_g f_c'/10)$, the member is considered a beam-column (21.4.1). The design of such members is governed by the requirements given below.

For the '89 Code, the first paragraph of Section 21.4.1 (former Section A.4.1) has been revised, without changing the intent of the original wording. It was felt that the original wording of Section A.4.1 could be misinterpreted to imply that members which did not meet the requirements of Sections 21.4.1.1 and 21.4.1.2 might be used, but did not have to comply with Section 21.4.

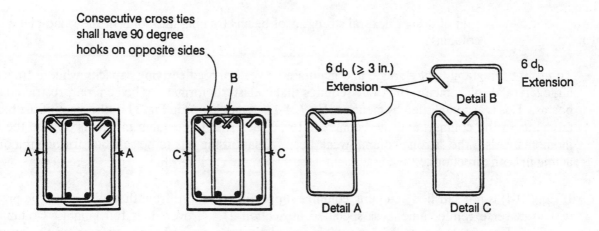

Fig. 31-4 Single- and Two-Piece Hoops

1. Limitations on section dimensions (21.4.1.1, 21.4.1.2):

   Shortest cross-sectional dimension $\geq$ 12 in.
   (measured on a straight line passing
   through the geometric centroid)

   $$\frac{\text{Shortest dimension}}{\text{Perpendicular dimension}} \geq 0.4$$

2. Limitations on longitudinal reinforcement (21.4.3.1):

   $$0.01 \leq \rho_g = A_s/A_g \leq 0.06$$

   Chapter 21 specifies a reduced upper limit for the reinforcement ratio in columns from the 8% of Chapter 10 of the Code to 6%. However, construction considerations will in most cases place the practical upper limit on the reinforcement ratio, ρ_g, near 4%. Convenience in detailing and placing reinforcement in beam-column connections makes it desirable to keep the column reinforcement low.

   The minimum reinforcement ratio is intended to provide for the effects of time-dependent deformations in concrete under axial loads as well as maintain a sizable difference between cracking and yield moments.

3. Flexural strength of columns relative to beams framing into a joint ("strong column-weak beam"provision) (21.4.2.2):

$$\Sigma M_e \geq \frac{6}{5} \Sigma M_g \qquad \text{Eq. (21-1)}$$

   where ΣM_e = sum of design flexural strengths of columns framing into joint. Column flexural strength shall be calculated for the factored axial force, consistent with the direction of lateral loading considered, which results in the lowest flexural strength.

   ΣM_g = sum of design flexural strengths of beams framing into joint in the same plane as the columns.

To ensure the stability of a frame and maintain its vertical load carrying capacity while it undergoes large lateral displacements, the Code requires that inelastic deformations be generally restricted to the beams. This is accomplished by means of Eq. (21-1). As indicated in Fig. 31-5, the signs of the bending moments in the columns and the beams are to be such that the column moments oppose the beam moments. Also, the "strong column-weak beam" relationship has to be satisfied for beam moments acting in both directions.

If Eq. (21-l) is not satisfied at a joint, columns supporting reactions from that joint are to be provided with transverse reinforcement, as specified in Section 21.4.4, over their full height. Columns not satisfying Eq. (21-1) are to be ignored in calculating the strength and stiffness of the structure. However, since such columns contribute to the stiffness of the structure before they suffer severe loss of strength due to plastic hinging, they should not be ignored if neglecting them results in unconservative estimates of design forces. This may occur in determining the design base shear or in calculating the effects of torsion in a structure. Columns not satisfying Eq. (21-1) should satisfy the minimum requirements for "members not proportioned to resist earthquake-induced forces", discussed under Section 21.8.

4. Restriction on lap splices (21.4.3.2):

Lap splices are to be used only within the center half of the column length and shall be designed as tension splices.

5. Welded splices or mechanical connectors in longitudinal reinforcement (21.4.3.2):

Welded splices or mechanical connectors may be used at any section of the column, provided

-they are used only on alternate longitudinal bars at a section
-the center-to-center distance between splices along the longitudinal axis of reinforcement $\geq$ 24 in.

6. Transverse reinforcement for confinement and shear (Section 21.4.4):

As in beams, transverse reinforcement in columns must provide confinement of the concrete core as well as shear resistance. In columns, however, the transverse reinforcement must all be in the form of closed hoops or continuous spiral reinforcement. Sufficient reinforcement should be provided to satisfy the requirement for confinement or shear, whichever is larger.

$$(\phi M_{nt}^+ + \phi M_{nb}^-) \geq \frac{6}{5}(\phi M_{nl}^+ + \phi M_{nr}^-)$$

$$(\phi M_{nt}^- + \phi M_{nb}^+) \geq \frac{6}{5}(\phi M_{nl}^- + \phi M_{nr}^+)$$

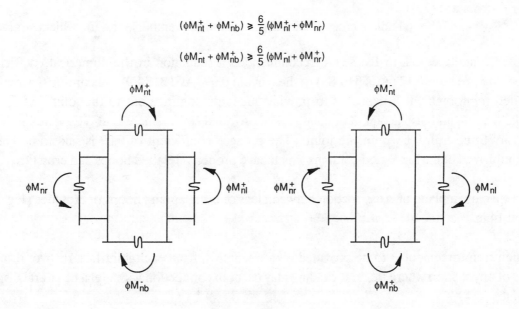

Subscripts l, r, t, and b stand for left support, right support, top of column, and bottom of column, respectively.

Fig. 31-5 "Strong Column-Weak Beam" Frame Requirements

Confinement requirements (Fig. 31-6) are

Volumetric ratio of spiral or circular hoop reinforcement

$$\rho_s \geq 0.12 \frac{f_c'}{f_{yh}} \qquad \text{Eq. (21-2)}$$

$$\geq 0.45 \left(\frac{A_g}{A_{ch}} - 1 \right) \frac{f_c'}{f_{yh}} \qquad \text{Eq. (10-5)}$$

where f_{yh} = specified yield strength of transverse reinforcement, in psi

A_{ch} = core area of column section, measured to the outside of transverse reinforcement, in sq. in.

For rectangular hoop reinforcement, total cross-sectional area within spacing s,

$$A_{sh} \geq 0.09 \, sh_c \frac{f_c'}{f_{yh}} \qquad \text{Eq. (21-4)}$$

$$\geq 0.3 \, sh_c \left(\frac{A_g}{A_{ch}} - 1 \right) \frac{f_c'}{f_{yh}} \qquad \text{Eq. (21-3)}$$

where h_c = cross-sectional dimension of column core, measured center-to-center of confining reinforcement, in.

s = spacing of transverse reinforcement measured along longitudinal axis of member, in., and

s_{max} = 1/4 × (smallest cross-sectional dimension of member), or 4 in., whichever is smaller

ACI-ASCE Committee 352, in Ref. 31.4, recommended a reduction in the numerical coefficient of Eq. (21-4) from the previous 0.12 (ACI 318-83) to the current 0.09 (ACI 318-89). According to Committee 352, the specified reinforcement is expected to provide adequate confinement to the joint during anticipated earthquake loading and displacement demands. The provided confinement is also expected to be sufficient for necessary force transfer within the joint. The reduced coefficient of 0.09 is said to be based on the observed improved behavior of tied columns which have properly detailed hoops and crossties.

Maximum spacing in plane of cross-section between legs of over-lapping hoops or crossties (Fig. 31-7) must not exceed 14 in.

Confinement reinforcement is to be provided over a length l_o from each joint face or over distances l_o on both sides of any section where flexural yielding may occur in connection with inelastic lateral displacements of the frame,

where $l_o \geq$ depth h of member

$\geq$ clear span of member/6

$\geq$ 18 in.

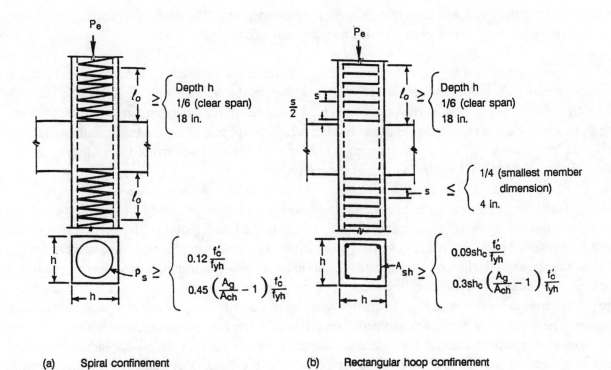

(a)   Spiral confinement
      reinforcement

(b)   Rectangular hoop confinement
      reinforcement A_{sh}

Fig. 31-6   Confinement Requirements at Column Ends

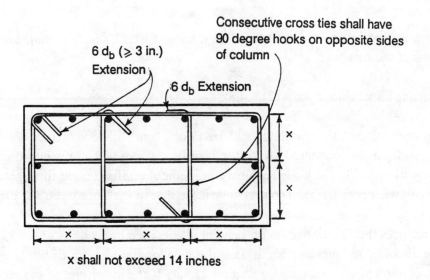

x shall not exceed 14 inches

Fig. 31-7   Transverse Reinforcement in Columns

Spirals represent the most efficient form of confinement reinforcement. The extension of such spirals into the beam-column joint, however, may cause some construction difficulties.

Rectangular hoops, when used in place of spirals, are considered less effective with respect to confinement of the concrete core. Their effectiveness is increased, however, by the use of supplementary crossties, each end of which has to engage a peripheral longitudinal bar. Crossties of the same bar size and spacing as the hoops may be used. Consecutive crossties are to be alternated end for end along the longitudinal reinforcement (Fig. 31-7). Crossties or legs of overlapping hoops are to be spaced no farther than 14 in. apart in a direction perpendicular to the longitudinal axis of the member.

In addition to satisfying confinement requirements, the transverse reinforcement in columns must resist the maximum shear associated with the formation of plastic hinges in the frame (Section 21.7.1.2). Although the "strong column-weak beam" provision governing the relative moment strengths of beams and columns is intended to have most of the inelastic deformation occur in the beams of a frame, the Code recognizes that hinging can occur in the columns. Thus, the shear reinforcement in columns is to be based on the condition that the probable moment strength, M_{pr} (calculated using $\varphi = 1.0$ and $f_s = 1.25f_y$) is developed at the ends of a column. The value of such probable moment strength is to be the maximum consistent with the possible factored axial compressive forces on the column. Moments associated with lateral displacements of the structure in both directions are to be considered, as indicated in Fig. 31-8. The axial load corresponding to the maximum moment capacity should then be used in computing the permissible shear stress in concrete, v_c.

In ACI 318-89, a sentence has been added to Section 21.7.1.2, pointing out that there is no need to calculate shear based on the full column moment strengths when the upper limit on moment is controlled by flexural members framing into the column. Where beams frame into opposite sides of a joint, the combined strength may be the sum of the negative moment strength of the beam on one side of the joint and the positive moment strength of the beam on the other side of the joint. Moment strengths must be determined using a strength reduction factor of 1.0 and reinforcing steel stress equal to at least $1.25f_y$. Distribution of the combined moment strength of the beams to the columns above and below the joint should be based on analysis.

The '89 Code also states that in no case shall the design shear force V_c be less than the factored shear determined by analysis of the structure.

7. Columns supporting discontinued walls (21.4.4.5):

Columns supporting discontinued shearwalls or stiff partitions tend to be subjected to large shear and compressive forces and can be expected to suffer significant inelastic deformations during strong earthquakes. In recognition of this, the Code requires confinement reinforcement throughout the height of such columns (Fig. 31-9) whenever the axial compressive force due to earthquake effects exceeds ($A_g f_c'/10$).

ACI 318-89 further requires that transverse reinforcement as specified in Sections 21.4.4.1 through 21.4.4.3 shall extend into the discontinued member for at least the development length of the largest longitudinal reinforcement in the column in accordance with Section 21.6.4. If the lower end of the column terminates on a wall, transverse reinforcement as specified above shall extend into the wall for at least the development length of the largest longitudinal reinforcement in the column at the point of termination. If the column terminates on a footing or mat, transverse reinforcement as specified shall extend at least 12 in. into the footing or mat. The extension of transverse reinforcement above and below the discontinuity is deemed necessary to provide confinement in areas of distribution of concentrated loads into a wall. At the foundation level, the footing pad or mat will provide confinement.

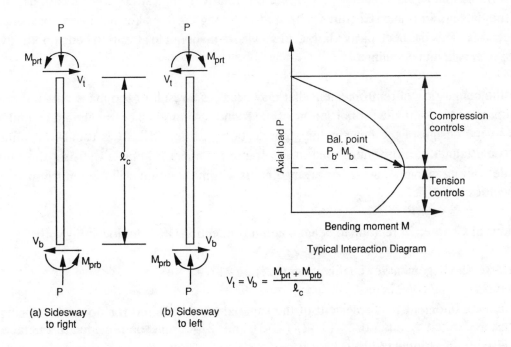

$$V_t = V_b = \frac{M_{prt} + M_{prb}}{\ell_c}$$

(a) Sidesway to right

(b) Sidesway to left

Fig. 31-8 Loading Cases for Design of Shear Reinforcement in Columns

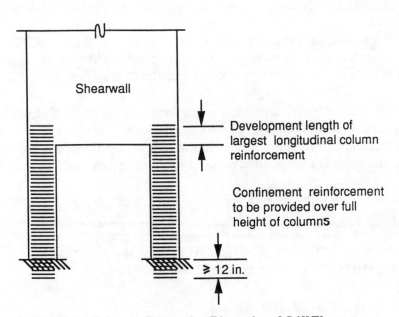

Fig. 31-9 Columns Supporting Discontinued Stiff Elements

21.6 JOINTS OF FRAMES

In conventional reinforced concrete buildings, the beam-column connections normally are not designed by the structural engineer. Detailing of bars within the joints is usually delegated to a draftsman or detailer. In earthquake-resistant frames, however, the design of beam-column connections requires as much attention as the design of the members themselves, since the integrity of the structure may well depend on the proper functioning of such connections. Beam-column joints represent regions of geometric and stiffness discontinuities in a frame and as such tend to be subjected to relatively high stress concentrations. A substantial portion of the damage in frame structures subjected to strong earthquakes has been observed to occur at these connections. This has been particularly evident where inadequate attention had apparently been given to their proper design and detailing.

Because of the congestion of reinforcement that may occur as a result of too many bars converging within the limited space of the joint, the proportioning of the frame columns and beams should be undertaken with due regard to the design of the beam-column connection. Little difficulty is usually encountered if the amount of longitudinal reinforcement used in the frame members is kept low. Also, the preparation of large-scale detailed drawings showing bar arrangements within the joints will help much in avoiding unexpected difficulties in the field.

The provisions of Chapter 21 relating to beam-column connections have to do mainly with:

1. Transverse reinforcement for confinement (Section 21.6.2):

    Minimum confinement reinforcement of the same amount required for potential hinging regions in columns, as defined by Eqs. (10-5), (21-2), (21-3), and (21-4), must be provided through beam-column joints around the column reinforcement.

    For joints confined on all four faces, a 50% reduction in the amount of confinement reinforcement is allowed. A member that frames into a face is considered to provide confinement to the joint if at least three-quarters of the face of the joint is covered by the framing member. The '89 Code further allows that, where a 50% reduction in the amount of confinement reinforcement is allowed, the spacing specified in Section 21.4.4.2(b) may be increased to 6 in. (Section 21.6.6.2).

    The transverse reinforcement in a beam-column connection is intended to provide adequate confinement of the concrete to ensure its ductile behavior and to allow it to maintain its vertical load-carrying capacity even after spalling of the outer shell. It also helps resist the shear forces transmitted by the framing members and improves the bond between steel and concrete within the connection.

    The minimum amount of confinement reinforcement, as given by Eqs. (10-5), (21-2), (21-3), and (21-4), must be provided through the joint regardless of the magnitude of the calculated shear force in the joint. The 50% reduction in the amount of confinement reinforcement allowed for joints having horizontal members framing into all four sides recognizes the beneficial effect provided by these members in resisting the bursting pressures generated within the joint.

2. Design for shear (21.6.1.1, 21.6.1.2, 21.6.3):

Shear force in a joint is to be calculated by assuming stress in tensile reinforcement of framing beams equal to 1.25 f_y. Shear strength of the connection is to be computed as

$$\varphi V_c = 20\sqrt{f_c'}\, A_j \text{ for joints confined on all four faces}$$
$$= 15\sqrt{f_c'}\, A_j \text{ for joints confined on three faces or on two opposite faces}$$
$$= 12\sqrt{f_c'}\, A_j \text{ for other joints}$$

(for normal weight concrete)

where φ = 0.85 for shear in joints

A_j = effective cross-sectional area within a joint, in a plane parallel to the plane of reinforcement generating shear in the joint. The joint depth is the overall depth of the column. Where a beam frames into a support of larger width, the effective width of the joint is not to exceed the smaller of:

(a) beam width plus the joint depth

(b) twice the smaller perpendicular distance from the longitudinal axis of the beam to the column side. See Fig. 31-10.

In no case is A_j greater than the column cross-sectional area.

In ACI 318-83, the definition of A_j in Section 21.0 did not agree with the definition of A_j in Section A.6.3.1. This has been corrected in ACI 318-89, in conjunction with revisions to Section 21.6.3.1.

The three-level shear strength provision for joints in Section 21.6.3 of ACI 318-89 is based on the recommendation of ACI Committee 352. Test data reviewed by Committee 352[31.4] indicated that the lower shear strength value given in Section A.6.3.1 of ACI 318-83 was unconservative when applied to corner joints.

For lightweight concrete, the nominal shear strength of the joint, shall be taken as no more than three-fourths of the values given above for normal weight concrete.

As indicated in Fig. 31-11, the design shear is based on the most critical combination of horizontal shears transmitted by the framing beams and columns. Tests have indicated that plastic hinging at the ends of beams, for deformations associated with response to strong earthquakes, impose strains in the flexural reinforcement well in excess of the yield strain. Because of the likelihood of strains in the tensile reinforcement going into the strain-hardening range, and to allow for the actual yield strength of the steel exceeding the specified value, the Code requires that the horizontal shear in the joint be determined by assuming the stress in the flexural tensile steel to be equal to 1.25 f_y.

Test results have indicated that the shear strength of joints is not too sensitive to the amount of transverse (shear) reinforcement. Based on these results, the 1983 edition of the ACI Code made the shear strength of beam-column connections a function only of the cross-sectional area of the joint, A_j, and f_c'.

When the factored shear in the joint exceeds the shear strength of the joint, the designer may either increase the column size or increase the depth of the beams. The former will increase the shear capacity of the joint section, while the latter will tend to reduce the required amount of flexural reinforcement in the beams, with an accompanying decrease in the shear transmitted to the joint.

3. Anchorage of longitudinal beam reinforcement within confined column core (21.6.1.3):

Beam longitudinal reinforcement terminated in a column is to be extended to the far face of the confined column core and anchored in tension according to Section 21.6.4 (discussed below) and in compression according to Chapter 12 of the Code.

4. Development length for reinforcement in tension (21.6.4):

For bar sizes #3 through #11 with standard 90-degree hooks (Fig. 31-12) in normal weight concrete,

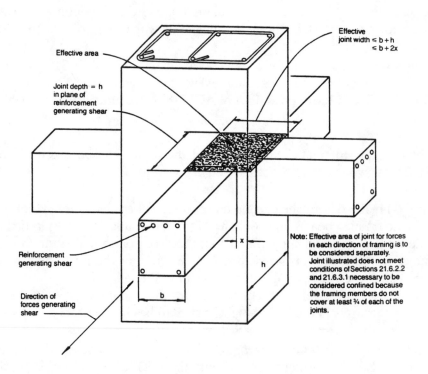

Fig. 31-10 Effective Area of Joint

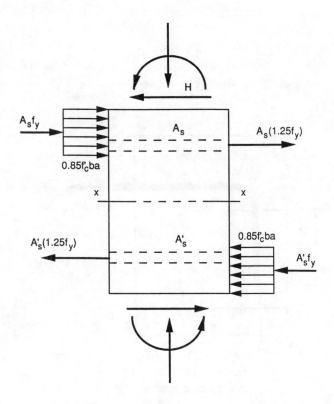

Fig. 31-11 Horizontal Shear in Beam-Column Connection

development length, $\ell_{dh} \geq f_y d_b / 65 \sqrt{f_c'}$

$\qquad \geq 8d_b$ (d_b is bar diameter)

$\qquad \geq 6$ in.

for the same bar sizes with 90-degree hooks embedded in lightweight concrete, ℓ_{dh} is to be at least 1.25 times the above values.

The 90-degree hook shall be located within the confined core of a column or other boundary element.

For straight bars of sizes #3 through #11,

$\ell_d = 2.5 \times \ell_{dh}$ specified for bars with 90-degree hooks) when the depth of concrete cast in one lift beneath the bar ≤ 12 in.

$\qquad = 3.5 \times \ell_{dh}$ specified for bars with 90-degree hooks) when the depth of concrete cast in one lift beneath the bar > 12 in.

Straight bars terminated at a joint shall pass through the confined core of a column or of a boundary member. Any portion of the straight embedment length not within the confined core shall be increased by a factor of 1.6.

31-23

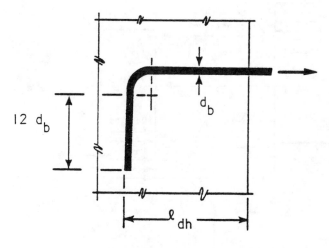

Note: Hook must be within
confined core

Fig. 31-12 Standard 90-Degree Hook

The expression for ℓ_{dh} given above already includes the coefficients 0.7 (for concrete cover) and 0.80 (for ties) that are normally applied to the required basic development length, ℓ_{hb}. This is because Chapter 21 requires that hooks be embedded in the confined core of the column or other boundary element. The expression for ℓ_{dh} also includes a factor of about 1.4, representing an increase over the development length required for conventional structures, to provide for the effect of load reversals.

Except in very large columns, it is usually not possible to develop the yield strength of a reinforcing bar from a framing beam within the width of a column. Where beam reinforcement can extend through a column, its capacity is developed by embedment in the column and within the compression zone of the beam on the far side of the connection (Fig. 31-1). Where no beam is present on the opposite side of a column, such as in exterior columns, the flexural reinforcement in a framing beam has to be developed within the confined region of the column. This is usually done by means of a standard 90-degree hook plus whatever extension is necessary to develop the bar, the development length being measured from the near face of the column (Fig. 31-11).

Chapter 21 makes no provision for the use of #14 and #18 bars because of a lack of information on the behavior of anchorages of such bars when subjected to load reversals simulating earthquake effects.

21.5 STRUCTURAL WALLS, DIAPHRAGMS, AND TRUSSES

21.7 SHEAR-STRENGTH REQUIREMENTS

When properly proportioned so that they possess adequate lateral stiffness to reduce interstory distortions due to earthquake-induced motions, structural walls (also called shearwalls) reduce the likelihood of damage to the nonstructural elements of a building. When used with rigid frames, walls form a system that combines the gravity-load-carrying efficiency of the rigid frame with the lateral-load-resisting efficiency of the structural wall.

Observations of the comparative performance of rigid-frame buildings and buildings stiffened by shearwalls during recent earthquakes have pointed to the consistently better performance of the latter. The performance of buildings stiffened by properly designed shearwalls has been better with respect to both safety and damage control. The need to ensure that critical facilities remain operational after a major tremor and the need to reduce economic losses from structural and nonstructural damage, in addition to the primary requirement of life safety, i.e., no collapse, has focused attention on the desirability of introducing greater lateral stiffness into earthquake-resistant multistory structures. Shearwalls, which have long been used in designing for wind resistance, offer a logical and efficient solution to the problem of lateral stiffening of multistory buildings.

Shearwalls are normally much stiffer than regular frame elements and are therefore subjected to correspondingly greater lateral forces during response to earthquake motions. Because of their relatively greater depth, the lateral deformation capacities of walls are limited, so that, for a given amount of lateral displacement, shearwalls tend to exhibit greater apparent distress than frame members. However, over a broad period range, a shearwall structure, which is substantially stiffer and hence has a shorter period than a frame structure, will suffer less lateral displacement than the frame, when subjected to the same ground motion intensity. Shearwalls with a height-to-depth ratio in excess of 2 behave essentially as vertical cantilever beams and should therefore be designed as flexural members, with their strength governed by flexure rather than by shear.

Primarily because of the greater stiffness of shearwall structures, but also because of earlier concerns about the deformation capacities of shearwalls, codes have specified larger design lateral forces for such structures.

Isolated shearwalls or individual walls connected to frames will tend to yield first at the base where the moment is the greatest. Coupled walls, i.e., two or more walls linked by short, rigidly-connected beams at the floor levels, on the other hand, have the desirable feature that significant energy dissipation through inelastic action in the coupling beams can be made to precede hinging at the bases of the walls.

The principal provisions of ACI Chapter 21 relating to structural walls (and diaphragms) are as follows (Fig. 31-13):

1. Reinforcement (21.5.2.1. 21.5.2.2):

    Walls and (diaphragms) are to be provided with shear reinforcement in two orthogonal directions in the plane of the wall. Minimum reinforcement ratios for both longitudinal and transverse directions,

    $$\rho_v = \frac{A_{sv}}{A_{cv}} = \rho_n \geq 0.0025$$

    with reinforcement continuous and distributed uniformly across the shear area where A_{cv} = net area of concrete section. i.e.. product of thickness and length of section in direction of shear considered

A_{sv} = projection on A_{cv} of area of shear reinforcement crossing the plane of A_{cv}, and

ρ_n = ratio of distributed shear reinforcement on a plane perpendicular to plane of A_{cv},

The spacing of reinforcement must not exceed 18 in.

At least two curtains of reinforcement - each having bars running in the longitudinal and transverse directions - are to be provided if the in-plane factored shear force assigned to the wall exceeds $2A_{cv}\sqrt{f_c'}$.

The use of two curtains of reinforcement in walls subjected to significant shears ($> 2A_{cv}\sqrt{f_c'}$) serves to reduce fragmentation and premature deterioration of the concrete under load reversals into the inelastic range. Distributing the reinforcement uniformly across the height and horizontal length of the wall helps control the width of inclined cracks.

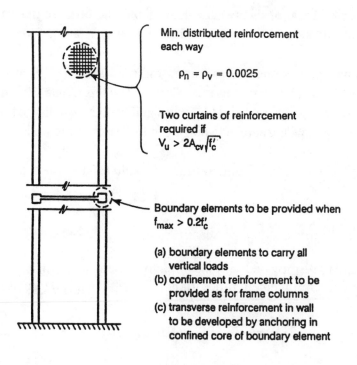

Min. distributed reinforcement each way

$$\rho_n = \rho_v = 0.0025$$

Two curtains of reinforcement required if

$V_u > 2A_{cv}\sqrt{f_c'}$

Boundary elements to be provided when $f_{max} > 0.2f_c'$

(a) boundary elements to carry all vertical loads
(b) confinement reinforcement to be provided as for frame columns
(c) transverse reinforcement in wall to be developed by anchoring in confined core of boundary element

Fig. 31-13 Shearwall Design and Detailing Requirements

It should be noted that the vertical reinforcement in the boundary elements (or reinforcement concentrated near the edges of the wall when no boundary elements are used) for resisting flexure in the wall is not to be included in determining satisfaction of the requirements for ρ_v or ρ_n.

b. Shear strength of walls (and diaphragms) (21.7.3):

For walls with height-to-horizontal length ratio, $h_w/\ell_w \geq 2.0$, the shear strength is to be determined from the expression

$$\varphi V_n = \varphi A_{cv}\left(2\sqrt{f_c'} + \rho_n f_y\right) \qquad\qquad \text{Eq. (21-6)}$$

where φ = 0.60, unless nominal shear strength provided exceeds the shear corresponding to development of nominal flexural capacity of wall

A_{cv} = as defined earlier

ρ_n = as defined earlier

h_w = height of entire wall or of segment of wall considered

ℓ_w = length of entire wall or of segment of wall considered in direction of shear force

For walls with $h_w/\ell_w < 2.0$, the shear strength may be determined from

$$\varphi V_n = \varphi A_{cv}\left(\alpha_c\sqrt{f_c'} + \rho_n f_y\right) \qquad\qquad \text{Eq. (21-7)}$$

where the coefficient α_c varies linearly from a value of 3.0 for $h_w/\ell_w = 1.5$ to 2.0 for $h_w/\ell_w = 2.0$.

Where a wall is divided into several segments by openings, the value of the ratio h_w/ℓ_w to be used in calculating V_n for any segment shall not be less than the corresponding ratio for the entire wall.

The nominal shear strength V_n of all wall segments or piers resisting a common lateral force shall not exceed $8 A_{cv}\sqrt{f_c'}$ where A_{cv} is the total cross-sectional area of the walls. The nominal shear strength of any individual vertical or horizontal segment of wall shall not exceed $10 A_{cp}\sqrt{f_c'}$ where A_{cp} is the cross-sectional area of the wall segment.

Chapter 21 allows calculation of the shear strength of any shearwall using a coefficient $\alpha_c = 2.0$. However, advantage can be taken of the greater observed shear strength of walls with low height-to-horizontal length (h_w/ℓ_w) ratios by using an α_c - value of up to 3.0 for $h_w/\ell_w = 1.5$ or less.

Chapter 21 limits the average nominal unit shear strength of structural walls to $8\sqrt{f_c'}$, with allowance for exceeding this average in any individual wall in a group of walls or wall segments, provided that the unit shear in the individual wall does not exceed $10\sqrt{f_c'}$. This upper bound on strength that may be developed in any individual segment is intended to limit the degree of shear redistribution among several connected wall segments. A wall segment refers to a part of wall bounded by openings or by an opening and an edge.

It is important to note that Section 9.3.4.1 of ACI 318-89 requires the use of a strength reduction factor φ for shear of 0.6 for all members (except joints) where the nominal shear strength is less than the shear

31-27

corresponding to the development of the nominal flexural strength of the member. In the case of beams, the design shears are obtained by assuming maximum probable moment strengths at the ends (21.7.1.1). Similarly, for a column, the design shears are determined not by applying load factors to shears obtained from a lateral load analysis, but from the consideration of maximum developable moments, consistent with the axial force on the column, occurring at the column ends. This approach to shear design is intended to ensure that even when flexural hinging occurs at member ends due to earthquake-induced deformations, no shear failure would develop. Under the above conditions, the Code allows the use of the normal strength reduction factor for shear of 0.85. When design shears are not based on the condition of maximum probable flexural strength being developed at member ends, the Code requires the use of a lower shear strength reduction factor to achieve the same result, i.e., prevention of premature shear failure.

In the case of shearwalls, a condition similar to that used for the shear design of beams and columns is not as readily established. This is primarily because the magnitude of the shear at the base of a wall (or at any level above) is influenced significantly by the forces and deformations beyond the particular level considered. Chapter 21 thus prescribes that the design shear force V_u for a shearwall shall be obtained from lateral load analysis of the structure containing the wall in accordance with the factored loads and combinations specified in Section 9.2 (21.7.1.3). Unlike the flexural behavior of beams and columns in a frame, which can be considered as close-coupled systems, i.e., with the forces and deformations in the members determined primarily by the displacements in the end joints, the state of flexural deformation at any section of a shearwall (a far-coupled system) is influenced significantly by the displacements at locations far removed from the section considered. Results of dynamic inelastic analyses of isolated shearwalls under earthquake loads also indicate that the base shear in such walls is strongly influenced by the higher modes of response.

A distribution of static lateral forces along the height of a wall, corresponding to the fundamental mode, such as is assumed by most current seismic codes, may produce flexural yielding at the base if the section at the base of the wall is designed for such yielding. However, other distributions of lateral forces, having a resultant closer to the base, can produce yielding at the base only if the magnitude of the resultant horizontal force, and hence the base shear, is increased. Results of research on isolated walls, which would also apply to frame-shearwall systems in which the frame is flexible relative to the wall, in fact indicate that for a wide range of wall properties and input motions, the resultant of the dynamic horizontal forces producing yielding at the base of the wall generally occurs well below the two-thirds-of-total height level associated with fundamental mode response. This would imply significantly larger base shears than those due to lateral forces distributed according to the fundamental mode response. Research on isolated walls indicates ratios of maximum dynamic shears to "fundamental mode shears" (i.e., shears associated with horizontal forces distributed according to the seismic codes) ranging from 1.3 to 4.0, the value of the ratio increasing with fundamental period.

3. Development length and splices (21.5.2.4, 21.5.3.5, 21.5.3.6):

All continuous reinforcement is to be anchored or spliced in accordance with the provisions for reinforcement in tension (21.6.4).

Where boundary elements are present, the transverse reinforcement in walls is to be anchored within the confined core of the boundary element to develop the yield stress in tension of the transverse reinforcement.

Actual forces in longitudinal bars of stiff members may exceed calculated forces. Because of this likelihood, and the importance of maintaining the flexural capacity of a wall, the Code requires that all continuous reinforcement be developed fully.

Similarly, the horizontal reinforcement in walls requiring boundary elements is called upon to function as web reinforcement. Because of this, the Code requires that such bars be fully anchored in the boundary elements (which act as flanges of vertical cantilever beams). Standard 90-degree hooks should be used whenever possible. Such hooks minimize the loss of bond that may otherwise result due to the occurrence of large transverse cracks in the boundary elements when subjected to large inelastic deformations.

Transverse reinforcement terminating at the edges of shearwalls without boundary elements is required to have a standard hook engaging the edge reinforcement, or the edge reinforcement must be enclosed in U-stirrups having the same size and spacing as, and spliced to, the transverse reinforcement. This requirement, now in ACI 318-89, need not apply when V_u in the plane of the wall is less than $A_{cv}\sqrt{f_c'}$. The addition of hooks or U-stirrups at the ends of transverse shearwall reinforcement provides anchorage so that the reinforcement will be effective in resisting shear forces. It will also tend to inhibit the buckling of the vertical edge reinforcement. In walls with low in-plane shear, the development of the horizontal reinforcement is not necessary.

4.   Boundary elements (21.5.3):

Boundary elements are to be provided, both along vertical boundaries of a wall and around the edges of openings, if any, when the maximum extreme-fiber stress in the wall due to factored forces including earthquake forces exceeds 0.2 f_c'. The boundary members may be discontinued when the calculated compressive stress becomes less than 0.15 f_c'.

Boundary elements need not be provided if the **entire** wall is reinforced in accordance with the provisions governing transverse reinforcement for members subjected to axial load and bending, as given by Eqs. (10-5), (21-2), (21-3), (21-4) and the related spacing requirements.

Boundary members of structural walls are to be designed to carry all the factored vertical loads on the wall, including self-weight and gravity loads tributary to the wall, as well as the vertical force required to resist the overturning moment due to factored earthquake loads. Such boundary elements are to be provided with confinement reinforcement in accordance with Eqs. (10-5), (21-2), (21-3), (21-4) and the related spacing requirements.

The Code uses a concrete stress of 0.2 f_c', calculated using a linearly elastic model based on gross sections of structural members and factored forces, as indicative of significant compression. Shearwalls subjected to compressive stresses exceeding this value are generally required to have boundary elements.

The condition assumed in requiring that boundary members be designed for all gravity loads as well as the vertical forces associated with overturning of the wall due to earthquake forces is illustrated in Fig. 31-14. This requirement assumes that the boundary member alone may have to carry all the vertical (compressive) forces at the critical wall section when the maximum horizontal earthquake force acts on the wall. Under load reversals, this condition imposes severe demands on the concrete in the boundary element. Hence the requirement for confinement reinforcement similar to those for members subjected to axial load and bending.

The design of the boundary element is carried out by considering it as an axially loaded column subjected to the factored compressive axial force at the critical section.

Diaphragms of reinforced concrete, such as floor slabs, that are designed to transmit horizontal forces through bending and shear in their plane, are treated in much the same manner as shearwalls.

Truss elements of reinforced concrete are also covered, although very briefly, in Chapter 21 of ACI 318-89. A major requirement for truss elements relates to the provision of special transverse reinforcement when the compressive stress exceeds $0.2 f_c'$.

21.8 FRAME MEMBERS NOT PROPORTIONED TO RESIST FORCES INDUCED BY EARTHQUAKE MOTIONS

Frame members that are not relied on to provide lateral resistance to earthquake-induced forces need not generally satisfy the stringent requirements governing lateral-load-resisting elements. This refers mainly to the requirements for transverse reinforcement for confinement and shear.

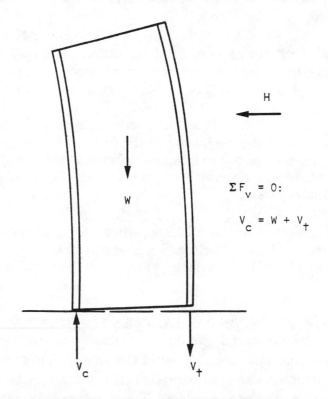

Fig. 31-14 Loading Condition Assumed for Design of Shearwall Boundary Element

A special requirement for non-lateral-load-resisting elements is that they be checked for adequacy with respect to a lateral displacement twice that calculated for the structure under the factored lateral forces. This requirement should enable the gravity-load system to maintain its vertical-load-carrying capacity, without reduction, under the specified lateral forces. Although elements of the gravity-load system need not be designed for moments related to the lateral forces, they may have to be provided with adequate confinement reinforcement in regions where plastic hinging can occur.

There had been some disagreement among engineers as to the intent of Section A.8 of ACI 318-83, and even more disagreement as to how the provisions were to be applied. Section 21.8 of ACI 318-89 offers specific requirements and has become more restrictive than the provisions in Section A.8.

The new Section 21.8.1 requires that frame members assumed not to contribute to lateral resistance shall be detailed according to Section 21.8.1.1 or Section 21.8.1.2, depending on the magnitude of moments induced in those members when subjected to twice the lateral displacement under the **factored** lateral forces.

Section 21.8.1.1 requires that when the induced moment exceeds the design moment strength of the frame member, members with factored gravity axial forces not exceeding $A_g f_c'/10$ must satisfy Sections 21.3.2.1 (minimum flexural reinforcement) and 21.7.1.1 (computation of design shear), and members with factored gravity axial forces exceeding $A_g f_c'/10$ must satisfy Sections 21.4.4 (transverse reinforcement), 21.5.2.1 (minimum horizontal and vertical reinforcement in shearwalls), and 21.7.1.2 (computation of design shear).

Section 21.8.1.2 specifies that when the induced moment does not exceed the design moment strength of the frame member, the member need only satisfy Section 21.3.2.1 (minimum flexural reinforcement).

For gravity-load frame members subjected to factored axial compressive forces exceeding $(A_g f_c'/10)$, the following requirements relating to transverse reinforcement have to be satisfied:

Maximum tie spacing, (over length ℓ_o from face of joint)
$$s_O \leq 8 \times \text{(diameter of smallest longitudinal bar)}$$
$$\leq 24 \times \text{tie diameter}$$
$$\leq \tfrac{1}{2} \text{ least cross}-\text{sectional dimension of column}$$

where, $\ell_o \geq 1/6$ clear height of column
$$\geq \text{maximum cross}-\text{sectional of column}$$
$$\geq 18 \text{ in.}$$

The first tie is to be located within a distance of $s_O/2$ from face of joint. Maximum tie spacing in any part of the column $= 2 s_O$.

21.9 REQUIREMENTS FOR FRAMES IN REGIONS OF MODERATE SEISMIC RISK

Although ACI Chapter 21 does not define "moderate seismic risk" in terms of a commonly accepted quantitative measure, it assumes that the probable ground motion intensity in such regions would be a fraction of that expected in a high seismic risk zone, to which the bulk of Chapter 21 is addressed. By the above description, an area of moderate seismic risk would correspond to Zone 2 as defined in UBC-88[31.2] or ANSI-88.[31.3]

For regions of moderate seismic risk, the provisions for the design of shearwalls given in the main body of the ACI Code are considered sufficient to provide the necessary toughness. The requirements of Chapter 21 for structures in moderate-risk areas relate mainly to frames.

The distinction between flexural members and columns based on an axial compressive force of $A_g f'_c/10$, as used in high seismic risk zones, also applies in regions of moderate seismicity (21.9.2).

For shear design of beams, columns or two-way slabs resisting earthquake effects, the magnitude of the design shear (21.9.3) should not be less than either

(a) The sum of the shear associated with the development of nominal moment strength at each restrained end and that due to factored gravity loads. This is similar to the corresponding requirements for high-risk zones illustrated in Fig. 31-3, except that the stress in the flexural reinforcement is taken as f_y rather than 1.25 f_y, or

(b) The maximum factored shear corresponding to the application of design gravity and earthquake forces, but with the earthquake effect taken at twice the calculated value. Thus, if the critical load combination consists of dead load (D) + live load (L) + earthquake effects (E), then the design shear is to be computed from

$$U = 0.75 [1.4D + 1.7L + 2(1.87E)]$$

Detailing requirements for beams (21.9.4) - The positive moment strength at the face of a joint shall not be less than 1/3 the negative moment capacity at the same section. (This compares with 1/2 for beams in areas of high seismic risk, Fig. 31-1.) The moment strength - positive or negative - at any section is not to be less than 1/5 the maximum moment strength at either end of the beam.

Stirrup spacing requirements are identical to those for beams in regions of high seismic risk (Fig. 31-2). However, closed hoops are not required within regions of potential plastic hinging.

Detailing requirements for columns (21.9.5) - The tie spacing requirements for columns are identical to those for gravity load carrying members as given in Section 21.8.2.2 (see above).

Detailing requirements for two-way slabs without beams (21.9.6) - It is worth noting that requirements for two-way slabs without beams are included in Chapter 21 for frames in regions of moderate seismic risk only. This suggests that the Code considers the use of properly designed two-way slabs without beams as acceptable components of the lateral-load-resisting system in regions of moderate seismic risk only.

The requirements for slabs without beams are illustrated in Figs. 31-15 and 31-16. The moment M_s in Fig. 31-15 is the portion of the factored slab moment balanced by the support moment. The factor γ_f represents the fraction of the unbalanced moment at a joint transferred by flexure, as defined in Chapter 13 of the Code, i.e.

$$\gamma_f = \frac{1}{1 + (2/3)\ \sqrt{b_1/b_2}}$$

Eq. (13-1)

where

b_1 = width of the critical section defined in Section 11.12.1.2 measured in the direction of the span for which moments are determined

b_2 = width of the critical section defined in Section 11.12.1.2 measured in the direction perpendicular to b_1

d = effective depth of slab

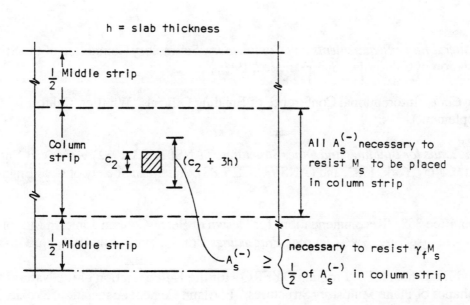

Fig. 31-15   Requirements Relating to Location of Reinforcement in Slabs without Beams in Regions of Moderate Seismic Risk

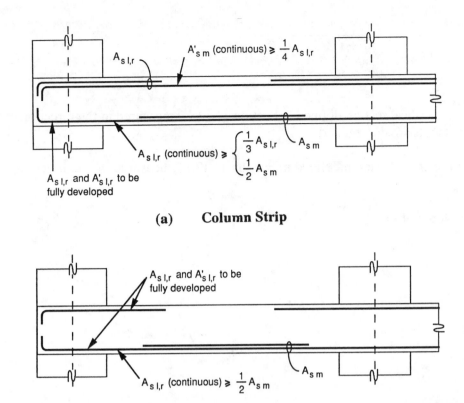

(a)   Column Strip

(b)   Middle Strip

Fig. 31-16   Requirements Relating to Arrangement of Reinforcement in Slabs without Beam in Regions of Moderate Seismic Risk

REFERENCES

31.1 *Recommended Lateral Force Requirements and Commentary,* Seismology Committee of the Structural Engineers Association of California, San Francisco, 1989.

31.2 *Uniform Building Code,* International Conference of Building Officials, Whittier, California, 1988 (including 1989 supplement).

31.3 *Minimum Design Loads for Buildings and Other Structures,* (ANSI A58.1-1982), American National Standards Institute, New York, 1982, and (ANSI/ASCE 7-88), American Society of Civil Engineers, New York, 1989.

31.4 ACI-ASCE Committee 352, "Recommendations for Design of Beam-Column Joints in Monolithic Reinforced Concrete Structures," *ACI Journal, Proceedings* Vol. 73, No. 7, July 1976, pp. 375-393.

31.5 Fugelso, L. E. and Derecho, A. T., "Program DYFRQ - for the Determination of the Natural Frequencies and Mode Shapes of Plane Multistory Structures," Portland Cement Association, Skokie, IL, 1975 (unpublished).

31.6 *STMFR, Analysis of Plane Multistory Frame-Shearwall Structures Under Lateral and Gravity Loads,*

31.7 Derecho, A. T., Fintel, M. and Ghosh, S. k., "Earthquake-Resistant Structures," Ch. 12, Handbook of Concrete Engineering, M. Fintel, ed., Van Nostrand Reinhold Company, New York, 2nd Ed., 1985, pp. 411-513.

31.8 *Design Handbook in Accordance with the Strength Design Method of ACI 318-83: Vol.2 - Columns,* Publication SP17A(85), American Concrete Institute, Detroit, 1985.

Example 31.1—Calculation of Seismic Design Forces for a 12-Story Frame-Shearwall Building and its Components

The computation of seismic design loads on a 12-story frame-shearwall building located in Uniform Building Code[31.2] Seismic Zone 4 is illustrated below. Also illustrated is the computation of seismic design forces for a number of typical members of the building. The design loads are calculated according to the 1988 edition of the Uniform Building Code.

A typical plan and elevation of the structure considered are shown in Figs. 31-17(a) and (b). The columns and shearwalls have constant cross sections throughout the height of the building,* the bases of the lowest story segments being assumed fixed. The beams and the slabs also have the same dimensions at all floor levels. Although the element dimensions in this example are within the practical range, the structure itself is a hypothetical one, and has been chosen mainly for illustrative purposes. Other pertinent design data are as follows:

Service loads-vertical:

Live load: 50 psf (UBC-88, Table 23-A)
          Additional average value to allow
          for heavier load on corridors = 25 psf
          Thus, total average live load = 75 psf

Superimposed   Average for partitions = 20 psf
dead load: Ceiling and mechanical = 10 psf
          Thus, total average superimposed
          dead load = 30 psf

Material properties:

Concrete: $f_c' = 4000$ psi, $w_c = 145$ pcf
Reinforcement: $f_y = 60,000$ psi

---

* The uniformity in member dimensions used in this example has been adopted mainly for simplicity.

Example 31.1—Continued

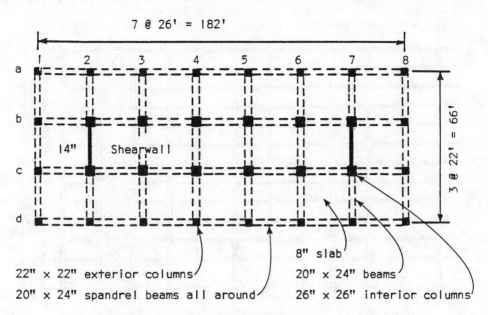

22" × 22" exterior columns

20" × 24" spandrel beams all around

8" slab

20" × 24" beams

26" × 26" interior columns

(a) Typical floor plan

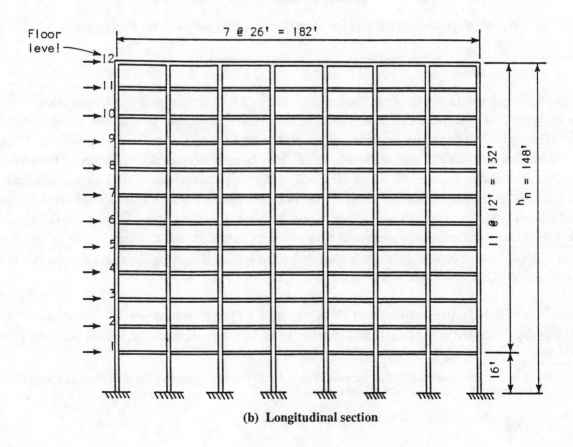

(b) Longitudinal section

Fig. 31-17 Example Building

Example 31.1—Continued

Calculations and Discussion

3 @ 22' = 66'

11 @ 12' = 132'

h_n = 148'

16'

Frame T-1

(4 Interior Frames)

Frame T-2

(2 Exterior Frames)

Frame T-3

(2 Interior Frames)

(c) Analytical model of building for lateral load analysis in transverse direction

Fig. 31-17 (Continued) Example Building

On the basis of the given data and the dimensions shown in Fig. 31-17, the weights of a typical floor[*] and the roof were estimated and are listed in Tables 31-1 and 31-2. The calculations of the base shear, V, for the transverse and longitudinal directions are shown at the bottom of Tables 31-1 and 31-2, respectively. For this example, the importance factor, I, and the soil factor, S, have been assigned values of unity. The period of the structure is calculated from Eq. 12-3 of UBC-88, using a C_t coefficient of 0.03 in the longitudinal direction and 0.02 in the transverse direction. Note that the value of R_w = 12 has been used for both directions because UBC-88 requires R_w = 12 for Special Moment Resisting Space Frames (SMRSF) which constitute the lateral load resisting system in the longitudinal direction and the same value of R_w is also required for a Dual System consisting of shearwalls and concrete SMRSF, which resists lateral loads in the transverse direction (UBC-88, Table No. 23-0.)

Calculations[**] of the undamped natural periods of vibration of the structure in the transverse direction, using the story weights listed in Table 31-1 and member stiffnesses based on gross concrete sections, gave a

---

[*] The weight of a typical floor includes that of all elements located between two imaginary parallel planes passing through the midheight of columns above and below the floor considered.

[**] Using the computer program described in Ref. 31.5.

Example 31.1—Continued

Calculations and Discussion

Table 31-1—Design Lateral Forces in Transverse (Short) Direction (for Entire Structure)

Floor Level (from base)	Seismic Forces					Wind Forces		
	Height h_x, ft	Story Weight w_x, kips	$w_x h_x$, ft-kips	Lateral Force F_x, kips	Story Shear ΣF_x, kips	Average Wind Pressure, psf	Lateral Force H_x, kips	Story Shear ΣH_x, kips
12 (Roof)	148	2100	311,000	239		23.7	30.5	
					239*			30.5
11	136	2200	299,000	160		23.7	52.3	
					399			82.8
10	124	2200	273,000	147		23.7	52.3	
					546			135.1
9	112	2200	246,000	132		23.7	52.3	
					678			187.4
8	100	2200	220,000	118		21.8	48.2	
					796			235.6
7	88	2200	194,000	104		20.0	44.1	
					900			279.7
6	76	2200	167,000	90		20.0	44.1	
					990			323.8
5	64	2200	141,000	76		19.7	43.5	
					1066			367.3
4	52	2200	114,000	61		18.2	40.1	
					1127			407.4
3	40	2200	88,000	47		16.4	36.2	
					1174			443.6
2	28	2200	61,600	33		14.6	32.2	
					1207			475.8
1	16	2200	35,200	19		12.9	52.1	
					1226			527.9

$$\Sigma \; = \; 26,300 \quad 2,148,700 \quad 1,226 \qquad \text{*Representing the sum } (F_t + F_{12})$$

Base shear, $V = \dfrac{Z1C}{R_w} W$

where $C = \dfrac{1.25\,S}{T^{2/3}}$ and $T = 0.02\,h_n^{3/4}$

In transverse direction, $h_n = 148$ ft, $T = 0.848$ sec., and $C = 1.39$ (with $S = 1$)

Thus, $V = \dfrac{(0.4)(1)(1.39)}{12}$ $(26,300) = 1226$ kips

$F_t = 0.07\,TV = (0.07)(0.848)(1226) = 72$ kips

Example 31.1—Continued

Calculations and Discussion

Table 31-2—Design Lateral Forces in Longitudinal Direction (for Entire Structure)

Floor Level (from base)	Height h_x, ft	Story Weight w_x, kips	$w_x h_x$, ft-kips	Lateral Force F_x, kips	Story Shear ΣF_x, kips	Average Wind Pressure, psf	Lateral Force H_x, kips	Story Shear ΣH_x, kips
			Seismic Forces			Wind Forces		
12 (Roof)	148	2100	311,000	205		23.7	11.3	
					205*			11.3
11	136	2200	299,000	118		23.7	19.3	
					323			30.6
10	124	2200	273,000	107		23.7	19.3	
					430			49.9
9	112	2200	246,000	97		23.7	19.3	
					527			69.2
8	100	2200	220,000	87		21.8	17.7	
					614			86.9
7	88	2200	193,000	76		20.0	16.3	
					690			103.2
6	76	2200	167,000	66		20.0	16.3	
					756			119.5
5	64	2200	141,000	55		19.7	16.1	
					811			135.6
4	52	2200	114,000	45		18.2	14.8	
					856			150.4
3	40	2200	88,000	35		16.4	13.3	
					891			163.7
2	28	2200	61,500	24		14.6	11.9	
					915			175.6
1	16	2200	35,200	14		12.9	19.2	
					929		194.8	194.8

$$\Sigma = 26,300 \qquad 2,148,700 \qquad 929 \qquad \text{*Representing the sum } (F_t+F_{12})$$

Base shear, $V = \dfrac{Z1C}{R_w} W$

where $C = \dfrac{1.25\,S}{T^{2/3}}$ and $T = 0.03\,h_n^{3/4}$

In longitudinal direction, $h_n = 148$ ft, $T = 1.273$ sec., and $C = 1.06$ (with $S = 1$)

Thus, $V = \dfrac{(0.4)(1)(1.06)}{12}\ (26,300) = 929$ kips

$F_t = 0.07\ TV = (0.07)\ (1.273)\ (929) = 83$ kips

Example 31.1—Continued

Calculations and Discussion

value for the fundamental period of 1.22 sec. compared to the T value of 0.48 sec. given by the approximate Eq. 12-3 of UBC-88. The calculated mode shapes as well as the corresponding periods of the first five modes of vibration of the structure in the transverse direction are shown in Fig. 31-18. Note that the mode shapes indicate only the relative displacements of the story masses (assumed concentrated at the floor levels). The top story displacement for each mode has been set equal to unity.

The lateral seismic design forces resulting from the distribution of the base shear in accordance with Eqs. 12-6, 12-7 and 12-8 of UBC-88 are listed in Tables 31-1 and 31-2. As an example, the seismic lateral force F_x at the 10th floor level in the transverse direction is given by

$$F_{10} = \frac{(V - F_t)w_x h_x}{\sum\limits_{i=1}^{n} w_i h_i} = \frac{(1226 - 72)(2200)(124)}{2,148,700} = 147 \text{ kips}$$

Also shown in the tables are the story shears corresponding to the distributed seismic forces.

For comparison, the wind forces and story shears corresponding to a basic wind speed of 70 mph and Exposure B, computed as prescribed in the Uniform Building Code [31.2], are shown for each direction in Tables 31-1 and 31-2.

Analysis of the structure in both directions under the respective seismic and seismic plus torsional[*] loads were carried out for three different cases,[**] using a plane frame computer program.[31-6] For the purpose of analyzing the structure in the transverse direction, the model shown in Fig. 31-17(c) was used. This model consists of three different frames linked by hinged rigid bars at the floor levels to impose equal horizontal displacements at these levels (this device is used to model the effect of floor slabs which may be assumed as rigid in their own planes). Frame T-1 represents the four identical interior frames along lines 3, 4, 5 and 6 which have been lumped together in this single frame, while Frame T-2 represents the two exterior frames along lines 1 and 8. The third frame, T-3, represents the two identical frame-shearwall systems along lines 2 and 7.

---

[*] UBC-88 Section 2312(e)5 requires that provision be made in design for the increased forces induced in resisting elements of the structural system resulting from torsion due to eccentricity between the center of application of the lateral forces and the center of rigidity of the lateral force-resisting system. Forces must not be decreased due to torsional effects. In addition, where the vertical resisting elements depend on diaphragm action for shear distribution at any level, the shear-resisting elements shall be capable of resisting an accidental torsional moment to be calculated as follows. To account for the uncertainties in locations of loads, the mass at each level shall be assumed to be displaced from the calculated center of mass in each direction a distance equal to five percent of the building dimension at that level perpendicular to the direction of the force under consideration (UBC-88 Section 2312(d)6).

[**] Case 1: 25% of design lateral loads on building excluding shearwalls.
Case 2: 100% of design lateral loads on building including shearwalls.
Case 3: 100% of design lateral loads plus torsional effects on building including shearwalls.

Example 31.1—Continued

Calculations and Discussion

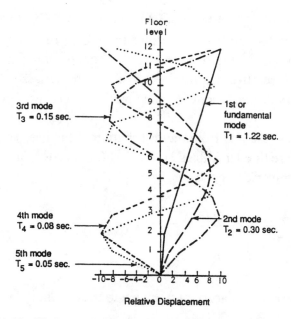

**Fig. 31-18   Undamped Natural Modes and Periods of Vibration
of Structure in Transverse Direction**

In the longitudinal direction, two linked frames, each similar to the frame shown in Fig. 31-17(b), were used to represent the two identical exterior frames L-1 along lines a and d and the two identical interior frames L-2 along lines b and c.

The lateral displacements due to seismic and wind forces, as listed in Table 31-3, are plotted in Fig. 31-19. Although the seismic forces used to obtain the curves of Fig 31-19 are fictitious, the results shown still serve to draw the distinction between wind and seismic forces, i.e., the fact that the former are external forces, the magnitudes of which are proportional to the exposed surface, while the latter are inertial forces depending primarily on the mass and stiffness properties of the structure. Thus, while the ratio of the total wind shear in the transverse direction to that in the longitudinal direction (see Tables 31-1 and 31-2) is about 2.7, the corresponding ratio for seismic shear is only 1.32. As a result of this and the smaller stiffness of the structure in the longitudinal direction, the displacement due to seismic forces in the longitudinal direction is significantly greater than that in the transverse direction, although the displacements due to wind in both directions are about the same. The typical deflected shapes associated with predominantly cantilever or flexure type structures (as in the transverse direction) and shear (open frame) type buildings (as in the longitudinal direction) are evident in Fig. 31-19. The average deflection indices, i.e., the ratio of the lateral displacement at the top to the total height of the structure, are 1/4209 for wind and 1/1344 for the code prescribed seismic loads in the transverse direction. The corresponding values in the longitudinal direction are 1/5567 for wind and 1/912 for seismic loads.

Example 31.1—Continued

Calculations and Discussion

Table 31-3 – Lateral Displacement due to Seismic and Wind Forces (in.)

Story Level	Transverse Direction		Longitudinal Direction	
	Wind	Seismic	Wind	Seismic
12	0.422	1.321	0.319	1.947
11	0.391	1.211	0.314	1.895
10	0.357	1.095	0.306	1.814
9	0.322	0.975	0.294	1.709
8	0.284	0.850	0.277	1.581
7	0.245	0.722	0.256	1.433
6	0.203	0.593	0.231	1.268
5	0.162	0.465	0.202	1.087
4	0.122	0.343	0.170	0.894
3	0.084	0.231	0.135	0.692
2	0.049	0.133	0.096	0.481
1	0.022	0.057	0.054	0.263

$$\frac{\delta_{nw}}{h_n} = \frac{0.422}{148 \times 12} = \frac{1}{4209}, \text{ transverse} \qquad \frac{\delta_{nw}}{h_n} = \frac{0.319}{148 \times 12} = \frac{1}{5567}, \text{ longitudinal}$$

$$\frac{\delta_{nE}}{h_n} = \frac{1.321}{148 \times 12} = \frac{1}{1344}, \text{ transverse} \qquad \frac{\delta_{nE}}{h_n} = \frac{1.947}{148 \times 12} = \frac{1}{912}, \text{ longitudinal}$$

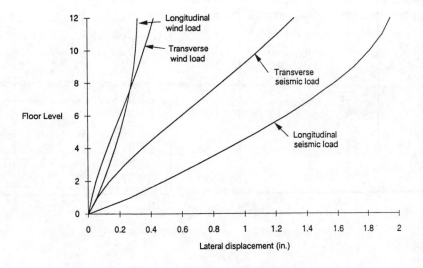

Fig. 31-19   Lateral Displacements under Seismic and Wind Loads

Example 31.1—Continued

Calculations and Discussion

An idea of the distribution of lateral loads among the different frames making up the structure in the transverse direction may be obtained from Table 31-4 which lists the portion of the total story shear at each level resisted by each of the three lumped frames. Note that at the 12th-story level, the lumped frame T-1 takes 115% of the total story shear. This reflects the fact that in frame-shearwall systems of average proportions, interaction between frame and wall under lateral loads results in the frame "supporting" the wall at the top, while at the base most of the horizonal shear is taken by the wall. Table 31-4 indicates that for the structure considered, the two frames with walls take 91% of the shear at the base in the transverse direction.

Table 31-5, which lists distribution of story shear due to seismic plus torsional loads at each level, also shows that, at the top level, interaction between frame and wall under lateral loads results in the frame "supporting" the wall, while at the base most of the horizonal shear is taken by the wall.

Table 31-4—Distribution of Horizontal Seismic Story Shears among the Three Transverse Frames Shown in Fig. 31-17(c).

Story Level	Frame T-1 (4 Interior Frames)		Frame T-2 (2 Exterior Frames)		Frame T-3 (2 Interior Frames with Shearwalls)		Total
	Story Shear, kips	% of Total	Story Shear, kips	% of Total	Story Shear, kips	% of Total	Story Shear, kips
12	275	115	116	48	-153	-64	238
11	209	52	93	23	96	24	398
10	236	43	103	19	205	37	544
9	247	36	108	16	321	47	675
8	258	32	113	14	424	53	793
7	264	29	115	12	519	58	896
6	264	26	115	11	608	61	985
5	256	24	111	10	694	65	1060
4	240	21	104	9	788	69	1130
3	211	17	91	7	877	74	1177
2	175	14	76	6	961	79	1210
1	73	5	28	2	1127	91	1226

Example 31.1—Continued

Calculations and Discussion

To illustrate the design of the typical beams on the sixth floor of an interior frame, the results of analysis of the structure in the transverse direction under seismic loads have been combined for the beams, using Eq. (9-1), Section 2609(c) and Eqs. (9-2) and (9-3), Section 2625(c)4 of UBC-88, with results obtained from gravity load analysis of a single-story bent including those beams. The results are listed in Table 31-6. Similar values for typical exterior and interior columns on the second floor of the same interior frame are shown in Table 31-7. Corresponding design forces for the shearwall section at the first floor level of Frame T-3 (Fig. 31-17(c)) are listed in Table 31-8. The last column in Table 31-8 lists the axial loads on the boundary elements (the 26 in. x 26 in. columns forming the flanges of the shearwalls) calculated in accordance with the UBC requirement that these be designed to carry all factored vertical loads on the wall, including gravity loads and vertical forces due to earthquake-induced overturning moments. The loading condition associated with this requirement is illustrated in Fig. 31-14. It should be pointed out that for buildings located in seismic Zones 2, 3 and 4 (i.e., moderate and high seismic risk areas), the detailing requirements for ductility provided in UBC-88, Section 2625 have to be met even when the design of a member is governed by wind loading rather than by seismic loads.

Table 31-5—Distribution of Horizontal Seismic plus Torsional Story Shears among the Eight Transverse Frames Shown in Fig. 31-17(a)

	Induced Story Shear due to Seismic + Torsional Forces, kips								Total Story Shear, kips
	Column Lines								
Level	1	2	3	4	5	6	7	8	
12	75	-93	77	72	66	60	-61	41	237
11	57	54	57	55	51	47	41	37	399
10	63	117	64	62	58	53	88	41	546
9	66	183	67	65	61	56	138	43	679
8	69	241	70	68	63	58	183	45	797
7	70	295	72	70	64	59	224	46	900
6	70	345	72	70	64	59	263	46	989
5	68	394	70	68	62	57	301	45	1065
4	64	447	66	64	58	53	342	42	1136
3	56	497	58	56	51	46	382	37	1183
2	47	543	48	47	42	38	419	31	1215
1	19	636	21	21	17	15	491	11	1231

Example 31.1—Continued

Calculations and Discussion

Table 31-6—Summary of Design Moments for Typical Beams on Sixth Floor of Interior Transverse Frames along Lines 3 through 6 (Fig. 31-17(a))

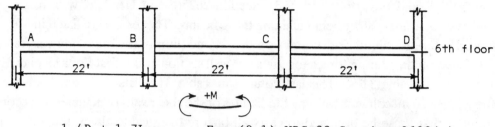

$$1.4D + 1.7L \qquad \text{Eq. (9-1) UBC-88 Section 2609(c)}$$
$$U = 1.4(D + L \pm E) \qquad \text{Eq. (9-2) UBC-88 Section 2625(c)4}$$
$$0.9D \pm 1.4E \qquad \text{Eq. (9-3) UBC-88 Section 2625(c)4}$$

Beam AB		A	Near Midspan	B
		Design Moment in ft-kips		
Eq. (9-1)		-127	+145	-203
Eq. (9-2)	Sidesway to right	+ 52	+139	-363
	Sidesway to left	-293	+136	- 22
Eq. (9-3)	Sidesway to right	+115	+ 67	-262
	Sidesway to left	-230	+ 65	+ 78
Beam BC		**B**	**Near Midspan**	**C**
Eq. (9-1)		-185	+127	-185
Eq. (9-2)	Sidesway to right	+ 23	+120	-374
	Sidesway to left	-374	+120	+ 23
Eq. (9-3)	Sidesway to right	+115	+ 58	-282
	Sidesway to left	-282	+ 58	+115

Example 31.1—Continued

Calculations and Discussion

Table 31-7—Summary of Design Moments and Axial Loads for Typical Columns on Second Floor of Interior Transverse Frames along Lines 3 through 6   (Fig. 31-17(a))

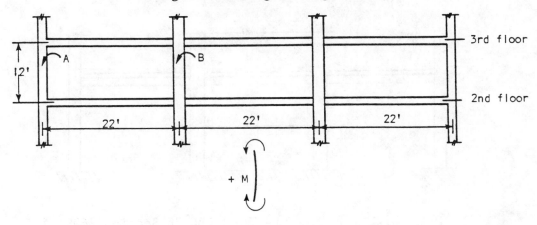

$$1.4D + 1.7L \qquad \text{Eq. (9-1) UBC-88 Section 2609(c)}$$
$$U = 1.4(D + L \pm E) \qquad \text{Eq. (9-2) UBC-88 Section 2625(c)4}$$
$$0.9D \pm 1.4E \qquad \text{Eq. (9-3) UBC-88 Section 2625(c)4}$$

		Exterior Column A			Interior Column B		
		Axial Load, kips	Moment, ft-kips		Axial Load, kips	Moment, ft-kips	
			Top	Bottom		Top	Bottom
Eq. (9-1)		- 451	- 64	+ 64	-1589	+ 9	- 9
Eq. (9-2)	Sidesway to right	- 770	- 4	- 11	-1511	+130	-158
	Sidesway to left	-1075	-116	+132	-1569	-113	+141
Eq. (9-3)	Sidesway to right	- 356	+ 27	- 43	- 815	+126	-154
	Sidesway to left	- 661	- 85	+100	- 874	-118	+146

Example 31.1—Continued

Calculations and Discussion

Table 31-8—Summary of Design Loads on Structural Wall Section at First Floor Level of Transverse Frame along Line 2 (or 7) (Fig. 31-17(a))

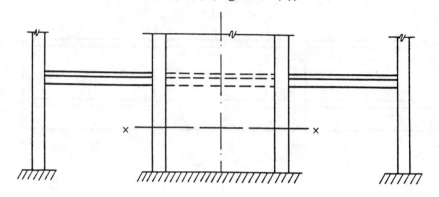

$$U = \begin{matrix} 1.4D + 1.7L \\ 1.4(D + L \pm E) \\ 0.9D \pm 1.4E \end{matrix}$$

Eq. (9-1) UBC-88 Section 2609(c)
Eq. (9-2) UBC-88 Section 2625(c)4
Eq. (9-3) UBC-88 Section 2625(c)4

Loading Condition	Design Forces Acting on Entire Structural Wall			
	Axial Load, kips	Bending (overturning) Moment, ft-kips	Horizontal Shear, kips	Axial Load* on Boundary Element, kips
Eq. (9-1)	-4926	Nominal	Nominal	-2463
Eq. (9-2)	-4808	46123	882	-4501
Eq. (9-3)	-2739	46123	882	-3466

*Based on loading condition illustrated in Fig. 31-14

The aim is to determine the flexural and shear reinforcement for the beam AB on the sixth floor of a typical interior transverse frame of the example building (Fig. 31-17(a)). The beam carries an unfactored dead load of 2.56 kips/ft of span and an unfactored reduced live load of 0.88 kip/ft. The design (factored) moments at column faces are as indicated below (see Table 31-6). The beam has dimensions of b = 20 in. and h = 24 in. (d = 21.5 in.). The slab is 8 in. thick. Use $f_c' = 4000$ psi and $f_y = 60,000$ psi.

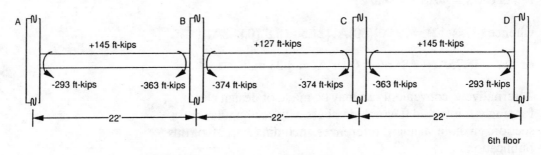

		Code
Calculations and Discussion		**Reference**

a. Check satisfaction of limitations on section dimensions.

$$\frac{\text{width}}{\text{depth}} = \frac{20}{21.5} = 0.93 > 0.3 \quad \text{O.K.}$$

21.3.1.3*
2625(d)1C*

width = 20 in. $\geq$ 10 in.

$\leq$ [width of supporting column + 1.5 $\times$ depth of beam]
$= 26 + 1.5\ (21.5) = 58$ in.

21.3.1.4
2625(d)1D

$$\frac{\ell_n}{d} = \frac{(20 \times 12) - 22}{21.5} = 10 > 4 \quad \text{O.K.}$$

21.3.1.2
2625(d)1B

b. Determine required flexural reinforcement.

1. Negative moment reinforcement at support B.

   Since the negative flexural reinforcement for both beams AB and BC at joint B will be provided by the same continuous bars, the larger negative moment at joint B will be used. Thus $M_u = 374$ ft-kips from Table 31-6. In the following calculations, the effect of any compressive reinforcement will be neglected.

---

* References starting with 2600 or higher are to UBC-88. All others are to ACI 318-89.

Example 31.2—Continued

Calculations and Discussion	Code Reference

$$a = \frac{A_s f_y}{0.85 f'_c b} = \frac{60 A_s}{(0.85)(4)(20)} = 0.882 A_s$$

$$M_u \leq \varphi M_n = \varphi A_s f_y [d - a/2]$$

whence $(374)(12) = (0.90)(60) A_s [21.5 - (0.5)(0.882 A_s)]$

or $A_s^2 - 48.75 A_s + 188.46 = 0$ or $A_s = 4.23$ in.^{2}

Alternatively, convenient use may be made of design charts for singly-reinforced flexural members with rectangular cross section, given in standard references, including Part 10 of this publication.

Use 5#9 bars, As = 5.00 in.^{2}. This gives a negative moment capacity at support B of φM_n = 434 ft-kips. Check satisfaction of limitations on reinforcement ratio.

$$\rho = \frac{A_s}{bd} = \frac{5.0}{(20)(21.5)} = 0.0116 > \rho_{min} = \frac{200}{f_y} = 0.0033$$

$$< \rho_{max} = 0.025 \quad \text{O.K.}$$

21.3.2.1
2625(d)2A

2. Negative moment reinforcement at support A.

$M_u = 293$ ft-kips

As at support B, a = 0.882As.

Substitution into $M_u = \varphi A_s f_y [d - a/2]$ yields $A_s = 3.24$ in.^{2}.

Use 5#8 bars, As = 3.95 in.^{2}. This gives a negative moment capacity at support A of φM_n = 351 ft-kips.

3. Positive moment reinforcement at support.

A positive moment capacity at the supports equal to at least 50% of the corresponding negative moment capacity is required.

21.3.2.2
2625(d)2B

$$\text{Min. } M_u^+ \text{ (support A)} = \frac{351}{2} = 175.5 \text{ ft-kips}$$

Example 31.2—Continued

Calculations and Discussion

Code Reference

which is more than M_{max}^+ at A (see Table 31-6) = 115 ft-kips

Min. M_u^+ (at support B for both spans AB and BC) = $\dfrac{434}{2}$

= 217 ft-kips

Note that the above required capacity at B is greater than the design positive moments near the midspans of both beams AB and BC.

4. Positive moment reinforcement at midspan—to be made continuous to supports.

(According to Section 21.3.2.2 (UBC-88 Section 2625(d)2B), neither the negative nor the positive-moment strength at any section along the member shall be less than one-fourth the maximum moment strength provided at the face of either joint.)

$$a = \frac{A_s f_y}{0.85 f_c' b} = \frac{60 A_s}{(0.85)(4)(20)} = 0.882 A_s$$

$$M_u^+ = (175.5)(12) = \varphi A_s f_y \left[d - a/2 \right]$$

yields A_s^+ required at A = 1.89 in². Similarly, corresponding to M_u^+ (required capacity) at support B = 217 ft-kips, A_s^+ required = 2.36 in².

Use 3#8 bars continuous through both spans. A_s = 2.37 in².
This provides a positive moment capacity of φM_n = 218 ft-kips.

$$\rho = \frac{2.37}{(20)(21.5)} = 0.0055 > \rho_{min} = \frac{200}{f_y} = 0.0033 \quad \text{O.K.}$$

10.5.1
2610(f)1

c. **Calculate required length of anchorage of flexural reinforcement in exterior column.**

Development length,
(plus standard 90 hook
located in confined
region of column)

$$\ell_{dh} \geq f_y d_b / 65 \sqrt{f_c'}$$
$$\geq 8 \, d_b$$
$$\geq 6 \text{ in.}$$

21.6.4.1
2625(g)4A

Example 31.2—Continued

Calculations and Discussion	Code Reference

According to Section 7.1.2 (UBC-88 Section 2607(b)2), a standard hook is defined as a 90-deg bend plus $12d_b$ extension at free end of a bar.

For the #8 (top) bars (bend radius $\geq 6d_b$)

7.2.1
2612(f)

$$\ell_{dh} \geq \begin{array}{l} (60{,}000)\,(1.00)/(65\,\sqrt{4000}) = 15 \text{ in.} \\ (8)(1.00) = 8 \text{ in.} \\ 6 \text{ in.} \end{array}$$

For the No. 8 (bottom) bars (bend radius $\geq 6d_b$)

7.2.1
2612(f)

$$\ell_{dh} \geq \begin{array}{l} (60{,}000)\,(1.00)/(65\,\sqrt{4000}) = 15 \text{ in.} \\ (8)(1.00) = 8 \text{ in.} \\ 6 \text{ in.} \end{array}$$

See Fig. 31-12 for detail of flexural reinforcement anchorage in exterior column. Note that the development length ℓ_{dh} is measured from the near face of the column to the far edge of the vertical 12-bar-diameter-extension (see sketch below). Section 21.6.1.3 (UBC-88 Section 2625(g)1c) requires that longitudinal reinforcement terminated in a column shall be extended to the far face of the confined column core and anchored in tension according to Section 21.6.4.1 (UBC-88 Section 2625(g)4A) and in compression according to Chapter 12 or UBC-88 Section 2612(f).

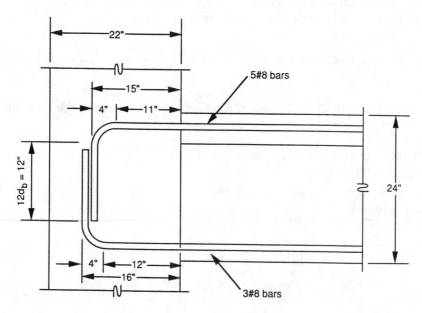

Example 31.2—Continued

Calculations and Discussion

Code Reference

d. Determine shear reinforcement requirements.

Design for shears corresponding to end moments obtained by assuming the stress in the tensile flexural reinforcement equal to 1.25 f_y and a strength reduction factor, $\varphi = 1.0$ (probable strength), plus factored tributary gravity loads according to Section 21.7.1.1 (Fig. 31-3), or unfactored tributary loads according to UBC-88 Section 2625(h)1A. The calculations below are based on unfactored tributary gravity loads, in accordance with UBC-88.

$$M_{pr} = 1.25\, A_s f_y \left[d - \frac{a}{2} \right]; \quad a = \frac{1.25 A_s f_y}{0.85\, f'_c b} = 1.103\, A_s$$

Table 31-9 shows values of design end shears corresponding to the two loading cases to be considered. In the table

$$w = (w_D + w_L) = (2.56) + (0.88) = 3.44 \text{ kips/ft.}$$

Chapter 21 (UBC-88 Section 2625(h)2) requires that the contribution of concrete to shear resistance, V_c, be neglected if the earthquake induced shear force (corresponding to the "probable flexural strengths" at beam ends calculated using $f_s = 1.25\, f_y$ and $\varphi = 1.0$) is greater than one-half of the total design shear and if the axial compressive force including earthquake effects is less than $A_g f'_c/20$.

For sideway to right, the shear at end B due to the plastic end moments in the beam (see Table 31-9) is calculated as follows:

$$M_{prl} = \frac{(2.37) \times 75,000}{12,000} \left[21.5 - \frac{1.103(2.37)}{2} \right] = 299 \text{ ft–kips}$$

$$M_{prr} = \frac{(5.00) \times 75,000}{12,000} \left[21.5 - \frac{1.103(5.00)}{2} \right] = 586 \text{ ft–kips}$$

$$V_B = \frac{299 + 586}{18.17} = 48.71 \text{ kips}$$

which is more than 50% ($V_e/2 = 39.98$ kips) of the total design shear, $V_e = 79.96$ kips. Therefore, the contribution of concrete to shear resistance can not be considered in determining shear reinforcement requirements.

Example 31.2—Continued

Calculations and Discussion

Table 31-9—Determination of Design Forces for Beam Spans

Loading	$V_e = \dfrac{M_{prA}^{\pm} + M_{prB}^{\mp}}{\ell_n} + \dfrac{w\ell_n}{2}$	
	A	B
A **B** 299 ft-kips 586 ft-kips w=2.56+0.88=3.44 kips/ft ℓ_n = 18.17 ft Sidesway to right	-17.46 kips	79.96 kips
A **B** 477 ft-kips 299 ft-kips w=3.44 kips/ft ℓ_n = 18.17 ft Sidesway to left	73.96 kips	-11.46 kips

Shear Diagram

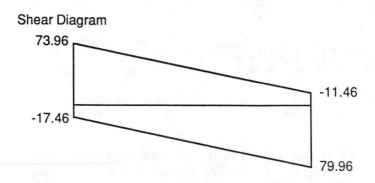

73.96

-11.46

-17.46

79.96

Example 31.2—Continued

Calculations and Discussion	**Code Reference**

At right end B, $V_e = 79.96$ kips.

Using $\varphi V_s = V_e = 79.96$ kips,

$$V_s = 94.1 \text{ kips} \leq 4\sqrt{f_c'}\, b_w d = 109 \text{ kips}$$

2611(f)4C
11.5.4.3

Required spacing of #3 closed stirrups (hoops) plus a #3 crosstie A_v (3 legs) = 0.33 in.^{2}.

$$s = \frac{A_v f_y d}{V_s} = \frac{(0.33)(60)(21.5)}{94.1} = 4.5 \text{ in.}$$

Maximum allowable hoop spacing within distance 2h = 2(24) = 48 in. from faces of supports.

$$s_{max} \leq \begin{cases} d/4 = 21.5/4 = 5.4 \text{ in.} \\ 8 \times (\text{dia. of smallest long. bar, which is \#8}) = 8(1.00) = 8 \text{ in.} \\ 24 \times (\text{dia. of hoop bars}) = 24(0.375) = 9 \text{ in.} \\ 12 \text{ in.} \end{cases}$$

21.3.3.2
2625(d)3B

Beyond distance 2h from the supports, maximum spacing of stirrups.

$$s_{max} = d/2 = 10.5 \text{ in.}$$

21.3.3.4
2625(d)3D

The first hoop shall be located not more than 2 in. from the column face.

Section 21.3.3.3 (UBC-88 Section 2625(d)3C) specifies that, where hoops are required, longitudinal bars on the perimeter shall have lateral support conforming to Section 7.10.5.3 (UBC-88 Section 2607(k)3C):

"Ties shall be arranged such that every corner and alternate longitudinal bar shall have lateral support provided by the corner of a tie with an included angle of not more than 135 degrees and a bar shall not be farther than 6 inches clear on each side along the tie from such a laterally supported bar."

Use #3 stirrups and hoops spaced as shown in Fig. 31-20 (see item g below). Where the loading is such that inelastic deformation may occur at intermediate points within the span, e.g., due to concentrated loads near midspan, the spacing of hoops will have to be determined in a manner similar to that used above for regions near supports, Section 21.3.3.1 (UBC-88 Section 2625(d)3A(ii)).

Example 31.2—Continued

Note that lap splices in longitudinal reinforcement should not be used
within joints or within a distance of 2h from the faces of supports,
Section 21.3.2.3 (UBC-88 Section 2625(d)2C). Where used outside
of these regions, such splices are to be confined over the length of the
lap by hoops or spirals with a maximum spacing or pitch of d/4 or 4 in.

e.   Negative reinforcement cutoff points.

For the purpose of determining cutoff points for the negative reinforcement,
a moment diagram corresponding to plastic end moments and 0.9 times the
dead load will be used. The cutoff point for two of the five #9 bars at the top,
near support B of beam AB, will be determined.

With the negative moment capacity of a section with 3#9 top bars
= 272 ft-kips (calculated using $f_s = f_y = 60$ ksi and $\varphi = 0.9$), the distance from
the face of the right support B to where the moment under the loading considered
equals 272 ft-kips is readily obtained by summing moments about Section a-a in
the sketch below, and equating these to −272 ft-kips.

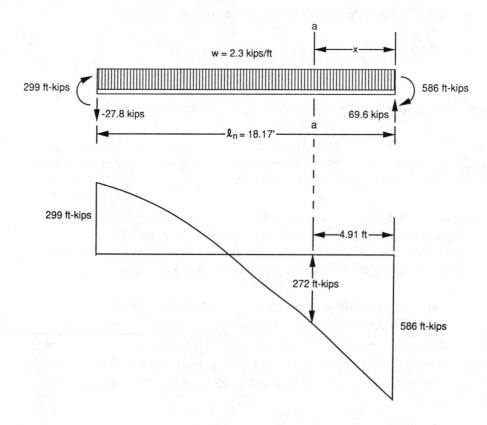

31-56

Example 31.2—Continued

Calculations and Discussion

Thus, $69.60x - 586 - 2.30 \frac{x^2}{2} = -272$

Solution of the above equation gives x = 4.91 ft. Hence two of the
5#9 top bars near support B may be cut off (noting that d = 21.5 in.
> $12d_b$ = 12 x 1.128 = 13.5 in.) at: x + d = 4.91 + 21.5/12 = 7.0 ft. from
face of right support B, Section 12.10.3 (UBC-88 Section 2612(k)3).

Since for a #9 top bar, ℓ_{dh} = 17 in., 3.5 ℓ_{dh} = 5 ft. < 7.0 ft, Section 21.6.4.2(b)
(UBC-88 Section 2625(g)4B(2)). So if 2#9 bars are cut off at 7.0 ft
from the face of support B, the two bars will still have the required
development length beyond the support face.

Of the 5#8 bars at the exterior support, two bars will also be cut
off at a similarly computed distance x + d = 4.43 + 21.5/12
= 6.22 ft ≈ 7ft away from the face of support A.

f. **Flexural reinforcement splices.**
    Lap splices of flexural reinforcement should not be placed within a joint,
    within a distance 2h from faces of supports or within regions of potential
    plastic hinging, Section 21.3.2.3 (UBC-88 Section 2625(d)2C)

    Note that all lap splices have to be confined by hoops or spirals
    with a maximum spacing or pitch of d/4 or 4 in. over the length
    of the lap, Section 21.3.2.3 (UBC-88 Section 2625(d)2C).

    1. Bottom bars, #8

        The bottom bars along most of the length of the beam may be
        subjected to maximum stress. The reinforcement area corresponding to
        the maximum positive moment near midspan of 145 ft-kips
        (Table 31-6), as obtained by analysis, is A_s = 1.55 in.^{2}
        The area of reinforcement provided corresponding to three
        #8 bars is A_s = 2.37 in.^{2}. Thus,

        $$\frac{A_s \text{ provided}}{A_s \text{ required}} = \frac{2.37}{1.55} = 1.5 < 2$$

        Since 100% of A_s will be spliced within the required length,
        use Class C splice according to UBC-88 Table No. 26-G

Example 31.2 —Continued

Calculations and Discussion	Code Reference

Please note that ACI 318-89 Section 12.15.1 does not include any reference to splices beyond Class B any more.

Required length of splice = $1.7\ell_d \geq 12$ in.

where $\ell_d = 0.04A_b f_y / \sqrt{f_c'} \geq 0.0004 d_b f_y$ 12.2.2

$\qquad = (0.04)(0.79)(60,000)/\sqrt{4000} \geq (0.0004)(1000)(60,000)$ 2612(c)

$\qquad = 30$ in. (governs) > 24 in.

Class C splice length = (1.7) (30) = 51 in.

2.   Top bars, #9, spliced to #8

Since the midspan portion of the span is always subject to a positive bending moment (see Table 31-6), splices in the top bars should be located at or near midspan.

Required length of Class A splice (top bars) = $1.4\ell_{db} \geq 12$ in., UBC Sections 2612(p)1 and 2612(c). Note that according to ACI 318-89 Sections 12.15.1, 12.2.2 and 12.2.4.1, the required splice length would have been $1.3\ell_{db} \geq 12$ in. Using the same expression given above (Section 12.2.2 or UBC Section 2612(c), one obtains $\ell_{db} = 38$ in.

Required splice length = 1.4(38) = 53 in.

g.   **Detail of beam—See Fig. 31-20.**

Example 31.2 —Continued

Calculations and Discussion

Code Reference

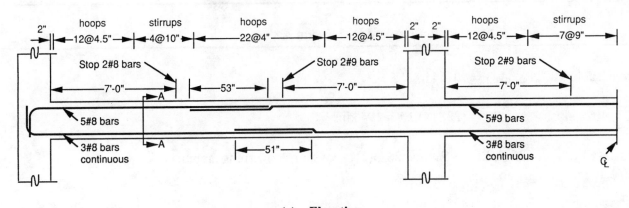

(a)   Elevation

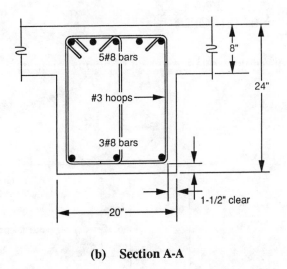

(b)   Section A-A

Fig. 31-20   Details of Beam Reinforcement

Example 31.3—Proportioning and Detailing of Columns of Building in Example 31.1

The aim is to design the transverse reinforcement for the exterior tied column on the second floor of a typical transverse interior frame, i.e., one of the frames making up Frame T-1 of Fig. 31-17(c). The column dimension has been established at 22 in. square and, on the basis of the different combinations of axial load and bending moment corresponding to the three loading conditions listed in Table 31-7, 8#8 bars arranged in a symmetrical pattern have been found adequate. Assume the same beam section framing into the column as considered in Example 31.2. Use $f'_c = 4000$ psi and $f_y = 60,000$ psi.

Calculations and Discussion

From Table 31-7, $P_{u\,(max)} = 1075$ kips

$P_u = 1075$ kips $> A_g f'_c/10 = (22)^2(4)/10 = 194$ kips

Thus ACI Chapter 21 (UBC-88 Section 2625(e)) provisions governing members subjected to bending and axial load apply.

a. **Check satisfaction of vertical reinforcement limitations and moment capacity requirements.**

    1.   Reinforcement ratio

        $0.01 \leq \rho \leq 0.06$ 21.4.3.1
                                                                          2625(e)3

        $\rho = \dfrac{A_{st}}{A_g} = \dfrac{8(0.79)}{(22)(22)} = 0.0131$ O.K.

    2.   Moment strength of columns relative to that of beams framing in the transverse direction.

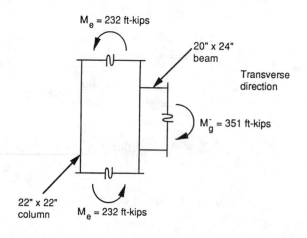

$M_e = 232$ ft-kips

20" x 24" beam

Transverse direction

$M_g^- = 351$ ft-kips

22" x 22" column $M_e = 232$ ft-kips

Example 31.3 – Continued

Calculations and Discussion

21.4.2.2
2625(e)2

$$M_e \text{ (columns)} \geq \frac{6}{5} \Sigma M_g \text{ (beams)}$$

From the preceding example, φM_n of the beam at A = 351 ft-kips, corresponding to sidesway to left.

From Table 31-7, maximum axial load on Column A at the 2nd floor level for sidesway to left, P_u = 1075 kips.

Using the PCA computer program PCACOL or interaction charts such as those given in ACI Publication SP 17A(85), see Part 11, the moment capacity of the column section corresponding to $P_u = \varphi P_n$ = 1075 kips is obtained as $\varphi M_n = M_e = 0.7(332) = 232$ ft-kips.

With the same size column above and below the beam, total moment capacity of columns = 2(232) = 464 ft-kips.

Thus, $\Sigma M_e = 464 > \frac{6}{5} M_g = (6)(351)/5 = 421$ ft-kips

Therefore, the lateral strength and stiffness of the column can be considered in determining the calculated strength and stiffness of the structure, Section 21.4.2.1 (UBC-88 Section 2625(e)2A).

3. Moment strength of columns relative to that of framing beams in longitudinal direction.

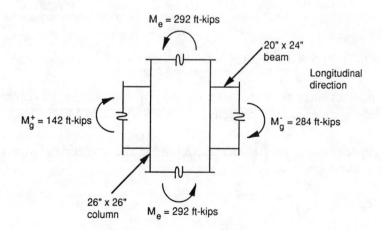

Example 31.3 – Continued

	Code
Calculations and Discussion	**Reference**

Since the columns considered here are located in the center portion of the exterior longitudinal frames, the axial forces due to seismic loads in the longitudinal direction are negligible. (Analysis of the longitudinal frames under seismic loads indicated practically zero axial forces in the exterior columns of the four transverse frames represented by Frame T-1 in Fig. 31-17(c).) Revaluating equations 9-1, 9-2 and 9-3 gives a critical axial load of $1.4D + 1.7L = 1.4 (565) + 1.7 (94) = 951$ kips.

From PCACOL or available interaction chart, the corresponding moment capacity of the column section with 8#8 bars is obtained as $\varphi M_n = M_e = 292$ ft-kips.

If we assume a ratio for the negative moment reinforcement of approximately 0.0075 $(200/f_y \le 0.0075 \le 0.025)$ in the beams of the exterior longitudinal frames ($b_w = 20$ in., $d = 21.5$ in.),

$$A_s = \rho b_w d = \text{approx. } (0.0075)(20)(21.5) = 3.23 \text{ in.}^2$$

Assume four 8#8 bars, $A_s = 3.14$ in.^{2}

Negative moment capacity of beam:

$$a \quad = \frac{A_s f_y}{0.85 f'_c\, b_w} = \frac{(3.14)(60)}{(0.85)(4)(20)} = 2.77 \text{ in.}$$

$$\varphi M_n^- = M_g = \varphi A_s f_y (d - a/2)$$
$$= (0.90)(3.16)(60)\left[(21.5 - 1.385)/12\right] = 284 \text{ ft−kips}$$

Assume a positive moment capacity of the beam on the opposite side of the column equal to one-half the negative moment capacity calculated above, or 142 ft-kips.

The total moment capacity of beams, framing into a joint in the longitudinal direction, for sidesway in either direction,

$$\Sigma M_g = 284 + 142 = 426 \text{ ft−kips}$$

$$\Sigma M_e = 2(292) = 584 \text{ ft−kips} > \frac{6}{5}\Sigma M_g = \frac{6}{5}(426) \qquad \text{(21.4.2.2)}$$
$$= 511 \text{ ft−kips} \quad \text{O.K.} \qquad \text{(25-1)}$$

Example 31.3 – Continued

Calculations and Discussion

b.   Orthogonal Effects

According to UBC-88 Section 2312(h)1, provision shall be made for the effects of earthquake forces acting in a direction other than the principal axes when a column forms part of two or more intersecting lateral force-resisting systems; except when the axial load in the column due to seismic forces acting in either direction is less than 20 percent of the column allowable axial load.

The axial load corresponding to seismic forces is equal to 109 kips. On the other hand, allowable axial load on the column is given by

$$P_O = 0.80\,\varphi\left[0.85f_c'(A_g - A_{st}) + A_{st}f_y)\right]$$

$$= 0.80 \times 0.70\left[0.85 \times 4(484 - (8)(0.79)) + (8)(0.79)(60)\right] = 1122 \text{ kips}$$

$$109 \text{ kips} < 0.2\,(1122) = 224 \text{ kips}$$

Therefore, orthogonal effects due to seismic loads need not be considered for this column.

c.   Determine transverse reinforcement requirements.

1.   Confinement reinforcement (see Fig. 31-7).

Transverse reinforcement for confinement is required over a distance ℓ_o from the column ends where

$\ell_o \geq$ Depth of Member = 22 in. (governs)   21.4.4.2
           1/6 Clear Height = $(10 \times 12)/6 = 20$ in.   2625(e)4D
           18 in.

Maximum allowable spacing of rectangular hoops
s_{max} = 4 in.   21.4.4.2
       > 1/4 smallest dimension of column = 22/4 = 5.5 in.   2625(e)4B

Example 31.3 – Continued

Required cross-sectional area of confinement reinforcement in the form of hoops

$$A_{sh} \geq \quad 0.12sh_c \frac{f'_c}{f_{yh}}$$

$$0.3sh_c \left[\frac{A_g}{A_{ch}} - 1\right] \frac{f'_c}{f_{yh}}$$

Eq. (25-4)
2625(e)4A
Eq. (25-3)
2625(e)4A

where

s	=	spacing of transverse reinforcement (in.)
h_c	=	cross-sectional dimension of column core, measured center-to-center of confining reinforcement (in.)
A_{ch}	=	core area of column section, measured outside to-outside of transverse reinforcement (in.^{2})
f_{yh}	=	specified yield strength of transverse reinforcement (psi)

Note that Eq.(21-3), Section 21.4.1.1 of ACI 318-89 is identical with Eq.(25-3) of UBC-88. Eq.(21-4), Section 21.4.4.1 of ACI 318-89 is the same as Eq.(25-4), except that the coefficient of 0.12 has been reduced to 0.09.

For a hoop spacing of 4 in., f_{yh} = 60,000 psi, and tentatively assuming #4 bar hoops (for the purpose of estimating h_c and A_{ch}), required cross-sectional area

$$A_{sh} \geq \quad (0.12)(4)(18.5)(4000)/60,000 = 0.59 \text{ in.}^2 \quad \text{(governs)}$$

$$(0.3)(4)(18.5)\left(\frac{484}{361} - 1\right)\frac{4000}{60,000} = 0.50 \text{ in}^2$$

#4 hoops with one crosstie Section 21.4.4.3 (UBC-88 Section 2625(e)4C), as shown in the sketch below, provide A_{sh} = 3(0.20) = 0.60 in.^{2}.

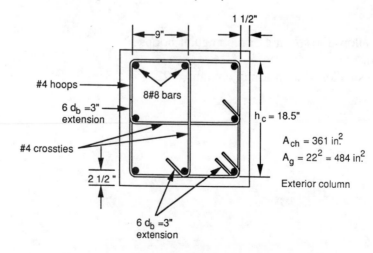

31-64

Example 31.3 – Continued

Calculations and Discussion

Code
Reference

2. Transverse reinforcement for shear.

As in the design of shear reinforcement for beams, the design shear in the columns is based not on the factored shear forces obtained from a lateral load analysis but rather on the nominal flexural strength provided in the columns. Section 21.7.1.1 (UBC-88 Section 2625(h)1B) requires that the column design shear forces shall be determined from the consideration of the maximum forces that can be developed at the faces of the joints, with the nominal moment strengths calculated for the factored axial compressive forces resulting in the largest moments acting at the joint faces, without strength reduction factors and assuming that the stress in tensile reinforcement is equal to at least 1.25 f_y.

The moment obtained from the balanced condition considering a tensile reinforcement stress of 1.25 f_y is $M_b = 599$ ft-kips (see Fig. 31-8).

However, Section 21.7.1.2 (UBC-88 Section 2625(h)1B) specifies that this moment need not exceed the moment that can be resisted by the flexural members framing into the joint calculated in accordance with UBC-88 Section 2625(h)1A ($M_n = 477$ ft-kips, see Table 31-9). Therefore, the column need only be designed to resist the maximum shear that can be transferred through the beam members.

Thus,

$$V_e = \frac{\frac{M_{n,beam}}{2}(top) + \frac{M_{n,beam}}{2}(bottom)}{h_n}$$

$$= \frac{\frac{477}{2} + \frac{477}{2}}{10} = 47.7 \text{ kips}$$

$$V_c = 2\sqrt{f_c'}\, bd \left(1 + \frac{N_u}{2000A_g}\right)$$

$$= \frac{2\sqrt{4000}\,(22)(19.5)}{1000} \left[1 + \frac{493,000}{2000 \times (22)^2}\right] = 82 \text{ kips}$$

$\varphi V_c > V_u$

Thus, the transverse reinforcement spacing over the distance $l_o = 22$ in. near the column ends is governed by the requirement for confinement rather than shear.

Maximum allowable spacing of shear reinforcement = d/2 or 9.7 in.

11.5.4
2611(f)

Example 31.3 – Continued

Calculations and Discussion	Code Reference

Use #4 hoops and cross-ties spaced at 4 in. within a distance of 22 in. from the column ends and #4 hoops spaced at 9 in. or less over the remainder of the column.

d.   Minimum length of lap splices of column vertical bars.

Section 21.4.3.2 (UBC-88 Section 2625(e)3B) limits the location of lap splices of column bars within the center half of the member length. Also, the splices are to be designed as Class A tension splices. Transverse reinforcement at 4 in. is to be provided over the full lap splice length in conformance with Section 21.4.4 (UBC-88 Section 2625(e)4).

Required length of splice = $1.0\ell_d$, where

$$\ell_d = 0.04A_b f_y / \sqrt{f_c'} \geq 0.0004 d_b f_y$$

$$= (0.04)(0.79)(60,000)/\sqrt{4000} \geq (0.0004)(1.000)(60,000)$$

$$= 30 \text{ in. (governs)} > 24 \text{ in.}$$

12.2.2
2625(c)

Thus required splice length = 1.0(30) = 30 in.

Use 30 in. lap splices

Example 31.3 – Continued

Calculations and Discussion

Code
Reference

e.   Detail of Column—See Fig. 31-21

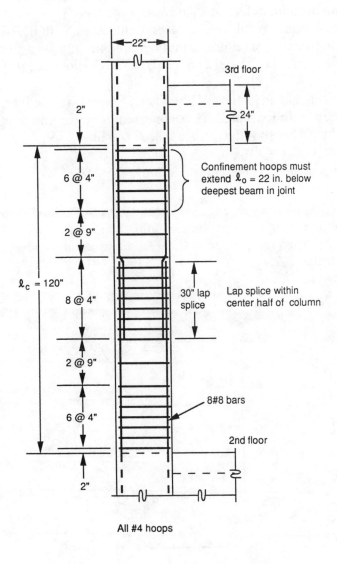

Fig. 31-21 Column Reinforcement Details

Example 31.4 - Proportioning and Detailing of Exterior Beam-Column Connection of Building in Example 31.1

The aim is to determine the transverse reinforcement and shear strength requirements for the exterior beam-column connection between the beam considered in Example 31.2 and the column of Example 31.3. Assume the joint to be located at the sixth floor level.

Calculations and Discussion

---

a.   Transverse reinforcement for confinement.

Chapter 21 and UBC-88 Section 2625 require the same amount of confinement reinforcement within the joint as for the length l_o at column ends, unless the joint is confined by beams framing into all vertical faces of the column. In the latter case, only one-half the amount of reinforcement required for joints not so confined need be provided, Sections 21.6.2.1 and 21.6.2.2 (UBC-88 Sections 2625(g)2A,2625(g)2B)

In the case of the beam-column joint considered here, beams frame into only three sides of the column. In Example 31.3, confinement requirements at column ends were satisfied by #4 hoops with crossties spaced at 4 in.

b.   Check shear strength of joint (transverse direction).

The shear across section x-x (see sketch below) of the joint is obtained as the difference between the tensile force from the top flexural reinforcement of the framing beam (stressed to 1.25f_y) and the horizontal shear from the column above, Section 21.6.1.1 (UBC-88 Section 2625(g)1A)

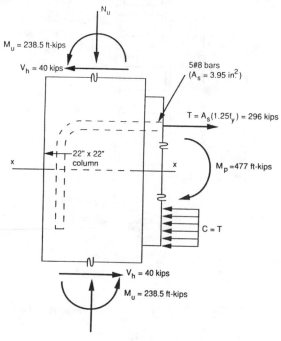

Example 31.4 – Continued

Calculations and Discussion	Code Reference

Tensile Force from Beam [A_s (5#8 bars) = 3.95 in^2]
= 3.95 (1.25)(60) = 296 kips

An estimate of the horizontal shear from the column, V_h, can be obtained by assuming that the beams in the adjoining floors are also deformed so that plastic hinges form at their junctions with the column, with M_{pr} (beam) = 477 ft-kips (see Table 31-9 for sidesway to left). By further assuming that the plastic moments in the beams are resisted equally by the columns above and below the joint, one obtains for the horizontal shear at the column ends.

$$V_h = \frac{M_{pr} \ (Beam)}{Story \ Height} = \frac{477}{12} = 40 \ kips$$

Thus, net shear at section x-x of the joint, V_u = 296 – 40 = 256 kips. Section 21.6.3.1 (UBC-88 Section 2625(g)3) makes the nominal shear strength of a joint a function only of the area of the joint cross section, A_j, and the degree of confinement by framing beams. For the joint confined on three faces considered here

$$\varphi V_c = \varphi 15 \sqrt{f_c'} \ A_j$$
$$= (0.85)(15)(\sqrt{4000}) \ (22)^2/1000$$
$$= 390 \ kips > V_u = 256 \ kips$$

21-6-3-1
2625(g)3A

Note that if the shear strength of the concrete in the joint as calculated above were inadequate, any adjustment would have to take the form of (since transverse reinforcement is considered not to have a significant effect on shear strength) either an increase in the column cross section (and hence A_j) or an increase in the beam depth (to reduce the amount of flexural reinforcement required and hence the tensile force T).

c. **Detail of joint—See Fig. 31-22.** (The detail should be checked for adequacy in the longitudinal direction).

NOTE: The use of crossties within the joint may cause some placement difficulties. To relieve the congestion, #6 hoops spaced at 4 in., but without crossties, may be considered as an alternative. Although the cross sectional area of confinement reinforcement provided by #6 hoops at 4 in. (A_{sh} = 0.88 in.^{2}) exceeds the required amount (0.59 in.^{2}), the requirement of Section 21.4.4.3 relating to a maximum spacing of 14 in. between crossties or legs of overlapping hoops (see Fig. 31-7) will not be satisfied. However, it is believed that this should not be a serious shortcoming in this case since the joint is restrained by beams on three sides.

Example 31.4 - Continued

Calculations and Discussion

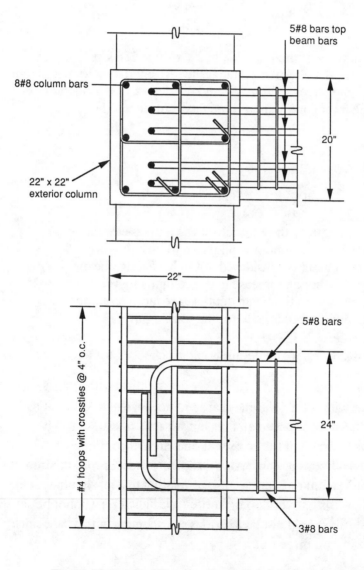

Fig. 31-22 Detail of Exterior Beam-Column Connection

Example 31.5 - Proportioning and Detailing of Interior Beam-Column Connection of Building in Example 31.1

The objective is to determine the transverse reinforcement and shear strength requirements for the interior beam-column connection at the sixth floor of the interior transverse frame considered in previous examples. The column is 26 in. square and is reinforced with 8#11 bars. The beams have dimensions of b = 20 in. and d = 21.5 in. and are reinforced as noted in Example 31.2 (see Fig. 31-20).

Calculations and Discussion	Code Reference

a.   Transverse reinforcement requirements (for confinement).

s_{max} = 4 in. (governs)   2625(e)4
        <1/4 in. smallest dimension of column = 26/4 = 6.5 in.   21.4.4

For the column cross section considered and assuming # 4 hoops,

h_c = 22.4 in., $A_{ch} = (22.9)^2 = 524$ in.^{2}

and $A_g = (26)^2 = 676$ in.^{2}

With a hoop spacing of 4 in., the required cross-sectional area of confinement reinforcement in the form of hoops,

$$0.12 s h_c \frac{f_c'}{f_{yh}} = (0.12)(4)(22.4)(4000)/(60,000)$$ Eq. (25-4)
                                                                                                                          2625(e)4A

$$= 0.72 \text{ in.}^2 \text{ (governs)}$$

$A_{sh} \geq$ Eq. (25-3)
                                                                                                                          2625(e)4A

$$0.3 s h_c \left(\frac{A_g}{A_{ch}} - 1\right)\frac{f_c'}{f_{yh}} = (0.3)(4)(22.4)\left(\frac{676}{524} - 1\right)\frac{4000}{60,000}$$

$$= 0.52 \text{ in.}^2$$

Note that Eq.(21-3), Section 21.4.4.1 of ACI 318-89 is identical with Eq. (25-3) of UBC-88, Eq. (21-4), Section 21.4.4.1 of ACI 318-89 is the same as in Eq. (25-4), except that the coefficient of 0.12 has been reduced to 0.09.

Since the joint is framed by beams (having widths (20 in.) ≥ 3/4 width of column) on all four sides, it is considered confined and a 50% reduction in the amount of confinement reinforcement indicated above is allowed. Thus A_{sh} (required) ≥ 0.36 in.^{2} Section 21.6.2.2 (UBC-88 Section 2625(g)2A).

Example 31.5 – Continued

	Code
Calculations and Discussion	**Reference**

#4 hoops with crossties spaced at 4 in. on center provide $A_{sh} = 0.60$ in.^{2}
(see Note at the end of Example 31.4).

b.   Check shear strength of joint.

Following the same procedure as used in Example 31.4, the forces
affecting the horizontal shear across a section near mid-depth
of the joint shown below are obtained.

Net shear across Section x-x = $T_1 + C_2 - V_h$ 21.6.3.1
= 375 + 177 – 74 = 478 kips = V_u 2625(g)3

Shear strength of joint, noting that the joint is confined on all faces,

$$\varphi V_c = \varphi 20 \sqrt{f_c'} \, A_j$$

$$= (0.85)(20)(\sqrt{4000})(26^2/1000) = 727 \text{ kips}$$ 21.6.3.1
$$> V_u = 478 \text{ kips} \quad \text{O.K.}$$ 2625(g)3A

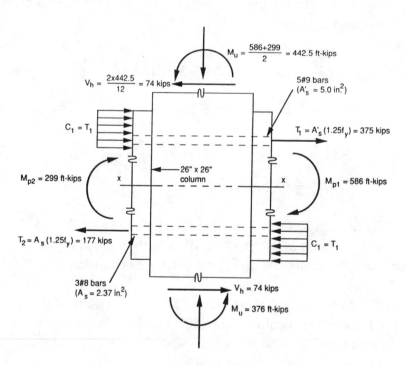

Example 31.6 - Proportioning and Detailing of Shearwall of Building in Example 31.1

The aim is to design the shearwall section at the first floor level of one of the identical frame-shearwall systems in Frame T-3 (Fig. 31-17(c)). The preliminary design, as shown in Fig. 31-17, is based on a 14 in. thick wall with 26 in. square vertical boundary elements, each of the latter being reinforced with 8 #11 bars.

Calculations and Discussion

Preliminary calculations indicated that the cross section of the shearwall at the lower floor levels needed to be increased. In the following, a 20 in. thick wall section with 32 in. x 50 in. boundary elements reinforced with 30#11 bars is investigated and other reinforcement requirements determined.

The design forces on the shearwall at the first floor level are listed in Table 31-8. Note that because the axis of the shearwall coincides with the centerline of the transverse frame of which it is a part, lateral loads do not induce any vertical (axial) forces on the wall.

The calculation of the maximum axial force on the boundary element corresponding to $1.4 [D + L \pm E]$, $P_u = 4501$ kips, shown in Table 31-8, involved the following steps:

At the base of the wall:

Dead Load, D = 3043 kips
Live Load, L = 391 kips

Moment at the base of the wall due to seismic load (from lateral load analysis of transverse frames), M_{base} = 32,945 ft-kips.

Referring to Fig. 31-14, and to UBC-88 Section 2625(c)4,

$$W = 1.4 (D + L + \pm E)$$
$$= 1.4 (3043 + 391 + 0) = 4808 \text{ kips}$$

$$Ha = 1.4 M_{base} = (1.4)(32945) = 46,123 \text{ ft}-\text{kips}$$

$$V_c = W/2 + Ha/\ell'_w = 4808/2 + 46,123/22 = 4501 \text{ kips}$$

Example 31.6 – Continued

Calculations and Discussion

a.   **Check if boundary elements are required**

Section 21.5.2.3 and UBC-88 Section 2625(f)3A require boundary elements to be provided if the maximum compressive extreme-fiber stress under factored forces exceeds $0.2f_c'$, unless the entire wall is reinforced to satisfy Sections 21.4.4.1 through 21.4.4.3 (UBC-88 Sections 2625(e)4A through 2625(e)4C) relating to confinement reinforcement .

It will be assumed that the wall will not be provided with confinement reinforcement over its entire section. For a homogeneous rectangular wall 26.17 ft. long (horizontally) and 20 in. (1.67 ft) thick,

$$I_{n.a.} = \frac{(1.67)(26.17)^3}{12} = 2694 \text{ ft}^4$$

$$A_g = (1.67)(26.17) = 43.7 \text{ ft}^2$$

Extreme-fiber compressive stress under $M_u = 46{,}123$ ft-kips and $P_u = 4808$ kips (see Table 31-8),

$$f_c = \frac{P_u}{A_g} + \frac{M_u \ell_w/2}{I_{n.a.}} = \frac{4808}{43.7} + \frac{(46{,}123)(26.17)/2}{2494}$$

$$= 352 \text{ ksf or } 2.44 \text{ ksi} > 0.2\, f_c' = (0.2)(4) = 0.8 \text{ ksi}$$

Therefore, boundary elements are required, subject to the confinement and special loading requirements specified in Chapter 21 and UBC-88 Section 2625.

b.   **Determine minimum longitudinal and transverse reinforcement requirements in the wall.**

1.   Check if two curtains of reinforcement are required.

Section 21.5.2.2 (UBC-88 Section 2625(f)2B) requires that two curtains of reinforcement be provided in a wall if the in-plane factored shear force assigned to the wall exceeds $2A_{cv}\sqrt{f_c'}$, where A_{cv} is the cross-sectional area bounded by the web thickness and the length of section in the

Example 31.6 – Continued

Code
Reference

Calculations and Discussion

direction of the shear force considered. From Table 31-8, the maximum factored shear force on the wall at the first floor level is $V_u = 882$ kips.

$$2A_{cv}\sqrt{f_c'} = (2)(20)(26.17 \times 12)(\sqrt{4000})/1000 = 794 \text{ kips}$$
$$< V_u = 882 \text{ kips}$$

Therefore, two curtains of reinforcement are required.

2. Required longitudinal and transverse reinforcement in wall.
    Minimum required reinforcement ratio.

21.5.2.1
2625(f)1A

$$\rho_v = \frac{A_{sv}}{A_{cv}} = \rho_n \geq 0.0025 \quad (\text{max. spacing} = 18 \text{ in.})$$

With A_{cv} (per foot of wall) $= (20)(12) = 240$ in.^{2}, required area of reinforcement in each direction per foot of wall $= (0.0025)(240) = 0.60$ in.^{2}/ft.

Required spacing of #5 bars (in two curtains, $A_s = 2\,(0.31) = 0.62$ in.^{2}),

$$s \text{ (required)} = \frac{2(0.31)}{0.60}\,(12) = 12.4 \text{ in.} < 18 \text{ in.}$$

c. **Determine reinforcement requirements for shear.**
    (Refer to Section 21.7 (UBC-88 Sections 2625(h)1C and 2625(h)3), for shear design of shearwalls.

    Assume two curtains of #5 bars spaced at 12 in. on center both ways. Shear strength of wall ($h_w/\ell_w = 148/26.17 = 5.66 > 2$),

21.7.3.2
2625(h)3

$$\varphi V_n = \varphi A_{cv}\,(2\sqrt{f_c'} + \rho_n f_y)$$

where $\varphi = 0.60$

$$A_{cv} = (20)(26.17 \times 12) = 6280 \text{ in.}^2$$
$$\rho_n = \frac{2\,(0.31)}{20\,(12)} = 0.00258$$

Thus $\quad \varphi V_n = (0.60)(6280)\big[2\sqrt{4000} + (0.00258)(60,000)\big]/1000$

$$= 3768.6\,\big[126.4 + 154.8\big]/1000 = 1060 \text{ kips} > V_u = 882 \text{ kips} \quad \text{O.K.}$$

Example 31.6 – Continued

Calculations and Discussion	**Code Reference**

Therefore, use two curtains of #5 bars spaced at 12 in. on center in both horizontal and vertical directions, Section 21.7.3.5 (UBC-88 Section 2625(h)3E).

d. Check adequacy of boundary element acting as a short column under factored vertical forces due to gravity and lateral loads.
see Fig. 31-14, Section 21.7.3.5 (UBC-88 Section 2625(f)3C).
From Table 31-8, maximum compressive axial load on boundary element,
$P_u = 4501$ kips

With boundary elements having dimensions 32 in. x 50 in. and reinforced with 30#11 bars,
$A_g = (32)(50) = 1600$ in.^{2}
$A_{st} = (30)(1.56) = 46.8$ in.^{2}
$\rho_{st} = 46.8/1600 = 0.0293$

$\rho_{min} = 0.01 < \rho_{st} < \rho_{max} = 0.06$ O.K.

21.4.3.1
2625(e)3A

Axial load capacity of boundary element acting as a short column

$$\varphi P_{n(max)} = 0.80\, \varphi \left[0.85 f_c' \left(A_g - A_{st} \right) + f_y A_{st} \right]$$
$$= (0.80)(0.70)[0.85)(4)(1600-46.8)+(60)(46.8)]$$
$$= (0.56)[5281+2808] = 4530\ \text{kips} > P_u = 4501\ \text{kips}\quad \text{O.K.}$$

10.3.5.2
2610(d)5A

e. Check adequacy of shearwall section at base under combined axial load and bending in the plane of the wall.

From Table 31-8, the following combinations of factored axial load and bending moment at the base of the wall are listed, corresponding to Eqs. (9-1), (9-2) and (9-3) of ACI 318-89, as modified by UBC 2625(c)4.

Eq. (9-1): $P_u = 4926$ kips, $M_u =$ small
Eq. (9-2): $P_u = 4808$ kips, $M_u = 46,123$ ft-kips
Eq. (9-3): $P_u = 2739$ kips, $M_u = 46,123$ ft-kips

2606(c)
2625(c)4

Example 31.6 – Continued

	Code
Calculations and Discussion	**Reference**

Figure 31-23 shows the $\varphi P_n - \varphi M_n$ interaction diagram (obtained using PCA computer program PCACOL) for a shearwall section having a 20-in. thick web reinforced with two curtains of reinforcement each having #5 horizontal and vertical bars spaced at 12 in. on center and 32 in. x 50 in. boundary elements reinforced with 30#11 vertical bars, with $f_c' = 4000$ psi and $f_y = 60,000$ psi . The design load combinations listed above are shown plotted in the figure. The point marked '1' represents the $P_u - M_u$ combination corresponding to Eq. (9-1), with similar notation for the other two combinations.

It is seen in Fig. 31-23 that the three design load combinations represent points inside the interaction diagram for the shearwall section considered. Therefore, the section is adequate with respect to combined bending and axial load.

f.   Determine lateral (confinement) reinforcement requirements for boundary element. (Fig. 31-24, item h below)

Maximum spacing, $s_{max} = 4$ in.   2625(e)4B
                   $< 1/4$ smallest dimension of boundary element   21.5.3.2
                   $= 32/4 = 8$ in.

1.   Required cross-sectional area of confinement reinforcement in short direction

$$\text{Ash} \geq \begin{cases} 0.12 s h_c \dfrac{f_c'}{f_{yh}} & \text{Eq. (25-4)} \\[2mm] & 2625(e)4A \\[2mm] 0.3 s h_c \left(\dfrac{A_g}{A_{ch}} - 1\right)\dfrac{f_c'}{f_{yh}} & \text{Eq. (25-3)} \\[2mm] & 2625(e)4A \end{cases}$$

Note that Eq. (21-3), Section 21.4.4.1 of ACI 318-89 is identical with Eq. (25-3) of UBC-88. Eq. (21-4), Section 21.4.4.1 of ACI 318-89 is the same as Eq. (25-4), except that the coefficent of 0.12 has been reduced to 0.09.

Assuming #5 hoops and crossties spaced at 4 in. on center and a distance from centerline of #11 vertical bars to face of column of 3 in.,

h_c (for short direction) $= 44 + 1.41 + 0.625 = 46.04$ in.
$A_{ch} = (46.04 + 0.625)(26 + 1.41 + 1.25) = 1337$ in.^{2}

Example 31.6 – Continued

Calculations and Discussion

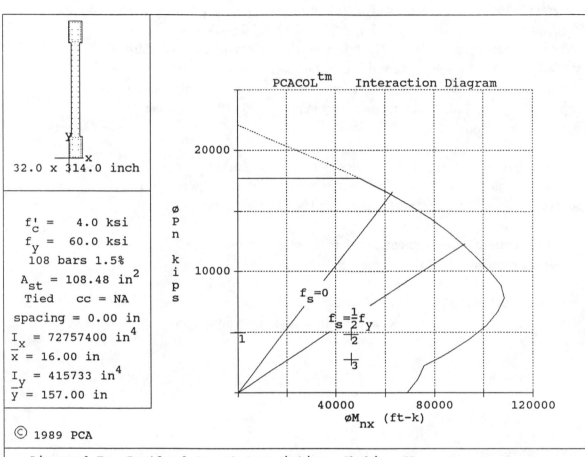

32.0 x 314.0 inch

f'_c = 4.0 ksi
f_y = 60.0 ksi
108 bars 1.5%
A_{st} = 108.48 in^2
Tied   cc = NA
spacing = 0.00 in
I_x = 72757400 in^4
$\overline{x}$ = 16.00 in
I_y = 415733 in^4
$\overline{y}$ = 157.00 in

© 1989 PCA

PCACOL^{tm} Interaction Diagram

$\emptyset P_n$ kips

20000

10000

$f_s=0$

$f_s=\frac{1}{2}f_y$

40000   80000   120000

$\emptyset M_{nx}$ (ft-k)

Project: Example 31-6   File name: C:\PCACOL\DATA\EX31-6.COL

Column Id: 20 in. shearwall   Material Properties:

Engineer: E_c = 3834 ksi ϵ_u = 0.003 in/in

Date: 12/1/89   Time: 16:07:19 f_c = 3.40 ksi E_s = 29000 ksi

Code: ACI 318-89 β_1 = 0.85

Version: 2.10   Stress Profile : Block

                                 Reduction: $\emptyset_c$ = 0.70 $\emptyset_b$ = 0.90

Slenderness not considered x-axis

Figure 31-23 Axial Load-Moment Interaction Diagram for Shearwall Section

Example 31.6 – Continued

	Code
Calculations and Discussion	Reference

h_c (for short direction) $= 44 + 1.41 + 0.625 = 46.04$ in.
$A_{ch} = (46.04 + 0.625)(26 + 1.41 + 1.25) = 1337$ in.^{2}

$$(0.12)(4)(46.04)(\frac{4}{60}) = 1.47 \text{ in.}^2 \text{ (governs)}$$

A_{sh} (required in short $\geq$ direction)

$$(0.3)(4)(46.04)\left(\frac{(32)(50)}{1337} - 1\right)\frac{4}{60} = 0.72 \text{ in.}^2$$

With three crossties (i.e., 5 legs, including outside hoop),

A_{sh} (provided) $= 5(0.31) = 1.55$ in.^{2} O.K.

2. Required cross-sectional area of confinement reinforcement in long direction

h_c (for long direction) $= 26 + 1.41 + 0.625 = 28.04$ in.
$A_{ch} = (46.04 + 0.625)(26 + 1.41 + 1.25) = 1337$ in^2

$$(0.12)(4)(28.04)(4/60) = 0.90 \text{ in.}^2 \text{ (governs)}$$

A_{sh} (required in long $\geq$ direction)

$$(0.3)(4)(28.04)(1.196 - 1)(4/60) = 0.44 \text{ in.}^2$$

With one crosstie (i.e., 3 legs, including outside hoop),

A_{sh} (provided) $= 0.93$ in^2

g. **Determine required development and splice lengths.**

Section 21.5.2.4 (UBC-88 Section 2625(f)2D) requires that all continuous reinforcement in shearwalls be anchored or spliced in accordance with the provisions for reinforcement in tension as given in Section 21.6.4 (UBC-88 Section 2625(g)4).

1. Lap splice for #11 vertical bars in boundary elements.
    (The use of mechanical connectors may be considered as an alternative to lap splices for these large bars.)

    Assuming that 50% or less of the vertical bars are spliced at any one location, a Class B splice may be used Section 12.15.2 (UBC-88 Section 2612(p)2 and Table No. 26-G).

    Required length of splice $= 1.3\,\ell_d$, where

Example 31.6 – Continued

Calculations and Discussion	Code Reference

$\ell_d = 0.40A_bf_y/\sqrt{f_c'} \geq 0.0004d_bf_y$

$= (0.04)(1.56)(60,000)/\sqrt{4000} \geq (0.0004)(1.41)(60,000)$

$= 59$ in. (governs) > 34 in.

12.2.2
2612(c)

Thus required splice length $= (1.3)(59) = 77$ in.

Note: splices must be staggered at least 24 in.

12.15.4.1
2625(d)C

2. Lap splice for #5 vertical bars in wall web.

Again assuming no more than 50% of bars spliced at any one level so that a Class B splice may be used, and using the same expression for ℓ_d above, $\ell_d = 15$ in. Hence required length of splice $= (1.3)(15) = 20$ in.

3. Development length for #5 horizontal bars in wall assuming no hooks are used within boundary element.

Since it is reasonable to assume that the depth of concrete cast in one lift beneath a horizontal bar will be greater than 12 in., the required factor of 3.5 to be applied to the development length, ℓ_{dh}, required for a 90 degree hooked bar will be used, Section 21.6.4.2 (UBC-88 Section 2625(g)4.B(2)).

$\ell_{dh} = f_yd_b/65\sqrt{f_c'} = (60,000)(0.625)/(65)/(\sqrt{4000})$

$= 9$ in. (governs)

$> 8d_b = (8)(0.625) = 5$ in.

> 6 in.

21.6.4.1
2625(g)4A

Thus required development length $\ell_d = 3.5(9) = 32$ in.

This length can be accommodated within the confined core of the boundary element so that no hooks are needed, as assumed.

No lap splices would be allowed for the #5 horizontal bars (full length bars will weigh approximately 25 lbs. and are easily installed).

Example 31.6 – Continued

Calculations and Discussion

h. Detail of structural wall—See Fig. 31-24.

It will be noted in Fig. 31-24 that the #5 vertical 'web' reinforcement required for shear resistance has been carried into the boundary elements. The Commentary to ACI Chapter 21 specifically states that the concentrated reinforcement provided at wall edges for bending shall not be included in determining shear reinforcement requirements. The area of vertical shear reinforcement located within the boundary elements could, if desired, be considered as contributing to axial load and bending capacity.

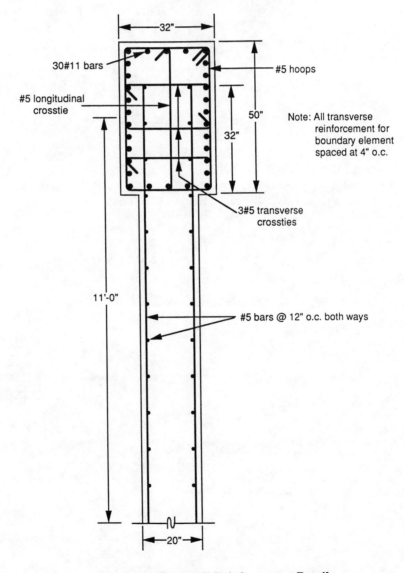

Fig. 31-24 Shearwall Reinforcement Details